AF837684

Topics in Geobiology

Volume 49

Series Editors

Neil H. Landman, Department of Paleontology, American Museum of Natural History, New York, NY, USA
Peter J. Harries, Department of Marine, Earth and Atmospheric Sciences, North Carolina State University, Raleigh, NC, USA

The Topics in Geobiology series covers the broad discipline of geobiology that is devoted to documenting life history of the Earth. A critical theme inherent in addressing this issue and one that is at the heart of the series is the interplay between the history of life and the changing environment. The series aims for high quality, scholarly volumes of original research as well as broad reviews.

Geobiology remains a vibrant as well as a rapidly advancing and dynamic field. Given this field's multidiscipline nature, it treats a broad spectrum of geologic, biologic, and geochemical themes all focused on documenting and understanding the fossil record and what it reveals about the evolutionary history of life. The Topics in Geobiology series was initiated to delve into how these numerous facets have influenced and controlled life on Earth.

Recent volumes have showcased specific taxonomic groups, major themes in the discipline, as well as approaches to improving our understanding of how life has evolved.

Taxonomic volumes focus on the biology and paleobiology of organisms—their ecology and mode of life—and, in addition, the fossil record—their phylogeny and evolutionary patterns—as well as their distribution in time and space.

Theme-based volumes, such as predator-prey relationships, biomineralization, paleobiogeography, and approaches to high-resolution stratigraphy, cover specific topics and how important elements are manifested in a wide range of organisms and how those dynamics have changed through the evolutionary history of life.

Comments or suggestions for future volumes are welcomed.

Neil H. Landman Department of Paleontology American Museum of Natural History New York, USA E-mail: landman@amnh.org

Peter J. Harries Department of Marine, Earth and Atmospheric Sciences North Carolina State University Raleigh, USA E-mail: pjharrie@ncsu.edu

More information about this series at http://www.springer.com/series/6623

Kenneth De Baets • John Warren Huntley
Editors

The Evolution and Fossil Record of Parasitism

Identification and Macroevolution of Parasites

Springer

Editors
Kenneth De Baets
GeoZentrum Nordbayern
Friedrich-Alexander-Universität
Erlangen-Nürnberg
Erlangen, Germany

John Warren Huntley
Department of Geological Sciences
University of Missouri
Columbia, MO, USA

ISSN 0275-0120
Topics in Geobiology
ISBN 978-3-030-42483-1 ISBN 978-3-030-42484-8 (eBook)
https://doi.org/10.1007/978-3-030-42484-8

© The Editor(s) (if applicable) and The Author(s) 2021
This work is subject to copyright. All rights are reserved by the Publisher, whether the whole or part of the material is concerned, specifically the rights of translation, reprinting, reuse of illustrations, recitation, broadcasting, reproduction on microfilms or in any other physical way, and transmission or information storage and retrieval, electronic adaptation, computer software, or by similar or dissimilar methodology now known or hereafter developed.
The use of general descriptive names, registered names, trademarks, service marks, etc. in this publication does not imply, even in the absence of a specific statement, that such names are exempt from the relevant protective laws and regulations and therefore free for general use.
The publisher, the authors, and the editors are safe to assume that the advice and information in this book are believed to be true and accurate at the date of publication. Neither the publisher nor the authors or the editors give a warranty, expressed or implied, with respect to the material contained herein or for any errors or omissions that may have been made. The publisher remains neutral with regard to jurisdictional claims in published maps and institutional affiliations.

This Springer imprint is published by the registered company Springer Nature Switzerland AG
The registered company address is: Gewerbestrasse 11, 6330 Cham, Switzerland

Foreword

The 'fossil record' and 'parasites' may seem strange bedfellows. Parasitism tends to be overlooked by palaeontologists even though it is one of the most successful ecological strategies and an important feature of the evolutionary history and palaeoecology of many groups. One measure of the importance of parasites is their diversity—they may account for more than 50% of all living species. Nonetheless, parasites are largely invisible today even if the same cannot be said of their effects. They are generally small and soft bodied, and certainly not strong candidates for fossilisation. However, their evolutionary history is intriguing and raises issues such as when parasitism originated and in which groups, how parasites coevolved with their hosts, how they impacted the ecology of ancient communities, and how interactions between parasite and host changed over time. The fossil record provides the only direct evidence of parasitism in the past, and it can contribute essential data to answering such questions.

Parasites, of their very nature, require exceptional conditions for fossilisation, and Konservat-Lagerstätten (conservation deposits) have proved important sources in recent decades. Malarial parasites and trypanosomatids, for example, have been discovered in the guts of biting insects in amber more than 100 million years old from the Cretaceous of Burma. Eggs and cysts of intestinal parasites are present in coprolites of late Palaeozoic elasmobranchs and Cretaceous dinosaurs, from aquatic and terrestrial settings, respectively. Giant flea-like insects have been found in Mesozoic lake sediments in Inner Mongolia and Liaoning Province in China. New examples of parasitism also continue to come to light in host fossils based on the galls, swellings and other malformations triggered by the parasite—but in this case the perpetrator is often difficult to identify and other factors may be at work.

Molecular data provide a major new line of evidence on the evolution of parasites, even though the nature of parasite genomes can present particular challenges. Gene sequences have allowed the phylogeny of different groups of parasites to be analysed where morphological data are limited. The phylogeny of the free-living relatives may also be informative. Phylogenies of parasites and of their host species, together with dates based on fossil occurrences, can yield estimates of divergence times (i.e. when they originated) based on molecular clocks. Such phylogenies reveal patterns of co-evolution and host switching. Fossils, however, are essential to calibrate clocks even though this is challenging due to the rare occurrence of parasites in the fossil record. Fossils may also provide more direct evidence of ancient parasite–host relationships. In rare cases they preserve associations that no longer persist, showing that initial hosts were different from those that are favoured today. The remarkable 430 million-year-old Silurian Herefordshire deposit in England, for example, which I have been investigating with colleagues for the last 25 years, has yielded a pentastomid crustacean which is clearly ectoparasitic on marine ostracods (examples are preserved attached to the ostracod carapace, and on eggs within it). Living pentastomids, in contrast, apart from a few exceptions that live on insects, are parasitic on terrestrial vertebrates where they invade the respiratory system.

A surprising range of parasites are represented in the fossil record but they have received relatively little attention to date, at least in terms of a major synthesis, an omission that is remedied here in grand style. The importance of parasitism is reflected in 25 contributions by more than 50 authors. All the important groups of parasites are reviewed: bacteria, protozoa, fungi, cnidarians, bivalves, gastropods, helminths, acanthocephalans, chelicerates, crustaceans and insects as well as fossil evidence on their impact on hosts: colonial organisms, crustaceans, trilobites, cephalopods, bivalves, echinoderms, vertebrates. The major issues covered include the ways in which parasites and their effects are fossilised, the evolution of host–parasite associations, the history of parasitism as recorded in the fossil record, and the utility of genomics and the molecular clock in revealing the course of host–parasite evolution. This welcome compilation will provide an indispensable platform for future work on this fascinating topic.

Yale University
New Haven, CT, USA

Derek E. G. Briggs

Reconstructions of known and lesser known fossil parasites made especially for this volume by (c) Franz Anthony (http://franzanth.com/): Cambrian pentastomid *Heymonsicambria scandica* (upper left) described by Walossek and Müller (1994: https://doi.org/10.1017/S0263593300006295), developing roundworm *Ascarites priscus* (upper centre) from the Lower Cretaceous described by Poinar and Boucot (2006: https://doi.org/10.1017/S0031182006000138), Cretaceous copepod *Kabatarina pattersoni* (upper right) described by Cressey and Boxshall (1989: https://doi.org/10.2307/1485466), putative parasitic isopod *Urda rostrata* (middle center) from the Jurassic described by Nagler et al. (2017: https://doi.org/10.1186/s12862-017-0915-1), Late Cretaceous engorged tick *Deinocroton draculi* (middle left) described by Peñalver et al. (2017: https://doi.org/10.1038/s41467-017-01550-z), Eocene bird louse *Megamenopon rasnitsyni* (middle right) described by Wappler et al. (2004: https://doi.org/10.1098/rsbl.2003.0158), developing acanthocephalan (lower left) from the Upper Cretaceous described Cardia et al. (2019: https://doi.org/10.1590/0001-3765201920170848) early nematode *Palaeonema phyticum* (lower center) described by Poinar et al. (2008: https://doi.org/10.1163/156854108783360159) from the Lower Devonian, and developing tapeworm from the Permian described by Dentzien-Dias et al. (2013: https://doi.org/10.1371/journal.pone.0055007)

Preface

Over 50% of living organisms have been interpreted to be parasitic, and most organisms host at least one species of parasite during their life. Their historical reputation is not the best for obvious reasons. This is even worse in the fossil record where parasites and their traces are considered to be too scarce to be helpful. Although some reviews have been published in the last decades, these often involve works of archaeological remains or focus on particular groups or time intervals. The fossil record is often still ignored by parasitologists and palaeontologists alike and only briefly mentioned in works focusing on parasite evolution and ecology. We felt we wanted to change this by highlighting the importance of the fossil record in reconstruction of the evolution of parasitism. During a research stay of John in Erlangen, we decided this could best be done in the form of two volumes dedicated to the evolution and fossil record of parasitism. The first volume covers the palaeobiological constraints on the fossil record, while the second volume focuses on novel techniques as well as the merit of co-evolutionary and pathological information linked with parasitic disease. Furthermore, we do not only discuss the origin of modern groups of parasites but also groups which are now extinct. We hope you as a reader will learn about how one can study parasites in the fossil record as well as what they can tell us about the evolution of parasitism. This journey started when looking for particular authors to contribute to this work—special thanks goes to all of them in accepting, contributing, and bearing with us during this endeavour. Special thanks also to the reviewers as well as Franz Anthony who designed the accompanying illustration featuring some more and some lesser known fossil parasites. Despite being two parasite aficionados, we have learned a lot about the parasite fossil record and evolution during the preparation and editing of this book. These volumes demonstrate that a lot of progress has been made, but also highlight areas where there is still some work to be done. This is particularly true regarding the prevalence of parasites and gaining a better understanding of their taphonomy and temporal and spatial trends. At least some of these things might benefit from novel techniques such a computed tomography and genomics. Both of us have been working on parasites for about 10 years and had great fun and effort to bring these chapters together. We therefore hope you will enjoy the ride through the fossil record of various groups of parasites, the usefulness in finding them or their traces still associated with their hosts, as well as new techniques to further constrain their evolutionary history.

Erlangen, Germany Kenneth De Baets
Columbia, MO, USA John Warren Huntley

Contents

Chapter 1
Parasites of Fossil Vertebrates: What We Know and What Can We Expect from the Fossil Record?

Tommy L. F. Leung

Abstract Parasites are ubiquitous in extant ecosystems and vertebrate animals often harbour rich parasite communities. However, the geological record of parasites is extremely sparse as their very nature means they are rarely fossilised. The few fossil parasites which have been described have provided interesting insights into the evolution of various parasite taxa, and the development of technology such as high-resolution computed tomography has made detecting signs of parasitism in the fossil record more practical. In this chapter, I will provide an overview of vertebrate-infecting macroparasites which have been described from fossils, and compare those fossil forms with their extant counterparts. I will also discuss what those fossils can tell us about the evolution of parasitism and the ecology of their hosts, the type of parasite fossils which may be associated with fossil vertebrates, and suggest some future research directions which combine aspects of palaeontology, ecology, and parasitology.

Keywords Parasitism · Parasites · Parasite communities · Helminths · Arthropods · Amber · Coprolite · Palaeoecology

1.1 Introduction

Parasites are ubiquitous in extant ecosystems, many of them play important ecosystem roles and form key parts of many food webs (Lafferty et al. 2006; Hatcher and Dunn 2011). As a life-style, parasitism has evolved independently multiple times in many disparate groups of organisms (Poulin and Morand 2004; Poulin 2007; Weinstein and Kuris 2016), and extant vertebrates are host to rich communities of

T. L. F. Leung (✉)
School of Environmental & Rural Sciences, University of New England, Armidale, NSW, Australia
e-mail: tleung6@une.edu.au

© The Author(s) 2021

K. De Baets, J. W. Huntley (eds.), *The Evolution and Fossil Record of Parasitism*, Topics in Geobiology 49, https://doi.org/10.1007/978-3-030-42484-8_1

parasites, both on their external surfaces as well as their various internal organs and tissues. These parasites vary in their taxonomy, life-cycle, infection site, and the effects that they have on their hosts. Many parasites are specialised to live on a specific part of the host's body, and the body of a vertebrate animal can be thought of as a collection of microhabitats, each inhabited by one or more specific parasite species (Loker and Hofkin 2015).

Despite their abundance in many extant ecosystems, parasites are exceedingly rare in the fossil record as most parasites are not conducive to being preserved as fossils. Many of them are small and have soft bodies, or do not leave traces of their presences on the host tissue in a way which will be fossilised (De Baets et al. 2021b). Additionally, most palaeontological studies are not focused on looking for traces of parasitism, so it is also possible that many fossil parasites are being overlooked (Littlewood and Donovan 2003). As such, little is known about the effects they have on their hosts or the ecological roles they played in various palaeoenvironments. However, there are a small collection of studies which have found evidence of parasites from fossil vertebrates (Boucot and Poinar 2011; De Baets and Littlewood 2015; Leung 2017). These studies provide vital glimpses into the evolution of parasitism throughout Earth's history.

The potential for detecting the signs of parasitism or parasites themselves in fossil samples has increased in recent years with technology such as computed tomography which allows well-preserved specimens to be examined for finer, microscopic details (e.g. Siveter et al. 2015; Robin et al. 2016; Poinar et al. 2017; Peñalver et al. 2017). However, an important step in furthering the study of fossil parasites is for palaeontologists to be aware of and recognise the potential for fossilised parasites. As such, any prospective palaeoparasitologists needs to be aware of what to expect and what to look for when it comes to potential fossil parasites. Since vertebrate taxa have relatively large body sizes and are host to a variety of different parasites, fossil vertebrates and their associated remains are great potential sources of parasite fossils. So how would a prospective parasite palaeontologist know what to look for when it comes to fossil parasites? How can they recognise a parasite when they see one? This is where having a general idea of the likely parasite community of a given fossil vertebrate taxa would be helpful, and there are methods for inferring potential parasites of extinct vertebrate taxa and anticipate their association with well-preserved vertebrate fossils.

The aim of this chapter is to provide an overview of vertebrate-infecting parasites which have been described from the fossil record, how they compare with their equivalents found in extant vertebrates, and how we can combine concepts and techniques from palaeontology and ecological parasitology to further develop research into this field. This chapter will be covering the types of macroparasites which have been found from fossil vertebrates that are pre-Pleistocene in age. For each group, I will provide (1) some background on the biology and ecology of their extant taxa, (2) what groups can be expected to host such parasites, (3) what types of fossils can be expect from such parasites, and (4) future research directions.

1.1.1 Cestodes (Tapeworms)

There are approximately 6000 described species of cestodes (Stunkard 1983). Better known as tapeworms, cestodes are found in all major groups of extant vertebrates (Littlewood et al. 2015). They have a complex life-cycle that is deeply linked with predator-prey relationships (see Mackiewicz 1988 for details). The adult cestode lives in the intestine of a vertebrate animals where it produces eggs via sexual reproduction. These eggs are released into the environment where they are ingested by an animal which functions as the intermediate host where the larva can grow and develop. This intermediate host is usually an arthropod, though for some families of cestodes, the intermediate host is a vertebrate animal, and may involve sequentially infecting two or more intermediate hosts. The life-cycle is completed when the infected intermediate host is eaten by the definitive vertebrate host (Mackiewicz 1988).

Being soft-bodied internal parasites, adult cestodes are not readily fossilised (Littlewood and Donovan 2003). The only known fossils for cestodes come in the form of their fossilised eggs (Zangerl and Case 1976; Dentzien-Dias et al. 2013; De Baets et al. 2015). While the sclerotized hooks on the cestode's scolex, the attachment organ anterior of the parasite, can potentially be fossilised under the right conditions, and putative fossils of such hooks in fossil cestode eggs have been reported by Dentzien-Dias et al. (2013). While Upeniece (2011) and De Baets et al. (2015) have suggested some fossil hook circlets associated with Devonian placoderm and acanthodians fossils might be cestode hooks, such fossilised hooks would be situated deep within the host's body and are unlikely to be found from an external examination of a fossil specimen. Upeniece (2001, 2011) has previously documented microscopic hooks present on placoderm fossils and suggested they might be cestode or monogenean hooks. Based on the arrangements of the hooks and where they were distributed on the placoderm, Leung (2017) suggested that it is more likely that they belong to monogeneans (see also De Baets et al. 2015, 2021a)—a group of ectoparasitic flatworms that will be discussed later in this chapter.

So far, there have been two published examples of fossilised cestode eggs, both of them from elasmobranch hosts (Zangerl and Case 1976; Dentzien-Dias et al. 2013), however, given the taxonomic range of extant vertebrates that act as definitive hosts for cestodes (Littlewood et al. 2015), there is a strong possibility that fossilised cestode eggs may be present in the coprolites or lower gastrointestinal tract of fossil vertebrates other than elasmobranchs. The findings of fossilised cestode eggs in elasmobranch coprolites supports molecular phylogeny studies which points to a very long coevolutionary history between elasmobranchs and many cestode lineages (Olson et al. 2010; Caira and Jensen 2014; Caira et al. 2014). Indeed a significant number of extant cestode species (28%) parasitise elasmobranch hosts (Littlewood et al. 2015), and the lineages that infect tetrapod and teleost hosts are nested within elasmobranch-infecting lineages (Caira et al. 2014). Fossils of cestode eggs can provide additional insight into when cestodes began colonising non-elasmobranch hosts, and act as useful calibration points for molecular phylogeny studies (De Baets et al. 2021a). Given fossils of other helminth eggs have previously been found as inclusions in coprolites (e.g. Poinar and Boucot 2006; Hugot et al. 2014; Chin 2021), the coprolites of many other vertebrates should be similarly examined for the presence of cestode eggs in a manner comparable to what has been achieved with subfossils and other archaeological remains (Gonçalves et al. 2003).

1.1.2 Trematodes (Flukes)

There are 18,000 described species of trematodes (Cribb et al. 2001). Like cestodes, trematodes are also internal parasites of vertebrates and have complex life-cycles (Cribb et al. 2003; Galaktionov and Dobrovolskij 2003). While most species have adult stages that live in the gastrointestinal tract of their definitive vertebrate host, there are some families of trematodes that have evolved to occupy other parts of the host's body such as the circulatory system, eyes, lungs, liver, bladder, and the connective tissue and muscles (Cribb et al. 2003; Poulin 2005). Additionally, extant trematodes in the Digenea group (which represents over 99.9% of all known living species) have an asexual reproduction stage (Cribb et al. 2003; Galaktionov and Dobrovolskij 2003) which results in the production of vast number of mobile larval stages which infect the next host in the life-cycle. Depending on the family, this is either a second intermediate host where they encyst, or the vertebrate definitive host where they will develop into sexually mature adults. This asexual reproduction stage in the intermediate host is absent in most other parasitic worms, and in most extant trematodes, asexual reproduction occurs in a mollusc host, which is usually a gastropod in most families (Cribb et al. 2003; Galaktionov and Dobrovolskij 2003). As with other parasitic worms, adult trematodes are soft-bodied internal parasites that do not usually fossilise even under the most ideal preservation conditions. However, they do produce environmentally resistant eggs that can potentially be fossilised with coprolites. So far, there has been one published example of fossilised trematode eggs, reported from dinosaur coprolites dating from the Early Cretaceous (Poinar and Boucot 2006). Given that trematodes are known from all major extant vertebrate groups (Cribb et al. 2003; Galaktionov and Dobrovolskij 2003; Littlewood et al. 2015), and that eggs from various lineages of trematode have been reported in more recent quaternary fossils and subfossil coprolites (e.g. Jouy-Avantin et al. 1999; Le Bailly and Bouchet 2010; Wood et al. 2013) examination of older coprolite samples may yield more trematode egg fossils.

A recent fossil find indicates that amber can preserve the larval stage of trematodes along with their host. Recently, Poinar et al. (2017) reported finding what appears to be the metacercaria stage of a trematode from a lizard preserved in Myanmar amber dated to the Early Cretaceous. The size and position of the fossil metacercaria is strikingly similar to that of some extant trematodes which use lizards (particularly Anoles) as the second intermediate host in their life-cycle (Poinar et al. 2017). This fossil provides some very useful insight into the evolutionary history of these trematodes, and its similarity in general morphology, host type, and host site to some extant trematodes indicates that the complex life-cycle of trematodes seen in extant taxa have existed since at least the Cretaceous.

Aside from their eggs and larval stages preserved *in situ* with their host in amber, larval trematodes may also leave other type of traces in the fossil record. The larval stages of some trematodes, namely those in the Gymnophallidae family, also infect bivalves as a part of their life-cycles and their presence can induce pitting, igloo-shaped concretions (e.g. Ituarte et al. 2005) and/or pearl formation in the shell which are more readily fossilised (e.g. Ruiz and Lindberg 1989; Ozanne and Harries 2002; see Huntley and De Baets 2015; Huntley et al. 2021). While shell pitting in bivalves can also be caused by other non-trematode factors (Leung 2017), the igloo-shaped concretions produced by gymnophallid larvae are quite distinct and is considered to be a reliable indicator of that particular trematode lineage (Huntley and De Baets 2015). A recent study places definitive evidence for the presence of igloo-shaped concretions in the Cretaceous (Rogers et al. 2018), and while similar traces have been found on the shells of Silurian bivalves (Liljedahl 1985), Huntley and De Baets (2015) cautioned against interpreting those as being caused by gymnophallid trematodes. Similarly, the presence of "blister pearls" (often represented as pits in steinkerns) of Early Devonian ammonoids might have been caused by parasites (De Baets et al. 2011), there is no conclusive evidence that they were caused by trematodes (De Baets et al. 2015). So while trematodes can leave potential trace fossils, fossilised eggs in coprolites or larval stages preserved in the intermediate hosts provide more definitive information on the likely lineage or taxonomic identity of the fossil trematode. Since trematodes are common in extant vertebrates with 18,000 known extant species, fossils can give insight into how and when they became so diverse and successful (De Baets et al. 2021a).

1.1.3 Nematodes (Roundworms)

Nematodes (commonly known as roundworms) are one of the most diverse and abundant animal phyla on earth, they inhabit a wide variety of ecological niches including parasitism (Poinar 1983). For the purpose of this chapter, I will be focusing on nematodes that infected fossil vertebrates, but for an extensive and detailed overview of parasitic nematodes in the fossil record, readers are referred to (De Baets et al. 2021a; Poinar 2015).

Parasitism has independently evolved at least 15 times in different nematode lineages, parasitising invertebrates, vertebrates, and plant hosts (Blaxter and Koutsovoulos 2015). Based on molecular phylogeny, nematodes have evolved to parasitise vertebrate animals on four separate occasions, and that they had arisen from arthropod-infecting taxa (Blaxter et al. 1998). A more recent molecular phylogenetic study proposed that Ascaridoidea—a diverse superfamily of nematodes found in all major vertebrate groups—has a common ancestor dating back to the Early Carboniferous (Li et al. 2018). Ascaridoidea is a particularly important group of vertebrate parasites, containing over 800 known species, many of which are widely recognised and studied because of their medical and economic importance (Anderson 2000). Much like cestodes and trematodes discussed above, the adult stage living in the gastrointestinal tract is very unlikely to become fossilised. However the environmentally-resistant eggs produced by the adult parasite can be preserved as inclusions in coprolites (e.g. Da Silva et al. 2014), survive treatment with strong acids and/or bases (e.g. Brinkkemper and van Haaster 2012; Dufour and Le Bailly 2013), and serve as a reliable indicator of the parasite's presence. Fossilised ascarid nematode eggs have so far been found in archosaur coprolites from the Early Cretaceous to Late Cretaceous (Poinar and Boucot 2006; Cardia et al. 2018, 2019a) and cynodont coprolites from Middle Triassic (Da Silva et al. 2014). Given the ubiquity of parasitic nematodes in extant vertebrate hosts (Anderson 2000), the presence of fossilised nematode eggs in the coprolites of phylogenetically distant groups as archosaurs and synapsids, and the result by Li et al. (2018) which proposed ascaridoids had originated in terrestrial hosts during the Early Carboniferous—close to the period when tetrapods began diversifying into various terrestrial habitats (Dunne et al. 2018)—it can be expected that most extinct vertebrate groups were host to ascaridoid nematodes, and their fossilised eggs may be present as inclusions in coprolites originating from many different tetrapods.

Another group of parasitic nematodes commonly found in the gastrointestinal tract of extant vertebrates are the Oxyurida—also known as "pinworms". Pinworms are found in the hindgut of many different invertebrate and vertebrate animals and are usually associated with hosts that have hindgut bacterial fermentation because of their bacterivorous diet (Adamson 1994). Fossils of pinworm eggs have been found in the coprolites attributed to therapsid cynodonts from the Middle Triassic (Hugot et al. 2014; Francischini et al. 2018), and given the taxonomic breadth of extant pinworm hosts, it can be expected that pinworm eggs may be present in coprolites of a wide range of terrestrial vertebrates. Various aspects of these pinworm biology also means they can provide valuable insight on the ecology and biology of their host. Francischini et al. (2018) pointed out that pinworm egg inclusions were found in "cylindrical" type coprolites, which were traditionally considered to be produced by carnivorous or omnivorous therapsids. However, the presences of the pinworm egg fossils overturn such an interpretation, as their presence implies that the originator of those coprolites included substantial amount of plant matter in their diet (Francischini et al. 2018). Also mentioned above, pinworms are restricted to infecting animals that use their hindgut as a site for bacterial fermentation, furthermore their eggs have limited dispersal ability and mostly rely upon social grooming or local contamination for transmission (Adamson 1994). Therefore, the very presence of fossilised pinworm eggs in coprolites can provide at least three key insights on the diet (mostly plant matter), physiology (hindgut fermentation) and behaviour (living in social groups or colonies) of the host taxa which may not be evident from examining the fossil of the host alone.

While all the known whole body fossils of animal-infecting parasitic nematodes have so far been from amber fossils of insect-infecting taxa (Poinar 2015), those fossils may nevertheless provide insight into nematodes that do infected fossil vertebrates. Some vertebrate-infecting nematodes with complex life-cycles are phylogenetically affiliated with and may have evolved from insect-infected lineages (Blaxter and Koutsovoulos 2015). Furthermore, many vertebrate-infecting nematodes use insects and other arthropods as intermediate hosts (Anderson 2000), thus much like the fossilised trematode metacercariae described by Poinar et al. (2017), some amber fossils may contain insects (and other small animals) which hosted the larval stage of vertebrate-infecting nematodes. Furthermore, some biting insects can act as vectors for parasitic nematodes—another superfamily of nematodes that commonly parasitises tetrapod vertebrates are the Filarioidea. Also known as filarials, the adult stage lives in the lymphatic or cardiovascular system of their vertebrate host, producing larval stages known as microfilarials which use blood-feeding arthropods as vectors to transmit to new hosts (Anderson 2000). Presently, there are three known species of filarials which have been described from fossils, all of which have been placed in the *Cascofilaria* genus; *C. baltica* which was found associated with a blackfly in Eocene age Baltic amber, *C. dominicana* which was found associated with a female mosquito in Dominican amber, and *C. parvus* which was also found associated with a mosquito host (Poinar 2015). The morphology of those fossil microfilarials resemble extant filarial species that infect mammals and amphibians (Poinar 2015). It is worth pointing out that there is some evidence to suggest human-infecting (and other mammal-infecting) filarial worms might have originated from bird-infecting filarials about 17–25 million years ago (Suh et al. 2016; Suh 2021), which overlaps with the age of Dominican amber which have preserved remains of tropical birds including their feathers and egg shells (Poinar 2010). Thus, parasites which are found in biting insects that are preserved in such amber deposits can provide vital insight into the origin and evolution of vector-transmitted parasites such as filarial nematodes.

1.1.4 Acanthocephalans (Thorny-Headed Worms)

Acanthocephalans, commonly known as thorny-headed worms, are a group of internal parasites with complex life cycles. There are about 1150 known living species, and all major groups of extant vertebrates have been found to host acanthocephalans, with the adult stage of this parasite living in the gastrointestinal tract of various vertebrate animals including fish, amphibians, reptiles, mammals, and birds (Kennedy 2006). The adult worm anchors itself to the gut wall of the vertebrate host using an eversible proboscis which is covered with hooks and spines (Miller and Dunagan 1985).

The typical life cycle of acanthocephalan involves two hosts, with the adult living in a vertebrate host, producing eggs which are released into the environment, and an arthropod intermediate host which become infected by acanthocephalan larvae when they ingest the parasite's eggs (Kennedy 2006). However, many species also incorporate additional vertebrates to act as paratenic (transport host) (e.g. Sinisalo and Valtonen 2003; Médoc et al. 2011)—the acanthocephalan larvae do not undergo further development in those hosts, however paratenic hosts can help the parasite complete its life-cycle as they serve as a bridge between arthropods and the acanthocephalan's final host, which are often macropredatory vertebrates which do not usually prey upon small arthropods (Nickol 1985; Kennedy 2006).

The adult stages of acanthocephalans are soft bodied and thus unlikely to be preserved as fossils, and while theoretically the hooks on the proboscis may potentially become fossilised under ideal circumstances, much like what has been discussed above for tapeworms, flukes, and roundworms, the stage most likely to be found as fossils would be eggs embedded in coprolites produced by their vertebrate hosts (Herlyn 2021).

Presently, the oldest known fossil example of acanthocephalans are fossilised eggs which were found as inclusions in coprolites that have been attributed to Late Cretaceous crocodyliformes (Cardia et al. 2019b). Much like the fossil pinworm eggs discussed above, they provide us with some interesting insight into the ecology of their host, since the presence of acanthocephalan eggs indicates the crocodyliformes that produced the coprolite had either consumed arthropods and/or vertebrate animals. This allows us to conclude that the coprolites most likely originated from one of the species of carnivorous (or at least omnivorous) crocodyliformes that had been found in that formation.

Furthermore fossils of acanthocephalans may also provide information on the evolution of Acanthocephala as a whole and their pattern of host-usage (Herlyn 2021). Presently, Acanthocephala is divided into four distinct classes, and each of those classes infect different suite of vertebrate animals from different environments as their final hosts (Verweyen et al. 2011). Finding evidence of acanthocephalan infection in fossil vertebrates, especially in lineages that left no living relatives, may help shed light how these parasites had evolved to infect the type of hosts that they do today.

1.1.5 Monogeneans

Monogeneans are a major group of parasitic flatworms which mostly live as ectoparasite of fish, with approximately 3000–4000 described species (Whittington 1998). Being soft-bodied animals they are highly unlikely leave behind fossils, however their attachment organ—the haptor—has microscopic, scleritised hooks which can potentially be fossilised under certain conditions (Littlewood and Donovan 2003). Fossilised monogenean hooks have previously been found associated with placoderms and acanthodians (Upeniece 2001, 2011), and their positioning on the fins, gill region, and abdomen of the host is comparable to that of extant monogeneans (Whittington 1998). Given their presence on those two phylogenetic distinct groups of extinct fish and the wide range of extant fish lineages that host monogeneans, it can be expected that most fossil fish from the Devonian period onward would have been parasitised by monogeneans.

1.1.6 Parasitic Copepods

Aside from monogeneans, one of the most commonly found ectoparasites of extant fishes are parasitic copepods (Boxshall 2005; Klompmaker and Boxshall 2015). Parasitic copepods make up approximately one-third of all known extant copepod diversity (Humes 1994; Ho 2001), and many of them are parasites of fish (Boxshall 2005) that attach themselves to the skin, gills, fins, or eyes of their hosts (Kearn 2004). Some parasitic copepods have highly derived morphology and are not easily recognisable as crustaceans, and some are considered as "mesoparasites" because the adult stages live partially embedded within the host's internal organ, while the rest of the body (usually the reproductive organs) protrude into the external environment (Kearn 2004).

Despite their ubiquity among extant fishes, there is currently only one documented example of a parasitic copepod in fossil fish (although some pathologies in fossil fish have been attributed to copepods, see Klompmaker and Boxshall 2015 for discussion), which was found within the gill chamber of an Early Cretaceous fish (Cressey and Patterson 1973). It was later described and named *Kabatarina pattersoni* and assigned to an extant family of parasitic copepod; Dichelesthiidae (Cressey and Boxshall 1989). This discovery raises two key points; (1) there may be extant parasitic copepod families that have had long histories and have relatives in the fossil record which have similar morphologies to their extant counterparts, and (2) well-preserved fish fossils presents opportunities for finding fossils of parasitic copepods (and indeed other parasites). This is a possibility which should be explored using micoCT scans, and much like monogeneans, given the ubiquity of parasitic copepods among extant fish (Kabata 1982), it can be expected that many extinct fish taxa would be host to parasitic copepods.

1.1.7 Parasitic Isopods

Parasitism has evolved in many crustacean taxa and aside from copepods, another group with many parasitic representatives are the isopods. Within Isopoda, there are three major groups of parasites: the Cymothoidae which are mostly ectoparasites of fish (Smit et al. 2014); Gnathiidae which are specialised ectoparasitic blood-feeders with a multi-stage developmental cycle (Smit and Davies 2004), and Epicaridea which are highly-derived parasites of crustaceans (Williams and Boyko 2012). Of those groups, the cymothoids and gnathiids are the two families which parasitise vertebrate hosts, mostly ray-finned fishes (Smit and Davies 2004; Smit et al. 2014).

A few recent studies on molecular phylogeny and fossils of parasitic isopods have provided some insight into the origin and evolution of parasitism in isopods. Smit et al. (2014) have suggested that based on the presence of fossil evidence for bopyrid (a very speciose family within Epicaridea) isopods in the Jurassic (Boucot and Poinar 2011; Klompmaker et al. 2014), fish-parasitising cymothoids most likely originated during the same time since the two group share a close common ancestry. This is supported by the recent discovery of some Middle to Late Jurassic fish fossils which are fossilised with isopods attached to their body (Haug et al. 2021; Nagler et al. 2016, 2017).

But as mentioned above, apart from the external body surfaces, cymothoids are also known to live in other parts of the host body such as the buccal cavity, gill chambers, and body cavity (Smit et al. 2014). A recent molecular phylogeny study indicates that the shift between living on the body surface and living in the buccal cavity or gill chambers might have independently occurred a number of times within the cymothoids (Hata et al. 2017). Furthermore, colonisation of the host body cavity has independently evolved in two separate lineages of cymothoids (Hata et al. 2017). Therefore, we should also expect that aside from being present on the skin and fins of fossil fish, there is also a possibility that some might be obscured within the host's mouth, gill chambers, and abdominal cavity, much like the parasitic copepod (Cressey and Patterson 1973) from the gill chamber of an Early Cretaceous fish fossil. As mentioned in relation to that example, microCT and similar techniques can potentially be used to examine well-preserved fish fossils for the presence of parasitic isopods (Nagler et al. 2017). While they are mainly parasites of teleost fish, cymothoids have also been documented from some elasmobranchs (Williams et al. 2010), and two disparate groups of extant marine reptiles—snakes (Saravanakumar et al. 2012) and turtles (Júnior et al. 2015). In the case of turtles, they were found around parts of the body that have thinner tissue and are more vascularized such as eyelids (Júnior et al. 2015). Given their capacity to infect those two extant marine reptile groups, it is possible that some Jurassic and Cretaceous marine reptiles may have also served as viable hosts for these parasitic isopods, thus expanding the range of potential hosts in the fossil record which may have hosted cymothoids.

However, a note of caution in regards to interpreting fossils of isopods as parasites when they are found in association with larger animals. Marine isopods are opportunistic feeders that frequently scavenge on vertebrate carcasses, and fossils of their scavenging activities have previously been found (Wilson et al. 2011; Klompmaker and Boxshall 2015; Robin et al. 2019). Thus before concluding any fossils of vertebrate-isopod association as a case of parasitism, one must ensure there is evidence to indicate that the host was alive and engaged in extended interaction with the isopod(s) when they became fossilised. Another possible method of establishing whether a fossil isopod was parasitic is by comparing its functional morphology and with that extant parasitic isopods, as parasitic species have specialised feeding and attachment structures that distinguish them from their free-living relatives (e.g. Nagler and Haug 2016).

1.1.8 Pentastomids (Tongue Worms)

Aside from copepods and isopods, another group of parasitic crustaceans which are known from fossils are the pentastomids. Also known as "tongue worms" due to the shape of the adult stage, the taxonomic affinity of pentastomids had been an enigma until they were finally recognised as highly-derived crustaceans (Lavrov et al. 2004). Extant pentastomids are respiratory tract parasites of tetrapods, mostly reptiles (Riley 1986), but the oldest known fossils of pentastomids are larval stages dating back from the Late Cambrian (Walossek and Müller 1994; Walossek et al. 1994; Klompmaker and Boxshall 2015), over 100 million years before the appearance of tetrapods, therefore it is unclear what those pentastomids parasitised (assuming that they were parasites at all). It has been suggested that those early pentastomids might have been parasites of basal chordates and led a more simple life-cycle than extant pentastomida (Sanders and Lee 2010), living in the gill cavities of their hosts (Walossek and Müller 1994).

A more recently described fossil suggests that the original hosts for pentastomids were in fact crustaceans (Siveter et al. 2015) and that they colonised vertebrate hosts much later in their evolutionary history. *Invavita piratica* lived during the mid-Silurian period and unlike the isolate larval microfossils found by Walossek and Müller (1994), the *I. piratica* specimen was interpreted as an adult which was found attached to its host—an ostracod (Siveter et al. 2015). This indicates that pentastomids might have originated as ectoparasites of crustaceans and incorporated vertebrates host and became internal parasites much later in their evolutionary history (but see Haug et al. 2021). So how and when did this switch in host and mode of parasitism occur?

Leung (2017) suggested it might have occurred through a process of "upward incorporation" (Parker et al. 2015)—the original crustacean host of pentastomids may have also been prey for the jawed fish that began appearing in the ocean during that period (Friedman and Sallan 2012; Klug et al. 2017), and what might have begun as an way of surviving their host's predation by transferring to the vertebrate predator subsequently became an obligate part of their life-cycle. In this scenario, as tetrapods evolved from those jawed fish and began colonising terrestrial biomes, the pentastomids coevolved with them, by which point they had evolved to live as internal parasites which allow them to survive in hosts which lived in terrestrial biomes. Further fossils of pentastomids can test the validity of this hypothesis, as well as provide more insights into how and when pentastomids transitioned from being ectoparasites of crustaceans to endoparasites of tetrapods, and how this might relate to the colonisation of land by tetrapods.

1.1.9 Ticks

There is abundant evidence for non-avian dinosaurs, pterosaurs, and non-mammalian synapsids being covered in fine integumentary features (Dhouailly et al. 2017)—which seems to be an ideal niche for ectoparasitic arthropods. Ticks are common ectoparasites of terrestrial vertebrates (Klompen et al. 1996), and while molecular phylogeny analysis placed the origin of ticks to the late Carboniferous (Mans et al. 2012), the oldest confidently identified fossil ticks are from the Cretaceous (e.g. Poinar and Buckley 2008; Estrada-Peña and de la Fuente 2018; Dunlop 2021).

Fossil ticks are usually found in isolation from host material, however, their morphology resembles that of extant ticks closely enough to infer their parasitic, hematophagous life-style, in some case can even be identified to extant genera (e.g. *Amblyomma birmitum*—Chitimia-Dobler et al. 2017). While only a few fossils of ticks have been described, they are likely to have been blood-feeding ectoparasites of many extinct terrestrial vertebrate taxa. This is supported by fossils that indicate such ticks were also hosts for vector-borne pathogens which are similar to those that infect extant vertebrates (Poinar 2019). A molecular phylogeny study found the divergence of hard and soft ticks to have occurred during the Early Permian, which indicates a Carboniferous origin for ticks (Mans et al. 2012). This was also during an important period in the evolution and diversification of terrestrial vertebrates and amniote animals (Clack 2002), therefore it is possible that ticks coevolved with terrestrial tetrapods.

Recent discoveries of dinosaur body parts or even entire hatchlings with intact plumage in amber (Xing et al. 2016, 2017) presented exceptional opportunities to investigate potential dinosaur ectoparasites such as ticks. Such amber fossils not only preserve the parasite in detail, but also *in situ* with host material, thus providing direct evidence for their parasite-host relationships. As an illustrative example, recently some tick fossils were described by Peñalver et al. (2017) from Middle Cretaceous Burmese amber. The ticks were found in association with loosely vaned pennaceous feathers which are found on penneraptoran dinosaurs, but not crown birds (Peñalver et al. 2017), thus providing unequivocal evidence of ectoparasites on non-avian theropod dinosaurs. Furthermore, some of the fossil ticks have morphology which indicates they belonged to families that no longer exist, such as *Deinocroton draculi* which was assigned to the family Deinocrotonidae (Peñalver et al. 2017). This indicates that there are families of ticks which have become co-extinct with their hosts, indeed this should be expected as many extant parasite groups had lineages that have become extinct with their hosts during various extinction events, indeed the potential co-extinction of parasites with their endangered hosts has recently emerged as a key issue in conservation biology (e.g. Campião et al. 2015; Strona 2015; Thompson et al. 2018).

1.1.10 Ectoparasitic Insects (Fleas and Lice)

While insects originated in the Devonian (Misof et al. 2014), evidence for insect feeding on the blood of vertebrates did not appear until the mid-Mesozoic (Lukashevich and Mostovski 2003). Nagler and Haug (2015) provided an extensive review on fossils of parasitic insects including those that are ectoparasites of vertebrates. There are at least seven orders of insects that have evolved to associate with vertebrates with their relationships ranging from commensalism to parasitism, with some feeding non-invasively on host secretion to those that actively feed on host tissue such as integumentary growth or blood (Waage 1979). It is worth noting that the major groups of ectoparasitic insects are usually associated with hosts that have integumentary features, it is possible that the evolution of integumentary structures in synapsids and dinosaurs might have provided suitable environments for the diversification of many ectoparasitic insect groups (Leung 2017), in a manner comparable to how angiosperm plants had coevolved with many herbivorous and pollinator insects (Labandeira et al. 1994; Grimaldi 1999). Of those, the insect orders that have evolved to parasitise vertebrates, the most successful groups are the lice (orders Phthiraptera) and the fleas (order Siphonaptera), both of which are composed wholly of ectoparasitic species, and both are common parasites of birds and mammals.

But while homeothermic terrestrial vertebrates had been interpreted to exist at least since the Triassic (Padian and Sues 2015), as mentioned above, both lice and fleas only began appearing in the fossil record after the K-Pg mass extinction event. Zhu et al. (2015) found that fleas diversified on mammals before they also colonised birds. A recent study on the molecular phylogeny of lice indicates that they radiated on birds and mammals after the K-Pg event, of which one lineage was exclusive to birds, and that the rapid diversification of birds and mammals after K-Pg was associated with the evolutionary radiation of their lice (Johnson et al. 2018). At the same time, another recent study indicates that there was a K-Pg mass extinction of stem birds, subsequently followed by a radiation of crown birds (Field et al. 2018). Together, these provide evidence that modern lice and fleas diversified alongside lineages of birds and mammals that survived the K-Pg extinction event. Much like other fossil insects, amber would be the most promising material to investigate for ectoparasitic insects, especially amber fossils that include some integumentary material or parts of the body from a potential host, as discussed above in relation to fossil ticks.

1.2 A Note of Caution Regarding Fossil Parasites

As discussed above, despite their diversity and abundance on extant birds and mammals, truly definitive fossils of lice and fleas did not appear until the Early Eocene (Wappler et al. 2004) and Miocene (Dittmar et al. 2015) respectively. While there have been some reports of "giant Jurassic fleas" from the mid-Mesozoic (Gao et al. 2012), their status as fleas or even as ectoparasites appears questionable. Dittmar et al. (2015, 2016) have pointed out a number of problems with the interpretation of those fossil insects as hematophagous ectoparasites. Indeed, that seems to be a recurring problem in the literature on various fossil insects which have been interpreted as having a parasitic life-style based on nothing more than conjecture (as discussed in Leung 2017). For example, the Strashilidae, a family of Jurassic insects, was initially interpreted as ectoparasites of pterosaurs or dinosaurs based on the morphology of mouthpart and hindlimbs (Ponomarenko 1976). This assumption was carried over in subsequent published studies on this family of insects (Rasnitsyn 1992; Vršanský et al. 2010) until newer fossil specimens led to a re-evaluation of its lifestyle and revealed it to be a sexually dimorphic aquatic insect (Huang et al.

2013). Furthermore, the morphological traits that were initially cited by Rasnitsyn (1992) as adaptations for parasitism—sucking mouthparts and prominent hindlimbs—are also found in a wide range of non-parasitic insects and should not be considered as definitive diagnostic features for ectoparasitism (Leung 2017). Similarly, the fossil larvae of Jurassic dipteran *Qiyia jurassica* was interpreted by Chen et al. (2014) as a hematophagous ectoparasite of aquatic amphibians on the basis of its sucking mouthpart and thoracic suckers, yet as Leung (2017) pointed out, the characteristics that Chen et al. (2014) interpreted as parasitic adaptations are also found on many non-parasitic freshwater insects. Most importantly, none of those fossils insects that have been interpreted as ectoparasites were ever found in association with their postulated hosts or host material. Misattribution of parasitism to extinct taxa is still an ongoing issue in the palaeontological literature. For example, Ponomarenko (1976) described the Early Cretaceous insect *Saurophthirus* as a kind of stem flea that lived on pterosaurs on the basis of a single fossil. But as Dittmar et al. (2015) and Zhu et al. (2015) pointed out, that interpretation is highly questionable and inconsistent with what is known about the morphology and phylogeny of crown fleas (Dittmar et al. 2016). Despite that, recently published studies on *Saurophthirus* has taken the "pterosaur ectoparasite" interpretation at face value, so much so that every aspect of its morphology has been interpreted by some workers as that of a hematophagous pterosaur ectoparasite (Rasnitsyn and Strelnikova 2017; Shcherbakov 2017) even though (1) no Saurophthiridae fossils have ever been found in association with pterosaur fossil material, (2) their morphological traits are also found on insects with non-parasitic life-styles, and (3) multiple aspects of the *Saurophthirus* fossil contradict this hypothesis (Rasnitsyn and Strelnikova 2018).

This should not be taken as a discouragement against searching for parasitic insects in the fossil record, but rather, a call for a more critical appraisal of fossil material and cultivating interdisciplinary research between palaeontologists, parasitologists and researchers from various other relevant fields. The lack of true fossil fleas and lice prior to the Cenozoic actually raises some intriguing questions about whether insects lived as ectoparasites of terrestrial homeotherms prior to the K-Pg extinction event, and if so (1) were they of the same order as those of extant ectoparasites, or did they belong to entirely different or potentially extinct groups? (2) Or perhaps the ectoparasitism niche was occupied by some other type of arthropods, and if so (3) what were the dominant ectoparasites of terrestrial homeotherms such as dinosaurs and mammals during the Mesozoic?

1.3 Framework for Inferring Parasite Communities of Extinct Vertebrate Groups

Based on fossil parasites which have been described thus far, it seems that most extinct vertebrates, at least those from the Mesozoic to recent, have similar or comparable parasites to their extant relatives (De Baets et al. 2015). While fossil parasites are rare and difficult to find, it is possible to anticipate the potential presence of certain parasite taxa in fossil taxa of vertebrates, based on the kind of parasites which are found on their extant relatives or modern ecological equivalents. This can be done by applying two concepts taken from palaeontology and ecology respectively: Extant Phylogenetic Bracketing and Ecological Fitting.

Extant Phylogenetic Bracketing (EPB; Bryant and Russell 1992; Witmer 1995) is often used for inferring the potential presence of certain soft tissues in extinct taxa, however, it can also be applied in a manner to inferring the potential parasite communities of extinct taxa. For example based on the ubiquity of tapeworms in extant vertebrates, it is possible to infer that many extinct fossil taxa bracketed within Vertebrata would had served as definitive hosts to the adult tapeworms (see Fig. 1.1). Similar, for monogeneans discussed above, since they are found in all extant groups of jawed fish and fossils of their hooks have been found on two wholly extinct groups (placoderms and acanthodians), it is possible to infer that many extinct taxa of fish could have potentially hosted monogeneans. By applying the EPB framework to parasite communities of fossil taxa, this would allow us to use their closest living relatives as a rough guide to the potential composition of their parasite communities.

But what if the extinct taxa are morphologically and ecologically very different from their closest living relatives (for example sphenacodontid synapsids, large theropod dinosaurs, sauropod dinosaurs, sauropterygians)? That is where the concept of "Ecological Fitting" (Agosta and Klemens 2008) would be useful. The term was coined by Janzen (1985) to describe an organism that is able to live and persist in a

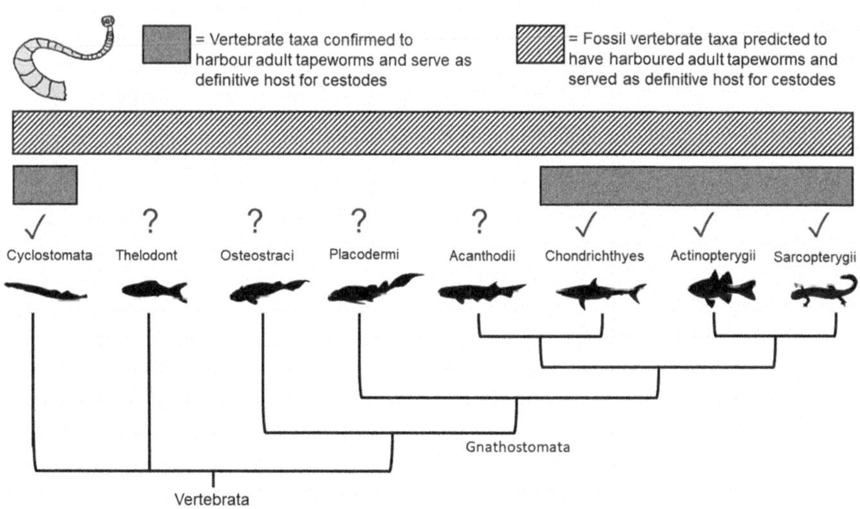

Fig. 1.1 An example of applying Extinct Phylogenetic Bracketing to fossil host taxa. Here, the focal parasites are the adult stage of cestodes (tapeworms). By looking at the extant groups which are known to act as definitive hosts for cestodes, it can be inferred that some wholly extinct vertebrate taxa may have also hosted adult cestodes in their gastrointestinal tract. Silhouette of the different hosts are from PhyloPic (http://phylopic.org/)

novel environment as a result of adaptations that have evolved under a different set of circumstances. As applied to parasites, it means some parasites may be found in host groups with which they did not have a long co-evolutionary history, and that taxonomically different hosts may share similar parasites because they share similar ecological niche and acquire similar parasite communities as a result of being exposed to a similar suite of infective stages (Hoberg and Brooks 2008).

This is broadly comparable to convergent evolution, where phylogenetically disparate taxa evolve similar adaptations under similar circumstances. In this case, such adaptations might have also predisposed them to being infected by similar parasites. For example, various studies have found associations between an animal's diet and the composition of its parasite fauna (Poulin and Morand 2004). Many parasites use trophic transmission, where the parasite is transmitted via the consumption of the infective stage in prey items as a mean of completing their life-cycles (Lafferty 1999), thus diet has a direct influence on the composition of their internal parasites. Indeed, parasites can be used to infer the diet of an animal and in contrast to stomach content, which can only provide a brief snapshot of what the animal had recently consumed, the presence of a parasite species is a lasting indicator of an animal's diet (e.g. Valtonen et al. 2010). In regards to fossil parasites of extinct vertebrate hosts, this line of inference can be inverted. The likely diet of a fossil animal might help us infer what kind of internal parasites we can expect to find associated with it.

In some cases, the presence of parasite larvae in prey species which are consumed by many different types of predators can facilitate host switch across phylogenetically distant taxa. For example, *Corynosoma australes* is an acanthocephalan commonly found in the gastrointestinal tract of pinnipeds, but they have also successfully parasitised Magellanic penguins (*Spheniscus magellanicus*) (Hernández-Orts et al. 2017). Despite being from completely separate branches of amniotes, due to the overlap in diet in both sea lions and penguins (in this case, fish), the latter acquired a parasite which would usually complete its life-cycle in a marine mammal host. Indeed, it seems that over long evolutionary time, endoparasitic helminths are particularly apt at switching between phylogenetically distant vertebrate definitive hosts and host-switching appear to be a common feature in their evolution (Poulin and Morand 2004). Therefore, extant species occupying the same ecological niche may provide a general approximation of the type of parasites that infect phylogenetically distant taxa that once occupied the same niche. For example, many of the marine reptiles that lived during the Mesozoic filled similar ecological niches to those occupied by extant marine mammals (Kelley and Motani 2015), and given that some of those extinct marine reptiles such as ichthyosaurs and plesiosaurs might have also been at least partially homeothermic (Bernard et al. 2010), there is a strong possibility that they shared similar parasite communities to some modern cetaceans.

Thus based on what is known about the ecology and host types of extant parasites, one can use information on the diet, habitat, morphology, physiology, phylogeny, and geological age of a given fossil taxon to infer what type of parasites were potentially included in (or conversely, excluded from) its parasite community (see

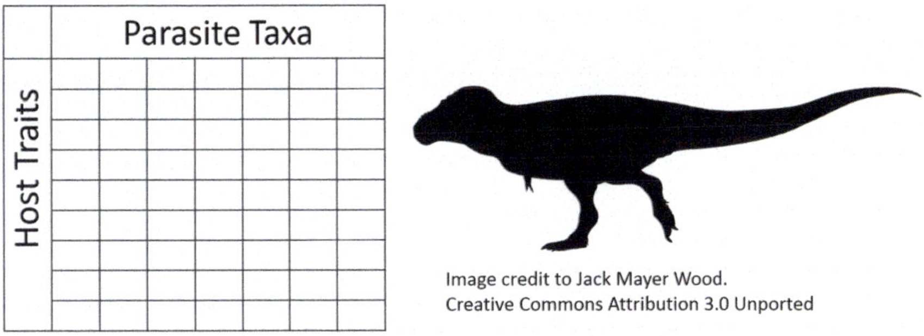

Tyrannosaurus rex

	Cestodes	Nematodes	Trematodes	Lice	Fleas	Ticks	Monogeneans	Parasitic Copepods	Parasitic Isopods
Diet: Carnivorous	✓	✓	✓	—	—	—	—	—	—
Habitat: Plains (terrestrial)	—	—	—	—	—	—	✗	✗	✗
Integumentary Cover: Absent or sparse (Bell et al. 2017)	—	—	—	✗	✗	—	—	—	—
Physiology: Homeothermic	—	—	—	✓	✓	—	—	—	—
Taxonomic Group: Archosaur	—	—	—	—	—	—	—	✗	—
Geological Age: Late Cretaceous	—	—	—	✗	✗	—	—	—	—

Fig. 1.2 A method for using host characteristics to infer the potential parasite community of a fossil taxa. Here *Tyrannosaurus rex* is used as an example, and the parasite taxa represented are based on the groups discussed in this chapter. Each row represents a given host characteristic, while each column represents a different parasite taxon. Check mark denotes that the characteristic predisposes the host to being infected with that given parasite type. Dash denotes the characteristic having neutral or no effect on the likelihood of harbouring such parasite type. Cross denotes the characteristic would exclude such parasite taxa as having occurred on/in the host. A single cross associated with any of the host characteristics will conclusively exclude the existence of said parasite taxon from the host taxon, even if other characteristics may theoretically predispose the host to it

Fig. 1.2). The focal host taxa can be as broad as an entire class or as narrow as a single species. The same also applies to the parasite taxa being examined which can be as broad as different phyla or as specific as different families within a given order (for example, comparing different families of animal-infecting nematodes which have different life-cycles and transmission pathways).

Through a combination of EPB and Ecological Fitting, one can deduce a list of potential parasites which are likely to have associated with a given extinct vertebrate taxa, and based on that, guide potential searches for fossils of said parasites whether they be in amber, coprolite, or compression fossils. Amber is an exceptional material that can preserve many organisms intact and in some cases, in their original ecological context (Poinar 1992), including some parasite-host relationships as shown through the examples discussed in this chapter. However, researchers must be mindful that amber is subjected to preservation biases (e.g. De Baets et al. 2021b; Kraemer et al. 2018), which provides an incomplete picture of the range of interactions which existed in a given palaeoenvironment. For coprolites, given the range of palaeoecology information that can be derived from studying the coprolites of vertebrates (Chin 2002, 2021; Bajdek et al. 2016), it would be advisable for researchers to also be on the lookout for fossils of parasite propagules while investigating such material.

1.4 Future Research Directions and Further Questions

Fossils can provide vital insights into the evolution of parasitism (De Baets and Littlewood 2015; De Baets et al. 2021b). Based on molecular phylogeny and fossils of parasites, it seems that major events in the history of parasite groups were often associated with diversification and extinction events of their vertebrate host taxa. Future fossil discoveries can address some key questions about evolution of parasites which are found on vertebrate hosts, and can in turn be used to help calibrate molecular studies on the phylogeny of various parasite groups. Many different taxa have independently evolved some kind of parasitic life style and often the parasitic representatives have heavily derived morphology that differ very significantly from their closest living non-parasitic relatives. Fossil parasites might provide insight into the transition between free-living and parasitism, both in terms of changes in their morphology and pattern of host or resource usage.

There are some fossil parasites that were discovered or described incidentally such as the case of Zangerl and Case (1976), where microfossils of what appears to be cestode eggs were found amidst the coprolitic material associated with the lower intestinal tract of a Carboniferous elasmobranch. Re-examination of older specimens using more recent technology such as microCT scans and other techniques (e.g., Maas 2013) may yield further discoveries. At the same time, microfossils of parasite materials such as helminth eggs or monogenean hooks may become inadvertently dislodged or destroyed during some preparation processes, thus any prospective palaeoparasitologists should make use of preparation and analysis methods that minimise disruption of the original material.

When examining, describing, and identifying potential fossil parasites, the study by Poinar et al. (2017) on the trematode metacercaria embedded in the leg of an amber-preserved lizard can be considered as an example of "best practice". The fossil they described was of the type of deposit/material (in this case, amber) that is known for preserving microscopic details of small organisms required for identifying parasites. They used the appropriate methodology for examining the fossil which allows them to closely examine and discern fine details required for identification (in this case, X-ray microcomputed tomography), and they subjected a comparable living specimen to the same examination method to check if they obtained similar or comparable results to what was found with the fossil.

Prospective palaeoparasitologists can take a pro-active approach to the study of fossil parasites by examining fossils of potential hosts for signs of parasitism based on what is known in extant parasite-host systems. As discussed above, extant fish are known to host copepods, isopods, and monogenean parasites on various parts of their body with some infecting their skin, fins, and eyes, while other are located under the gill covers or within the body cavity. Therefore, well-preserved fish fossils should be examined at those parts for remains of potential parasites using a combination of microscopy and microCT scans. Additionally, some host taxa and characteristics are associated with certain types of parasites. Among extant terrestrial vertebrates, integumentary covering such as feathers or fur are associated with parasitic arthropods as they provide shelter for ectoparasites, therefore, well-preserved fossil feathers and fur may also be a source for ectoparasite fossils which may be revealed through careful examination, as suggested by Peñalver et al. (2017). Likewise examining faecal samples is common practice in parasitology to determine the parasite communities of various animals, this can also be applied to coprolites which should be examined for parasite egg inclusions, as suggested by Chin (2021) and Francischini et al. (2018).

Conversely, fossil parasites can also be used to infer aspects of their host's ecology which might not be evident solely through their fossil remains. For example, as discussed in the section pertaining to nematodes, the presence of pinworm eggs as inclusions in coprolite can provide valuable insight into the ecology, physiology, and even social behaviour of the animal from which the coprolite originated. In a more general sense, because some parasites have complex life-cycles and are transmitted via ingestion of specific prey animals, the presence of a parasite species (or their eggs in host faecal matter) can be an indicator of the host's diet. Thus, vertebrate palaeontologists should also be investigating fossil parasites as their presence can reveal important information about the animals that they lived on/in.

Beyond providing a better understanding of how parasitism evolved, fossil parasites can also provide a more complete picture of palaeoecosystems and the ecology of extinct animals. Further research on fossil parasites will provide us with a new understanding and appreciation for the roles that parasites played in ancient ecosystems.

Acknowledgements I would like to thank Kenneth De Baets for inviting me to contribute to this volume and his helpful comments and suggestions on the initial drafts of this chapter. The silhouettes for Figs. 1.1 and 1.2 were obtained from PhyloPic (http://phylopic.org/).

References

Adamson M (1994) Evolutionary patterns in life histories of Oxyurida. Int J Parasitol 24(8):1167–1177

Agosta SJ, Klemens JA (2008) Ecological fitting by phenotypically flexible genotypes: implications for species associations, community assembly and evolution. Ecol Lett 11(11):1123–1134

Anderson RC (2000) Nematode parasites of vertebrates: their development and transmission, 2nd edn. CABI Publishing, New York, NY

Bajdek P, Qvarnström M, Owocki K, Sulej T, Sennikov AG, Golubev VK, Niedźwiedzki G (2016) Microbiota and food residues including possible evidence of pre-mammalian hair in Upper Permian coprolites from Russia. Lethaia, 49(4): 455–477

Bell PR, Campione NE, Persons WS IV, Currie PJ, Larson PL, Tanke DH, Bakker RT (2017) Tyrannosauroid integument reveals conflicting patterns of gigantism and feather evolution. Biol Lett 13(6):20170092

Bernard A, Lécuyer C, Vincent P, Amiot R, Bardet N, Buffetaut E, Cuny G, Fourel F, Martineau F, Mazin J-M, Prieur A (2010) Regulation of body temperature by some Mesozoic marine reptiles. Science 328(5984):1379–1382

Blaxter M, Koutsovoulos G (2015) The evolution of parasitism in Nematoda. Parasitology 142(S1):S26–S39

Blaxter ML, De Ley P, Garey JR, Liu LX, Scheldeman P, Vierstraete A, Vanfleteren JR, Mackey LY, Dorris M, Frisse LM, Vida JT, Thomas WK (1998) A molecular evolutionary framework for the phylum Nematoda. Nature 392(6671):71

Boucot AJ, Poinar GO Jr (2011) Fossil behavior compendium. CRC Press, London

Boxshall G (2005) Copepoda. In: Rohde K (ed) Marine parasitology. CSIRO Publishing, Collingwood, VIC, pp 123–138

Brinkkemper O, van Haaster H (2012) Eggs of intestinal parasites whipworm (Trichuris) and mawworm (Ascaris): non-pollen palynomorphs in archaeological samples. Rev Palaeobot Palynol 186:16–21

Bryant HN, Russell AP (1992) The role of phylogenetic analysis in the inference of unpreserved attributes of extinct taxa. Philos Trans R Soc Lond B 337:405–418

Caira JN, Jensen K (2014) A digest of elasmobranch tapeworms. J Parasitol 100(4):373–391

Caira JN, Jensen K, Waeschenbach A, Olson PD, Littlewood DTJ (2014) Orders out of chaos–molecular phylogenetics reveals the complexity of shark and stingray tapeworm relationships. Int J Parasitol 44(1):55–73

Campião KM, de Aquino Ribas AC, Cornell SJ, Begon M, Tavares LER (2015) Estimates of coextinction risk: how anuran parasites respond to the extinction of their hosts. Int J Parasitol 45(14):885–889

Cardia DFF, Bertini RJ, Camossi LG, Letizio LA (2018) The first record of Ascaridoidea eggs discovered in Crocodyliformes hosts from the Upper Cretaceous of Brazil. Revista Brasileira de Paleontologia 21(3):238–244

Cardia DF, Bertini RJ, Camossi LG, Letizio LA (2019a) Two new species of ascaridoid nematodes in Brazilian Crocodylomorpha from the upper cretaceous. Parasitol Int 72:101947. https://doi.org/10.1016/j.parint.2019.101947

Cardia DF, Bertini RJ, Camossi LG, Letizio LA (2019b) First record of Acanthocephala parasites eggs in coprolites preliminary assigned to Crocodyliformes from the Adamantina Formation (Bauru Group, Upper Cretaceous), São Paulo, Brazil. An Acad Bras Cienc 91(S2):e20170848

Chen J, Wang B, Engel MS, Wappler T, Jarzembowski EA, Zhang H, Wang X, Zheng X, Rust J (2014) Extreme adaptations for aquatic ectoparasitism in a Jurassic fly larva. elife 3:e02844

Chin K (2002) Analyses of coprolites produced by carnivorous vertebrates. Paleontol Soc Pap 8:43–50

Chin K (2021) Gastrointestinal parasites of ancient non-human vertebrates: evidence from coprolites and other materials. In: De Baets K, Huntley JW (eds) The evolution and fossil record of parasitism: Coevolution and paleoparasitological techniques. Topics in Geobiology 50

Chitimia-Dobler L, De Araujo BC, Ruthensteiner B, Pfeffer T, Dunlop JA (2017) Amblyomma birmitum a new species of hard tick in Burmese amber. Parasitology 144(11):1441–1448
Clack JA (2002) Gaining ground: the origin and evolution of tetrapods. Indiana University Press, Bloomington, IN
Cressey R, Boxshall G (1989) *Kabatarina pattersoni*, a fossil parasitic copepod (Dichelesthiidae) from a Lower Cretaceous fish. Micropaleontology 35:150–167
Cressey R, Patterson C (1973) Fossil parasitic copepods from a Lower Cretaceous fish. Science 180:1283–1285
Cribb TH, Bray RA, Littlewood DTJ, Pichelin SP, Herniou EA (2001) The digenea. In: Littlewood DTJ, Bray RA (eds) Interrelationships of the platyhelminthes. Taylor and Francis, London, pp 168–185
Cribb TH, Bray RA, Olson PD, Timothy D, Littlewood J (2003) Life cycle evolution in the Digenea: a new perspective from phylogeny. Adv Parasitol 54:197–254
Da Silva PA, Borba VH, Dutra JM, Leles D, Da-Rosa AA, Ferreira LF, Araujo A (2014) A new ascarid species in cynodont coprolite dated of 240 million years. An Acad Bras Cienc 86:265–269
De Baets K, Klug C, Korn D (2011) Devonian pearls and ammonoid-endoparasite co-evolution. Acta Palaeontol Pol 56(1):159–180
De Baets K, Littlewood DTJ (2015) The importance of fossils in understanding the evolution of parasites and their vectors. Adv Parasitol 90:1–51
De Baets K, Dentzien-Dias P, Upeniece I, Verneau O, Donoghue PC (2015) Constraining the deep origin of parasitic flatworms and host-interactions with fossil evidence. Adv Parasitol 90:93–135
De Baets K, Dentzien-Dias P, Harrison GWM, Littlewood DTJ, Parry LA (2021a) Fossil Constraints on the Timescale of Parasitic Helminth Evolution. In: De Baets K, Huntley JW (eds) The evolution and fossil record of parasitism: Identification and macroevolution of parasites. Topics in Geobiology 49
De Baets K, Huntley JW, Klompmaker AA, Schiffbauer JD, Muscente AD (2021b) The fossil record of parasitism: Its extent and taphonomic constraints. In: De Baets K, Huntley JW (eds) The evolution and fossil record of parasitism: Coevolution and paleoparasitological techniques. Topics in Geobiology 50
Dentzien-Dias PC, Poinar G Jr, de Figueiredo AEQ, Pacheco ACL, Horn BL, Schultz CL (2013) Tapeworm eggs in a 270 million-year-old shark coprolite. PLoS One 8(1):e55007
Dhouailly D, Godefroit P, Martin T, Nonchev S, Caraguel F, Oftedal O (2017) Getting to the root of scales, feather and hair: as deep as odontodes? Exp Dermatol 28:503. https://doi.org/10.1111/exd.13391
Dittmar K, Zhu Q, Hastriter MW, Whiting MF (2015) Evolutionary history of Siphonaptera: fossils, origins, vectors. In: Morand S, Krasnov BR, Littlewood DTJ (eds) Parasite diversity and diversification: evolutionary ecology meets phylogenetics. Cambridge University Press, Cambridge, pp 230–245
Dittmar K, Zhu Q, Hastriter MW, Whiting MF (2016) On the probability of dinosaur fleas. BMC Evol Biol 16(1):9
Dufour B, Le Bailly M (2013) Testing new parasite egg extraction methods in paleoparasitology and an attempt at quantification. Int J Paleopathol 3(3):199–203
Dunne EM, Close RA, Button DJ, Brocklehurst N, Cashmore DD, Lloyd GT, Butler RJ (2018) Diversity change during the rise of tetrapods and the impact of the 'Carboniferous rainforest collapse'. Proc R Soc B 285(1872):20172730
Dunlop J (2021) Chelicerates as parasites. In: De Baets K, Huntley JW (eds) The evolution and fossil record of parasitism: Identification and macroevolution of parasites. Topics in Geobiology 49
Estrada-Peña A, de la Fuente J (2018) The fossil record and the origin of ticks revisited. Exp Appl Acarol 75(2):255–261
Field DJ, Bercovici A, Berv JS, Dunn R, Fastovsky DE, Lyson TR, Vajda V, Gauthier JA (2018) Early evolution of modern birds structured by global forest collapse at the end-Cretaceous mass extinction. Curr Biol 28(11):1825–1831

Francischini H, Dentzien-Dias P, Schultz CL (2018) A fresh look at ancient dungs: the Brazilian Triassic coprolites revisited. Lethaia 51:389–405. https://doi.org/10.1111/let.12251
Friedman M, Sallan LC (2012) Five hundred million years of extinction and recovery: a Phanerozoic survey of large-scale diversity patterns in fishes. Palaeontology 55(4):707–742
Galaktionov KV, Dobrovolskij AA (2003) The biology and evolution of trematodes. Kluwer Academic Publishers, Dordrecht
Gao TP, Shih CK, Xu X, Wang S, Ren D (2012) Mid-Mesozoic flea-like ectoparasites of feathered or haired vertebrates. Curr Biol 22(8):732–735
Gonçalves MLC, Araújo A, Ferreira LF (2003) Human intestinal parasites in the past: new findings and a review. Mem Inst Oswaldo Cruz 98:103–118
Grimaldi D (1999) The co-radiations of pollinating insects and angiosperms in the Cretaceous. Ann Mo Bot Gard 86:373–406
Hata H, Sogabe A, Tada S, Nishimoto R, Nakano R, Kohya N, Takeshima H, Kawanishi R (2017) Molecular phylogeny of obligate fish parasites of the family Cymothoidae (Isopoda, Crustacea): evolution of the attachment mode to host fish and the habitat shift from saline water to freshwater. Mar Biol 164:105. https://doi.org/10.1007/s00227-017-3138-5
Hatcher MJ, Dunn AM (2011) Parasites in ecological communities: from interactions to ecosystems. Cambridge University Press, New York, NY
Herlyn H (2021) Thorny-headed Worms (Acanthocephala): Jaw-less Members of Jaw-bearing Worms that Parasitize Jawed Arthropods and Jaw-bearing Vertebrates. In: De Baets K, Huntley JW (eds) The evolution and fossil record of parasitism: Identification and macroevolution of parasites. Topics in Geobiology 49.
Hernández-Orts JS, Brandão M, Georgieva S, Raga JA, Crespo EA, Luque JL, Aznar FJ (2017) From mammals back to birds: host-switch of the acanthocephalan Corynosoma australe from pinnipeds to the Magellanic penguin Spheniscus magellanicus. PLoS One 12(10):e0183809
Ho JS (2001) Why do symbiotic copepods matter? Hydrobiologia 453(1):1–7
Hoberg EP, Brooks DR (2008) A macroevolutionary mosaic: episodic host-switching, geographical colonization and diversification in complex host–parasite systems. J Biogeogr 35(9):1533–1550
Huang D, Nel A, Cai C, Lin Q, Engel MS (2013) Amphibious flies and paedomorphism in the Jurassic period. Nature 495(7439):94–97
Haug JT, Haug C, Nagler C (2021) Evolutionary history of crustaceans as parasites. In: De Baets K, Huntley JW (eds) The evolution and fossil record of parasitism: Identification and macroevolution of parasites. Topics in Geobiology 49
Hugot JP, Gardner SL, Borba V, Araujo P, Leles D, Da-Rosa ÁAS, Dutra J, Ferreira LF, Araújo A (2014) Did the dinosaurs have pinworms? Discovery of a 240 million year old oxyurid nematode parasite egg sheds light on the early origin of nematode parasitism in vertebrates. Parasit Vectors 7:486. https://doi.org/10.1186/s13071-014-0486-6
Humes A (1994) How many copepods? Hydrobiologia 292(293):1–7
Huntley JW, De Baets K (2015) Trace fossil evidence of trematode—bivalve parasite—host interactions in deep time. Adv Parasitol 90:201–231
Huntley JW, De Baets K, Scarponi D, Linehan LC, Epa YR, Jacobs GS, Todd JA (2021) Bivalve Mollusks as Hosts in the Fossil Record. In: De Baets, K., Huntley, J. W. (eds). The Evolution and Fossil Record Of Parasitism: Coevolution and Paleoparasitological Techniques. Topics in Geobiology 50
Ituarte C, Cremonte F, Zelaya DG (2005) Parasite-mediated shell alterations in Recent and Holocene sub-Antarctic bivalves: the parasite as modeler of host reaction. Invertebr Biol 124(3):220–229
Janzen DH (1985) On ecological fitting. Oikos 45(3):308–310
Johnson KP, Nguyen NP, Sweet AD, Boyd BM, Warnow T, Allen JM (2018) Simultaneous radiation of bird and mammal lice following the K-Pg boundary. Biol Lett 14(5):20180141. https://doi.org/10.1098/rsbl.2018.0141

Jouy-Avantin F, Combes C, Miskovsky JC, Moné H (1999) Helminth eggs in animal coprolites from a Middle Pleistocene site in Europe. J Parasitol 85:376–379
Júnior JCR, Pfaller JB, Corbetta R, Veríssimo L (2015) Parasitic isopods associated with sea turtles nesting in Brazil. J Mar Biol Assoc U K 95(5):973–981
Kabata Z (1982) Copepoda (Crustacea) parasitic on fishes: problems and perspectives. Adv Parasitol 19:1–71
Kearn GC (2004) Leeches, lice and lamprey. A natural history of skin and gill parasites of fishes. Springer, Dordrecht
Kelley NP, Motani R (2015) Trophic convergence drives morphological convergence in marine tetrapods. Biol Lett 11(1):20140709. https://doi.org/10.1098/rsbl.2014.0709
Kennedy CR (2006) Ecology of the acanthocephala. Cambridge University Press, New York, NY
Klompen JSH, Black W, Keirans JE, Oliver JH Jr (1996) Evolution of ticks. Annu Rev Entomol 41(1):141–161
Klompmaker AA, Boxshall GA (2015) Fossil crustaceans as parasites and hosts. Adv Parasitol 90:233–289
Klompmaker AA, Artal P, Van Bakel BW, Fraaije RH, Jagt JW (2014) Parasites in the fossil record: a Cretaceous fauna with isopod-infested decapod crustaceans, infestation patterns through time, and a new ichnotaxon. PLoS One 9:e92551. https://doi.org/10.1371/journal.pone.0092551
Klug C, Frey L, Pohle A, De Baets K, Korn D (2017) Palaeozoic evolution of animal mouthparts. Bull Geosci 92(4):511–524. https://doi.org/10.3140/bull.geosci.1648
Kraemer MMS, Delclòs X, Clapham ME, Arillo A, Peris D, Jäger P, Stebner F, Peñalver E (2018) Arthropods in modern resins reveal if amber accurately recorded forest arthropod communities. Proc Natl Acad Sci 115:6739-6744
Labandeira CC, Dilcher DL, Davis DR, Wagner DL (1994) Ninety-seven million years of angiosperm-insect association: paleobiological insights into the meaning of coevolution. Proc Natl Acad Sci 91(25):12278–12282
Lafferty KD (1999) The evolution of trophic transmission. Parasitol Today 15:111–115
Lafferty KD, Dobson AP, Kuris AM (2006) Parasites dominate food web links. Proc Natl Acad Sci 103(30):11211–11216
Lavrov DV, Brown WM, Boore JL (2004) Phylogenetic position of the Pentastomida and (pan) crustacean relationships. Proc R Soc Lond B Biol Sci 271(1538):537–544
Le Bailly M, Bouchet F (2010) Ancient dicrocoeliosis: occurrence, distribution and migration. Acta Trop 115(3):175–180
Leung TLF (2017) Fossils of parasites: what can the fossil record tell us about the evolution of parasitism? Biol Rev 92(1):410–430
Li L, Lü L, Nadler SA, Gibson DI, Zhang LP, Chen HX, Zhao WT, Guo YN (2018) Molecular phylogeny and dating reveal a terrestrial origin in the early carboniferous for ascaridoid nematodes. Syst Biol 67:888–900
Liljedahl L (1985) Ecological aspects of a silicified bivalve fauna from the Silurian of Gotland. Lethaia 18(1):53–66
Littlewood DTJ, Donovan SK (2003) Feature: fossil parasites: a case of identity. Geol Today 19(4):136–142
Littlewood DTJ, Bray RA, Waeschenbach A (2015) Phylogenetic patterns of diversity in cestodes and trematodes. In: Morand S, Krasnov BR, Littlewood DTJ (eds) Parasite diversity and diversification: evolutionary ecology meets phylogenetics. Cambridge University Press, Cambridge, pp 304–319
Loker E, Hofkin B (2015) Parasitology: a conceptual approach. Garland Science, New York, NY
Lukashevich ED, Mostovski MB (2003) Hematophagous insects in the fossil record. Paleontol J 37:153–161
Maas A (2013) Gastrotricha, cycloneuralia and gnathifera: the fossil record. In: Schmidt-Rhaesa A (ed) Handbook of zoology. Gastrotricha, Cycloneuralia and Gnathifera. 1, Nematomorpha, Priapulida, Kinorhyncha, Loricifera. de Gruyter, Berlin, pp 11–28
Mackiewicz JS (1988) Cestode transmission patterns. J Parasitol 74:60–71

Mans BJ, de Klerk D, Pienaar R, de Castro MH, Latif AA (2012) The mitochondrial genomes of *Nuttalliella namaqua* (Ixodoidea: Nuttalliellidae) and *Argas africolumbae* (Ixodoidae: Argasidae): estimation of divergence dates for the major tick lineages and reconstruction of ancestral blood-feeding characters. PLoS One 7(11):e49461

Médoc V, Rigaud T, Motreuil S, Perrot-Minnot MJ, Bollache L (2011) Paratenic hosts as regular transmission route in the acanthocephalan Pomphorhynchus laevis: potential implications for food webs. Naturwissenschaften 98(10):825

Miller DM, Dunagan TT (1985) Functional morphology. In: Crompton DWT, Nickol BB (eds) Biology of the acanthocephala. Cambridge University Press, Cambridge, pp 73–123

Misof B, Liu S, Meusemann K, Peters RS, Donath A, Mayer C, Frandsen PB, Ware J, Flouri T, Beutel RG, Niehuis O, Peterson M, Izquierdo-Carrasco F, Wappler T, Rust J, Aberer AJ, Aspöck U, Aspöck H, Bartel D, Blanke A, Berger S, Böhm A, Buckley TR, Calcott B, Chen J, Friedrich F, Fukui M, Fujita M, Greve C, Grobe P, Gu S, Huang Y, Jermiin L, Kawahara AY, Krogmann L, Kubiak M, Lanfear R, Letsch H, Li Y, Li Z, Li J, Lu H, Machida R, Mashimo Y, Kapli P, McKenna DD, Meng G, Nakagaki Y, Navarrete-Heredia JL, Ott M, Ou Y, Pass G, Podsiadlowski L, Pohl H, von Reumont B, Schutte K, Sekiya K, Shimizu S, Slipinksi A, Stamatakis A, Song W, Su X, Szucsich NU, Tan M, Tan X, Tang M, Uchifune T, Walzl MG, Wiegmann BM, Wilbrandt J, Wipfler B, Wong TKF, Wu Q, Wu G, Xie Y, Yang S, Yang Q, Yeates DK, Yoshizawa K, Zhang Q, Zhang R, Zhang W, Zhang Y, Zhao J, Zhou C, Zhou L, Ziesmann T, Zou S, Li Y, Xu X, Zhang Y, Yang H, Wang J, Wang J, Kjer KM, Zhou X (2014) Phylogenomics resolves the timing and pattern of insect evolution. Science 346(6210):763–767

Nagler C, Haug JT (2015) From fossil parasitoids to vectors: insects as parasites and hosts. Adv Parasitol 90:137–200

Nagler C, Haug JT (2016) Functional morphology of parasitic isopods: understanding morphological adaptations of attachment and feeding structures in Nerocila as a pre-requisite for reconstructing the evolution of Cymothoidae. PeerJ 4:e2188. https://doi.org/10.7717/peerj.2188

Nagler C, Haug C, Resch UKJ, Haug JT (2016) 150 million years old isopods on fishes: a possible case of palaeo-parasitism. Bull Geosci 91(1):1–12. https://doi.org/10.3140/bull.geosci.1586

Nagler C, Hyžný M, Haug JT (2017) 168 million years old "marine lice" and the evolution of parasitism within isopods. BMC Evol Biol 17(1):76. https://doi.org/10.1186/s12862-017-0915-1

Nickol BB (1985) Epizootiology. In: Crompton DWT, Nickol BB (eds) Biology of the acanthocephala. Cambridge University Press, Cambridge, pp 307–346

Olson PD, Caira JN, Jensen K, Overstreet RM, Palm HW, Beveridge I (2010) Evolution of the trypanorhynch tapeworms: parasite phylogeny supports independent lineages of sharks and rays. Int J Parasitol 40(2):223–242

Ozanne CR, Harries PJ (2002) Role of predation and parasitism in the extinction of the inoceramid bivalves: an evaluation. Lethaia 35(1):1–19

Padian K, Sues H-S (2015) The age of transformation: the Triassic period and the rise of today's land vertebrate fauna. In: Dial KP, Shubin N, Brainerd EL (eds) Great transformations: major events in the history of vertebrate life. University of Chicago Press, Chicago, IL, pp 351–374

Parker GA, Ball MA, Chubb JC (2015) Evolution of complex life cycles in trophically transmitted helminths. I. Host incorporation and trophic ascent. J Evol Biol 28(2):267–291

Peñalver E, Arillo A, Delclòs X, Peris D, Grimaldi DA, Anderson SR, Nascimbene PC, Pérez-de la Fuente R (2017) Ticks parasitised feathered dinosaurs as revealed by Cretaceous amber assemblages. Nat Commun 8(1):1924. https://doi.org/10.1038/s41467-017-01550-z

Poinar G (2019) Vertebrate pathogens vectored by ancient hematophagous arthropods. Hist Biol 32(7):888-901. https://doi.org/10.1080/08912963.2018.1545018

Poinar G, Boucot AJ (2006) Evidence of intestinal parasites of dinosaurs. Parasitology 133(2):245–249

Poinar GO, Buckley R (2008) Compluriscutula vetulum (Acari: Ixodida: Ixodidae), a new genus and species of hard tick from Lower Cretaceous Burmese amber. Proc Entomol Soc Wash 110(2):445–450

Poinar GO Jr (1983) The natural history of nematodes. Prentice Hall Inc, Upper Saddle River, NJ

Poinar GO Jr (1992) Life in amber. Stanford University Press, Stanford, CA
Poinar G Jr (2010) Palaeoecological perspectives in Dominican amber. Annales de la Société entomologique de France 46(1–2):23–52
Poinar GO Jr (2015) The geological record of parasitic nematode evolution. Adv Parasitol 90:53–92
Poinar G Jr, Philbrick KA, Cohn MJ, Turner RT, Iwaniec UT, Wunderlich J (2017) X-ray micro-computed tomography reveals putative trematode metacercaria in a 100 million year-old lizard (Squamata: Agamidae). Cretac Res 80:27–30
Ponomarenko AG (1976) A new insect from the Cretaceous of Transbaikalia Ussr a possible parasite of pterosaurians. Paleontol Zh 3:102–106
Poulin R (2005) Evolutionary trends in body size of parasitic flatworms. Biol J Linn Soc 85(2):181–189
Poulin R (2007) Evolutionary ecology of parasites. Princeton University Press, Princeton, NJ
Poulin R, Morand S (2004) Parasite biodiversity. Smithsonian Books, Washington, DC
Rasnitsyn AP (1992) *Strashila incredibilis*, a new enigmatic mecopteroid insect with possible siphonapteran affinities from the Upper Jurassic of Siberia. Psyche 99(4):323–333
Rasnitsyn AP, Strelnikova OD (2017) Tracheal system and biology of the Early Cretaceous Saurophthirus longipes Ponomarenko, 1976 (Insecta,? Aphaniptera, Saurophthiroidea stat. nov.). Paleontol J 51(2):171–182
Rasnitsyn AP, Strelnikova OD (2018) Digestive system of the early cretaceous insect Saurophthirus longipes Ponomarenko (Insecta,? Aphaniptera, Saurophthiroidea). Paleontol J 52(2):146–154
Riley J (1986) The biology of pentastomids. Adv Parasitol 25:45–128
Robin N, Béthoux O, Sidorchuk E, Cui Y, Li Y, Germain D, King A, Berenguer F, Ren D (2016) A Carboniferous mite on an insect reveals the antiquity of an inconspicuous interaction. Curr Biol 26(10):1376–1382
Robin N, Marramà G, Vonk R, Kriwet J, Carnevale G (2019) Eocene isopods on electric rays: tracking ancient biological interactions from a complex fossil record. Palaeontology 62:287-303. https://doi.org/10.1111/pala.12398
Rogers RR, Curry Rogers KA, Bagley BC, Goodin JJ, Hartman JH, Thole JT, Zatoń M (2018) Pushing the record of trematode parasitism of bivalves upstream and back to the Cretaceous. Geology 46(5):431–434
Ruiz GM, Lindberg DR (1989) A fossil record for trematodes: extent and potential uses. Lethaia 22(4):431–438
Sanders KL, Lee MS (2010) Arthropod molecular divergence times and the Cambrian origin of pentastomids. Syst Biodivers 8:63–74
Saravanakumar A, Balasubramanian T, Raja K, Trilles JP (2012) A massive infestation of sea snakes by cymothoid isopods. Parasitol Res 110(6):2529–2531
Shcherbakov DE (2017) Cretaceous Saurophthiridae (Aphaniptera) as pupiparous pre-fleas of diving pterosaurs. Paleontol J 51(2):183–185
Sinisalo T, Valtonen ET (2003) Corynosoma acanthocephalans in their paratenic fish hosts in the northern Baltic Sea. Parasite 10(3):227–233
Siveter DJ, Briggs DE, Siveter DJ, Sutton MD (2015) A 425-million-year-old Silurian pentastomid parasitic on ostracods. Curr Biol 25(12):1632–1637
Smit NJ, Davies AJ (2004) The curious life-style of the parasitic stages of gnathiid isopods. Adv Parasitol 58:289–391
Smit NJ, Bruce NL, Hadfield KA (2014) Global diversity of fish parasitic isopod crustaceans of the family Cymothoidae. Int J Parasitol Parasit Wildl 3(2):188–197
Strona G (2015) Past, present and future of host–parasite co-extinctions. Int J Parasitol Parasit Wildl 4(3):431–441
Stunkard HW (1983) Evolution and systematics. In: Arme C, Pappas PW (eds) Biology of the eucestoda, vol 1. Academic Press, New York, NY, pp 1–25
Suh A, Witt CC, Menger J, Sadanandan KR, Podsiadlowski L, Gerth M, Weigert A, McGuire JA, Mudge J, Edwards SV, Rheindt FE (2016) Ancient horizontal transfers of retrotransposons

between birds and ancestors of human pathogenic nematodes. Nat Commun 7:11396. https://doi.org/10.1038/ncomms11396

Suh A (2021) Horizontal Transfer of Transposons as Genomic Fossils of Host-Parasite Interactions. In: De Baets K, Huntley JW (eds) The Evolution and Fossil Record of Parasitism: Coevolution and Paleoparasitological Techniques. Topics in Geobiology 50

Thompson RA, Lymbery AJ, Godfrey SS (2018) Parasites at risk–insights from an endangered marsupial. Trends Parasitol 34(1):12–22

Upeniece I (2001) The unique fossil assemblage from the Lode quarry (Upper Devonian, Latvia). Foss Rec 4(1):101–119

Upeniece I (2011) Palaeoecology and juvenile individuals of the Devonian placoderm and acanthodian fishes from Lode site, Latvia. Dissertation, University of Latvia. Available at http://www.lu.lv/fileadmin/user_upload/lu_portal/zinas/Upeniece_Ieva_-_2011_-_promocijas_darba_kopsavilkums.pdf. Accessed 15 May 2018

Valtonen ET, Marcogliese DJ, Julkunen M (2010) Vertebrate diets derived from trophically transmitted fish parasites in the Bothnian Bay. Oecologia 162(1):139. https://doi.org/10.1007/s00442-009-1451-5

Verweyen L, Klimpel S, Palm HW (2011) Molecular phylogeny of the Acanthocephala (class Palaeacanthocephala) with a paraphyletic assemblage of the orders Polymorphida and Echinorhynchida. PLoS One 6(12):e28285. https://dx.doi.org/10.1371%2Fjournal.pone.0028285

Vršanský P, Ren D, Shih C (2010) Nakridletia ord. n.–enigmatic insect parasites support sociality and endothermy of pterosaurs. Amba Projekty 8:1–16

Waage JK (1979) The evolution of insect/vertebrate associations. Biol J Linn Soc 12(3):187–224

Walossek D, Müller KJ (1994) Pentastomid parasites from the Lower Palaeozoic of Sweden. Earth Environ Sci Trans R Soc Edin 85(1):1–37

Walossek D, Repetski JE, Müller KJ (1994) An exceptionally preserved parasitic arthropod, Heymonsicambria taylori n. sp. (Arthropoda incertae sedis: Pentastomida), from Cambrian–Ordovician boundary beds of Newfoundland, Canada. Can J Earth Sci 31(11):1664–1671

Wappler T, Smith VS, Dalgleish RC (2004) Scratching an ancient itch: an Eocene bird louse fossil. Proc R Soc Lond B Biol Sci 271(Suppl 5):S255–S258

Weinstein SB, Kuris AM (2016) Independent origins of parasitism in Animalia. Biol Lett 12(7):20160324. https://doi.org/10.1098/rsbl.2016.0324

Whittington ID (1998) Diversity "down under": monogeneans in the Antipodes (Australia) with a prediction of monogenean biodiversity worldwide. Int J Parasitol 28(10):1481–1493

Williams JD, Boyko CB (2012) The global diversity of parasitic isopods associated with crustacean hosts (Isopoda: Bopyroidea and Cryptoniscoidea). PLoS One 7(4):e35350. https://doi.org/10.1371/journal.pone.0035350

Williams EH, Bunkley-Williams L, Ebert DA (2010) An accidental attachment of Elthusa raynaudii (Isopoda, Cymothoidae) in Etmopterus sp.(Squaliformes, Etmopteridae). Acta Parasitol 55(1):99–101

Wilson GD, Paterson JR, Kear BP (2011) Fossil isopods associated with a fish skeleton from the Lower Cretaceous of Queensland, Australia–direct evidence of a scavenging lifestyle in Mesozoic Cymothoida. Palaeontology 54(5):1053–1068

Witmer LM (1995) The extant phylogenetic bracket and the importance of reconstructing soft tissues in fossils. In: Thomason J (ed) Functional morphology in vertebrate paleontology. Cambridge University Press, New York, NY, pp 19–33

Wood JR, Wilmshurst JM, Rawlence NJ, Bonner KI, Worthy TH, Kinsella JM, Cooper A (2013) A megafauna's microfauna: gastrointestinal parasites of New Zealand's extinct moa (Aves: Dinornithiformes). PLoS One 8(2):e57315. https://doi.org/10.1371/journal.pone.0057315

Xing L, McKellar RC, Xu X, Li G, Bai M, Persons WS IV, Miyashita T, Benton MJ, Zhang J, Wolfe AP, Yi Q, Tseng K, Ran H, Currie PJ (2016) A feathered dinosaur tail with primitive plumage trapped in mid-Cretaceous amber. Curr Biol 26(24):3352–3360

Xing L, O'Connor JK, McKellar RC, Chiappe LM, Tseng K, Li G, Bai M (2017) A mid-Cretaceous enantiornithine (Aves) hatchling preserved in Burmese amber with unusual plumage. Gondwana Res 49:264–277

Zangerl R, Case GR (1976) *Cobelodus aculeatus* (Cope) an anacanthous shark from Pennsylvanian Black Shales of North America. Palaeontogr Abt A 154:107–157

Zhu Q, Hastriter MW, Whiting MF, Dittmar K (2015) Fleas (Siphonaptera) are Cretaceous, and evolved with Theria. Mol Phylogenet Evol 90:129–139

Chapter 2
Fossil Record of Viruses, Parasitic Bacteria and Parasitic Protozoa

George Poinar

Abstract Fossil evidence of ancient pathogens is rare and limited mostly to specimens associated with arthropod vectors in amber. In various fossil vectors is evidence of cytoplasmic and nuclear polyhedrosis viruses, pathogenic spirochetes, rickettsia, actinomycetes and protozoa. Also present in amber are examples of fossil virus-like tumors and indirect evidence of iridoviruses, polydnaviruses, pathogenic bacteria and protozoans. Many of these pathogens resemble those being transmitted today to a range of vertebrate hosts by arthropod vectors. Based on these findings, it is clear that invertebrate and vertebrate viruses, as well as a range of pathogenic bacteria and protozoa, were well established by the mid-Cretaceous.

Keywords Fossils · Polyhedrosis viruses · Iridoviridae · Polydnaviruses · Pathogenic bacteria · Pathogenic protozoa · Cretaceous · Cenozoic

2.1 Introduction

Insects evolved some 400 million years ago and probably became infected with viruses bacteria and protozoa soon after. Over time, these microbes developed various symbiotic associations with invertebrates and vertebrates (De Baets and Littlewood 2015). One phase of the evolution of microbes was their association with hematophagous arthropods which resulted in the introduction of viral, bacterial and protozoan parasites into vertebrates. In searching for records of fossil microbial parasites, it is useful to carefully examine fossils in amber, especially hematophagous arthropods. Due to the excellent preservation qualities of amber, it is one of the most suitable fossil sources for studying pathogenic microbes. This is due to the ability of resin to infiltrate, fix, and dehydrate the inclusions. During these processes, oxygen is restricted and associated decay-causing microorganisms are destroyed. Terpenoids and resin acids are responsible for tissue fixation, while

G. Poinar (✉)
Department of Integrative Biology, Oregon State University, Corvallis, OR, USA
e-mail: poinarg@science.oregonstate.edu

© The Author(s) 2021
K. De Baets, J. W. Huntley (eds.), *The Evolution and Fossil Record of Parasitism*, Topics in Geobiology 49, https://doi.org/10.1007/978-3-030-42484-8_2

sugars and terpines replace water by inert dehydration (Poinar and Hess 1985). During these processes, arthropod tissues are often partially cleared, thus revealing associated microorganisms.

The oldest amber with abundant identifiable insects is 120–130 million year old Lebanese amber (Poinar and Milki 2001). The oldest evidence of vertebrate pathogens occurs in 100 million year-old Myanmar amber. In this amber is evidence of cytoplasmic and nuclear polyhedrosis viruses as well as pathogenic spirochetes, rickettsia, actinomycetes and protozoa in bloodsucking arthropods and indirect evidence of polydnaviruses. In Cenozoic amber are examples of virus-like tumors and a range of pathogenic bacteria. In both Myanmar and Dominican amber is evidence of symbiotic and parasitic protozoa, often inside their arthropod vectors. Based on these findings, it is clear that invertebrate and vertebrate viruses as well as a range of pathogenic bacteria and protozoa were well established by the mid-Cretaceous. By the mid-Cenozoic, these infective agents had spread to various host groups, many of which remain infected today. Ages of the various amber deposits with samples discussed below are 15–45 million years ago for Dominican (Iturralde-Vinent and MacPhee 1996; Cepek in Schlee 1990); 22.5–26 million years ago for Mexican (Berggren and van Couvering 1974); 45–55 million years ago for Baltic (Wolfe et al. 2016); 76.5–79.5 million years ago for Canadian (Eberth and Hamblin 1993) and 97–110 million years ago for Myanmar (Burmese) (Cruickhank and Ko 2003; Shi et al. 2012).

2.2 Virus Fossils

Viruses are quite ancient and based on molecular studies, may date back to the origin of life on earth (Foreteer 2006). And since virus particles are so small, fossil evidence of them is quite rare. Presently, most studies on viral evolution are based on "molecular fossils" that explores endogenous retroviruses (EVRs) that have become established in their host's genome. If these viruses were the result of a single infection event, then phylogenetic screening of related hosts can be used to predict when such infections occurred, thus providing putative dates when the virus was incorporated into the host genome. Using these genomic methods, retrovirus infections have been traced back millions of years (Katzouakis 2014; Lee et al. 2013; Suh 2021).

Other branches of paleovirology involve searching for ancient remains that show physical signs or symptoms of ancient viruses such as in amber fossils. In the present work, fossil evidence of viruses is discussed under "direct evidence" and "indirect evidence." Direct evidence is when polyhedra (inclusion bodies) or virions are recovered in well-preserved fossils. Historic evidence of viruses in humans is usually based on genetic evidence in Pleistocene samples. If the specimens are well preserved, then the viruses can be identified from their remains by genomic techniques, which is how most ancient viruses infecting humans have been found (Duggan et al. 2016), thus providing information on widespread epidemics that occurred during historic times, such as discovering evidence of smallpox in Africa at 3700 BC (Schatmayr 2014).

Indirect evidence of virus infections is when some modification of an organism, such as color or form that has been shown to be the result of an extant virus infection is recovered in the fossil state. Also included in this category are fossil organisms whose present day descendants are known to harbor genomic viruses such as polydnaviruses.

2.2.1 *Direct Evidence of Fossil Viral Infections*

Several groups of insect pathogenic viruses are known as occluded viruses since their virions are embedded in a crystaline protein lattice called a polyhedron or inclusion body. Some insect polyhedra are large enough to be seen with the light microscope and their identification can be based on their shape, size, location and dimensions of the embedded virions. Two viral groups that form relatively large polyhedra visible under the light microscope (Poinar and Thomas 1984) are the cypoviruses (cytoplasmic polyhedrosis viruses or CPV such as Reoviridae) and the nuclear polyhedrosis viruses (NPV such as Baculoviridae). Fossil examples of both of these viral types have been identified in Myanmar amber (Poinar and Poinar 2005), dated to the mid-Cretaceous (Cruickhank and Ko 2003; Shi et al. 2012).

In the midgut cells of a small adult biting midge (Diptera; Ceratopogonidae) (Fig. 2.1) in Myanmar amber were numerous polyhedra of a cypovirus (Figs. 2.2 and 2.3). The polyhedra are irregular (mostly hexagonal and cuboidal) and range

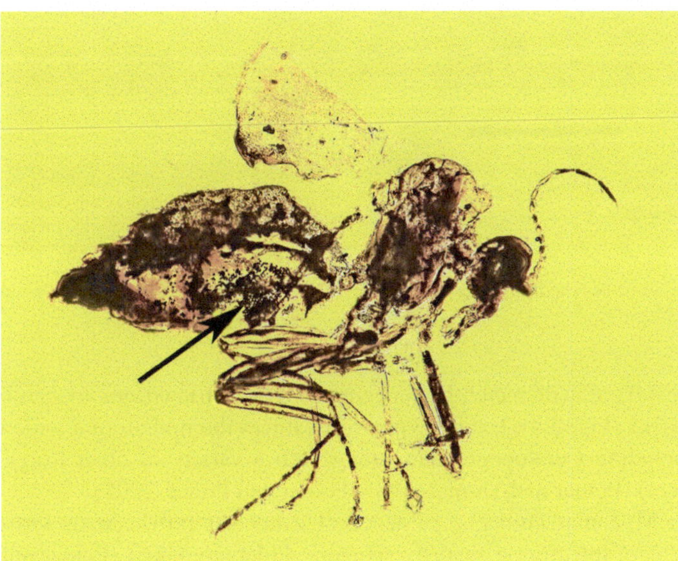

Fig. 2.1 Biting midge (Diptera: Ceratopogonidae) in 100 million years ago Myanmar amber. Arrow shows location of cypovirus polyhedra (CPV) in midgut. Scale bar = 22 μm

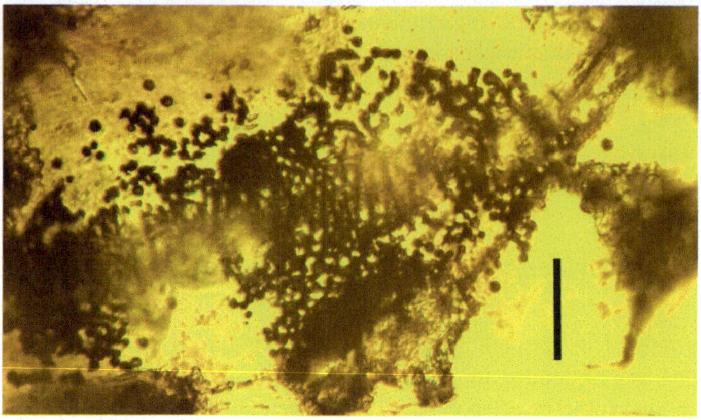

Fig. 2.2 Group of polyhedra with cypoviruses (CPV) in the midgut of the biting midge in Fig. 2.1. Scale bar = 33 μm

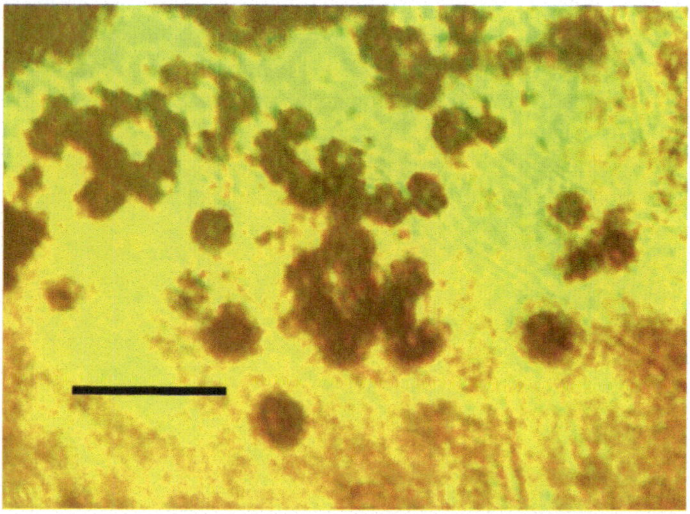

Fig. 2.3 Detail of polyhedra of a cypovirus (CPV) in the midgut of the biting midge in Fig. 2.1. Scale bar = 12 μm

from 0.2 to 3.0 μm in diameter. In some, putative isometric virions are barely visible as dark specks (Fig. 2.4). Extant cypoviruses infect the midgut of a wide range of insects, including Ceratopogonidae, along with a variety of other lower Diptera (Nematocera) (Poinar and Thomas 1984; Poinar and Poinar 2005).

Also in Myanmar amber is evidence of a nuclear polyhedrosis virus (NPV) (Baculoviridae) infecting a sand fly (Diptera: Phlebotomidae) (Poinar and Poinar 2005) (Figs. 2.5 and 2.6). The inclusion bodies (polyhedra) are irregular and in some, putative rod-shaped virions are barely visible as dark rods (Fig. 2.7). Nuclear

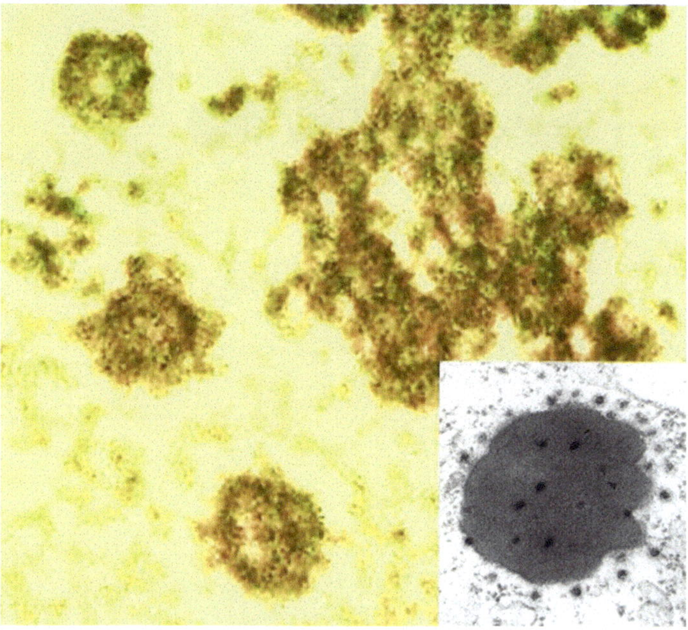

Fig. 2.4 Isometric cypovirus virions (black specks) in polyhedra of the biting midge in Fig. 2.1. ×9000. Insert shows a polyhedrin of an extant cypovirus with embedded isometric virions. ×82,000

Fig. 2.5 Myanmar amber sand fly (Diptera: Phlebotomidae) containing inclusion bodies of a nuclear polyhedrosis virus (NPV). Scale bar = 425 μm

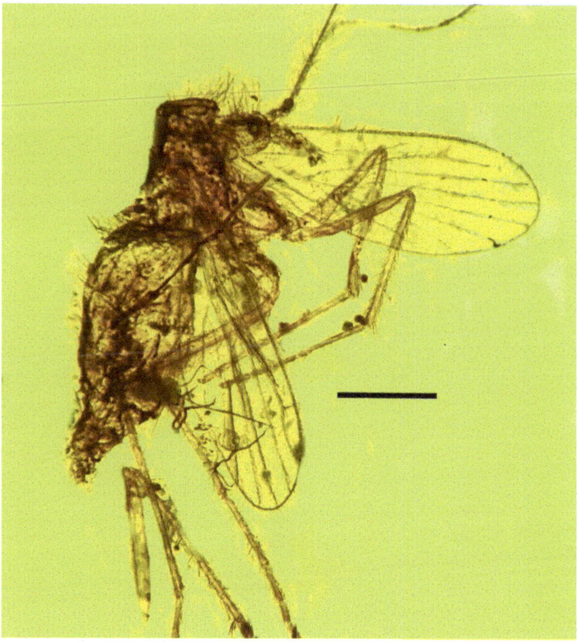

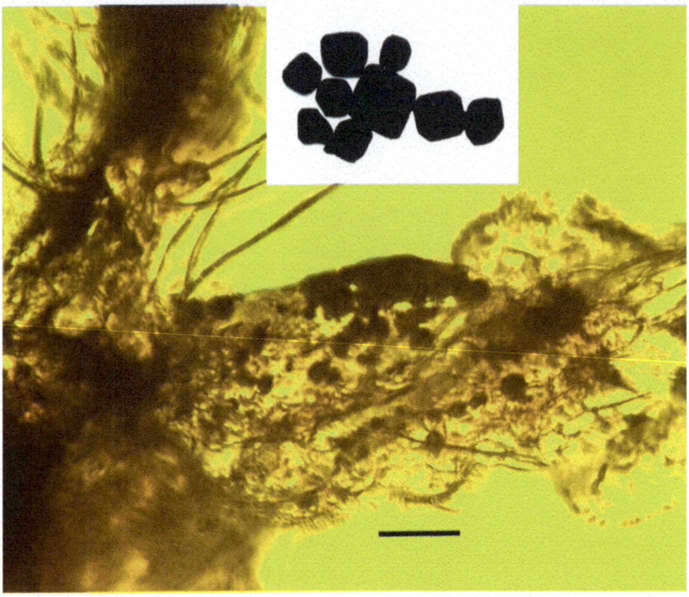

Fig. 2.6 Detail of NPV polyhedra in the proboscis of Myanmar amber sand fly in Fig. 2.5. Scale bar = 6 μm. Insert shows the range of shapes of extant NPV polyhedra. ×2000

Fig. 2.7 Rod-shaped virions (arrowhead) of an NPV in a polyhedron in the Myanmar amber sand fly shown in Fig. 2.5. ×15,000. Insert shows a section of a polyhedron of an extant nuclear polyhedrosis virus with rod-shaped nucleocapsids. ×80,000

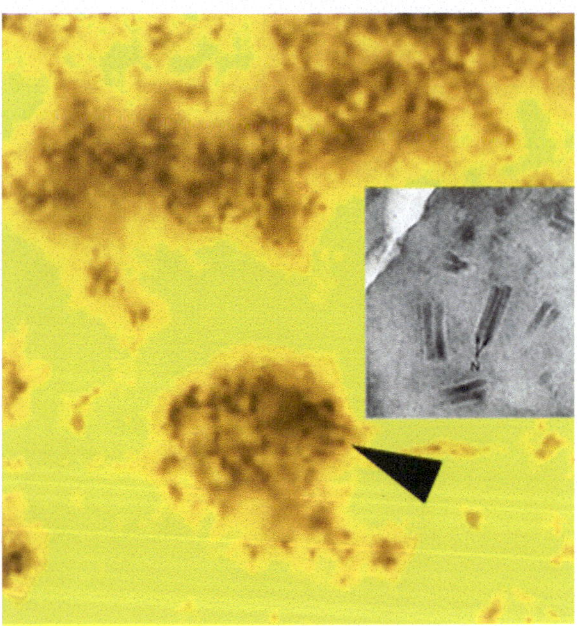

polyhedrosis viruses have been reported from extant lower Diptera (Nematocera) (Poinar and Poinar 2005). These two cases show that cypoviruses and NPV viruses were infecting lower Diptera by the mid-Cretaceous.

2.2.2 *Indirect Evidence of Virus Infections*

2.2.2.1 Parasitic Wasps with Polydnaviruses

In the mid-1970s, researchers first observed that virus particles from the calyx fluid of a parasitic wasp could infect cells of their insect hosts (Poinar et al. 1976) (Fig. 2.8). These enveloped viruses turned out to be a unique group now known as polydnaviruses that are transmitted by female parasitic wasps in several subfamilies of the families Braconidae and Ichneumonidae (Hymenoptera) (Labandeira and Li 2021). The viral genomes are integrated into the wasp genome and virus replication occurs in the reproductive system of the female wasp. During oviposition, the virus

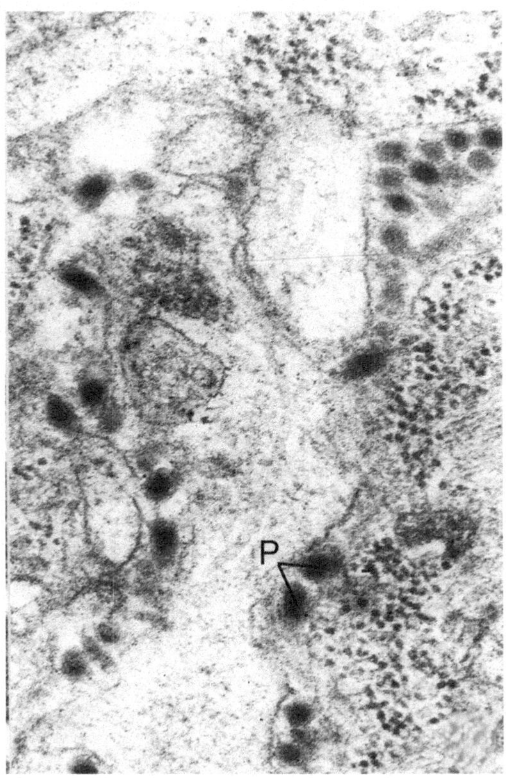

Fig. 2.8 Particles (P) of a polydnavirus originating from the calyx of the Chelonine braconid *Phanerotoma flavitestacea* and developing in the tissue of its lepidopterous host, *Amyelois transitella* (Pyralidae). ×66,000

Fig. 2.9 Adult female wasp, *Microctonus* sp. (Hymenoptera: Braconidae: Euphorinae) in mid-Cenozoic Dominican amber that could be carrying the genome of a *Bracovirus* in its reproductive system. Scale bar = 0.7 mm

is introduced and replicates in cells of the potential host, thus preventing the host immune system from encapsulating the wasp eggs or developing larvae. Polydnaviruses in braconid wasps (Hymenoptera: Braconidae) are placed in the genus *Bracovirus* and polydnaviruses in ichneumonid wasps (Hymenoptera: Ichneumonidae) in the genus *Ichnovirus*. A female euphorine wasp in 20–30 million years Dominican amber (Fig. 2.9) suggests that polydnaviruses were already incorporated into the wasp genome by the mid-Cenozoic based on the law of behavioral fixity (the behavior, ecology and climatic preferences of fossil organisms will be similar to that found in their present day descendants at the generic and often family level) (Boucot 1990).

Species of *Bracovirus* are thought to have evolved from a nudivirus some 100 million years ago (Herniou et al. 2013). This date corresponds with the discovery of a 100 million years ago old pupa of a euphorine braconid wasp attached to its weevil host in mid-Cretaceous Myanmar amber (Poinar and Shaw 2016; Poinar et al. 2016) (Fig. 2.10). Virus-like particles have been shown to occur in the extant braconid, *Microctonus aethiopoides* (Hymenoptera: Braconidae: Euphorinae) that parasitizes adult weevils today (Barratt et al. 1999).

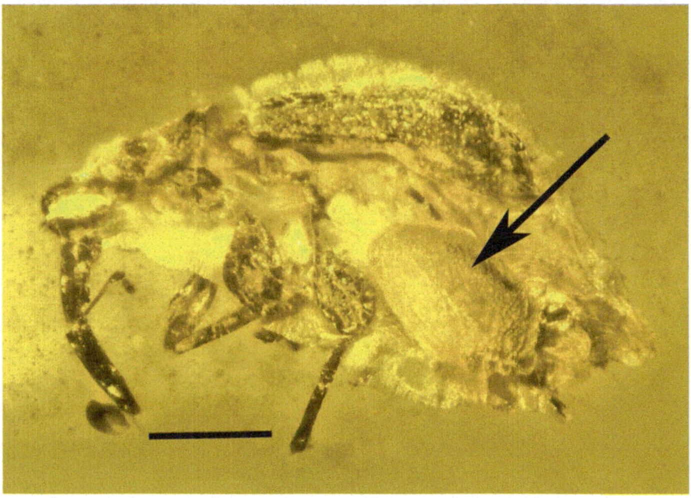

Fig. 2.10 The pupa (arrow) of a braconid wasp (Hymenoptera: Braconidae: Euphorinae) attached to its weevil host, *Habropezus plaisiommus* (Coleoptera: Curculionoidea: Ithyceridae) in mid Cretaceous Myanmar amber. Scale bar = 0.8 mm

2.2.2.2 Tumors in Lepidoptera

Tumors in insects are relatively rare (Harker 1963), yet are often associated with virus infections (Bird 1949; Neilson and Elgee 1968). It is very likely that in the examples presented below, viruses are the cause of the tumors since there are no signs of fungi, protozoa, bacteria or other parasites associated with these growths.

The first example is a caterpillar in Mexican amber that contains numerous tumors in its body cavity (Figs. 2.11 and 2.12). The specimen originated from mines in the Simojovel area in the state of Chiapas, Mexico, which are dated from 22.5 to 26 million years (Berggren and Van Couvering 1974). The fossil tumors resemble those described by Federley (1926) in living caterpillars of the genus *Pygaera* Ochsenheimer (Lepidoptera: Notodontidae). No causal agent of the tumors in these extant caterpillars was reported.

The second fossil tumor was found in a Dominican amber moth of the family Gracillariidae (Lepidoptera). Neoplastic tissue has grown out between the abdominal segments on the lateral-ventral side of the body. Several neoplastic lobes are present, with the largest greater than the width of the moth (Fig. 2.13). A detailed view of the surface of the medium-sized lobe showed it to be covered with an anastomosed network of various sized ridges (Fig. 2.14). It is quite possible that this condition was the result of a virus infection since there is no evidence of fungi, protozoa, bacteria or other parasites associated with the growth.

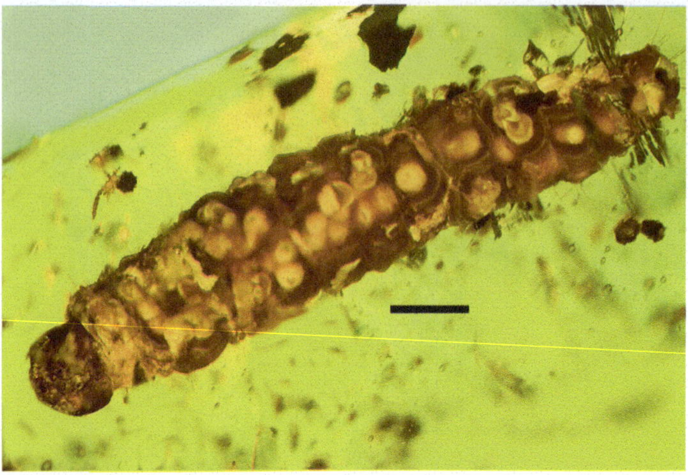

Fig. 2.11 Mexican amber caterpillar (Lepidoptera) with white, spherical tumors inside its body cavity. Scale bar = 1.0 mm

Fig. 2.12 Detail of white, spherical tumors inside caterpillar shown in Fig. 2.11. Scale bar = 380 μm

2.2.2.3 Iridoviridae

Members of this family of double-stranded DNA viruses have icosahedral virions that are usually arranged in paracrystalline arrays in the host tissue. Such a viral configuration often imparts an iridescent color to the tissues of infected hosts. Iridoviruses are widespread in both vertebrates and invertebrates and among the latter group are terrestrial isopods (Hess and Poinar 1985). When such infections

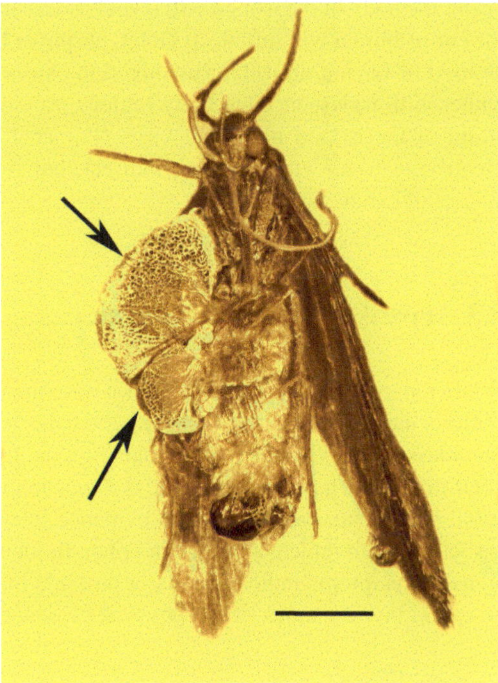

Fig. 2.13 Moth (Lepidoptera: Gelichiidae) in Dominican amber with tumors arising from body. Arrows show two large tumor lobes. Scale bar = 770 μm

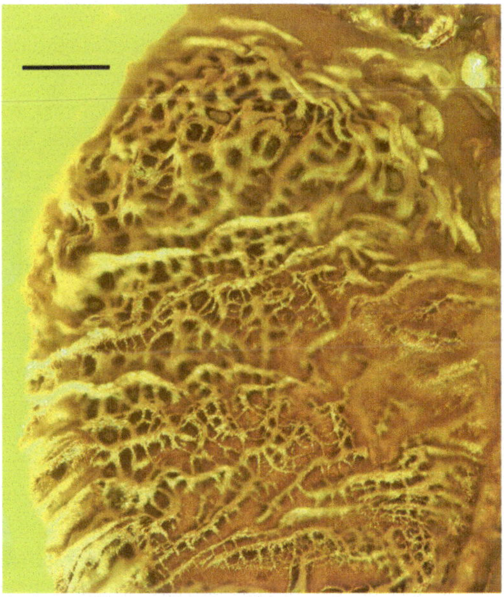

Fig. 2.14 Detail of a tumor lobe on the moth shown in Fig. 2.13. Scale bar = 90 μm

occur, the body of the isopod will reveal iridescent spots. When infections are heavy, the entire animal will turn deep violet, purple or blue. An isopod thus colored can be assured of having an iridovirus infection. An isopod in mid-Cretaceous Myanmar amber with iridescent blue body regions provides indirect evidence of an ancient strain of the isopod iridescent virus (Poinar 2014c). This is another example of identifying fossil viruses using indirect, but reliable evidence based on detailed studies of extant cases.

2.3 Fossil Pathogenic Bacteria

While reports of fossil bacteria are not rare, the ubiquitous presence of saprophytic bacteria makes it difficult to separate the latter from pathogenic or symbiotic bacteria unless other evidence is available (e.g., the location of the infected cells and the condition of the host). Again, there is direct and indirect evidence of the presence of fossil pathogenic bacteria. Direct evidence involves finding fossil bacteria closely associated with infected animal and plant tissue while indirect evidence is based on signs or symptoms indicating that a possible bacterial infection is present—based on extant investigations, when physical evidence of bacteria is lacking.

2.3.1 Direct Evidence of Fossil Pathogenic Bacteria

There are several examples of direct evidence of pathogenic bacteria in amber. One is a *Bacillus* sporangium inside the pseudocoel of the fossil mycetophagous nematode, *Oligaphelenchus atrebora* (Fig. 2.15) in Mexican amber. This bacterium was most likely a parasite that developed in the body cavity of the nematode (Poinar 1977). Members of *Bacillus* sp. are known to infect present day nematodes (Dollfus 1946).

Another example of direct fossil evidence involves insect-parasitic heterorhabditid nematodes. Several infective stages of *Proheterorhabditis burmanicus* were preserved as they emerged from their beetle host in mid-Cretaceous Myanmar amber. Extant species of the related genus *Heterorhabditis* carry and release spores of the luminescent bacterium, *Photorhabdus luminescens*, into the body cavity of insects. The bacteria multiply and kill the insects within 24 h. The nematodes then feed on the combined bacteria and decomposing host tissue and during development, newly formed infective stage juvenile nematodes acquire bacterial cells in the lumen of their alimentary tract. These juvenile nematodes then introduce the bacteria into new developmental hosts to continue the cycle. Bacterial cells that had emerged from an infected fossil beetle parasitized by *Proheterorhabditis burmanicus* are adjacent to the infected host (Poinar 2011) (Fig. 2.16).

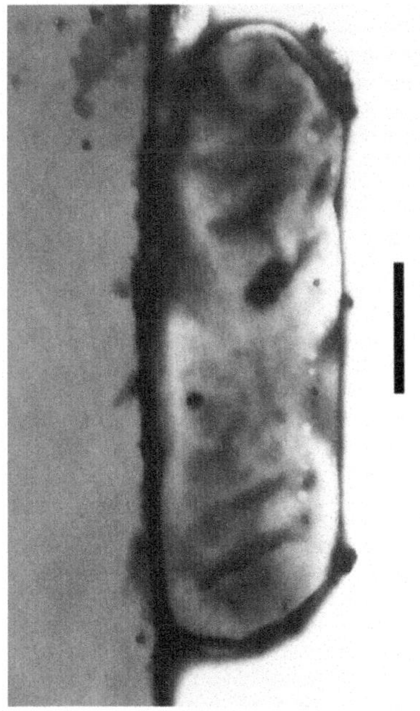

Fig. 2.15 A *Bacillus* sporangium attached to the pseudocoel of the nematode *Oligaphelenchus atrebora* in Mexican amber. Scale bar = 420 μm

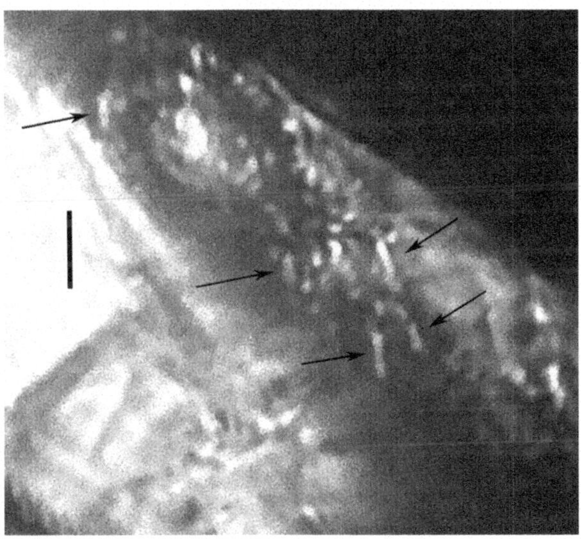

Fig. 2.16 Rod-shaped bacteria (arrows) from the body cavity of a beetle that was parasitized by the nematode *Proheterorhabditis burmanicus*, in mid-Cretaceous Myanmar amber. Scale bar = 8 μm

Other examples of direct evidence of fossil pathogenic bacteria are spirochetes of *Palaeoborrelia dominicana* (Spirochaetales) infecting a larva *Amblyomma* sp. tick (Ixodidae) in Dominican amber (Fig. 2.17). The size and shape of the fossil spirochetes (Fig. 2.18) are similar to those of extant *Borrelia* species, including *Borrelia burgdorferi*, the causal agent of Lyme disease (Poinar 2014a; Johnson et al. 1984).

Fig. 2.17 Larva *Amblyomma* sp. (Ixodidae) tick in Dominican amber infected with the spirochete *Palaeoborrelia dominicana*. Scale bar = 200 μm

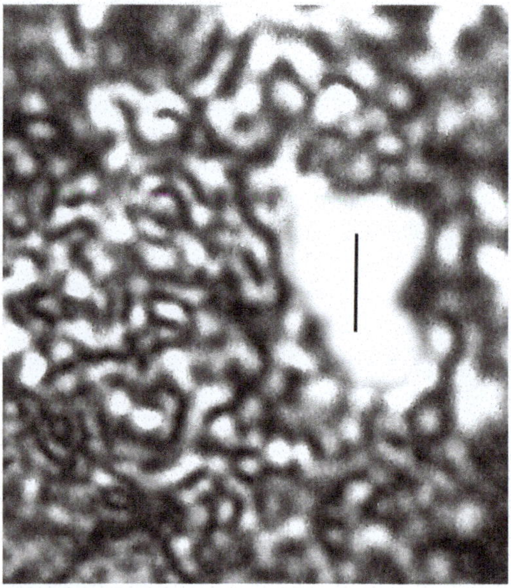

Fig. 2.18 Spirochetes of *Palaeoborrelia dominicana* infecting the larval *Amblyomma* sp. tick shown in Fig. 2.17. Scale bar = 6.4 μm

Spirochetes, as well as an endospore of a possible filamentous intestinal bacterium, were reported from a fossil termite, *Mastotermes electrodominicus*, in Dominican amber, however whether these bacteria represented pathogens or symbiotes is unclear (Wier et al. 2002).

Ticks are known to vector a number of rickettsial pathogens and evidence in amber shows that this association dates back at least 100 million years. The gastric

Fig. 2.19 An ixodid tick larva, *Cornupalpatum burmanicum*, in mid-Cretaceous Myanmar amber. Scale bar = 128 μm

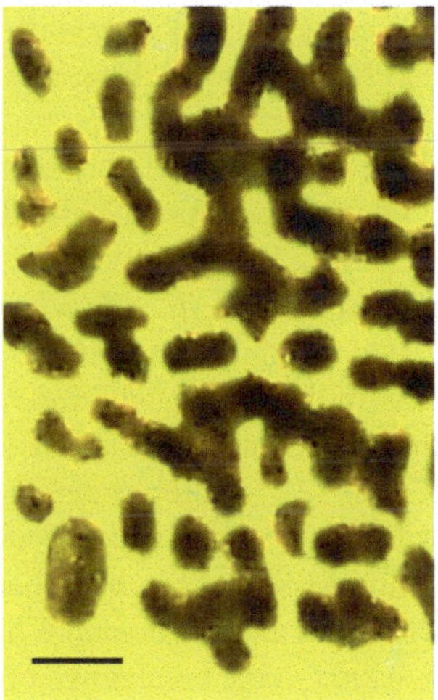

Fig. 2.20 Short rods of *Palaeorickettsia protera* in the larva of *Cornupalpatum burmanicum* shown in Fig. 2.19. Scale bar = 2.0 μm

ceca of the tick, *Cornupalpatum burmanicum* (Ixodida: Ixodidae) in mid-Cretaceous Myanmar amber (Poinar and Brown 2003) (Fig. 2.19) was infected with coccoid, diplococcoid and short rod-shaped cells (Fig. 2.20). These rickettsial-like cells were described as *Palaeorickettsia protera*. Their features closely resemble those of extant rickettsia, which are known to occur in the body of extant ticks (Poinar

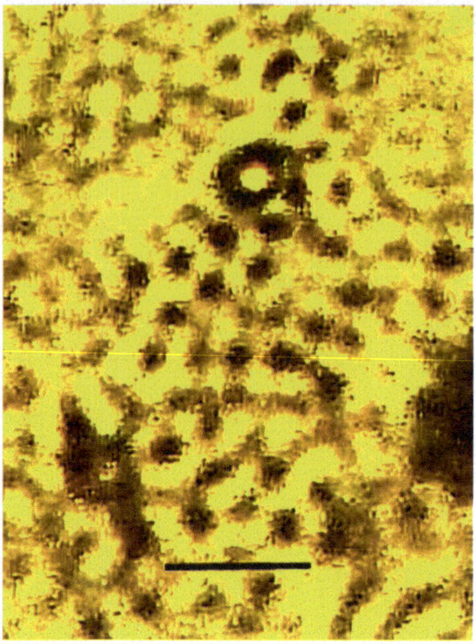

Fig. 2.21 Ovoid cells of *Palaeorickettsia protera* in the larva of *Cornupalpatum burmanicum* shown in Fig. 2.19. The cells are surrounded by a lighter-colored slime layer (halo). Such cells represent infective stages. Scale bar = 3.8 μm

2014a). In some areas of the Cretaceous tick's hemocoel, ovoid cells surrounded by an apparent lighter-colored slime layer (haloes) were present (Fig. 2.21). In extant ticks, such "halo" cells are considered to represent infective stages of rickettsial pathogens.

Species in the reduviid subfamily Triatomidae (Hemiptera: Reduviidae) are notorious blood-suckers, requiring vertebrate blood to complete their development. Extant species harbor parasitic nocardiform actinobacteria that are thought to assist in the breakdown of vertebrate blood (Poinar 2011). These are strange bacteria that produce fugacious mycelia that break up into non-motile rod-shaped or coccoidal elements. While some are considered symbiotic, others, like the horse parasite *Rhodococcus equi*, are pathogenic.

One of these actinomycetes was discovered in a fecal droplet issuing from the anus of the fossil bug, *Triatoma dominicana* in Dominican amber (Poinar 2005c). In the droplet were numerous coccoid elements with associated mycelial fragments (Fig. 2.22), thus representing the first fossil record of a nocardioform organism that was described as *Paleorhodococcus dominicanus* (Poinar 2011). Species of the genus *Rhodococcus* have been reported from extant reduviid bugs in the genera *Rhodnius* and *Triatoma*. While *P. dominicus* could have been symbiotic if it was assisting in the breakdown of blood components, it could also have been infectious.

Bacteria have also been found in fossil fleas in amber. Coccobacilli were discovered in the rectum (Fig. 2.23) and on the tip of the beak (Fig. 2.24) of the flea, *Atopopsyllus cionus* (Siphonaptera: Pulicidae) in Dominican amber. This flea had previously been identified as *Rhopalopsyllus* sp. (Poinar 2014b), but further investigations revealed that it was a new genus, *Atopopsyllus* (Poinar 2015). The size of the coccobacilli (1.0–4.0 μm) fell within the size range of vertebrate pathogenic bacteria and the short rods and nearly spherical cells of the fossil coccobacilli are

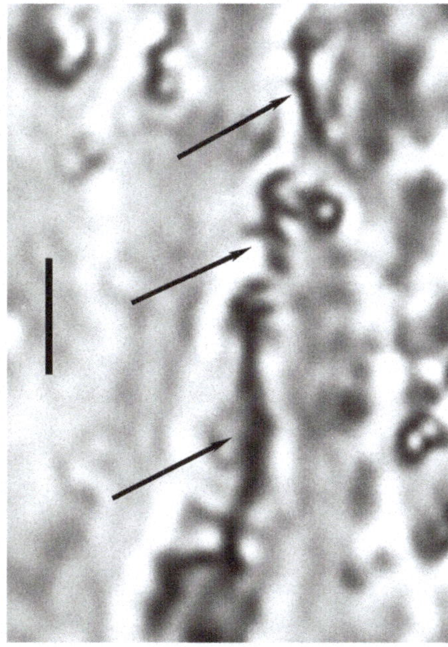

Fig. 2.22 Filaments with elementary branching (arrows) of the Actinobacterium, *Paleorhodococcus dominicanus*, in an anal droplet of the Dominican amber kissing bug, *Triatoma dominicana*. Scale bar = 7 μm

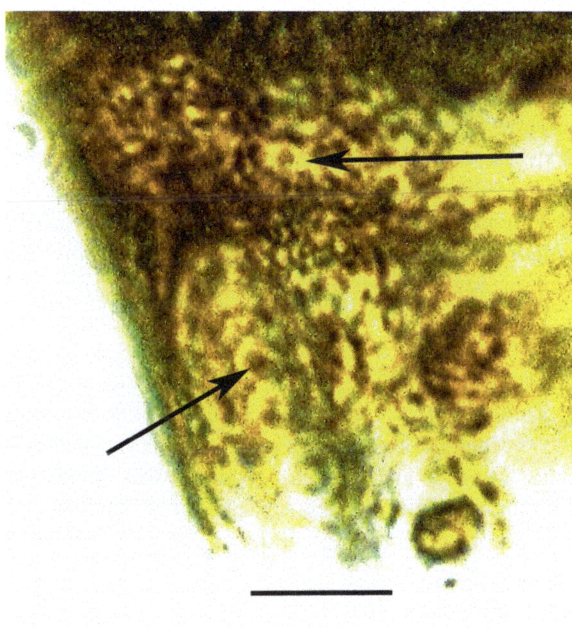

Fig. 2.23 Clusters of coccobacilli (arrows) in the rectum of the flea, *Atopopsyllus cionus*, in Dominican amber. Scale bar = 7 μm

characteristic of the modern genus *Yersinia* (Poinar 2015). *Yersinia pestis* is the causal agent of plague, a flea-transmitted disease that has decimated human populations throughout history (Perry and Fetherston 1997). It is not possible to determine if the bacteria associated with the fossil flea is related to *Yersinia*, however it was

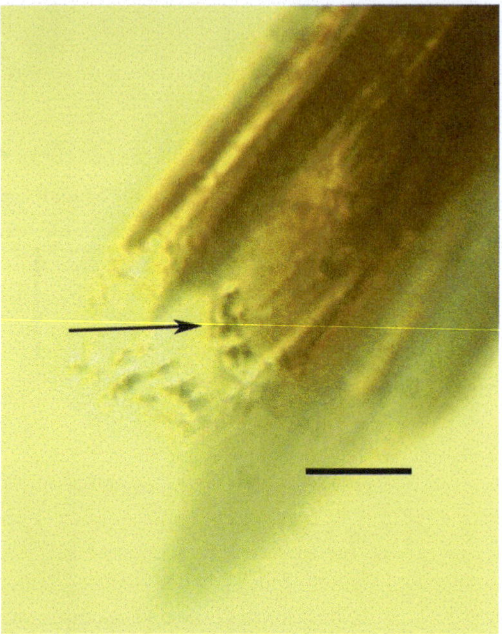

Fig. 2.24 A group of coccobacilli (arrow) adhering to the tip of the beak of the flea, *Atopopsyllus cionus*, in Dominican amber. Scale bar = 35 μm

very likely transmitted to hosts of the fossil flea. Some reports on the ancient history of *Y. pestis* infecting humans have used ancient DNA obtained from teeth and bones to characterize strains that caused Black Death during the fourteenth to seventeenth centuries. While these reports are too young to be classified as fossils, they demonstrate the use of molecular genetic techniques for tracing the history of various human pathogens (Bos et al. 2011). Bacteria have also been detected in fossil teeth, as demonstrated by Fostowicz-Frelik and Frelik (2010), who reported filamentous actinomycetes in a North American Eocene rabbit tooth. Actinomycetes were also recognized in a Miocene *Sivapithecus* (Hershkovitz et al. 1997).

Discovering bacteria in coprolites (fossil feces) is to be expected but whether these bacteria are saprophytic or parasitic is difficult to determine. An example is the Permian actinomycete, *Palaeostromatus diairetus*, that was recovered inside fish coprolites. Colonies consisted of groups of substrate mycelia with emerging spore chairs separated by elongate connectives (Fig. 2.25). This actinomycete was probably part of the gut flora and whether it was innocuous, beneficial or pathogenic is difficult to determine (Dentzien-Dias et al. 2016).

Actinobacteria also infect plants and one such species with coccoidal cells was found infecting the petals of the mid-Cretaceous fossil flower, *Eoëpigynia burmensis*, in Myanmar amber (Poinar et al. 2007) (Fig. 2.26). It was not possible to determine if these fossils belonged to the family Micrococcaceae or Microbacteriaceae, both of which contain species that infect plant surfaces. Actinobacteria were also reported from silicified wood of a Late Cretaceous *Laurinoxylon* and an Early Cretaceous *Protocupressinoxylon* in Romania. Iamandei (2003) concluded that the presence of filamentous and coccoid bacteria in xylem elements indicated a pathogenic condition.

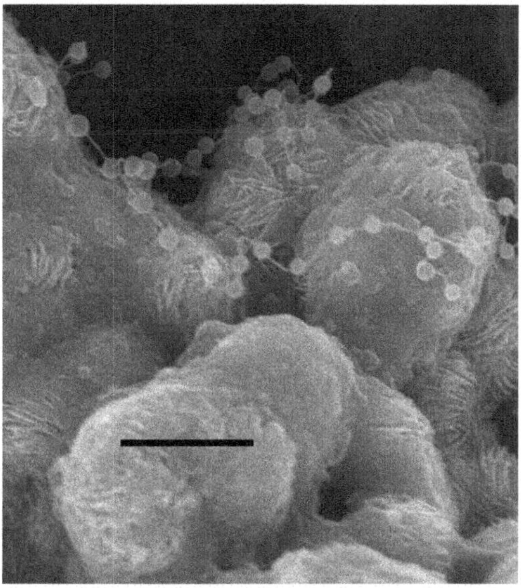

Fig. 2.25 Spore chains of the Permian actinomycete, *Palaeostromatus diairetus* arising from verrucose colonies of substrate mycelium on a coprolite. Scale bar = 4.4 μm. Photo courtesy of CEME-Sul FURG

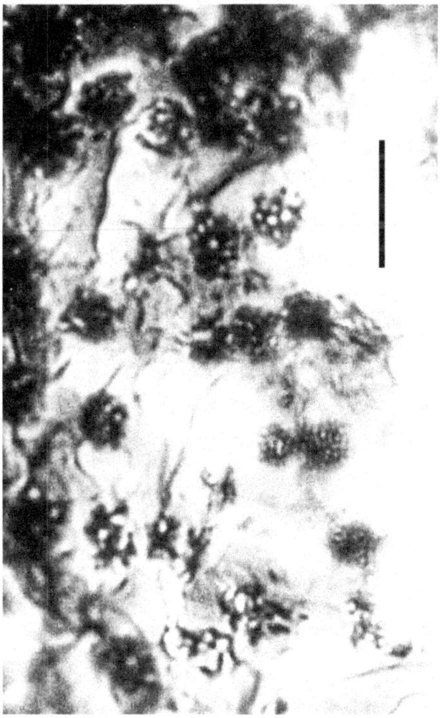

Fig. 2.26 Coccoidal cells of an actinobacterium infecting the petals of the mid-Cretaceous fossil flower, *Eoëpigynia burmensis*, in Myanmar amber. Scale bar = 12 μm

2.3.2 Indirect Evidence of Fossil Symbiotic-Pathogenic Bacteria

Indirect evidence of fossil pathogenic bacteria has been reported in the middle Cambrian Burgess Shale onychophoran, *Aysheaia pedunculata*. Dark areas on the body are thought to represent bacterial infections (Whittington 1985). Other indirect examples are fossil fish with luminescent organs occupied by luminescent bacteria in present day descendants, suggesting both a symbiotic and parasitic association (Boucot and Poinar 2010).

Insects producing flatus in amber are an indication of alimentary tract bacteria, some of which could be pathogenic (Poinar 2010a, b). There are numerous bacteria in the alimentary tract of honeybees, some of which are symbiotic and others pathogenic. One pathogenic species that is a common member of the gut flora is *Frischella perrara* that causes scabs on gut epithelium (Engel et al. 2016).

Gut bacteria also occur in stingless bees. A member of the extinct stingless bee genus *Proplebeia* shows the presence of gut bacteria on the basis of flatulence in Dominican amber (Fig. 2.27). Extant stingless bees closely related to the fossil species carry *Bacillus* spores and cells in their alimentary tract (Poinar 2010c). These bacteria are deposited on pollen in the nest and are thought to preserve this food. Attempts were made to determine if some bacterial spores inside the amber *Proplebeia* bees might still be viable. Pieces of Dominican amber containing worker *Proplebeia* bees were subjected to a period of ethyl bromide fumigation and then crushed in a sterile hood. The crushed amber pieces were then placed on culture dishes. Out of 40 culture plates, two were positive for *Bacillus* sp., however it was not possible to determine if the isolated bacteria were truly ancient or contaminants (Poinar and Poinar 1994). In 1994, Cano et al. recovered *Bacillus* DNA from the

Fig. 2.27 Flatus (arrowhead) released from a stingless bee in Dominican amber. Scale bar = 0.6 mm

abdomen of *Proplebia dominicana* preserved in 25–40 million year old amber from the Dominican Republic. Cano and Borucki (1995) claim to have cultured a strain of *Bacillus schadericus* from *P. dominicana* in Dominican amber. These studies corroborate an earlier work by Galippe (1920) who claimed to have revived bacteria (some of which appear to be *Bacillus* spp.) from 40 to 50 million years ago Baltic amber, although none were fully characterized or named. However, more recently, Hamamoto and Horikoshi (1994) reported that they isolated, cultured and characterized a bacterium from Baltic amber that had a 99.6% homology with that of *B. subtilis*. And if Dombrowski's report (1963) on reviving Permian *Bacillus circulans* from the Kali and Zechstein salt deposits in Germany can be verified, it would show the environmental persistence of *Bacillus* spores. While such an extended period of dormancy for bacterial spores appears incredulous, Hornek et al. (1994) documented bacterial spore survival for long periods in outer space, adding credence to their ability to survive in the ground encased in amber for millions of years.

2.4 Protozoan Fossils

Protozoa, a group of single-celled eukaryotic organisms formely assigned to the Kingdom Protista, are an ancient group with a fossil record extending back to the Archean Eon, some 2 billion years ago (Baker 1965). Protozoa probably originated in the sea where they formed associations with various organisms. One of their first habitats with marine invertebrates and vertebrates probably was in the alimentary tract. Freshwater and terrestrial habitats would have been explored soon afterwards. Some protozoa became permanent residents with their "hosts" and established mutualistic and parasitic associations. Eventually, life cycles were established that involved the transfer of protozoa between hematophagous arthropods and vertebrates. Fossil examples of a range of associations between parasitic protozoa and animals are presented below.

2.4.1 Direct Evidence of Fossil Protozoan Parasites

2.4.1.1 Fossil Trypanosome Parasites

One of the earliest fossil records of a protozoan parasite was in the gut lumen of a phlebotomine sand fly entrapped in mid-Cretaceous Myanmar amber (Fig. 2.28) (Poinar 2004). In the midgut lumen of the sand fly, *Palaeomyia burmitis*, were various stages of the leishmanial parasite, *Paleoleishmania proterus* (Poinar and Poinar 2004a, b) (Fig. 2.29). Aside from normal promastigotes, elongate nectomonad promastigotes (Fig. 2.30) and short procyclic promastigotes (Fig. 2.31) were also present. Vertebrate infective amastigotes occurred in the proboscis of the fossil sand fly (Fig. 2.32) and a view into the thoracic lumen of the fossil sand fly revealed nucleated vertebrate blood cells. Based on morphological characteristics of the blood

Fig. 2.28 The Burmese amber sand fly, *Palaeomyia burmitis*, that was vectoring the leishmanial parasite, *Paleoleishmania proterus*. Scale bar = 400 μm

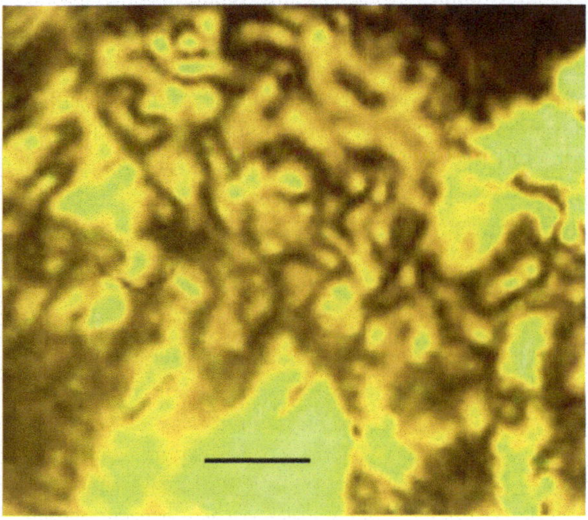

Fig. 2.29 Various stages of the leishmanial parasite, *Paleoleishmania proterus*, in the midgut lumen of the sand fly, *Palaeomyia burmitis*, in Burmese amber. Scale bar = 6.0 μm

cells, the vertebrate host was reptilian and *Paleoleishmania proterus* was considered to be an ancestor of the genus *Sauroleishmania* that infects lizards and snakes today (Poinar and Poinar 2004a, b).

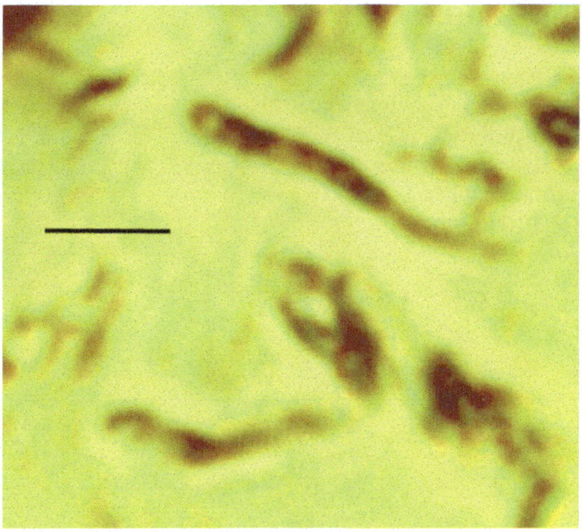

Fig. 2.30 Normal and elongate nectomonad promastigotes of *Paleoleishmania proterus* in the midgut lumen of the sand fly, *Palaeomyia burmitis*, in Burmese amber. Scale bar = 3.4 μm

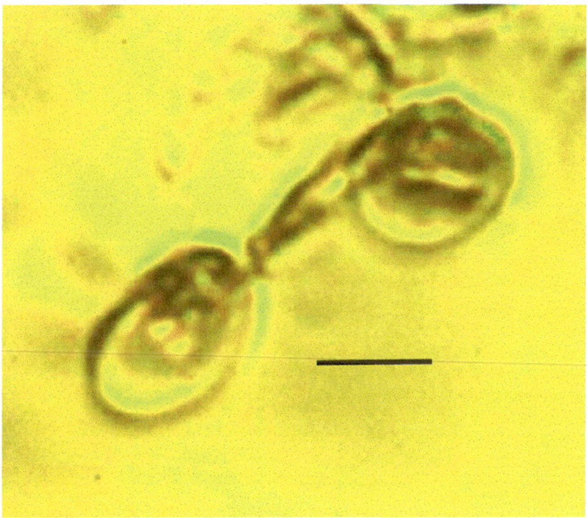

Fig. 2.31 Short procyclic promastigotes of *Paleoleishmania proterus* in the midgut lumen of the sand fly, *Palaeomyia burmitis*, in Burmese amber. Scale bar = 5.0 μm

Other flies also vector trypanosomes and one biting midge, *Leptoconops nosopheris* (Diptera: Ceratopogonidae), in mid-Cretaceous Myanmar amber (Fig. 2.33) had trypanosomes of *Paleotrypanosoma burmanicus* in its midgut. Trypanosomes also occurred in the head, salivary glands and salivary ducts of *L. nosopheris* and some flagellates were in a salivary droplet that collected on the tip of the proboscis (Fig. 2.34) (Poinar 2008). The vertebrate host of *Paleotrypanosoma burmanicus* was probably a reptile (Auezova et al. 1990).

Kissing bugs (Hemiptera: Reduviidae), well known for their bloodsucking habits, transmit a trypanosome to humans that causes Chagas disease (*Trypanosoma*

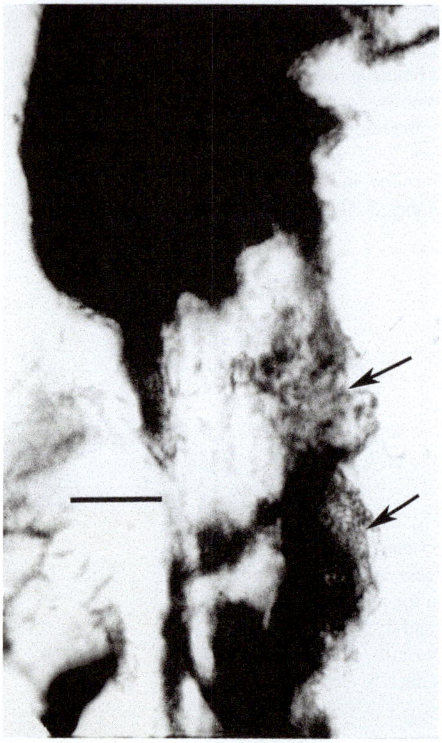

Fig. 2.32 Vertebrate infective amastigotes (arrows) of *Paleoleishmania proterus* in the proboscis of *Palaeomyia burmitis* in Burmese amber. Scale bar = 10 μm

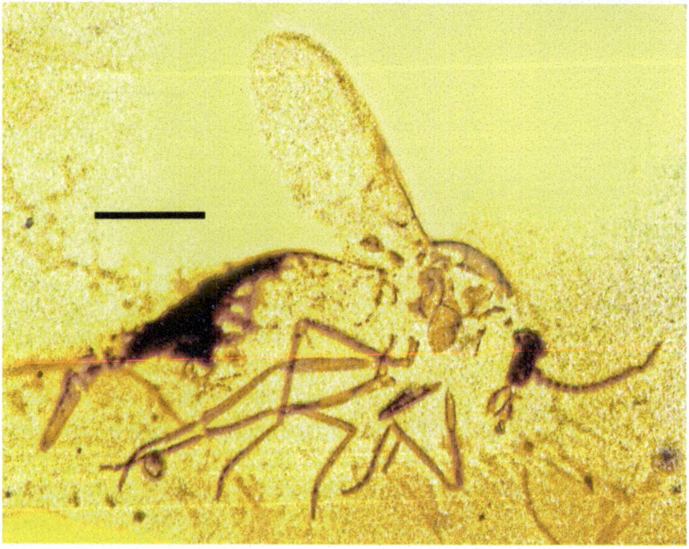

Fig. 2.33 *Leptoconops nosopheris* (Diptera: Ceratopogonidae) in Burmese amber. Scale bar = 260 μm

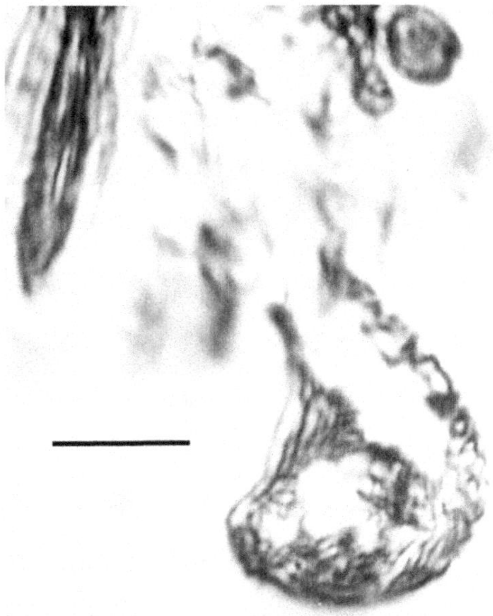

Fig. 2.34 Trypanosomes of *Paleotrypanosoma burmanicus* in a salivary droplet of *Leptoconops nosopheris* (Diptera: Ceratopogonidae) in Burmese amber. Scale bar = 15 μm

cruzi) in the American tropics. A fifth instar nymph of *Triatoma dominicana* in Dominican amber (Fig. 2.35) was adjacent to fecal droplets that contained metatrypanosomes of *Trypanosoma antiquus* (Fig. 2.36). Mammalian hairs adjacent to the fecal droplets indicate that the vertebrate host was a bat. Extant triatomines are known to vector trypanosomes to New World bats (Hoare 1972).

Other types of flagellates have been discovered in fossil triatomines. Epimastigotes of a *Blastocrithidia* sp. were detected in an anal droplet of *Panstrongylus hispaniolae* in Dominican amber (Figs. 2.37 and 2.38) (Poinar 2013b). The fossil flagellates are similar in size and shape to the epimastigotes of *Blastocrithidia triatomae* that were described from the intestine of *Triatoma infestans* (Cerisola et al. 1971). *Blastocrithidia* can be pathogenic in triatomes when populations built up in the midgut (Schaub 1994).

The primitive mid-Cretaceous triatomine, *Paleotriatoma metaxytaxa* (Fig. 2.39) in Burmese amber is considered a transitional species bridging the gap between the invertebrate feeding Reduviinae and the vertebrate feeding Triatominae (Poinar 2018). This Cretaceous triatomine was carrying trypomastigote stages of a trypanosome in its hindgut lumen (Fig. 2.40). These stages correspond to those that normally occur in the blood stream of infected vertebrates today, suggesting that the trypanosomes in *Paleotriatoma* were digenetic forms vectored to vertebrates.

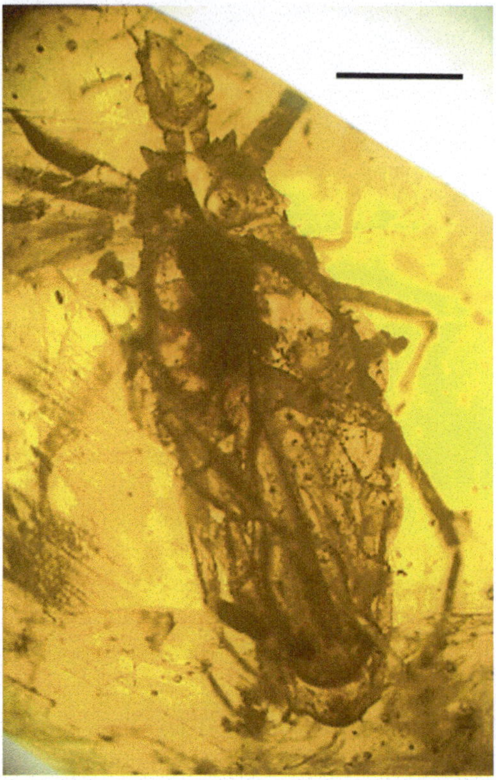

Fig. 2.35 A fifth instar nymph of *Triatoma dominicana* in Dominican amber. Scale bar = 662 μm

2.4.1.2 Fossil Malaria Parasites

There are numerous types of malaria today in a wide range of vertebrates and vectored by a wide range of arthropods, however very few species have left a fossil record. Malaria is the most notorious protozoan disease of humans today and continues to cause immense suffering and deaths worldwide.

The oldest record of malaria vectored by a biting midge involved a *Protoculicoides* sp. in Early Cretaceous Myanmar amber (Poinar and Telford 2005) (Fig. 2.41). Developing oocysts and sporozoites of the malarial parasite, *Paleohaemoproteus burmacis*, were inside the abdomen of the biting midge (Fig. 2.42). The development of *Paleohaemoproteus burmacis* was probably very similar to species of the extant genus *Haemoproteus* that are vectored by biting midges to birds and reptiles today (Garnham 1966).

Discovering oocysts, sporozoites, ookinetes and microgametocytes of *Plasmodium dominicana* in the Dominican amber mosquito, *Culex malariager* (Fig. 2.43) showed that *Plasmodium* malaria was established in the Americas some 20–30 million years ago (Poinar 2005a, b). Two of the oocysts had ruptured (Fig. 2.44) and were releasing sporozoites (Fig. 2.45), some of which had reached the salivary glands before the mosquito was entrapped. Based on the structure of

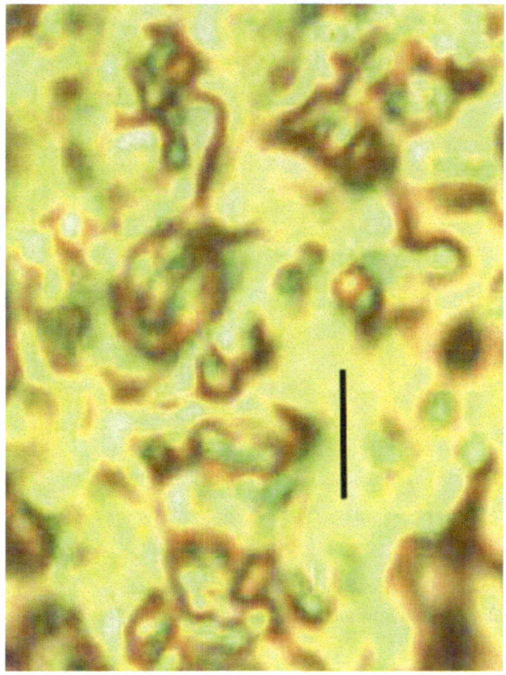

Fig. 2.36 Fecal droplets with metatrypanosomes of *Trypanosoma antiquus* in Dominican amber. Scale bar = 3.5 µm

Fig. 2.37 *Panstrongylus hispaniolae* in Dominican amber. Scale bar = 3.2 mm

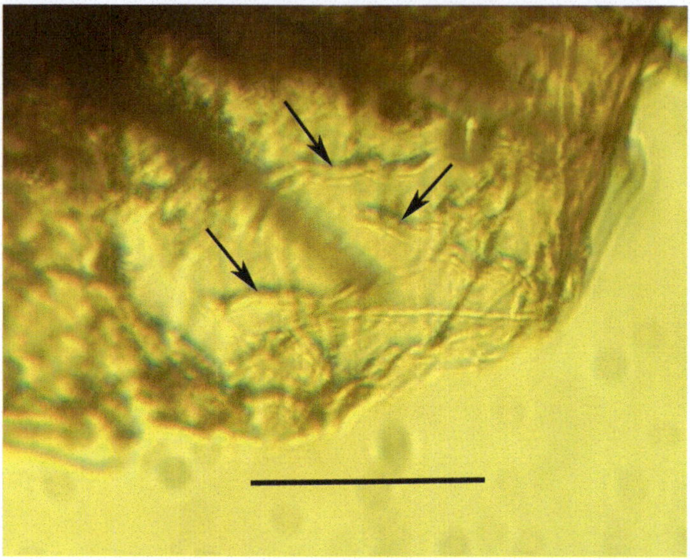

Fig. 2.38 Epimastigotes (arrows) of *Blastocrithidia* sp. in an anal droplet of *Panstrongylus hispaniolae* in Dominican amber. Scale bar = 77 μm

Fig. 2.39 Burmese amber triatomine bug, *Paleotriatoma metaxytaxa*. Scale bar = 1.6 mm

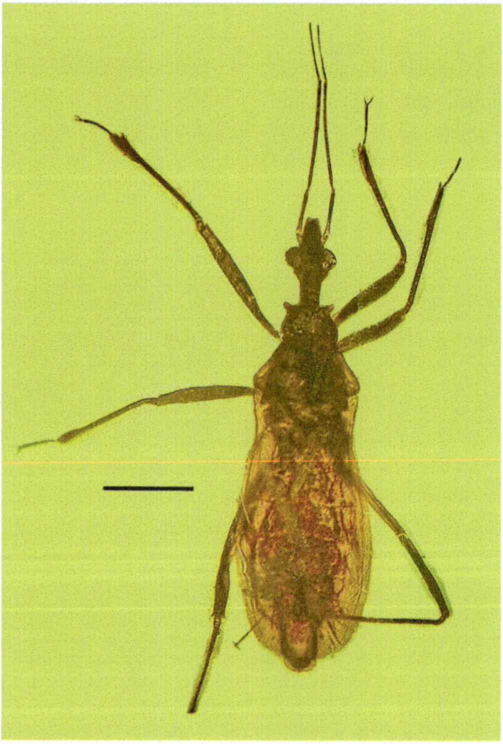

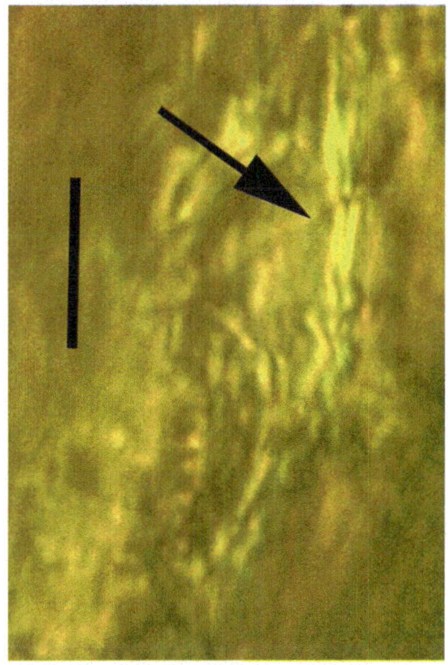

Fig. 2.40 Trypomastigote stages (arrow) of trypanosomes in the hindgut lumen of *Paleotriatoma metaxytaxa*. Scale bar = 30 μm

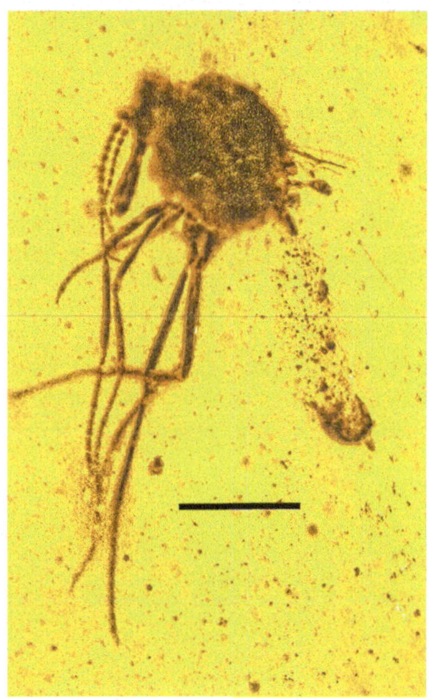

Fig. 2.41 A biting midge, *Protoculicoides* sp., in Burmese amber. Scale bar = 400 μm

these developing stages and the *Culex* vector, it was concluded that *P. dominicana* parasitized birds.

Another malarial parasite was discovered in the Dominican amber bat fly, *Enischnomyia stegosoma* (Diptera: Streblidae) (Fig. 2.46) (Poinar and Brown 2012). Two ovoid oocysts that contained short, stubby sporozoites of the malarial

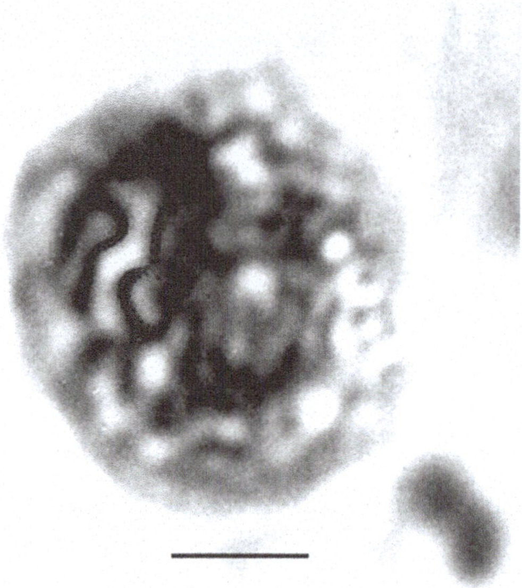

Fig. 2.42 Developing oocysts and sporozoites of the malarial parasite, *Paleohaemoproteus burmacis*, in the abdomen of *Protoculicoides* sp. Scale bar = 6.0 μm

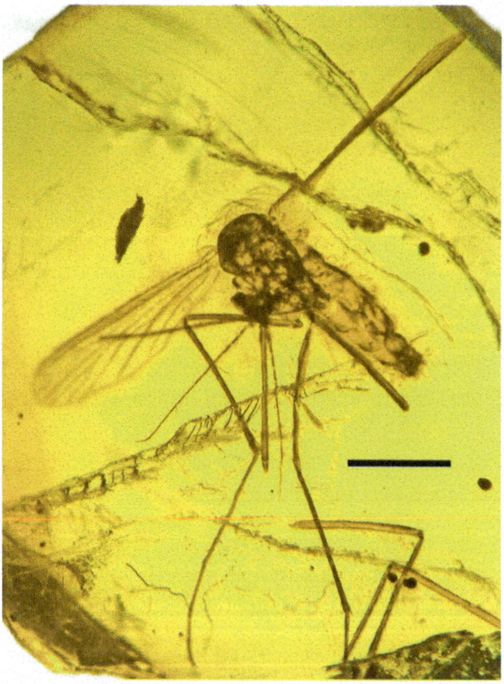

Fig. 2.43 *Culex malariager*, a vector of *Plasmodium dominicana* in Dominican amber. Scale bar = 1.2 mm

Fig. 2.44 Oocyst of *Plasmodium dominicana* in the body cavity of *Culex malariager*. Scale bar = 34 μm

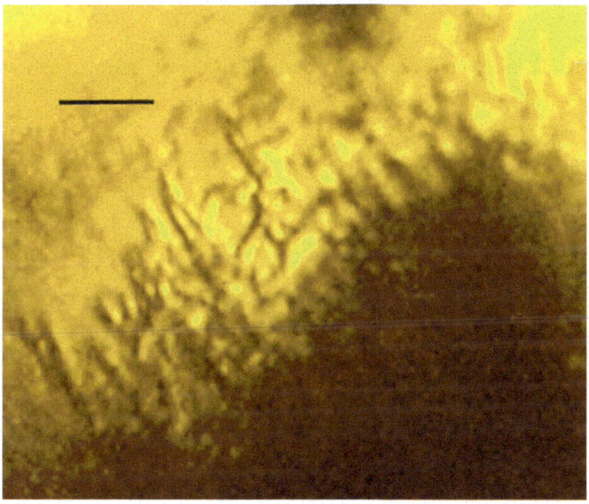

Fig. 2.45 Sporozoites of *Plasmodium dominicana* being released from a ruptured oocyst. Scale bar = 8.0 μm

parasite, *Vetufebrus ovatus*, were attached to the midgut wall of the host fly (Fig. 2.47). Mature sporozoites were present in salivary ducts and secretions of the fossil fly, indicating that Neotropical bat flies were vectoring malaria to bats in the mid-Cenozoic (Poinar 2012a, b).

Fig. 2.46 Dominican amber bat fly, *Enischnomyia stegosoma* (Diptera: Streblidae) Scale bar = 245 µm

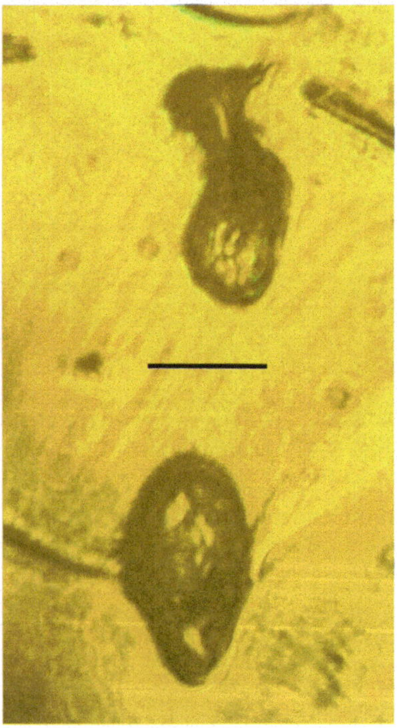

Fig. 2.47 Ovoid oocysts with short, stubby sporozoites of the malarial parasite, *Vetufebrus ovatus*, attached to the midgut wall of *Enischnomyia stegosoma*. Scale bar = 5 µm

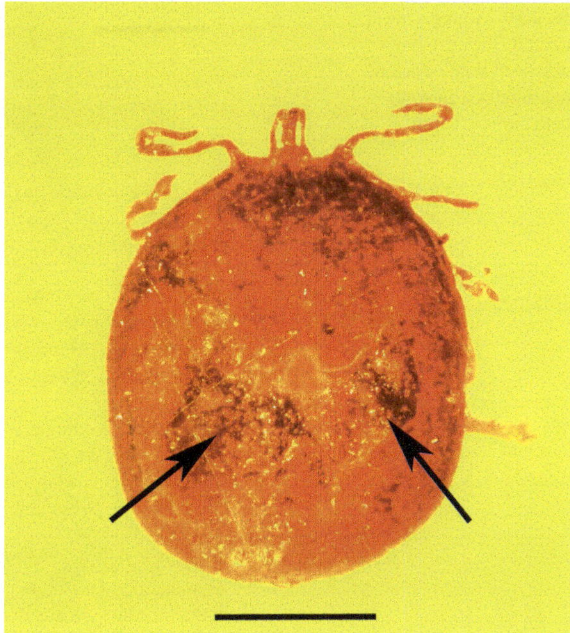

Fig. 2.48 Engorged nymphal *Amblyomma* tick with two body injuries (arrows) in Dominican amber. Scale bar = 960 μm

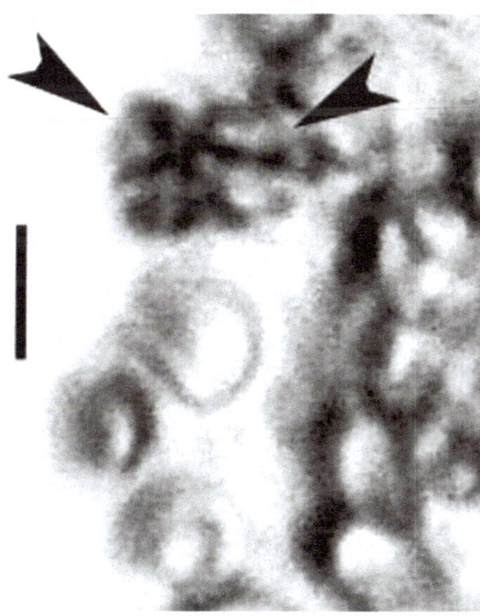

Fig. 2.49 Vertebrate erythrocytes (arrowheads) infected with the piroplasm, *Paleohaimatus calabresi*, in Dominican amber. Scale bar = 8.0 μm

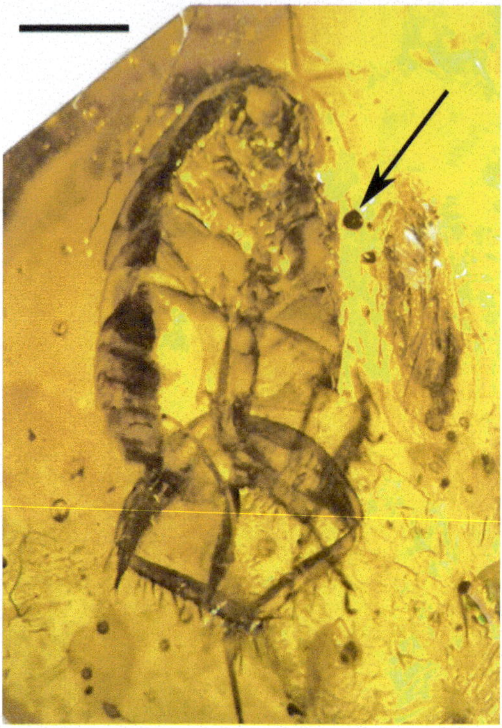

Fig. 2.50 A gregarine-infected Burmese amber cockroach with adjacent gregarine cyst (arrow). Scale bar = 1.1 mm

2.4.1.3 Other Fossil Records of Protozoan Parasites

Fossil Piroplasmid Parasites

Ticks also vector parasitic protozoa and one engorged nymphal *Amblyomma* tick in Dominican amber was recently found to vector a representative of the Order Piroplasmida. The tick was in a pool of mammalian blood that had flowed out of a wound on its back (Fig. 2.48). Some of the vertebrate erythrocytes contained the developmental stages of the piroplasm, *Paleohaimatus calabresi* (Fig. 2.49). Internally, the tick showed developmental stages of *P. calabresi* in its hemocoel, gut lumen and gut epithelial cells (Fig. 2.36). Based on features of the erythrocytes, it was concluded that the tick had been feeding on a monkey and was plucked off by a companion during a grooming operation. Thus ticks were vectoring piroplasmas to vertebrates in the mid-Cenozoic (Poinar 2017).

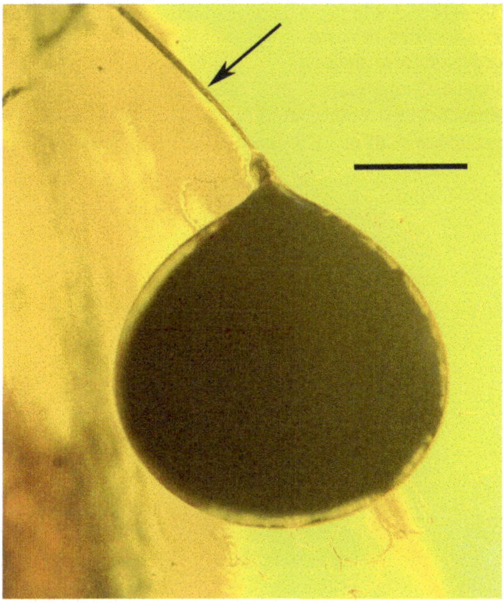

Fig. 2.51 Oocyst of *Primigregarina burmanica*, a cockroach parasite in Burmese amber. Scale bar = 80 μm

Fossil Gregarine Parasites

Gregarines are a very complex group of protozoans that evolved as monoxenous parasites of a wide variety of marine and terrestrial invertebrates (Desportes and Schrevel 2013). Among the insects that are parasitized are cockroaches, which may have been one of the original host groups. Discovering trophozoites and oocysts of *Primigregarina burmanica* from a mid-Cretaceous Myanmar amber cockroach shows that the basic developmental stages of extant gregarines were firmly established some 100 million years ago (Poinar 2010a, b, 2013b) (Figs. 2.50 and 2.51).

2.4.2 *Indirect Evidence of Fossil Protozoan Parasites*

2.4.2.1 Fossil Parasites in Coprolites

Protozoans cause a number of alimentary tract infections in all vertebrates; however, the fossil presence of such infections are difficult to find in situ. Reptiles were one of the early hosts of intestinal protozoan parasites (Keymer 1981) as was made evident by the discovery of mature cysts of *Entamoebites antiquus* in a dinosaur coprolite from the Early Cretaceous Bernissart Iguanodon shaft in Belgium (Poinar and Boucot 2006) (Fig. 2.52). The fossil cysts resembled species of the widespread extant genus *Entamoeba* that infect amphibians, reptiles, birds, and mammals.

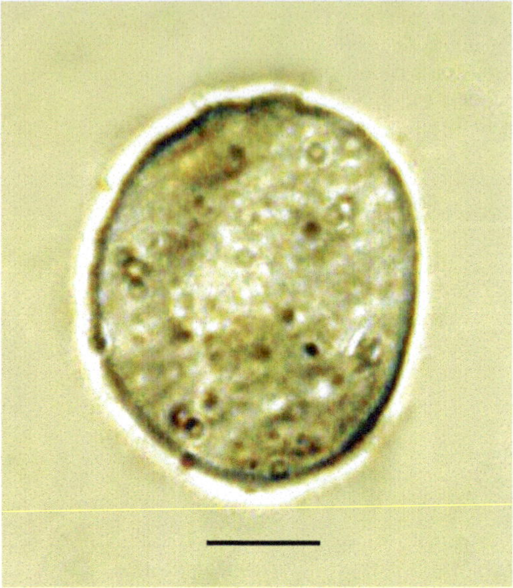

Fig. 2.52 Cyst of *Entamoebites antiquus* recovered from an Early Cretaceous dinosaur coprolite at the Bernissart Iguanodon shaft in Belgium. Scale bar = 4.0 μm

2.5 Conclusions

Based on direct evidence of fossil pathogenic viruses, bacteria and protozoa in hosts and arthropod vectors embedded in amber, it is obvious that these agents have been present since at least the mid-Cretaceous. The majority of the findings reported here are due to the preservative qualities of amber (De Baets et al. 2021; Poinar and Hess 1985).

Determining that the viruses, bacteria and protozoa reported here involved pathogenic and not other symbiotic associations was based on their comparisons with present day pathogens and the law of behavior fixity. The latter states that the behavior of fossil organisms will be similar to that of their present day descendants at the generic level (Boucot 1990; Boucot and Poinar 2010; Poinar and Poinar 1999).

The association between organisms are typically expressed with the following terms: Inquilinism, where neither organism suffers or benefits from the relationship; commensalism, where one organism benefits while the other organism is not harmed; mutualism, where both organisms benefit and neither is harmed and; parasitism, where one organism benefits by taking nourishment from the other organism (now a host) that is harmed (Poinar 1983). These basic definitions are very simplified since associations between two organisms are often much more intricate and complicated.

In regards to polydnaviruses in parasitic wasps, it is generally agreed that since the viral genome is incorporated into that of the wasp, there is no detriment to the wasp (Hess et al. 1980). However, this scenario changes dramatically after the wasp introduces the virus into an insect host. Not only does the virus enter many of the host's tissues, but it interferes with the defense responses of the host, rendering it susceptible to the parasite, resulting in its demise.

A similar scenario occurs with many pathogenic bacteria that are vectored by arthropods. While inside the vector, the bacteria cause no obvious damage. After being introduced by the vector into a vertebrate host, however, the bacteria can have disastrous effects. Variable factors include the nutrition or stage of the host and whether the host is already infected with other microbial agents. In reference to the latter condition, a detailed examination of an insect thought to have a single virus infection revealed that it was actually infected with five separate viruses (Hess et al. 1979), all of which could be detrimental and produce a synergistic effect on the host.

Acknowledgments The author thanks Roberta Poinar for comments on the manuscript and CEME-Sul FURG for permission to use the photograph in Fig. 2.25.

References

Auezova G, Brushko Z, Kubykin R (1990) Feeding of biting midges (Leptoconopidae) on reptiles. In: Proceeding of the 2nd International Congress of Dipterology, Bratislava, Czechoslovakia, September 1990, p 12

Baker R (1965) The evolution of parasitic protozoa. In: Taylor AER (ed) Evolution of parasites. Blackwell Scientific, Oxford, pp 1–27

Barratt BI, Evans AA, Stoltz DB, Vinson SB, Easingwood R (1999) Virus-like particles in the ovaries of *Microctonus aethiopoides* Loan (Hymenoptera: Braconidae), a parasitoid of adult weevils (Coleoptera: Curculionidae). J Invertebr Pathol 73:182–188

Berggren WA, Van Couvering JA (1974) The late Neogene and Palaeogene. Palaeoclimat Palaeoecol 16:1–216

Bird FT (1949) Tumours associated with a virus infection in an insect. Nature 4150:777–778

Bos KI, Schuenemann VJ, Golding GB et al (2011) A draft genome of *Yersinia pestis* from victims of the Black Death. Nature 478(7370):506–510

Boucot AJ (1990) Evolutionary paleobiology of behavior and coevolution. Elsevier, Amsterdam, p 735

Boucot AJ, Poinar GO Jr (2010) Fossil behavior compendium. CRC Press, Boca Raton, FL, pp 1–424

Cano RJ, Borucki MK (1995) Revival and identification of bacterial spores in 25- to 40-million-year-old Dominican amber. Science 268(5213):1060–1064

Cano RJ, Borucki MK, Higby Schweitzer M, Poinar HN, Poinar GO Jr, Pollard KJ (1994) *Bacillus* DNA in fossil bees: an ancient symbiosis? J App Environ Microbiol 60:2164–2167

Cerisola JA, del Prado CE, Rohwedder R, Bozzini JP (1971) *Blastocrithidia triatomae* n. sp. found in *Triatoma infestans* from Argentina. J Protozool 18(3):503–506

Cruickhank RD, Ko K (2003) Geology of an amber locality in the Hukawng valley, northern Myanmar. J Asian Earth Sci 21:441–455

De Baets K, Littlewood DTJ (2015) The importance of fossils in understanding the evolution of parasites and their vectors. Adv Parasitol 90:1–51

De Baets K, Huntley JW, Klompmaker AA, Schiffbauer JD, Muscente AD (2021) The fossil record of parasitism: Its extent and taphonomic constraints. In: De Baets K, Huntley JW (eds) The evolution and fossil record of parasitism: Coevolution and paleoparasitological techniques. Topics in Geobiology 50

Dentzien-Dias P, Poinar G, Francischini H (2016) A new actinomycete from a Guadalupian vertebrate coprolite from Brazil. Hist Biol 29:770–776
Desportes I, Schrevel J (2013) The gregarines: the early branching apicomplexa. EJ Brill, Leiden. vols 1 and 2
Dollfus RP (1946) Parasites des Helminthes. Paul Lechevalier, Paris, pp 1–482
Dombrowski H (1963) Bacteria from Palaeozoic deposits. Ann N Y Acad Sci 108:453–460
Duggan AT, Perdomo MF, Piombino-Mascali D et al (2016) 17th Century Variola Virus reveals the recent history of Smallpox. Curr Biol 26:3407–3412
Eberth DA, Hamblin AP (1993) Tectonic, stratigraphic, and sedimentological significance of a regional discontinuity in the Judith River Group (Belly River Wedge) of southern Alberta, Saskatchewan and northern Montana. Can J Earth Sci 30:174–200
Engel P, Kwong WK, McFrederick Q et al (2016) The bee microbiome: impact on bee health and model for evolution and ecology of host-microbe interactions. MBio 7:2164–2215
Federley H (1926) Sex-limited hereditary cancer in lepidopterous larvae. Hereditas 22:193–216
Foreteer P (2006) The origin of viruses and their possible role in major evolutionary transition. Virus Res 117:5–16
Fostowicz-Frelik Ł, Frelik GJ (2010) Earliest record of dental pathogen discovered in a North American Eocene rabbit. PALAIOS 25:818–822
Galippe V (1920) Recherches sur la resistance des microzymes à l'action du temps et sur leur survivance dans l'ambre. Comp Rendu Acad Sci (Paris) 170:856–858
Garnham PCC (1966) Malaria parasites and other haemosporidia. Blackwell Scientific Publications, Oxford
Hamamoto T, Horikoshi K (1994) Characterization of a bacterium isolated from amber. Biodivers Conserv 3:567–572
Harker JE (1963) Tumors. In: Steinhaus EI (ed) Insect pathology. An advanced treatise, vol 1. Academic Press, New York, NY, pp 191–214
Herniou EA, Huguet E, Thézé J, Bézier A, Periquet G, Drezen J-M (2013) When parasitic wasps hijacked viruses: genomic and functional evolution of polydnaviruses. Philos Trans R Soc B 368:20130051. https://doi.org/10.1098/rstb.2013.0051
Hershkovitz IH, Kelly J, Latimer B, Rothschild BM, Simpson S, Polak I, Rosenberg M (1997) Oral bacteria in Miocene *Sivapithecus*. Journal of Human Evolution 33: 507–512
Hess RT, Poinar GO Jr (1985) Iridoviruses infecting terrestrial isopods and nematodes. In: Current topics in microbiology and immunology, vol 116. Springer, Berlin, pp 49–76
Hess RT, Summers MD, Falcon LA (1979) A mixed virus infection in midgut cells of *Autographa californica* and *Trichoplusia ni* larvae. J Ultrastruct Res 65:253–265
Hess RT, Poinar GO Jr, Etzel L, Merritt CC (1980) Calyx particle morphology of *Bathyplectes anurus* and *B. curculionis* (Hymenoptera: Ichneumonidae). Acta Zool (Stockh) 61:111–116
Hoare C (1972) The trypanosomes of mammals. Blackwell Scientific Publications, Oxford
Hornek G, Bucker H, Reitz G (1994) Long-term survival of bacterial spores in space. Adv Space Res 14:41–45
Iamandei E (2003) Fossil bacteria and fungi in some Cretaceous woods and their paleoecological implications. Rev Roumaine Géol 47:31–41
Iturralde-Vinent MA, MacPhee RDE (1996) Age and paleogeographic origin of Dominican amber. Science 273:1850–1852
Johnson RC, Schmid GP, Hyde FW, Steigerwalt AG, Brenner DJ (1984) *Borrelia burgdorferi* sp. nov.: etiologic agent of Lyme disease. Int J Syst Evol Microbiol 34:496–497
Katzouakis A (2014) Paleovirology: inferring viral evolution from host genome sequence data. Phil Trans R Soc B 368:20120493. https://doi.org/10.1098/rstb.2012.0493
Keymer IF (1981) Protozoa In: Cooper JE, Jackson OF (eds) Diseases of the Reptilia, Vol. 1. Academic Press, New York, pp 233–290
Labandeira CC, Li L (2021) The History of Insect Parasitism and the Mid-Mesozoic Parasitoid Revolution. In: De Baets K, Huntley JW (eds) The Evolution and Fossil Record Of Parasitism: Identification and Macroevolution of Parasites. Topics in Geobiology 49

Lee A, Nolan A, Watson J, Tristem M (2013) Identification of an ancient endogenous retrovirus, predating the divergence of the placental mammals. Philos Trans R Soc B 368:20120503. https://doi.org/10.1098/rstb.2012.0503

Neilson MN, Elgee DE (1968) Tumorlike bodies in virus-infected and noninfected adults of the spruce sawfly, *Diprion hercyniae*. J Invertebr Pathol 10:70–75

Perry RD, Fetherston JD (1997) *Yersinia pestis*--etiologic agent of plague. Clin Microbiol Rev 10:35–66

Poinar GO Jr (1977) Fossil nematodes from Mexican amber. Nematologica 23:232–238

Poinar GO Jr (1983) The natural history of nematodes. Prentice Hall, New York, NY. 323 p

Poinar GO Jr (2004) *Palaeomyia burmitis* (Diptera: Phlebotomidae), a new genus and species of Cretaceous sand flies with evidence of blood-sucking habits. Proc Entomol Soc Wash 106:598–605

Poinar GO Jr (2005a) *Culex malariager*, n. sp. (Diptera: Culicidae) from Dominican amber: the first mosquito vector of *Plasmodium*. Proc Entomol Soc Wash 107:548–553

Poinar GO Jr (2005b) *Plasmodium dominicana* n. sp. (Plasmodiidae: Haemospororida) from Tertiary Dominican amber. Syst Parasitol 61:47–52

Poinar GO Jr (2005c) *Triatoma dominicana* sp. n., (Hemiptera: Reduviidae:Triatominae), and *Trypanosoma antiquus* sp. n. (Stercoraria: Trypanosomatidae), the first fossil evidence of a Triatomine-Trypanosomatid vector association. Vector Borne Zoonot Dis 5:72–81

Poinar GO Jr (2008) *Leptoconops nosopheris* sp. n. (Diptera: Ceratopogonidae) and *Paleotrypanosoma burmanicus* gen. n., sp. n. (Kinetoplastida: Trypanosomatidae), a biting midge-trypanosome vector association from the Early Cretaceous. Mem Inst Oswaldo Cruz 103:468–471

Poinar GO Jr (2010a) *Primigregarina burmanica* n. gen., n. sp., an Early Cretaceous gregarine (Apicomplexa: Eugregarinorida) parasite of a cockroach (Insecta: Blattodea). In: Boucot AJ, Poinar GO Jr (eds) Fossil behavior compendium. CRC Press, Boca Raton, FL, pp 54–56

Poinar GO Jr (2010b) Fossil flatus: indirect evidence of intestinal microbes. In: Boucot AJ, Poinar GO Jr (eds) Fossil behavior compendium. CRC Press, Boca Raton, FL, pp 22–25

Poinar GO Jr (2010c) Notes on the origins and evolution of Bacillus in relation to insect parasitism. CRC Press, Boca Raton, FL, pp 68–71

Poinar G Jr (2011) *Paleorhodococcus dominicanus* n. gen., n. sp. (Actinobacteria) in a fecal droplet of *Triatoma dominicana* (Hemiptera: Reduviidae: Triatominae) in Dominican amber. Hist Biol 24:219–221

Poinar GO Jr (2012a) *Vetufebrus ovatus* n. gen. n. sp. (Hemosporida: Plasmodiidae) vectored by a streblid bat fly (Diptera: Streblidae) in Dominican amber. Parasit Vectors 4:229–233

Poinar GO Jr (2012b) (Haemospororida: Plasmodiidae) vectored by a streblid bat fly (Diptera: Streblidae) in Dominican amber. Parasit Vectors 4:229–233

Poinar GO Jr (2013a) *Panstrongylus hispaniolae* sp. n. (Hemiptera: Reduviidae: Triatominae), a new fossil in Dominican amber. Parasites and Vectors 4: 229–233. of gut flagellates. Palaeodiversity 6:1–8

Poinar GO Jr (2013b) Fossil gregarines in Dominican and Burmese amber: examples of accelerated development? Palaeodiversity 5:1–6

Poinar G Jr (2014a) Spirochete-like cells in a Dominican amber *Amblyomma* tick (Arachnida: Ixodidae). Hist Biol 27:565–570

Poinar G Jr (2014b) Rickettsial-like cells in the Cretaceous tick, *Cornupalpatum burmanicum* (Ixodida: Ixodidae). Cretac Res 52:623–627

Poinar G Jr (2014c) Evolutionary history of terrestrial pathogens and endoparasites as revealed in fossils and subfossils. Adv Biol 2014:1–29. https://doi.org/10.1155/2014/181353

Poinar G Jr (2015) A new genus of fleas with associated microorganisms in Dominican amber. J Med Entomol 52:1234–1240

Poinar G Jr (2017) Fossilized mammalian erythrocytes associated with a tick reveal ancient Piroplasms. J Med Entomol 54:895–900

Poinar G Jr (2018) A primitive triatomine bug, *Paleotriatoma metaxytaxa* gen. et sp. nov. (Hemiptera: Reduviidae: Triatominae), in mid-Cretaceous amber from northern Myanmar. Cretac Res 93:90–97

Poinar G Jr, Brown A (2003) A new genus of hard ticks in Cretaceous Burmese amber (Acari: Ixodida: Ixodidae). Systematic Parasitol 54: 199–205

Poinar G Jr, Boucot AJ (2006) Evidence of intestinal parasites of dinosaurs. Parasitology 133:245–249

Poinar G Jr, Brown A (2012) The first fossil streblid bat fly, *Enischomyia stegosoma* n. g, n. sp. (Diptera: Hippoboscoidea: Streblidae). Syst Parasitol 81:79–86

Poinar GO Jr, Hess R (1985) Preservative qualities of recent and fossil resins: Electron micrograph studies on tissue preserved in Baltic amber. J Balt Stud 16:222–230

Poinar G Jr, Milki R (2001) Lebanese amber. The oldest insect ecosystem in fossilized resin. Oregon State University Press, Corvallis, OR, pp 1–96

Poinar G Jr, Poinar R (1994) The quest for life in amber. Addison Wesley Publishing Company, New York, NY, pp 1–219

Poinar GO Jr, Poinar R (1999) The amber forest. Princeton University Press, Princeton, NJ, pp 1–270

Poinar G Jr, Poinar R (2004a) *Paleoleishmania proterus* n. gen., n. sp., (Trypanosomatidae: Kinetoplastida) from Cretaceous Burmese amber. Protist 155:305–310

Poinar G Jr, Poinar R (2004b) Evidence of vector-borne disease of early Cretaceous reptiles. Vector Borne Zoonot Dis 4(4):281–284

Poinar GO Jr, Poinar R (2005) Fossil evidence of insect pathogens. J Invertebr Pathol 89:243–250

Poinar G Jr, Shaw SR (2016) Endoparasitism of a Cretaceous adult weevil by a euphorine wasp (Hymenoptera: Braconidae). Neues Jahrbuch Geologie Paläontologie 282:109–113

Poinar GO Jr, Thomas GM (1984) Laboratory guide to insect pathogens and parasites. Plenum Press, New York, NY, pp 1–392

Poinar GO Jr, Hess R, Caltagirone LE (1976) Virus-like particles in the calyx of *Phanerotoma flavitestacea* (Hymenoptera: Braconidae) and their transfer into host tissues. Acta Zool (Stockh) 57:161–165

Poinar G Jr, Chambers KL, Buckley R (2007) *Eoëpigynia burmensis* gen. and sp. nov., an Early Cretaceous eudicot flower (Angiospermae) in Burmese amber. J Bot Res Inst Texas 1:91–96

Poinar GO Jr, Brown AE, Legalov AA (2016) A new weevil tribe, Mekorhamphini trib. Nov. (Coleoptera, Ithyceridae) with two new genera in Burmese amber. Biol Bull Bogdan Chmelnitskiy Melitopol State Pedagogical University 6:157–163

Poinar G, Telford SR (2005) *Paleohaemoproteus burmacis* gen. n., sp. n. (Haemospororida: Plasmodiidae) from an Early Cretaceous biting midge (Diptera: Ceratopogonidae). Parasitology 131:79–84

Schatmayr HG (2014) Viruses and paleoparasitology. In: Ferreira LF, Reinhard KJ, Araúyo A (eds) Foundations of Paleoparasitology. Editora Fiocruz, Rio de Janeiro, pp 199–203

Schaub GA (1994) Pathogenicity of trypanosomatids on insects. Parasitol Today 10(12):463–468

Schlee D (1990) Das Bernstein-Kabinett. Stuttgarter Beiträge zur Naturkunde C 28:1–100

Shi G, Grimaldi DA, Harlow GE, Wang J, Wang J, Yang M, Lei W, Li Q, Li X (2012) Age constraint on Burmese amber based on U-Pb dating of zircons. Cretac Res 37:155–163

Suh A (2021) Horizontal transfer of transposons as genomic fossils of host-parasite interactions. In: De Baets K, Huntley J W (eds) The evolution and fossil record of paraistism: Coevolution and paleoparasitological techniques. Topics in Geobiology 50

Whittington HB (1985) The burgess shale. Yale University Press, New Haven, CN, pp 1–168

Wier A, Dolan M, Grimaldi D, Guerrero R, Wagensberg J, Margulis L (2002) Spirochete and protist symbionts of a termite (*Mastotermes electrodominicus*) in Miocene amber. Proc Natl Acad Sci 99:1410–1413

Wolfe AP, McKellar RC, Tappert R, Sodhi RNS, Muehlenbachs K (2016) Bitterfeld amber is not Baltic amber: three geochemical tests and further constraints on the botanical affinities of succinite. Rev Palaeobot Palynol 225:21–32

Chapter 3
Fungi as Parasites: A Conspectus of the Fossil Record

Carla J. Harper and Michael Krings

Abstract Fungal parasites are important drivers in ecosystem dynamics today that can have far-reaching effects on the performance and community structure of other organisms. Knowledge of the fossil record and evolution of fungal parasitism is therefore a key component of our understanding of the complexity and functioning of ancient ecosystems. However, the fossil record of fungi as parasites remains exceedingly incomplete for several reasons. This chapter provides selected fossil examples of (putative) fungal parasites in association with land plants, algae, other fungi, and animals, and elucidates the inherent problems that often render interpretation of even the most exquisite fungal fossils difficult. Of all the potential levels of fungal interaction, parasitism is perhaps the most difficult to demonstrate in the fossil record. Different lines of evidence obtained from both the host and fungus are required to safely discriminate parasitic fungi from saprotrophs and even mutualists when examined in fossils.

Keywords Chert · Disease symptom · Host response · Interaction · Mycoparasitism · Preservation · Rhynie chert

C. J. Harper (✉)
Botany Department, School of Natural Sciences, Trinity College Dublin, Dublin 2, Ireland

SNSB-Bayerische Staatssammlung für Paläontologie und Geologie, Munich, Germany

Department of Ecology and Evolutionary Biology, University of Kansas, Lawrence, KS, USA

Natural History Museum and Biodiversity Institute, University of Kansas, Lawrence, KS, USA
e-mail: charper@tcd.ie

M. Krings
SNSB-Bayerische Staatssammlung für Paläontologie und Geologie, Munich, Germany

Department of Ecology and Evolutionary Biology, University of Kansas, Lawrence, KS, USA

Natural History Museum and Biodiversity Institute, University of Kansas, Lawrence, KS, USA

Department für Geo- und Umweltwissenschaften, Paläontologie und Geobiologie, Ludwig-Maximilians-Universität, Munich, Germany
e-mail: krings@snsb.de

© The Author(s) 2021

K. De Baets, J. W. Huntley (eds.), *The Evolution and Fossil Record of Parasitism*, Topics in Geobiology 49, https://doi.org/10.1007/978-3-030-42484-8_3

3.1 Introduction

Fungi occur in next to every ecosystem of the world, colonizing numerous substrates and performing multiple functions (Treseder and Lennon 2015; Dighton and White 2017). As carbon heterotrophs, they have, by necessity, evolved several different nutritional modes and mastered various levels of cooperation with, and exploitation of, other organisms to acquire carbon (Lewis 1973). Many fungi degrade complex organic compounds such as lignin and cellulose and, through this recycling, are important in returning minerals to the soil and CO_2 to the atmosphere (Hatakka 2005; Baldrian and Valášková 2008). Others partner with certain types of algae and cyanobacteria to form lichens (Nash 2008), while still others enter into mycorrhizal associations with non-vascular and vascular land plants (Brundrett and Tedersoo 2018). Fungi have also evolved mutualistic associations with animals; some even thrive within the animal, in anaerobic environments (Orpin and Joblin 1997; Dollhofer et al. 2015). On the other hand, what has been termed the "dark side" of the fungal Kingdom (Taylor et al. 2015) is that, as parasites and pathogens, fungi negatively affect the performance of other microorganisms, plants, animals, and even humans and are causative agents of many diseases (e.g., Sharon and Schlezinger 2013; Köhler et al. 2015; Hall and Noverr 2017; Möller and Stukenbrock 2017).

Parasitic fungi live and derive the majority of their nutrients at the expense of other organisms that are alive at the time of infection (Deverall 1969; Zelmer 1998). Biotrophic parasitic relationships represent physiologically balanced systems, in which the parasite coexists with its host for an extended period of time, whereas necrotrophic parasites kill host tissue and then feed saprotrophically on the dead remains (Glazebrook 2005; Delaye et al. 2013). However, it is known today that, while this subdivision is generally accurate, the actual situation is more complex because many fungi behave as both biotrophs and necrotrophs, depending on the conditions in which they find themselves or the stages of their life cycles (Glazebrook 2005).

The origin of the true fungi is estimated at between 660 Ma and up to 2.6 Ga ago based on molecular clock data and some paleontological evidence (for details on early fungal fossils, see Krings et al. 2017c; Loron et al. 2019; Bonneville et al. 2020), and the divergence of the fungal-animal lineage from the plant lineage at between 780 Ma and up to 2.5 Ga ago (e.g., Altermann and Schopf 1995; Martin et al. 2003; Taylor and Berbee 2006; Blair 2009; Lücking et al. 2009; Sharpe et al. 2015; Bengtson et al. 2017; Berbee et al. 2017). The nutritional mode of the common ancestor of the true fungi remains elusive. However, early-diverging branches of the fungal stem lineage include the Aphelida, which are parasites of planktonic algae (Letcher et al. 2013, 2017; Karpov et al. 2014, 2017), and the animal-endoparasitic Cryptomycota and Microsporidia (Keeling and Fast 2002; James et al. 2006; Jones et al. 2011a, b; Vávra and Lukeš 2013; Han and Weiss 2017; Bass et al. 2018), suggesting that the evolutionary arms race of fungi as parasites of other organisms is of ancient origin (Anderson et al. 2010). Unfortunately, none of these early-diverging lineages (except possibly Aphelida; see Krings and Kerp 2019) have been documented in the fossil record, due probably to the simple fact that it is very difficult to recognize these life forms as fossils because of their diminutive size (e.g., Garcia 2002). Moreover, they lack structural features that could be used to safely identify them, with one possible exception. Xenomas (or xenoparasitic complexes; see Chatton 1920; Weissenberg 1968) are tumor-like growths caused by a variety of parasitic protists and fungi, including Microsporidia. They can occur on numerous organisms such as oligochaetes, insects, and crustaceans; however, they are predominantly found on fishes (Matos et al. 2003; Lom and Dyková 2005; Weiss and Becnel 2014). It might therefore be worth looking for such abnormal growths also in well preserved fossil fish specimens, although it will most likely be very difficult to positively identify the actual causative agent(s) (e.g., Petit 2010; Petit and Khalloufi 2012).

Fungal parasites can have profound influence on the performance and community structure of other organisms, and thus are important players in ecosystem functioning today (e.g., Marcogliese 2004; Skerratt et al. 2007; Sime-Ngando 2012; Frenken et al. 2017). Knowledge of their fossil record, evolution, and the roles they played in biological and ecological processes in the past is therefore a key component of our understanding of the complexity and functioning of ancient ecosystems. However, the fossil record of fungi as parasites remains incomplete and is often difficult to interpret for several reasons as explained in the following section (Fig. 3.1a, b).

3.2 Identifying Fungal Parasitism in the Fossil Record

3.2.1 Finding Fossil Fungi

The success of finding fungi from the geologic past heavily relies on the way the fossils are preserved (Krings et al. 2012), even more so if nutritional modes and interactions with other organisms are to be resolved, too. Cherts certainly represent the most important sources of evidence of fossil fungi (Taylor et al. 2015). Chert deposits occur at various points in geologic time and typically represent an extremely dense microcrystalline or cryptocrystalline type of sedimentary rock (Laschet 1984; Hesse 1989). Some cherts may be fossiliferous and demonstrate not only three-dimensional and structural preservation of the organisms (sometimes even in situ), but often also details of individual cells and subcellular structures. As a result of faithful fossil preservation, cherts provide an ideal matrix from which to extract information about fungi and fungal interactions with other organisms. Moreover, cherts provide the only source of direct evidence of the fungal world within the context of ecosystem complexity, versatility, and dynamics. Foremost among the chert deposits yielding evidence of fungi is the Lower Devonian (~410 Ma) Rhynie chert from Aberdeenshire, Scotland (Trewin and Kerp 2017; Garwood et al. 2020), that contains representatives of all major fungal lineages except Basidiomycota (Taylor et al. 2004; Krings et al. 2017a). The Rhynie chert is perhaps best known

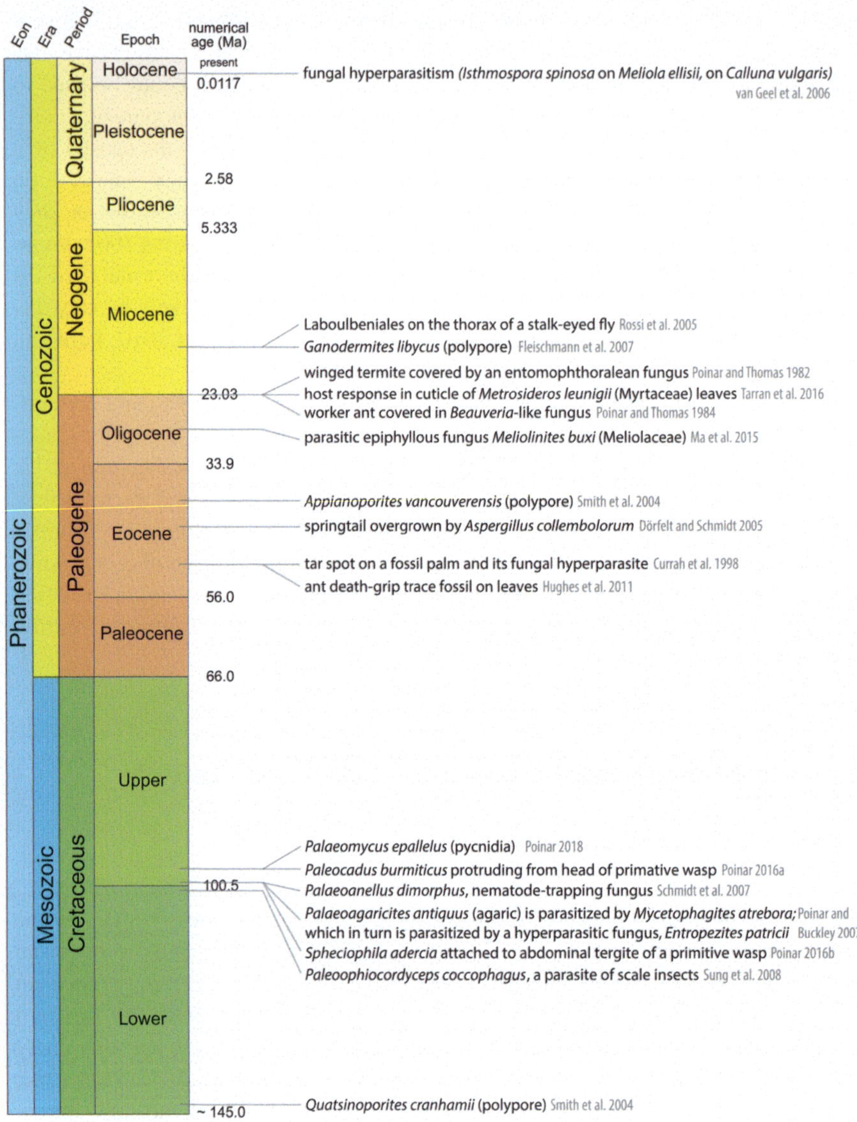

Fig. 3.1 (**a**) Selected fossils of fungi as parasites plotted on the Cretaceous to Holocene stratigraphic chart. Stratigraphic chart based on Cohen et al. (2013). (**b**). Selected fossils of fungi as parasites plotted on the Devonian to Jurassic stratigraphic chart. Stratigraphic chart based on Cohen et al. (2013)

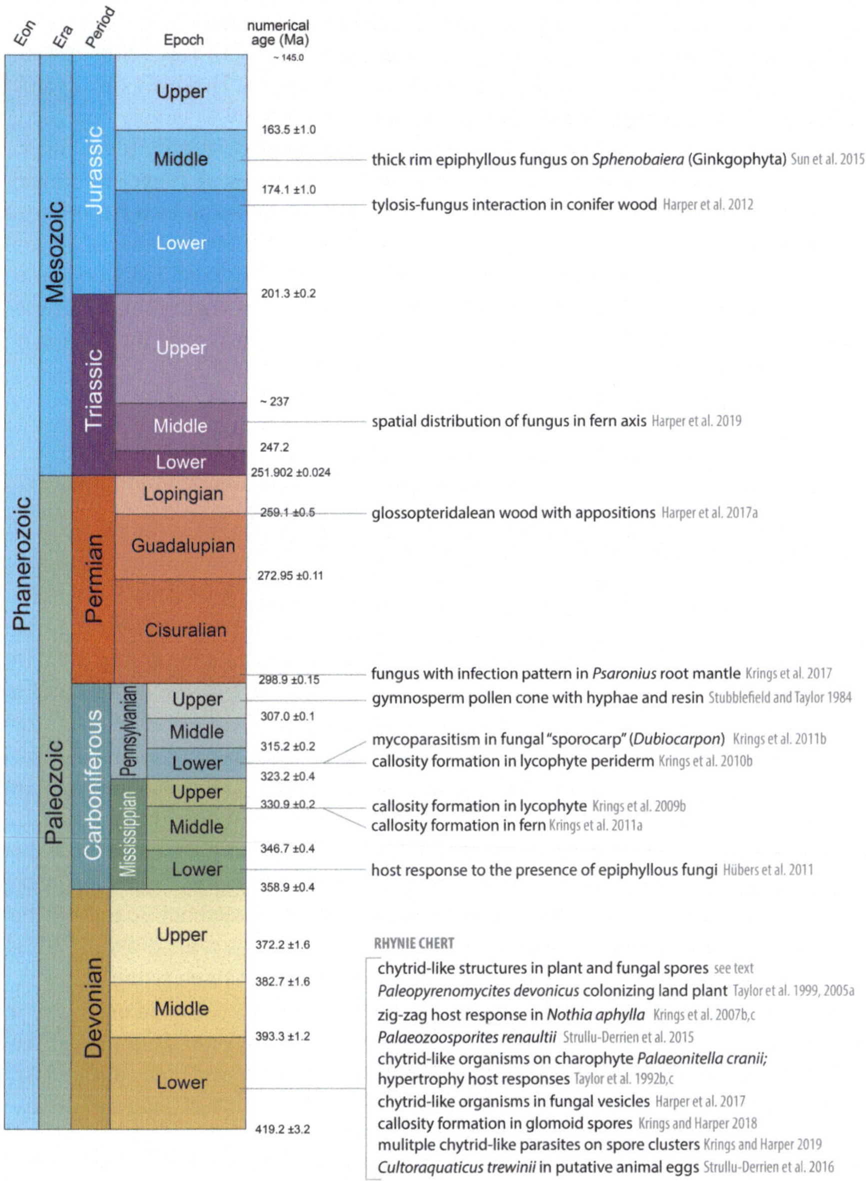

Fig. 3.1 (continued)

among mycologists for several specimens of fungi that are exquisitely preserved in situ together with their host organisms and demonstrate the existence of different types of fungal associations and interactions, including arbuscular mycorrhizas (Taylor et al. 1995, 2005b; Brundrett et al. 2018; Walker et al. 2018), and parasitism of land plants, algae, other fungi, and possibly animals (see below).

Permineralized peat and coal balls are two other rock matrices that may contain abundant fossil evidence of fungi (e.g., Williamson 1878, 1880, 1883; Stubblefield and Taylor 1988; Harper et al. 2015, 2016; Slater et al. 2015). However, the organic remains in these matrices are usually compacted and partially to largely degraded, and hence render it more difficult, but not impossible, to safely identify fungal interactions (e.g., Knoll 1985; Krings et al. 2014). Various types of fungi and fungal interactions, as well as indirect evidence of fungal activities such as microborings and chemical traces (Golubic et al. 1975; Marynowski et al. 2013), have also been exquisitely preserved by other preservation modes, including silicified wood, animal hard parts, and amber (Taylor et al. 2015). However, the ecological configuration of the community in which these organisms lived is often less completely known.

3.2.2 *Tracing Fungal Parasitism in the Fossil Record*

Parasitic fungi exploit carbon sources that are, by definition, alive at the time of infection. However, fossils represent snap-shots in time, with no long-term and follow-up studies available to determine the condition of the host before and after fungal colonization. This raises the question how, if at all, parasitic fungi can be recognized as fossils and distinguished from epi-/endophytes and saprotrophs? In other words, how can we tell if a fossil containing evidence of the presence of a fungus was (part of) a living organism at the time of fungal colonization, and how can we determine whether the fungus thrived at the host's expense?

If the host organism is preserved in pristine condition, then this could mean that it was alive at the time of fungal colonization. Conversely, a host that is tattered, fragmented, and shows tissue disruption and disintegration might have been in the process of decay when colonized. Nevertheless, it is not normally possible to determine whether fragmentation and tissue destruction were initiated before or after the fungus colonized the host, or are preservation artifacts (Krings et al. 2009b, 2010b). Somewhat more reliable is perhaps the spatial distribution of fungi within the host. A fungal distribution pattern within an intact host that reflects forced entry and a consistent pathway of colonization (e.g., along the vascular bundles in plants; see Harper et al. 2019), or is restricted to certain body parts or tissue types of the host, is suggestive of colonization of a living host, while colonization of dead and decaying matter more likely results in random fungal distribution.

Structural features suggestive of parasitism and pathogenicity in fossil organisms include disease symptoms and host reactions such as cell and tissue alteration or local necroses (e.g., Mendgen et al. 1996; Pearce 1996). In rare instances, the fungal perpetrator and the disease symptom/host response even co-occur, providing additional lines of direct evidence (Taylor et al. 1992b, c; Krings and Harper 2018). For example, a common type of host response to fungal intrusion is the formation of callosities, which are inwardly directed projections consisting of newly synthesized cell wall material that encase the parasite's penetration device, and thus may reduce or inhibit nutrient extraction from the host (Akai 1959; Aist 1976, 1977). Callosities have been observed in fossil plants belonging to several different lineages, including lycophytes (Krings et al. 2009b, 2010b), sphenophytes (Taylor et al. 2012), ferns (Krings et al. 2011a), and gymnosperms (Stubblefield et al. 1984), but have also been recorded in fossil fungi (Hass et al. 1994). Several of the latter records even provide evidence of a biotrophic relationship, in which the parasite was contained to a certain extent by the callosities, but was still able to grow and extract nutrients, while the host remained demonstrably viable for an extended period of time while being parasitized (Krings and Harper 2018). However, not all parasites elicit host responses, and it may therefore be difficult, if not impossible, to determine the nutritional modes of asymptomatic fossil fungi associated with intact hosts. For example, commonly present within structurally preserved plants throughout the Phanerozoic are small fungal reproductive units (e.g., spores, sporangia, cleistothecia, pycnidia) and mycelia that are randomly distributed; no evidence of host responses has been found (Magnus 1903; Stubblefield and Taylor 1986; LePage et al. 1994; García Massini et al. 2012; Klymiuk et al. 2013). Some of these fungal remains, including ascomycotan hyphae, pseudothecia, pycnidia, and hyphomycetous spores, have nonetheless been interpreted as parasites because their (presumed) modern equivalents are parasites of plants (LePage et al. 1994; García Massini et al. 2012). Finally, many of the host responses known in extant organisms (e.g., chemical responses; see Swain 1977; Langenheim 1994) cannot be identified in fossils or are easily mistaken for natural decay (e.g., necroses; see Van Loon et al. 2006).

Evidence of fungal parasitism in ancient ecosystems also occurs in the form of fungal structures that are found as detached fossils (i.e. with no information on the host available), but that can be directly compared to modern fungal taxa known to be parasites. For example, polyporous fungi or polypores (Basidiomycota) today thrive as saprotrophs in decaying wood or as parasites and perpetrators of diseases in conifers and hardwoods (Blanchette 1991; Ryvarden 1991; Schwarze et al. 2000). The Cretaceous and Cenozoic record of these fungi is quite extensive and consists primarily of basidiocarps (conks) that usually can be assigned to modern families and genera with some confidence based on morphology and spore structure (e.g., Smith et al. 2004; Fleischmann et al. 2007). The inventory of fossil polypores suggests that these fungi were widely distributed and diverse in Neogene and Quaternary forest paleoecosystems and significant in delignification processes and as pathogens of woody plants.

Of all the potential levels of interaction between fungi and other organisms, parasitism is perhaps the most difficult to demonstrate in the fossil record. Without a combination of different lines of evidence obtained from both the host and fungus, this type of interaction cannot be discriminated from saprotrophism and even mutualism when examined in fossils (Taylor et al. 2009). Excellent examples illustrating this problem occur in the form of structurally preserved remains of Lepidodendrales (arborescent lycophytes) from the Carboniferous of central Europe that contain diverse assemblages of fungal mycelia and reproductive units (Krings et al. 2007a, 2009b, 2010b, 2011d; summarized in Fig. 3.2). Some of these fungi have been interpreted as parasites based on morphology and distribution (Fig. 3.2j) or the presence of host responses (Fig. 3.2f), while others were probably mutualists (mycorrhizal

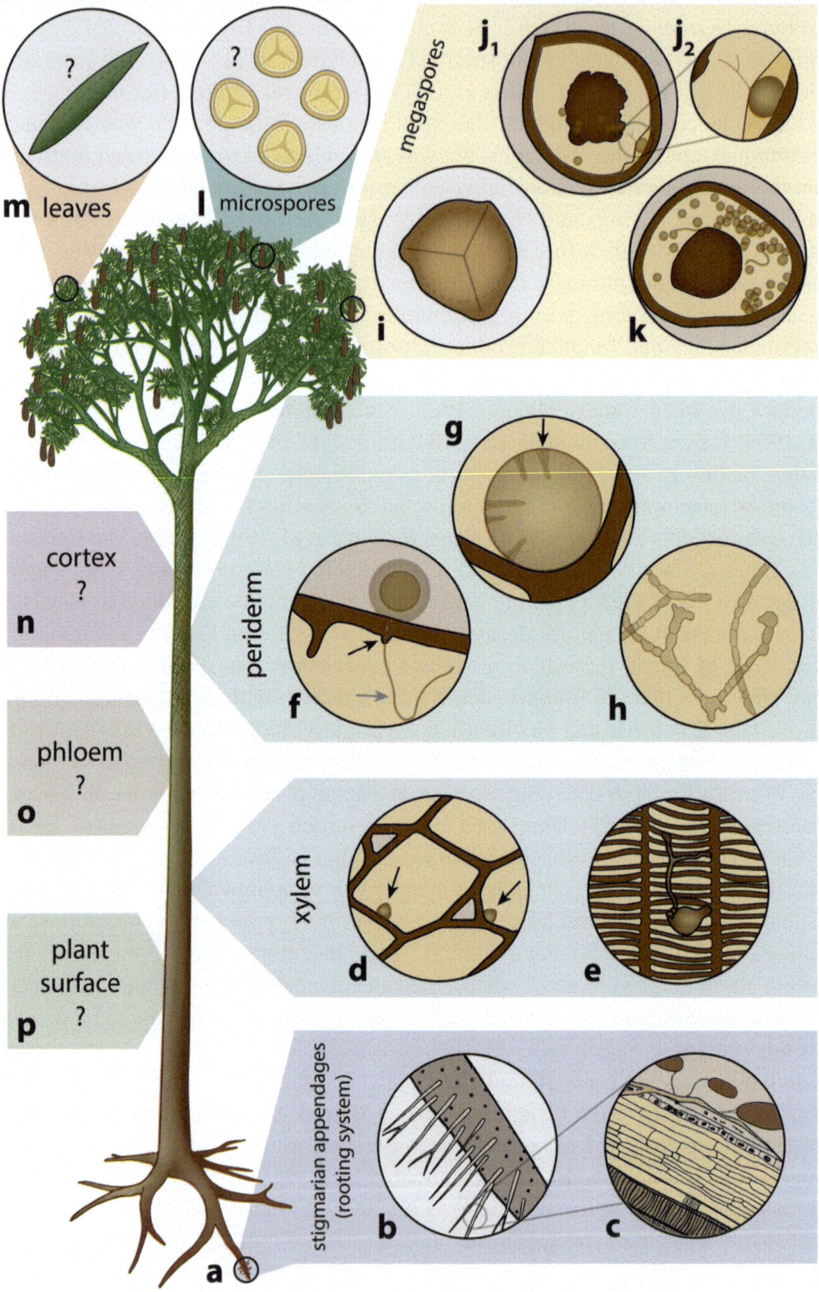

Fig. 3.2 Synopsis of documented evidence of fungi associated with Carboniferous arborescent lycophytes (Lepidodendrales). Some of these fungi, especially those eliciting host responses, were probably parasites, whereas others were mutualists or saprotrophs. (**a**) Stigmarian appendages

fungi; Fig. 3.2a–c) because their morphology parallels that seen in present-day mycorrhizal fungi (Glomeromycota). However, the nutritional modes of most fungal remains associated with Carboniferous Lepidodendrales remain elusive.

3.3 Fossils of Fungi as Parasites

Fossil evidence of fungi has been documented throughout the Phanerozoic (Taylor et al. 2015), but it is fungal associations with land plants and other fungi from the Early Devonian, Carboniferous, Triassic, and Cenozoic that have to date been examined more systematically. As a result, there are several well-documented examples of fungal parasites of plants and other fungi from these periods of geologic time. On the other hand, the fossil record of parasitic fungi of animals is scanty throughout for several reasons (see Sect. 3.3.4 below). The purpose of this chapter is to portray the fossil record of fungi as parasites. In the sections below, we have not attempted to provide exhaustive coverage, but rather have selected fossil examples of (putative) fungal parasites in association with plants (i.e. land plants and a few algae), other fungi, and animals, with a slight emphasis on the Rhynie chert, and have elucidated the inherent problems that often render interpretation of even the most exquisite fungal fossils difficult.

3.3.1 Fungal Parasites of Land Plants

Land plants today are exposed to a wide variety of different fungi, many of which are parasites (Cannon and Hawksworth 1995). Although one can only speculate about the events during the terrestrialization of plants some 515–485 Ma ago (Morris et al. 2018), the conquest of the terrestrial realm has likely been profoundly influenced by interactions with saprotrophic, parasitic, and mutualistic fungi

Fig. 3.2 (continued) (below-ground rooting system). (**b**) Higher magnification of Fig. 3.3a, focusing on appendages. (**c**) Higher magnification of Fig. 3.3b, showing mycorrhizal association in cortex of appendage; based on fig. 2a in Krings et al. 2011d. (**d**) Chytrid-like sporangia (arrows) attached to tracheid walls; based on pl. II, 1 in Krings et al. 2009b. (**e**) Pear-shaped sporangium with narrow, aseptate subtending hypha growing along inner surface of tracheid; based on fig. 1j in Krings et al. 2007a. (**f**) Putative chytrid resting sporangium with primary rhizoidal axis (gray arrow) and early stage in callosity development (black arrow); note another, more prominent callosity to left; based on fig. 4j in Krings et al. 2010b. (**g**) Glomeromycotan spore with callosities (arrow); based on pl. III, 10 in Krings et al. 2009b. (**h**) Septate, irregularly swollen hyphae; based on figs 1e, f in Krings et al. 2010b. (**i**) Surface view of megaspore. (**j_1**) Megaspore with large chytrid zoosporangium between wall layers; based on pl. I, 7 in Krings et al. 2009b. (**j_2**) Higher magnification of chytrid, showing rhizoidal system extending into spore lumen; based on pl. I, 8 in Krings et al. 2009b. (**k**) Megaspore containing numerous spheres (?zoosporangia of a polycentric chytrid) and tenuous hyphae; based on pl. I, 1 in Krings et al. 2009b. (**l–p**) Lepidodendralean organs, tissues, and reproductive structures for which there is currently no documented evidence of associated fungi

(Chisholm et al. 2006). To become effective as a plant parasite, a fungus must access the plant interior, either by actively penetrating the surface or by finding, recognizing, and entering through wounds or natural openings such as stomata (Hoch and Staples 1991; Knogge 1996; Mendgen et al. 1996; Szabo and Bushnell 2001). Alternatively, a fungus on the plant surface may produce substances that kill (parts of) the host and subsequently feed on the decaying tissue (Berrocal-Lobo et al. 2002; Dean et al. 2012). Fungal parasites today are found on all parts of plants, including roots, stems, leaves, reproductive structures, and pollen grains (Money 2016). Interestingly, there are also some 400 species of plants that parasitize fungi and exploit them as their principle source of carbon (Leake 2005; Merckx 2013), but that's another story.

3.3.1.1 Early Land Plants

Several (putative) fungal parasitic interactions with early land plants have been reported from the Lower Devonian Rhynie chert, including chytrid-like fungi interpreted as parasites that are associated with the spores of several early land plants (Fig. 3.3a) (Kidston and Lang 1921; Harvey et al. 1969; Illman 1984; Taylor et al. 1992a), and *Paleopyrenomycites devonicus*, a perithecial ascomycete colonizing the land plant *Asteroxylon mackiei* (Taylor et al. 1999, 2005a). Although no host response has been found, Taylor et al. (2005a) submit that *P. devonicus* colonized *A. mackiei* while it was alive based on the fact that the perithecia often occur within the substomatal chambers of the host plant, with the ostioles directly beneath the stomata to facilitate spore dissemination (Fig. 3.3b). Moreover, some of the perithecia contain remains of other fungi believed to represent mycoparasites (Taylor et al. 2005a: fig. 41). Another example of fungal parasitism has been described in rhizomes of the land plant *Nothia aphylla* (Krings et al. 2007b, c), but this time the fungal intruders elicit host responses in the form of characteristic cell and tissue alterations. A hypodermal zigzag line composed of secondarily thickened cell walls characterizes heavily infected rhizomes (Fig. 3.3c). This line marks the outer

Fig. 3.3 (continued) from ray (r), with fungal hyphae (white arrows) penetrating through tylosis (Jurassic); University of Kansas paleobotanical collection slide TS-GIX-SB-036-01; scale bar = 20 μm. (**g**) Fungus extending into, and subsequently forming coralloid branching systems within, lumen of *Psaronius* root mantle cell (Permian); pl. IV, fig. 3 in Krings et al. 2017b; scale bar = 20 μm. (**h**) Ascoma surrounded by incompletely thickened ring (arrowheads) formed by host leaf cuticle (Jurassic); pl. II, fig. 15 in Sun et al. 2015; scale bar = 100 μm. (**i**) Higher magnification of Fig. 3.2h, focusing on thickened rim; scale bar = 20 μm. (**j**) Chytrid-like inclusions in pollen grain of *Striatopodocarpites multistriatus* (Permian); fig. 2D in Aggarwal et al. 2015; scale bar = 20 μm. (**k**) Angiosperm leaf portion with 5 pycnidia (arrows) of *Palaeomycus epallelus* in Myanmar amber (Cretaceous); fig. 1 in Poinar 2018; scale bar = 1 cm. (**l**) *Milleromyces rhyniensis* chytrid zoosporangium extending through cell surface (arrowhead) of the charophyte *Palaeonitella cranii* (Devonian); color version of fig. 10 in Taylor et al. 1992b; scale bar=10 μm. (**m**) Two *Palaeonitella cranii* cells showing extensive enlargement (hypertrophy host response; h) when compared with normal cells at base (Devonian); color version of fig. 26 in Taylor et al. 1992b; scale bar = 100 μm. (**n**) Longitudinal section of normal cells (n) of *Palaeonitella cranii* (Devonian); color version of fig. 1 in Taylor et al. 1992b; scale bar = 100 μm

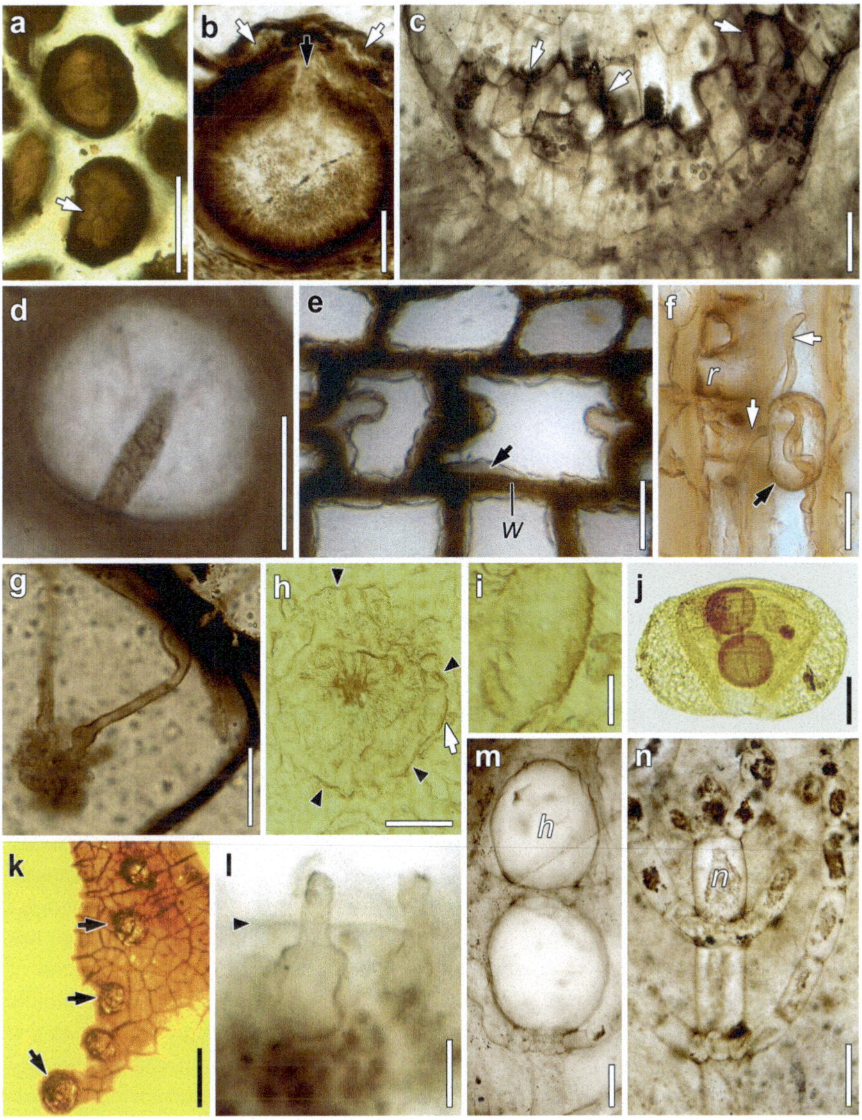

Fig. 3.3 Fossils of fungus-land plant and fungus-alga (putatively) parasitic interactions (detailed explanations in the text). (**a**) Endobiotic fungi (arrow) in spores of *Horneophyton lignieri* (Devonian); Munich collection, slide SNSB-BSPG 1964 XX 24; scale bar = 50 μm. (**b**) *Paleopyrenomycites devonicus* perithecium with ostiole (black arrow) in sub-stomatal chamber; white arrows indicate guard cells (Devonian); color version of fig. 7 in Taylor et al. 2005a; scale bar = 100 μm. (**c**) Zig-zag host response (arrows) in *Nothia aphylla* (Devonian); color version of fig. 2e in Krings et al. 2007c; scale bar = 100 μm. (**d**) Callosity extending into lumen of *Botryopteris antiqua* cortical cell (Carboniferous); color version of fig. 1n in Krings et al. 2011a; scale bar = 20 μm. (**e**) Glossopteridalean cell walls (w) with appositions and wall swellings (arrow) (Permian); fig. 2E in Harper et al. 2017a. (**f**) Conifer wood with tylosis (black arrow) extending

boundary of cells containing one of the intrusive fungi, and hence probably represents a host response effective in separating infected from uninfected tissues. Moreover, several of the infected rhizomes contain peripheral regions that are devoid of cells. Krings et al. (2007c) suggest that this tissue degradation may have been effective as a defense mechanism based on the fact that, in some extant plants, phytopathogenic microorganisms are deterred by programmed cell death around the infected areas that inhibit the microbes from spreading (Hammond-Kosack and Jones 1996; Veronese et al. 2003; Glazebrook 2005; Anderson et al. 2010).

Clusters of globose vesicles attached to branching hyphae characterize *Palaeozoosporites renaultii*, another fungus in *Asteroxylon mackiei* (Strullu-Derrien et al. 2015). These authors suggest that *P. renaultii* was a parasite with affinities to the Chytridiomycota, and report, but do not illustrate, a host response in the form of secondarily thickened cell walls. We hold the opinion that *P. renaultii* represents a cluster of glomoid spores; however, Strullu-Derrien et al. (2015) reject affinities to the Glomeromycota because "hyphal structures…narrow progressively as they branch", which is in fact a common morphology within the Glomeromycota (Walker et al. 2018).

3.3.1.2 Plant Structural Alterations in Response to Fungal Intrusion

The previous section provided examples of structural defense mechanisms effective in slowing down or deterring "unwanted" fungal colonization or spreading that were in place in early land plants by the Devonian. A little later, in the Carboniferous, vascular plants showed host responses against fungal intrusion in the form of callosities (also called appositions, lignotubers, or papillae, among other terms; see Stubblefield et al. 1984) that closely resemble defenses employed by plants today (Akai 1959; Pearce 1996; Schwarze et al. 2000; Schulze-Lefert 2004). One example of callosity formation occurs in a rachis of the filicalean fern *Botryopteris antiqua* (Krings et al. 2011a) from the Mississippian of France (Fig. 3.3d), while another has been reported in a lycophyte (*Lepidodendron* sp.), also from the Mississippian of France (Krings et al. 2009b: pl. II, figs 12–17). The latter specimen even contains two different types of callosities, namely a narrow form that does not show evidence of a penetration canal, and a larger form that may be straight or curved and usually contains a central penetration canal. The presence of two different types of callosities may be evidence that this host recognized two different intruders. Although putative chytrid zoosporangia occur in the same tissue samples as the callosities, they have not been found in organic connection, and thus cannot be positively linked to one another. Other documented evidence of callosity formation in Carboniferous plants includes lycophyte periderm from the Pennsylvanian of Great Britain (Krings et al. 2010b: fig. 4J–M) and sphenophyte rootles from the Pennsylvanian of France (Taylor et al. 2012: pl. I, fig. 1, pl. II, fig. 9, pl. III, figs 1–3). Finally, the gymnosperm pollen cone *Lasiostrobus polysacci* from the Carboniferous of North America contains septate fungal hyphae in the cortex and microsporophylls (Stubblefield et al. 1984). The host cells are sometimes accompanied by opaque matter interpreted as resin that might represent a host response. In addition, on the inner surface of the cells are swellings suggestive of some type of wall apposition. Although fungi today frequently target the nutritional density in reproductive structures of plants (Vujanovic et al. 2009), the preceding is one of the few persuasive fossil examples of this fungal strategy.

3.3.1.3 Host Responses in Woody Plants

Woody plants have a long evolutionary and antagonist history with fungi (Schwarze et al. 2000; Vacher et al. 2008). Wood-degrading fungi encompass a heterogeneous assemblage of basidiomycetes and ascomycetes, and constitute one of the major drivers of carbon cycling in forest ecosystems today (Lindahl et al. 2002; van der Heijden et al. 2008). Some investigators have suggested that wood-rotting fungi begin their life cycle as parasites, but then, once the host is dead, switch to saprotrophism (Garrett 1970; Lewis 1973). Evidence of decay attributable to fungi is frequently encountered in fossil wood; however, studies focusing on fossil fungal wood degradation are rare (see Harper et al. 2016 for a review; Wan et al. 2017), and documented examples of (partially) decayed fossil wood containing well preserved fungal remains are even rarer. Harper et al. (2017a) report on decaying glossopterid wood from the Permian of Antarctica that contain fungal remains, symptoms of white pocket-rot decay, arthropod remains, and host-responses in the form of appositions (Fig. 3.3e). Appositions that occur at sites of infection or attempted penetration by a fungus (Pearce 1996) are composed of material and components not normally present in cell walls (e.g., phenolic compounds, callose, silicon) plus normal cell wall components, especially suberin (Pearce and Holloway 1984), that can partially to fully occlude cell lumina to contain or prevent further spreading of the intruder (Aist 1976, 1983). The lumina of some of the tracheids in the glossopterid wood are completely sealed by some opaque matter (Harper et al. 2017a: fig. 2F, J), while the cell walls of other tracheids are swollen and partially occlude the lumen (Harper et al. 2017a: fig. 2I). Both types of cell lumen occlusion might represent strategies of passive defense against antagonistic fungal expansion within the wood. Conspicuous swellings in extant wood have been interpreted as a reaction or barrier zone to penetration by delignifying fungi (Schwarze and Baum 2000). Similar appositions have also been documented in other Permian woods from Antarctica (Stubblefield et al. 1985; Stubblefield and Taylor 1986; Weaver et al. 1997). Other structures in fossil woods believed to represent host responses to fungi include ergastic substances and resin (Stubblefield et al. 1985; Gnaedinger et al. 2015). However, it is difficult to specifically attribute these formations to fungal parasitism because they are also known to be produced in response to damages caused by fire or mechanical injury (e.g., Shrimpton 1973; Blanchette and Biggs 1992). In rare cases, such as in a Jurassic conifer wood from Antarctica, the fungus is in direct contact with a possible host response in the form of tyloses (Fig. 3.3f) (Harper et al. 2012). Tyloses are bladder- or sac-like outgrowths (protoplasmic bulges) on parenchyma cells that extend into adjacent conducting cells (tracheids, vessels) via pits in the cell walls and, in this way, can block the dead conducting cells to counteract the spreading of phytopathogenic organisms. They may be filled with various substances (tannins, gums, resins, etc.) as a result of physical damage or parasitic activity (Collins and Parke 2008; Feng et al. 2013; for an extensive review, see De Micco et al. 2016). Harper et al. (2012) hypothesize that the Jurassic tyloses formed a physical barrier to prevent the fungus from spreading. The morphology and pattern of colonization suggest that the fossil shares similarities with various extant Ascomycota, including sap-stain, blue-stain, and dark-stain fungi that are pathogens of various conifers (see Ballard et al. 1982; Hessburg and Hansen 1987). Last, in some extant gymnosperms, including *Pinus*, the number of resin ducts in the xylem may increase as a result of a fungal infection (Martín-Rodrigues et al. 2013). This is certainly a structural feature that can also be recognized in fossil wood. In addition, certain types of tissue disruption are caused by parasitic plants invading stem tissue, but may also be the result of fungal infection (Gomes and Fernandes 1994; do Amaral and Ceccantini 2011). However, no evidence of such tissue disruptions in fossil wood has been produced to date, which may be due in part to the fact that most investigators of fossil wood lack a search image for such structures, or perhaps attribute the disruptions to a different cause.

3.3.1.4 Host Plant Preservation and Fungal Distribution

While the evidence used to infer fungal parasitism in the fossils surveyed in the preceding sections largely consists of fossilized host responses, there is one example of a fossil fungus-land plant interaction that deserves special mention because in this case host plant preservation and fungal distribution within the host have been used to infer the nutritional mode of the fungus (Barthel et al. 2010; Krings et al. 2017b). This fungus occurs in a silicified Early Permian *Psaronius* root mantle from Germany, and displays a consistent pattern of host cell colonization that includes the formation of swellings effective in pushing a hyphal tip through the host cell wall and multi-branched structures remotely resembling arbuscules and certain haustoria that probably served in nutrient extraction or exchange (Fig. 3.3g). The different tissues of the host root mantle, including the fragile root aerenchyma, are exquisitely preserved, suggesting that the roots were intact, and thus probably alive at the time of fossilization. Moreover, the strictly intracellular growth pattern of the fungus seems implausible for a saprotroph that extends through moribund or dead and decaying plant tissue. However, the fungus did apparently not trigger any host response or disease symptom, suggesting it may have been a harmless endophyte or mild parasite, which extracted some nutrients, but not in an amount sufficient enough to cause notable damage. It is also possible, however, that the fungus was well adapted to its mode of life, rendering it "invisible" to the immune response of the plant, although this is virtually impossible to test in fossils.

3.3.1.5 Epiphyllous Fungi

The cuticle, a waxy coating of all aerial plant parts before secondary growth (Pollard et al. 2008), is the first line of physical defense and barrier against pathogenic fungi (Martin 1964; Serrano et al. 2014). However, many fungi have evolved strategies to breach this barrier, pass into the interior of the plant, and spread out (Kolattukudy 1985; Nicholson and Epstein 1991), while others grow on the plant surface and locally penetrate the cuticle to extract nutrients from the underlying tissues (Mendgen and Deising 1993; Tucker and Talbot 2001). Still other fungi reside on the plant surface without ever entering the host (e.g., Hongsanan et al. 2016). Fungi that grow on leaves are termed epiphyllous, regardless of whether they are parasites or just surface residents. Since leaf cuticles often survive fossilization and diagenesis relatively unaltered, they can be freed from the surrounding rock matrix and cleared through chemical maceration processes and studied in transmitted light (Kerp 1990; Kerp and Krings 1999). Fossil leaf cuticles provide information on epidermal anatomy, including cell pattern and stomatal morphology, but may also contain information on leaf-associated fungi. There are numerous reports of fossil epiphyllous fungi, mostly microthyriaceous types, for which details of the fungus and the host are known (e.g., Dilcher 1965; Elsik 1978; Phipps and Rember 2004; Limaye et al. 2007; Bannister et al. 2016). For the most part, the nutritional modes of these fungi remain unknown; some authors indicate there are morphological similarities to modern plant pathogens such as *Asterina*, *Vizella*, and *Trichothyrina*, thus inferring the nutritional mode as parasitism (Ellis 1977; Phipps 2007; Khan et al. 2015). Evidence suggestive of a host response to the presence of an epiphyllous fungus in the form of a rim of thickened cuticle has been described in a Jurassic *Sphenobaiera* (Ginkgophyta) leaf from China (Fig. 3.3h, i) (Sun et al. 2015: pl. II, fig. 15). Other examples of cuticle alterations interpreted as a host response include *Metrosideros leunigii* (Myrtaceae) leaves from the Eocene-Oligocene of Australia that appear to have produced cuticle thickenings to divert the growth of the hyphae of a fungal parasite (Tarran et al. 2016: fig. 8A). Another interesting epiphyllous fungus, *Meliolinites buxi* (Meliolaceae), occurs on the cuticles of Oligocene *Buxus* leaves from China (Ma et al. 2015: fig. 3A–H). These authors offer the hypothesis that *M. buxi* is a parasite based on the thickening and twisting of epidermal cell walls in the host leaf, along with the parasitic life style of the extant Meliolaceae. The earliest fossil evidence in plant cuticles of a host response to the presence of epiphyllous fungi occurs in the form of impressions of rosette-like fungal thalli on a dispersed plant cuticle of unknown systematic affinity from the Carboniferous of Germany (Hübers et al. 2011). The host reaction occurs in the form of extensive cutinizations around the thallus margins. The thalli are interpreted as hyphopodia or some other epiphyllous structure of a parasitic fungus that facilitated host attachment and penetration.

3.3.1.6 Dispersed Remains and Plant Pathogens

The dispersed microfossil record yields a plethora of information on fungi and fungal interactions in the geologic past (Kalgutkar and Jansonius 2000). For example, small 'spherules' that are sometimes attached to the outer surface or occurring within the body of Carboniferous to Cenozoic land plant spores and pollen grains obtained through palynological sampling (Fig. 3.3j) (e.g., Daugherty 1941: pg. 43; Phipps et al. 2000: pl. 2, figs 5 and 6; Aggarwal et al. 2015: figs 2–4) are mostly interpreted as remains of fungi and compared to modern pollen-colonizing Chytridiomycota. On the other hand, Mesozoic and Cenozoic non-pollen palynomorph (NPP) assemblages frequently contain spores of microthyriaceous and other fungi that are morphologically similar to the spores produced by certain present-day fungal parasites and pathogens (e.g., Höhnel 1924; Van Geel and Anderson 1988; Kalgutkar and Jansonius 2000; Van Geel 2002; Singh and Tripathi 2010; Kürschner et al. 2015; Schumilovskikh et al. 2015; Roth and Lorscheitter 2016).

There are certain fungal plant pathogens such as rusts (Pucciniomycetes), smuts (Ustilaginomycetes), and leaf spot diseases (e.g., *Alternaria, Cercospora*) (Agrios 2005) that are widespread today but rare or absent in the fossil record (reviewed in Taylor et al. 2015). This is surprising since innumerable fossils of leaves are available, and one would expect to find at least some showing evidence of these fungi in the form of lesions or galls (Callow and Ling 1978). However, there is likely a collection bias for undamaged leaves, thus probably discarding leaves with fungal remains (Taylor and Krings 2010; Krings et al. 2012). While no convincing evidence of fossil rusts has been documented, there are several reports of dispersed spores which are similar in morphology to extant *Puccinia*, *Gymnosporangium*, and *Uromyces* (Bradley 1931; Wolf 1969; Kalgutkar and Jansonius 2000). Documented evidence of fossil smuts is in a similar situation. Most of the reports of fossil smuts have later been dismissed or remain inconclusive. For example, fossils interpreted as spore clusters similar to *Ustilago* have been reported in degrading plant tissue from Deccan Intertrappean cherts (Cretaceous) from India (Kapgate 2016). However, none of the specimens figured display features of sufficient clarity to allow assignment to any group of fungi with confidence. The dispersed spore type *Ustilago deccanii* from the same beds was initially reported as a spore of a smut fungus (Chitaley and Yawale 1976, 1978), but has subsequently been transferred to *Inapertisporites*, a taxon used for fossil amerospores of Fungi Imperfecti (Kalgutkar and Jansonius 2000). No information is available on the nutritional modes of these fossil fungi. In yet another case, small spores in *Saururus tuckerae* anthers from the Eocene of North America were initially identified as a smut fungus (Currah and Stockey 1991; LePage et al. 1994), but are now thought to represent minute pollen grains produced by the flower (Smith and Stockey 2007). Another important plant pathogen today are the leaf spot fungi (Agrios 2005). There are numerous reports of specks and dots on Mesozoic and Cenozoic plant remains (surveyed in Tiffney and Barghoorn 1974), and even one report of a putative Paleozoic leaf spot (Wang 1997). More recently, Poinar (2018) reported well preserved pycnidia, formally described as *Palaeomycus epallelus*, on an angiosperm leaf preserved in mid-Cretaceous Myanmar amber (Fig. 3.3k). The author suggests that, although there are no modern equivalents to the pycnidia, they are most similar to leaf spot-producing members of the coelomyceteous fungi. Because the taxonomy and identification of modern rusts, smuts, and other fungal disease causative agents in plants is largely dependent on micromorphological characters of the spores, careful palynological preparations and/or examination of permineralized angiosperms are likely the keys to more accurately resolving the fossil history of these pathogens.

Another approach that has been employed to better understand the geologic history of fungus-plant parasitic interactions is to look for the host plants. For example, the Erysiphales (Ascomycota), or powdery mildews, produce cleistothecia with very characteristic appendages and are associated with specific angiosperm hosts (Braun 1987). Consequently, the presence of certain angiosperm hosts during the Late Cretaceous has been used as indirect evidence of the initial radiation of this group of fungi (Takamatsu et al. 2010; Takamatsu 2013). However, we feel that this approach, although interesting, is also problematic (see De Baets and Littlewood 2015). In the absence of a fossil record, how do we know if the ancestors of present-day (hyper-)host-specific fungal parasites parasitized the ancestors of the present-day hosts and elicited the same disease symptoms? Nevertheless, it should be possible to identify members of the Erysiphales on the surface of fossil leaf cuticles since many of the reproductive structures are highly ornamented.

3.3.2 Fungal Parasites of Algae

Algae are critical elements in modern aquatic ecosystems, not only in producing oxygen for other aquatic life, but also in serving as primary producers of organic matter at the base of the food chain (Round 1981). Some are pivotal in the biology of aquatic animals, while others are major structural contributors to the formation of reefs (Coates and Jackson 1987; Weiss and Martindale 2017). The fossil record of algae is extensive and dates back to the Late (perhaps even Middle) Proterozoic (Coniglio and James 1985; Graham and Wilcox 2000; Butterfield 2015; Bengtson et al. 2017). Fungal parasitism of algae today is common, and some fungi enter into complex relationships with their algal hosts (e.g., Kohlmeyer 1979; Kohlmeyer and Kohlmeyer 1979; Gachon et al. 2010). One interesting example consists of endolithic microscopic algae inhabiting coral skeletons as a convenient shelter and endolithic fungi colonizing the corals primarily for food and feeding on both the coral polyps and the endolithic algae (for details, see Le Campion-Alsumard et al. 1995; Golubic et al. 2005). Parasitic fungi can have a profound impact on freshwater and/or marine phytoplankton and algal populations (Ibelings et al. 2004; Kagami et al. 2007; Wang and Johnson 2009; Gleason et al. 2011); however, documented evidence of fungal parasites of fossil algae is very rare. This dearth of evidence is due probably to the fact that the most common modes of preservation of fossil algae (e.g., as cysts, calcareous skeletons, or thallus impressions) are not conducive to the preservation in recognizable form of microbial parasites associated with these organisms in vivo. It is also possible that some researchers have noted fungi occurring with their algae fossils, but did not bother to address them because they were merely interested in the algae and/or lacked the expertise to adequately describe fungal remains.

Perhaps the best fossils of fungal parasites of algae come from the Lower Devonian Rhynie chert, together with the earliest evidence of hypertrophy (i.e. increase in cell size as a result of an external stimulus) in the fossil record. Chytrid-like organisms are common parasites of the Rhynie chert charophyte *Palaeonitella cranii* (Taylor et al. 1992b, c). One of these organisms, *Milleromyces rhyniensis*, is characterized by an endobiotic zoosporangium extending out from the charophyte cell wall (Fig. 3.3l). At the base of the zoosporangium is a rhizoidal system. Other chytrid-like organisms associated with *P. cranii* include *Lyonomyces pyriformis* and *Krispiromyces discoides*, which differ from one another in thallus morphology. Both *M. rhyniensis* and *K. discoides* are associated with hypertrophic host cells (Fig. 3.3m), which grow to approximately five times the diameter of normal cells (Fig. 3.3n), and thus prove that colonization occurred while the host was alive. This same pattern in cell increase in response to chytrid parasitism has been reported in the modern genus *Chara* (Karling 1928), a distant relative of *P. cranii*.

3.3.3 *Fungal Parasites of Other Fungi*

The term mycoparasitism is used to describe the interfungal interrelationships of a fungus parasite and a fungus host (Barnett 1963; Jeffries and Young 1994). There are numerous examples of mycoparasitism in the fossil record, the majority of which come from the Lower Devonian Rhynie chert.

3.3.3.1 Rhynie Chert Interfungal Interactions

Rhynie chert evidence of interfungal associations ranges from fungal mycelia and reproductive units in the lumen of other fungal reproductive units (Kidston and Lang 1921; Krings et al. 2009a, 2010a, 2015, 2016), to fungal hyphae enveloping and subsequently penetrating fungal vesicles (Krings and Taylor 2014b), to fungal reproductive units developing in glomeromycotan vesicles (Fig. 3.4a) (Harper et al. 2017b). Moreover, numerous monocentric and polycentric chytrid-like organisms have been described as intruders of fungal hyphae and spores. Most of these organisms consist of epibiotic sporangia and rhizoidal systems extending into the host spore lumen (Fig. 3.4b) (Taylor et al. 1992a; Krings and Taylor 2014a; Krings and Harper 2019). Other chytrid-like intruders of fungal spores are found between particular wall layers of fungal spores or occupying the spore lumen (Hass et al. 1994). Unfortunately, the majority of fungi associated with other fungi in the Rhynie chert cannot be identified as to their nutritional mode because there is no observable host response. Moreover, larger sample sets of specimens displaying consistent patterns of such associations are relatively rare. Spectacular exceptions include one particular type of glomoid spore located in degrading land plant axes that contains evidence of (simultaneous) colonization by three different intrusive fungi (Krings and Harper 2018). In this study, the authors report massive callosities that occur around the intrusion filaments of a chytrid-like parasite with epibiotic sporangia, while at the same time the penetration hyphae of another fungus can extend into the same spore without triggering a host response. Moreover, several of the spores show large numbers of short, inwardly directed projections, each consisting of a short hyphal branch of a third fungus encased in a callosity of host wall material (Fig. 3.4c). It should be noted that it is exceedingly rare to find the causative agent preserved in direct contact with the callosity (or any host response); typically, only evidence of the host response is found (Fig. 3.4d).

There are several reports of fungal mycelia and reproductive units occurring on the surface of or within other fungi also from geologically younger deposits (e.g., White and Taylor 1989: pl. 2, figs 1, 2, 6; 1991: pl. II, figs 1 and 3, pl. III, figs 1–5; Taylor et al. 1994: pl. I, fig. 2; Taylor et al. 2005c: fig. 7; Krings et al. 2011c: fig. 1L; Krings and Taylor 2012: pl. I, fig. 4; Harper et al. 2015: pl. 1, figs 13 and 14; White et al. 2018: pl. IV, figs 3–6). However, as with the Rhynie chert specimens, there is mostly no direct evidence of parasitism in the form of a host response.

3.3.3.2 Fossil Fungal "Sporocarps"

One group of fungal fossils that have received considerable attention are the so-called "sporocarps", enigmatic structures, mostly from the Carboniferous and Triassic, that are composed of a walled cavity enveloped in an investment of interlacing or tightly compacted hyphae (surveyed in Krings et al. 2013; Taylor et al. 2015). In some specimens, the cavity is empty, but more often contains one to several spheres that have been the basis for several hypotheses regarding the affinities of these structures. One interprets them as cleistothecia and suggests affinities with the Ascomycota based on specimens containing one large sphere believed to represent an ascus that in turn contains several smaller spheres interpreted as ascospores (Stubblefield and Taylor 1983; Stubblefield et al. 1983). An alternative interpretation views the large sphere as a zygospore, and the entire structure accordingly as a reproductive structure (i.e. a mantled zygosporangium) of a member of the zygomycetous fungi (Pirozynski 1976; Taylor and White 1989). If this latter interpretation is accurate, then the smaller spheres found within the large sphere in some specimens would represent some type of mycoparasite. There is an increasing body of circumstantial evidence to corroborate the latter hypothesis. For example, a "sporocarp" specimen from the Carboniferous of Great Britain contains not only spherical structures, but also hyphae forming appressorium-like swellings at the contact region with host walls (Fig. 3.4e, f) (Krings et al. 2011b).

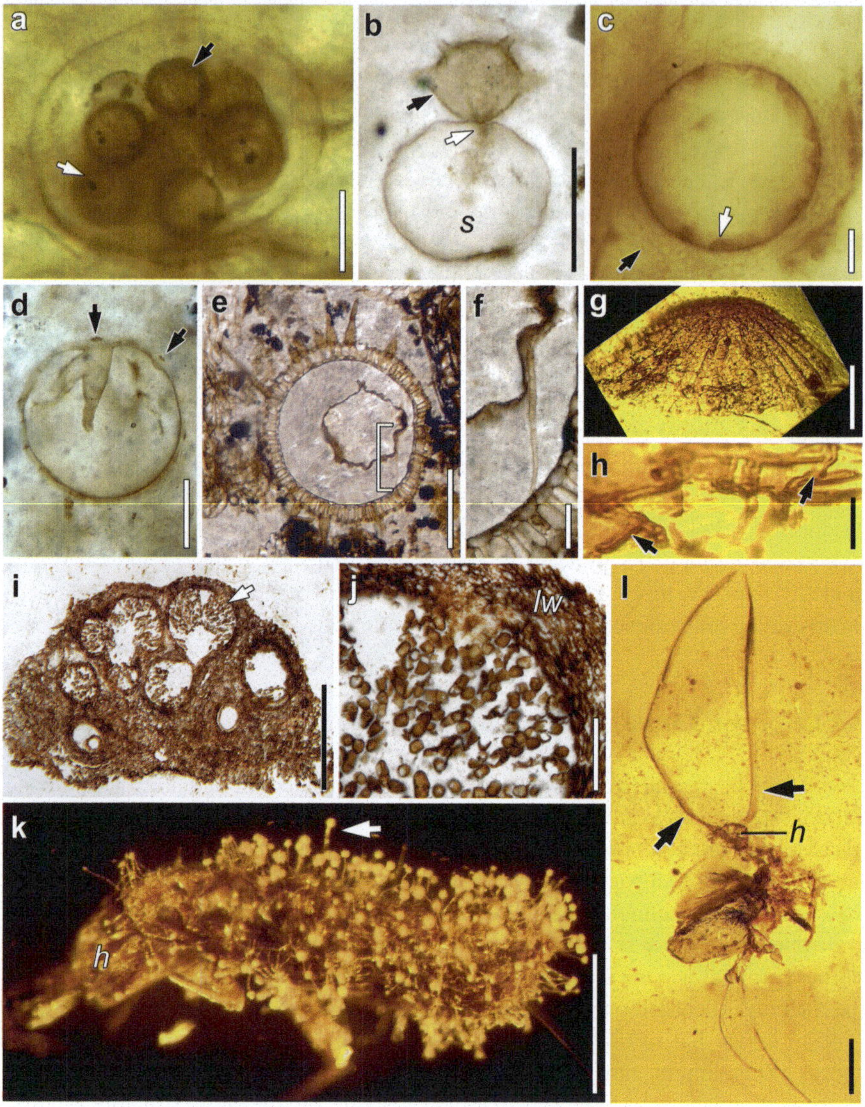

Fig. 3.4 Fossils of fungus-fungus and fungus-animal (putatively) parasitic interactions. (**a**) Thick-walled (black arrow) and smaller thinner-walled propagules (white arrow) of fungal intruder in glomeromycotan vesicle (Devonian); fig. 2, 5 in Harper et al. 2017b; scale bar = 20 μm. (**b**) *Illmanomyces corniger* zoosporangium (black arrow) with four discharge tubes and endobiotic rhizoidal system (white arrow) extending into lumen of host spore (s) (Devonian); color version of fig. 2f, g in Krings and Taylor 2014a; scale bar = 100 μm. (**c**) Glomeromycotan spore enveloped in a compact, multi-layered hyphal sheath and showing numerous penetration sites (white arrow) of another, parasitic fungus (black arrow) (Devonian); fig. 3F in Krings and Harper 2018; scale bar = 10 μm. (**d**) Massive, branched callosity in a glomeromycotan spore (Devonian); note parasite penetration sites on spore surface (arrows); Munich collection, slide SNSB-BSPG 2017 XXXII 1; scale bar = 50 μm. (**e**) Overview of fungal sporocarp *Dubiocarpon* sp. with mycoparasite (bracket) (Carboniferous); color version of fig. 1a in Krings et al. 2011b; scale bar = 250 μm. (**f**) Detail of

3.3.3.3 Hyperparasitism

A parasite that infects a host which is parasitizing a third organism is termed a hyperparasite (Parratt and Laine 2016). One of the oldest gilled mushrooms, *Palaeoagaricites antiquus*, comes from Myanmar amber that is Early Cretaceous in age (Fig. 3.4g) (Poinar and Buckley 2007). This fossil consists of a portion of a cap, 2.2 mm in diameter. Especially interesting is that this fossil agaric is parasitized by a mycoparasite, *Mycetophagites atrebora*, which in turn is parasitized by a hyperparasitic fungus, *Entropezites patricii* (Fig. 3.4h). A structurally preserved ascomycete, *Paleoserenomyces allenbyensis* from the Eocene Princeton chert, occurs in leaf tissue of the palm *Uhlia allenbyensis* (Fig. 3.4i) (Currah et al. 1998). The fungus is composed of multiple locules lined by thin-walled hyphae. Present in some of the locules are globose ascomata interpreted as hyperparasites. The hyperparasite, formally described as *Cryptodidymosphaerites princetonensis* (Fig. 3.4j), shares structural details with certain present-day species in *Didymosphaeria*, a plant pathogen of multiple hosts that is included in the Pleosporales (Aptroot 1995). Last, an interesting report by Van Geel et al. (2006), which demonstrated the potential of examining multiple host–parasite interactions among organisms in the fossil record, noted the subfossil occurrence of *Isthmospora spinosa*, a hyperparasite of various genera within the Meliolaceae. Specimens were recovered from a Holocene bog and demonstrated the tripartite relationship between the host *Calluna vulgaris* (heather) and the parasitic fungus *Meliola ellisii*, which is in turn parasitized by *I. spinosa*.

3.3.4 *Fungal Parasites of Animals*

There is a diverse suite of fungi today that thrive as facultative and/or obligatory parasites of invertebrates and vertebrates (Ainsworth et al. 1973); some are also the causative agents of mild to severe diseases (Góralska and Błaszkowska 2015). Although there are fungal infections that elicit abnormal growth or tissue

Fig. 3.4 (continued) Fig. 3.3e, focusing on bracketed area, showing hypha of a parasite entering sporocarp cavity and extending to sac-like structure; color version of fig. 1C in Krings et al. 2011b; scale bar = 50 µm. (**g**) Pileus of *Palaeoagaricites antiquus* preserved in amber and covered with the mycelium of the mycoparasite *Mycetophagites atrebora* (Cretaceous); fig. 1a in Poinar and Buckley 2007; scale bar = 500 µm. (**h**) Mycoparasite *Mycetophagites atrebora* parasitized by the hyperparasite *Entropezites patricii* (arrows) (Cretaceous; fig. 1C in Poinar and Buckley 2007; scale bar = 20 µm. (**i**) Section of stroma of *Paleoserenomyces allenbyensis* with endoparasite (arrow) in locules (Eocene); color version of fig. 13 in Currah et al. 1998; scale bar = 500 µm. (**j**) Section of *Paleoserenomyces allenbyensis* locule (lw) containing asci of endoparasite *Cryptodidymosphaerites princetonensis* (Eocene); color version of fig. 15 in Currah et al. 1998; scale bar = 50 µm. (**k**) Conidial heads of *Aspergillus collembolorum* (arrow) attached to the surface of a springtail, head = h (Eocene); fig. 2 in Dörfelt and Schmidt 2005; scale bar = 500 µm. (**l**) *Paleoophiocordyceps coccophagus* showing two synnemata (arrows) arising from head (h) of a scale insect (Cretaceous); fig. 1 in Sung et al. 2008; scale bar = 500 µm destruction in hard parts such as bones, tests, and shells (Meyers 1990; Stewart 1993; Cook et al. 2003), most host responses and disease symptoms (mycoses) that can develop in animals in response to fungal infections occur in the soft parts of the body, in non-mineralized tissues that are readily degraded by bacteria and other microorganisms. Because soft tissue preservation in animals is exceedingly rare in the fossil record (Allison and Briggs 1993), there is only a narrow chance for finding fossil evidence of fungal parasitism and pathogenicity in animals.

3.3.4.1 Rhynie Chert

Although the documented record of animals from the Rhynie chert is quite extensive and diverse (e.g., Anderson and Trewin 2003; Dunlop and Garwood 2017), there is only a single report to date of an ostensible interaction of fungi with animals. The co-occurrence of a chytrid-like organism, *Cultoraquaticus trewinii*, with peculiar spherules interpreted as resting eggs of the branchiopod crustacean *Lepidocaris rhyniensis*, is purported to represent compelling evidence of a role for chytrids in a mycoloop (Kagami et al. 2014) that transferred nutrients obtained from a substrate to the crustacean (Strullu-Derrien et al. 2016). The spherules, which are of varying diameters and bear spines of varying lengths (Strullu-Derrien et al. 2016: fig. 4D, E, G, H, J, K), are compared to the resting eggs of the modern *Linderiella santarosae* (Anostraca) (Thiéry and Fugate 1994).

3.3.4.2 Amber Inclusions

Specimens enshrined in amber dominate the fossil record of parasitic and pathogenic fungi on insect hosts (Boucot and Poinar 2010). One report describes *Paleocadus burmiticus*, a member of the Eccrinales, which were previously thought to be zygomycetous fungi but are today considered members of the Mesomycetozoea (Opisthokonta), producing two types of sporangiospores on different thalli that protrude from a primitive wasp preserved in Cretaceous amber from Myanmar (Poinar 2016a). Present-day Eccrinales do not infect members of the Hymenoptera, suggesting a wider host range during the Mesozoic. Geologically younger (Eocene) Baltic amber has also yielded exquisite examples of insect colonization by fungi, including a springtail overgrown by conidiophores of the fossil fungus *Aspergillus collembolorum* (Fig. 3.4k) (Dörfelt and Schmidt 2005). The authors suggest that *A. collembolorum* was a facultative parasite because modern *Aspergillus* species usually are facultative parasites or saprotrophs. Another example of fossil *Aspergillus* comes from Dominican amber and occurs in the form of well-preserved tufts of catenulate chains of conidia covering the surface of the abdomen of a fly (Thomas and Poinar 1988). Although the authors do not comment on the nutritional mode of the fungus, it is likely that, similar to *A. collembolorum*, it was a facultative parasite. Another example of fungal parasitism of animals in Dominican amber is a winged termite covered by an entomophthoralean fungus (Poinar and Thomas 1982). The authors conclude that the fungus was parasitic based on the presence of conidia budding along with smaller secondary conidia adjacent to the main mycelial mat, a characteristic of modern Entomophthorales (Prasertphon 1963). In addition, a fossil member of the Laboulbeniales, which are obligate ectoparasites, has been discovered on the thorax of a fossil stalk-eyed fly (*Prosphyracephala succini*) in Eocene Baltic amber (Rossi et al. 2005). An enigmatic fossil from Myanmar amber, *Speciophila adercia*, also attributed to the Laboulbeniales, consists of a thallus with numerous perithecia and antheridia that is attached to the abdominal tergite of a primitive wasp (Poinar 2016b). This author suggests that *S. adercia* belongs to an extinct lineage because there are no other extant thallus-forming ectoparasitic fungi. Finally, an example of a special form of fungal parasitism, predation (carnivory), has also been fossilized in amber (Schmidt et al. 2007). Several specimens of a fungus that used hyphal rings as trapping devices occur in Late Cretaceous amber from France together with the fungus' prey, small nematodes. The fossil nematode-trapping fungus cannot be assigned to any recent taxon of carnivorous fungi, but rather suggests that different groups occupied this ecological niche in the Cretaceous and that trapping devices evolved independently multiple times in the course of Earth history. Predatory fungi catch microorganisms using a remarkable array of trapping devices; however, their primary ecosystem function appears to be that of wood decay, and hence they are saprotrophs that attack other organisms as sources of nitrogen to supplement a primarily carbohydrate (woody) diet (Barron 2003).

3.3.4.3 Cordycipitaceae Interactions with Arthropods

Fungi in the family Cordycipitaceae (Ascomycota) enter into several types of fascinating parasitic interrelationships with insects and other arthropods that usually, but not always, result in the death of the arthropod host (Sung et al. 2007). For example, *Ophiocordyceps unilateralis* enters ants and eventually takes control over the host's brain activities (commonly named 'zombie' infection) and manipulates its behavior. The manipulated ant, which becomes a so-called parasite-extended phenotype (Hughes 2014), is forced by the fungus to move to so-called death locations (usually plant parts) that represent ideal spots for fungal spore dispersal (de Bekker et al. 2014, 2015; Shang et al. 2015). Arrived at its death location, the fungus forces the ant to bite into the substrate (e.g., a leaf or small plant axis) and remain in that position until death arrives. These bites leave a characteristic scar in the plant known as the 'death-grip' (Anderson et al. 2009). Although it is exceedingly rare to capture multiple stages of fungal life histories in the fossil record, there are three documented examples of fossils displaying stages of the Cordycipitaceae life cycle. One includes an ant preserved in Dominican amber that is covered in a fungus morphologically similar to certain present-day species in *Beauveria* (Cordycipitaceae), which are obligate endoparasites (Poinar and Thomas 1984). The sexual stages (or teleomorphs) of *Beauveria*, where known, are species of *Cordyceps* (Rehner et al. 2011). The second example is a spectacular specimen of a fungus formally described as *Paleoophiocordyceps coccophagus*, which is a parasite of Cretaceous scale insects (Fig. 3.41) (Sung et al. 2008). This fossil provides the oldest compelling evidence of animal parasitism by fungi, and is characterized by several synnemata emerging from the head of the host. However, there is no evidence indicative of host behavior manipulation. The third example is a trace fossil that occurs in the form of well-preserved death-grip scars in Eocene (48 Ma) angiosperm leaves from the Messel pit in Germany (Hughes et al. 2011b). The characteristic scars are remarkably similar to the death-grip scars produced by some present-day fungus-infected carpenter ants (Hughes et al. 2011a).

3.3.4.4 Dinosaurs

Some of the first animals that come to mind when we think of life in the geologic past are dinosaurs. There are limited reports of pathogenic fungi associated with dinosaur remains. One possible example is a *Tyrannosaurus rex* mandible that shows multiple erosive lesions (Rega and Brochu 2001; Wolff et al. 2009; but see Rothschild and Martin 2006; Watson and Rothschild 2021 for alternative interpretations). The authors postulate that such abscesses may eventually form large, localized caseous masses which would be susceptible to fungal parasitic infections, similar to those seen in modern crocodilians (Huchzermeyer 2003). Another interesting example includes fossilized Cretaceous sauropod dung from India that contains an array of fungal remains, including forms that are known to be plant parasites (e.g., *Colletotrichum*), thus indicating that some sauropods ate the leaves of fungus-infected plants as food (Kar et al. 2004). Other sauropod coprolites from India contain several different fungal remains characteristic of present-day plant pathogens, including *Colletotrichum*-like acervuli (leaf spot, red rot disease), *Erysiphe*- and *Uncinula*-like cleistothecia (powdery mildew), and black spot-producing microthyriaceous ascostromata (Sharma et al. 2005). A single specimen of an infected grass spikelet with similarities to extant *Claviceps purpurea* (commonly known as ergot) has been identified in Cretaceous Myanmar amber (Poinar 2015). The authors suggest that, if ergot-infected grass was ingested by herbivore animals, they may have felt the effects of the psychotropic compounds produced by the fungus in a similar way to that seen in livestock animals today (Bove 1970). Finally, numerous large coprolites from three different horizons (spanning at least 6 Ma) within the Upper Cretaceous Two Medicine Formation of Montana have revealed that some herbivorous dinosaurs sometimes consumed large amounts of fungus-infected wood (Chin 2007). This author states that decaying wood would have provided an assortment of nutritious foods, including cellulose from the wood, fungi, other microbes, and detritivorous invertebrates.

Some investigators even have suggested that fungal parasites may have contributed to the demise of the dinosaurs at the end of the Cretaceous. Several dinosaur eggs have been reported with mycelium-like structures or endolithic fungi in the eggshells (Gong et al. 2008). Based on fungal morphology and the areas in the shells in which the fungi occur, it has been hypothesized that the fungi were parasitic and invaded the eggs before they became lithified. A very provocative hypothesis regarding the extinction of dinosaurs relating to fungal infections by Casadevall

(2005) suggests that the purported unprecedented accumulation of fungal spores after the global plant die-off at the end of the Cretaceous would have altered the usual balance of power, namely by delivering such massive concentrations of spores into the lungs of the dinosaurs that their immune defenses were overwhelmed. If the body temperature of dinosaurs was lower than that of mammals, then these reptiles might well have been susceptible to fungal infection, perhaps giving an advantage to early mammals to survive the KT-extinction event. While there are numerous censures surrounding this hypothesis, including the debate as to whether the spores that accumulated indeed are fungal in origin (Hochuli 2016) and the ongoing discussion of whether dinosaurs were warm- or cold-blooded (e.g., Grady et al. 2014), it nevertheless opens an interesting and thought-provoking new perspective on the end of the age of dinosaurs.

3.4 Concluding Remarks

Deciphering the roles of fungi colonizing other organisms in natural environments today is challenging because of the difficulties in making field observations (Jeffries 1995). Even more challenging is the analysis of fungal relationships from the geologic past. The lack of information that can be used to safely assign fungal fossils systematically is one of the principle problems exacerbating the assessment of fossil fungal associations (Krings et al. 2016). As we have tried to exemplify in this chapter, however, there are also persistent uncertainties with regard to determining the fungal nutritional modes that are connected to the inherent limitations of the fossil record (Fig. 3.5). Based on the examples of (assumed) fungal parasitism from the fossil record presented throughout the sections of this chapter, certain patterns nonetheless begin to emerge:

1. Sublime preservation (e.g., in amber or chert) is a precondition to identify direct features such as fungal mycelia spreading along consistent pathways in intact plant tissue, and indirect evidence such as host responses or death-grip scars in fossils that can be attributed to fungal parasitism.
2. Host responses currently represent the most reliable fossil evidence in support of fungal infection of a living host, albeit not necessarily of a parasitic nutritional mode of the intruder.
3. In the absence of host responses, the presence of certain structural features regularly seen in extant fungal parasites (e.g., haustoria, endobiotic rhizoidal systems, and holdfasts) can provide hints at parasitism in fossil fungi (see Karling 1932).
4. Fossils that can be attributed to present-day parasitic fungal families and genera with confidence are suggestive of fossil parasitism even if information on the hosts is not available. Unfortunately, fungal fossils older than Cretaceous cannot normally be attributed to modern families and genera with confidence.

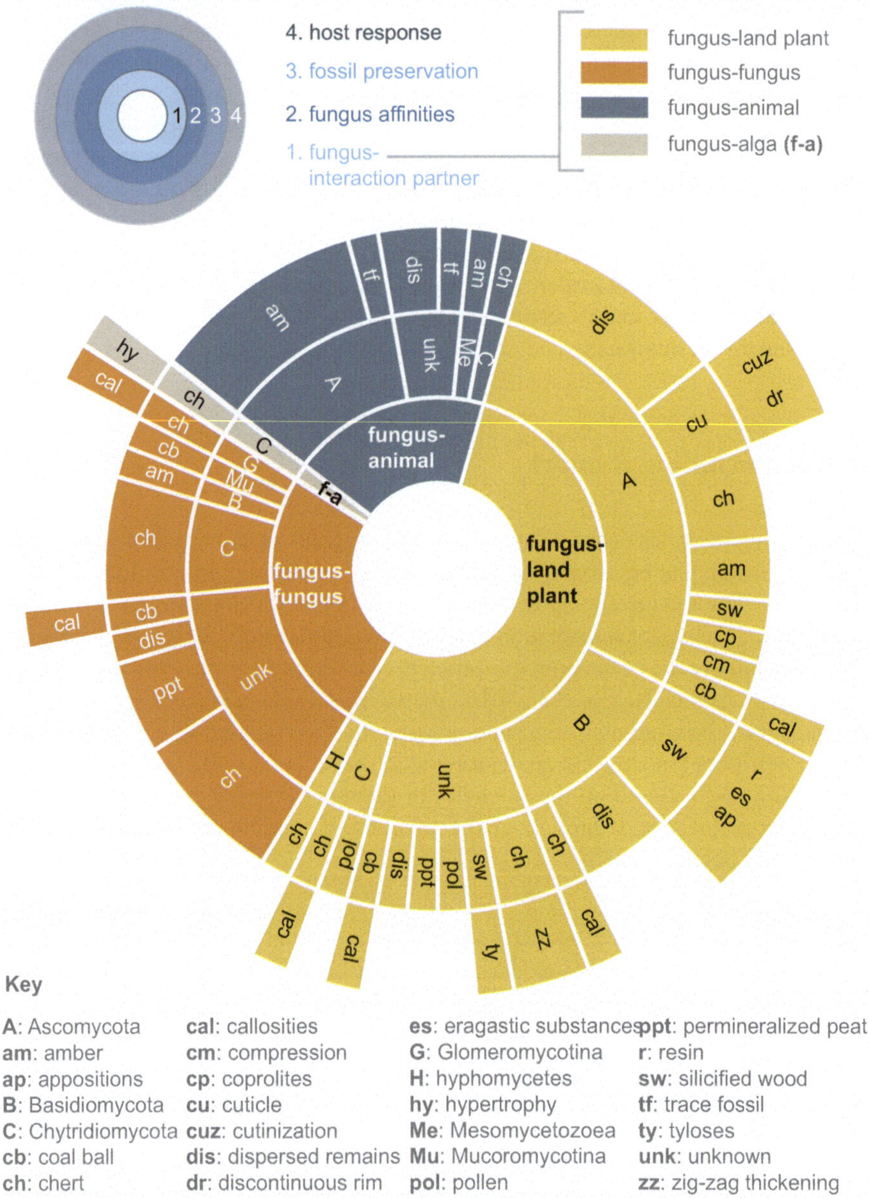

Fig. 3.5 Graphical synopsis of the information contained in the chapter, showing proportional abundance of (1) fossil hosts containing fungal parasites, (2) suggested affinities of the fungal parasites, (3) mode of fossil preservation, and (4) host response(s) if present. We acknowledge that the figure contains a prominent bias towards the paleobotanical evidence

The sobering truth is that, in most of the fossil record, we will never be able trace parasitic fungal interactions, simply because either the fungi are not preserved or the structural features required in determining the nutritional modes of the fungi cannot be resolved. On the other hand, we must not overhear the clarion call for concerted research efforts aimed at integrating other research fields such as geochemistry or biomarker analysis into paleomycology, because they might provide some of the information that cannot be obtained by using traditional paleontological techniques. More than anything, collaborative and synergistic research efforts are needed between neo-mycologists, pathologists, and paleontologists to accurately document fossil fungi and their many different interactions with other ecosystem components so that these fossils can be placed in a greater context such as (paleo-) ecosystem functioning and/or phylogenetic analyses.

Acknowledgements We acknowledge financial support from the Alexander von Humboldt-Foundation (3.1-USA/1160852 STP to C.J.H.), and the National Science Foundation (DEB-1441604 subcontract S1696A-A to M.K.). We gratefully acknowledge H. Kerp and H. Hass (both Münster, Germany), G.O. Poinar and R.A. Stockey (both Corvallis, OR, USA), and A.R. Schmidt (Göttingen, Germany) for providing images, A.-L. Decombeix (Montpellier, France) for fruitful discussions, as well as N. Dotzler, H. Martin, and S. Sónyi (all Munich, Germany) for technical assistance, and K. De Baets (Erlangen, Germany) for insightful comments on the manuscript.

References

Aggarwal N, Krings M, Jha N, Taylor TN (2015) Unusual spheroidal inclusions in Late Permian gymnosperm pollen grains from southern India revisited: evidence of a fungal nature. Grana 54:174–183

Agrios GN (2005) Plant pathology, 5th edn. Academic Press, Amsterdam

Ainsworth GC, Sparrow FK, Sussman AS (eds) (1973) The fungi, an advanced treatise, vol IV A: A taxonomic review with keys: Ascomycetes and Fungi Imperfecti. Academic, New York, NY

Aist JR (1976) Papillae and related wound plugs of plant cells. Annu Rev Phytopathol 14:145–163

Aist JR (1977) Mechanically induced wall appositions of plant cells can prevent penetration by a parasitic fungus. Science 197:568–570

Aist JR (1983) Structural responses as resistance mechanisms. In: Bailey JA, Deverall BJ (eds) The dynamics of host defence. Academic, Sydney

Akai S (1959) Histology of defense in plants. In: Horsfall JG, Dimond AE (eds) Plant pathology, an advanced treatise, vol. I: the diseased plant. Academic, New York, NY

Allison PA, Briggs DEG (1993) Exceptional fossil record: distribution of soft-tissue preservation through the Phanerozoic. Geology 21:527–530

Altermann W, Schopf JW (1995) Microfossils from the Neoarchean Campbell Group, Griqualand West Sequence of the Transvaal Supergroup, and their paleoenvironmental and evolutionary implications. Precambrian Res 75:65–90

Anderson JP, Gleason CA, Foley RC, Thrall PH, Burdon JB, Singh KB (2010) Plants versus pathogens: an evolutionary arms race. Funct Plant Biol 37:499–512

Anderson LI, Trewin NH (2003) An Early Devonian arthropod fauna from the Windyfield cherts, Aberdeenshire, Scotland. Palaeontology 46:467–509

Anderson SB, Gerritsma S, Yusah KM, Mayntz D, Hywel-Jones NL, Billen J, Boomsma JJ, Hughes DP (2009) The life of a dead ant: the expression of an adaptive extended phenotype. Am Nat 174:424–433

Aptroot A (1995) A monograph of *Didymosphaeria*. Stud Mycol 37:1–160

Baldrian P, Valášková V (2008) Degradation of cellulose by basidiomycetous fungi. FEMS Microbiol Rev 32:501–521

Ballard RG, Walsh MA, Cole WE (1982) Blue-stain fungi in xylem of lodgepole pine: a light-microscope study on extent of hyphal distribution. Can J Bot 60:2335–2341

Bannister JM, Conran JG, Lee DE (2016) Life on the phylloplane: eocene epiphyllous fungi from Pikopiko Fossil Forest, Southland, New Zealand. New Zealand J Bot 54:412–432

Barnett HL (1963) The nature of mycoparasitism by fungi. Ann Rev Microbiol 17:1–14

Berrocal-Lobo M, Molina A, Solano R (2002) Constitutive expression of *ETHYLENE-RESPONSE-FACTOR1* in *Arabidopsis* confers resistance to several necrotrophic fungi. Plant J 29:23–32

Barron GL (2003) Predatory fungi, wood decay, and the carbon cycle. Biodiversity 4:3–9

Barthel M, Krings M, Rössler R (2010) Die schwarzen Psaronien von Manebach, ihre Epiphyten, Parasiten und Pilze. Semana 25:41–60

Bass D, Czech L, Williams BAP, Berney C, Dunthorn M, Mahé F, Torruella G, Steniford GD, Williams TA (2018) Clarifying the relationships between Microsporidia and Cryptomycota. J Eukaryot Microbiol 65:1–10

Bengtson S, Rasmussen B, Ivarsson M, Muhling J, Broman C, Marone F, Stampanoni M, Bekker A (2017) Fungus-like mycelial fossils in 2.4 billion-year-old vesicular basalt. Nature Ecol Evol 1:0141

Berbee ML, James TY, Strullu-Derrien C (2017) Early diverging fungi: diversity and impact at the dawn of terrestrial life. Annu Rev Microbiol 71:41–60

Blair JE (2009) Fungi. In: Hedges SB, Kumar S (eds) The timetree of life. Oxford University Press, Oxford

Blanchette RA (1991) Delignification by wood-decay fungi. Annu Rev Phytopathol 29:381–398

Blanchette RA, Biggs AR (eds) (1992) Defense mechanisms of woody plants against fungi. Springer-Verlag, Berlin

Bonneville S, Delpomdor F, Préat A, Chevalier C, Araki T, Kazemian M, Steele A, Schreiber A, Wirth R, Benning LG (2020) Molecular identification of fungi microfossils in a Neoproterozoic shale rock. Sci Adv 6(4):eaax7599

Boucot AJ, Poinar GO (2010) Fossil behavior compendium. CRC Press, Boca Raton, FL

Bove FJ (1970) The story of ergot. S. Karger, New York, NY

Bradley WH (1931) Origin and microfossils of the oil shale of the Green River Formation of Colorado and Utah. US Geol Surv Prof Paper 168:58

Braun U (1987) A monograph on the Erysiphales (Powdery Mildews). J. Cramer, Berlin

Brundrett MC, Tedersoo L (2018) Evolutionary history of mycorrhizal symbioses and global host plant diversity. New Phytol 220:1108–1115

Brundrett MC, Walker C, Harper CJ, Krings M (2018) Fossils of arbuscular mycorrhizal fungi give insights into the history of a successful partnership with plants. In: Krings M, Harper CJ, Cúneo NR, Rothwell GW (eds) Transformative paleobotany: papers to commemorate the life and legacy of Thomas N. Taylor. Elsevier, Cambridge, MA

Butterfield NJ (2015) Early evolution of the Eukaryota. Palaeontology 58:5–17

Callow JA, Ling IT (1978) Histology of neoplasms and chlorotic lesions in maize seedlings following the injection of sporidia of *Ustilago maydis* (DC) Corda. Physiol Plant Pathol 3:489–490

Cannon PF, Hawksworth DL (1995) The diversity of fungi associated with vascular plants: the known, the unknown and the need to bridge the knowledge gap. In: Andrews JH, Tommerup IC (eds) Advances in plant pathology, vol. 11. Academic, London

Casadevall A (2005) Fungal virulence, vertebrate endothermy, and dinosaur extinction: is there a connection? Fung Gen Biol 42:98–106

Chatton E (1920) Un complexe xéno-parasitaire morphologique et physiologique *Neresheimeria paradoxa* chez *Fritillaria pellucida*. CR Acad Sci Paris 171:55–57

Chin K (2007) The paleobiological implications of herbivorous dinosaur coprolites form the Upper Cretaceous Two Medicine Formation of Montana: Why eat wood? Palaios 22:554–566

Chisholm ST, Coaker G, Day B, Staskawicz BJ (2006) Host-microbe interactions: shaping the evolution of the plant immune response. Cell 124:803–814

Chitaley SD, Yawale NR (1976) Fungal remains from the Deccan Intertrappean beds of Mohgaonkalan, India. Proc Indian Sci Cong, Part 3, Sect VI Botany 63:52

Chitaley SD, Yawale NR (1978) Fungal remains from the Deccan Intertrappean beds of Mohgaonkalan, India. Botanique 7:189–194

Coates AG, Jackson JBC (1987) Clonal growth, algal symbiosis, and reef formation by corals. Paleobiology 13:363–378

Cohen KM, Finney SC, Gibbard PL, Fan J-X (2013) The ICS International Chronostratigraphic chart (2013; updated). Episodes 36:199–204

Collins B, Parke J (2008) Spatial and temporal aspects of tylosis formation in tanoak inoculated with *Phytophora ramorum*. In: Frankel SJ, Kliejunas JT, Palmieri KM (eds) Proceedings of the sudden oak death third science symposium, US Department of Agriculture General Technology Report PSW-GTR-214

Coniglio M, James NP (1985) Calcified algae as sediment contributors to Early Paleozoic limestones: evidence from deep-water sediments of the Cow Head Group, Western Newfoundland. J Sed Petrol 55:746–754

Cook MI, Beissinger SR, Toranzos GA, Rodriguez RA, Arendt WJ (2003) Trans-shell infection by pathogenic micro-organisms reduces the shelf life of non-incubated bird's eggs: a constraint on the onset of incubation? Proc Roy Soc Biol Sci B 270:2233–2240

Currah RS, Stockey RA (1991) A fossil smut fungus from the anther of an Eocene angiosperm. Nature 350:698–699

Currah RS, Stockey RA, LePage BA (1998) An Eocene tar spot on a fossil palm and its fungal hyperparasite. Mycologia 90:667–673

Daugherty LH (1941) The Upper Triassic flora of Arizona. Carnegie Institute of Washington Publishers, Washington, DC

Dean R, Van Kan JA, Pretorius ZA, Hammond-Kosack KE, Di Pietro A, Spanu PD, Rudd JJ, Dickman M, Kahmann R, Ellis J, Foster GD (2012) The top 10 fungal pathogens in molecular plant pathology. Mol Plant Pathol 13:414–430

De Baets K, Littlewood DTJ (2015) The importance of fossils in understanding the evolution of parasites and their vectors. Adv Parasitol 90:1–51

De Bekker C, Ohm RA, Loreto RG, Sebastian A, Albert I, Merrow M, Brachmann A, Hughes DP (2015) Gene expression during zombie ant biting behavior reflects the complexity underlying fungal parasitic behavioral manipulation. BMC Genomics 16:620

De Bekker C, Quevillon LE, Smith PB, Fleming KR, Ghosh D, Patterson AD, Hughes DP (2014) Species-specific ant brain manipulation by a specialized fungal parasite. BMC Evol Biol 14:166

Delaye L, García-Guzmán G, Heil M (2013) Endophytes versus biotrophic and necrotrophic pathogens—are fungal lifestyles evolutionarily stable traits? Fung Div 60:125–135

De Micco V, Balzano A, Wheeler EA, Baas P (2016) Tyloses and gums: a review of structure, function, and occurrence of vessel occlusions. IAWA J 37:186–205

Deverall BJ (1969) Fungal parasitism (Institute of Biology's Studies in Biology no. 17). Edward Arnold, London

Dilcher DL (1965) Epiphyllous fungi from Eocene deposits in western Tennessee, USA. Palaeontographica B116:1–54

Dighton J, White JF (eds) (2017) The fungal community: its organization and role in the ecosystem, 4th edn. CRC Press, Boca Raton, FL

Do Amaral MM, Ceccantini G (2011) The endoparasite *Pilostyles ulei* (Apodanthaceae–Cucurbitales) influences wood structure in three host species of *Mimosa*. IAWA J 32:1–13

Dollhofer V, Podmirseg S, Callaghan T, Griffith GW, Fliegerová K (2015) Anaerobic fungi and their potential for biogas production. In: Guebitz GM, Bauer A, Bochmann G, Gronauer A, Weiss S (eds) Biogas science and technology. Springer International Publishing, Cham

Dörfelt H, Schmidt AR (2005) A fossil *Aspergillus* from Baltic amber. Mycol Res 109:956–960
Dunlop JA, Garwood RJ (2017) Terrestrial invertebrates in the Rhynie chert ecosystem. Proc Trans Roy Soc B 373:20160493
Ellis JP (1977) The genera *Trichothyrina* and *Actinopeltis* in Britain. Trans Brit Mycol Soc 68:145–155
Elsik WC (1978) Classification and geologic history of the microthyriaceous fungi. In: Bharadwaj DC, Lele KM, Kar RK (eds) 4th international palynological conference, Lucknow, proceedings, vol 1. Birbal Sahni Institute of Palaeobotany, Lucknow
Feng Z, Wang J, Rößler R, Kerp H, Wei HB (2013) Complete tylosis formation in a latest Permian conifer stem. Ann Bot 111:1075–1081
Fleischmann A, Krings M, Mayr H, Agerer R (2007) Structurally preserved polypores from the Neogene of North Africa: *Ganodermites libycus* gen. et sp. nov. (Polyporales, Ganodermataceae). Rev Palaeobot Palynol 145:159–172
Frenken T, Alacid E, Berger SA, Bourne EC, Gerphagnon M, Grossart HP, Gsell AS, Ibelings BW, Kagami M, Küpper FC, Letcher PM, Loyau A, Miki T, Nejstgaard JC, Rasconi S, Reñé A, Rohrlack T, Rojas-Jimenez K, Schmeller DS, Van de Waal DB, Van den Wyngaert S, Van Donk E, Wolinska J, Wurzbacher C, Agha R (2017) Integrating chytrid fungal parasites into plankton ecology: research gaps and needs. Environ Microbiol 19:3802–3822
Gachon CM, Sime-Ngando T, Strittmatter M, Chambouvet A, Kim GH (2010) Algal diseases: spotlight on a black box. Trends Plant Sci 15:633–640
Garcia LS (2002) Laboratory identification of the Microsporidia. J Clinical Microbiol 40:1892–1901
García Massini JL, Channing A, Guido DM, Zamuner AB (2012) First report of fungi and fungus-like organisms from Mesozoic hot springs. Palaios 27:55–62
Garrett SD (1970) Pathogenic root infecting fungi. Cambridge University Press, Cambridge
Garwood RJ, Oliver H, Spencer ART (2020) An introduction to the Rhynie chert. Geol Mag 157:47–64
Glazebrook J (2005) Contrasting mechanisms of defense against biotrophic and nectrotrophic pathogens. Annu Rev Phytopathol 43:205–227
Gleason FH, Küpper FC, Amon JP, Picard K, Gachon CMM, Marano AV, Sime-Ngando T, Lilje O (2011) Zoosporic true fungi in marine ecosystems: a review. Marine Freshwater Res 62:383–393
Gnaedinger S, García Massini JL, Bechis F, Zavattieri AM (2015) Coniferous woods and wood-decaying fungi from the el Freno Formation (Lower Jurassic), Neuquen Basin, Mendoza Province, Argentina. Ameghiniana 52:447–467
Golubic S, Perkins RD, Lukas KJ (1975) Boring microorganisms and microborings in carbonate substrates. In: Frey RW (ed) The study of trace fossils. Springer, Berlin
Golubic S, Radtke G, Le Campion-Alsumard T (2005) Endolithic fungi in marine ecosystems. Trends Microbiol 13:229–235
Gomes AL, Fernandes GW (1994) Influence of parasitism by *Pilostyles ingae* (Rafflesiaceae) on its host plant, *Mimosa naguirei* (Leguminosae). Ann Bot 74:205–208
Gong Y-M, Xu R, Hu B (2008) Endolithic fungi: a possible killer for the mass extinction of Cretaceous dinosaurs. Sci China Ser D: Earth Sci 51:801–807
Góralska K, Błaszkowska J (2015) Parasites and fungi as risk factors for human and animal health. Ann Parasitol 61:207–220
Grady JM, Enquist BJ, Dettweiler-Robinson E, Wright NA, Smith FA (2014) Evidence for mesothermy in dinosaurs. Science 344:1268–1272
Graham LE, Wilcox LW (2000) Algae. Prentice Hall, Upper Saddle River NJ
Hall RA, Noverr MC (2017) Fungal interaction with the fungal animal parasites man host: exploring the spectrum of symbiosis. Curr Op Microbiol 40:58–64
Hammond-Kosack KE, Jones JDG (1996) Resistance gene-dependent plant defense responses. Plant Cell 8:1773–1791
Han B, Weiss LM (2017) Microsporidia: obligate intracellular pathogens within the fungal kingdom. Microbiol Spectr 5(2). https://doi.org/10.1128/microbiolspec.FUNK-0018-2016

Harper CJ, Bomfleur B, Decombeix A-L, Taylor EL, Taylor TN, Krings M (2012) Tylosis formation and fungal interactions in an Early Jurassic conifer from northern Victoria Land, Antarctica. Rev Palaeobot Palynol 175:25–31
Harper CJ, Taylor TN, Krings M, Taylor EL (2015) Fungi associated with *Glossopteris* (Glossopteridales) leaves from the Permian of Antarctica: a preliminary report. Zitteliana A 55:107–114
Harper CJ, Taylor TN, Krings M, Taylor EL (2016) Structurally preserved fungi from Antarctica: diversity and interactions in late Palaeozoic and Mesozoic polar forest ecosystems. Antarct Sci 28:153–173
Harper CJ, Decombeix A-L, Taylor EL, Taylor TN, Krings M (2017a) Fungal decay in Permian glossopteridalean stem and root wood from Antarctica. IAWA J 38:29–48
Harper CJ, Krings M, Dotzler N, Taylor EL, Taylor TN (2017b) Deciphering interfungal relationships in the 410-million-yr-old Rhynie chert: morphology and development of vesicle-colonizing microfungi. Geobios 50:9–22
Harper CJ, Galtier J, Taylor TN, Taylor EL, Rößler R, Krings M (2019) Distribution of fungal endophytes in a Triassic fern stem. Earth Environ Sci Trans Roy Soc Edinburgh 108:387–398
Harvey R, Lyon AG, Lewis PN (1969) A fossil fungus from Rhynie chert. Trans Br Mycol Soc 53:155–156
Hass H, Taylor TN, Remy W (1994) Fungi from the Lower Devonian Rhynie chert: mycoparasitism. Amer J Bot 81:29–37
Hatakka A (2005) Biodegradation of lignin. In: Steinbüchel A, Hofrichter M (eds) Biopolymers online. Weinheim, Wiley-VCH Verlag, pp 129–145. https://doi.org/10.1002/3527600035.bpol1005
Hessburg PF, Hansen EM (1987) Pathological anatomy of black stain root disease of Douglas-fir. Can J Bot 65:962–971
Hesse R (1989) Silica diagenesis: origin of inorganic and replacement cherts. Earth Sci Rev 26:253–284
Hoch HC, Staples RC (1991) Signaling for infection structure formation in fungi. In: Cole GT, Hoch HC (eds) The fungal spore and disease initiation in plants and animals. Springer, Boston, MA
Hochuli PA (2016) Interpretation of "fungal spikes" in Permian-Triassic boundary sections. Glob Plan Change 144:48–50
von Höhnel F (1924) Studien über Hyphomyzeten. Zbl Bakteriol Parasitol A 2(60):1–26
Hongsanan S, Sánchez-Ramírez S, Crous PW, Ariyawansa HA, Zhao RL, Hyde KD (2016) The evolution of fungal epiphytes. Mycosphere 7:1690–1712
Hübers M, Bomfleur B, Krings M, Kerp H (2011) An Early Carboniferous leaf-colonizing fungus. N Jb Geol Paläontol, Abh 261:77–82
Huchzermeyer FW (2003) Crocodiles: biology, husbandry and diseases. CABI Publishing, Wallingford, Oxfordshire
Hughes DP (2014) On the origins of parasite-extended phenotypes. Int Comp Biol 54:210–217
Hughes DP, Andersen SB, Hywel-Jones NL, Himaman W, Billen J, Boomsma JJ (2011a) Behavioral mechanisms and morphological symptoms of zombie ants dying from fungal infection. BMC Ecol 11:13
Hughes DP, Wappler T, Labandeira CC (2011b) Ancient death-grip leaf scars reveal ant-fungal parasitism. Biol Lett 7:67–70
Ibelings BW, de Bruin A, Kagami M, Rijkeboer M, Brehm M, Van Donk E (2004) Host parasite interactions between freshwater phytoplankton and the chytrid fungi (Chytridiomycota). J Phycol 40:437–453
Illman WI (1984) Zoosporic fungal bodies in the spores of the Devonian fossil vascular plant, *Horneophyton*. Mycologia 76:545–547
James TY, Kauff F, Schoch CL, Matheny PB, Hoftetter V, Cox CJ, Celio G, Gueidan C, Fraker E, Miadlikowska J, Lumbsch HT, Rauhut A, Reeb V, Arnold AE, Amtoft A, Stajich JE, Hosaka K, Sung G-H, Johnson D, O'Rourke B, Crockett M, Binder M, Curtis JM, Slot JC, Wang Z,

Wilson AW, Schüßler A, Longcore JE, O'Donnell K, Mozley-Standridge S, Porter D, Letcher PM, Powell MJ, Taylor JW, White MM, Griffith GW, Davies DR, Humber RA, Morton JB, Sugiyama J, Rossman AY, Rogers JD, Pfister DH, Hewitt D, Hansen K, Hambleton S, Shoemaker RA, Kohlmeyer J, Volkmann-Kohlmeyer B, Spotts RA, Serdani M, Crous PW, Hughes KW, Matsuura K, Langer E, Langer G, Untereiner WA, Lücking R, Büdel B, Geiser DM, Aptroot A, Diederich P, Schmitt I, Schultz M, Yahr R, Hibbett DS, Lutzoni F, McLaughlin DJ, Spatafora JW, Vilgalys R (2006) Reconstructing the early evolution of Fungi using a six-gene phylogeny. Nature 443:818–822

Jeffries P (1995) Biology and ecology of mycoparasitism. Can J Bot 73(Suppl 1):S1284–S1290

Jeffries P, Young TWK (1994) Interfungal parasitic relationships. CABI, Wallingford

Jones MDM, Forn I, Gadelha C, Egan MJ, Bass D, Massana R, Richards TA (2011a) Discovery of novel intermediate forms redefines the fungal tree of life. Nature 474:200–203

Jones MDM, Richards TA, Hawksworth DL, Bass D (2011b) Validation and justification of the new phylum name *Cryptomycota* phyl. nov. IMA Fungus 2:173–175

Kagami M, de Bruin A, Ibelings BW, Van Donk E (2007) Parasitic chytrids: their effects on phytoplankton communities and food-web dynamics. Hydrobiologia 578:113–129

Kagami M, Miki T, Takimoto G (2014) Mycoloop: chytrids in aquatic food webs. Frontiers in Microbiology, 5, 166

Kalgutkar RM, Jansonius J (2000) Synopsis of fossil fungal spores, mycelia and fructifications. Amer Assoc Strat Palynol Found, Dallas, TX

Kapgate VD (2016) Fossil fungi from Deccan Intertrappean cherts of Madhya Pradesh, India. Int J Life Sci Spec Issue A6:117–120

Karling JS (1928) Studies in the Chytridiales III. A parasitic chytrid causing cell hypertrophy in *Chara*. Amer J Bot 15:485–496

Karling JS (1932) Studies in the Chytridiales VII. The organization of the chytrid thallus. Amer J Bot 19:41–74

Kar RK, Sharma N, Kar R (2004) Occurrence of fossil fungi in dinosaur dung and its implication on food habit. Cur Sci 87:1053–1056

Karpov SA, Mamkaeva MA, Benzerara K, Moreira D, López-García P (2014) Molecular phylogeny and ultrastructure of *Aphelidium* aff. *melosirae* (Aphelida, Opisthosporidia). Protist 165:512–526

Karpov SA, Torruella G, Moreira D, Mamkaeva MA, López-García P (2017) Molecular phylogeny of *Paraphelidium letcheri* sp. nov. (Aphelida, Opisthosporidia). J Eukaryot Microbiol 64:573–578

Keeling PJ, Fast NM (2002) Microsporidia: biology and evolution of highly reduced intracellular parasites. Annu Rev Microbiol 56:93–116

Kerp H (1990) The study of fossil gymnosperms by means of cuticular analysis. Palaios 5:548–569

Kerp H, Krings M (1999) Light microscopy of cuticles. In: Jones TP, Rowe NP (eds) Fossil plants and spores: modern techniques. Spec Publ Geol Soc, London

Khan MA, Bera S, Ghosh R, Spicer RA, Spicer TEV (2015) Leaf cuticular morphology of some angiosperm taxa from the Siwalik sediments (middle Miocene to lower Pleistocene) of Arunachal Pradesh, eastern Himalaya: Systematic and paleaoclimatic implications. Rev Palaeobot Palynol 214:9–26

Kidston R, Lang WH (1921) On Old Red Sandstone plants showing structure, from the Rhynie Chert Bed, Aberdeenshire. Part V. The Thallophyta occurring in the peat-bed; the succession of the plants throughout a vertical section of the bed, and the conditions of accumulation and preservation of the deposit. Trans Roy Soc Edinburgh 52:855–902

Klymiuk AA, Taylor TN, Taylor EL, Krings M (2013) Paleomycology of the Princeton Chert. I. Fossil hyphomycetes associated with the early Eocene aquatic angiosperm, *Eorhiza arnoldii*. Mycologia 105:521–529

Knogge W (1996) Fungal infection of plants. Plant Cell 8:1711–1722

Knoll AH (1985) Exceptional preservation of photosynthetic organisms in silicide carbonates and silicified peats. Phil Trans Roy Soc London B 311:111–122

Köhler JR, Casadevall A, Perfect J (2015) The spectrum of fungi that infects humans. Cold Spring Harb Perspect Med 5:a019273
Kohlmeyer J (1979) Marine fungal pathogens among Ascomycetes and Deuteromycetes. Experientia 35:437–439
Kohlmeyer J, Kohlmeyer L (1979) Marine mycology: the higher fungi. Academic, New York, NY
Kolattukudy PE (1985) Enzymatic penetration of the plant cuticle by fungal pathogens. Annu Rev Phytopathol 23:223–250
Krings M, Harper CJ (2018) Deciphering interfungal relationships in the 410 million-year-old Rhynie chert: glomoid spores under attack. Geobios 51:151–160
Krings M, Harper CJ (2019) Fungal intruders of enigmatic propagule clusters occurring in microbial mats from the Lower Devonian Rhynie chert. PalZ 93:135–149
Krings M, Kerp H (2019) A tiny parasite of unicellular microorganisms from the Lower Devonian Rhynie and Windyfield cherts, Scotland. Rev Palaeobot Palynol 271:104106
Krings M, Taylor TN (2012) Fungal reproductive units enveloped in a hyphal mantle from the Lower Pennsylvanian of Great Britain, and their relevance to our understanding of Carboniferous fungal "sporocarps.". Rev Palaeobot Palynol 175:1–9
Krings M, Taylor TN (2014a) An unusual fossil microfungus with suggested affinities to the Chytridiomycota from the Lower Devonian Rhynie chert. Nova Hedw 99:403–412
Krings M, Taylor TN (2014b) Deciphering interfungal relationships in the 410-million-yr-old Rhynie chert: an intricate interaction between two mycelial fungi. Symbiosis 64:53–61
Krings M, Dotzler N, Taylor TN, Galtier J (2007a) A microfungal assemblage in *Lepidodendron* from the Upper Visean (Carboniferous) of central France. CR Palevol 6:431–436
Krings M, Taylor TN, Hass H, Kerp H, Dotzler HEJ (2007b) An alternative mode of early land plant colonization by putative endomycorrhizal fungi. Plant Signal Behav 2:125–126
Krings M, Taylor TN, Hass H, Kerp H, Dotzler HEJ (2007c) Fungal endophytes in a 400-million-yr-old land plant: infection pathways, spatial distribution, and host responses. New Phytol 174:648–657
Krings M, Dotzler N, Taylor TN (2009a) *Globicultrix nugax* nov. gen. et nov. spec. (Chytridiomycota), an intrusive microfungus in fungal spores from the Rhynie chert. Zitteliana A 48(49):165–170
Krings M, Dotzler N, Galtier J, Taylor TN (2009b) Microfungi from the upper Visean (Mississippian) of central France: Chytridiomycota and chytrid-like remains of uncertain affinity. Rev Palaeobot Palynol 156:319–328
Krings M, Dotzler N, Longcore JE, Taylor TN (2010a) An unusual microfungus in a fungal spore from the Lower Devonian Rhynie chert. Palaeontology 53:753–759
Krings M, Dotzler N, Taylor TN, Galtier J (2010b) A fungal community in plant tissue from the Lower Coal Measures (Langsettian, Lower Pennsylvanian) of Great Britain. Bull Geosci 85:679–690
Krings M, Dotzler N, Galtier J, Taylor TN (2011a) Oldest fossil basidiomycete clamp connections. Mycoscience 52:18–23
Krings M, Dotzler N, Taylor TN (2011b) Mycoparasitism in *Dubiocarpon*, a fungal sporocarp from the Carboniferous. N Jb Geol Paläontol Abh 262:241–245
Krings M, Taylor TN, Dotzler N, Galtier J (2011c) Fungal remains in cordaite (Cordaitales) leaves from the Upper Pennsylvanian of central France. Bull Geosci 86:777–784
Krings M, Taylor TN, Taylor EL, Dotzler N, Walker C (2011d) Arbuscular mycorrhizal-like fungi in Carboniferous arborescent lycopsids. New Phytol 191:311–314
Krings M, Taylor TN, Dotzler N (2012) Fungal endophytes as a driving force in land plant evolution: evidence from the fossil record. In: Southworth D (ed) Biocomplexity of plant–fungal interaction. Wiley, Ames, IA
Krings M, Taylor TN, Dotzler N (2013) Fossil evidence of the zygomyceteous fungi. Persoonia 30:1–10

Krings M, Taylor TN, Dotzler N (2014) Microorganisms associated with the seed fern *Lyginopteris oldhamia* (Binney) H. Potonié (Lyginopteridales) from the Carboniferous of Great Britain. Palaeontographica B290:109–125

Krings M, Taylor TN, Kerp H, Walker C (2015) Deciphering interfungal relationships in the 410-million-yr-old Rhynie chert: Sporocarp formation in glomeromycotan spores. Geobios 48:449–458

Krings M, Taylor TN, Dotzler N, Harper CJ (2016) Morphology and ontogenetic development of *Zwergimyces vestitus*, a fungal reproductive unit enveloped in a hyphal mantle from the Lower Devonian Rhynie chert. Rev Palaeobot Palynol 228:47–56

Krings M, Harper CJ, Taylor EL (2017a) Fungi and fungal interactions in the Rhynie chert: a review of the evidence, with the description of *Perexiflasca tayloriana* gen. et sp. nov. Phil Trans Roy Soc B 373:20160500

Krings M, Harper CJ, White JF, Barthel M, Heinrichs J, Taylor EL, Taylor TN (2017b) Fungi in a *Psaronius* root mantle from the Rotliegend (Asselian, Lower Permian/Cisuralian) of Thuringia, Germany. Rev Palaeobot Palynol 239:14–30

Krings M, Taylor TN, Harper CJ (2017c) Early Fungi. In: Dighton J, White JF (eds) The fungal community: its organization and role in the ecosystem. CRC Press, Boca Raton, FL

Kürschner H, Schumilovskikh L, Djamali M, de Beaulieu J-L (2015) A late Holocene subfossil record of *Sphagnum squarrosum* Crome (Sphagnopsida, Bryophyta) from NW Iran. Nova Hedw 100:373–381

Langenheim JH (1994) Higher plant terpenoids: a phytocentric overview of their ecological roles. J Chem Ecol 20:1223–1280

Laschet C (1984) On the origin of cherts. Facies 10:257–289

Le Campion-Alsumard T, Golubic S, Hutchings P (1995) Microbial endoliths in skeletons of live and dead corals, *Porites lobata* (Moorea, French Polynesia). Mar Ecol Prog Ser 117:149–157

Leake JR (2005) Plants parasitic on fungi: Unearthing the fungi in myco-heterotrophs and debunking the 'saprophytic' plant myth. Mycologist 19:113–122

LePage BA, Currah RS, Stockey RA (1994) The fossil fungi of the Princeton chert. Int J Plant Sci 155:828–836

Letcher PM, Lopez S, Schmieder R, Lee PA, Behnke C, Powell MJ, McBride RC (2013) Characterization of *Amoeboaphelidium protococcarum*, an algal parasite new to the Cryptomycota isolated from an outdoor algal pond used for the production of biofuel. PLoS One 8(2):e56232

Letcher PM, Powell MJ, Lee PA, Lopez S, Burnett M (2017) Molecular phylogeny and ultrastructure of *Aphelidium desmodesmi*, a new species in Aphelida (Opisthosporidia). J Eukaryot Microbiol 64:655–667

Lewis DH (1973) Concepts in fungal nutrition and the origin of biotrophy. Biol Rev 48:261–278

Limaye RB, Kumaran KPN, Nair KM, Padmalal D (2007) Non-pollen palynomorphs as potential palaeoenvironmental indicators in the Late Quaternary sediments of the west coast of India. Curr Sci 92:1370–1382

Lindahl BO, Taylor AFS, Finlay RD (2002) Defining nutritional constraints on carbon cycling in boreal forests—towards a less 'phytocentric' perspective. Plant Soil 242:123–135

Lom J, Dyková I (2005) Microsporidian xenomas in fish seen in wider perspective. Folia Parasitol 52:69–81

Loron CC, François C, Rainbird RH, Turner EC, Borensztaijn S, Javaux EJ (2019) Early fungi from the Proterozoic Era in Arctic Canada. Nature 570:232–235

Lücking R, Huhndorf S, Pfister DH, Plata ER, Lumbsch HT (2009) Fungi evolved right on track. Mycologia 101:810–822

Ma F-J, Sun B-N, Wang Q-J, Dong J-L, Yang G-L, Yang Y (2015) A new species of *Meliolinites* associated with *Buxus* leaves from the Oligocene of Guangxi, southern China. Mycologia 107:505–511

Magnus P (1903) Ein von F.W. Oliver nachgewiesener fossiler parasitischer Pilz. Ber Deutsch Bot Ges 21:248–250

Marcogliese DJ (2004) Parasites: small players with crucial roles in the ecological theater. EcoHealth 1:151–164

Martin JT (1964) Role of cuticle in the defense against plant disease. Annu Rev Phytopathol 2:81–100

Martin W, Rotte C, Hoffmeister M, Theissen U, Gelius-Dietrich G, Ahr S, Henze K (2003) Early cell evolution, eukaryotes, anoxia, sulfide, oxygen, fungi first (?), and a tree of genomes revisited. IUBMB Life 55:193–204

Martín-Rodrigues N, Espinel S, Sanchez-Zabala J, Ortíz A, González-Murua C, Duñabeitia MK (2013) Spatial and temporal dynamics of the colonization of *Pinus radiata* by *Fusarium circinatum*, of conidiophore development in the pith and of traumatic resin duct formation. New Phytol 198:1215–1227

Marynowski L, Smolarek J, Bechtel A, Philippe M, Kurkiewicz S, Simoneit BRT (2013) Perylene as an indicator of conifer fossil wood degradation by wood-degrading fungi. Org Geochem 59:143–151

Matos E, Corral L, Azevedo C (2003) Ultrastructural details of the xenoma of *Loma myrophis* (phylum Microsporidia) and extrusion of the polar tube during autoinfection. Dis Aquat Org 54:203–207

Mendgen K, Deising H (1993) Infection structures of fungal plant pathogens—a cytological and physiological evulation. New Phytol 124:193–213

Mendgen K, Hahn M, Deising H (1996) Morphogenesis and mechanisms of penetration of plant pathogenic fungi. Annu Rev Phytopathol 34:367–386

Merckx VSFT (ed) (2013) Mycoheterotrophy. The biology of plants living on fungi. Springer Science+Business Media, New York

Meyers JD (1990) Fungal infections in bone marrow transplant patients. Seminars Oncol 17:10–13

Möller M, Stukenbrock EH (2017) Evolution and genome architecture in fungal plant pathogens. Nature Rev Microbiol 15:756–771

Money NP (2016) Fungi: a very short introduction. Oxford University Press, Oxford, MA

Morris JL, Puttick MN, Clark JW, Edwards D, Kenrick P, Pressel S, Wellman CH, Yang Z, Schneider H, Donoghue CJ (2018) The timescale of early land plant evolution. Proc Natl Acad Sci USA 115:E2274–E2283

Nash TH (Ed) (2008) Lichen biology. 2nd edition. Cambridge University Press, Cambridge, UK

Nicholson RL, Epstein L (1991) Adhesion of fungi to the plant surface: prerequisite for pathogenesis. In: Cole GT, Hoch HC (eds) The fungal spore and disease initiation in plants and animals. Plenum Press, New York, NY

Orpin CG, Joblin KN (1997) The rumen anaerobic fungi. In: Hobson PN, Stewart CS (eds) The rumen microbial ecosystem. Springer, Dordrecht

Parratt SR, Laine AL (2016) The role of hyperparasitism in microbial pathogen ecology and evolution. ISME J 10:1815–1822

Pearce RB (1996) Antimicrobial defences in the wood of living trees. New Phytol 132:203–233

Pearce RB, Holloway PJ (1984) Suberin in the sapwood of oak (*Quercus robur* L.): its composition from a compartmentalization barrier and its occurrence in tyloses in undecayed wood. Physiol Plant Pathol 24:71–81

Petit G (2010) Skin nodules in fossil fishes from Monte Bolca (Eocene, Northern Italy). Geodiversitas 32:157–163

Petit G, Khalloufi B (2012) Paleopathology of a fossil fish from the Solnhofen Lagerstätte (Upper Jurassic, southern Germany). Int J Paleopathol 2:42–44

Phipps CJ (2007) *Entopeltacites remberi* sp. nov. from the Miocene of Clarkia, Idaho. Rev Palaeobot Palynol 145:193–200

Phipps CJ, Rember WC (2004) Epiphyllous fungi from the Miocene of Clarkia, Idaho: reproductive structures. Rev Palaeobot Palynol 129:67–79

Phipps CJ, Axsmith BJ, Taylor TN, Taylor EL (2000) *Gleichenipteris antarcticus* gen. et sp. nov. from the Triassic of Antarctica. Rev Palaeobot Palynol 108:75–83

Pirozynski KA (1976) Fossil fungi. Annu Rev Phytopathol 14:237–246

Poinar GO (2015) One hundred million year old ergot: psychotropic compounds in the Cretaceous? Palaeodiversity 8:13–19

Poinar GO (2016a) A mid-Cretaceous Eccrinales infesting a primitive wasp in Myanmar amber. Fung Biol 120:1537–1539

Poinar GO (2016b) A mid-Cretaceous ectoparasitic fungus, *Spheciophila adercia* gen et sp. nov., attached to a wasp in Myanmar amber. Fungal Genom Biol 6:145

Poinar GO (2018) A mid-Cretaceous pycnidia, *Palaeomycus epallelus* gen. et sp. nov., in Myanmar amber. Hist Biol 32:234–237

Poinar GO, Buckley R (2007) Evidence of mycoparasitism and hypermycoparasitism in Early Cretaceous amber. Mycol Res 111:503–506

Poinar GO, Thomas GM (1982) An entomophthoralean fungus from Dominican amber. Mycologia 74:332–334

Poinar GO, Thomas GM (1984) An entomogenous fungus from Dominican amber. Experientia 40:578–579

Pollard M, Beisson F, Li Y, Ohlrogge JB (2008) Building lipid barriers: biosynthesis of cutin and suberin. Trends Plant Sci 13:236–246

Prasertphon S (1963) Conidial formation in *Entomophthora* species with *E. muscae*-like conidia. Mycologia 68:1–29

Rega EA, Brochu C (2001) Paleopathology of a mature *Tyrannosaurus rex*. J Vert Paleontol 21:(Suppl 3):92A

Rehner SA, Minnis AM, Sung G-H, Luangsa-ard JJ, Devotto L, Humber RA (2011) Phylogeny and systematics of the anamorphic, entomopathogenic genus *Beauveria*. Mycologia 103:1055–1073

Roth L, Lorscheitter ML (2016) Fungi, algae, and other palynomorphs in sedimentary profiles collected from two forests in the northernmost coastal plain from Rio Grande do Sul, southern Brazil. Brazil J Bot 39:1135–1143

Rothschild BM, Martin, LD (2006) Skeletal impact of disease: bulletin 33 (Vol. 33). New Mexico Museum of Natural History and Science

Rossi W, Kotrba M, Triebel D (2005) A new species of *Stigmatomyces* from Baltic amber, the first fossil record of *Laboulbeniomycetes*. Mycol Res 109:271–274

Round FE (1981) The ecology of algae. Cambridge University Press, Cambridge, MA

Ryvarden L (1991) Genera of polypores. Nomenclature and taxonomy (Synopsis fungorum 5, fungiflora). Grønlands Grafiske A/S, Oslo

Schmidt AR, Dörfelt H, Perrichot V (2007) Carnivorous fungi from Cretaceous amber. Science 318:1743

Schulze-Lefert P (2004) Knocking on the heaven's wall: pathogenesis of and resistance to biotrophic fungi at the cell wall. Curr Op Plant Biol 7:377–383

Schumilovskikh LS, Schlütz F, Achterberg I, Kvitkina A, Bauerochse A, Leuschner HH (2015) Pollen as nutrient source in Holocene ombrotrophic bogs. Rev Palaeobot Palynol 221:171–178

Schwarze FWMR, Baum S (2000) Mechanisms of reaction zone penetration by decay fungi in wood of beech (*Fagus sylvatica*). New Phytol 146:129–140

Schwarze FWMR, Engels J, Mattheck C (2000) Fungal strategies of wood decay in trees. Springer-Verlag, Berlin

Serrano M, Coluccia F, Torres M, L'Haridon F, Métraux J-P (2014) The cuticle and plant defense to pathogens. Front Plant Sci 5:274

Shang Y, Feng P, Wang C (2015) Fungi that infect insects: altering host behavior and beyond. PLoS Pathog 11:e1005037

Sharon A, Schlezinger N (2013) Fungi infecting plants and animals: killers, non-killers, and cell death. PLoS Pathog 9:e1003517

Sharma N, Kar RK, Agarwal A, Kar R (2005) Fungi in dinosaurian (*Isisaurus*) coprolites from the Lameta Formation (Maastrichtian) and its reflection on food habit and environment. Micropaleontology 51:73–82

Sharpe SC, Eme L, Brown MW, Roger AJ (2015) Timing the origins of multicellular eukaryotes through phylogenetic and relaxed molecular clock analysis. In: Ruiz-Trillo IR, Nedelecu AM (eds) Evolutionary transitions to multicellular life. Springer, Dordrecht, NL

Shrimpton DM (1973) Extractives associated with wound response of lodgepole pine attacked by the mountain pine beetle and associated microorganisms. Can J Bot 51:527–534

Sime-Ngando T (2012) Phytoplankton chytridiomycosis: fungal parasites of phytoplankton and their imprints on the food web dynamics. Front Microbiol 3:361. https://doi.org/10.3389/fmicb.2012.00361

Singh H, Tripathi SKM (2010) Fungal remains from the Early Paleogene sub-surface sediments of Karakha, Barmer Distinct, western India. Geophytology 39:9–15

Skerratt LF, Berger L, Speare R, Cashins S, McDonald KR, Phillott AD, Hines HB, Kenyon N (2007) Spread of chytridiomycosis has caused the rapid global decline and extinction of frogs. EcoHealth 4:125–134

Slater BJ, McLoughlin S, Hilton J (2015) A high-latitude Gondwanan lagerstätte: the Permian permineralised peat biota of the Prince Charles Mountains, Antarctica. Gondwana Res 27:1446–1473

Smith SY, Currah RS, Stockey RA (2004) Cretaceous and Eocene poroid hymenophores from Vancouver Island, British Columbia. Mycologia 96:180–186

Smith SY, Stockey RA (2007) Establishing a fossil record for the perianthless Piperales: *Saururus tuckerae* sp. nov. (Saururaceae) from the Middle Eocene Princeton Chert. Amer J Bot 94:1642–1657

Stewart JE (1993) Infectious disease of marine crustaceans. In: Couch JA, Fournie JW (eds) Pathbiology of marine and estuarine organisms. CRC Press, Boca Raton, FL

Strullu-Derrien C, Wawrzyniak Z, Goral T, Kenrick P (2015) Fungal colonization of the rooting system of the early land plant *Asteroxylon mackiei* from the 407-myr-old Rhynie chert (Scotland, UK). Bot J Linn Soc 179:201–213

Strullu-Derrien C, Goral T, Longcore JE, Olesen J, Kenrick P, Edgecombe GD (2016) A new chytridiomycete fungus intermixed with crustacean resting eggs in a 407-million-year-old continental freshwater environment. PLoS One 11:e0167301

Stubblefield SP, Taylor TN (1983) Studies of paleozoic fungi. I. The structure and organization of *Traquairia* (Ascomycota). Amer J Bot 70:387–399

Stubblefield SP, Taylor TN (1986) Wood decay in silicified gymnosperms from Antarctica. Bot Gaz 147:116–125

Stubblefield SP, Taylor TN (1988) Recent advances in paleomycology. New Phytol 108:3–25

Stubblefield SP, Taylor TN, Miller CE, Cole GT (1983) Studies in Carboniferous fungi. II. The structure and organization of *Mycocarpon*, *Sporocarpon*, *Dubiocarpon*, and *Coleocarpon* (Ascomycotina). Amer J Bot 70:1482–1498

Stubblefield SP, Taylor TN, Miller CE, Cole GT (1984) Studies in Paleozoic fungi. III. Fungal parasitism in a Pennsylvanian gymnosperm. Amer J Bot 71:1275–1284

Stubblefield SP, Taylor TN, Beck CB (1985) Studies of Paleozoic fungi. IV. Wood-decaying fungi in *Callixylon newberryi* from the Upper Devonian. Amer J Bot 72:1765–1774

Sun C, Taylor TN, Na Y, Li T, Krings M (2015) Unusual preservation of a microthyriaceous fungus (Ascomycota) on *Sphenobaiera* (ginkgophyte foliage) from the Middle Jurassic of China. Rev Palaeobot Palynol 223:21–30

Sung G-H, Hywel-Jones NL, Sung J-M, Luangsa-ard JJ, Shrestha B, Spatafora JW (2007) Phylogenetic classification of *Cordyceps* and the clavicipitaceous fungi. Stud Mycol 57:5–59

Sung G-H, Poinar GO, Spatafora JW (2008) The oldest fossil evidence of animal parasitism by fungi supports a Cretaceous diversification of fungal–arthropod symbioses. Mol Phylogen Evol 49:495–502

Swain T (1977) Secondary compounds as protective agents. Ann Rev Plant Physiol 28:479–501

Szabo LJ, Bushnell WR (2001) Hidden robbers: the role of fungal haustoria in parasitism of plants. Proc Natl Acad Sci USA 98:7654–7655

Takamatsu S, Niinomi S, Harada M, Havrylenko M (2010) Molecular phylogenetic analyses reveal a close evolutionary relationship between *Podosphaera* (Erysiphales: Erysiphaceae) and its rosaceous hosts. Persoonia 24:38–48

Takamatsu S (2013) Origin and evolution of the powdery mildews (Ascomycota, Erysiphales). Mycoscience 54:75–86

Tarran M, Wilson PG, Hill RS (2016) Oldest record of *Metrosideros* (Myrtaceae): fossil flowers, fruits, and leaves form Australia. Amer J Bot 103:754–768

Taylor JW, Berbee ML (2006) Dating divergences in the fungal tree of life: review and new analyses. Mycologia 98:838–849

Taylor TN, Krings M (2010) Paleomycology: the rediscovery of the obvious. Palaios 25:283–286

Taylor TN, White JF (1989) Fossil fungi (Endogonaceae) from the Triassic of Antarctica. Amer J Bot 76:389–396

Taylor TN, Remy W, Hass H (1992a) Fungi from the Lower Devonian Rhynie chert: Chytridiomycetes. Amer J Bot 79:1233–1241

Taylor TN, Hass H, Remy W (1992b) Devonian fungi: interactions with the green alga *Palaeonitella*. Mycologia 84:901–910

Taylor TN, Remy W, Hass H (1992c) Parasitism in a 400-million-year-old green alga. Nature 357:493–494

Taylor TN, Galtier J, Axsmith BJ (1994) Fungi from the Lower Carboniferous of central France. Rev Palaeobot Palynol 83:253–260

Taylor TN, Remy W, Hass H, Kerp H (1995) Fossil arbuscular mycorrhiza from the Early Devonian. Mycologia 87:560–573

Taylor TN, Hass H, Kerp H (1999) The oldest fossil ascomycetes. Nature 399:648

Taylor TN, Klavins SD, Krings M, Taylor EL, Kerp H, Hass H (2004) Fungi from the Rhynie chert: a view from the dark side. Trans R Soc Edinburgh, Earth Sci 94:457–473

Taylor TN, Hass H, Kerp H, Krings M, Hanlin RT (2005a) Perithecial ascomycetes from the 400 million years old Rhynie chert: an example of ancestral polymorphism. Mycologia 97:269–285

Taylor TN, Kerp H, Hass H (2005b) Life history biology of early land plants: dedciphering the gametophyte phase. Proc Natl Acad Sci USA 102:5892–5897

Taylor TN, Krings M, Klavins SD, Taylor EL (2005c) *Protoascon missouriensis*, a complex fossil microfungus revisited. Mycologia 97:725–729

Taylor TN, Taylor EL, Krings M (2009) Palaeobotany: the biology and evolution of fossil plants. Elsevier, Burlington, MA

Taylor TN, Krings M, Galtier J, Dotzler N (2012) Fungal endophytes in *Astromyelon*-type (Sphenophyta, Equisetales, Calamitaceae) roots from the Upper Pennsylvanian of France. Rev Palaeobot Palynol 171:9–18

Taylor TN, Krings M, Taylor EL (2015) Fossil fungi, 1st edn. Elsevier, Burlington, MA

Thiéry A, Fugate M (1994) A new American fairy shrimp, *Linderiella santarosae* (Crustacea, Anostraca, Linderiellidae), from vernal pools of California. Proc Biol Soc Wash 107:641–656

Thomas GM, Poinar GO (1988) A fossil *Aspergillus* from Eocene Dominican amber. J Paleontol 62:141–143

Tiffney BH, Barghoorn ES (1974) The fossil record of the fungi. Occas Pap Farlow Herb Cryptog Bot 7:1–42

Treseder KK, Lennon JT (2015) Fungal traits that drive ecosystem dynamics on land. Microbiol Mol Biol Rev 79:243–262

Trewin NH, Kerp H (2017) The Rhynie and Windyfield cherts, Early Devonian, Rhynie, Scotland. In: Fraser NC, Sues H-D (eds) Terrestrial conservation Lagerstätten—windows into the evolution of life on land. Dunedin Academic Press, Edinburgh

Tucker SL, Talbot NJ (2001) Surface attachment and pre-penetration stage development plant pathogenic fungi. Annu Rev Phytopathol 39:385–417

Vacher C, Piou D, Desprez-Loustau ML (2008) Architecture of an antagonistic tree/fungus network: the asymmetric influence of past evolutionary history. PLoS One 3:e1740

Vávra J, Lukeš J (2013) Microsporidia and 'the art of living together'. Adv Parasitol 82:253–319

Van Der Heijden MG, Bardgett RD, Van Straalen NM (2008) The unseen majority: soil microbes as drivers of plant diversity and productivity in terrestrial ecosystems. Ecol Lett 11:296–310
Van Geel B (2002) Non-pollen palynomorphs. In: Smol JP, HJB B, Last WM, Bradley RS, Alverson K (eds) Tracking environmental change using lake sediments. developments in paleoenvironmental research, vol 3. Springer, Dordrecht
Van Geel B, Andersen ST (1988) Fossil ascospores of the parasitic fungus *Ustulina deusta* in Eemian deposits in Denmark. Rev Palaeobot Palynol 56:89–93
Van Geel B, Aptroot A, Mauquoy D (2006) Sub-fossil evidence for fungal hyperparasitism (*Isthmospora spinosa* on *Meliola ellisii*, on *Calluna vulgaris*) in a Holocene intermediate ombrotrophic bog in northern-England. Rev Palaeobot Palynol 141:121–126
Van Loon LC, Rep M, Pieterse CMJ (2006) Significance of inducible defense-related proteins in infected plants. Annu Rev Phytopathol 44:135–162
Veronese P, Ruiz MT, Coca MA, Hernandez-Lopez A, Lee H, Ibeas JI, Darnsz B, Pardo JM, Hasegawa PM, Bressan RA, Narasimhan ML (2003) In defense against pathogens. Both plant sentinels and foot soldiers need to know the enemy. Plant Physiol 131:1580–1590
Vujanovic V, St-Arnaud M, Neumann P-J (2009) Susceptibility of cones and seeds to fungal infection in a pine (*Pinus* spp.) collection. Forest Pathol 30:305–320
Walker C, Harper CJ, Brundrett MC, Krings M (2018) Looking for arbuscular mycorrhizal fungi (AMF) in the fossil record: an illustrated guide. In: Krings M, Harper CJ, Cúneo NR, Rothwell GW (eds) Transformative paleobotany: papers to commemorate the life and legacy of Thomas N. Taylor. Elsevier, Cambridge, MA
Wan M, Yang W, He X, Liu L, Wang J (2017) First record of fossil basidiomycete clamp connections in cordaitalean stems from the Asselian-Sakmarian (lower Permian) of Shanxi Province, North China. Palaeogeogr Palaeoclimatol Palaeoecol 466:353–360
Wang Z-Q (1997) Permian *Supaia* fronds and an associated *Autunia* fructification from Shanxi, China. Palaeontology 40:245–277
Wang G, Johnson ZI (2009) Impact of parasitic fungi on the diversity and functional ecology of marine phytoplankton. In: Kersey WT, Munger SP (eds) Marine phytoplankton (Oceanography and Ocean Engineering). Nova Science Publishers Inc., New York
Watson V, Rothschild B (2021) Deep origin of parasitic disease in vertebrates. In: De Baets K., Huntley JW (eds) The evolution and fossil record of Parasitism: Coevolution and paleoparasitological techniques. Topics in Geobiology 50
Weaver L, McLoughlin S, Drinnan AN (1997) Fossil woods from the Upper Permian Bainmedart Coal Measures, northern Prince Charles Mountains, East Antarctica. AGSO J Aust Geol Geophys 16:655–676
Weiss LM, Becnel JJ (2014) Microsporidia: pathogens of opportunity. Wiley, Oxford
Weiss A, Martindale RC (2017) Crustose coralline algae increased framework and diversity on ancient coral reefs. PLoS One 12:e0181637
Weissenberg R (1968) Intracellular development of the microsporidian *Glugea anomala* Moniez in hypertrophying migratory cells of the fish *Gasterosteus aculeatus* L., an example of the formation of "xenoma tumors". J Protozool 15:44–57
White JF, Taylor TN (1989) Triassic fungi with suggested affinities to the Endogonales (Zygomycotina). Rev Palaeobot Palynol 61:53–61
White JF, Kingsley K, Harper CJ, Verma SK, Brindisi L, Chen Q, Chang X, Micci A, Bergen M (2018) Reactive oxygen defense against cellular endoparasites and the origin if eukaryotes. In: Krings M, Harper CJ, Cúneo NR, Rothwell GW (eds) Transformative paleobotany: papers to commemorate the life and legacy of Thomas N. Taylor. Elsevier, Cambridge, MA
Williamson WC (1878) On the organization of the fossil plants of the coal-measures. Part IX. Phil Trans Roy Soc London B Biol Sci 169:319–364
Williamson WC (1880) On the organization of the fossil plants of the coal-measures. Part X. Including an examination of the supposed radiolarians of the Carboniferous rocks. Phil Trans Roy Soc London B Biol Sci 171:493–539

Williamson WC (1883) On the organization of the fossil plants of the coal-measures: part XII. Phil Trans Roy Soc London B Biol Sci 174:459–475
Wolf FA (1969) A rust and an alga in Eocene sediment from western Kentucky. J Elisha Mitchell Sci Soc 85:57–58
Wolff EDS, Salisbury SW, Horner JR, Varricchio DJ (2009) Common avian infection plagued the tyrant dinosaurs. PLoS One 4:e7288
Zelmer DA (1998) An evolutionary definition of parasitism. Int J Parasitol 28:531–533

Chapter 4
Evolution, Origins and Diversification of Parasitic Cnidarians

Beth Okamura and Alexander Gruhl

Abstract Parasitism has evolved in cnidarians on multiple occasions but only one clade—the Myxozoa—has undergone substantial radiation. We briefly review minor parasitic clades that exploit pelagic hosts and then focus on the comparative biology and evolution of the highly speciose Myxozoa and its monotypic sister taxon, *Polypodium hydriforme*, which collectively form the Endocnidozoa. Cnidarian features that may have facilitated the evolution of endoparasitism are highlighted before considering endocnidozoan origins, life cycle evolution and potential early hosts. We review the fossil evidence and evaluate existing inferences based on molecular clock and cophylogenetic analyses. Finally, we consider patterns of adaptation and diversification and stress how poor sampling might preclude adequate understanding of endocnidozoan diversity.

Keywords Myxozoa · *Polypodium* · Adaptations to parasitism · Life-cycle evolution · Cnidarian origins · Fossil record · Host acquisition · Molecular clock analysis · Co-phylogenetic analysis · Unknown diversity

4.1 Introduction

Cnidarians are generally regarded as a phylum of predatory free-living animals that occur as benthic polyps and pelagic medusae in the world's oceans. They include some of the most iconic residents of marine environments, such as corals, sea anemones and jellyfish. Cnidarians are characterised by relatively simple body-plans, formed entirely from two tissue layers (the ectoderm and endoderm), and by their stinging cells or nematocytes. Nematocytes are unique to Cnidaria and function primarily for prey capture and defense. Phylogenetic analyses identify cnidarians as

B. Okamura (✉)
Department of Life Sciences, Natural History Museum, London, UK
e-mail: b.okamura@nhm.ac.uk

A. Gruhl
Department of Symbiosis, Max Planck Institute for Marine Microbiology, Bremen, Germany

© The Author(s) 2021

K. De Baets, J. W. Huntley (eds.), *The Evolution and Fossil Record of Parasitism*, Topics in Geobiology 49, https://doi.org/10.1007/978-3-030-42484-8_4

early diverging metazoans and sister to Bilateria, with the Ctenophora and Porifera variously placed as earlier diverging sister lineages to Metazoa (e.g. Ryan et al. 2013; Simion et al. 2017; Whelan et al. 2017). Accordingly, cnidarians have a convincing fossil record dating from the early Cambrian (e.g. Dong et al. 2013) and, as discussed later, with probable representation even earlier in the Ediacaran Period.

Although mainly viewed as marine animals, a few cnidarians have invaded freshwater habitats (Jankowski et al. 2008), including the model organism *Hydra*. In addition, parasitic lifestyles have been adopted on several occasions by different cnidarian lineages. One parasitic group in particular—the Myxozoa—has undergone extensive radiation as endoparasites with complex life cycles, exploiting invertebrate and vertebrate hosts. According to the most recent estimate, myxozoans represent some 20% (2596/14,355) of all described cnidarian species (Okamura et al. 2018), a proportion expected to rise further in view of extensive undersampling.

This chapter reviews the evolution, origins and diversification of parasitic cnidarians. We first describe the variety of cnidarian parasites known to date and highlight cnidarian features that may be generally conducive for adopting parasitic lifestyles. We then focus on the major clade of parasitic cnidarians, the Endocnidozoa, which contains the diverse Myxozoa and the monotypic *Polypodium hydriforme*. This leads us to consider more explicitly pathways to endoparasitism, origins and early hosts, and patterns and drivers of diversification within the Endocnidozoa.

4.2 Parasitic Cnidarians Other than Endocnidozoans

According to current views of cnidarian systematics (Fig. 4.2; Kayal et al. 2018) parasitic forms have evolved at least twice in Anthozoa (Rodríguez et al. 2014) and perhaps twice or more in Hydrozoa (Bentlage et al. 2018; Table 4.1). In all cases parasitic stages are associated with pelagic animal hosts. They have been described in the distantly-related burrowing anemone families, Edwardsiidae and Haloclavidae (Rodríguez et al. 2014), and in families belonging to the hydrozoan orders Narcomedusae and Anthoathecata (Collins et al. 2008; Bentlage et al. 2018). Infection is likely generally to occur via the larval (planula) stage (Boero and Bouillon 2005) with parasites then undergoing further development (Table 4.1). For example, larvae of the anthozoan *Peachia* develop to polyps on their medusa hosts, which then drop off to take up benthic existence. Polyp stages of hydrozoan narcomedusae develop as endoparasites in medusa and polychaete hosts prior to assuming life as free-living medusae. Other hydrozoans develop as ectoparasitic colonies on fish, copepods and pteropods during the polyp phase of the life cycle. It is argued that the anthozoan *Edwardsiella lineata* develops in the digestive cavity of ctenophores as a novel life history stage. The latter is inferred on the basis of a unique combination of features (no cilia or tentacles but possessing a pharynx, retractor muscles and mesenteries), tissue remodelling (including apoptosis), and a clear shift

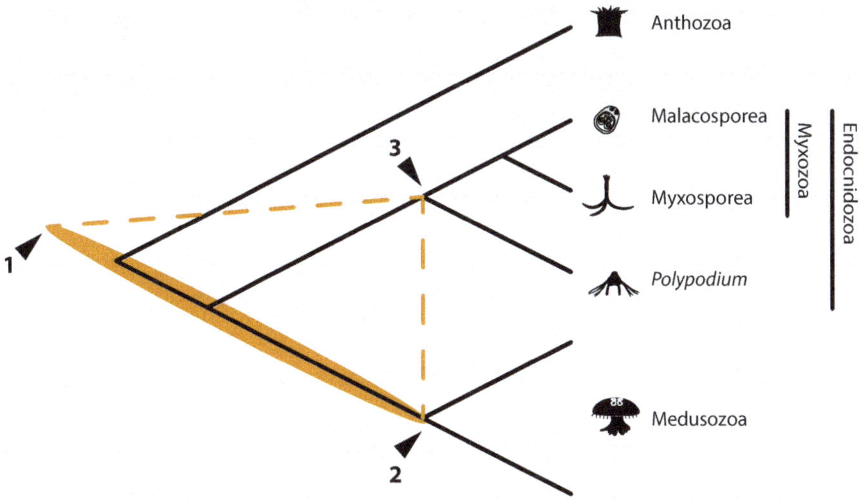

Fig. 4.2 Cnidarian phylogenetic tree with mapped time constraints for the evolution of Endocnidozoa. (1) Maximum age of crown Cnidaria (~900 Ma); (2) Minimum age of crown Medusozoa (505 Ma); (3) Minimum age of crown Endocnidozoa (not calibrated by fossils)

in the ecological niche between the parasitic and free-living life history stages (Reitzel et al. 2006).

The number of cnidarian species currently recognised to include parasitic stages (Table 4.1) is probably underestimated (Appeltans et al. 2012). Parasitic narcomedusans are particularly likely to be poorly known as their hosts are relatively infrequently sampled open ocean animals (often from the deep sea), they may be overlooked due to their inconspicuous nature, and they may occur at low prevalences of infection. It is possible that parasitic stages may be linked in future with narcomedusan species previously thought to be entirely free-living. Nor would it be surprising if further entirely new species with parasitic stages are detected, especially in poorly sampled habitats such as the deep sea and polar regions (Okamura et al. 2018). However, it may also be the case that our current understanding is compromised. For example, taxa described long ago as distinct genera may belong to a common genus (e.g. the pteropod-infecting taxa *Perigonella*, *Pandea* and *Kinetocodium*; Table 4.1). Alternatively, taxa currently recognised as different could be part of the same life cycle (e.g. if a broad range of hosts is exploited or distinct parasitic stages have evolved).

Despite caveats regarding sampling effort and taxonomic uncertainties, parasitic cnidarians other than endocnidozoans have apparently undergone little radiation and they are not associated with markedly long branches in phylogenetic trees (e.g. Rodríguez et al. 2014; Bentlage et al. 2018). In all cases the number of described species within lineages is ≤11. Thus even if we assume that all species of e.g. *Edwardsiella* incorporate a parasitic phase they would only amount to some 0.08% (11/14,355) of currently described cnidarian species diversity. The

Table 4.1 Details for parasitic cnidarians (excluding Endocnidozoa) including higher taxonomy (Class, Order, Family according to WoRMS Editorial Board 2018), genus (and number of described species according to WoRMS Editorial Board 2018), hosts, examples of lifestyles, and inferred life history stages of parasitic forms (numbered references as superscripts). Note that the life cycles of many cnidarians are poorly known (Collins et al. 2008) thus it is unclear whether all species within the listed genera include parasitic stages. Also note that the narcomedusan families Cuninidae and Solmarisidae are polyphyletic ([P]) and recent phylogenetic analysis (Bentlage et al. 2018) indicates that parasitic members of both families are each other's closest relatives

Higher taxonomy	Genus	Hosts	Example lifestyles	Life history stages	References
Anthozoa, Actiniaria, Edwardsiidae	*Edwardsiella* (11)	Ctenophores	*E. lineata* positioned along pharynx or in ciliated region near esophagus as vermiform stages with oral end inside digestive cavity feeding on pre-digested material by ciliary currents[1]	Novel stage[1]	1. Reitzel et al. (2006)
Anthozoa, Actiniaria, Haloclavidae	*Peachia* (11)	Hydrozoan and scyphozoan medusae	*P. quinquecapitata* initially parasitic in gastrovascular system (feeding on pre-digested material like *E. lineata*[1]) then moving to and replacing gonad of medusa host[2]; *P. parasitica* attached by expanded mouth to subumbrella or tissues of host[3]	Larvae and pre-adults (larvae mature to adult anemones that drop off host)[2]	2. Spaulding (1972); 3. McDermott et al. (1982)
Hydrozoa, Narcomedusae, Cuninidae[P]	*Cunina* (11)	Hydrozoan medusae	Parasitic in gastrovascular system[4,5], also attached to and apparently replacing gonad[5]	Larvae[4]; polyps[5]	4. Boero and Bouillon (2005); 5. Bentlage et al. (2018)

(continued)

Table 4.1 (continued)

Higher taxonomy	Genus	Hosts	Example lifestyles	Life history stages	References
Hydrozoa, Narcomedusae, Solmarisidae[P]	*Pegantha* (5)	Polychaetes (e.g. *Tomopteris*)	Polyps of *P. martagon* attached to peritoneum of *Tomopteris* and inferred to absorb nutrients from coelomic fluid (as there is no mouth), budded medusae released from polyps into coelomic cavity of host[5]	Larvae[4]; polyps + early medusae[5]	4. Boero and Bouillon (2005); 5. Bentlage et al. (2018)
Hydrozoa, Anthoathecata, Pandeidae	*Hydrichthys* (6)	Fish, copepod	Attached to and eroding fish surface, tentacle-less polyps bend and mouth sucks in blood and tissues[4,6]; also attached to copepods parasitic on mesopelagic lanternfish[7]	Polyps[6]	4. Boero and Bouillon (2005); 6. Boero et al. (1991); 7. Moser and Taylor (1978)
	Larsonia (1)	Fish	Attached to and eroding fish surface, tentacle-less polyps bend and mouth sucks in blood and tissues[6]	Polyps[6]	6. Boero et al. 1991
	Perigonella (1)	Pteropods	*P. sulfurea* polyps attach to pteropod shells and feed on epithelia and embryos[4]	Polyps[4]	4. Boero and Bouillon (2005)
	Pandea (4)	Pteropods	*P. conica* polyps attach to pteropod shells and feed on epithelia and embryos[4]	Polyps[4]	4. Boero and Bouillon (2005)
Hydrozoa, Anthoathecata, Incertae sedis	*Kinetocodium* (1)	Pteropods	*K. danae* polyps attach to pteropod shells and feed on epithelia and embryos[4]	Polyps[4]	4. Boero and Bouillon (2005)

non-endocnidozoan lineages of cnidarians that incorporate parasitic stages in their life history therefore appear to have evolved fairly recently and to have undergone modest to minimal radiation.

4.3 The Endocnidozoa

As the name implies, the Endocnidozoa is comprised of endoparasitic cnidarians. This recently recognised clade incorporates the sister taxa *Polypodium hydriforme* (henceforth referred to as *Polypodium*) and the diverse Myxozoa (Collins 2009; Chang et al. 2015; Kayal et al. 2018). For a considerable time period long-branch attraction obscured phylogenetic placement of both Myxozoa and *Polypodium* in molecular phylogenetic analyses (Zrzavý and Hypša 2003; Foox and Siddall 2015; Okamura and Gruhl 2015). However, phylogenomic (Chang et al. 2015; Kayal et al. 2018) and some morphological (e.g. Siddall et al. 1995; and see below) evidence currently places these as sister taxa comprising the Endocnidozoa, which itself is sister to the Medusozoa (Fig. 4.1).

4.3.1 General Biology

Polypodium's one-host life cycle (Fig. 4.1c) includes a free-living adult phase and parasitic larval stages in acipenseriform fish (sturgeon and paddlefish). Myxozoans have complex parasitic life cycles and require both invertebrate and vertebrate hosts for development (Fig. 4.1a, b). Invertebrate hosts include freshwater (phylactolaemate) bryozoans (exploited by the Malacosporea) and marine and freshwater oligochaetes and polychaetes (exploited by the Myxosporea) (Fiala et al. 2015a). Myxosporean infections (including spore production) reported in octopus (Yokoyama and Masuda 2001) and in a monogenean infecting fish (a case of hyperparasitism) (Freeman and Shinn 2011) suggest that other invertebrate hosts may at least occasionally be exploited. By far the greatest number of recognised vertebrate hosts of myxozoans are fish (including representatives of both subclasses of primitive cartilaginous fishes and a broad range of derived bony fish; Lom and Dyková 2006; Kodádková et al. 2015), but myxozoans also infect reptiles (turtles and tortoises), waterfowl (ducks), small mammals (shrews and probably moles) (Lom and Dyková 2006; Hallett et al. 2015) and all orders of amphibians (Hartigan et al. 2016).

The free-living *Polypodium* stage emerges from spawned eggs of acipenseriform fish as chains or stolons of budded but connected tentaculate individuals. The stolons fragment into individual buds that take up benthic life, actively feeding and undergoing growth and fission during summer months (Fig. 4.1). Reproductively mature individuals produce a specialised multicellular stage derived from gonadal tissue (Raikova 1994, 2008) that enables infection following direct contact with larval fish. Post-invasion infection dynamics are unknown until fish become

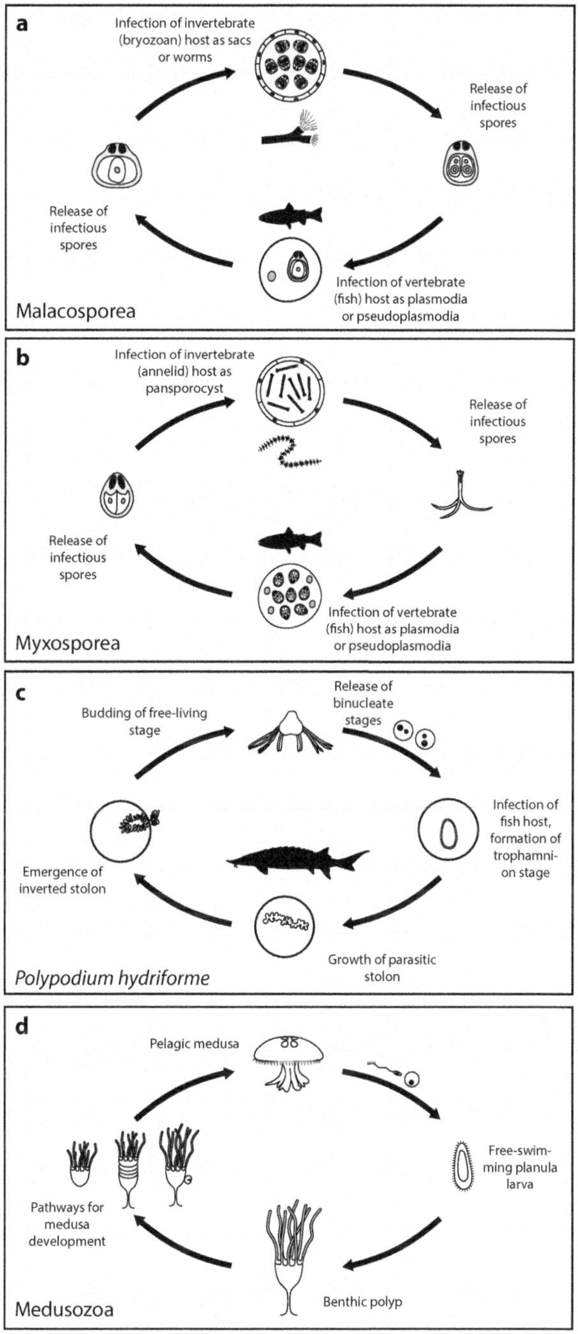

Fig. 4.1 Life cycles of endocnidozoans and free-living cnidarians: (**a**) Malacosporea; (**b**) Myxosporea; (**c**) *Polypodium hydriforme*; (**d**) Medusozoa

reproductively mature, at which stage the development of *Polypodium* within fish eggs has been characterised (Raikova 1994). Larvae and budding stolons inside the eggs have inverted germ layers, an inner ectoderm and outer endoderm—a condition that is reversed prior to emerging from fish eggs. Stolons are liberated from eggs in the oviducts of spawning fish (Raikova 2002). *Polypodium* infections have been recorded in 78% of mature female sterlet in the Volga and Kama Rivers and up to 100% of eggs per female can apparently be infected (Raikova 1994)—a cause for concern given potential impacts on caviar production. The lineage is currently regarded as monotypic although sequence divergence in housekeeping genes has recently been revealed between North American and Russian isolates (Hartigan et al. unpublished data).

Myxozoans exploit invertebrate and vertebrate hosts that act as definitive and intermediate hosts, respectively (Fig. 4.1). Multicellular spores released from hosts into the environment to achieve transmission are non-feeding and metabolically inactive. The early-diverging major myxozoan clades, the Malacosporea and Myxosporea, differ in invertebrate host use and morphological complexity (Fiala et al. 2015a; Gruhl 2015). In their freshwater bryozoan hosts, malacosporean sporogonic (spore-producing) stages develop in the host coelomic cavity as sacs (~300–700 μm in diameter) and 'myxoworms' (up to ~3 mm in length) that exhibit clearly recognisable epithelial layers and muscle systems (the latter only in myxoworms) (Feist et al. 2015; Gruhl 2015). Multicellular spores that achieve transmission to fish are produced within the hollow spaces of sacs and myxoworms. In contrast, malacosporean sporogonic stages that develop in fish kidney (called pseudoplasmodia) are comprised of a single cell (the so-called primary cell) within which multicellular spores infectious to bryozoans are produced. Malacosporeans exploit a broad range of freshwater bryozoans (Hartikainen et al. 2014) and infections have so far been detected in fish hosts in the families Salmonidae, Cyprinidae and Percidae (Grabner and El-Matbouli 2010; Bartošová-Sojková et al. 2014; Naldoni et al. 2019), however some fish in these families may be accidental hosts because spore development has not always been demonstrated.

Myxosporean sporogonic stages in annelids, called pansporocysts, are comparable to malacosporean sacs, but have an outer lining that is made up of only eight cells (El-Matbouli and Hoffmann 1998). These cells are extremely thin and show hardly any epithelial characteristics. Pansporocysts are very small (10–100 μm range) and develop in the epidermis, gut epithelium, and coelomic cavities of annelid hosts (Lom and Dyková 1997; Gruhl 2015). Myxosporean sporogonic stages in vertebrate hosts are either plasmodia (unicellular, multinucleate forms) or pseudoplasmodia (unicellular, uninucleate forms) within which multicellular spores develop. Spores produced by malacosporeans in invertebrate and vertebrate hosts (malacospores and fish malacospores, respectively) are morphologically similar and short-lived (e.g. <24 h; de Kinkelin et al. 2002). Spores produced by myxosporeans (myxospores produced in fish hosts and actinospores produced in annelid hosts) show variable morphologies some of which can be useful taxonomically (Atkinson et al. 2015; Fiala et al. 2015b). Histozoic myxosporeans develop in small intercellular spaces within epithelia or connective tissues, whereas coelozoic species develop in the lumen of organs (e.g. bile duct, urinary bladder). The most common infection sites include gills, skin, fins, eyes, kidney, intestine, liver and gall bladder, nervous system, cartilage, musculature, and swim bladder (Molnár and Eszterbauer 2015). Myxosporean plasmodia and pseudoplasmodia are generally tiny, but some are mm to cm in dimensions. Annelid hosts of myxosporeans are mostly unknown—infections so far have been detected in 7 (Alexander et al. 2015) of approximately 120 families (Fauchald and Rouse 1997; Erséus 2005). Fish are widely exploited by myxosporeans with infections reported in many families of both cartilaginous and bony fishes.

Myxozoans are causative agents of several diseases impacting aquaculture and wild fish populations. These include whirling disease, proliferative kidney disease and enteronecrosis in salmonids and pharyngeal myxosporidiosis in carp (Feist and Longshaw 2006; Jones et al. 2015).

4.3.2 Comparative Development and Body Plans

There are noteworthy similarities and differences in morphology and life history stages between *Polypodium* and myxozoans consistent with the retention of a free-living stage in the life cycle of the former and the entirely endoparasitic life cycle of the latter (Table 4.2). In particular, myxozoans have lost many organs (e.g. no recognisable gut or gonads) and certain tissues (e.g. nerves, gonad) although proper epithelia characterise the early diverging malacosporeans. *Polypodium* and myxoworms both develop independent, sub-epidermal muscle systems (Raikova et al. 2007; Gruhl and Okamura 2012) in contrast to the typical cnidarian epitheliomuscular cells. Cilia, centrioles, and cnidocils are absent in myxozoans but present in *Polypodium*. Pinpointing the cells that act as gametes requires observing meiosis and the brief and transitory process of fusion and there are conflicting interpretations of these events in myxozoans (reviewed in Feist et al. 2015; Okamura et al. 2015b). Fusion may occur within spores that develop in myxozoan invertebrate hosts (resulting in self-fertilisation) or in fish hosts after cells released from spores invade and proliferate within fish (also potentially involving self-fertilisation). In *Polypodium* fusion has been inferred to happen at some time after fish hosts are invaded and to involve self-fertilisation of 'blastomeres' that develop within a nurse cell (the trophamnion) which becomes polyploid (Raikova 2008). This trophamnion-blastomere complex arises from the original binucleate cells that invade fish hosts. The complex is initiated by engulfment of one nucleus in the binucleate cell, resulting in a cell-in-cell organisation (Raikova 1994). It is apparent that fish hosts of both *Polypodium* and myxozoans support early stages of development. In addition, both *Polypodium* and myxozoans have similar nematocysts (organelles within nematocytes and referred to as polar capsules in myxozoans) that are used for attachment and are characterised by hollow tubes and an absence of spines. Such nematocysts, called 'atrichous isorhizae' are also found in free-living cnidarians (Okamura et al. 2015b). *Polypodium* additionally possesses nematocysts that penetrate prey.

Table 4.2 Similarities and differences in morphological and life history stages between *Polypodium* and myxozoans

Feature	*Polypodium* (reference/s)	Myxozoans (reference/s)
Cilia/flagella	Flagellated endodermal cells (lining gut) (Raikova 2008)	Absent (Lom 1990; Feist et al. 2015)
Centrioles	Present (Raikova 2008)	Absent (Lom 1990; Canning et al. 2000; Feist et al. 2015)
Gametes	Absence of 'true' egg and sperm cells; 'egg' inferred to be small cell enclosed within another cell that will become a trophamnion (forming cell-in-cell complex); this complex results from 2nd meiotic division with trophamnion arising from polar body; diploidy proposed to be restored by fusion of blastomeres (Raikova 1994, 2002, 2008)	Not evident or recognisable; conflicting interpretations of which cells represent gametes during spore development; meiosis during sporogony in invertebrate hosts results in production of spores inferred to be haploid; when fusion (diploidy) is achieved is unknown (Feist et al. 2015; Okamura et al. 2015a)
Nematocysts	Atrichous isorhizae (enable attachment of walking tentacles); holotrichous isorhizae (penetrants for prey capture) (Raikova 1990; Ibragimov and Raikova 2004)	Nematocyst homologues (polar capsules) equated to atrichous isorhizae; function in attachment to hosts (Siddall et al. 1995)
Cnidocil (a modified cilium)	Present in unusual apical location above nematocyst lid (Raikova 1990)	Absent; apical plug-like structure similar to that in medusozoans in equivalent position (Canning and Okamura 2004; Okamura et al. 2015a)
Mitochondria with tubular cristae	Present (Raikova 2008)	Present but variability of this trait suggests it is not particularly noteworthy (Canning and Okamura 2004)
Epithelia	Present (Raikova 1994)	Present in sporogonic stages of malacosporeans developing in invertebrate hosts (Canning et al. 2000; Gruhl and Okamura 2015)
Independent muscles (non-epitheliomuscular cells)	Present: Muscles sub-epidermal, underlying the ectoderm (Raikova et al. 2007)	Only present in vermiform malacosporeans: 4 longitudinal, sub-epidermal muscle blocks with chiral pattern (Gruhl and Okamura 2012)
Digestive tract	Present (Raikova 1994)	Absent (Gruhl and Okamura 2015)
Nervous system	Present (Raikova and Raikova 2016)	Absent (Gruhl and Okamura 2015)
Gonads	Two sequentially developing gonads (Type I and II); the first possesses gonoducts and degenerates (Raikova 1994, 2002, 2008)	Absent (Gruhl and Okamura 2015)

(continued)

Table 4.2 (continued)

Feature	*Polypodium* (reference/s)	Myxozoans (reference/s)
Transmission stage	Modified gonad that attaches to larval acipenseriform fish (Raikova 1994, 2008)	Multicellular spores released into the water column (Canning and Okamura 2004; Feist and Longshaw 2006)
Early presporogonic stages	Binucleate cells that leave gonad attached to larval fish (Raikova 2002, 2008)	Invertebrate hosts: amoeboid uninucleate cells in malacosporeans; binucleate cells in myxosporeans (Feist et al. 2015) Vertebrate hosts: primary cell with enclosed secondary cell/s in malacosoreans and in myxosporeans, respectively (Feist et al. 2015)
Cell-within-cell development and proliferation	Trophic cell (referred to as 'trophamnion' and acting as 'nurse cell') surrounds developing larvae in fish eggs and becomes polyploid (Raikova 1994, 2008)	Characterises sporogonic stages with primary and secondary cells in myxosporeans and in malacosporeans developing in fish hosts (but not in invertebrate hosts); primary cells become multinucleate forming plasmodia in myxosporeans but are uninucleate in pseudoplasmodia (Feist et al. 2015)
Intracellular parasitism	Present: larval stages develop in eggs of acipenseriform fishes	Present in some myxozoans (Canning and Okamura 2004; Sitjà-Bobadilla et al. 2015)

The free-living *Polypodium* body plan is relatively consistent with that of other free-living cnidarians with an external ectoderm and internal endoderm (gastrodermis). We postulate that the independent sub-epidermal muscles may have evolved to facilitate unrestricted reversal of ectoderm and endoderm positions when these germ layers invert during formation of the mature stolon. Two Type I gonads inferred to be female (with 'oviducts' opening into the gastric cavity) develop, but mature eggs have not been observed and the gonadal complexes degenerate (Raikova 2008). Four Type II gonads (without gonoducts) support gametogenesis which is described as first resembling spermatogenesis but subsequently becoming more similar to oogenesis (entailing 'reorientation of male into female gonad'; Raikova 2008). Fertilisation would be achieved at a much later stage given Raikova's conclusion that the binucleate cells (produced by Type II gonads) that go on to invade fish are haploid. Cleavage and early morula-like embryonic stages are observed within fish eggs that may not develop for years (e.g. sterlet and beluga require up to 10 and 16 years to become reproductive, respectively) indicating prolonged arrested development of *Polypodium* stages following infection of juvenile fish (Raikova 1994). It is unclear whether early invading binucleate cells multiply within fish and thus eventually contribute to high percentages of infected eggs, however Raikova (2002) notes that in sections of ovary from young fish, cells resembling binucleate stages were repeatedly observed. Nourishment from the fish egg yolk is provided indirectly by the surrounding trophamnion whilst larvae develop to the adult stolon. Further yolk incorporated during the inversion of germ layers just prior to egg spawning supports several days of free-living existence after the stolon is released and undergoes fragmentation in the environment.

The endo- and ectodermal nature of myxozoan epithelia remains unknown, but, if homologous with that of *Polypodium*, the outer epithelium of malacosporean sacs and worms and the wall of myxosporean pansporocysts in invertebrate hosts would be endodermal (see also below discussion). In contrast, sporogonic stages developing in fish hosts (plasmodia of myxosporeans and pseudoplasmodia of malacosporeans) are characterised by development within an outer cell—an arrangement similar to the trophamnion-enclosed larval stages of *Polypodium* (Table 4.2).

Endocnidozoans have repeatedly evolved stages that retain multiple copies of nuclear content within a single cell. These include cell-in-cell complexes of myxozoans, binucleate stages of *Polypodium*, syncytial plasmodia of myxosporeans, and polyploidy in the trophamnion stage of *Polypodium* (the latter inferred in an early cytophotometry study (Raikova 1965)). All cell-in-cell myxozoan stages appear to originate from engulfment of one cell by another forming cell complexes (Feist et al. 2015). These have mistakenly been described as binucleate stages similar to those of *Polypodium* (Raikova 1994, 2008; Holzer et al. 2018). However, Morris (2012) showed that the cell membrane surrounding engulfed myxozoan cells becomes indistinct (suggesting a binucleate condition), but this membrane reappears later in sporoplasm development. Binucleate cells in *Polypodium* are regarded as long-lived haploid stages that have not subsequently divided after the second meiotic division (i.e. cytokinesis is postponed; Raikova 2008).

4.4 Evolution and Life Cycles of Endocnidozoans

4.4.1 Preadaptations to Parasitism

Cnidarians possess several traits that may predispose them for endoparasitic lifestyles (Okamura et al. 2015b). Their diploblastic body plan is manifested by the extensive development of external and internal epithelial layers across which resource capture, uptake and excretion are performed. An inherent capacity for uptake of dissolved organic material across these surfaces (e.g. Grover et al. 2008) may particularly facilitate adopting endoparasitic lifestyles (Okamura et al. 2015b). These combined cnidarian features are similarly likely to support the intimate relationships cnidarians have repeatedly evolved with endosymbionts (Kayal et al. 2018) based on nutrient exchange (e.g. in sea anemones, corals, green *Hydra*, and stalked jellyfish).

Nematocysts are triggered to discharge filaments in response to specific chemical or mechanical environmental stimuli. These filaments function variously in prey capture or defense, achieved by penetration and injection of venoms or digestive enzymes, attachment, and entanglement. Nematocysts have been co-opted in endoparasitic cnidarians. For example, the filaments discharged from myxozoan polar capsules anchor transmission stages (spores) to host surfaces to enable the initiation of infection (Kallert et al. 2015). Cnidarians secrete, either via nematocysts or standard secretory pathways, a wide variety of compounds such as toxins and antimicrobial peptides as well as cytolytic, proteolytic and other digestive enzymes (Balasubramanian et al. 2012; Dunlap et al. 2013). These substances have a high potential of becoming co-opted for host-parasite interactions such as host invasion, tissue modification or defence against host immune system elements. Furthermore, soft-bodied cnidarians have evolved specific epithelial defence mechanisms manifested in a high structural and molecular diversity of the glycocalyx, the apical membrane surface layer (Bosch 2016). The glycocalyx forms the outer layer of myxozoan spores, but also covers endoparasitic stages possibly functioning in immune evasion or other host-parasite interactions (Gruhl and Okamura 2015).

Cnidarians are characterised by a high degree of asexual reproduction in the form of e.g. fission, fragmentation and budding (Fautin 2002). The ability to proliferate asexually is a hallmark of many parasites, enabling them to exploit host resources, to produce new stages, to overwhelm host immune responses, and to facilitate horizontal transmission (Poulin 2007). Examples of such asexual replication in endoparasitic cnidarian stages include evidence for fission (Fig. 3.2c, d; in Okamura et al. 2015a) and budding (Okamura 1996; McGurk et al. 2006) in sac-forming myxozoans in the body cavity of bryozoan hosts, and budding of medusae from polyp-like endoparasitic narcomedusans within the body cavity of polychaete hosts (Bentlage et al. 2018). Parasitic polyp stages also bud off medusae in the ring canal system of parental medusae (Bigelow 1909). However, the proliferation of early stages of endocnidozoan infections and of parasitic larval narcomedusans is achieved by binucleate cells and cell-within-cell stages (Bigelow 1909; Feist et al. 2015) that, as far as we are aware, have no ready homology with cell complexes in free-living cnidarians.

There is clearly substantial plasticity in cnidarian life histories and the ability to develop novel stages (Cartwright and Nawrocki 2010; Okamura et al. 2015b). For example, medusa stages have been lost in many hydrozoans, including in *Hydra*, with progenesis enabling sexual reproduction (Boero et al. 1992) and some medusozoans incorporate parasitic stages inferred to be specialised polyps (Table 4.1). Novel stages of cnidarians can undergo dormancy, regeneration and transdifferentiation (reviewed in Lai and Aboobaker 2017)—processes that are intimately related to asexual reproduction. For example, planulae or polyps shrink and persist as cysts and dormant hydrorhizae with low metabolic costs during stressful periods. When favourable conditions return, cell proliferation and morphogenesis ensue (Boero et al. 1992). Polyp stages of the hydrozoan genus, *Turritopsis*, can be reformed from regressed tissues of sexual stages (medusa) via reverse ontogeny—a process mediated by proliferation of interstitial stem cells and cell transdifferentiation (Piraino et al. 2004). Medusa buds of the hydrozoan, *Sarsia tubulosa*, transform back to polyp buds when exposed to different temperatures (Werner 1963). This phenomenon could be more common in cnidarians than previously anticipated (e.g. He et al.

2015). These features may have aided precursors of parasitic cnidarians initially to survive within hosts. For example, cryptic dormant stages of myxozoans are viable and present in dormant propagules produced by invertebrate hosts (freshwater bryozoans) (Abd-Elfattah et al. 2014). Such capacities of regeneration and transdifferentiation may have facilitated the evolution of novel stages during transition to parasitism.

4.4.2 Life Cycle Speculations

Sexual reproduction, evidenced by meiosis, occurs when myxozoans develop in their invertebrate hosts. This suggests that trophic stages that develop in invertebrates are extremely morphologically simplified stages that may be equivalent to the free-living, sexual stages of *Polypodium* (Okamura et al. 2015b). Although the typical medusa umbrella is lacking, the free-living stage of *Polypodium* has been inferred to be a modified medusa (Raikova 1994; Raikova and Raikova 2016). This inference is primarily based on the assumption that sexual reproduction invariably is expressed in the medusa stage. The developing stolon enclosed within fish eggs therefore is proposed to correspond to a polyp stage. However, molecular phylogenetic placement and life cycle plasticity arguments (see above) would equally suggest that the free-living stage could be equivalent to a polyp stage. Regardless of the unresolved nature of *Polypodium's* free-living stage (medusa vs. polyp), it is achieved by larval development in fish eggs, albeit this is delayed until fish reach sexual maturity. In contrast, myxozoans incorporate a second life cycle phase that develops in vertebrate intermediate hosts. The lack of a biphasic life cycle in *Polypodium* would suggest this could have been absent in the common ancestor of *Polypodium* and Myxozoa and thus that myxozoan stages in vertebrate hosts are novel.

4.5 Origins and Fossil Records of Endocnidozoa and Their Recognised Major Host Groups

The fossil record of endoparasites is poor for many reasons (De Baets and Littlewood 2015; De Baets et al. 2021a; Leung 2017, 2021). Although endocnidozoan parasites are diverse and abundant and have probably existed on earth for hundreds of millions of years, there is so far neither a direct nor indirect fossil record for this group. The fossilization potential of within-host life-cycle stages of endocnidozoans is arguably low. Such stages are soft, microscopic and most often reside within soft host tissues which themselves are only rarely preserved. However, mature myxozoan spores, especially myxospores (myxosporean spores produced in vertebrate hosts) can withstand adverse conditions like desiccation, extreme temperatures, chemical treatments, and gut passage (El-Matbouli and Hoffmann 1991; Hedrick et al. 2008). Their outer layer consists of a hardened, cuticle-like, extracellular material secreted by the underlying valve cells. This material may be elaborated as ridges and other surface features across the spore surface (reviewed in Gruhl and Okamura 2015). In some species a pronounced cytoskeleton inside the valve cells adds further mechanical reinforcement, and chitin has been identified in the periphery of myxospores (Munoz et al. 1999). Myxospores in mud have been shown to remain viable for several months (El-Matbouli and Hoffmann 1991). Thus, there is a good chance that these could be present along with other palynomorphs in sedimentary deposits or in peat. Such preservation has been shown for other parasite remains (e.g. nematode and trematode eggs), but recovery may be heavily influenced by the extraction method used (Dufour and Le Bailly 2013). It is unclear, however, how easily such remains could be identified as myxospores. Although spore shape and surface structure are of some use, the fossilization potential of the most important diagnostic myxozoan characters, the polar capsules, is unclear. Clusters of nematocysts (for example so-called "nematocyst batteries" in jellyfish) are recognisable in fossil cnidarians with soft-tissue preservation (Han et al. 2016), but the individual nematocysts are, to our knowledge, hardly ever retained. As of yet, no experimental studies on the taphonomy of myxospores exist. Such data would inform on the potential recognition of fossilized myxozoan spores.

Indirect fossil evidence of endocnidozoan parasitism, i.e. pathological changes in their hosts, could theoretically be recognised, particularly if such pathologies occur in skeletal tissues. Soft tissue pathologies such as cysts or swollen organs require exceptional preservation, but even in such cases it will still be difficult to unambiguously identify myxozoan infection. This is because infections by many other organisms can produce similar disease symptoms, as exemplified by fossilized fish skin nodules (Petit 2010; Petit and Khalloufi 2012) from the Monte Bolca and Solnhofen deposits. In addition, many diagnostic techniques, such as histological staining, are inapplicable to fossilized material.

Because of the inherent restrictions of the fossil record, current reconstructions of endocnidozoan evolutionary history have to use other sources of information (reviewed in De Baets and Littlewood 2015; Martínez-Aquino 2016; Warnock and Engelstädter 2021). These can include extant parasite and host phylogenies, the fossil record of closely related free-living cnidarian taxa as well as that of hosts, and paleo-environmental data. Such data can be analysed and combined in different ways. Ancestral character states (such as host preference) can be reconstructed using standard phylogenetic techniques. Phylogenetic bracketing, for example, postulates the occurrence of the parasite in the last common ancestor of all recent hosts (except for those that were clearly acquired by recent host switching). Molecular clock analyses can also provide age estimates for origins of groups that lack fossil data when based on both a well-resolved phylogeny and a reliable fossil calibration of a wide set of nodes (Benton et al. 2009; Parham et al. 2012). A common practice for age estimates of a taxon is to use the oldest reliable fossil as a minimum age estimate and either the earliest molecular clock or the stratigraphic maximum estimate as a (soft) maximum age estimate (De Baets et al. 2015; Parham et al. 2012).

Molecular clock analyses, however, can be particularly challenging for parasitic groups for many reasons (Warnock and Engelstädter 2021). First, evolutionary rates have repeatedly been shown to be increased in parasite lineages (Bromham 2009; Bromham et al. 2013; Paterson et al. 2010). In addition, reconstructions and calibrations of inferred parasite-host associations are only reliable if host switching is very rare. The latter can be detected independent of molecular clock analyses, by co-phylogenetic analyses, which, however have certain pitfalls (de Vienne et al. 2013; see also discussion in later section). For example, parasites, including many endocnidozoans, can often exploit a range of hosts and this introduces complications to co-phylogenetic analyses that assume strict host use (Banks and Paterson 2005). Finally, any inferences of parasite evolution, including molecular clock methods and host-parasite cophylogenies, can be highly compromised when parasite diversities are poorly known. As we argue here and elsewhere (Okamura et al. 2018) endocnidozoan diversities are certainly grossly underestimated and pivotal taxa may be overlooked due to lack of sampling in many environments.

An additional and important conceptual point to stress in any discussion of evidence for fossil remains is the appreciation that evolution is a continuous process. Thus, at some period endocnidozoans will have diverged from an entirely free-living common ancestor. However, whether this divergence was simultaneously linked with adoption of parasitism or whether parasitism evolved subsequently is an open question. This question is of course relevant to a timeframe when fossils of parasites might be expected. Later we will review how initial and further hosts may have been adopted in simple and complex parasite life cycles, but complications arising from the continuous nature of evolution should be born in mind as we discuss the origins of endocnidozoans and of their potential early hosts in this section.

Below we review the current state of knowledge of the origins of cnidarians and of hosts known to be used by endocnidozoans based on the fossil record and other evidence (e.g. molecular clock analyses). We also assess the potential for recognizing myxozoan infections in host fossil material. We then critically evaluate conclusions of a recent study of the origin and evolution of endocnidozoans deriving from molecular clock and cophylogenetic analyses.

4.5.1 Cnidarian Origins and Fossil Record

Phylogenomic analyses of multi-gene datasets across the animal kingdom have in recent decades resolved many questions. Basal metazoan relationships, however, including those of cnidarians, remain especially controversial and several equally well supported hypotheses exist (discussed in Dohrmann and Wörheide 2013; Dunn et al. 2014; Halanych 2015). The two most prominent groups of hypotheses mainly differ in the placement of Placozoa and Ctenophora: either united with Cnidaria to form the "Diploblastica" (e.g. Schierwater et al. 2009) or as basal metazoan branches leaving Cnidaria as sister to bilaterians (e.g. Simion et al. 2017; Whelan et al. 2017). Although most molecular clock analyses predict both metazoan and bilaterian origins in the Neoproterozoic (900–600 Ma) (e.g. Dohrmann and Wörheide 2017), reliable fossil evidence from these times is extremely scarce (reviewed in Cunningham et al. 2017). Both the oldest reliable crown-bilaterian and crown-cnidarian fossils, *Kimberella* (555 Ma) and *Corumbella* (543 Ma), respectively, are dated to the uppermost Ediacaran (Grazhdankin 2004; Warren et al. 2012). Other Ediacaran fossils, including the classic Vendobionta (soft-bodied Ediacaran macrofossils like *Dickinsonia*) and the Weng'an and Lantian biota are difficult to interpret as to both their exact age and affinities.

Extant Cnidaria comprise the taxa Anthozoa, Staurozoa, Endocnidozoa, Scyphozoa, Cubozoa and Hydrozoa. Phylogenetic relationships among these taxa have been subject to extensive debate. The most recent phylogenomic analyses (Kayal et al. 2018) support a sister group-relationship between Anthozoa and the clade comprising Medusozoa and Endocnidozoa (Fig. 4.2). Previously regarded as sister group to Medusozoa, Kayal et al. (2018) recover Staurozoa within Medusozoa, as sister to a clade uniting Cubozoa and Hydrozoa. The earliest generally accepted cnidarian polyps, *Corumbella* and *Paraconularia*, are from the uppermost Ediacaran or earliest Cambrian (<550/540 Ma) (Van Iten et al. 2014). Several Neoproterozoic cnidarian candidates remain questionable, including *Haootia*, a 560 Ma impression from the Ediacaran of Newfoundland interpreted as a staurozoan polyp. *Lantianella* bears similarity to a scyphozoan medusa, but the Lantian Formation is not dated precisely, with estimates ranging from 590 to 635 Ma (Wan et al. 2016). According to molecular clock analyses the origin of cnidarians and their initial diversification most likely date to the Cryogenian or even earlier (Park et al. 2012; dos Reis et al. 2015; Cunningham et al. 2017; Dohrmann and Wörheide 2017). Given its phylogenetic position, the endocnidozoan stem lineage must have diverged at some point between the basal branching of Cnidaria and the emergence of crown medusozoans (Fig. 4.2). The earliest unambiguous cnidarian medusae are from the Middle Cambrian Burgess Shale (Epoch 3, 505 Ma) (Young and Hagadorn 2010 and references therein). Stem medusozoans, perhaps still lacking the medusa stage could of course be older, but recognising these forms could be highly challenging. Molecular clock analyses date crown medusozoans to the late Cryogenian to early Ediacaran (~700–600 Ma) (Park et al. 2012; Dohrmann and Wörheide 2017; Holzer et al. 2018). Holzer et al. (2018) infer ages of 651 Ma (601–700 Ma) and 588 Ma (540–642 Ma) for the origins of Endocnidozoa and Myxozoa, respectively.

4.5.2 Vertebrate Origins and Fossil Record

Vertebrates, especially "fish", are the most speciose and best studied endocnidozoan hosts, and they have a relatively extensive fossil record (Friedman and Sallan 2012). A number of soft-bodied potential vertebrates are known from Cambrian strata. A particular problem in the interpretation of these fossils, however, is the lack of crown-vertebrate characters. Among the more convincing fossil vertebrates are *Metaspriggina* from the Burgess Shale and *Haikouichtys* from Chengjiang. These fossils show definitive chordate characters, including gill slits, fins, and metameric musculature. The provenance of many other proposed vertebrate fossils is more controversial. For example, *Haikouella* or *Yunnanozoon* could be fossil hemichordates, acranians (vertebrates lacking skulls), stem deuterostomes or stem vertebrates (Donoghue and Keating 2014). Several attempts have been made to date the vertebrate phylogeny using molecular clocks. Most studies, however, focus on the gnathostomes for which more fossil calibration points exist, leaving the origin of the vertebrates relatively uncertain. Dohrmann and Wörheide (2017) place the deuterostome origin before the Marinoan glaciation and chordate and vertebrate origins into Cambrian and Ordovician, respectively. The origin of (crown) vertebrates has been variously placed at 490 (504–476) Ma (Delsuc et al. 2018), 460–533 Ma (dos Reis et al. 2015) and 480 Ma (Holzer et al. 2018).

Recent craniate vertebrates include the jawless lampreys and hagfishes, the chondrichthyans (cartilaginous fishes), and the osteichthyans (bony fishes), the latter comprising sarcopterygians and actinopterygians. The Agnatha (or Cyclostomata) (jawless fishes) are currently considered to be monophyletic, forming the sister group to Gnathostomata (vertebrates with jaws, including most present-day fish). Because of the complete lack of hard skeletons the origin of agnathans is difficult to trace; the earliest reliable fossils are from the Devonian and Carboniferous. Molecular clock studies converge on ~460 Ma for the gnathostome divergence (Irisarri et al. 2017; dos Reis et al. 2015; Broughton et al. 2013).

Two further exclusively fossil groups are of relevance for early vertebrate relationships: the conodonts and the ostracoderms. Conodonts are known by their characteristic tooth-like skeletal elements from the late Cambrian until the Late Triassic when they apparently went extinct. Currently conodonts are mostly considered to branch at the base of vertebrates, either as stem agnathans or stem gnathostomes (Goudemand et al. 2011; Murdock et al. 2013; Turner et al. 2010). The ostracoderms (Placodermi, Osteostraci, Acanthodii, with origins estimated from 467 to 295 Ma) are interpreted as a paraphyletic assemblage of gnathostome stem-lineage taxa that mostly lacked jaws (Janvier 2001). The earliest lobe-fin fossils are from 423 Ma (Brazeau and Friedman 2015), marking the putative chondrichthyan/osteichthyan (crown-gnathostome) divergence. Chondrichthyan-like scales are known from as early as the late Ordovician (458 Ma; Andreev et al. 2016). Vertebrates with chondrichthyan body fossils, however, do not appear in the fossil record before 400 Ma (Brazeau and Friedman 2015). The first definitive actinopterygian, *Cheirolepis*, is from the early Middle Devonian (~390 Ma), although isolated scales of ray-fins were present from ~427 Ma (Friedman 2015). Molecular clock analyses suggest crown Actinopterygii to have been present by 400 Ma (Broughton et al. 2013), 384 Ma (Near et al. 2012), or 320 Ma (Irisarri et al. 2017) and ages for teleosts to range from 300 to 200 Ma. Recent Actinopterygii include the three sequentially branching taxa Cladistia (bichirs and ropefish), Chondrostei (sturgeon and paddlefish), and Holostei (gars and bowfin). The most derived actinopterygians, the Teleostei, includes the majority of the ~30,000 actinopterygian species (Friedman 2015). Only scattered actinopterygian fossils are known from the Devonian indicating relatively cryptic evolution until several radiations took place after the Devonian/

Carboniferous boundary. Further teleost diversification occurred in the Triassic, including transitions from marine to freshwater and back. As a result the majority of recent marine actinopterygians derive from freshwater ancestors (Carrete Vega and Wiens 2012). These radiations are of potential significance, especially for the diversification of myxosporeans, which show a distinct subdivision into freshwater and marine clades (Fiala et al. 2015a, b; Holzer et al. 2018). Cophylogenetic analyses that take these habitat changes into account may better resolve myxozoan radiations and could additionally serve to explain switches of invertebrate hosts. Moreover, node calibrations for fish radiations following invasions of freshwater or marine habitats could be directly transferred to the corresponding parasite groups. For example, the hypothesis could be examined that the marine myxosporean clade diverged from a freshwater clade that infected oligochaetes following the first transition of teleosts to the marine environment.

Myxozoans infect a range of organs and tissues of their vertebrate hosts causing diverse disease symptoms. The site of sporogony usually demonstrates the greatest pathology and is often quite specific (Molnár and Eszterbauer 2015). The fossilization potential of myxozoan stages in lumina and soft tissues is probably very low. Fish immune responses can, however, lead to encapsulation of parasites by dense connective or cartilaginous tissues, which may have a higher probability of preservation. Furthermore, a common reaction against myxozoan infection is the formation of granulomata with melanomacrophage centres—dense accumulations of immune cells containing melanin (Sitjà-Bobadilla et al. 2015; Steinel and Bolnick 2017), a substance which has been characterised from well-preserved fossil vertebrates (Colleary et al. 2015). Cysts occurring in gills, skin or muscle tissue are the most common form of tissue alteration and may be recognisable in fossils with exceptional soft-tissue preservation, but recognition of spores (see above) within these structures is crucial to unequivocally link to myxozoan infection. Myxozoan infection symptoms most likely to be preserved in the fossil record would be expected to result from species that infect cartilage and cause skeletal deformation. This phenomenon is best studied in salmonid whirling disease caused by *Myxobolus cerebralis*, which infects cartilage prior to ossification leading to malformations of spine, skull, jaw or fin rays (Sarker et al. 2015).

Due to their high phosphate content and rapid fossilization potential, vertebrate faeces could provide a further means of detecting myxozoans via analyses of coprolite contents. Coprolites can sometimes offer exceptionally well-preserved soft tissue remains comparable to those of a Konservat-Lagerstätte (Qvarnström et al. 2016). Accordingly, coprolites have been found to contain parasite remains (Poinar and Boucot 2006; Hunt et al. 2012; Dentzien-Dias et al. 2013; Hugot et al. 2014; Brachaniec et al. 2015; Dentzien-Dias et al. 2018; Chin 2021) along with preserved hair, feathers, muscles, bones, chitinous exoskeletons, bacteria, and fungi. Coprolites could therefore have a high potential to yield myxozoan spores. These spores may derive from species infecting sites where spores are released into the digestive tract (e.g. infections in the bile ducts, gall bladder or intestine), or they could derive from infected prey. The latter possibility is supported by the detection of myxozoan DNA in faeces collected from fish-eating birds (cormorants) (Briscoe et al. unpublished data; Hartikainen et al. 2016) and evidence for viable spores excreted from great blue herons (Koel et al. 2010). Not all coprolites have equal preservation potential. Shark coprolites are often well-preserved (Hunt et al. 2012) and have yielded tapeworm eggs (e.g. Dentzien-Dias et al. 2013). Smaller actinopterygian coprolites have also been successfully analyzed for microfossils (e.g. from the Lake Messel deposit; Richter and Baszio 2001; Richter and Wedmann 2005).

4.5.3 Lophotrochozoan Origins and Fossil Record

The most important invertebrate host groups of myxozoans are annelids and phylactolaemate bryozoans, both of which are members of the large superphylum Lophotrochozoa. The internal relationships of lophotrochozoans have proven notoriously difficult to resolve. Lophotrochozoan stem-group members are among the classic fossils of the Cambrian Lagerstätten (Sirius Passet, Burgess Shale, Chengjiang). Many of these show character combinations that have led to varying assignments as stems of extant phyla (e.g. molluscs, annelids, brachiopods).

4.5.4 Annelid Origins and Fossil Record

The traditional annelid phylogeny with subdivision into polychaetes and clitellates has undergone major changes given recent phylogenomic data (Struck et al. 2011, 2015; Weigert et al. 2014; Weigert and Bleidorn 2016). Polychaetes now have to be considered paraphyletic with Clitellata being an ingroup of a large taxon embracing most of the classical "sedentarian" polychaetes. Thus, the ancestral annelids were polychaete-like and marine. Clitellates (which include the oligochaetes), in contrast, originated in freshwater or terrestrial habitats as shown by adaptations such as direct development inside a cocoon and reduction of palps and parapodia. Recent marine clitellates (mostly tubificids) have clearly invaded the sea secondarily. Inferred basal splits within Annelida are still somewhat unstable and lack robust support, but generally suggest a motile or errant ancestor. The earliest annelid fossils have traditionally been traditionally reported from the Cambrian Sirius Passet formation (Conway Morris and Peel 2008; De Baets et al. 2021b; Vinther et al. 2011). Later Cambrian annelid fossils have been found in the Burgess Shale (Conway Morris 1979) and slightly younger ones from the Chengjiang Lagerstätten (Liu et al. 2015). These animals were in the mm-cm size range and had homonomous segmentation, parapodia and palps (Parry et al. 2014, 2015). If Sipunculida, which are also reported from Chengjiang (Huang et al. 2004), are really an annelid ingroup, the initial annelid radiation must have happened in the Early Cambrian or before. Although Clitellata are now consistently placed within polychaetous taxa, their sister group is not yet identified unambiguously (Weigert and Bleidorn 2016). Some of the oldest clitellate fossils are leech cocoons from the Triassic (Manum et al. 1991; Bomfleur et al. 2012; De Baets et al. 2021b) suggesting a late Paleozoic origin of this group. Most modern polychaete groups first appear in the Carboniferous, but some are already known from the Devonian (Parry et al. 2014).

Few studies have investigated myxozoan infections in annelids (e.g. El-Matbouli and Hoffmann 1998; Bartholomew et al. 1997). Pathological effects on annelid hosts include tissue damage, reduced fecundity and hypertrophic growth (Elwell et al. 2009; Alexander et al. 2015)—as occurs in infected phylactolaemate bryozoans (see below). In general, myxozoan stages are unlikely to be preserved in fossil annelids, however, soft-body features are clearly recognisable in several exceptional preservations (e.g. Briggs and Bartels 2010; Timm et al. 2016). A particularly stunning example is the preservation of spermatozoa—structures comparable in size to myxozoan spores—from Eocene clitellate cocoons (Bomfleur et al. 2015). Considering the vast diversity of annelids, it is clear that only a minor proportion of recent annelid taxa is currently known to serve as hosts of myxozoans. This could argue for a late acquisition of this host group, but might also reflect serious undersampling, as definitive hosts of the nearly 2200 myxosporean species described from fish are yet to be resolved (Eszterbauer et al. 2015).

4.5.5 Bryozoan Origins and Fossil Record

Recent Bryozoa comprise the taxa Phylactolaemata, Stenolaemata and Gymnolaemata. Phylactolaemata is considered sister to the other two groups (Waeschenbach et al. 2012) and includes <100 described species (Massard and Geimer 2008). In contrast to the vast majority of bryozoans, phylactolaemates are uncalcified and occur exclusively in fresh water. Relationships amongst phylactolaemates are mostly unresolved in molecular phylogenetic analyses (Waeschenbach et al. 2012; Hartikainen et al. 2013b). The oldest bryozoan fossils occur in the lowest Ordovician (Xia et al. 2007). The diversity of the calcified stenolaemates present at that time was already high (six major groups, of which only the cyclostomes survived the later Permian and Triassic extinctions), suggesting an earlier Cambrian radiation. However, neither calcified nor soft-bodied bryozoans have been found so far in the Cambrian (Taylor and Waeschenbach 2015). The lack of bryozoan soft body fossils hampers reconstruction of the origin and history of the uncalcified Phylactolaemata. Stem members of this group theoretically should have co-occurred with stenolaemates and ctenostomatous gymnolaemates. The oldest chitinous statoblasts (asexual propagules and resting buds of phylactolaemates) are found in the Permian (Vinogradow 1996). Statoblasts are interpreted as clear adaptations to fresh water, indicating that this lifestyle had evolved by then. Earlier stem phylactolaemates are likely to have been marine. Other recent marine bryozoan groups (the cheilostomes and cyclostomes) diversified in the mid-Mesozoic (Taylor and Waeschenbach 2015).

So far phylactolaemates are the only known bryozoan hosts of myxozoans. Hartikainen et al. (2013a) have shown that malacosporeans can affect phylactolaemate host morphology when the development of spore-producing infectious stages results in larger zooids, malformed statoblasts and reduced statoblast production. Recognition of such effects in fossil material would require good preservation, lack of post-preservation deformation, and comparative assessment of many individuals. Several extinct bryozoan groups could be promising candidates as hosts of ancient myxozoans.

A predisposing trait of phylactolaemates as myxozoan hosts is the relatively large and confluent colony-wide coelomic cavity. The continuous action of cilia lining this coelomic cavity ensures that metabolites are distributed throughout the colony. In addition the unobstructed coelomic space allows malacosporean sacs and myxoworms to move freely amongst the zooids of a colony, with the circulation of sacs being assisted by ciliary beating. Cyclostome and gymnolaemate zooids, in contrast, are separated by walls, which have communication pores that are either very small (cyclostomes) or filled with tissue (gymnolaemates), thus preventing myxozoan stages from migrating between zooids. Furthermore, nutrient transfer in cyclostomes and gymnolaemates is achieved largely via tissue connections (the funicular system)—leaving the coelomic fluids with relatively low concentrations of nutrients. The Paleozoic stenolaemate *Corynotrypida*, however, completely lacked interzooidal walls. In addition, some representatives of other Paleozoic stenolaemate groups (Cystoporata, Esthonioporata, Trepostomata and Cryptostomata) possessed zooids with larger communication pores while others (the free-walled stenolaemates) lacked calcified exterior walls and thus are assumed have had confluent hypostegal coelomic cavities (Boardman 1998; Ernst and Schäfer 2006). These ancient bryozoan taxa could, thus potentially have served as malacosporean or stem myxozoan hosts.

4.5.6 Other Potential Ancient Invertebrate Hosts

Discoveries of myxosporeans in octopus (a species of *Kudoa*; Yokoyama and Masuda 2001) and as hyperparasites in three monogenean species (*Myxidium* or *Myxidium*-like species) and in two digenean species (*Fabespora vermicola* and *Fabespora* sp.; Freeman and Shinn 2011) suggest that a greater range of invertebrate hosts may be routinely used. Notably *Kudoa*, *Myxidium* and *Fabespora* are all derived lineages within the Myxosporea (Fiala 2006; Freeman and Shinn 2011). These host findings are, however, not straightforward. Thus, at least in some cases monogenean 'hosts' could be incidentally exploited by myxozoans infecting fish and *Kudoa* infection has only been observed in a single octopus. It is conceivable that the latter could have developed if infection was transmitted from infected fish prey (perhaps if injury was sustained in catching and subduing the fish). In addition, the spores that develop in both octopus and the platyhelminth species are typical myxospores (Freeman and Shinn 2011). This implies that these invertebrates may either be used as alternative hosts to fish or have been adopted as novel hosts via host switching. There is no evidence so far of a further range of invertebrate hosts associated with actinospore- or malacospore-producing stages of myxozoans.

Annelids and phylactolaemate bryozoans share traits that might predispose them to be parasitised. These include collecting small particles for ingestion by screening relatively large environmental samples (via suspension or deposit feeding) and large body cavities in which parasites can develop. Other invertebrate groups with such traits (for example bivalves, echinoderms or phoronids), might act or have acted as hosts but have yet to be detected. Infection of fish may also have predisposed myxosporean hyperparasitism of platyhelminths enabling diversification in host use over time.

4.5.7 *Summary of Origins and Ancient Hosts*

Due to the complete lack of fossils, we have no means of calibrating the age of the last common ancestor of endocnidozoans. We can, however, constrain the divergence of the endocnidozoan stem lineage by using both the maximum age estimate for crown Cnidaria and the minimum age estimate for Medusozoa (505 Ma) (Fig. 4.2). Endocnidozoan characters, however, could have evolved at any point along the endocnidozoan stem lineage—between their divergence from the rest of the Cnidaria and the presence of the last common ancestor of all living endocnidozoans. Thus we cannot currently assess the traits of endocnidozoan stem lineage members, including whether they were parasitic or which hosts they parasitised. Molecular clock estimates so far undertaken propose ages of 651 Ma (601–700 Ma) for Endocnidozoa and 588 Ma (540–642 Ma) for Myxozoa (Holzer et al. 2018) and divergence of Malacosporea and Myxosporea at 540 ± 73 Ma (Kodádková et al. 2015). Although the crown groups of the main hosts of Recent endocnidozoans are mostly estimated to be slightly younger than endocnidozoans (see above) it is entirely possible that stem group members acted as hosts.

4.6 Inferring Endocnidozoan Origins and Acquisition of Early Hosts

The lack of a fossil record and the possibility of host switching over time may highly constrain our understanding of endocnidozoan origins and what hosts were acquired when. Any insights, however flawed they may be, must be gained by evaluating data on extant taxa. Here we consider in general how hosts may be acquired and, by extension, host acquisition by endocnidozoans. We then go on to examine more closely some of the pitfalls of inferring parasite origins and patterns of host use over time on the basis of molecular clock and cophylogenetic analyses.

4.6.1 The Process of Host Acquisition

Adoption of a parasitic life style would have required frequent proximity of future hosts and parasites leading to enhanced fitness of parasite precursors. Transitions to parasitism could be based on precursors depending on future hosts for e.g. food or dispersal. Alternatively, precursors may have been regularly ingested by future hosts (Poulin 2007; Schmid-Hempel 2011). Precursor pre-adaptations may have facilitated benefitting from such an association (Poulin 2007). For example, as outlined above, cnidarian traits such as uptake of dissolved organic matter across relatively extensive epithelial surfaces, the ability to persist as dormant stages, regeneration, and nematocysts may have variously facilitated host exploitation, survival in adverse host environments and attachment to hosts.

Exploitation of fish hosts by both *Polypodium* and myxozoans along with their sister relationship status suggest that their common ancestor could have exploited fish (Fig. 4.3; Scenario 1—gain of fish host during A, gain of invertebrate host during B; Scenario 2—gain of fish host during A, gain of bryozoan host at D, gain of annelid host at E). It can be further argued that fish hosts were exploited either by a

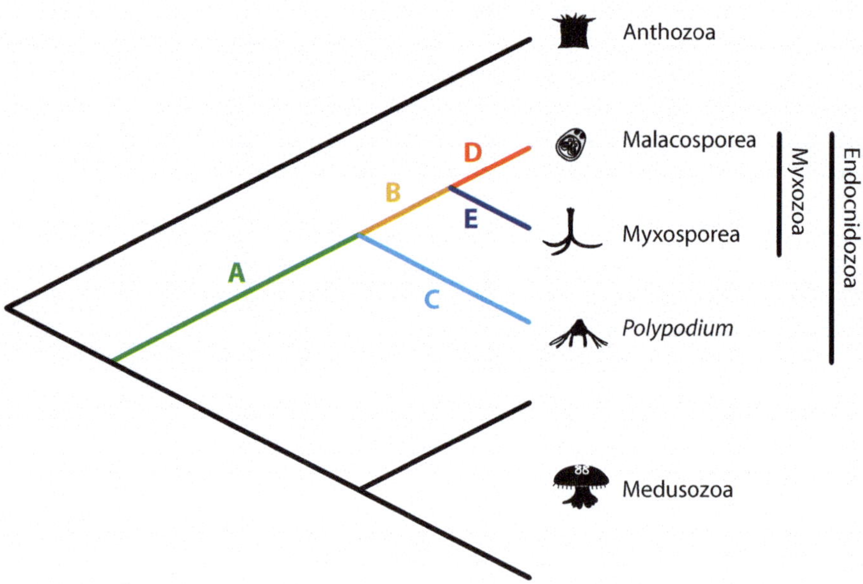

Fig. 4.3 Cnidarian phylogeny with A–E delineating time periods during which various hosts were acquired. Note that because we lack fossil evidence, host acquisition could have occurred at nodes or at any subsequent time period prior to later nodes or the present-day. Scenarios for host acquisition are: (1) Gain of fish host during A, gain of invertebrate host during B; (2) Gain of fish host during A, gain of bryozoan host at D, gain of annelid host at E; (3) Gain of both fish and invertebrate hosts during A, loss of invertebrate host at C; (4) Gain of invertebrate host during A, gain of vertebrate host at B, switching from invertebrate to vertebrate host at C; (5) Gain of invertebrate host at B, gain of fish host at C; (6) Gain of fish host at C; gain of bryozoan host at B

larval or novel pre-adult stage of this ancestor because sexual reproduction occurs in the free-living stage of *Polypodium* and when myxozoans are exploiting invertebrate hosts. Alternative more complicated scenarios include loss of primary invertebrate hosts in *Polypodium* (a scenario that would require the re-acquisition of an adult free-living stage in *Polypodium*) (Fig. 4.3; Scenario 3—gain of both fish and invertebrate hosts during A, loss of invertebrate host at C), switching from vertebrate to invertebrate host in *Polypodium* (Fig. 4.3; Scenario 4—gain of invertebrate host during A, gain of vertebrate host at B, switching from invertebrate to vertebrate host at C) and independent transitions to parasitism in *Polypodium* and Myxozoa (Fig. 4.3; Scenario 5—gain of invertebrate host at B, gain of fish host at C). The latter scenario implies a unique transition resulting in the single host life cycle of *Polypodium* and another unique transition resulting in the initial single host life cycle of myxozoans. Precedence for this scenario is provided by the multiple independent origins of parasitism across the animal kingdom (Weinstein and Kuris 2016) and also within taxa, including within the Cnidaria (as described above) and the Nematoda (Blaxter and Koutsovoulos 2014). Flexibility in cnidarian life histories and the capacity to produce novel stages are features that would support the more complicated scenarios described here, but we must also appreciate that unknown host switching events potentially complicate any interpretations about the identity of ancestral hosts.

The evolution of complex life cycles is linked with the great radiation of myxozoans and was achieved by the incorporation of a second host. For helminths it has been argued that secondary hosts may be acquired when parasites evolve to exploit predators or prey of the definitive first host (Choisy et al. 2003; Parker et al. 2003). For example, frequently ingested original hosts may become intermediate hosts when new, larger hosts are exploited by 'upward incorporation'—an outcome that may be associated with increased parasite fecundity, increased probability of finding a sexual partner (Brown et al. 2001; Parker et al. 2015) or a decrease in inbreeding because of multiple infections of larger hosts (Rauch et al. 2005). Alternatively, 'downward incorporation' could occur when prey of the original host frequently ingest parasite propagules and become intermediate hosts, thus enhancing transmission to the original host (Parker et al. 2003). However, these scenarios are based on helminth life cycles that initially involved sexual reproduction in the first hosts.

If fish (or their ancestors) were first hosts of ancestral larval myxozoan stages then invertebrates could have been adopted as hosts of adult forms by a kind of downward incorporation (Fig. 4.3; Scenarios 1, 2). This would require release of larval forms in sufficient numbers from fish (or ancestral fish) hosts that they were frequently consumed by invertebrates, with sexual reproduction then being undertaken in invertebrate hosts. Alternatively, if a stem lophotrochozoan (some ancestral precursor to bryozoans and annelids) served as first host of ancestral myxozoans then upward incorporation may have enabled e.g. a stem chordate or vertebrate to be adopted as a secondary host (Fig. 4.3; Scenarios 4, 5). The retention of primitive features in malacosporeans (e.g. recognisable epithelia, musculature) suggests that ancestral invertebrate hosts may have been more similar to present day freshwater bryozoans than annelids. Subsequently, annelids (or their ancestors) may have been incorporated as invertebrate hosts by the common ancestor of the more derived Myxosporea via host-switching. It is also possible that first hosts were acquired much later and that present-day patterns of host use are directly representative of host group acquisition. In this case, an argument could be made against annelids (or their relatively recent ancestors) as first hosts because this scenario would require the unlikely re-acquisition of primitive traits in the malacosporeans. However, if *Polypodium* and myxozoans acquired parasitism independently and relatively recently (Fig. 4.3; Scenario 6) then bryozoans or fish as first myxozoan hosts seem equally feasible scenarios.

4.6.2 Molecular Clock and Cophylogenetic Investigations

4.6.2.1 Some General Pitfalls

Investigations of parasite origins are inherently difficult due to their extremely poor fossil record. Some researchers have therefore adopted the use of molecular clock and cophylogenetic analyses involving parasites and their hosts in order to infer when parasites may have originated. A general constraint in these approaches is the central assumption that parasite origins can be imputed from host origins while at best they can only be constrained by these (De Baets et al. 2015). In addition, there are a number of other assumptions typically involved in such analyses that should be appreciated. For example, patterns of host use by parasite taxa that we recognise today may have arisen from a complex history of host switching (De Baets et al. 2015; Warnock and Engelstädter 2021). Myxozoans clearly do undergo host switching and indeed the acquisition of new hosts in their complex life cycles provides a case example. Furthermore, identifying first hosts based on those used in the present day may be erroneous. Presently known hosts (or their ancestors) may have been acquired at any time during the potentially very long history of lineages observed today. We appreciate that it can be tempting to propose that early-diverging parasites associated in the present day with early-diverging hosts represent ancient parasite-host associations. For instance, Kodádková et al. (2015) make this argument for the early-diverging myxosporeans *Bipteria* and *Ceratomyxa* in cartilaginous fish hosts. However, species of both *Bipteria* (albeit these may be currently mis-assigned) and *Ceratomyxa* also exploit teleost fish hosts (Lom and Dyková 2006; Adriano and Okamura 2017), an observation suggesting that cartilaginous fishes could have been adopted much later when teleost-infecting lineages switched hosts. Presentation of such alternative scenarios would be helpful to promote more balanced interpretations of results.

Extinctions may also obscure inferences about parasite origins and patterns of diversification by influencing the present-day distributions of parasites. For example, if parasitism first evolved when stem endocnidozoans began to exploit a stem lophotrochozoan or craniate host in the late Cryogenian, then extinctions over time (of hosts or parasites) could explain gaps in the distribution of parasites across the range of present-day hosts (De Baets et al. 2015). Any inferences concerning the origins and evolutionary histories of parasitic lineages should therefore acknowledge the possibility of a potentially rich complexity of host-parasite interactions that may have been impacted over time by gradual and mass extinctions. Indeed, it is fascinating to contemplate what parasites may have been lost during mass extinction events and what host lineages may have survived and subsequently diversified having lost those parasites (van Dijk and De Baets 2021). It should be noted that there is a growing appreciation that co-extinction of parasites (affiliate taxa) along with taxa they depend on (hosts) may be the most common form of species loss (Dunn et al. 2009; Strona 2015). However, as far as we are aware, the modelling approaches developed so far to estimate parasite extinctions are constrained to estimating the number of extinctions of affiliate species as a function of host extinctions retrospectively, based on knowledge of present-day patterns (Colwell et al. 2012), or by modelling dynamics of digitally evolving organisms (reviewed in Strona 2015). Parasite extinctions have also been inferred by detection of ancient DNA (e.g. host-specific heterokoid nematodes in coprolites of moa; Boast et al. 2018), but this technique cannot be applied to reveal extinctions over deep time.

In addition, limited sampling of the diversity of many parasites will constrain insights about patterns of host use. Current undersampling of both host and parasite diversities may thus compromise current views, for example that all major myxozoan clades are characterised by different invertebrate host groups. This limitation is particularly relevant when cophylogenetic analyses attempt to characterise patterns of parasite co-evolution relative to focal host groups. Such analyses require reliable phylogenetic trees—otherwise congruence of phylogenies may be compromised by poor phylogenetic resolution.

4.6.2.2 Endocnidozoan Origins and Host Use Over Time

The first attempt to characterise the timing and divergence of endocnidozoans (Holzer et al. 2018) provides a useful platform for interpreting endocnidozoan evolution as well as to illustrate how inferences may be compromised by the various pitfalls of such analyses. We variously elaborate on these issues below.

To estimate divergence times, Holzer et al. (2018) analysed concatenated alignments of six protein-coding genes in endocnidozoans, their hosts, and other metazoans (10 myxozoans, *Polypodium* and 127 other metazoan taxa from Erwin et al. 2011). Molecular clock analyses estimated divergence times of endocnidozoans and myxozoans to be in the late Cryogenian at 651 Ma (700–601 Ma) and the Ediacaran at 588 Ma (642–540 Ma), respectively. As outlined earlier, such divergence times are consistent with other molecular clock analyses which predict cnidarian origins in the Cryogenian (720–625 Ma) and the presence of crown cnidarians in the uppermost Ediacaran or earliest Cambrian. However, one of the main conclusions, that endocnidozoans diverged when invertebrates were incorporated as first hosts in the Cryogenian, is not well supported. Parasitism of stem hosts other than invertebrates is also possible. Although crown group vertebrates are estimated to be much younger, the vertebrate stem lineage extends well back into the early Cambrian or even Ediacaran. The tunicate-vertebrate divergence, for instance, is resolved at 575 and 550 Ma (by Dohrmann and Wörheide 2017 and dos Reis et al. 2015, respectively), close to the Cryogenian estimate for the inferred endocnidozoan divergence. This provides evidence that fish ancestors were likely present and could equally have been first hosts. The scenario that very ancient endocnidozoans may have been free-living is also not recognised. So far there is no fossil evidence to demonstrate when and in what hosts endocnidozoans or myxozoans first appeared. Further processes that are not considered include host switching and extinction. Myxozoans clearly do undergo host switching and presently known hosts may have been acquired at any time since they have evolved. However, under Holzer et al.'s scenario, myxozoans have consistently only exploited those host lineages in which they are currently parasites.

Cophylogenetic studies conducted by Holzer et al. (2018) are employed along with their molecular clock analyses to argue that invertebrates were adopted as first hosts. However, the inferences proposed as a result of these analyses are potentially problematic. Thus, Holzer et al. (2018) infer that invertebrate host and parasite phylogenies are congruent 'down to the most basal branches' and highlight that all major myxozoan clades are associated with different invertebrate acquisition events (phylactolaemates, oligochaetes and polychaetes) (apart from *Sphaerospora* whose invertebrate hosts are unknown). Yet all major clades (including *Sphaerospora*) are also associated with fish hosts and, as argued above, fish (or their ancestors) cannot necessarily be excluded as first hosts. Furthermore, relationships among phylactolaemate bryozoans are largely unresolved in molecular phylogenetic analyses (most genera and families in this small group comprise a polytomy; Waeschenbach et al. 2012), and recent studies reveal new higher level taxa in undersampled environments (e.g. Amazonia; Wood and Okamura 2017). Thus the congruence of phylogenies may be anomalous and compromised by both limited sampling and poor phylogenetic resolution of invertebrate hosts. Finally, their multiple examples of congruent cophylogenies of myxozoans and fish hosts could arise if fish served as primary hosts followed by host switching, rather than being multiply acquired as secondary hosts as proposed. Quite apart from these specific issues are more general concerns about what can be inferred by such cophylogenetic investigations. These include the problem of non-independence of phylogenies using PARAFIT as done by Holzer et al. (2018) (Felsenstein 1985; de Vienne et al. 2013), testing for congruence on the basis of estimated phylogenies without accounting for uncertainty in the inference (Warnock and Engelstädter 2021) and underestimation of the high probability of host-shift speciations (de Vienne et al. 2013).

4.6.2.3 The Endocnidozoa

Despite their molecular clock evidence for a common origin of Myxozoa/*Polypodium* in the late Cryogenian, Holzer et al. (2018) propose that *Polypodium* and myxozoans are actually not sister taxa and that Endocnidozoa is thus an invalid taxon. They suggest that the apparently contradictory evidence for the Devonian appearance of acipenseriform fish hosts of *Polypodium* (419–395 Ma) and their molecular clock estimate of a much earlier origin for the divergence of *Polypodium*/Myxozoa (651 Ma) is a result of long branch attraction. This view ignores the possibilities of host switching or prolonged existence as free-living forms. Furthermore, none of the specific arguments raised in support of these scenarios are convincing. These include: large differences in genome sizes of *Polypodium* and the myxosporean *Kudoa iwatai*; differences in gene content in myxozoans and *Polypodium*; comparative differences in orthologous genes of *K. iwatai*, *Polypodium* and free-living cnidarians; putative differences in stages that invade hosts ('binucleate cells'), and; contrasting diversities of *Polypodium* and myxozoans. We consider these in order as follows:

- Genome sizes are notoriously variable and they are often reduced in parasitic organisms (e.g. Tsai et al. 2013; Poulin and Randhawa 2015). The larger genome size of *Polypodium* may thus reflect inclusion of a free-living stage in the life cycle. Furthermore, the genome size of the myxozoan *Thelohanellus kitauei* is 8× larger (188.5 Mb) than that of *K. iwatai* (and is some 33% of the genome size of *Polypodium*).
- Described differences in gene content between *Polypodium* and myxozoans (e.g. reductions of genes related to development, cell differentiation and cell-cell communication in myxozoans) can be explained by retention of a free-living stage in *Polypodium*. In addition, the comparison so far only includes myxosporeans, which show a higher degree of simplification, whereas the morphologically more complex malacosporeans may have retained more of these genes.
- The highlighted lower overlap in exclusive orthologous genes (OGs) between *K. iwatai* and *Polypodium* than between *K. iwatai* and two free-living cnidarians (*Nematostella vectensis* and *Hydra magnipapillata*) is based on relatively low numbers (24 with *Polypodium*, 44 with *N. vectensis*, 31 with *H. magnipapillata*). Such variation could reflect any number of processes (e.g. patterns of gene loss, gene assembly, contamination (e.g. see Kayal et al. 2018), etc.).
- Binucleate cells in *Polypodium* are described as 'a larval, diploid stage' that invades fish hosts and those in myxozoans as haploid cells that invade annelid hosts. Both the descriptions and juxtapositions are flawed. *Polypodium* binucleate cells that invade fish are haploid (Raikova 1994, 2008; and see earlier discussion) and myxozoan stages that invade both fish and annelid hosts are cell-in-cell complexes (Feist et al. 2015; and see earlier discussion). Furthermore, in malacosporeans unicellular haploid stages invade bryozoan hosts.
- There are precedents for extensive radiation in one parasitic lineage and little in its sister taxon within parasite clades. For example, the Aspidogastrea (with some 80 species) is sister to the Digenea (with >10,000 species) (Cribb et al. 2003). Such patterns of radiations are likely explained by unique, lineage-specific key innovations (see section 4.7.2). The argument that different patterns of diversification signify independently evolved lineages is therefore not necessarily convincing. In addition, the monotypic status of *Polypodium* may partly be an

artefact arising from lack of investigation of parasites in eggs of non-commercial fish. New evidence (divergence in housekeeping genes; Hartigan et al. unpublished data) also demonstrates cryptic species within acipenseriform hosts.

In summary, we propose that various shared aspects of development and life history that we highlight above (e.g. sub-epidermal muscles, development within cells, cell-mediated invasion of fish hosts) are consistent with the repeated identification of *Polypodium* and myxozoans as sister taxa in phylogenomic analyses (Chang et al. 2015; Kayal et al. 2018). In contrast, none of the arguments for the demise of Endocnidozoa forwarded by Holzer et al. (2018) are well supported. We therefore conclude that the available evidence continues to support *Polypodium* and Myxozoa as sister taxa forming the Endocnidozoa.

4.6.3 Scenarios of Endocnidozoan Evolution and Recommendations for Future Study

We now have evidence in the form of molecular clock analysis (Holzer et al. 2018) that endocnidozoans are an ancient lineage that originated during a period when ancestors to all their currently recognised hosts likely were also present. Parasitism could have evolved at that time (or later) when first hosts were acquired that belonged to groups currently recognised as endocnidozoan hosts (or their precursors). Alternatively, first hosts may have belonged to an unrecognised or extinct group. Extrapolation from the life cycle of *Polypodium* suggests that early endocnidozoans may have been free-living benthic animals that used tentacles for food capture and to gain purchase to and move across the substratum using nematocysts and muscles. In view of the phylogenetic placement of Endocnidozoa, this suggests a beguiling scenario of a transitional form that shares a benthic habitat with attached basal anthozoans and an unattached mobile adult form with its sister medosozoans. Inferring patterns of host acquisition over time (Fig. 4.3) are difficult in view of the many issues we discuss above, including undersampling, lack of fossil evidence, extinctions, and plasticity in cnidarian life cycles. However, if these various issues are not problematic, we propose that the more convincing scenarios are that fish or their precursors served as first hosts (Fig. 4.3; Scenarios 1, 2) or that parasitism was acquired independently in lineages leading to *Polypodium* and Myxozoa (Fig. 4.3; Scenario 6). The scenario of invertebrates as first hosts acquired by both lineages (Fig. 4.3; Scenario 4) is more unlikely because it requires re-acquisition of a complex free-living adult stage in *Polypodium*. Because parasitism generally entails a transition to host dependency associated with specialized and reductive evolution (e.g. simplified metabolism, loss of digestive tracts, expansive uptake surfaces) reversion back to the more complex features of free-living ancestors is often regarded as nearly impossible—in keeping with Dollo's law (a complex trait cannot re-evolve in the same form) (Cruickshank and Paterson 2006). There is evidence for such reversions in taxa that are not strongly modified for parasitism (e.g. mites and strongyloid nematodes; Cruickshank and Paterson 2006). Reversion to a free-living form has also been inferred for a diplomonad flagellate via acquisition of bacterial genes coding for prey-degrading enzymes needed by a free-living phagotroph (Xu et al. 2016). We contend that reversibility is unlikely for highly modified metazoans like cestodes or myxozoans.

De Baets et al. (2015) outlined various recommendations for future work on discerning the origins and patterns of host acquisition by parasite taxa, and proposed that the most conservative and least circular approach would be to use robust relaxed molecular clock estimates for the origin of free-living ancestors and the earliest certain appearance of parasites in the fossil record. It is just conceivable that fossil myxozoan spores may be preserved along with the soft and small remains of e.g. embryos in ancient deposits (such as the Doushantuo Formation) and these would be extremely valuable as calibrations in molecular clock analyses to constrain age estimates. Furthermore, as myxozoan diversity and ecology becomes better understood, other calibrations, such as unique radiations associated with geological events whose timing is well constrained, may be adopted (De Baets et al. 2016; Ho et al. 2015; Warnock and Engelstädter 2021). An exciting possibility here would be successful sequencing of ancient DNA from chitinised myxosporean spores preserved in sediment cores. Calibration of the molecular clock is the most important factor influencing divergence dates (Inoue et al. 2010; dos Reis et al. 2015), thus incorporating myxozoan-specific calibration points would greatly improve future molecular clock investigations. This could also help to control for their notoriously rapid rates of molecular evolution that may remain problematic even in analyses that accommodate rate variation (Warnock and Engelstädter 2021). We also note that improved understanding of host use would enable greater confidence in patterns revealed by any phylogenetic analyses. Future studies of endocnidozoan origins and host acquisition should be judicious in selecting data for analysis and consider how conclusions may be compromised and constrained by current knowledge of endocnidozoan diversity and patterns of host use in the present-day.

4.7 Adaptation and Diversification of Endocnidozoans

4.7.1 Adaptations to a Parasitic Life Style

Many endoparasites exhibit morphologies that are simplified in comparison to those of their free-living relatives. These may reflect processes such as selection against features that are no longer functional, increased allocation of resources for reproduction, or adaptation to confined spaces within hosts (Okamura et al. 2015b). Malacosporeans exhibit considerable tissue loss, but this morphological simplification is taken to the extreme in myxosporeans which develop exclusively as amorphous plasmodia and pseudoplasmodia. Myxozoans have also undergone miniaturisation, perhaps as a consequence of adapting for life in confined host spaces, such as body cavities, blood vessels and between or within cells (Okamura et al. 2015b). Simplification along with miniaturisation has enabled myxozoans to converge on patterns of host exploitation similar to those of protists (Okamura et al. 2015a), involving extensive multiplication as unicellular forms within hosts. In myxozoans these forms comprise invasive cells and cell-in-cell complexes that multiply within hosts whilst journeying to target infection sites, as well as sacs, myxoworms, pansporocysts, pseudoplasmodia and plasmodia that support spore development. Free-living cnidarians produce relatively few and large eggs and numerous tiny sperm (anisogamy). Even though there is some confusion over which myxozoan cells are gametes, it is clear that both *Polypodium* and myxozoans are isogamous (Raikova 2008; Okamura et al. 2015b). This may be adaptive for parasitism if self-fertilisation is undertaken because finding a mate is difficult within hosts (Okamura et al. 2015b). Other adaptations to parasitism in *Polypodium* include inversion of germ layers and a polyploid trophamnion (Raikova 1994) functioning as a nurse cell during larval development and the multinucleate plasmodia that support sporogony in myxosporeans.

Parasites, however, may also evolve innovations involving complex and specialised designs. Examples include arrays of hooks and spines for host attachment in the scolex of cestodes and the proboscis of acanthocephalans, and the outer syncytial tegument specialised for absorption, secretion and protection in trematodes and cestodes (Ruppert et al. 2004). Multicellular spores of myxozoans and the modified detachable gonads (gonophores) of *Polypodium* are both specialised endocnidozoan innovations for transmission using anchoring nematocysts (Ibragimov and Raikova 2004). In addition, myxosporean spores display further morphological and structural innovations in the form of inflatable caudal processes in actinospores and hardening of the outer wall of myxospores. Caudal processes enable actinospores to be carried away from the benthic habitats of worm hosts, presumably facilitating transmission to fish hosts. Hardening confers environmental resistance and hence spore longevity, presumably facilitating eventual ingestion of myxospores by worms. Myxospores also notably vary in form with some morphotypes arising from convergence. The drivers and functional significance of variation in myxospore morphology remain obscure. Similar actinospore morphotypes are produced in species that develop different myxospore morphotypes suggesting that myxospores have undergone greater morphological diversification than actinospores (Fiala et al. 2015a). However, it must be stressed that actinospores are very poorly sampled (see below).

4.7.2 Patterns of Diversification

A meta-analysis study has revealed that parasite species richness is strongly correlated with that of their hosts—richer host clades harbour richer parasite assemblages (Kamiya et al. 2014). This pattern is reflected in the invertebrate host diversities of the relatively species-poor malacosporeans and the highly speciose myxosporeans. Malacosporeans exploit freshwater bryozoans belonging to the

Class Phylactolaemata, a notably depauperate group of 74 extant species (Massard and Geimer 2008), whilst myxosporeans exploit freshwater and marine oligochaetes and polychaetes which comprise 3500 and 8000 extant species, respectively (Ruppert et al. 2004). Coincident with myxosporean radiations in polychaete and oligochaete radiation is residency in marine and freshwater environments, respectively (Fiala et al. 2015a). The relatively limited radiation of malacosporeans (despite their apparent capacity to infect diverse fish families) suggests that invertebrate host diversity may primarily explain the contrasting diversification patterns of malacosporeans and myxosporeans. Congruence in molecular phylogenies of some myxosporeans and fish provides evidence for co-diversification with certain fish (Holzer et al. 2018).

In the present day the highly speciose myxosporeans infect vertebrate host gills, skin, fins, eyes, kidney, intestine, liver, gall bladder, nervous system, cartilage, muscles, swimbladder and gonad to produce infectious spores (Feist and Longshaw 2006; Molnár and Eszterbauer 2015) whilst the malacosporeans exploit fish kidney for spore production. The acquisition of plasmodia suited for sporulation in organs and tissues may have been critical for myxosporeans to exploit tissues (Fiala et al. 2015a). Phylogenetic analyses have provided evidence that patterns of 'tissue tropism' (site-preference) are lineage-specific and that the ability to exploit particular host environments may support radiations in at least some myxosporeans (e.g. Eszterbauer 2004; Holzer et al. 2004; Heiniger et al. 2013).

Several other traits have been identified as potential drivers of the spectacular myxosporean radiation or as constraints on malacosporeans to undergo radiation (Fiala et al. 2015a). These include: acquisition of hardened, environmentally resistant spore valves in myxosporeans; improved uptake of metabolites by syncytial myxosporean stages (vs. epithelial malacosporean stages), and; incorporation of additional vertebrate host groups (e.g. amphibians, waterfowl and shrews) by myxosporeans. Some non-fish hosts (e.g. shrews) are entirely terrestrial thus terrestrial oligochaetes are highly likely to serve as primary hosts with trophic transmission achieved when infected worms are ingested. This raises the spectre of a diversity of fully terrestrialised myxozoans with life cycles involving vertebrate hosts that eat earthworms.

A current hindrance to understanding the significance of host diversity and environments for myxozoan diversification is our considerable ignorance of parasite diversity in general and of myxozoans in particular (Okamura et al. 2018). Even in relatively well-sampled regions, novel myxozoan diversity is being detected by eDNA sampling (Hartikainen et al. 2016). Particularly poorly sampled regions include the deep sea, polar and tropical environments. Ongoing research suggests that extrapolation of diversity in tropical fish may lead to predictions of myxozoan species richness that rivals or exceeds that of their free-living cnidarian relatives. For instance, the high host specificity of many ceratomyxids (Gunter et al. 2010; Heiniger and Adlard 2013) suggests that Australia's coral reef fish will be exploited by over 1500 species of *Ceratomyxa* of which <1% have been described (Queensland Museum Network 2010). Recall that the currently described 2596 myxozoans represent some 20% of cnidarian species diversity already. While new free-living cnidarians species can also be expected (Appeltans et al. 2012) the hidden nature of endoparasites is likely to make myxozoans particularly undersampled. Further impediments to understanding myxozoan diversification include: limited study of invertebrates as parasite hosts; a research focus on economically important parasites; inapparency of many myxozoan infections; and difficulties in sampling and identification (including low infection prevalences, lack of expertise, and requirement of microscopy and molecular sequence data) (Okamura et al. 2018).

Finally, it is clear that endocnidozoans are characterised by rapid rates of molecular evolution. How this trait may relate to diversification remains unclear. We predict that this issue will begin to be resolved with the development and analysis of new transcriptome, genome and proteome datasets with larger taxonomic coverage.

4.8 Conclusions

Cnidarians have evolved parasitism on multiple occasions but only in the Myxozoa has this entailed substantial radiation. There is evidence that myxozoan diversification is linked with host diversity and it is likely that various cnidarian traits served as preadaptations for assuming parasitic lifestyles. However, our chapter raises more questions than answers. Was parasitism adopted as an ancient lifestyle in the cnidarians when the Endocnidozoa first appeared in the Ediacaran or was there a substantial period when early endocnidozoans were free-living? Does undersampling and our current lack of knowledge of endocnidozoan diversity and host use greatly compromise insights on host groups and co-phylogenetic analyses? To what extent might host switching preclude inferences about ancient hosts and how have mass extinctions driven patterns of endocnidozoan diversity over time? The holy grail of gaining insights into endocnidozoan stem lineages may begin to be addressed if relevant data can be gathered. Such data include fossil evidence that would strengthen molecular clock analyses. Thus, systematic investigation of host pathologies and attempts to retrieve remains of myxozoan spores in palynological samples or from coprolites could be fruitful, particularly from sites with exceptional soft-tissue preservation. Investigations of myxozoan spore taphonomy would be complementary and informative for such research. Meanwhile, more convincing cophylogenetic analyses require better understanding of myxozoan diversity and better resolution of host phylogenies. This could be gained by systematic sampling of further potential host groups, eDNA sampling (especially from severely undersampled habitats), screening of existing archival databases, and incorporation of more comprehensive data in molecular phylogenies of host groups. We expect that advances will be made on at least some of these fronts and could thus help to reveal how endocnidozoans have evolved and diversified to become one of the major cnidarian clades in the present day. We also expect that further work will provide a better understanding of the origins and diversification of the other more limited radiations of parasitic cnidarians.

Acknowledgements We thank Joe Ryan and David Plachetszki for illuminating discussion and insights about patterns of myxozoan evolution over time. We thank Kenneth De Baets for inviting us to develop our chapter. Comments from Steve Feist and Kenneth De Baets helped us to improve our chapter. Insights relevant to the development of this chapter have benefitted by support for research involving myxozoans from the Natural Environment Research Council grants (GR9/04271; GR3/11068; GR3/09956; NER/A/S/1999/00075; NER/B/S/2000/00336; NER/S/A/2004/12399; NE/019227/1), The Leverhulme Trust (a Research Fellowship awarded to B. Okamura), the Biotechnology and Biological Sciences Research Council (BB/F003242/1); a DAAD postdoctoral fellowship to AG, and a Marie-Curie Fellowship (272772, Myxozoa Evo Devo).

References

Abd-Elfattah A, Fontes I, Kumar G, Soliman H, Hartikainen H, Okamura B, El-Matbouli M (2014) Vertical transmission of *Tetracapsuloides bryosalmonae* (Myxozoa), the causative agent of salmonid proliferative kidney disease. Parasitology 141:482–490

Adriano EA, Okamura B (2017) Motility, morphology and phylogeny of the plasmodial worm, *Ceratomyxa vermiformis* n. sp. (Cnidaria: Myxozoa: Myxosporea). Parasitology 144:158–168

Alexander JD, Kerans BL, El-Matbouli M, Hallett SL, Stevens L (2015) Annelid-Myxosporean interactions. In: Okamura B, Gruhl A, Bartholomew JL (eds) Myxozoan evolution, ecology and development. Springer International Publishing, Cham, pp 217–234

Andreev P, Coates MI, Karatajūtė-Talimaa V, Shelton RM, Cooper PR, Wang N-Z, Sansom IJ (2016) The systematics of the Mongolepidida (Chondrichthyes) and the Ordovician origins of the clade. PeerJ 4:e1850

Appeltans W et al (2012) The magnitude of global marine species biodiversity. Curr Biol 22:2189–2202

Atkinson SD, Bartošová-Sojková P, Whipps CM, Bartholomew JL (2015) Approaches for characterising myxozoan species. In: Okamura B, Gruhl A, Bartholomew J (eds) Myxozoan evolution, ecology and development. Springer International Publishing, Cham, pp 111–123

Balasubramanian PG, Beckmann A, Warnken U, Schnölzer M, Schüler A, Bornberg-Bauer E, Holstein TW, Özbek S (2012) Proteome of the *Hydra* nematocyst. J Biol Chem 287:9672–9681

Banks JC, Paterson AM (2005) Multi-host parasite species in cophylogenetic studies. Int J Parasitol 35:741–746

Bartholomew J, Whipple M, Stevens D, Fryer JL (1997) The life cycle of *Ceratomyxa shasta*, a myxosporean parasite of salmonids, requires a freshwater polychaete as an alternate host. J Parasitol 83:859–868

Bartošová-Sojková P, Hrabcová M, Pecková H, Patra S, Kodádková A, Jurajda P, Tyml T, Holzer AS (2014) Hidden diversity and evolutionary trends in malacosporean parasites (Cnidaria: Myxozoa) using molecular phylogenetics. Int J Parasitol 44:565–577

Bentlage B, Osborn KJ, Lindsay DJ, Hopcroft RR, Raskoff KA, Collins AG (2018) Loss of metagenesis and evolution of a parasitic life style in a group of open-ocean jellyfish. Mol Phylogenet Evol 124:50–59

Benton MJ, Donoghue PCJ, Asher R (2009) Calibrating and constraining molecular clocks. In: Hedges SB, Kumar S (eds) The timetree of life. Oxford University Press, Oxford, pp 35–86

Bigelow HB (1909) Reports on the scientific results of the expedition to the eastern tropical Pacific, in charge of Alexander Agassiz, by the U.S. Fish Commission Steamer 'Albatross', from October, 1904, to March, 1905, Lieut. Commander L.M. Garrett, U.S.N., commanding. XVI. The Medusae. Memoirs of the Museum of Comparative Zoology at Harvard University 37:1–243

Blaxter ML, Koutsovoulos G (2014) The evolution of parasitism in Nematoda. Parasitology 142(Suppl 1):S26–S39

Boardman RS (1998) Reflections on the morphology, anatomy, evolution, and classification of the class Stenolaemata (Bryozoa). Smithson Contrib Paleobiol 86:1–60
Boast AP, Weyrich LS, Wood JR, Metcalf JL, Knight R, Cooper A (2018) Coprolites reveal ecological interactions lost with the extinction of New Zealand birds. Proc Natl Acad Sci U S A 115:1546–1551
Boero F, Bouillon J (2005) Cnidaria and Ctenophora (cnidarians and comb jellies). In: Rohde K (ed) Marine parasitology. CABI Publishing, Wallingford, UK, pp 177–182
Boero F, Bouillon J, Gravili C (1991) The life cycle of *Hydrichthys mirus* (Cnidaria: Hydrozoa: Anthomedusae: Pandeidae). Zool J Linnean Soc 101:189–199
Boero F, Bouillon J, Piraino S (1992) On the origins and evolution of hydromedusan life cycles (Cnidaria, Hydrozoa). In: Dallai R (ed) Sex origin and evolution, Symposia and monographs U.Z.I, vol 6, pp 59–68
Bomfleur B, Kerp H, Taylor TN, Moestrup Ø, Taylor EL (2012) Triassic leech cocoon from Antarctica contains fossil bell animal. Proc Natl Acad Sci USA 109:20971–20974. https://doi.org/10.1073/pnas.1218879109
Bomfleur B, Mörs T, Ferraguti M, Reguero MA, McLoughlin S (2015) Fossilized spermatozoa preserved in a 50-Myr-old annelid cocoon from Antarctica. Biol Lett 11:20150431
Bosch TCG (2016) Emergence of immune system components in cnidarians. In: Ratcliffe MJH (Editor in Chief) Encyclopedia of immunobiology, vol 1. Academic, Oxford, pp 397–406
Brachaniec T, Niedźwiedzki R, Surmik D, Krzykawski T, Szopa K, Gorzelak P, Salamon MA (2015) Coprolites of marine vertebrate predators from the Lower Triassic of southern Poland. Palaeogeogr Palaeoclimatol Palaeoecol 435:118–126
Brazeau MD, Friedman M (2015) The origin and early phylogenetic history of jawed vertebrates. Nature 520:490–497
Briggs DEG, Bartels C (2010) Annelids from the Lower Devonian Hunsrück Slate (Lower Emsian, Rhenish Massif, Germany). Palaeontology 53:215–232
Bromham L (2009) Why do species vary in their rate of molecular evolution? Biol Lett 5:401–404
Bromham L, Cowman PF, Lanfear R (2013) Parasitic plants have increased rates of molecular evolution across all three genomes. BMC Evol Biol 13:1
Broughton RE, Betancur-R R, Li C, Arratia G, Ortí G (2013) Multi-locus phylogenetic analysis reveals the pattern and tempo of bony fish evolution. PLOS Curr Tree Life. https://doi.org/10.1371/currents.tol.2ca8041495ffafd0c92756e75247483e
Brown SP, Renaud F, Guégan JF, Thomas F (2001) Evolution of trophic transmission in parasites: the need to reach a mating place? J Evol Biol 14:815–820
Canning EU, Okamura B (2004) Biodiversity and evolution of the Myxozoa. Adv Parasitol 56:43–131
Canning EU, Curry A, Feist SW, Longshaw M, Okamura B (2000) A new class and order of myxozoans to accommodate parasites of bryozoans with ultrastructural observations on *Tetracapsula bryosalmonae* (PKX organism). J Euk Microbiol 47:456–468
Carrete Vega G, Wiens JJ (2012) Why are there so few fish in the sea? Proc R Soc B Biol Sci 279:2323–2329
Cartwright P, Nawrocki AM (2010) Character evolution in Hydrozoa (phylum Cnidaria). Integr Comp Biol 50:456–472
Chang ES, Neuhof M, Rubinstein ND, Diamant A, Philippe H, Huchon D, Cartwright P (2015) Genomic insights into the evolutionary origin of Myxozoa within Cnidaria. Proc Natl Acad Sci U S A 112:14912–14917
Chin K (2021) Gastrointestinal parasites of ancient non-human vertebrates: evidence from coprolites and other materials. In: De Baets K, Huntley JW (eds) The evolution and fossil record of parasitism: Coevolution and paleoparasitological techniques. Topics in Geobiology 50
Choisy M, Brown SP, Lafferty KD, Thomas F (2003) Evolution of trophic transmission in parasites: why add intermediate hosts? Am Nat 162:172–181
Colleary C, Dolocan A, Gardner J, Singh S, Wuttke M, Rabenstein R, Habersetzer J, Schaal S, Feseha M, Clemens M, Jacobs BF, Currano ED, Jacobs LL, Sylvestersen RL, Gabbott SE,

Vinther J (2015) Chemical, experimental, and morphological evidence for diagenetically altered melanin in exceptionally preserved fossils. Proc Natl Acad Sci U S A 112:12592–12597

Collins AG (2009) Recent insights into cnidarian phylogeny. In: Lang MA et al (eds) Proceedings of the Smithsonian marine science symposium, Smithsonian contributions to the marine sciences, vol 38. Smithsonian Institution, Washington, DC, pp 139–149

Collins AG, Bentlage B, Lindner A, Lindsay D, Haddock SHD, Jarms G, Norenburg JL, Jankowski T, Cartwright P (2008) Phylogenetics of Trachylina (Cnidaria: Hydrozoa) with new insights on the evolution of some problematical taxa. J Mar Biol Assoc UK 88:1673–1685

Colwell RK, Dunn RR, Harris NC (2012) Coextinction and persistence of dependent species in a changing world. Annu Rev Ecol Evol Syst 43:183–203

Conway Morris S (1979) Middle Cambrian Polychaetes from the Burgess Shale of British Columbia. Philos Trans R Soc London B Biol Sci 285:227–274

Conway Morris S, Peel JS (2008) The earliest annelids: lower Cambrian polychaetes from the Sirius Passet Lagerstätte, Peary Land, North Greenland. Acta Palaeontol Pol 53:137–148

Cribb TH, Bray RA, Olson PD, Littlewood DT (2003) Life cycle evolution in the Digenea: a new perspective from phylogeny. Adv Parasitol 54:197–254

Cruickshank RH, Paterson AM (2006) The great escape: do parasites break Dollo's law? Trends Parasitol 22:509–515

Cunningham JA, Liu AG, Bengtson S, Donoghue PCJ (2017) The origin of animals: can molecular clocks and the fossil record be reconciled? BioEssays 39:1–12

De Baets K, Littlewood DTJ (2015) The importance of fossils in understanding the evolution of parasites and their vectors. Adv Parasitol 90:1–51

De Baets K, Dentzien-Dias P, Upeniece I, Verneau O, Donoghue PCJ (2015) Constraining the deep origin of parasitic flatworms and host-interactions with fossil evidence. Adv Parasitol 90:93–135

De Baets K, Antonelli A, Donoghue PCJ (2016) Tectonic blocks and molecular clocks. Philos Trans R Soc B 371:20160098

De Baets K, Huntley JW, Klompmaker AA, Schiffbauer JD, Muscente AD (2021a) The fossil record of parasitism: Its extent and taphonomic constraints. In: De Baets K, Huntley JW (eds) The evolution and fossil record of parasitism: Coevolution and paleoparasitological techniques. Topics in Geobiology 50

De Baets K, Dentzien-Dias P, Harrison, GWM, Littlewood, DTJ, Parry LA (2021b) Fossil constraints on the timescale of parasitic helminth evolution. In: De Baets K, Huntley JW (eds) The evolution and fossil record of parasitism: Identification and Macroevolution of Parasites. Topics in Geobiology 49

de Kinkelin P, Gay M, Forman S (2002) The persistenceof infectivity of *Tetracapsula bryosalmonae*-infected water for rainbow trout, *Oncorhynchus mykiss* (Walbaum). J Fish Dis 25:477–482

de Vienne DM, Refrégier G, López-Villavicencio M, Tellier A, Hood ME, Giraud T (2013) Cospeciation vs host-shift speciation: methods for testing, evidence from natural associations and relation to coevolution. New Phytol 198:347–385

Delsuc F, Philippe H, Tsagkogeorga G, Simion P, Tilak M-K, Turon X, López-Legentil S, Piette J, Lemaire P, Douzery EJP (2018) A phylogenomic framework and timescale for comparative studies of tunicates. BMC Biol 16:39

Dentzien-Dias PC, Poinar G Jr, de Figueiredo AEQ, Pacheco ACL, Horn BLD, Schultz CL (2013) Tapeworm eggs in a 270 million-year-old shark coprolite. PLoS One 8:e55007

Dentzien-Dias P, Carrillo-Briceño JD, Francischini H, Sánchez R (2018) Paleoecological and taphonomical aspects of the Late Miocene vertebrate coprolites (Urumaco formation) of Venezuela. Palaeogeogr Palaeoclimatol Palaeoecol 490:590–603

Dohrmann M, Wörheide G (2013) Novel scenarios of early animal evolution—is it time to rewrite textbooks? Integr Comp Biol 53:503–511

Dohrmann M, Wörheide G (2017) Dating early animal evolution using phylogenomic data. Sci Rep 7:3599

Dong X-P, Cunningham JA, Bengtson S, Thomas C-W, Liu J-B, Stampanoni M, Donoghue PCJ (2013) Embryos, polyps and medusae of the early Cambrian scyphozoan *Olivooides*. Proc R Soc B 280:20130071

Donoghue PCJ, Keating JN (2014) Early vertebrate evolution. Palaeontology 57:879–893

dos Reis M, Thawornwattana Y, Angelis K, Telford MJ, Donoghue PCJ, Yang Z (2015) Uncertainty in the timing of origin of animals and the limits of precision in molecular timescales. Curr Biol 25:2939–2950

Dufour B, Le Bailly M (2013) Testing new parasite egg extraction methods in paleoparasitology and an attempt at quantification. Int J Paleopathol 3:199–203

Dunlap WC, Starcevic A, Baranasic D, Diminic J, Zucko J, Gacesa R, van Oppen MJH, Hranueli D, Cullum J, Long PF (2013) KEGG orthology-based annotation of the predicted proteome of *Acropora digitifera*: ZoophyteBase - an open access and searchable database of a coral genome. BMC Genomics 14:509

Dunn RR, Harris NC, Colwell RK, Koh LP, Sodhi NS (2009) The sixth mass coextinction: are most endangered species parasites and mutualists? Proc R Soc B 276:3037–3045

Dunn CW, Giribet G, Edgecombe GD, Hejnol A (2014) Animal phylogeny and its evolutionary implications. Annu Rev Ecol Evol Syst 45:371–395

El-Matbouli M, Hoffmann RW (1991) Effects of freezing, aging, and passage through the alimentary canal of predatory animals on the viability of *Myxobolus cerebralis* spores. J Aquat Anim Health 3:260–262

El-Matbouli M, Hoffmann RW (1998) Light and electron microscopic studies on the chronological development of *Myxobolus cerebralis* to the actinosporean stage in *Tubifex tubifex*. Int J Parasitol 28:195–217

Elwell LC, Kerans BL, Zickovich J (2009) Host–parasite interactions and competition between tubificid species in a benthic community. Freshw Biol 54:1616–1628

Ernst A, Schäfer P (2006) Palaeozoic vs. post-Palaezoic Stenolaemata: phylogenetic relationship or morphological convergence? Courier Forsch Inst Senckenberg 257:49–63

Erséus C (2005) Phylogeny of oligochaetous Clitellata. Hydrobiologia 535/536:357–372

Erwin DH, Laflamme M, Tweedt SM, Sperling EA, Pisani D, Peterson KJ (2011) The Cambrian conundrum: early divergence and later ecological success in the early history of animals. Science 334:1091–1097

Eszterbauer E (2004) Genetic relationship among gill-infecting Myxobolus species (Myxosporea) of cyprinids: molecular evidence of importance of tissue-specificity. Dis Aquat Org 58:35–40

Eszterbauer E, Atkinson S, Diamant A, Morris D, El-Matbouli M, Hartikainen H (2015) Myxozoan life cycles: practical approaches and insights. In: Okamura B, Gruhl A, Bartholomew J (eds) Myxozoan evolution, ecology and development. Springer International Publishing, Cham, pp 175–198

Fauchald K, Rouse G (1997) Polychaete systematics: past and present. Zool Scripta 26:71–138

Fautin DG (2002) Reproduction of Cnidaria. Can J Zool 80:1735–1754

Feist SW, Longshaw M (2006) Phylum Myxozoa. In: Fish diseases and disorders, Protozoan and metazoan infections, vol 1, 2nd edn. CABI Publishing, Wallingford, UK, pp 230–296

Feist SW, Morris DJ, Alama-Bermejo G, Holzer AS (2015) Cellular processes in myxozoans. In: Okamura B, Gruhl A, Bartholomew JL (eds) Myxozoan evolution, ecology and development. Springer International Publishing, Cham, pp 139–154

Felsenstein J (1985) Phylogenies and the comparative method. Am Nat 125:1–15

Fiala I (2006) The phylogeny of Myxosporea (Myxozoa) based on small subunit ribosomal RNA gene analysis. Int J Parasitol 36:1521–1534

Fiala I, Bartošová-Sojková P, Okamura B, Hartikainen H (2015a) Adaptive radiation and evolution within the Myxozoa. In: Okamura B, Gruhl A, Bartholomew J (eds) Myxozoan evolution, ecology and development. Springer International Publishing, Cham, pp 69–84

Fiala I, Bartošová-Sojková P, Whipps CM (2015b) Classification and phylogenetics of Myxozoa. In: Okamura B, Gruhl A, Bartholomew J (eds) Myxozoan evolution, ecology and development. Springer International Publishing, Cham, pp 85–110

Foox J, Siddall ME (2015) The road to Cnidaria: history of phylogeny of the Myxozoa. J Parasitol 101:269–274

Freeman MA, Shinn AP (2011) Myxosporean hyperparasites of gill monogeneans are basal to the Multivalvulida. Parasit Vectors 4:220

Friedman M (2015) The early evolution of ray-finned fishes. Palaeontology 58:213–228

Friedman M, Sallan LC (2012) Five hundred million years of extinction and recovery: a phanerozoic survey of large-scale diversity patterns in fishes. Palaeontology 55:707–742

Goudemand N, Orchard MJ, Urdy S, Bucher H, Tafforeau P (2011) Synchrotron-aided reconstruction of the conodont feeding apparatus and implications for the mouth of the first vertebrates. Proc Natl Acad Sci U S A 108:8720–8724

Grabner DS, El-Matbouli M (2010) Experimental transmission of malacosporean parasites from bryozoans to common carp (*Cyprinus carpio*) and minnow (*Phoxinus phoxinus*). Parasitology 137:629–639

Grazhdankin D (2004) Patterns of distribution in the Ediacaran biotas: facies versus biogeography and evolution. Paleobiology 30:03–221

Grover R, Marguer JF, Allemand D, Ferrier-Pagès C (2008) Uptake of dissolved free amino acids by the scleractinian coral *Stylophora pistillata*. J Exp Biol 211:860–865

Gruhl A (2015) Myxozoa. In: Wanninger A (ed) Evolutionary developmental biology of invertebrates 1: introduction, non-Bilateria, Acoelomorpha, Xenoturbellida, Chaetognatha. Springer, Wien, pp 165–177

Gruhl A, Okamura B (2012) Development and myogenesis of the vermiform *Buddenbrockia* (Myxozoa) and implications for cnidarian body. EvoDevo 3:10

Gruhl A, Okamura B (2015) Tissue characteristics and development of Myxozoa. In: Okamura B, Gruhl A, Bartholomew JL (eds) Myxozoan evolution, ecology and development. Springer International Publishing, Cham, pp 155–174

Gunter NL, Burger MAA, Adlard RD (2010) Morphometric and molecular characterisation of four new *Ceratomyxa* species (Myxosporea: Bivalvulida: Ceratomyxidae) from fishes off Lizard Island, Australia. Folia Parasitol 57:1–10

Halanych KM (2015) The ctenophore lineage is older than sponges? That cannot be right! Or can it? J Exp Biol 218:592–597

Hallett SL, Atkinson SD, Bartholomew JL, Székely C (2015) Myxozoans exploiting homeotherms. In: Okamura B, Gruhl A, Bartholomew J (eds) Myxozoan evolution, ecology and development. Springer International Publishing, Cham, pp 125–135

Han J, Hu S, Cartwright P, Zhao F, Ou Q, Kubota S, Wang X, Yang X (2016) The earliest pelagic jellyfish with rhopalia from Cambrian Chengjiang Lagerstätte. Palaeogeogr Palaeoclimatol Palaeoecol 449:166–173

Hartigan A, Wilkinson M, Gower DJ, Streicher JW, Holzer AS, Okamura B (2016) Myxozoan infections of caecilians demonstrate broad host specificity and indicate a link with human activity. Int J Parasitol 46:375–381

Hartikainen H, Fontes I, Okamura B (2013a) Parasitism and phenotypic change in colonial hosts. Parasitology 140:1403–1412

Hartikainen H, Waeschenbach A, Wöss E, Wood T, Okamura B (2013b) Divergence and species discrimination in freshwater bryozoans (Bryozoa: Phylactolaemata). Zool J Linnean Soc 68:61–80

Hartikainen H, Gruhl AG, Okamura B (2014) Diversification and repeated morphological transitions in endoparasitic cnidarians (Myxozoa: Malacosporea). Mol Phylogenet Evol 76:261–269

Hartikainen H, Bass D, Briscoe AG, Knipe H, Green AJ, Okamura B (2016) Assessing myxozoan presence and diversity using environmental DNA. Int J Parasitol 46(12): 781–792

He J, Zheng L, Zhang W, Lin Y (2015) Life cycle reversal in *Aurelia* sp.1 (Cnidaria, Scyphozoa). PLoS One 10:e0145314

Hedrick RP, McDowell TS, Mukkatira K, MacConnell E, Petri B (2008) Effects of freezing, drying, ultraviolet irradiation, chlorine, and quaternary ammonium treatments on the infectivity of myxospores of *Myxobolus cerebralis* for *Tubifex tubifex*. J Aquat Anim Health 20:116–125

Heiniger H, Adlard RD (2013) Molecular identification of cryptic species of *Ceratomyxa* Thélohan, 1892 (Myxosporea: Bivalvulida) including the description of eight novel species from apogonid fishes (Perciformes: Apogonidae) from Australian waters. Acta Parasitol 58:342–360

Heiniger H, Cribb TH, Adlard RD (2013) Intra-specific variation of *Kudoa* spp. (Myxosporea: Multivalvulida) from apogonid fishes (Perciformes), including the description of two new species, *K. cheilodipteri* n. sp. and *K. cookii* n. sp., from Australian waters. Syst Parasitol 84:193–215

Ho SYW, Tong KJ, Foster CSP, Ritchie AM, Lo N, Crisp MD (2015) Biogeographic calibrations for the molecular clock. Biol Lett 11:20150194

Holzer AS, Sommerville C, Wootten R (2004) Molecular relationships and phylogeny in a community of myxosporeans and actinosporeans based on their 18S rDNA sequences. Int J Parasitol 34:1099–1111

Holzer AS, Bartošová-Sojková P, Born-Torrijos A, Lövy A, Hartigan A, Fiala I (2018) The joint evolution of the Myxozoa and their alternate hosts: a cnidarian recipe for success and vast biodiversity. Mol Ecol 27:1651–1666

Huang D-Y, Chen J, Vannier J, Saiz Salinas JI (2004) Early Cambrian sipunculan worms from Southwest China. Proc R Soc B Biol Sci 271:1671–1676

Hugot J-P, Gardner SL, Borba V, Araujo P, Leles D, Stock Da-Rosa ÁA, Dutra J, Ferreira LF, Araújo A (2014) Discovery of a 240 million year old nematode parasite egg in a cynodont coprolite sheds light on the early origin of pinworms in vertebrates. Parasit Vectors 7:486

Hunt AP, Milàn J, Lucas SG, Spielmann JA (2012) Vertebrate coprolites. N M Mus Nat Hist Sci Bull 57:5–24

Ibragimov A, Raikova E (2004) Nematocysts of *Polypodium hydriforme*, a cnidarian parasite of acipenseriform fishes. Hydrobiologia 530:165–171

Inoue J, Donoghue PCJ, Yang Z (2010) The impact of the representation of fossil calibrations on Bayesian estimation of species divergence times. Syst Biol 59:74–89

Irisarri I, Baurain D, Brinkmann H, Delsuc F, Sire J-Y, Kupfer A, Petersen J, Jarek M, Meyer A, Vences M, Philippe H (2017) Phylotranscriptomic consolidation of the jawed vertebrate timetree. Nat Ecol Evol 1:1370–1378

Jankowski T, Collins AG, Campbell R (2008) Global diversity of inland water cnidarians. Hydrobiologia 595:35–40

Janvier P (2001) Ostracoderms and the shaping of the gnathostome characters. In: Ahlberg PE (ed) Major events in early vertebrate evolution. Taylor and Francis, London, pp 172–186

Jones SRM, Bartholomew JL, Zhang JY (2015) Mitigating myxozoan disease impacts on wild fish populatins. In: Okamura B, Gruhl A, Bartholomew JL (eds) Myxozoan evolution, ecology and development. Springer International Publishing, Cham, pp 139–154

Kallert DM, Grabner DS, Yokoyama H, El-Matbouli M, Eszterbauer E (2015) Transmission of myxozoans to vertebrate hosts. In: Okamura B, Gruhl A, Bartholomew JL (eds) Myxozoan evolution, ecology and development. Springer International Publishing, Cham, pp 235–250

Kamiya T, O'Dwyer K, Nakagawa S, Poulin R (2014) Host diversity drives parasite diversity: meta-analytical insights into patterns and causal mechanisms. Ecography 37:689–697

Kayal E, Bentlage B, Pankey MS, Ohdera AH, Medina M, Plachetzki DC, Collins AG, Ryan JF (2018) Phylogenomics provides a robust topology of the major cnidarian lineages and insights on the origins of key organismal traits. BMC Evol Biol 18(1)

Kodádková A, Bartošová-Sojková P, Holzer AS, Fiala I (2015) *Bipteria vetusta* n. sp. – an old parasite in an old host: tracing the origin of myxosporean parasitism in vertebrates. Int J Parasitol 45:269–276

Koel TM, Kerans BL, Barras SC, Hanson KC, Wood JS (2010) Avian piscivores as vectors for *Myxobolus cerebralis* in the Greater Yellowstone ecosystem. Trans Am Fish Soc 139:976–988

Lai AG, Aboobaker AA (2017) EvoRegen in animals: time to uncover deep conservation or convergence of adult stem cell evolution and regenerative processes. Dev Biol 433:118–131

Leung TLF (2017) Fossils of parasites: what can the fossil record tell us about the evolution of parasitism? Biol Rev 92:410–430

Leung TLF (2021) Parasites of fossil vertebrates: what we know and what can we expect from the fossil record? In: De Baets K, Huntley JW (eds) The evolution and fossil Record of Parasitism: Identification and macroevolution of parasites. Topics in Geobiology 49

Liu J, Ou Q, Han J, Li J, Wu Y, Jiao G, He T (2015) Lower Cambrian polychaete from China sheds light on early annelid evolution. Sci Nat 102:34

Lom J (1990) Phylum Myxozoa. In: Margulis L, Corliss JO, Melkonian M, Chapman DJ (eds) Handbook of Protoctista. Jones and Bartlett, Boston, pp 36–52

Lom J, Dyková I (1997) Ultrastructural features of the actinosporean phase of Myxosporea (Phylum Myxozoa): a comparative study. Acta Protozool 36:83–103

Lom J, Dyková I (2006) Myxozoan genera: definition and notes on taxonomy, life-cycle terminology and pathogenic species. Folia Parasitol 53:1–36

Manum S, Bose M, Sawyer R (1991) Clitellate cocoons in fresh-water deposits since the Triassic. Zool Scripta 20:347–366

Martínez-Aquino A (2016) Phylogenetic framework for coevolutionary studies: a compass for exploring jungles of tangled trees. Curr Zool 62:393–403

Massard JA, Geimer G (2008) Global diversity of bryozoans (Bryozoa or Ectoprocta) in freshwater: an update. Bull Soc Nat Luxemb 109:139–148

McDermott JJ, Zubkoff PL, Lin AL (1982) The occurrence of the anemone *Peachia parasitica* as a symbionts in the scyphozoan *Cyanea capillata* in the lower Chesapeake Bay. Estuaries 5:319–321

McGurk C, Morris DJ, Auchinachie NA, Adams A (2006) Development of *Tetracapsuloides bryosalmonae* (Myxozoa: Malacosporea) in bryozoan hosts (as examined by light microscopy) and quantitation of infective dose to rainbow trout (*Oncorhynchus mykiss*). Vet Parasitol 135:249–257

Molnár K, Eszterbauer E (2015) Specificity of infection sites in vertebrate hosts. In: Okamura B, Gruhl A, Bartholomew JL (eds) Myxozoan evolution, ecology and development. Springer International Publishing, Cham, pp 295–313

Morris DJ (2012) A new model for myxosporean (Myxozoa) development explains the endogenous budding phenomenon, the nature of cell within cell life stages and evolution of parasitism from a cnidarian ancestor. Int J Parasitol 42:829–840

Moser M, Taylor S (1978) Effects of the copepod *Cardiodectes medusaeus* on the laternfish *Stenobrachius leucopsarus* with notes on hypercastration by the hydroid *Hydrichthys* sp. Can J Zool 56:2372–2376

Munoz P, Palenzuela O, Alvarez-Pellitero P, Sitjà-Bobadilla A (1999) Comparative studies on carbohydrates of several myxosporean parasites of fish using lectin histochemical methods. Folia Parasitol 46:241–247

Murdock DJE, Dong X-P, Repetski JE, Marone F, Stampanoni M, Donoghue PCJ (2013) The origin of conodonts and of vertebrate mineralized skeletons. Nature 502:546–549

Naldoni J, Adriano E, Hartigan A, Sayer C, Okamura B (2019) Malacosporean myxozoans exploit a diversity of fish hosts. Parasitology 146:1–11. https://doi.org/10.1017/S0031182019000246

Near TJ, Eytan RI, Dornburg A, Kuhn KL, Moore JA, Davis MP, Wainwright PC, Friedman M, Smith WL (2012) Resolution of ray-finned fish phylogeny and timing of diversification. Proc Natl Acad Sci U S A 109:13698–13703

Okamura B (1996) Occurrence, prevalence, and effects of the myxozoan *Tetracapsula bryozoides* Canning, Okamura and Curry, 1996, parasitic in the freshwater bryozoan *Cristatella mucedo* Cuvier (Bryozoa, Phylactolaemata). Folia Parasitol 43:262–266

Okamura B, Gruhl A (2015) Myxozoan affinities and route to endoparasitism. In: Okamura B, Gruhl A, Bartholomew JL (eds) Myxozoan evolution, ecology and development. Springer International Publishing, Cham, pp 23–44

Okamura B, Gruhl A, Bartholomew JL (2015a) An introduction to myxozoan evolution, ecology and development. In: Okamura B, Gruhl A, Bartholomew JL (eds) Myxozoan evolution, ecology and development. Springer International Publishing, Cham, pp 1–20

Okamura B, Gruhl A, Reft AI (2015b) Cnidarians origins of the Myxozoa. In: Okamura B, Gruhl A, Bartholomew JL (eds) Myxozoan evolution, ecology and development. Springer International Publishing, Cham, pp 45–68

Okamura B, Hartigan H, Naldoni N (2018) Extensive uncharted biodiversity: the parasite dimension. Integr Comp Biol 58:1132–1145

Parham JF, Donoghue PCJ, Bell CJ, Calway TD, Head JJ, Holroyd PA, Inoue JG, Irmis RB, Joyce WG, Ksepka DT, Patané JSL, Smith ND, Tarver JE, vanTuinen M, Yang Z, Angielczyk KD, Greenwood JM, Hipsley CA, Jacobs L, Makovicky PJ, Müller J, Smith KT, Theodor JM, Warnock RCM, Benton MJ (2012) Best practices for justifying fossil calibrations. Syst Biol 61:346–359. https://doi.org/10.1093/sysbio/syr107

Park E, Hwang D-S, Lee J-S, Song J-I, Seo T-K, Won Y-J (2012) Estimation of divergence times in cnidarian evolution based on mitochondrial protein-coding genes and the fossil record. Mol Phylogenet Evol 62:329–345

Parker GA, Chubb JC, Roberts GN, Michaud M, Milinski M (2003) Evolution of complex life cycles in helminth parasites. Nature 425:480–484

Parker GA, Ball MA, Chubb JC (2015) Evolution of complex life cycles in trophically transmitted helminths. I. Host incorporation and trophic ascent. J Evol Biol 28:267–291

Parry L, Tanner A, Vinther J (2014) The origin of annelids. Palaeontology 57:1091–1103

Parry L, Vinther J, Edgecombe GD (2015) Cambrian stem-group annelids and a metameric origin of the annelid head. Biol Lett 11:20150763

Paterson S, Vogwill T, Buckling A, Benmayor R, Spiers AJ, Thomson NR, Quail M, Smith F, Walker D, Libberton B, Fenton A, Hall N, Brockhurst MA (2010) Antagonistic coevolution accelerates molecular evolution. Nature 464:275–278

Petit G (2010) Skin nodules in fossil fishes from Monte Bolca (Eocene, Northern Italy). Geodiversitas 32:157–163

Petit G, Khalloufi B (2012) Paleopathology of a fossil fish from the Solnhofen Lagerstätte (Upper Jurassic, southern Germany). Int J Paleopathol 2:42–44

Piraino S, De Vito D, Schmich J, Bouillon J, Boero F (2004) Reverse development in Cnidaria. Can J Zool 82:1748–1754

Poinar G, Boucot AJ (2006) Evidence of intestinal parasites of dinosaurs. Parasitology 133:245–249. https://doi.org/10.1017/S0031182006000138

Poulin R (2007) Evolutionary ecology of parasites, 2nd edn. Princeton University Press, Princeton, NJ

Poulin R, Randhawa HS (2015) Evolution of parasitism along convergent lines: from ecology to genomics. Parasitology, 142(S1):S6–S15

Queensland Museum Network (2010). http://www.qm.qld.gov.au/Research/Projects/Projects/Biodiversity/Parasitology/Internal+parasites+of+coral+reef+fishes#.WgG49lu0Pcs. Accessed 7 Nov 2017

Qvarnström M, Niedźwiedzki G, Žigaitė Ž (2016) Vertebrate coprolites (fossil faeces): an underexplored Konservat-Lagerstätte. Earth-Sci Rev 162:44–57

Raikova E (1965) A cytophotometric study of the DNA content in the cell nuclei of *Polypodium hydriforme* Ussov (Coelenterata) at various stages of its life cycle. Zhurnal Obchei Biologii 26:546–552. [in Russian, English summary]

Raikova EV (1990) Fine structure of the nematocytes of *Polypodium hydriforme* Ussov (Cnidaria). Zool Scripta 19:1–11

Raikova E (1994) Life cycle, cytology, and morphology of *Polypodium hydriforme*, a coelenterate parasite of the eggs of Acipenseriform fishes. J Parasitol 80:1–22

Raikova EV (2002) *Polypodium hydriforme* infection in the eggs of acipenseriform fishes. J Appl Ichthyo 18:405–415

Raikova EV (2008) Cytological peculiarities of *Polypodium hydriforme* (Cnidaria). J Mar Biol Assoc UK 88:1695–1702

Raikova EV, Raikova OI (2016) Nervous system immunohistochemistry of the parasitic cnidarian *Polypodium hydriforme* at its free-living stage. Zoology 119:143–152

Raikova EV, Ibragimov AY, Raikova OI (2007) Muscular system of a peculiar parasitic cnidarian *Polypodium hydriforme*: a phalloidin fluorescence study. Tiss Cell 39:79–87

Rauch G, Kalbe M, Reusch TBH (2005) How a complex life cycle can improve a parasite's sex life. J Evol Biol 18:1069–1075
Reitzel AM, Sullivan JC, Finnerty JR (2006) Qualitative shift to indirect development in the parasitic sea anemone *Edwardsiella lineata*. Integr Comp Biol 46:827–837
Richter G, Baszio S (2001) Traces of a limnic food web in the Eocene Lake Messel—a preliminary report based on fish coprolite analyses. Palaeogeogr Palaeoclimatol Palaeoecol 166:345–368
Richter G, Wedmann S (2005) Ecology of the Eocene Lake Messel revealed by analysis of small fish coprolites and sediments from a drilling core. Palaeogeogr Palaeoclimtol Palaeoecol 223:147–161
Rodríguez E, Barbeitos MS, Brugler MR, Crowley LM, Grajales A et al (2014) Hidden among sea anemones: the first comprehensive phylogenetic reconstruction of the order Actiniaria (Cnidaria, Anthozoa, Hexacorallia) reveals a novel group of hexacorals. PLoS One 9(5):e96998
Ruppert EE, Fox RS, Barnes RD (2004) Invertebrate zoology. A functional evolutionary approach, 7th edn. Thomson Brooks/Cole, Belmont
Ryan JF et al (2013) The genome of the ctenophore *Mnemiopsis leidyi* and its implications for cell type evolution. Science 342:1242592
Sarker S, Kallert DM, Hedrick RP, El-Matbouli M (2015) Whirling disease revisited: pathogenesis, parasite biology and disease intervention. Dis Aquat Org 114:155–175
Schierwater B, Eitel M, Jakob W, Osigus HJ, Hadrys H, Dellaporta SL, Kolokotronis SO, DeSalle R (2009) Concatenated analysis sheds light on early metazoan evolution and fuels a modern "urmetazoon" hypothesis. PLoS Biol 7:36–44
Schmid-Hempel P (2011) Evolutionary parasitology. The integrated study of infections, immunology, ecology, and genetics. Oxford University Press, Oxford
Siddall ME, Martin DS, Bridge D, Desser SS, Cone DK (1995) The demise of a phylum of protists: phylogeny of Myxozoa and other parasitic Cnidaria. J Parasitol 81:961–967
Simion P, Philippe H, Baurain D, Jager M, Richter DJ, Di Franco A, Roure B, Satoh N, Quéinnec É, Ereskovsky A, Lapébie P, Corre E, Delsuc F, King N, Wörheide G, Manuel M (2017) A large and consistent phylogenomic dataset supports sponges as the sister group to all other animals. Curr Biol 27:958–967
Sitjà-Bobadilla A, Schmidt-Posthaus H, Wahli T, Holland JW, Secombes CJ (2015) Fish immune responses to Myxozoa. In: Okamura B, Gruhl A, Bartholomew JL (eds) Myxozoan evolution, ecology and development. Springer International Publishing, Cham, pp 253–280
Spaulding JG (1972) The life cycle of *Peachia quinquecapitata*, an anemone parasitic on medusae during its larval development. Biol Bull 143:440–453
Steinel NC, Bolnick DI (2017) Melanomacrophage centers as a histological indicator of immune function in fish and other poikilotherms. Front Immunol 8:827
Strona G (2015) Past, present and future of host–parasite coextinctions. Int J Parasitol Parasites Wildl 4:431–441
Struck TH, Paul C, Hill N, Hartmann S, Hösel C, Kube M, Lieb B, Meyer A, Tiedemann R, Purschke G, Bleidorn C (2011) Phylogenomic analyses unravel annelid evolution. Nature 471:95–98
Struck TH, Golombek A, Weigert A, Franke FA, Westheide W, Purschke G, Bleidorn C, Halanych KM (2015) The evolution of annelids reveals two adaptive routes to the interstitial realm. Curr Biol 25:1993–1999
Taylor PD, Waeschenbach A (2015) Phylogeny and diversification of bryozoans. Palaeontology 58:585–599
Timm T, Vinn O, Buscalioni ÁD (2016) Soft-bodied annelids (Oligochaeta) from the Lower Cretaceous (La Huérguina formation) of the Las Hoyas Konservat-Lagerstätte, Spain. N Jb Geol Paläont (Abh) 280:315–324
Tsai IJ, Zarowiecki M, Holroyd N, Garciarrubio A, Sanchez-Flores A, Brooks KL, Tracey A, Bobes RJ, Fragoso G, Sciutto E, Aslett M, Beasley H, Bennett HM, Cai J, Camicia F, Clark R, Cucher M, De Silva N, Day TA, Deplazes P, Estrada K, Fernández C, Holland PWH, Hou J, Hu S, Huckvale T, Hung SS, Kamenetzky L, Keane JA, Kiss F, Koziol U, Lambert O, Liu K, Luo X, Luo Y, Macchiaroli N, Nichol S, Paps J, Parkinson J, Pouchkina-Stantcheva N, Riddiford N, Rosenzvit M, Salinas G, Wasmuth JD, Zamanian M, Zheng Y, Cai X, Soberón X, Olson

PD, Laclette JP, Brehm K, Berriman M (2013) The genomes of four tapeworm species reveal adaptations to parasitism. Nature 496:57–63

Turner S, Burrow CJ, Schultze HP, Blieck A, Reif WE, Rexroad CB, Bultynck P, Nowlan GS (2010) False teeth: conodont–vertebrate phylogenetic relationships revisited. Geodiversitas 32:545–594

van Dijk J, De Baets K (2021) Biodiversity and host-parasite (co)extinction. In: De Baets K, Huntley JW (eds) The evolution and fossil record of parasitism: Coevolution and paleoparasitological techniques. Topics in Geobiology 50

Van Iten H, Marques AC, de Moraes Leme J, Pacheco MLAF, Simões MG (2014) Origin and early diversification of the phylum Cnidaria Verrill: major developments in the analysis of the taxon's Proterozoic-Cambrian history. Palaeontology 57:677–690

Vinogradow AV (1996) New fossil freshwater bryozoans from the Asiatic part of Russia and Kazakhstan. Paleontol J 30:284–292

Vinther J, Eibye-Jacobsen D, Harper DAT (2011) An early Cambrian stem polychaete with pygidial cirri. Biol Lett 7:929–932

Waeschenbach A, Taylor PD, Littlewood DTJ (2012) A molecular phylogeny of bryozoans. Mol Phylogenet Evol 62:718–735

Wan B, Yuan X, Chen Z, Guan C, Pang K, Tang Q, Xiao S (2016) Systematic description of putative animal fossils from the early Ediacaran Lantian formation of South China. Palaeontology 59:515–532

Warnock RCM, Engelstädter J (2021) The molecular clock as tool for understanding host-parasite evolution. In: De Baets K, Huntley JW (eds) The evolution and fossil record of parasitism: Coevolution and paleoparasitological techniques. Topics in Geobiology 50

Warren LV, Pacheco MLAF, Fairchild TR, Simoes MG et al (2012) The dawn of animal skeletogenesis: ultrastructural analysis of the Ediacaran metazoan *Corumbella werneri*. Geology 40:691–694

Weigert A, Bleidorn C (2016) Current status of annelid phylogeny. Org Divers Evol 16:345–362

Weigert A, Helm C, Meyer M, Nickel B, Arendt D, Hausdorf B, Santos SR, Halanych KM, Purschke G, Bleidorn C, Struck TH (2014) Illuminating the base of the annelid tree using transcriptomics. Mol Biol Evol 31:1391–1401

Weinstein SB, Kuris AM (2016) Independent origins of parasitism in Animalia. Biol Lett 12:20160324

Werner B (1963) Effect of some environmental factors on differentiation and determination in marine Hydrozoa, with a note on their evolutionary significance. Ann N Y Acad Sc 105:461–488

Whelan NV, Kocot KM, Moroz TP, Mukherjee K, Williams P, Paulay G, Moroz LL, Halanych KM (2017) Ctenophore relationships and their placement as the sister group to all other animals. Nat Ecol Evol 1:1737–1746

Wood TS, Okamura B (2017) New species, genera, families, and range extensions of freshwater bryozoans in Brazil: the tip of the iceberg? Zootaxa 4306:383–400

WoRMS Editorial Board (2018) World Register of Marine Species. http://www.marinespecies.org. Accessed 7 Apr 2018

Xia F-S, Zhang S-G, Wang Z-Z (2007) The oldest bryozoans: new evidence from the Late Tremadocian (Early Ordovician) of East Yangtze Gorges in China. J Paleontol 81:1308–1326

Xu F, Jerlström-Hultqvist J, Kolisko M, Simpson AG, Roger AJ, Svärd SG, Andersson JO (2016) On the reversibility of parasitism: adaptation to a free-living lifestyle via gene acquisitions in the diplomonad *Trepomonas* sp. PC1. BMC Biol 14:62

Yokoyama H, Masuda K (2001) *Kudoa* sp. (Myxozoa) causing a post-mortemmyoliquefaction of North-Pacific giant octopus *Paroctopus dofleini* (Cephalopoda: Octopodidae). Bull Eur Assoc Fish Pathol 21:266–268

Young GA, Hagadorn JW (2010) The fossil record of cnidarian medusae. Palaeoworld 19:212–221

Zrzavý J, Hypša V (2003) Myxozoa, *Polypodium*, and the origin of the Bilateria: the phylogenetic position of "Endocnidozoa" in the light of the rediscovery of *Buddenbrockia*. Cladistics 19:164–169

Chapter 5
Evolutionary History of Bivalves as Parasites

Aleksandra Skawina

Abstract Bivalves commonly associate with other organisms, however, examples of true parasitic associations are described only for members of the marine superfamily Galeommatoidea (two species of approximately 500 known species, facultative parasitism) and the larvae of members of the freshwater order Unionida (almost all of 958 known species, obligate parasitism). The evolution toward such a close relationship required establishing a close association with the host's body, resulting in being enclosed within its tissues. Clear adaptations to the host species are observed in both groups. Most galeommatoideans live in soft sediments and are associated with other benthic organisms or their burrows—settlement in a burrow or on/within the host's body protects these little bivalves, while life activity of the host possibly ensures oxygenated water currents with a food source for the bivalve. However, a few recent examples of bivalve settlement within the body cavity of crabs (believed as accidental, nevertheless bivalves feed in the crab's hemocoel), or in the oesophagus of holothurians (common, presumably nutrition from the host is possible) indicate possible pathways for an evolutionary transition from free- or commensal-living to a parasitic lifestyle. Unionoids, large freshwater bivalves, are characterized by their tiny larvae that parasitize fish. This close relationship primarily benefits bivalves through enhanced dispersal abilities, but fish tissues may also serve as a source of nutrients for the larvae. Parasitic association likely established when close and common contact of both associates could happen, thus one may hypothesize a fish that lived close to the bottom of lakes and rivers (including durophagous species) as a likely host at the beginning of their co-evolution. Accidental contact of the larvae with the body of fishes (during predation on bivalves or caused by anti-sinking mucous or the larval threads tangled with fish) could result in increased bivalve

Electronic supplementary materials: The online version of this chapter (https://doi.org/10.1007/978-3-030-42484-8_5) contains supplementary material, which is available to authorized users.

A. Skawina (✉)
Department of Animal Physiology, Faculty of Biology, University of Warsaw, Warszawa, Poland
e-mail: a.skawina@uw.edu.pl

© The Author(s) 2021

K. De Baets, J. W. Huntley (eds.), *The Evolution and Fossil Record of Parasitism*, Topics in Geobiology 49, https://doi.org/10.1007/978-3-030-42484-8_5

dispersal. Subsequently, firmer attachment on fish tissues was acquired, followed by encapsulation of the larva within the host epithelium. This might have allowed for the feeding on host tissues, but required developing resistance to the host's immune system, which might have further strengthened their association.

Keywords Bivalvia · Galeommatoidea · Unionida · Parasitism · Evolution

5.1 Introduction

Parasitism has evolved independently 223 times in nearly half of the generally recognized animal phyla. The majority (83%) of unique origins are today described at or below the family level (Weinstein and Kuris 2016). Bivalves commonly are the commensals of other invertebrates (Boss 1965a). Although some authors indicate even a symbiotic relationship between some bivalves and their hosts (e.g. some galeommatoid bivalves according to Boss 1965a; Goto et al. 2014, 2016), it is often difficult to detect what benefit, if any, the host species get (Boss 1965a). On the other hand, possible damage of the host tissues or its skeleton, the influence on physiology or food competition are commonly described (e.g. Meyers and Millemann 1977; Scott and Risk 1988; Haag et al. 1993; Lafferty 1993; Strayer and Smith 1996; Ricciardi et al. 1998; Jansen et al. 2001; Dodd et al. 2006; Bagur et al. 2013; Poulter et al. 2018). Nevertheless true parasitic associations seem to be rare in bivalves, with the exceptions of the parasitising larvae characterizing members of the freshwater order Unionida (obligate parasites; e.g. Bauer 2001a), and facultative parasitic associations described for two species within the marine superfamily Galeommatoidea (members of which are generally known to establish obligate commensal relations with their hosts; Goto et al. 2007, 2012, 2018; Bristow et al. 2010a; Bhaduri et al. 2017; Fig. 5.1). Furthermore, most of the known obligate commensal or parasitic bivalve species appear to have a high degree of specificity to their hosts (Boss 1965a; Bauer 2001b; Sato et al. 2011; Goto et al. 2012, 2018; Patterson 2018), however parasitic larvae of members of the freshwater order Unionida are diverse in this aspect and some are highly specialized, while the others often are able to utilize a broader host species range (Bauer 2001b).

5.1.1 Bivalves in Relations with Other Organisms

Bivalves often settle on the hard surfaces of other organisms, especially in soft bottom environments—commonly on the shells of the other mollusks, on alive and dead corals or on the external skeletons of crustaceans (e.g. Bromley and Heinberg 2006). They commonly actively choose a living host individual, and their molts, dead skeletons or empty shells (Merrill 1960; Morton 1972; Lewandowski 1976; Morton 1988; Scott 1988; Kleemann 1996; Owada 2009; Li and O'Foighil 2012; Fig. 5.1). They often co-inhabit burrows of other invertebrates (Morton 1972; Mikkelsen and Bieler 1989; Morton and Scott 1989; Goto et al. 2011, 2018; Bhaduri et al. 2017). Some commensal clams have strong host-habitat preference and settle

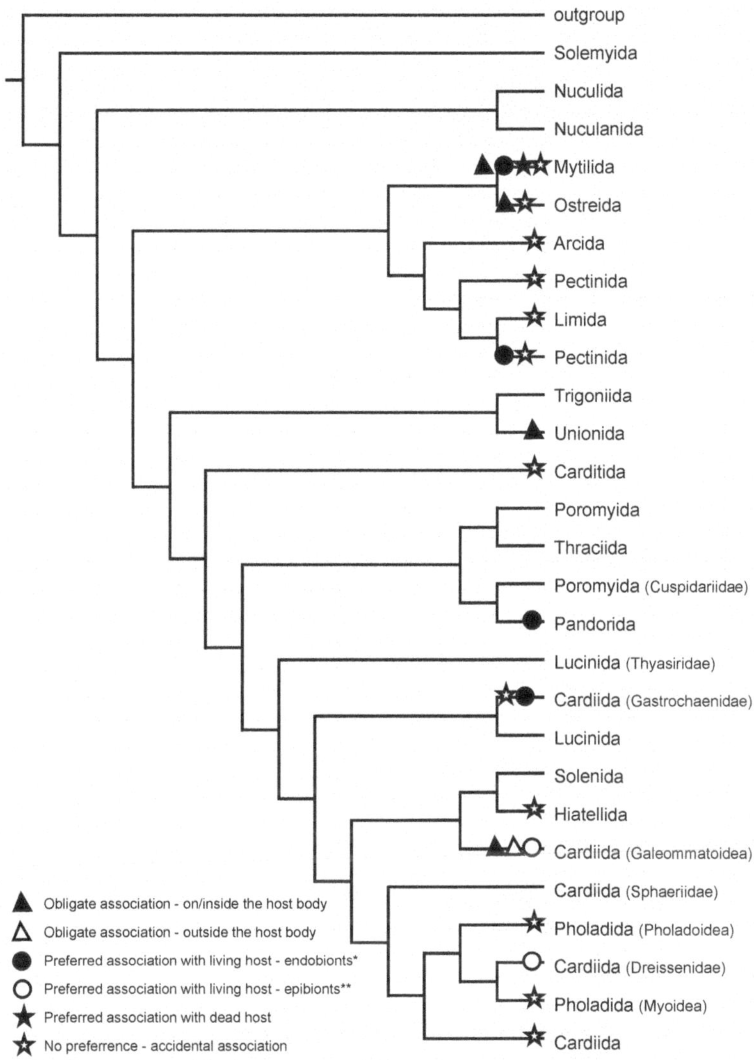

Fig. 5.1 Phylogenetic tree of Bivalvia after Combosch et al. (2017), simplified. Lineages in which different kinds of association with other organisms were developed (literature data) are indicated. *—endobionts *sensu* biocorroders, bioeroders, covered by tissues; **—epibionts *sensu* settlement on the skeleton; based on electronic supplementary materials

directly on the selected surface of exoskeleton or soft body of their hosts (sometimes being then covered by host tissues, e.g. on echiurids, or inside sponges); the other attach within the branchial cavity of host crabs or inside the esophagus of holothurians (Morton 1972; Lafferty 1993; Kato 1998; Boyko and Mikkelsen 2002; Bristow et al. 2010a; Tsubaki and Kato 2014; Bhaduri et al. 2017; Goto and Ishikawa 2019). Yet, it is described at least for two marine species, that a commensal bivalve becomes possibly endoparasitic, by being enclosed in the host's body and presumably feeding on the host's tissues (Bristow et al. 2010a; Bhaduri et al. 2017).

Boring clams like pholadids can excavate a variety of substrates, including corals and shells of the other mollusks (Lauckner 1983). For example *Diplothyra curta* is known to bore the shells of oysters (*Crassostrea virginica*; Turner 1954; Galtsoff 1964), while *Penitella conradi* bores in the shells of both the bivalves and snails *Mytilus californianus*, *Pododesmus* sp., *Haliotis* spp. and *Astraea* (Turner 1955). The mytilid genus *Lithophaga* contains a variety of species that bore limestone, dead and living corals, as well as the shells of living mollusks, i.e. oysters (Ogle 1976; Morton and Scott 1980; Scott 1986; Brickner et al. 1993; Kleemann 1996; Mauna et al. 2005; Scaps et al. 2008; Bagur et al. 2013). Varying degrees of bivalve specialization to host species are observed. For example, *L. mucronata* is host specific to the coral genus *Montipora* and *L. simplex* larval metamorphosis is induced by living corals of only two species (*Astreopora myriophthalma, Goniastrea pectinata*)—most commonly the parent host coral (Morton and Scott 1980; Mokady et al. 1992). These clams burrow in their substrates by chemical means (Morton and Scott 1980). The mytilid *Mytilus edulis* was found by Wiborg (1946) according to Lauckner (1983) to have settled within the mantle cavity of *Modiolus modiolus*—this location may indicate a parasitic relationship; however these authors believe it was accidental settlement. Adult *Mytilus edulis* were also described as settling on the gills of decapods—*Paralithodes camtschaticus* and *Carcinus meanas* (Jansen et al. 1998; Poulter et al. 2018). Most likely it is a case of accidental colonization, with negative outcomes for the crab. The bivalves were presumably not feeding directly on their hosts' tissues, however they may have negatively altered the crabs' respiratory function (Poulter et al. 2018). Other mytilids, members of the genus *Musculus*, and a pandorid *Mytilimeria nuttalli* often live covered by the tissues of a variety ascidians species (Roberts and Breen 1985; Cañete and Rocha 2013). The pteriid *Vulsella vulsella* is an obligate sponge dwelling bivalve within the order Ostreida, however it is interpreted as a case of a filtering-mutualism, not a parasitic relation (Tsubaki and Kato 2014). The hiatellid *Hiatella arctica* was also described as an epibiont in the branchial chamber of *Chionoecetes bairdi* and (together with *M. edulis*, *M. trossulus* and pectinid *Heteranomia scuamula*) from the carapax surface of *Lithodes maia*, *Paralithodes camtschaticus* and *Hemigrapsus sanguineus* crabs (Dick et al. 1998; Jansen et al. 1998; Isaeva et al. 2001; Dvoretsky and Dvoretsky 2008) with no preferred surface for settlement (live or dead) thus might be best interpreted as accidental associates.

5.1.2 *Galeommatoidea*

Galeommatoidea (= Erycinacea, = Leptonacea) bivalves are a diverse group of marine, globally distributed, small-bodied bivalves (often less than 1 cm), known since the Cretaceous and living in shallow-water environments (Li et al. 2012, 2016). Some species are free-living, typically restricted to hard bottom environments (Wear 1966; Booth 1979; Chanley and Chanley 2010; Li et al. 2012; Bhaduri et al. 2017), while most species have been reported from soft sediments, living their lives as commensals or even parasites of other benthic organisms (Appukuttan

1972; Bristow et al. 2010a; Goto et al. 2012; Bhaduri et al. 2017). They are associated with a number of benthic and burrowing invertebrate phyla, e.g. Arthropoda, Echinodermata, Annelida, Porifera, Brachiopoda, among others (Boss 1965a, b; Wear 1966; Narchi 1969; Morton 1972; Mikkelsen and Bieler 1989; Kato 1998; Sato et al. 2011; Goto et al. 2012, 2016; Li et al. 2012). Li et al. (2016) indicated, that ancestral galeommatoideans were sediment-dwelling commensals with one major free-living radiation into hard-bottom environments, coupled with losses of commensalism. Commensals either live in or around host burrows, or attach by the byssus threads directly to the host body surface (Morton and Scott 1989). Some do attach inside the shell carried by their host—hermit crabs (e.g. *Curvemysella paula*; Goto et al. 2007), inside the holothurian host esophagus (e.g. *Entovalva* spp.; Kato 1998), or inside the gill chamber of crabs (e.g. *Kurtiella pedroana*; Lafferty 1993; Boyko and Mikkelsen 2002; Bhaduri et al. 2017). Presumably they benefit from their habitat by both the anti-predator shelters and oxygen and food rich areas, e.g. in result of the water currents pumped through the tunnel by their host (Morton and Scott 1989; Boyko and Mikkelsen 2002; Goto et al. 2007). Goto et al. (2018) proposed, that evolution from sediment-dwelling to ectocommensal (or even endocommensal) lifestyle required a stage of a burrow-wall-commensal lifestyle.

Most species of the galeommatoid commensal bivalves are host specific (Table 5.1). Sato et al. (2011) summarized that among 29 ectocommensal species recorded from Japan, Korea and China, 23 were reported to attach directly on the body or shell of only one host species. Goto et al. (2018) tested if the obligate ectocommensal *Parabonia squilina* collected from abdomen of mantis shrimp *Lysiosquilla scabricauda* would reattach after removal—the bivalves crawled to the host and attached by the byssal threads once their foot reached the host pleon. Morton (1972) widely discussed the possible functional interpretation of the location of *Pseudopythina subsinuata* on its host stomatopod *Oratosquilla oratorio*—the bivalve is always found attached (and returns if detached) by its byssus to the lateral undersurface of the thorax of the host, possibly regarding the respiratory currents and cleaning behavior of the stomatopod. Similar observations were illustrated by Dall (1899) (in Morton 1972) for *Pseudopythina rugifera* attached to the undersurface of the mud shrimp *Upogebia pugetensis* (also *Naeromya rugifera* prefers this locality; Li and O'Foighil 2012). *Scintillona brissae* is always being found attached to the anus area of the echinoderm *Brissus latecarinatus* (Morton and Scott 1989), while *Waldo arthuri* chooses oral surface of the echinoderm *Brisaster latifrons* (Valentich-Scott et al. 2013). Wear (1966) describes that *Arthritica bifurca* attaches usually to the outer surface of polychaete *Pectinaria australis* tube—at the head end of it. *Sagamiscintilla thalassemicola* crawles to the inside of the proboscis gutter of the echiuran *Anelassorhynchus* sp., which is immediately closed and the bivalve is completely enclosed by the host tissues. Similarly, all the four members of the genus *Entovalva* live inside the esophagus of their holothurian hosts (Kato 1998; Bristow et al. 2010a). Such close relation may be granted by higher fitness of offspring, if the commensal bivalve evolves specialized adaptations to a specific host (cf. Reznick and Ghalambor 2001).

On the other hand some individual clam species may associate with multiple host species. Among them there are species associating with closely related hosts e.g. *Pseudopythina macrophtalmensis* is an associate of six species of *Macrophtalmus* crab (Sato et al. 2011), or *Pseudopythina subsinuata* live in a close commensal association with the stomatopod crustaceans—members of the genera *Oratosquilla*, *Harpiosquilla* or *Squilla* (Appukuttan 1972; Morton 1972). Others associate with even strikingly different hosts. For example, *Naeromya rugifera* associates with *Upogebia pugettensis* shrimp or the polychaete sea mouse *Aphrodita* spp. (Li and O'Foighil 2012). The latter example provides proof for the ecophenotypic plasticity of *N. rugifera*, whose shell differs morphologically depending on the host (but no host-specific genetic structuring was recorded). This finding is confirmed by the results of Sato et al. (2011), who found that *Koreamya arcuata* associated with two species of *Lingula* having different shell morphologies (differing mostly by the size of the adult shell), but DNA sequences of COI and ITS1 confirmed genetic similarity of both morphotypes. Another galaeommatid fits in this category—*Kurtiella pedroana* (formerly *Mysella*), which associates with different genera of mole crabs (*Blepharipoda occidentalis*, *Emerita analoga*) and hermit crab (*Isocheles pilosus*) settles on their exoskeletons, or—more commonly—inside their gill chambers, attached by their byssal threads, however it also is able to live free (Lafferty 1993; Boyko and Mikkelsen 2002; Carpenter 2005; Bhaduri et al. 2017; Table 5.1). Recently Bhaduri et al. (2017) described an unusual location for this species—inside the body of a mole crab. These authors discovered several adult clams (and some veligers) that were living directly in the hemocoel of *E. analoga*. These clams lacked byssal threads and were smaller than external bivalves. It is not clear, if the crabs' hemocoel is an accidental locality for the settlement of *K. pedroana*, neither how the bivalves infected this locality (likely as veligers), nor how they obtain oxygen or food (possibly from hemolymph of crab). It is ambiguous if they can reproduce. This bivalve is a simultaneous hermaphrodite with a female-only phase (Boyko and Mikkelsen 2002), thus enclosed veligers could be an offspring of enclosed bivalves, but it is not clear if enclosed veligers can leave the crab. The answer is presumably not for both reproduction and larvae evacuation, which is interpreted as a likely dead-end for the bivalve in the evolution of a host-parasite relationship. Furthermore no wound reactions or pathologies were associated with presence of internal bivalves, suggesting no immune reaction of the crabs to bivalves in their hemocoel. This example may be considered as a possible transition from a commensal to a possible endoparasitic lifestyle.

Similarly, although all other members of *Entovalva* genus are endocommensals of holothurians, *Entovalva nhatrangensis*, an associate in the esophagus of two holothurian species from Vietnam may be considered not as a commensal, but as a possible parasite (Bristow et al. 2010a). It is regarded as a most modified member of the genus by having its outer surface heavily folded, ensuring by this a large surface of contact with its host's tissues. This folded surface is presumed to serve as absorption/excretion organ. Its foot is curved and not movable, which ensures firmer contact with the host. The byssus is secreted through the gaps between folds and glue the bivalve to the host tissues. *E. nhatrangensis* is confirmed to reproduce in

Table 5.1 The list of examples of galeommatoid bivalves and their host organisms

Galeommatoid bivalve		Host animal				
Family	Species	No. of host species	Species	Taxon	Body/burrow/free living	References
Galeommatidae	*Divariscintilla luteocrinita*	1	*Lysiosquilla scabricauda*	Arthropoda	Burrow-wall	Goto et al. (2018)
	Divariscintilla octotentaculata	1	*Lysiosquilla scabricauda*	Arthropoda	Burrow-wall	
	Divariscintilla toyohiwakensis	1	*Bigelowina phalangium*	Arthropoda	Burrow-wall	
	Divariscintilla troglodytes	1	*Lysiosquilla scabricauda*	Arthropoda	Burrow-wall	Mikkelsen and Bieler (1989), Goto et al. (2018)
	Divariscintilla yoyo	1	*Lysiosquilla scabricauda*	Arthropoda	Burrow-wall	
	Divariscintilla aff. yoyo	1	*Lysiosquilla scabricauda*	Arthropoda	Burrow-wall	Goto et al. (2018)
	Pythina deshayesiana		–	–	Free-living	Bernard et al. (1993)
	Sagamiscintilla thalassemicola	1	*Anelassorhynchus mucosus*	Echiura	Body surface, enclosed by host tissues	Habe (1962), Goto and Ishikawa (2019)
	Scintilla (Lactemiles) strangei		–	–	Free-living	Li et al. (2012)
	Scintillona bellerophon	1	*Leptosynapta clarki*	Echinodermata	Body	
	Scintillona brissae	1	*Brissus latecarinatus*	Echinodermata	Body surface (close to anus)	Morton and Scott (1989)
	Scintillona zelandica	1	*Taeniogyrus dendyi*	Echinodermata	Body surface	Morton (1957)

(continued)

Table 5.1 (continued)

Galeommatoid bivalve		Host animal				
Family	Species	No. of host species	Species	Taxon	Body/burrow/free living	References
Lasaeidae	*Arthritica bifurca*	1	*Pectinaria australis*	Annelida	Outer tube surface (head-end)	Wear (1966)
			–	–	Free-living	Ponder (1965)
	Arthritica crassiformis	1	*Barnea similis*	Mollusca	Border of shell	Chanley and Chanley (2010)
	Barrimysia siphonosomae	1	*Siphonosoma cumanense*	Sipuncula	Burrow wall	Morton and Scott (1989)
	Borniola lepida		–	–	Free-living	Li et al. (2012)
	Curvemysella paula	2	*Diogenes edwardsii*	Arthropoda	Inside shell of hermit crab	Morton and Scott (1989), Goto et al. (2007), Morton (2017)
			Spiropagurus spiriger	Arthropoda	Inside shell of hermit crab	Goto et al. (2007)
	Devonia perrieri	1	*Leptosynapta inhaerens*	Echinodermata	Body surface	Bristow et al. (2010b)
	Devonia semperi (=*Entovalva semperi*)	1	*Protankyra bidenta*	Echinodermata	Body surface	Morton and Scott (1989), Goto et al. (2012)
	Entovalva amboinensis	1	*Patinapta laevis*	Echinodermata	Inside the esophagus (pouch in wall)	Sparck 1931 in Kato (1998)
	Entovalva lessonothuriae	1	*Holothuria (Lessonothuria) pardalis*	Echinodermata	Inside the esophagus	Kato (1998)
	Entovalva mirabilis	1	*Patinapta crosslandii*	Echinodermata	Inside the esophagus	Voeltzkow 1890 in Kato (1998)

Entovalva nhatrangensis	2	*Holothuria spinifera*	Echinodermata	Inside the esophagus (*possible parasite*)	Bristow et al. (2010a)
		Holothuria leucospilota	Echinodermata	Inside the esophagus	Bristow et al. (2010a)
Kellia sp.		–	–	Free-living	Li et al. (2012)
Kurtiella (formerly—*Mysella*) *pedroana*	3	*Blepharipoda occidentalis*	Arthropoda	Body (branchial chamber, outside surface)	Scott (1987), Lafferty (1993), Boyko and Mikkelsen (2002)
		Emerita analoga	Arthropoda	Body (lateroventral surface, branchial chamber, hemocoel) (parasite in hemocoel)	Bhaduri et al. (2017)
		Isocheles pilosus	Arthropoda		Carpenter (2005)
		–	–	Free-living (not preferred)	Scott (1987)
Kurtiella tumida (formerly *Rochefortia (Mysella) tumida)*	1	*Mesochaetopterus taylori*	Annelida		Li et al. (2012)
Mysella (Montacutona) compacta		Nestling in coral crevices; attached to the byssus of *Septifer virgatus*	Cnidaria; Mollusca	Coral crevices, byssus, free-living (?) on rocks	Morton and Scott (1989)
Mysella cuneata	1	*Phascolion strombi*	Sipuncula	Burrow walls	Gage (1968)
Mysella (Montacutona) mutsuwanensis	1	*Cerianthus filiformis*	Cnidaria	Tube of the burrowing sea anemone	Morton and Scott (1989)
Mysella vitrea	1	*Trypacea australiensis*	Arthropoda	Burrows (facultative)	Kerr and Corfield (1998)

(continued)

Table 5.1 (continued)

Galeommatoid bivalve		Host animal				
Family	Species	No. of host species	Species	Taxon	Body/burrow/free living	References
	Neaeromya rugifera	2	*Upogebia pugettensis*	Arthropoda	Body (ventral surface of abdomen)	Li and O'Foighil (2012), Li et al. (2012)
			Aphrodita spp.	Annelida	Body (ventral surface)	
	Parabornia squilina	1	*Lysiosquilla scabricauda*	Arthropoda	Body (abdomen—lateral part of pleonal sternite)	Boss (1965b), Goto et al. (2018)
	Pseudopythina (Borniopsis) ariakensis	1	*Protankyra bidentata*	Echinodermata	Burrow wall	Morton and Scott (1989)
	Pseudopythina macrophthalmensis	At least 6	*Macrophthalmus* spp.	Arthropoda	Body (lateral margins of the carapace and appendages)	Morton and Scott (1989), Sato et al. (2011)
	Pseudopythina nodosa	1	*Sipunculus nudus*	Sipuncula	Body surface	Morton and Scott (1989)
	Pseudopythina ochestostomae	2	*Ochestostoma erythrogrammon*	Echiura	Burrow wall	Morton and Scott (1989)
			Listriolobius sorbillans	Echiura	Burrow wall, body	Goto et al. (2012)
	Pseudopythina rugifera	2	*Upogebia pugetensis*	Arthropoda	Body (lateral parts of underthorax and abdomen)	Dall 1899 in (Morton 1972)
			Aphrodita refulgida	Annelida	Body (lateral parts)	Narchi (1969)
	Pseudopythina subsinuata	6	*Oratosquilla perpensa*	Arthropoda	Body (lateral parts of underthorax and abdomen)	Morton (1972)

			Oratosquilla oratorio	Arthropoda	Body (lateral parts of underthorax)	Morton (1972), Morton and Scott (1989)
			Oratosquilla interrupta	Arthropoda	Body (lateral parts of underthorax and abdomen)	Morton (1972)
			Harpiosquilla japonica	Arthropoda	Body (lateral parts of underthorax)	
			Squilla raphidea	Arthropoda	Body (lateral parts ventral)	Appukuttan (1972)
			Squilla nepa	Arthropoda	Body (lateral parts ventral)	
	Pseudopythina (Borniopsis) tsurumaru	2	*Protankyra bidentata*	Echinodermata	Burrow wall	Morton and Scott (1989)
			Hexapus anfraetus	Arthropoda	Body surface (occasionally)	
	Tellimya fujitaniana	1	*Echinocardium cordatum*	Echinodermata	Posterior body surface (juv) & burrow wall (adult)	Goto et al. (2016)
	Waldo arthuri	1	*Brisaster latifrons*	Echinodermata	(Oral surface)	Valentich-Scott et al. (2013)
	Waldo sp.	1	*Brisaster latifrons*	Echinodermata	Body	Li et al. (2012)
Montacutidae	*Koreamya arcuata*	3	*Lingula anatina*	Brachiopoda	Body (shell outside)	Lützen et al. (2009), Sato et al. (2011)
			Lingula translucida	Brachiopoda	Body (shell outside)	Sato et al. (2011)
			Lingula adamsi	Brachiopoda	Body (shell outside)	
	Koreamya setoucniensis	1	*Lingula anatina*	Brachiopoda	Body (shell outside)	Goto et al. (2014)
Basterotiidae	*Basterotia* sp.	1	*Ikedosoma gogoshimense*	Echiura	Burrow wall	Goto et al. (2011)

the host—the bivalve is a hermaphrodite, producing sperm as a young adult and becomes a female when older (Bristow et al. 2010a). Although it has not been determined to be a parasite or a commensal (*E. nhatrangensis* had diatoms in their gills, suggesting active filtration from the sea water), the example of *Entovalva* may indicate an evolutionary sequence of steps towards parasitism—one may suppose that being a commensal enclosed in the host body may facilitate the transition to parasitic mode of life, involving feeding on the surrounding host tissues.

Li et al. (2016) and Goto et al. (2012) provided a molecular phylogeny of Galeommatoidea indicating that many commensal lineages lack clade-specific host fidelity; they suggest host shifts between distantly related taxa. Nevertheless the same host species are known to be shared by some congeneric galeommatoidean species, which suggests the possibility of within-host speciation in these lineages (Mikkelsen and Bieler 1989; Goto et al. 2014, 2016, 2018).

5.1.3 Unionida

The freshwater pearly mussels, naiads, unionoids, members of the order Unionida, are characterized by their unique life cycle. Not only do their larvae receive parental care, developing in the females' gills, but they also must parasitize aquatic vertebrates, most often fish, to complete their development and to metamorphose into juvenile mussels (e.g. Kat 1984; Bauer 1994, 2001a; Graf and Cummings 2006; Barnhart et al. 2008; Strayer 2008; Patterson 2018).

All Recent unionoids are restricted to freshwater (e.g. Bauer 2001a; Graf 2013). They are relatively large (often over a dozen cm), long-lived (tens to a hundred years) pearly mussels, spending their lives as semi-infaunal burrowers in streams and lakes of all continents, except Antarctica (Bauer 2001a, c, d; Graf and Cummings 2007; Bogan 2008; Bogan and Roe 2008). Their fossil record can be confidently traced back to the Triassic (Good 1998; Watters 2001; Skawina and Dzik 2011; Van Damme et al. 2015), however today their diversity is estimated to be at least 840 (up to 958) species and is declining worldwide (Graf and Cummings 2007, 2021; Bogan and Roe 2008; Patterson et al. 2018). Adults are filter feeders through the activity of the cilia present on their paired gills, which results in water currents and the transportation of mucus with trapped food particles to the mouth of the animal. Gills of these mussels serve not only as a food-capturing organ (and gaseous exchange organ)—unionoids brood their larvae in the females' modified demibranchs—so-called marsupia (e.g. Lefevre and Curtis 1910a, Tankersley and Dimock 1992). It is believed that this is an adaptation to freshwater environments, where all the lotic habitats (flowing waters like streams or rivers; e.g. Pennak 1971) finish in marine conditions. Larval brooding and elimination of the typical free-swimming veliger prevent larvae from being transported to the sea (Graf and O'Foighil 2000; Wächtler et al. 2001; Graf and Cummings 2006; Graf 2013; Haag 2013). Because unionoids are rather large and heavy, their dispersal relies mostly on their minute larvae, which parasitise fish. Larvae (most commonly glochidia) must come in contact with a host soon after being released into the water otherwise they die (they usually survive from a few days to weeks; e.g. Murphy 1942; Jansen et al. 2001 and works there cited). Nevertheless their successful infection of the host commonly depends on the time of suspension in the water column, as sedimentation on bottom deposits limits their chances of host attachment (glochidia do not actively swim; Paling 1968). This is the most critical period in the life of unionids. Contact with fish tissue results in larval attachment and quick encapsulation by host tissues (in a few, up to a dozen hours; Lefevre and Curtis 1910b; Nezlin et al. 1994; Rogers-Lowery and Dimock 2006). In most species larvae are not growing on their host, however they do feed and produce storage granules (Wächtler et al. 2001). After some time (days to months) larvae metamorphose and actively detach from their host to start a juvenile period of life as fairly mobile deposit to filter feeding infaunal bivalves (Yeager et al. 1994; Jansen et al. 2001; Reis et al. 2014; Araujo et al. 2018). Later (months to years) juveniles emerge to the depositional surface, mature and spend the rest of their life as adult filter feeding semi-infaunal burrowers (Bauer 2001c).

5.2 Functional Anatomy of Adult Mussels and the Larvae

5.2.1 *Adults*

Adult unionoids usually produce a thick pearly shell. The thin mantle underlies it and secretes it. The mantle forms incurrent and excurrent apertures at the posterior end of the mussel (Fig. 5.2). The incurrent aperture has ciliated receptors on papillae

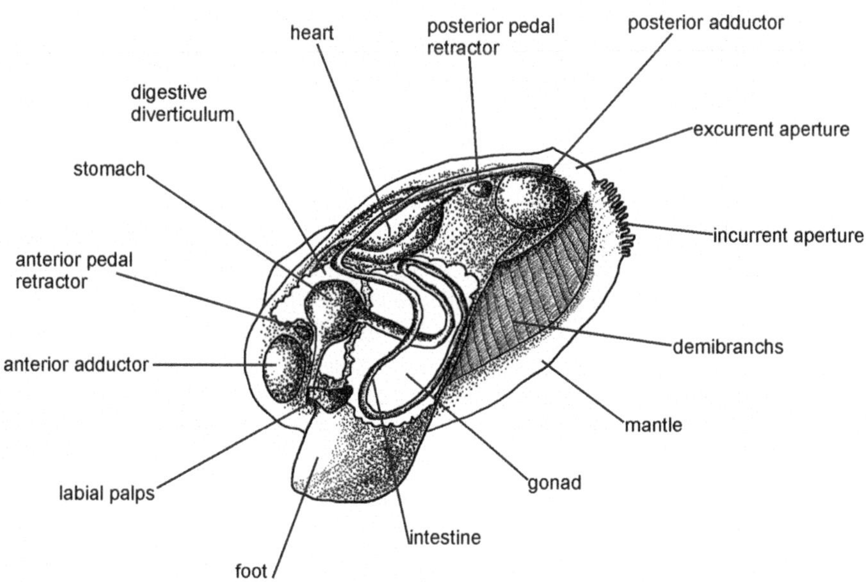

Fig. 5.2 Soft tissue anatomy of an adult freshwater mussel

(Petit et al. 1978; McElwain and Bullard 2014). In Lampsilini (Unionidae) the mantle forms mimetic flaps—lures (attracting fish for transmission of the glochidia; Zanatta and Murphy 2006, Barnhart et al. 2008). Within the mantle cavity, the gills (two paired demibranchs; in females modified into marsupia) are placed on each side of the visceral mass and a large foot. Adult Unionida use the foot for burrowing or for locomotion over short distances (Amyot and Downing 1998; Schwalb and Push 2007), while juveniles also use their ciliated foot for food collection (Gatenby et al. 1996; Schartum et al. 2017; Araujo et al. 2018). Adult unionoids are considered filter feeders. Their eulamellibranch gills are ciliated—the activity of these cilia results in water currents. The food particles that are suspended in the water are captured on the gills, trapped in the mucus to be later sorted and transported by labial palps to the mouth of animal (Allen 1921; Galbraith et al. 2009). The esophagus connects the mouth with the stomach and digestive gland, which surrounds the stomach in a visceral mass (McElwain and Bullard 2014). Food is transported into the blind-ended ciliated tubules of the digestive diverticulum where it is digested, then the non-digested particles pass through the stomach, are compacted by the crystalline style sac, and removed through the intestine (Owen 1955; McElwain and Bullard 2014). Particles not ingested by the digestive system are removed as "pseudofeces". Ciliary activity draws fresh water through incurrent aperture into the mantle cavity and then into the watertubes of the gills. Next, the water comes to the suprabranchial chamber and leaves the mantle cavity with the waste materials and "pseudofeces" throughout the excurrent aperture. During this passage through the gills haemolymph is oxygenated and carbon dioxide is released. Gills of the females serve as brooding chambers—the marsupium. Unionoids are usually dioecious, only some species are hermaphroditic, or occasionally hermaphroditic (Bauer 2001b; Watters 2007; Patterson 2018). Males release sperm into the water as spermatozoa, which can survive there up to 48 h (Ishibashi et al. 2000). The female draws the spermatozoa with the water to the mantle cavity and to the suprabranchial chamber, where fertilization occurs. The embryos and larvae are then brooded in the marsupium. The brooding negatively influences females' respiration (Richard et al. 1991; Tankersley and Dimock 1993a) and also filtration rate—due to functional changes in the ciliature of the marsupial demibranchs (Tankersley and Dimock 1993b; Tankersley 1996). Most species have marsupia in the outer demibranchs (species-rich family Unionidae), but e.g. Margaritiferidae and several species of Unionidae are tetragenous (which is interpreted as a derived condition; Graf and O'Foighil 2000), and members of families Hyriidae, Iridinidae, Mycetopodidae and Etheriidae are endobranchous (Graf and O'Foighil 2000; Bauer 2001b; Graf and Cummings 2006; Fig. 5.3). The larvae are numerous, small (0.06–0.4 mm), and must parasitize fish (with some exceptions) to complete metamorphosis (Kat 1984; Bauer 2001a; Watters and O'Dee 1998; Wächtler et al. 2001; Barnhart et al. 2008).

The nervous system consists of four ganglia, each with several nerve fibers extending throughout the body. Bilobed pedal ganglion innervates both the ventral pedal foot musculature and the lateral portions of the visceral mass. Paired cerebral ganglia are responsible for innervation of anterior adductor, retractor, foot and mantle edge, thus are required for complex motor functions. Visceral ganglion fibers

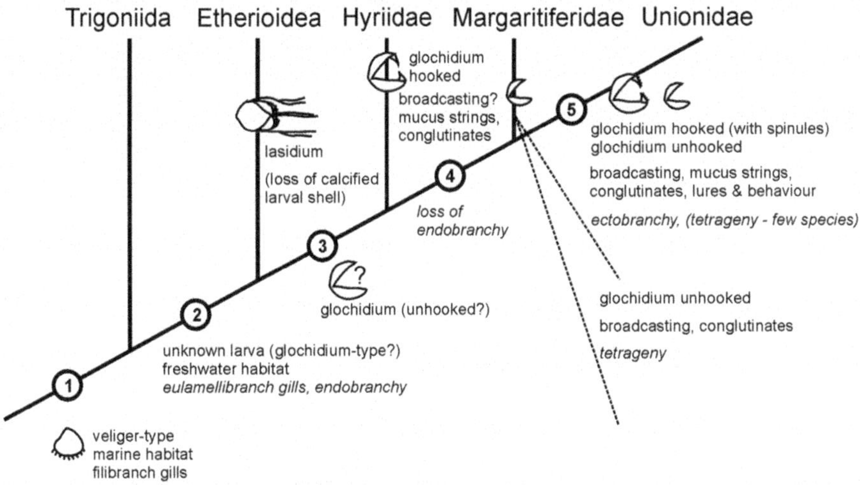

Fig. 5.3 A schematic reconstruction of scenario of evolution of parasitic and brooding (in italic) characters within Unionida presented on one of the recently published topology (e.g. Pfeiffer et al. 2019). 1. marine ancestor, with veliger larva; 2. freshwater habitat, brooding in internal pair of eulamellibranch demibranchs, unknown larva—possibly a kind of glochidium; in lineage to Etherioidea loss of calcified larval shell and development of lasidium; 3. Glochidium, possibly unhooked; development of hooks in Hyriidae; 4. Loss of endobranchy; brooding in all four demibranchs (Margaritiferidae); 5. Development of hooked (with spinules) glochidia; brooding in external demibranchs; development of diverse adaptations to host attraction, including lures & behaviour

extend to the base of the inner demibranchs and are responsible for closing the shell valves (McElwain and Bullard 2014).

The large anterior and posterior adductors are responsible for closing the shell. Other muscles attached to the shell are: anterior and posterior pedal retractors (which retract the foot), pedal protractor attached at the anterior area of the shell, and a set of tiny umbonal muscles playing role of a pedal elevator, attached within the umbonal cavity of the shell (Cox 1969b; Graf and Cummings 2006; Skawina and Dzik 2011; Carter et al. 2012).

5.2.2 *Larvae*

A glochidium is the most common type of larva within Unionida (Fig. 5.3). It is found in families Unionidae, Margaritiferidae and Hyriidae, thus in about 90% extant species within this order (Wächtler et al. 2001; Graf and Cummings 2007; Pfeiffer et al. 2019). The shell of the glochidium is slightly calcified and may carry hooks (Fig. 5.4) and denticles/spinules on them (e.g. Wächtler et al. 2001; Pimpão et al. 2012; Malchus and Sartori 2013). Their size, depending on species, vary

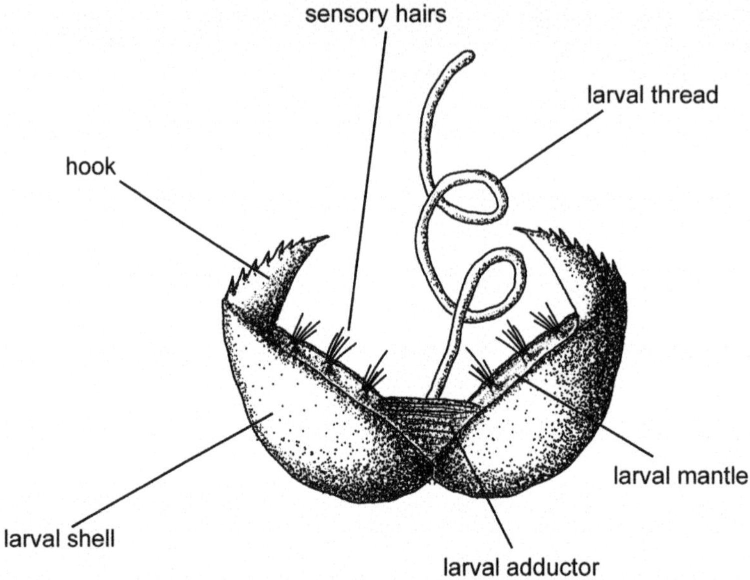

Fig. 5.4 Hooked glochidium morphology

between 60–400 μm in diameter (Wächtler et al. 2001; Pimpão et al. 2012). Pfeiffer and Graf (2015) listed five primary types of glochidia: unhooked; hooked with basal spines; S-shaped hooked; axe-head shaped; and bilaterally asymmetrical, while Wächtler et al. (2001) described them as "triangular, spherical or hatch-shaped". Hooked glochidia are usually larger and easily attach to both the gills and the skin (or fins; Davis and Fuller 1981) of a broad spectrum of fish, for a shorter time, while hookless glochidia are usually small and found on the gills of the host fish, where they encapsulate for a longer period of time (Bauer 1994, 2001b; Jansen et al. 2001; Barnhart et al. 2008). The valves of a glochidium are closed due to contraction of a single larval adductor (Lefevre and Curtis 1910b). On the soft body of a glochidium, sensory tufts/hairs are present (possible chemo and mechanoreceptors) and ciliary fields (possibly for respiration). Cells of larval mantle have microvilli, which are believed to gather nutrition from the host tissue after encapsulation (Nezlin et al. 1994; Wächtler et al. 2001).

Lasidium, the second type of larvae, is found in Mycetopodidae, Iridinidae and Etheriidae (Bogan and Hoeh 2000; Wächtler et al. 2001; Graf and Cummings 2006; Fig. 5.3). According to Wächtler et al. (2001), a lasidium is characterized by the lack of a calcified shell. Their size is about 85–330 μm in diameter (Fryer 1961; Wächtler et al. 2001; Graf and Cummings 2006). They attach to the host by anterior ciliated lobes and posterior adhesive structures, with hooks and sensory hairs (Fryer 1961; Wächtler et al. 2001; Barnhart et al. 2008; Silva-Souza et al. 2011). In most cases it forms a cyst in the fish epithelium. Depending on the species, it may parasitize a variety of fish or have a narrow host spectrum. The lasidium of *Mutela* however is not cyst-forming, but transforms into a subsequent larva stage, rooted in its host tissues (commonly fins of a cyprinid *Barbus altanialis radcliffi*) by a fork-like, long (up to 3 mm), nutrient transmitting stalk, a haustorium (Fryer 1959, 1961; Wächtler et al. 2001).

5.3 Life History, with the Review of the Adaptations to Attracting the Host

5.3.1 Life History

Unionoid bivalves live in lakes and rivers on all the continents, except Antarctica (Graf and Cummings 2007; Bogan 2008; Bogan and Roe 2008; Bolotov et al. 2016; Lopes-Lima et al. 2017). Their lack on this continent is believed as a result of climate limitation (Pugh and Scott 2002), possibly together with host limitation (no freshwater fish fauna lives in Antarctic freshwater today; Eastman 1993). Additionally, although most Antarctic lakes are today inhabited by various invertebrates (tardigrades, rotifers, crustaceans; McInnes and Pugh 1998; Pugh and McInnes 1998; Boxshall and Defaye 2008; Segers 2008), no freshwater mollusk is currently known from Antarctica (Pugh and Scott 2002). Antarctica has been an isolated glaciated continent for about the last 34–36 million years (DeConto and Pollard 2003; Carter et al. 2017), previously it was a part of Gondwana (together with Africa, Australia and South America), and before that—up to Middle Mesozoic it was a part of Pangea with all the rest of the continents (e.g. Eastman 1993; Jokat et al. 2003; Eagles and König 2008; Seton et al. 2012). Before isolation and glaciation freshwater habitats were well developed there, with numerous groups of freshwater fish (the possible hosts) that are preserved as fossils there today (Eastman 1993). However no unionoid fossil has so far been published from Antarctic freshwater fossil sites (to my best knowledge the only published freshwater bivalve fossils are Pliocene sphaeriids; Ashworth and Preece 2003). Even though one may suppose there is no reason for their absence in Antarctic freshwaters since the middle Mesozoic, until glaciation (Thompson and Stilwell 2010; Graf et al. 2015; Van Damme et al. 2015).

Unionoids are large (commonly over a dozen cm) and long lived (tens to a hundred years; Bauer 1992, 2001b; Mutvei and Westermark 2001). Juveniles spend the first years of their life completely burrowed within aerated sediments, feeding on particles present in interstitial water, until they emerge to the sediment surface at the size of about 1 cm to live as semi-infaunal burrowing filter feeders (Yeager et al. 1994; Gatenby et al. 1997; Bauer 2001b; Jansen et al. 2001; Reis and Araujo 2016; Araujo et al. 2018). After they mature (after several years), they were observed to group for spawning (Amyot and Downing 1998) and males release the spermatozoa to the water (Fig. 5.5).

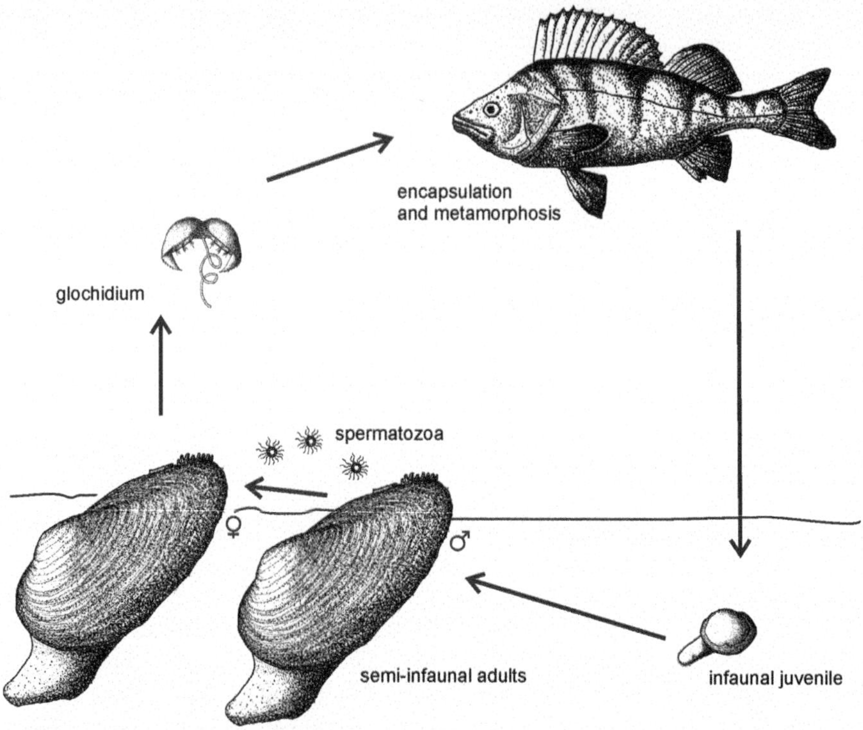

Fig. 5.5 Life cycle of freshwater mussels

After fertilization in females' suprabranchial chamber, females brood the larvae within marsupia of the modified gills—filled marsupia are more swollen thus in several species sexual dimorphism occurs in shape of the shell (e.g. *Epioblasma obliquata obliquata*; Patterson 2018). The duration of brooding period is short (weeks to several months) in tachytictic species (short-term brooders), or long (months, usually overwinter, with supplying larvae with nutrients) in bradytictic species (long-term brooders, considered a derived character; Wood 1974a; Graf and O'Foighil 2000; Bauer 2001b; Wächtler et al. 2001; Gascho Landis and Stoeckel 2016). Almost all known species of unionoids produce larvae, which temporarily must parasitize on fish to metamorphose into a juvenile and drop off to the sediments (Kat 1984). The exceptions include direct development (a few North and South American species; Lefevre and Curtis 1911; Parodiz and Bonetto 1963; Watters 1992; Dickinson and Sietman 2008), or parasites of amphibians (Watters and O'Dee 1998; Barnhart et al. 2008).

5.3.2 Adaptations to Host Infection

Larvae of Unionida do not actively swim (Paling 1968). Irrespective of the fact if they are host generalists or host specialists, unionoids have evolved diverse adaptations for parasitizing their hosts. These include: morphological modification of larvae (size, hooks, threads); physiological adaptations (e.g. ability to resist host immune response and larvae probably facilitate encapsulation); various adult strategies for host infection (morphological and behavioral). These adaptations have recently been reviewed in detail by Barnhart et al. (2008) and Haag (2012).

Broadcasting the larvae implies releasing free larvae to the water with no strategy to attract the host (Barnhart et al. 2008; Fig. 5.3). This strategy is considered to be the more primitive strategy by Haag (2012). It occurs in both the Margaritiferidae and some Unionidae, and is associated with huge mortality of the offspring (Bauer 2001b; Haag 2013). Soler et al. (2018) observed, that *Margaritifera auricularia* (kept in laboratory) in addition to massive broadcasting of free glochidia, releases “small white masses composed of larval material, including glochidia”, These masses remain close to the female and may act as a host attraction (conglutinates?), or are a result of respiratory stress (Aldridge and McIvor 2003; Haag and Warren 2003). Haag (2012) noticed that bivalves with a broadcasting strategy sometimes release larvae entangled in mucus strings or that larva have a larval thread, possibly to facilitate being suspended in water and to increase the chances for attachment to the host (Wood 1974b; found also in Hyriidae, Klunzinger et al. 2013). *Margaritifera margaritifera* and *Anodonta implicata*, however, are known to enhance the chances for host infection by synchronizing the timing of reproduction with the presence of their host (Davenport and Warmuth 1965; Barnhart et al. 2008). Several other species of *Anodonta* can recognize the chemical cues when host fish are near and only then release glochidia (Jokela and Palokangas 1993; Haag and Warren 2000; Bauer 2001b). Vicentini (2005) reported an unusual behavior of *Unio crassus* broadcasting larvae by vigorously spurting out glochidia, causing the water to splash.

Hyriidae and many Unionidae, together with mentioned above Margaritiferidae produce so-called conglutinates. Lefevre and Curtis (1912) who introduced this term, defined conglutinates as masses of eggs (and later larvae) molded into the shape of the watertubes in marsupia. These packages of larvae are formed in the female’s marsupia and often mimic preferred food items of the host (Kat 1984; Haag et al. 1995; Bauer 2001b; Walker et al. 2001; Barnhart et al. 2008; Haag 2013). They also prolong the time of survivorship for glochidia. Mussels that produce conglutinates often produce less glochidia than broadcasting ones (Barnhart et al. 2008). These egg aggregations vary morphologically from irregular and fragile that break during release (in Margaritiferidae), through leaf-like fragile and easily breaking ones (in most conglutinate producing Unionidae), to the reinforced by unfertilized, structural eggs that keep the shape of the aggregation (in tribes Pleurobemini and Lampsilini, Unionidae). The latter conglutinates often mimic the prey of the host (Barnhart et al. 2008).

Active host attraction is present in numerous members of tribe Lampsilini (Unionidae).

Within this tribe some species (e.g. *Lampsilis* spp.) attract the host with mantle lures that resemble hosts' prey. After the fish touches (attacks) the mantle, the mussel releases the glochidia (Haag et al. 1999; Haag and Warren 2000; Bauer 2001b; Zanatta and Murphy 2006; Barnhart et al. 2008). Female *Epioblasma* trap the head of small fishes, which were attracted by the exposed mantle, by the quick closure of the valves of their shell. While the fish is trapped, the mussel releases free glochidia directly into its mouth, which enhances chances of infestation (Zanatta and Murphy 2006; Barnhart et al. 2008). However, fish often die as their head may be crushed, so Barnhart et al. (2008) hypothesized that this strategy might aid in attracting other fish using trapped fish as a lure.

5.4 Phylogeny of the Unionida

5.4.1 Origin of Unionida

Bivalves originated in the oceans not later than the Early Cambrian (Jell 1980; Cope 1996; Pojeta 2000; Parkhaev 2008; Elicki and Gürsu 2009; Bieler et al. 2013). Since then they have invaded freshwaters multiple times (Gray 1988; Graf 2013). Bivalves that are classified within the order Unionida have been present in freshwater habitats at least since the Late Triassic (Reeside 1927; Good 1998; Watters 2001; Graf and Cummings 2006; Skawina and Dzik 2011; Van Damme et al. 2015), and presumably even earlier (Yates et al. 2012; Fig. 5.6). Unionoids are accepted as a monophyletic lineage, which is supported by several characters including their restriction to freshwater, brooding larvae in marsupia, and obligatory parasitic larvae of fish (with a few exceptions; Hoeh et al. 1998, 2001; Graf and Cummings 2006; Whelan et al. 2011). Members of the marine order Trigoniida, which is only represented today by the relict Australian genus *Neotrigonia*, are considered to be the closest relatives and the sistergroup of Unionida (Hoeh et al. 1998, 2001; Giribet and Wheeler 2002; Graf and Cummings 2006; Combosch et al. 2017; Lopes-Lima et al. 2017; Fig. 5.7). Newell and Boyd (1975) trace back the split between Unionida and their closest relatives, the mostly brackish-water trigoniids classified in the family Pachycardiidae, to the Triassic.

5.4.2 Classification and Diversity

Members of the order Unionida are traditionally classified in six families: Unionidae (674–753 species, and most widespread worldwide), Margaritiferidae (12–17 species, mostly Holarctic), Hyriidae (71–92 species, Neotropical and Australasian),

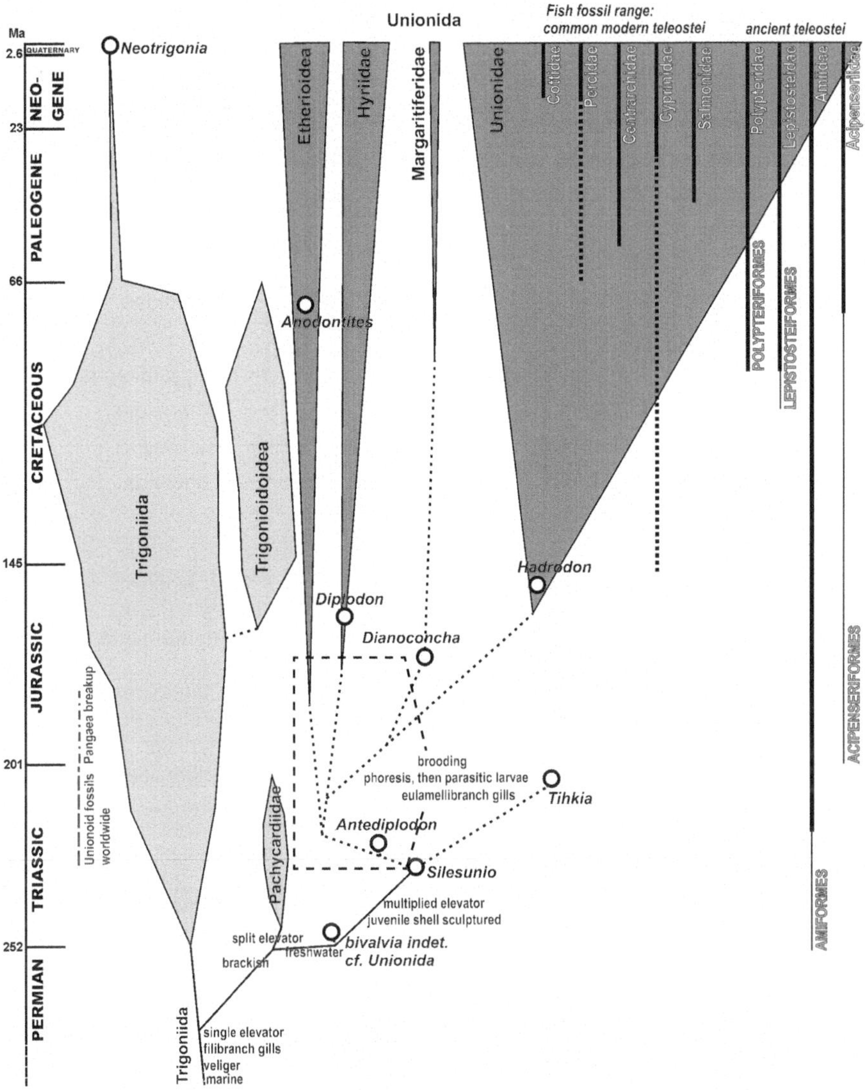

Fig. 5.6 The proposed scenario of evolutionary transformations leading from the trigonioids to the main lineages of the unionoids. Brooding, eulamellibranch gills and phoresis, followed by development of a parasitic relationship with host appeared after the origin of unionoids and before the Middle Jurassic. The time range of crown unionoid lineages is based on divergence time estimates (after Bolotov et al. 2016; Bolotov et al. 2017b; Huang et al. 2018), with examples of fossil occurrences indicated. The recent species diversity of Unionida follows estimations of Graf (2013). The fossil range of common modern and ancient teleost fish, the possible hosts, are indicated (bold lines—ranges of the families; thin lines—ranges of orders). The diversity of the Trigoniida on the level of known genera is assessed according to Newell and Boyd (1975; not in scale to Unionida). Dotted lines indicate possible but unclear estimates

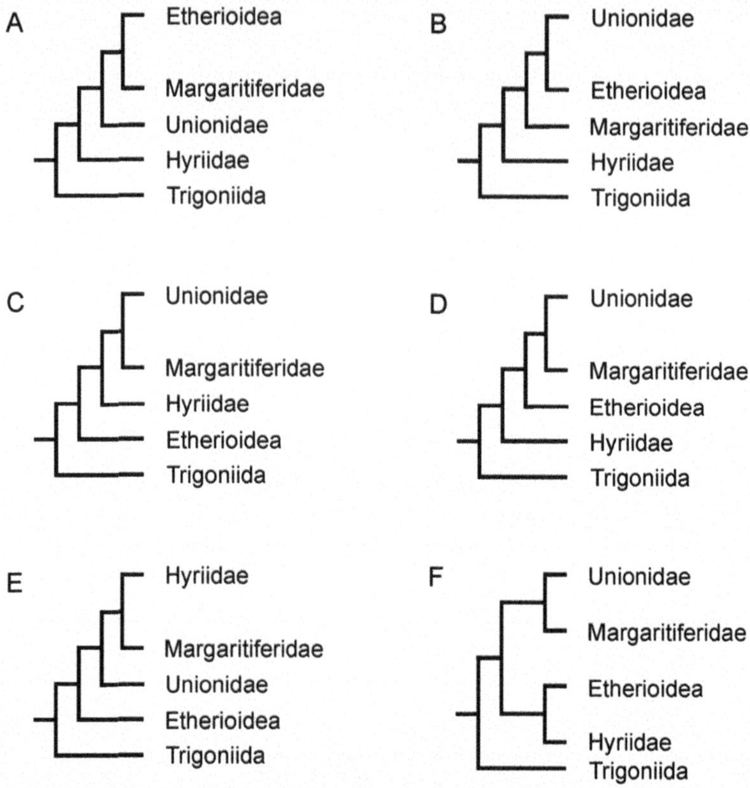

Fig. 5.7 Recently published phylogenetic hypotheses of Palaeoheterodonta (Unionida + Trigoniida), based on molecular, or molecular and morphological data, simplified to the family level (with superfamily Etherioidea = Etheriidae + Mycetopodidae + Iridinidae). (**a**) Combosch et al. (2017): H3, COI, 16S, 18S, 28S; Pfeiffer et al. (2018): COI, 28S; (**b**) Whelan et al. (2011): COI, 28S (maximum likelihood); Lopes-Lima et al. (2017): COI, 28S; (**c**) Graf and O'Foighil (2000): COI, morphology; Bolotov et al. (2017): COI, 16S, 28S; Guerra et al. (2017): mtDNA; Huang et al. (2018): mt DNA; Huang et al. (2019): mtDNA; Pfeiffer et al. (2019): multi-locus nuclear DNA via anchored hybrid enrichment probe (maximum likelihood); (**d**) Hoeh et al. (2001): COI, morphology; Graf et al. (2015): COI, 16S, 28S; Pfeiffer and Graf (2015): COI, 16S, 28S; Bolotov et al. (2016): 16S, 18S, 28S, D3; Huang et al. (2018): COI, 16S, 18S, 28S, H3; Huang et al. (2019): COI, 28S, 16S; (**e**) Whelan et al. (2011): COI, 28S (maximum parsimony); Pfeiffer et al. (2019): multi-locus nuclear DNA via anchored hybrid enrichment probe (coalescent-based ASTRAL); (**f**) Graf and Cummings (2006): COI, 28S, morphology

Etheriidae (4 species, Tropical), Mycetopodidae (36–53 species, Neotropical) and Iridinidae (39–43 species, Afrotropical; Wächtler et al. 2001; Graf and Cummings 2007, Bogan 2008; Bogan and Roe 2008; Graf 2013). Yet, currently Graf and Cummings (2021) estimated total diversity of Recent unionoid to 958 species. The most species-rich family Unionidae is today composed of four to seven subfamilies (e.g. summary by Pfeiffer et al. 2019), according to Whelan et al. (2011): Tropical African and Asian Parreysiinae, mostly Holarctic Unioninae, Indotropical

Rectidentinae, Nearctic and Neotropical Ambleminae (comprising Lampsilini and Pleurobemini in which advanced strategies of host attraction evolved), together with mostly Holarctic paraphyletic assemblage of species "Gonideinae" (Whelan et al. 2011; Bolotov et al. 2017a, b; Lopes-Lima et al. 2017; Huang et al. 2019; Pfeiffer et al. 2019). According to Bolotov et al. (2017a) the most species rich family Unionidae possibly originated during the Middle Jurassic in an evolutionary hotspot in Southeast and East Asia, then colonized North America (not earlier than in mid-Cretaceous), and later Europe and India (since Paleocene).

5.4.3 Phylogeny

The phylogeny of Unionida is under active research, and no consensus about the phylogenetic relationships among families is reached yet (Fig. 5.7). Nevertheless recently Combosch et al. (2017) for bivalvia (including unionoids) and Lopes-Lima et al. (2017) for Unionidae (and the other families of the order) proposed a phylogenetic scenario at a family level on the basis of molecular datasets. They both confirm trigoniids as a sister group to monophyletic unionoids, and Hyriidae as sister group to all the other unionoid families (Fig. 5.7a, b). Similar results were obtained by e.g. Hoeh et al. (2001), Whelan et al. (2011), Graf et al. (2015), Pfeiffer and Graf (2015), Huang et al. (2018, 2019), although they differ in reconstructing phylogenetic relations between the remaining five families (Fig. 5.7a, b, d). Most authors however place Unionidae as sister group to Margaritiferidae (Fig. 5.7c, d, f). Above reconstructions require that families that have glochidia (Hyriidae, Unionidae and Margaritiferidae) are paraphyletic, and a glochidium is an ancestral larva for lasidium. The alternative is that glochidia-bearing families are monophyletic as well as lasidia-bearing families (superfamily Etherioidea). This alternative was erected (on the basis of morphology and life history) by Parodiz and Bonetto (1963), who directly excluded a possibility that such different larvae could be derived from each other. This scenario was recently supported by several authors on the basis of analyses of the mitochondrial genome (Bolotov et al. 2017b; Guerra et al. 2017; Pfeiffer et al. 2019) which in unionoids (like in some other bivalves) may come from both sexes of the parents (DUI: doubly uniparental inheritance) depending on the method.

There are however differences depending on the method (Whelan et al. 2011; Huang et al. 2018, 2019; Fig. 5.7c, e). Also, Combosch et al. (2017) noticed that assessing the main divisions within the Unionida is sensitive to methodologies and needs more data.

The molecular clock—the idea of dating the separation of two taxa on the basis of their molecular sequences (more precisely the number of substitutions which accumulated between their DNA sequences) provides a powerful tool in estimations of phylogeny of taxa (Warnock and Engelstädter 2021). Nevertheless, because molecular clocks vary between evolutionary lineages, such phylogeny usually requires the comparison of clock-based data with the other sources of time information, e.g. fossils (Pulquerio and Nichols 2007). Recently, several attempts at the estimation of divergence of evolutionary traits of unionoid lineages were performed

Table 5.2 The molecular clock estimates indicating the latest possible date of origin of main lineages of Unionida

Family	Estimation of earliest origin of crown group (**fossil-calibrated: bold**)	References
Margaritiferidae	~175 Ma	Graf et al. (2015)
	93 Ma	Bolotov et al. (2016)
	~80 Ma	Bolotov et al. (2017b)
	88 Ma	Huang et al. (2018)
Unionidae	~175 Ma	Graf et al. (2015)
	125 Ma	Bolotov et al. (2016)
	177 Ma	Bolotov et al. (2017a)
	158 Ma (Oriental species)	Bolotov et al. (2017b)
	166 Ma	Huang et al. (2018)
Hyriidae	194 Ma	Graf et al. (2015)
	209 Ma	Bolotov et al. (2016)
	225 Ma	Santos-Neto et al. (2016)
	~110 Ma	Bolotov et al. (2017b)
	178 Ma	Huang et al. (2018)
Etherioidea (Iridinidae + Etheriidae + Mycetopodidae)	~165 Ma	Graf et al. (2015)
	186 Ma	Bolotov et al. (2016)
	~100 Ma	Bolotov et al. (2017b)
	251 Ma	Huang et al. (2018)

using this method (Table 5.2). Depending on the data used (different molecular markers and their number—both nuclear and mitochondrial, the number of specimens from each family used for obtaining molecular material) time estimations vary for indicating the origin of the crown group of six unionoid families. The origin of Margaritiferidae most commonly is assessed as in Late Cretaceous (~80–93 Ma; Bolotov et al. 2016, 2017b; Huang et al. 2018) however estimations twice as old have been made (however, this is not fossil-calibrated; Graf et al. 2015). Unionidae, which have recently received increased attention, are estimated to appear between the Early Jurassic and Early Cretaceous (most commonly 177–158 Ma; Bolotov et al. 2017a, b; Huang et al. 2018). Hyriidae most likely appeared between Late Triassic and Early Jurassic (178–225 Ma; Graf et al. 2015; Bolotov et al. 2016; Santos-Neto et al. 2016; Huang et al. 2018). Members of the superfamily Etherioidea presumably originated as early as Early Triassic (Huang et al. 2018) to Late Cretaceous (Bolotov et al. 2017b), with Early and Middle Jurassic estimations (Graf et al. 2015; Bolotov et al. 2016).

5.5 Evolution of Parasitism

Little direct fossil evidence is available to constrain the evolution of parasitism between unionoids and fishes. According to Boucot and Poinar (2010) the best evidence to interpret fossil behavior (including the parasite-host relation) requires both preservation of an organism while doing something allowing for a functional interpretation (e.g. direct association with its hosts). Fossil data like this exist

(Boucot and Poinar 2010; De Baets and Littlewood 2015; Leung 2017; De Baets et al. 2021) but are comparatively rare (e.g. nematodes protruding from insects; Poinar 2015). Unfortunately, mussel-fish interactions still lack such evidence. This might however be a matter of time—entirely soft-bodied parasites or their sclerotized hooks have been found associated with gills of their fish hosts as well as shipworm with a strongly reduced shells (Cressey and Patterson 1973; Cressey and Boxshall 1989; Upeniece 2001; De Baets et al. 2015; Robin et al. 2018). The fossilization potential might depend on the microenvironment building in the gills during decay (e.g. decaying tissue provides phosphates, drop of pH secures stability of calcium phosphate in common phosphatization of soft tissues) as well as on microbial activity (Martill and Harper 1990; Briggs and Kear 1993; Briggs et al. 1993; Wilby and Whyte 1995; Briggs and Wilby 1996; Wilby and Briggs 1997; Briggs 2003; Briggs and McMahon 2016; Purnell et al. 2018). According to Purnell et al. (2018) processes of decay, maturation, and mineralization are controlled by diverse factors that vary spatially and temporally in the way they act during these processes. As a result, characters of extraordinarily preserved fossils may not entirely correspond to what simple experimental decay data predict (there is a need to better understand the fossilization process); also the effects of processes of preservation may be selective to tissue types and may impose on the results of simple decay experiments. Various traits (size/morphology/habitat) could however be indicative/suggestive of a parasitic lifestyle.

5.5.1 Parental Care

The fossil record of unionoids ranges back to the Triassic, however little is known about their early evolution, because their shells are of rather generalized morphology (e.g. McMichael 1957; Good 1998; Watters 2001; Skawina and Dzik 2011; Van Damme et al. 2015). During this time all the present continents formed the supercontinent of Pangaea (Dubiel et al. 1991; Seton et al. 2012)—this suggests a possible worldwide dispersal of the early unionoids. In the Late Triassic of Poland, fossil unionoids occur that preserved characters important for unionoid taxonomy: characteristic juvenile sculpture on the shell (post-larval), muscle attachment scars and phosphatized gills (Skawina and Dzik 2011, Skawina 2010). This is the oldest data on the anatomy of the organ, which is today crucial for unionoid reproduction although it isn't necessary for developing a parasitic life mode. The Late Triassic unionoids had filibranch gill anatomy, but no evidence is available regarding possible brooding of larvae in these bivalves (Skawina and Dzik 2011). Although today in freshwater habitats bivalves commonly exhibit parental care (a convergence in Unionida, Sphaeridae, Cyrenidae and some Dreissenidae; Graf and O'Foighil 2000; Morton and Puljas 2012; Graf 2013; Mansur et al. 2019), none of these bivalves have filibranch grade gills. All of them brood their offspring in modified eulamellibranch demibranchs (Unionida, Cyrenidae, *Congeria* [Dreissenidae]), or first within the demibranchs, then in the mantle cavity (*Rheodreissena* [Dreissenidae]).

Filibranch gills are present in both fossil (Klug et al. 2005) and extant trigoniids (Tevesz 1975; Morton 1987), which is widely accepted as the sister group and hypothesized ancestor to freshwater unionoids (Newell and Boyd 1975; Hoeh et al. 1998, 2001; Waller 1998; Giribet and Wheeler 2002; Graf and Cummings 2006; Combosch et al. 2017; Lopes-Lima et al. 2017). Regarding anatomical similarities, members of both orders share similar ctenidial ciliary patterns (Tevesz 1975) and have chitinous gill supports, which are reinforced with calcium phosphate concretions (Ridewood 1903; Atkins 1938; Morton 1987; Silverman et al. 1987a, b, 1989; Whyte 1991; Wilby and Whyte 1995; Hinzmann et al. 2015).

The presence of a veliger in *Neotrigonia margaritacea* (a member of the only extant genus of trigoniids) has still not been confirmed but is presumed by most authors (Tevesz 1975; O'Foighil and Graf 2000; Graf and Cummings 2006). Atypical for marine bivalves (e.g. Mackie 1984)—*Neotrigonia margaritacea* likely lacks a typical veliger larva and its prodissoconch morphology indicates a non-planktotrophic pre-juvenile ontogeny (O'Foighil and Graf 2000), suggesting a somewhat derived evolutionary status. O'Foighil (1986) noticed, that parental care of young among bivalves is associated with a flattened prodissoconch. *Neotrigonia margaritacea* has an inflated prodissoconch, which suggests that it is a broadcast spawner, but it is unknown whether its pre-juvenile development is benthic or pelagic (O'Foighil and Graf 2000).

Richard et al. (1991), followed by Watters (2001) and Haag (2012), noticed that the presence of calcium concretions in mineralized gills of fossil unionoids from the Jurassic onwards may suggest parental care, as today these concretions serve as a source of calcium for glochidial shells (Silverman et al. 1985, 1987a, b; Machado and Lopes-Lima 2011; Hinzmann et al. 2015). One may indicate that such structures are today present both in trigoniid and unionoid gills (e.g. Atkins 1938), as well as are believed to exist also in fossil members of both orders, facilitating their gill mineralization (documented by specimens with phosphatized gills from the Triassic, Jurassic and Cretaceous; Richard et al. 1991; Simone and Mezzalira 1993; Klug et al. 2005; Skawina 2010, Whyte 1991; Wilby and Whyte 1995; Skawina, personal observations on Rhaetian *Tihkia silesiaca*, specimen ZPAL V33/305). Both the trigoniids and the oldest unionoids share filibranch grade gill anatomy, which lacks the tissue fusion of the gill filaments seen in eulamellibranch gills (which should be preserved, if it was present, e.g. Simone and Mezzalira 1993; Skawina 2010; Knight et al. 2014). In filibranch gills, filaments are connected instead only through the association of ciliary discs (e.g. Ridewood 1903; Morton 1987) making incubation of larvae in suprabranchial cavity or in interlamellar spaces serving as a marsupium impossible (Mackie 1984). Nevertheless Mackie (1984) noticed that marine bivalves with filibranch gills may incubate embryos within their mantle cavity (infrabranchial cavity, e.g. *Philobrya munita*; Morton 1978), which does not exclude a similar intermediate status in the early unionoids.

It is possible, that early unionoids established brooding not only for protection of their larvae (e.g. against water currents, osmotic and nutrient poor environment, Gray 1988; Graf and O'Foighil 2000; Graf and Cummings 2006; Graf 2013; Graf et al. 2015), but also to ease embryonic shell formation by the close contact with maternal calcium concretions inside their gills, while spending their lives in a habitat with low availability of environmental calcium (Silverman et al. 1985, 1987a, b; Richard et al. 1991; Machado and Lopes-Lima 2011; Hinzmann et al. 2015). As far as the advantages of maternal care in unionoids seems to be resolved, there are still open questions such as when did unionoids evolve eulamellibranch gills allowing them to incubate larvae in interlamellar spaces, and independently—from when onwards did their larvae became parasites?

5.5.2 *Parasitic Larvae*

All the extant members of the Unionida have parasitic larvae (glochidium or lasidium—sometimes with haustorium; e.g. Kat 1984; Wächtler et al. 2001; Graf and Cummings 2006; Pfeiffer et al. 2019; Fig. 5.3), brooded in the female gills—in marsupium, until they are released to parasitize their host. It is not clear when this type of larvae and their parasitism arose. The oldest fossil glochidia are found as late as the Miocene (Gross and Piller 2018), Pleistocene (Brodniewicz 1968), and some younger deposits (Aldridge and Horne 1998). If the behavior of the extant mussels has been similar in the past (Boucot and Poinar 2010; Nagler and Haug 2015), one may expect finding fossils with glochidia while encapsulated on fish gills, skin or fins. This could provide direct fossil evidence of the evolution of this relationship. The preservation of fossilized soft tissues of fishes (most often muscles, but including the gills or skin) is not rare in several so-called Konservat-Lagerstätte (Wilby and Briggs 1997). The detailed preservation of soft tissue of fish in marine Jurassic Solnhofen Lagerstätte (Germany) allowed identification of pathologies—skin lesions (Petit and Khalloufi 2012); such lesions are also recognizable even at the cellular level on mineralized fish from the marine Jurassic Osteno Lagerstätte (Italy; Pinna 1985). Unfortunately freshwater sites like this are exceptional (e.g. review by Cavin 2017), but in brackish deposits of the Romualdo Member of the Santana Formation of Brazil (Cretaceous) three-dimensionally preserved fishes (most commonly *Rhacolepis* sp.) with mineralized soft tissues occur (e.g. Martill 1988; Martill and Harper 1990; Wilby and Martill 1992; Maldanis et al. 2016; Osés et al. 2017). On the gills of *Cladocyclus* sp. from this Formation tiny fossil parasitic copepods are preserved (Cressey and Patterson 1973; Cressey and Boxshall 1989). Bruno and Hessel (2006) report, that *Cladocyclus* sp. could be tolerant to variations in salinity and thus could be contaminated with marine parasites as copepods, thus this does not exclude possibility of carrying glochidia as well. Glochidia of extant *Glebula rotundata* (Lampsilini, Unionidae) were found on two estuarine fish species (*Trinectes maculatus* and *Anchoa mitchilli*; Parker et al. 1984). Although there is no morphological evidence of glochidiosis in mineralized gills of *Rhacolepis* sp. or *Cladocyclus* sp., as well as to my knowledge no unionoid fossils were discovered from this formation, the resolution of preservation together with the freshwater depositional environment are promising (Martill and Harper 1990). Additionally, in the underlying lacustrine deposits of the Crato Formation, the freshwater fish *Dastilbe* is commonly found (sometimes with mineralized soft tissues; Davis and Martill 1999), as well as other freshwater fish (e.g. amiiforms, semionotiforms; Brito et al. 2008). Bruno and Hessel (2006) list *Pseudohyria*, in freshwater layers (however they do not picture it)—a representative of Trigonioidoidea (Unionida or Trigoniida member, according to Carter et al. 2011). Recently a possible member of Unionida was also discovered in deposits of this Formation (da Silva et al. 2020).

5.5.3 Origin and Evolution of Unionoid Parasitism

5.5.3.1 Phoresis

The great advantage of dispersal and upstream transport using the fish might have been an important selective force during the early evolution of unionoids (e.g. Graf 1997; Watters 2001; Barnhart et al. 2008; Strayer 2008). Although all the extant members of this order possess both parasitic larvae and the eulamellibranch gills (Wächtler et al. 2001; Graf and Cummings 2006, Fig. 5.3), one may expect these characters already existed in their last common ancestor. So probably after the origin of the Unionida but before the Middle Jurassic, when advanced unionoids already existed (based on their characteristic, double-looped or V/W shaped [according to Zieritz et al. 2015] umbonal sculpture; Watters 2001; Skawina and Dzik 2011; Bolotov et al. 2016, Fig. 5.6). The material from the early Late Triassic (specimen ZPAL AbIII/2210, Fig. 2a in Skawina and Dzik 2011) preserve juvenile shell morphology which indicates an early postlarval development similar to extant unionoids, but unfortunately the preservation of the area of larval shell does not allow discussion about the presence of glochidium (Skawina and Dzik 2011). According to Malchus and Sartori (2013), the early ontogenetic shells (prodissoconch = glochidia) from fossil unionoids are currently unknown, also because juvenile part of the shell frequently suffers from abrasion (e.g. McMichael 1957; McMichael and Hiscock 1958).

According to Nagler and Haug (2015), who discussed direct and indirect evidence which allow parasitism to be inferred in the fossil record, among others, that phylogenetic inference may give some indications. Following this rationale, the genera of unionoids, which live today, possibly had a parasitic larvae at their origin. Thus estimations on the basis of molecular clock, or direct fossil evidences of the origin of Recent genera might cautiously set the minimum time constraints (e.g. *Margaritifera* [Margaritiferidae] about 70 Ma; *Parreysia* [Unionidae]—45 Ma; *Diplodon* [Hyriidae] about 130 Ma; *Etheria* [Etheriidae]—150 Ma; Graf et al. 2015; Bolotov et al. 2016; Santos-Neto et al. 2016; Bolotov et al. 2017b; Huang et al. 2018). The maximum age assessment could possibly be related to the origin of Recent families of the crown unionoid group (Table 5.2), as all the glochidia-bearing and lasidia-bearing unionoids were recently confirmed to be reciprocally monophyletic by mitogenomic phylogenetic reconstructions (Guerra et al. 2017; Huang et al. 2018, 2019; Pfeiffer et al. 2019). Thus, presumably one may assume, that stem-group unionoids, or even their marine/brackish waters ancestors, had a character which eased acquiring a parasitic relation with fish after they invaded freshwaters, which led to the independent development of recent lasidia and glochidia larvae.

On the other hand, numerous previous morphological and molecular phylogenies have consistently supported that glochidia-bearing mussels are paraphyletic and that glochidia are ancestral larvae for lasidia (e.g. Bogan and Hoeh 2000; Hoeh et al. 2001; Graf and Cummings 2006; Whelan et al. 2011; Graf et al. 2015; Pfeiffer and Graf 2015; Bolotov et al. 2016, 2017a; Lopes-Lima et al. 2017; Huang et al. 2018, 2019). Assuming that a parasitic relation requires a unique set of characters, one may hypothesize that unionoids obtained this ability once and early in their evolution, followed by this ability becoming ubiquitous and a selective force shaping their further evolution (e.g. Watters 2001; Barnhart et al. 2008)—then modification of the larva type appeared, which might support the latter hypothesis.

Unionoids are today typically large and heavy with limited mobility, thus their dispersal relies on their minute larvae that attach to the fish (Schwalb and Push 2007; Schwalb et al. 2011b; Graf 2013; Patterson 2018). If the size of an adult is a sufficient clue for a phoretic relationship similarly large mussels (about at least 8 cm in length), members of the genus *Tihkia*, were discovered by the end of Triassic: Carnian of Manda Beds of Tanzania and Ntawere Formation of Zambia (Cox 1969a; Peecook et al. 2017), Carnian/Norian of Maleri Formation of the Hyderabad State and the region of Tihki, Vindhya Pradesh, India (Sahni and Tewari 1958); early Rhaetian of the Upper Silesia, Poland (Skawina and Dzik 2011). On the other hand, Bauer (2001d) suggested the possibility that the development of parasitism, with the cost of the high mortality of larvae that occurs during transition to the host (Jansen et al. 2001), might have been a selective factor for a larger size (and thus fertility).

Watters (2001) suggested that ancestral unionoids probably had fewer, large larvae, similar to larvae of the extant Sphaeridae. Findings of Tevesz (1975) may support this proposition as the trigoniid *Neotrigonia gemma*, sectioned by him, had "large, white, irregularshaped eggs", but exhibited presumably external fertilization. The ancestral larvae of unionoids—presumably a kind of veliger—might have been released either freely, or entangled in mucous packages. Larvae might be predated by fish and this could have been the start of the contact with the fish body, as today sticklebacks (*Gasterosteus aculeatus*) actively feed on large, individual glochidia of *Anodonta cygnea* (Dartnall and Walkey 1979). Nevertheless this is a rare observation—glochidia are usually not a part of a fish diet today (Jansen et al. 2001).

Glochidia of several extant species have adhesive larval threads, which can act as an attachment organ (Wood 1974b). Wächtler et al. (2001) report that in European species this feature varies in closely related species like *Anodonta anatina*, *A. cygnea* (present) and *Pseudoanodonta complanata* (absent), or *Unio tumidus*, *U. pictorum* (present) and *U. crassus* (absent). It is also absent in *Margaritifera margaritifera*. They presume it is related to the lentic environments (streams, rivers). Nevertheless threads of many glochidia are capable of forming a floating web, increasing their chances of host infestation and limiting the possibility of larvae sinking to the sediments (Wood 1974b). Barnhart et al. (2008) and Haag (2012) found this structure in many species within family Unionidae, but not in tribe Lampsilini—a group that developed more specialized host infection strategies. Mucous threads secreted through the exhalent siphon of the freshwater cyrenid (Cardiida) *Corbicula fluminea* are known to facilitate its downstream or even interstream transportation because they act as a draglines, when tangled to the fish or bird (Prezant and Chalermwat 1984; Kimura et al. 2004; Haag 2012). If such entanglement may be an advantage of upstream dispersal, the relation between fish and glochidium may have arose initially as a phoretic relationship (Watters 2001; Barnhart et al. 2008; Haag 2012). Strayer (2008) reports an example of rapid dispersal of *Lampsilis cardium*, into the Potomac basin as a consequence of the introduction of its host, the smallmouth bass (*Micropterus dolomieu*). Actual dispersal due to phoresis is related to dispersal abilities of host fish. Unionoids today often parasitize small benthic fishes, with limited mobility, so their dispersal may be low (Watters 1992; Haag and Warren 2003; McLain and Ross 2005; Blazek and Gelnar 2006; Strayer 2008; Schwalb et al. 2011b; Vaughn 2012). Dispersal of mussels associated with mobile or migratory species may be higher (Sepkoski and Rex 1974; Schwalb et al. 2011a; Lellis et al. 2013; Terui et al. 2014). Still, many species of unionoids are capable of parasitizing several species of fish, ranging from mobile to mobility limited forms (e.g. Blazek and Gelnar 2006; Van Snik Gray et al. 2009; St. John White et al. 2017).

Barnhart et al. (2008) and Haag (2012) propose that unionoids might have routinely dispersed on fish, being solely attached by the larval threads—before acquiring a firmer attachment, by clamping the valves on the fish tissue. The latter could cause harm to the fish, initiating a wound reaction (which facilitates encapsulation), but could also have facilitated obtaining the nutrition from tissues of their host (Watters 2001). The crucial step for developing a parasitic mode of life is resisting the host responses—both behavioral and cellular.

5.5.3.2 Encapsulation

Close contact with an object results in the closing of the glochidium valves, thus larvae often attach to unsuitable hosts and even to inanimate objects (Lefevre and Curtis 1910b; Wood 1974b; Neves et al. 1985; Jansen et al. 2001). After their valves are closed, glochidia remain attached to the object for several hours, so they are not able to change the place of attachment whether or not they reached their host (Lefevre and Curtis 1910b; Wood 1974a). Within this time encapsulation by host epithelium tissues must occur, providing both the mechanical protection and nutrients for larvae (Karna and Millemann 1978; Wächtler et al. 2001; Rogers-Lowery and Dimock 2006; Reis et al. 2014). Mussel larva sometimes can infest and induce the formation of a cyst even in an unsuitable host. As a result, most often the host epithelium, which forms the capsule, is destroyed and the glochidia fall off before they complete metamorphosis (Waller and Mitchell 1989; Watters and O'dee 1996; Rogers-Lowery and Dimock 2006; Barnhart et al. 2008).

Jansen et al. (2001) report the observations of Pflugfelder (1951) that fish are able to detach glochidia by rubbing their body against objects. The solution of this problem would be quick encapsulation in host epithelium. Host epithelial cells called keratocytes quickly cover the glochidium, instead of cell proliferation (Nezlin et al. 1994; Dodd

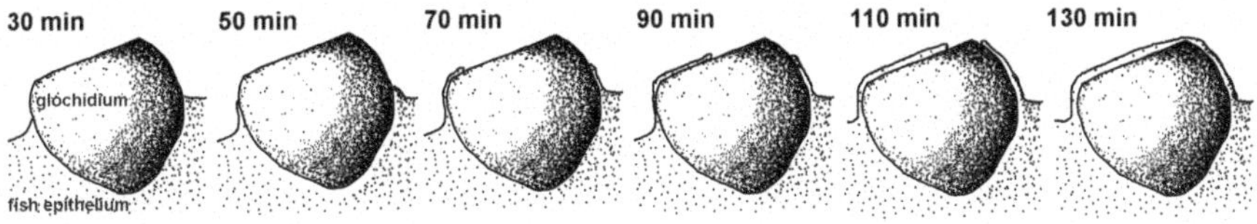

Fig. 5.8 Encapsulation of a glochidium by migrating epithelial cells of the fish. Drawn on the basis of figure 1 in Rogers-Lowery and Dimock (2006), with the permission to use material from the publication of the University of Chicago Press. The sequence of pictures starts 30 min after attachment of glochidium

et al. 2005; Rogers-Lowery and Dimock 2006; Barnhart et al. 2008). Today, glochidia are completely enclosed by single or multiple layers of host cells within 2–36 h (Lefevre and Curtis 1910b; Nezlin et al. 1994; Jansen et al. 2001; Araujo et al. 2002; Howerth and Keller 2006; Rogers-Lowery and Dimock 2006; Fig. 5.8). This time may be affected by the size of glochidium (e.g. Lefevre and Curtis 1910b) as well as the water temperature (Marwaha et al. 2017) and the place of attachment (Jansen et al. 2001).

Once glochidia are encapsulated, the fish is still able to get rid of them, due to immunological reactions, especially if the fish is an unsuitable host (has natural immunity) or is a suitable host that developed acquired resistance against mussel glochidia (has acquired immunity; Rogers and Dimock 2003; Dodd et al. 2005, 2006; Watters 2007). This rejection may occur at different times during larval development—most commonly during encapsulation and shortly after, but also up to a dozen weeks after (Arey 1932a; Hove and Neves 1994; Watters and O'Dee 1996; Haag and Warren 1997; Haag et al. 1999; Jansen et al. 2001; Rogers-Lowery and Dimock 2006; Barnhart et al. 2008). Arey (1932a) and Waller and Mitchell (1989) noticed that capsules formed on an unsuitable host with acquired immunity differ from those on a suitable host by being enlarged and irregular, prior to destruction of glochidia by concentrated phagocytes and granulocytes. Gills of the fish are the critical organ for gas exchange, ion homeostasis, and acid-base balance. They are covered by a thin epithelium—a monolayer of cells that acts as a boundary between the extracellular fluids and the surrounding water (Evans et al. 2005). As long as this organ is exposed to the water with a large surface of contact, it is likely place for entry of pathogenic organisms. Yet, this tissue can have an immune response (Campos-Perez et al. 2000; Olsen et al. 2011). Once glochidia attach to the fish gill epithelium, tissue response includes inflammation, migration of different host cells to the area of larvae attachment, as well as the appearance of necrotic cells (filled with connective fibers) at the edges of the growth of capsules and occasionally epithelial cell hyperplasia were observed (Waller and Mitchell 1989; Rogers-Lowery and Dimock 2006). Both the inflammatory responses and reduced presence of eosinophilic granular cells in and around the glochidial capsule are reported from more suitable hosts (Waller and Mitchell 1989; Fast et al. 2006).

An inflammatory response in fish is characterized by an extensive proliferation of fibrous connective tissue elements (e.g. observed as a result of the attachment of copepod *Lernaea cruciate* to skin of serranid fish, *Morone chrysops*; Joy and Jones 1973). Healing of the wounds in fish consist of rapid spreading of the epithelium cells and later their proliferation and differentiation (Quilhac and Sire 1999). Additionally Rogers-Lowery and Dimock (2006) documented that cells in fish epithelium are in constant motion. It is reasonable, that a combination of these characters might have facilitated quick encapsulation of glochidia once the parasitic relation was developed. Yet, Rogers-Lowery and Dimock (2006) highlighted that modern glochidia must additionally stimulate host tissue to form a capsule, because glochidia of different mussel species and similar morphology are not encapsulated in the same rate on the same fish species.

5.5.3.3 Suitable Host

The magnitude of fish tissue reactions is amplified in unsuitable or resistant hosts, compared to naïve suitable hosts, resulting in the shedding the larvae (Waller and Mitchell 1989; Watters and O'Dee 1996; Dodd et al. 2006). Thus the transformation success of glochidium depends on if and how often the host suitable fish was infected before (review by Jansen et al. 2001; Rogers and Dimock 2003; Dodd et al. 2005). Usually smaller (younger) fishes are more heavily infested than larger (Tedla and Fernando 1969; Hastie and Young 2001), but Rogers and Dimock (2003) found a positive correlation between the size and the number of attached larvae, during laboratory infestation experiments (nevertheless smaller fishes were more often infested by fewer glochidia). Similarly Hastie and Young (2001) described even greater initial infection load of *Margaritifera margaritifera* glochidia on older salmonid fish (with larger gill surface areas/greater ventilation rates) but finally young ones were much more susceptible. The metamorphosis success on older and larger fish appeared lower than on fish younger and smaller even in the wild (Bauer 1987 in Strayer 2008). In some cases host infection by one species of mussel may even result in cross-resistance of fish to the other mussel genera and even subfamilies and might involve both specific and nonspecific mechanisms, as found by Dodd et al. (2005). Yet these studies show conflicting results related with the degree of resistance (discussed by Jansen et al. 2001; Strayer 2008). Nevertheless, Dodd et al. (2006) noticed a progressive loss of immunity to glochidia of *Lampsilis reeveiana* in the largemouth bass (*Micropterus salmoides*) about a half year past developing the immune response. Little is known about the mechanisms involved in the specificity of the association between mussels and their hosts, however Rogers-Lowery et al. (2007) identified that fish antibodies recognize glochidial granular structures located between the larval mantle and shell, together with the mantle cells in relation of bluegill sunfish (*Lepomis macrochirus*) and *Utterbackia imbecillis*. Dubansky et al. (2011) gave insight into a poorly understood physiological mechanism regulating this host-ectoparasite interaction. Cortisol is the main corticosteroid hormone in vertebrates, influencing a variety of physiological and behavioral actions including stress reactions, during which its concentration in plasma rises (Mommsen et al. 1999; Fürtbauer and Heistermann 2016). Dubansky et al. (2011) showed a wide range of actions elicited by the cortisol on fish, including its immunosuppressant properties, enhancing glochidial metamorphosis probably by improving the retention of attached glochidia of *U. imbecillis* on gills of *L. macrochirus*. It is also clear that glochidia encystment has a physiological effect on their host (e.g. higher energetic costs of movement; Slavík et al. 2017). Douda et al. (2017) reported reduction in the body mass, condition factor and changes in several physiological parameters measured in plasma of European chub (*Squalius cephalus*) infested with glochidia of *Sinanodonta woodiana*. Filipsson et al. (2017) recently noticed, that juvenile brown trout (*Salmo trutta*) infested with glochidia of *Margaritifera margaritifera* had an increased metabolic rate and haematocrit values, compared with the control group. This is possibly an effect of a compensatory response to the potential respiratory stress caused by the encapsulation of glochidia on its gills. Nevertheless, it is believed that positive selection for fish to resist glochidial infection is weaker, than strong positive selection of larvae, to develop resistance to its host (Haag 2012). To survive, the glochidium must resist the host's immune response (Rogers and Dimock 2003; Dodd et al. 2005, 2006; Watters 2007). Mussels that possess a more resistant phenotype will be positively selected for (Graf 1997; Barnhart et al. 2008). High levels of glochidial infestation may cause a severe load to fish gills, enough to kill the fish host (Howerth and Keller 2006), but in nature overcrowding of glochidia on host (which may primarily lead to increased larval mortality) is rare, due to temporal and spatial differences in glochidial distribution on their hosts (Hastie and Young 2001; Jansen et al. 2001; Blazek and Gelnar 2006). Nevertheless, lower levels of infestation may change hosts' physiology (e.g. by stress reaction) enough to decrease glochidium chances for metamorphosis (Dubansky et al. 2011). Strayer (2008) discussed epidemiological models of mussel-host dynamics to illuminate the importance of the immune response of the fish and mussel competition for hosts. He concluded that there must be a density limit to the number of glochidia that successfully metamorphose on an individual host.

5.5.3.4 Duration of Encapsulation and Metamorphosis

Larvae of most species do not grow during encapsulation (with the exceptions of Margaritiferidae, Iridinidae, and a dozen of North American species listed by Patterson (2018); Wächtler et al. 2001; Barnhart et al. 2008), however stable isotope analysis proved they obtain nutrients from their host (Wächtler et al. 2001; Fritts et al. 2013; Denic et al. 2015) possibly due to the close presence of fish blood capillaries, which are involved in formation of the cyst. The food is gathered from the host by the dense microvilli of the larval mantle cells and the larval mantle also produces storage granules (Wächtler et al. 2001). An additional source of nutrients comes from the larval adductor and mantle remnant that are degenerated during metamorphosis (Arey 1932b). Encapsulation lasts from days to months, often depending on the glochidium size (Bauer 1994, 2001d; Wächtler et al. 2001), which is species specific. Smaller glochidia grow during encapsulation. They are usually unhooked, and need more time to complete metamorphosis when compared to large, hooked ones (respectively 1300–1600 days compared to 240–544 days in similar temperatures; Bauer 1994; Barnhart et al. 2008). Meaning they are exposed to the host immune response for a longer time, so they must be closely adapted to the host species. They commonly parasitize one, or a few closely related, fish species while mussel species with large glochidia (and a shorter time of encapsulation) are host generalists—commonly successfully developing on over a dozen host fish species (Bauer 1994, 2001d; Hastie and Young 2001; Barnhart et al. 2008; Strayer 2008). The duration of encapsulation is variable within species and among species of mussels (O'Brien and Box 1999; O'Dee and Watters 2000; McLeod et al. 2017), as well as between hosts (Yeager and Saylor 1995; Haag and Warren 1997; O'Brien and Box 1999; Taeubert et al. 2013). It is suggested that the most significant external factor is the water temperature, which influences both the metabolism of the mussel and the fish and leads to slower metamorphosis if it decreases (Bauer 1994; Jansen et al. 2001; Van Snik Gray et al. 2009; Taeubert et al. 2013, 2014).

Metamorphosis during encapsulation includes two main stages: (1) disintegration of larval adductor muscle and formation of a mushroom-like body by the larval mantle cells; (2) de novo development of two post-larval adductor muscles; formation of stomach, digestive gland and intestine; formation of large ciliated foot with foot retractor; larval mantle cells replaced by post-larval mantle cells; formation of nervous system and rudimentary gills; and finally the disappearance of the mushroom-like body (Waller and Mitchell 1989; Wächtler et al. 2001; Fisher and Dimock 2002).

During encapsulation fish are dispersing the glochidia, possibly to new places (Sepkoski and Rex 1974; Graf 1997; Graf and O'Foighil 2000; Wächtler et al. 2001; Strayer 2008). If metamorphosis is completed, juveniles actively break the wall of cyst (which become thinner in this time; Waller and Mitchell 1989) and drop off as free-living mussels (Karna and Millemann 1978; Jansen et al. 2001). Juveniles migrate into sediments (where they feed on sediment pore water by employing a pedal feeding strategy), until they emerge at a size of about 1 cm as filter feeders (Yeager et al. 1994; Bauer 2001c; Schwalb and Push 2007; Schartum et al. 2017; Araujo et al. 2018).

5.5.3.5 In Search of the Host of Early Unionids

Irrespective if unionoids used fish as their hosts or just in a phoretic relationship, between Late Triassic and Middle Jurassic, they must have developed this relationship with the fishes they could encounter. Today unionoids usually parasitize small, young fishes of the most common species in their environments. Related mussel species often parasitize related fish species (Strayer 2008). The global geographical distribution of mussel families (Graf and Cummings 2007; Bogan 2008; Bogan and Roe 2008; Lopes-Lima et al. 2017) is today correlated with the distribution of their likely fish hosts' families (Berra 2001). Today unionoid diversity is primarily related to the fish diversity, not to the drainage area (Watters 1992; Modesto et al. 2018). Nevertheless it appears that the mussel populations are not just controlled by the presence of their host, as geographic ranges of mussel species are smaller than that of their hosts (Strayer 1983, 2008; Bauer et al. 1991; Watters 1992). Today the most common host species belong to the modern families of teleostei (class: Actinopterygii) like Percidae, Cottidae, Cyprinidae, Centrarchidae, and Salmonidae

Table 5.3 The fossil record and molecular clock estimates indicating the latest possible date of origin of phyletic freshwater common modern teleostei lineages

Order	Family	Earliest occurrence (Mya)	References (fossil data indicated with asterisk)
Scorpaeniformes	Cottidae	8–3.3 ~15 23.0–5.3 25.8	Yokoyama and Goto (2005) Betancur-R et al. (2017) Jordan and Gilbert (1919)*, Knope (2013)* http://timetree.org
Perciformes	Percidae	~50 ~40 66–58 28.4–23.0 95	Betancur-R et al. (2013) Betancur-R et al. (2017) Collette and Bănărescu (1977)* Lopatin (1996)* http://timetree.org
	Centrarchidae	33 ~40 55.8–50.3 35	Near et al. (2007) Betancur-R et al. (2017) Divay and Murray (2016)* http://timetree.org
	Percidae (North American darters)	34	Near et al. (2011)
Cypriniformes	Cyprinidae	~60 42.38 56–55(?), 31 152–145 118	Betancur-R et al. (2017) Wang et al. (2012) Cavender (1991)* Fürsich and Pandey (1999)* http://timetree.org
Salmoniformes	Salmonidae	59.1 106 54.02 ~40 58.9 47.8–41.3 69	Crête-Lafrenière et al. (2012) Shedko et al. (2013) Ma et al. (2016) Betancur-R et al. (2017) Lecaudey et al. (2018) Kemp (1985)* http://timetree.org

(reviewed by Strayer 2008). Although the fossil record of the modern teleostei range back to the Late Cretaceous (Benton 2005; Broughton et al. 2013; Betancur-R et al. 2017), the families of the common extant hosts appeared much later—ca. 60 to ca. 3 Mya (Paleocene-Pliocene; Table 5.3, Fig. 5.6). Watters (2001) and Haag (2012) interpret that a great radiation in freshwater teleostei, which started at the beginning of the Mesozoic (Benton 2005; Broughton et al. 2013), might have facilitated diversification of the unionoid fauna. Extant unionoids however do metamorphose also using members of the ancient groups of the subclasses Actinopteri and Cladistia (orders Acipenseriformes, Amiiformes, Lepistosteiformes and Polypteriformes; Araujo and Ramos 2001; Wächtler et al. 2001). The fish fossil record documents their Mesozoic origin; they probably appeared in Late Triassic (Amiiformes), Jurassic (Acipenseriformes) and in late Early Cretaceous (Polypteriformes, Lepistosteiformes). The molecular clock methods date divergence of the modern families of these ancient groups to at least Late Cretaceous or post-Mesozoic (e.g. Broughton et al. 2013; Betancur-R et al. 2017), however the fossil record might support an older origin—since the Late Cretaceous, with the exception of much older

Table 5.4 The fossil record and molecular clock estimates indicating the latest possible date of origin of ancient fish orders and families that serve as hosts to the extant unionids

Order Family	Earliest occurrence (Mya)	References (fossil data indicated with asterisk)
Polypteriformes	Albian 112–99	Broughton et al. (2013)
	Cenomanian 100–94	Gayet et al. (2002)* Grandstaff et al. (2012)*
Polypteridae	Rupelian ~30	Betancur-R et al. (2017)
	Cenomanian 100–94	Grandstaff et al. (2012)*
	Coniacian 87	http://timetree.org
Lepistosteiformes	Carnian 237–227	Goodwin et al. (1999)*
	Albian 113–100	Gayet et al. (2002)*
	Cenomanian 100–94	López-Arbarello (2012)*
Lepisosteidae	Burdigalian ~20	Betancur-R et al. (2017)
	Cenomanian 100–94	López-Arbarello (2012)*
	Thanethian 57	http://timetree.org
Amiiformes	Changhsingian 254	Wang and Jin (2007)*
Amiidae	Ladinian ~240	Liu et al. (2003)*
	Norian 222	Bogan et al. (2013)*
Acipenseriformes	Induan 252	Nielsen (1955)*
	Hettangian 200	Bemis et al. (1997)*
Acipenseriidae	Bartonian ~40	Betancur-R et al. (2017)
	Campanian 80–75	Grande and Hilton (2006)*
	Oxfordian 158	http://timetree.org

(Late Triassic) Amiidae (Liu et al. 2003; Grande and Hilton 2006; Grandstaff et al. 2012; López-Arbarello 2012; Table 5.4). Thus they possibly could have served as host fishes together with the modern teleostei. Several authors (e.g. Dubiel et al. 1991; Watters 2001; Dzik and Sulej 2007; Skrzycki 2015) noticed co-occurrence of the early unionoids with the widespread lung fish, Dipnoi, which, since the Carboniferous, lived in freshwaters and could serve as hosts. The evidence of the shell damage (presumably in Late Triassic, Middle-Late Jurassic), which can be attributed to the predation by a lungfish, may support their somewhat close relation (Gorzelak et al. 2010; Van Damme et al. 2015, Qvarnström et al. 2017). Barnhart et al. (2008), Haag (2013) and Patterson (2018) have recently noticed that some extant unionids (*Ellipsaria*, *Leptodea*, *Potamilus* and *Truncilla* [Lampsilini, Unionidae]) evolved unique strategies to infest a freshwater drum (*Aplodinotus grunniens*) which feeds on them. These strategies include: broadcasting the larvae close to the river bottom, where the drum feeds; attracting the drum to induce the ingestion of larvae by gaping widely the gravid females; or gravid females being ingested by the drum. One may assume, that the latter "strategy" could have been easily introduced in early unionoids, although all of these strategies could have been present then. This also could be an evolutionary selecting pressure to develop huge amounts of small larvae (with their low cost), which may cause an unusually high infestation of their hosts (Coker et al. 1921; Haag 2013). Other molluscivorous fish were also reported from Mesozoic freshwaters like hybodontid sharks (*Hybodus* spp.) and neopterygians (e.g. *Lepidotes* [Lepistosteiformes]; Kear and Godthelp 2008; Bermúdez-Rochas 2009; Bermudez-Rochas et al. 2013; Van Damme et al. 2015; Cavin 2017) whose diet in combination with close contact with early unionoids could have facilitated the establishment of parasite-host associations. Other freshwater fish, e.g. non-durophagous neopterygians like *Pliodetes* (Lepistosteiformes), or *Semionotus* (Semionotiformes) as well as freshwater members of ancient Palaeonisciformes also co-occurred with early unionoids (Dzik and Sulej 2007; Lucas and Tanner 2007; Cavin 2017; Gibson 2018), thus might also have served as their early associates.

5.6 Conclusions

Many bivalves live in associations with the other organisms, but only members of the marine superfamily Galeommatoidea and larvae of the freshwater order Unionida are known to live in facultative (Galeommatoidea) or obligate (Unionida) parasitising relationships. In both cases, but for different reasons, these relationships require close association with the host body followed by covering the bivalve by the host tissues.

In the case of galeommatoideans, it is proposed, that the close association with their hosts evolved from the stage of a free living ancestor, through a burrow-wall commensal, to a commensal that settles on a precise area of a host body, sometimes being covered by the soft tissues of the host, or settling inside the gill chambers of the host (depending on both the bivalve and host species). This may be corroborated by recent findings of interpreted accidental larval settlement within the body cavity of the host-crab in case of *Kurtiella pedroana*, leading to feeding and presumably reproduction attempt in the crab's hemocoel, or living in the body and presumably parasitising the holothurian by *Entovalva nhatrangensis*.

Their freshwater habitat, parasitising larvae, and maternal care characterize unionoids. Because of their large adult size and limited mobility, their dispersal abilities rely on their minute larvae (most commonly the glochidium) and their close association with fish hosts. To enhance the larval survival chances, females brood them in their own modified gills (interpreted as protection against the chemical parameters of freshwaters, ensuring calcium for the larval shell, and a protection against water currents, which are accompanied by mantle morphology and behavior that lure potential host fish, differently developed by different lineages). Unionoids appeared in Pangean freshwaters by the Late Triassic. At this time gills of the bivalves were not yet modified (which does not exclude the possibility of maternal brooding), but presumably in the Middle Jurassic and surely by the Cretaceous they were, allowing larval incubation. Middle Jurassic unionoids of modern lineages probably already inhabited freshwaters on Laurasia (Unionidae, later Margaritiferidae) and Gondwana (Hyriidae, Etherioidea). Thus, at the latest during the Middle Jurassic, one may suppose the presence of both characters—derived gills with marsupia and parasitic larvae for dispersion in their last common ancestor.

The parasitic association could be established with the fish that co-occurred in freshwaters and which possibly were common and lived close to the bottom of ancient lakes or rivers (e.g. durophagous fish). In the context of this close proximity (possibly including predation), the phoretic relationship was possibly established, which enhanced the dispersal abilities of the bivalve associates (e.g. via larval mucous threads, which presumably were tangled on fish, but could also serve as anti-sinking floating web or dragline). Then a firmer attachment was acquired, by clamping the bivalve larval valves on the fish tissue, which started a wound reaction in fish tissue and resulted in quick larval encapsulation due to host epithelium cell migration. Further development of mechanisms, which allowed resistance to the host immunity, strengthened this association and opened the possibility of nutrition on fish tissues. On the one hand, larval feeding on host tissues enabled reducing the size of larvae, which may lower the cost of females producing a huge amount of small larvae (which could cause high infestation of the host increasing the probability of survival and dispersal of the offspring). On the other hand, the small size of larvae requires a longer timespan spent in the environment of fish tissues in order to gather nutrition for growth to a size that ensures safe metamorphosis and thus being exposed to immune responses for a longer time than larger larvae. In consequence a small larva must have developed a close adaptation to the host species, which in addition to their small size might hamper their fossil preservation potential.

Little direct fossil evidence is known to constrain the transition from free-living to a parasitic lifestyle in bivalves from both parasitic groups (Galeommatoidea, Unionida). Based on knowledge on the reasons or consequences of establishing a parasitic relationship from living species and indirect fossil record, at least two major evolutionary steps towards parasitism can be described: close contact with the host body, followed by the enclosing of the bivalve within the host's tissues. Detailed considerations about the mechanisms and scenario of such relations originating still need more direct data.

Acknowledgments I am grateful to the Editors, Kenneth De Baets and John Warren Huntley for their immense effort in editing and reviewing this chapter. I am thankful to John M. Pfeiffer for the kind discussions about evolution of larva types in Unionida, to Andrzej Kołodziejczyk who proof-read the draft, to Maria Cristina Dreher Mansur for sharing her observations on lasidium morphology and to Marcello Guimares Simões for sharing his knowledge on bivalves from Crato Formation. I am grateful to the University of Chicago Press for granting the permission to use material from the publication of Rogers-Lowery and Dimock (2006) as the model for a new illustration in this work.

References

Aldridge DC, Horne DC (1998) Fossil glochidia (Bivalvia, Unionidae): identification and value in palaeoenvironmental reconstructions. J Micropalaeontol 17:179–182

Aldridge DC, McIvor AL (2003) Gill evacuation and release of glochidia by *Unio pictorum* and *Unio tumidus* (Bivalvia: Unionidae) under thermal and hypoxic stress. J Molluscan Stud 69:55–59

Allen WR (1921) Studies of the biology of freshwater mussels. Experimental studies of the food relations of certain Unionidae. Biol Bull 40:210–241
Amyot J-P, Downing JA (1998) Locomotion in *Elliptio complanata* (Mollusca: Unionidae): a reproductive function? Freshw Biol 39:351–358
Appukuttan KK (1972) *Pseudopythina subsinuata* (Lischke) a commensal bivalve of *Squilla nepa* (Latricelle) and *Squilla raphidea* Fabricius. J Mar Biol Assoc India 14:412–415
Araujo R, Camara N, Ramos MA (2002) Glochidium metamorphosis in the endangered freshwater mussel *Margaritifera auricularia* (Spengler, 1793): a histological and scanning electron microscopy study. J Morphol 254:259–265
Araujo R, Campos M, Feo C, Varela C, Soler J, Ondina P (2018) Who wins in the weaning process? Juvenile feeding morphology of two freshwater mussel species. J Morphol 279:4–16
Araujo R, Ramos MA (2001) Life-history data on the virtually unknown *Margaritifera auricularia*. In: Bauer G, Wächtler K (eds) Ecology and evolution of the freshwater mussels Unionoida. Springer, Berlin, pp. 143–152
Arey LB (1932a) A microscopical study of glochidial immunity. J Morphol 53:367–379
Arey LB (1932b) The nutrition of glochidia during metamorphosis. A microscopical study of the sources and manner of utilization of the nutritive substances. J Morphol 53:201–221
Ashworth AC, Preece RC (2003) The first freshwater molluscs from Antarctica. J Molluscan Stud 69:89–92
Atkins D (1938) On the ciliary mechanisms and interrelationships of Lamellibranchs. Part VII: laterofrontal cilia of the gill filaments and their phylogenetic value. Q J Microsc Sci 80:345–436
Bagur M, Richardson CA, Gutiérrez JL, Arribas LP, Doldan MS, Palomo MG (2013) Age, growth and mortality in four populations of the boring bivalve *Lithophaga patagonica* from Argentina. J Sea Res 81:49–56
Barnhart MC, Haag WR, Roston WN (2008) Adaptations to host infection and larval parasitism in Unionoida. J N Am Benthol Soc 27:370–394
Bauer G (1992) Variation in the life span and size of the freshwater pearl mussel. J Anim Ecol 61:425–436
Bauer G (1994) The adaptive value of offspring size among freshwater mussels (Bivalvia; Unionoidea). J Anim Ecol 63:933–944
Bauer G (2001a) Characterization of the Unionoida (= Naiads). In: Bauer G, Wächtler K (eds) Ecology and evolution of the freshwater mussels Unionoida. Springer, Berlin, pp. 3–4
Bauer G (2001b) Life-history variation on different taxonomic levels of naiads. In: Bauer G, Wächtler K (eds) Ecology and evolution of the freshwater mussels Unionoida. Springer, Berlin, pp. 83–91
Bauer G (2001c) Factors affecting naiad occurrence and abundance. In: Bauer G, Wächtler K (eds) Ecology and evolution of the freshwater mussels Unionoida. Springer, Berlin, pp. 155–162
Bauer G (2001d) Framework and driving forces for the evolution of naiad life histories. In: Bauer G, Wächtler K (eds) Ecology and evolution of the freshwater mussels Unionoida. Springer, Berlin, pp. 233–255
Bauer G, Hochwald S, Silkenat W (1991) Spatial distribution of freshwater mussels: the role of host fish and metabolic rate. Freshw Biol 26:377–386
Bemis WE, Findeis EK, Grande L (1997) An overview of Acipenseriformes. Environ Biol Fish 48:25–71
Benton MJ (2005) Vertebrate palaeontology, 3rd edn. Blackwell Science Ltd, Malden
Bermúdez-Rochas DD (2009) New hybodont shark assemblage from the Early Cretaceous of the Basque-Cantabrian Basin. Geobios 42:675–686
Bermudez-Rochas DD, Delvene G, Ruiz-Omenaca JI (2013) Evidence of predation in Early Cretaceous unionoid bivalves from freshwater sediments in the Cameros Basin, Spain. Lethaia 46:57–70
Bernard FR, Cai Y-Y, Morton B (1993) A catalogue of the living marine bivalve molluscs of China. Hong Kong University Press, Hong Kong
Berra TM (2001) Freshwater fish distribution. Academic, San Diego

Betancur-R R, Broughton RE, Wiley EO, Carpenter K, López JA, Li C, Holcroft NI, Arcila D, Sanciangco M, Cureton II JC, Zhang F, Buser T, Campbell MA, Ballesteros JA, Roa-Varon A, Willis S, Borden WC, Rowley T, Reneau PC, Hough DJ, Lu G, Grande T, Arratia G, Ortí G (2013) The Tree of Life and a new classification of bony fishes. PLOS Currents 5: ecurrents.tol.53ba26640df0ccaee75bb165c8c26288. https://doi.org/10.1371/currents.tol.53ba26640df0caee75bb165c8c26288

Betancur-R R, Wiley EO, Arratia G, Acero A, Bailly N, Miya M, Lecointre G, Ortí G (2017) Phylogenetic classification of bony fishes. BMC Evol Biol 17:162. https://doi.org/10.1186/s12862-017-0958-3

Bhaduri RN, Valentich-Scott P, Hilgers MS, Singh R, Hickman ME, Lafferty KD (2017) Facultative parasitism by the bivalve *Kurtiella pedroana* in the mole crab *Emerita analoga*. J Parasitol 103:646–651

Bieler R, Mikkelsen PM, Giribet G (2013) Bivalvia—a discussion of known unknowns*. Am Malacol Bull 31:123–133

Blazek R, Gelnar M (2006) Temporal and spatial distribution of glochidial larval stages of European unionid mussels (Mollusca: Unionidae) on host fishes. Folia Parasitol 53:98–106

Bogan AE (2008) Global diversity of freshwater mussels (Mollusca, Bivalvia) in freshwater. Hydrobiologia 595:139–147

Bogan AE, Hoeh WR (2000) On becoming cemented: evolutionary relationships among the genera in the freshwater bivalve family Etheriidae (Bivalvia: Unionoida). Geol Soc Spec Publ 177:159–168

Bogan AE, Roe KJ (2008) Freshwater bivalve (Unioniformes) diversity, systematics, and evolution: status and future directions. J N Am Benthol Soc 27:349–369

Bogan S, Taverne L, Agnolin F (2013) First Triassic and oldest record of a South American amiiform fish: *Caturus* sp. from the Los Menucos Group (lower Upper Triassic), Río Negro province, Argentina. Geol Belg 16:191–195

Bolotov IN, Vikhrev IV, Bespalaya YV, Gofarov MY, Kondakov AV, Konopleva ES, Bolotov NN, Lyubas AA (2016) Multi-locus fossil-calibrated phylogeny, biogeography and a subgeneric revision of the Margaritiferidae (Mollusca: Bivalvia: Unionoida). Mol Phylogenet Evol 103:104–121

Bolotov IN, Kondakov AV, Vikhrev IV, Aksenova OV, Bespalaya YV, Gofarov MY, Kolosova YS, Konopleva ES, Spitsyn VM, Tanmuangpak K, Tumpeesuwan S (2017a) Ancient river inference explains exceptional oriental freshwater mussel radiations. Sci Rep 7:2135. https://doi.org/10.1038/s41598-017-02312-z

Bolotov IN, Vikhrev IV, Kondakov AV, Konopleva ES, Gofarov MY, Aksenova OV, Tumpeesuwan S (2017b) New taxa of freshwater mussels (Unionidae) from a species-rich but overlooked evolutionary hotspot in Southeast Asia. Sci Rep 7:11573. https://doi.org/10.1038/s41598-017-11957-9

Booth JD (1979) Common bivalve larvae from New Zealand: Leptonacea. N Z J Mar Freshw Res 13:241–254

Boss KJ (1965a) Symbiotic erycinacean bivalves. Malacologia 3:183–195

Boss KJ (1965b) A new mollusk (Bivalvia, Erycinidae) commensal on the stomatopod crustacean *Lysiosquilla*. Am Mus Novit 2215:1–12

Boucot AJ, Poinar GOJ (2010) Fossil behavior compendium. CRC Press, Boca Raton, FL

Boxshall GA, Defaye D (2008) Global diversity of copepods (Crustacea: Copepoda) in freshwater. Hydrobiologia 595:195–207

Boyko CB, Mikkelsen PM (2002) Anatomy and biology of *Mysella pedroana* (Mollusca: Bivalvia: Galeommatoidea), and its commensal relationship with *Blepharipoda occidentalis* (Crustacea: Anomura: Albuneidae). Zool Anzeig 241:149–160

Brickner I, Kramarsky-Winter E, Mokady O, Loya Y (1993) Speciation in the coral-boring bivalve *Lithophaga purpurea*: evidence from ecological, biochemical and SEM analysis. Mar Ecol Prog Ser 101:139–145

Briggs DEG (2003) The role of decay and mineralization in the preservation of soft-bodied fossils. Annu Rev Earth Planet Sci 31:275–301

Briggs DEG, Kear AJ (1993) Fossilization of soft tissue in the laboratory. Science 259:1439–1442
Briggs DEG, Kear AJ, Martill DM, Wilby PR (1993) Phosphatization of soft-tissue in experiments and fossils. J Geol Soc 150:1035–1038
Briggs DEG, McMahon S (2016) The role of experiments in investigating the taphonomy of exceptional preservation. Palaeontology 59:1–11
Briggs DEG, Wilby PR (1996) The role of the calcium carbonate-calcium phosphate switch in the mineralization of soft-bodied fossils. J Geol Soc 153:665–668
Bristow GA, Berland B, Schander C, Vo DT (2010a) A new endosymbiotic bivalve (Heterodonta: Galeommatoidea), from Pacific holothurians. J Parasitol 96:532–534
Bristow GA, Johannessen P, Berland B, Schander C (2010b) *Devonia perrieri* (Malard, 1903) (Bivalvia) found in situ on the holothurian *Leptosynapta inhaerens* (Müller, 1776) in Norway. Fauna Norv 30:25–27
Brito P, Yabumoto Y, Grande L (2008) New amiid fish (Halecomorphi: Amiiformes) from the Lower Cretaceous Crato Formation, Araripe Basin, Northeast Brazil. J Vertebr Paleontol 28:1007–1014
Brodniewicz I (1968) On glochidia of the genera *Unio* and *Anodonta* from the Quaternary freshwater sediments of Poland. Acta Palaeontol Pol 13:619–628
Bromley RG, Heinberg C (2006) Attachment strategies of organisms on hard substrates: a palaeontological view. Palaeogeogr Palaeoclimatol Palaeoecol 232:429–453
Broughton RE, Betancur-R R, Li C, Arratia G, Ortí G (2013) Multi-locus phylogenetic analysis reveals the pattern and tempo of bony fish evolution. PLoS Curr 5:ecurrents.tol.2ca8041495ffafd8041490c8092756e75247483e. https://doi.org/10.1371/currents.tol.2ca8041495ffafd0c92756e75247483e
Bruno AP, Hessel MH (2006) Registros paleontológicos do Cretáceo marinho na Bacia do Araripe. Estud Geol 16:30–49
Campos-Perez JJ, Ward M, Grabowski PS, Ellis AE, Secombes CJ (2000) The gills are an important site of iNOS expression in rainbow trout *Oncorhynchus mykiss* after challenge with the gram-positive pathogen *Renibacterium salmoninarum*. Immunology 99:153–161
Cañete JI, Rocha R (2013) *Modiolarca lateralis* (Pteryomorphia: Mytilidae): bivalve associated to six species of ascidians from Bocas del Toro, Panama. Lat Am J Aquat Res 41:1030–1035
Carpenter SM (2005) *Mysella pedroana*, a commensal bivalve (Lasaeidae) on two decapod crustacean hosts. Nautilus 119:105–108
Carter A, Riley TR, Hillenbrand C-D, Rittner M (2017) Widespread Antarctic glaciation during the Late Eocene. Earth Planet Sci Lett 458:49–57
Carter JG, Altaba CR, Anderson LC, Araujo R, Biakov AS, Bogan AE, Campbell DC, Campbell M, Chen J, Cope JCW, Delvene G, Dijkstra HH, Zong-jie F, Gavrilova VA, Goncharova I, Guzhov AV, Harries PJ, Hartman JH, Hautmann M, Hoeh WR, Hylleberg J, Jiang B, Johnston P, Kirkendale L, Kleemann K, Koppka J, Kříž J, Machado D, Malchus N, Márquez-Aliaga A, Masse JP, Middelfart PU, Mitchell S, Nevesskaja LA, Özer S, Pojeta JJ, Polubotko IV, Pons JM, Popov S, Sánchez T, Sartori AF, Scott RW, Sey II, Signorelli JH, Silantiev VV, Skelton PW, Steuber T, Waterhouse JB, Wingard GL, Yancey T (2011) A synoptical classification of the bivalvia (Mollusca). Paleontological Contributions 2011(4):1–47
Carter JG, Harries PJ, Malchus N, Sartori AF, Anderson LC, Bieler R, Bogan AE, Coan EV, Cope JCW, Cragg SM, Garcia-March JR, Hylleberg J, Kelley P, Kleemann K, Kriz J, McRoberts C, Mikkelsen PM, Pojeta J, Temkin I, Yancey T, Zieritz A (2012) Part N, Revised, Volume 1, Chapter 31: Illustrated Glossary of the Bivalvia. Treatise Online 48:1–209, 327 fig.
Cavender TM (1991) The fossil record of the Cyprinidae. In: Winfield IJ, Nelson JS (eds) Cyprinid fishes: systematics, biology and exploitation. Springer, Dordrecht, pp. 34–54
Cavin L (2017) Freshwater fishes: 250 million years of evolutionary history. Oxford, London
Chanley P, Chanley M (2010) Reproductive biology of *Arthritica crassiformis* and *A. bifurca*, two commensal bivalve molluscs (Leptonacea). N Z J Mar Freshw Res 14:31–43
Coker RE, Shira AF, Clark HW, Howard AD (1921) Natural history and propagation of fresh-water mussels. Bull Bureau Fish 37:79–181

Collette BB, Bănărescu P (1977) Systematics and zoogeography of the fishes of the family Percidae. J Fish Res Board Can 34:1450–1463

Combosch DJ, Collins TM, Glover EA, Graf DL, Harper EM, Healy JM, Kawauchi GY, Lemer S, McIntyre E, Strong EE, Taylor JD, Zardus JD, Mikkelsen PM, Giribet G, Bieler R (2017) A family-level Tree of Life for bivalves based on a Sanger-sequencing approach. Mol Phylogenet Evol 107:191–208

Cope JCW (1996) The early evolution of the Bivalvia. In: Taylor JD (ed) Origin and evolutionary radiation of the Mollusca. Oxford University Press, Oxford, pp. 361–370

Cox CB (1969a) Two new dicynodonts from the Triassic Ntawere Formation, Zambia. Bull Br Mus (Nat Hist) Geol 17:255–294

Cox LR (1969b) General features of the bivalvia. In: Moore RC (ed) Treatise on invertebrate paleontology, part N, Mollusca, 6. Bivalvia. Geological Society of America and University of Kansas Press, Lawrence, pp. N2–N129

Cressey R, Boxshall G (1989) *Kabatarina pattersoni*, a fossil parasitic copepod (Dichelesthiidae) from a Lower Cretaceous fish. Micropaleontology 35:150–167

Cressey R, Patterson C (1973) Fossil parasitic copepods from a Lower Cretaceous fish. Science 180:1283–1285

Crête-Lafrenière A, Weir LK, Bernatchez L (2012) Framing the Salmonidae family phylogenetic portrait: a more complete picture from increased taxon sampling. PLoS One 7:e46662. https://doi.org/10.1371/journal.pone.0046662

Dartnall HJG, Walkey M (1979) The distribution of glochidia of the swan mussel, *Anodonta cygnea* (Mollusca) on the three-spined stickleback *Gasterosteus aculeatus* (Pisces). J Zool 189:31–37

Davenport D, Warmuth M (1965) Notes on the relationship between the freshwater mussel *Anodonta implicata* Say and the alewife *Pomolobus pseudoharengus* (Wilson). Limnol Oceanogr 10:R74–R78

Davis GM, Fuller SLH (1981) Genetic relationships among recent Unionacea of North America. Malacologia 20:217–253

Davis SP, Martill DM (1999) The gonorynchiform fish *Dastilbe* from the Lower Cretaceous of Brazil. Palaeontology 42:715–740

da Silva VR, Varejão FG, Matos SA, Fürsich FT, Skawina A, Schneider S, Simões MG (2020) *Cratonaia novaolindensis* gen. et sp. nov. (Unionida, Silesunionoidea) from the Aptian of Brazil (Araripe Basin), and its implications for the early evolution of freshwater mussels. Cretaceous Research, 107:104275. https://doi.org/10.1016/j.cretres.2019.104275

De Baets K, Dentzien-Dias P, Upeniece I, Verneau O, Donoghue PCJ (2015) Constraining the deep origin of parasitic flatworms and host-interactions with fossil evidence. Topics in Geobiology 90:93–135

De Baets K, Littlewood DTJ (2015) The importance of fossils in understanding the evolution of parasites and their vectors. Advances in Parasitology 90:1–51

De Baets K, Huntley JW, Klompmaker AA, Schiffbauer JD, Muscente AD (2021) The fossil record of parasitism: Its extent and taphonomic constraints. In: De Baets K, Huntley JW (eds) The evolution and fossil record of parasitism: Coevolution and paleoparasitological techniques. Topics in Geobiology 50

DeConto RM, Pollard D (2003) Rapid Cenozoic glaciation of Antarctica induced by declining atmospheric CO2. Nature 421:245–249

Denic M, Taeubert JE, Geist J (2015) Trophic relationships between the larvae of two freshwater mussels and their fish hosts. Invertebr Biol 134:129–135

Dick MH, Donaldson WE, Vining IW (1998) Epibionts of the tanner crab *Chionoecetes bairdi* in the region of Kodiak Island, Alaska. J Crustac Biol 18:519–527

Dickinson D, Sietman BE (2008) Recent observation of metamorphosis without parasitism in *Utterbackia imbecillis*. Ellipsaria 10:7–8

Divay JD, Murray AM (2016) An early Eocene fish fauna from the Bitter Creek area of the Wasatch Formation of southwestern Wyoming, USA. J Vertebr Paleontol 36:e1196211. https://doi.org/10.1080/02724634.2016.1196211

Dodd BJ, Barnhart MC, Rogers-Lowery CL, Fobian TB, Dimock RV Jr (2005) Cross-resistance of largemouth bass to glochidia of unionid mussels. J Parasitol 91:1064–1072
Dodd BJ, Barnhart MC, Rogers-Lowery CL, Fobian TB, Dimock RV Jr (2006) Persistence of host response against glochidia larvae in *Micropterus salmoides*. Fish Shellfish Immunol 21:473–484
Douda K, Velisek J, Kolářová J, Rylková K, Slavík O, Horký P, Langrová I (2017) Direct impact of invasive bivalve (*Sinanodonta woodiana*) parasitism on freshwater fish physiology: evidence and implications. Biol Invasions 19:989–999
Dubansky B, Whitaker B, Galvez F (2011) Influence of cortisol on the attachment and metamorphosis of larval *Utterbackia imbecillis* on bluegill sunfish (*Lepomis macrochirus*). Biol Bull 220:97–106
Dubiel R, Parrish JT, Good S (1991) The Pangean megamonsoon-evidence from the Upper Triassic Chinle formation. Palaios 6:347–370
Dvoretsky AG, Dvoretsky VG (2008) Epifauna associated with the northern stone crab *Lithodes maia* in the Barents Sea. Polar Biol 31:1149–1152
Dzik J, Sulej T (2007) A review of the early Late Triassic Krasiejów biota from Silesia, Poland. Palaeontol Pol 64:3–27
Eagles G, König M (2008) A model of plate kinematics in Gondwana breakup. Geophys J Int 173:703–717
Eastman JT (1993) Antarctic fish biology: evolution in a unique unvironment. Academic, San Diego
Elicki O, Gürsu S (2009) First record of *Pojetaia runnegari* Jell, 1980 and *Fordilla* Barrande, 1881 from the Middle East (Taurus Mountains, Turkey) and critical review of Cambrian bivalves. Paläontol Z 83:267–291
Evans DH, Piermarini PM, Choe KP (2005) The multifunctional fish gill: dominant site of gas exchange, osmoregulation, acid-base regulation, and excretion of nitrogenous waste. Physiol Rev 85:97–177
Fast MD, Muise DM, Easy RE, Ross NW, Johnson SC (2006) The effects of *Lepeophtheirus salmonis* infections on the stress response and immunological status of Atlantic salmon (*Salmo salar*). Fish Shellfish Immunol 21:228–241
Filipsson K, Brijs J, Näslund J, Wengström N, Adamsson M, Závorka L, Österling EM, Höjesjö J (2017) Encystment of parasitic freshwater pearl mussel (*Margaritifera margaritifera*) larvae coincides with increased metabolic rate and haematocrit in juvenile brown trout (*Salmo trutta*). Parasitol Res 116:1353–1360
Fisher GR, Dimock RJ (2002) Morphological and molecular changes during metamorphosis in *Utterbackia imbecilis* (Bivalvia: Unionidae). J Molluscan Stud 68:159–164
Fritts MW, Fritts AK, Carleton SA, Bringolf RB (2013) Shifts in stable-isotope signatures confirm parasitic relationship of freshwater mussel glochidia attached to host fish. J Molluscan Stud 79:163–167
Fryer G (1959) Development in a mutelid lamellibranch. Nature 183:1342–1343
Fryer G (1961) The developmental history of *Mutela bourguignati* (Ancey) Bourguignat (Mollusca: Bivalvia). Philos Trans R Soc B 244:259–298
Fürsich FT, Pandey DK (1999) Genesis and environmental significance of Upper Cretaceous shell concentrations from the Cauvery Basin, southern India. Palaeogeogr Palaeoclimatol Palaeoecol 145:119–139
Fürtbauer I, Heistermann M (2016) Cortisol coregulation in fish. Sci Rep 6:30334. https://doi.org/10.1038/srep30334
Gage J (1968) The mode of life of *Mysella cuneata*, a bivalve 'commensal' with *Phascolion strombi* (Sipunculoidea). Can J Zool 46:919–934
Galbraith HS, Frazier SE, Allison B, Vaughn CC (2009) Comparison of gill surface morphology across a guild of suspension-feeding unionid bivalves. J Molluscan Stud 75:103–107
Galtsoff PS (1964) The American oyster *Crassostrea virginica* Gmelin Fishery Bulletin, USA. Fish and Wildlife Service, Washington

Gascho Landis AM, Stoeckel JA (2016) Multi-stage disruption of freshwater mussel reproduction by high suspended solids in short- and long-term brooders. Freshw Biol 61:229–238

Gatenby CM, Neves RJ, Parker BC (1996) Influence of sediment and algal food on cultured juvenile freshwater mussels. J N Am Benthol Soc 15:597–609

Gatenby CM, Parker BC, Neves RJ (1997) Growth and survival of juvenile rainbow mussels, *Villosa iris* (Lea, 1829) (Bivalvia: Unionidae), reared on algal diets and sediment. Am Malacol Bull 14:57–66

Gayet M, Meunier FJ, Werner C (2002) Diversification in Polypteriformes and special comparison with the Lepisosteiformes. Palaeontology 45:361–376

Gibson SZ (2018) A new species of *Lasalichthys* (Actinopterygii, Redfieldiiformes) from the Upper Triassic Dockum Group of Howard County, Texas, with revisions to the genera *Lasalichthys* and *Synorichthys*. J Vertebr Paleontol 38:e1513009. https://doi.org/10.1080/02724634.2018.1513009

Giribet G, Wheeler W (2002) On bivalve phylogeny: a high-level analysis of the Bivalvia (Mollusca) based on combined morphology and DNA sequence data. Invertebr Biol 121:271–324

Good SC (1998) Freshwater bivalve fauna of the Late Triassic (Carnian-Norian) Chinle, Dockum, and Dolores Formations of the Southwest United States. In: Johnston PA, Haggart JW (eds) Bivalves: an eon of evolution – paleobiological studies honoring Norman D. Newell. University of Calgary Press, Calgary, 223-249

Goodwin MB, Clemens WA, Hutchison JH, Wood CB, Zavada MS, Kemp A, Duffin CJ, Schaff CR (1999) Mesozoic continental vertebrates with associated palynostratigraphic dates from the northwestern Ethiopian plateau. J Vertebr Paleontol 19:728–741

Gorzelak P, Niedźwiedzki G, Skawina A (2010) Pathologies of non-marine bivalve shells from the Late Triassic of Poland. Lethaia 43:285–289

Goto R, Hamamura Y, Kato M (2007) Obligate commensalism of *Curvemysella paula* (Bivalvia: Galeommatidae) with hermit crabs. Mar Biol 151:1615–1622

Goto R, Hamamura Y, Kato M (2011) Morphological and ecological adaptation of *Basterotia bivalves* (Galeommatoidea: Sportellidae) to symbiotic association with burrowing echiuran worms. Zool Sci 28:225–234

Goto R, Harrison TA, Ó Foighil D (2018) Within-host speciation events in yoyo clams, obligate commensals with mantis shrimps, including one that involves a change in microhabitat and a loss of specialized traits. Biol J Linn Soc 124:504–517

Goto R, Ishikawa H (2019) An unusual habitat for bivalves: rediscovery of the enigmatic commensal clam *Sagamiscintilla thalassemicola* (Habe, 1962) (Bivalvia: Galeommatoidea) from spoon worm's spoon. Mar Biodivers 49:1553–1558

Goto R, Ishikawa H, Hamamura Y (2016) Symbiotic association of the bivalve *Tellimya fujitaniana* (Galeommatoidea) with the heart urchin *Echinocardium cordatum* (Spatangoida) in the Northwestern Pacific. Zool Sci 33:434–440

Goto R, Ishikawa H, Hamamura Y, Sato S, Kato M (2014) Evolution of symbiosis with *Lingula* (Brachiopoda) in the bivalve superfamily Galeommatoidea (Heterodonta), with description of a new species of *Koreamya*. J Molluscan Stud 80:148–160

Goto R, Kawakita A, Ishikawa H, Hamamura Y, Kato M (2012) Molecular phylogeny of the bivalve superfamily Galeommatoidea (Heterodonta, Veneroida) reveals dynamic evolution of symbiotic lifestyle and interphylum host switching. BMC Evol Biol 12:172. https://doi.org/10.1186/1471-2148-12-172

Graf D (1997) Sympatric speciation of freshwater mussels (Bivalvia: Unionoidea): a model. Am Malacol Bull 14:35–40

Graf DL (2013) Patterns of freshwater bivalve global diversity and the state of phylogenetic studies on the Unionoida, Sphaeriidae, and Cyrenidae. Am Malacol Bull 31:135–153

Graf DL, Cummings KS (2006) Palaeoheterodont diversity (Mollusca: Trigonioida + Unionoida): what we know and what we wish we knew about freshwater mussel evolution. Zool J Linnean Soc 148:343–394

Graf DL, Cummings KS (2007) Review of the systematics and global diversity of freshwater mussel species (Bivalvia: Unionoida). J Molluscan Stud 73:291–314

Graf DL, Cummings KS (2021) A 'big data' approach to global freshwater mussel diversity (Bivalvia: Unionoida), with an updated checklist of genera and species. J Molluscan Stud 87: eyaa034

Graf DL, Jones H, Geneva AJ, Pfeiffer JM 3rd, Klunzinger MW (2015) Molecular phylogenetic analysis supports a Gondwanan origin of the Hyriidae (Mollusca: Bivalvia: Unionida) and the paraphyly of Australasian taxa. Mol Phylogenet Evol 85:1–9

Graf DL, O'Foighil D (2000) The evolution of brooding characters among the freshwater pearly mussels (Bivalvia: Unionoidea) of North America. J Molluscan Stud 66:157–170

Grande L, Hilton EJ (2006) An exquisitely preserved skeleton representing a primitive sturgeon from the Upper Cretaceous Judith River formation of Montana (Acipenseriformes: Acipenseridae: n. gen. and sp.). J Paleontol 80(S65):1–39

Grandstaff BS, Smith JB, Lamanna MC, Lacovara KJ, Abdel-Ghani MS (2012) *Bawitius*, gen. nov., a giant polypterid (Osteichthyes, Actinopterygii) from the Upper Cretaceous Bahariya formation of Egypt. J Vertebr Paleontol 32:17–26

Gray J (1988) Evolution of the freshwater ecosystem: the fossil record. Palaeogeogr Palaeoclimatol Palaeoecol 62:1–214

Gross M, Piller WE (2018) Fossil glochidia (Bivalvia: Unionida: Hyriidae) from the middle Miocene of western Amazonia (Peru). J Syst Palaeontol:1–12

Guerra D, Plazzi F, Stewart DT, Bogan AE, Hoeh WR, Breton S (2017) Evolution of sex-dependent mtDNA transmission in freshwater mussels (Bivalvia: Unionida). Sci Rep 7:1551. https://doi.org/10.1038/s41598-017-01708-1

Haag WH (2012) North American freshwater mussels: natural history, ecology and conservation. Cambridge University Press, Cambridge

Haag WR (2013) The role of fecundity and reproductive effort in defining life-history strategies of North American freshwater mussels. Biol Rev 88:745–766

Haag WR, Berg DJ, Garton DW, Farris JL (1993) Reduced survival and fitness in native bivalves in response to fouling by the introduced zebra mussel (*Dreissena polymorpha*) in western lake Erie. Can J Fish Aquat Sci 50:13–19

Haag WR, Butler RS, Hartfield PD (1995) An extraordinary reproductive strategy in freshwater bivalves: prey mimicry to facilitate larval dispersal. Freshw Biol 34:471–476

Haag WR, Warren ML Jr (1997) Host fishes and reproductive biology of 6 freshwater mussel species from the Mobile Basin, USA. J N Am Benthol Soc 16:576–585

Haag WR, Warren ML Jr (2000) Effects of light and presence of fish on lure display and larval release behaviours in two species of freshwater mussels. Anim Behav 60:879–886

Haag WR, Warren ML Jr (2003) Host fishes and infection strategies of freshwater mussels in large Mobile Basin streams, USA. J N Am Benthol Soc 22:78–91

Haag WR, Warren ML Jr, Shillingsford M (1999) Host fishes and host-attracting behavior of *Lampsilis altilis* and *Villosa vibex* (Bivalvia: Unionidae). Am Midl Nat 141:149–157

Habe T (1962) *Achasmea thalassemicola* sp. nov., a new commensal bivalve found in an echiuroid, *Thalassema mucosum* Ikeda. Venus 22:117–119

Hastie LC, Young MR (2001) Freshwater pearl mussel (*Margaritifera margaritifera*) glochidiosis in wild and farmed salmonid stocks in Scotland. Hydrobiologia 445:109–119

Hinzmann MF, Lopes-Lima M, Bobos I, Ferreira J, Domingues B, Machado J (2015) Morphological and chemical characterization of mineral concretions in the freshwater bivalve *Anodonta cygnea* (Unionidae). J Morphol 276:65–76

Hoeh WR, Black MB, Gustafson R, Bogan AE, Lutz RA, Vrijenhoek RC (1998) Testing alternative hypotheses of *Neotrigonia* (Bivalvia: Trigonioida) phylogenetic relationships using Cytochrome C oxidase subunit I DNA sequences. Malacologia 40:267–278

Hoeh WR, Bogan AE, Heard WH (2001) A phylogenetic perspective on the evolution of morphological and reproductive characteristics in the Unionoida. In: Bauer G, Wächtler K (eds) Ecology and evolution of the freshwater mussels Unionoida. Springer, Berlin, pp. 257-280

Hove MC, Neves JR (1994) Life history of the endangered James spinymussel *Pleurobema collina* (Conrad 1837) (Mollusca: Unionidae). Am Malacol Bull 11:29–40

Howerth EW, Keller AE (2006) Experimentally induced glochidiosis in smallmouth bass (*Micropterus dolomieu*). Vet Pathol 43:1004–1007

Huang X-C, Su J-H, Ouyang J-X, Ouyang S, Zhou C-H, Wu X-P (2019) Towards a global phylogeny of freshwater mussels (Bivalvia: Unionida): species delimitation of Chinese taxa, mitochondrial phylogenomics, and diversification patterns. Mol Phylogenet Evol 130:45–59

Huang XC, Wu RW, An CT, Xie GL, Su JH, Ouyang S, Zhou CH, Wu XP (2018) Reclassification of *Lamprotula rochechouartii* as *Margaritifera rochechouartii* comb. nov. (Bivalvia: Margaritiferidae) revealed by time-calibrated multi-locus phylogenetic analyses and mitochondrial phylogenomics of Unionoida. Mol Phylogenet Evol 120:297–306

Ishibashi R, Komaru A, Kondo T (2000) Sperm sphere in unionid mussels (Bivalvia: Unionidae). Zool Sci 17:947–950

Isaeva V, Kulikova V, Kasyanov V (2001) Bivalve molluscs, *Mytilus trossulus* and *Hiatella arctica*, as facultative epibionts of the crab, *Hemigrapsus sanguineus*, infested by the rhizocephalan, *Sacculina polygenea*. J Mar Biol Assoc U K 81:891–892

Jansen P, Mackenzie K, Hemmingsen W (1998) Some parasites and commensals of red king crabs, *Paralithodes camtschaticus* (Tilesius), in the Barents Sea. Bull Eur Assoc Fish Pathol 18:46–49

Jansen W, Bauer G, Zahner-Meike E (2001) Glochidial mortality in freshwater mussels. In: Bauer G, Wächtler K (eds) Ecology and evolution of the freshwater mussels Unionoida. Springer, Berlin, pp. 185-211

Jell PA (1980) Earliest known pelecypod on Earth—a new Early Cambrian genus from Australia. Alcheringa 4:233–239

Jokat W, Boebel T, König M, Meyer U (2003) Timing and geometry of early Gondwana breakup. J Geophys Res Solid Earth 108: 2428. https://doi.org/10.1029/2002JB001802

Jokela J, Palokangas P (1993) Reproductive tactics in *Anodonta* clams: parental host recognition. Anim Behav 46:618–620

Jordan DS, Gilbert JZ (1919) Fossil fishes of the Miocene (Monterey) formations. Leland Stanford Junior University Publications, University Series, Redwood City, pp 13–60

Joy JE, Jones LP (1973) Observations on the inflammatory response within the dermis of a white bass, *Morone chrysops* (Rafinesque), infected with *Lernaea cruciata* (Copepoda: Caligidea). J Fish Biol 5:21–23

Karna DW, Millemann RE (1978) Glochidiosis of salmonid fishes. III. Comparative susceptibility to natural infection with *Margaritifera margaritifera* (L.) (Pelecypoda: Margaritanidae) and associated histopathology. J Parasitol 64:528–537

Kat PW (1984) Parasitism and the Unionacea (Bivalvia). Biol Rev 59:189–207

Kato M (1998) Morphological and ecological adaptations in montacutid bivalves endo- and ecto-symbiotic with holothurians. Can J Zool 76:1403–1410

Kear BP, Godthelp H (2008) Inferred vertebrate bite marks on an Early Cretaceous unionoid bivalve from Lightning Ridge, New South Wales, Australia. Alcheringa 32:65–71

Kemp D (1985) The Selsey division (Bracklesham Group) at Lee-on-the-Solent, Gosport, (Hants). Tertiary Res 7:35–44

Kerr G, Corfield J (1998) Association betweenthe ghost shrimp *Trypacea australiensis* Dana 1852 (Crustacea: Decapoda) and a small deposit feeding bivalve *Mysella vitrea* Laserson 1952 (Mollusca: Letonidae). Mar Freshw Res 49:801–806

Kimura T, Soutome Y, Sekiguchi H (2004) Larval development of the brackish water clam *Corbicula japonica* (Bivalvia: Corbiculidae), with special reference to morphological comparison with concurrent tidal flat bivalves. Venus 63:33–48

Kleemann K (1996) Biocorrosion by bivalves. Mar Ecol 17:145–158

Klug C, Hagdorn H, Montenari M (2005) Phosphatized soft-tissue in Triassic bivalves. Palaeontology 48:833–852

Klunzinger MW, Thomson GJ, Beatty SJ, Morgan DL, Lymbery AJ (2013) Morphological and morphometrical description of the glochidia of *Westralunio carteri* Iredale, 1934 (Bivalvia: Unionoida: Hyriidae). Molluscan Res 33:104–109

Knight RI, Morris NJ, Todd JA, Howard LE, Ball AD, Jagt J (2014) Exceptional preservation of a novel gill grade in large Cretaceous inoceramids: systematic and palaeobiological implications. Palaeontology 57:37–54

Knope ML (2013) Phylogenetics of the marine sculpins (Teleostei: Cottidae) of the North American Pacific Coast. Mol Phylogenet Evol 66:341–349

Lafferty KD (1993) Ecology and parasites of the spiny sand crab, *Blepharipoda occidentalis*. Am Zool 33:19A

Lauckner G (1983) Diseases of Mollusca: bivalvia. In: Kinne O (ed) Diseases of marine animals. Biologische Anstalt Helgoland, Hamburg

Lecaudey LA, Schliewen UK, Osinov AG, Taylor EB, Bernatchez L, Weiss SJ (2018) Inferring phylogenetic structure, hybridization and divergence times within Salmoninae (Teleostei: Salmonidae) using RAD-sequencing. Mol Phylogenet Evol 124:82–99

Lefevre G, Curtis WC (1910a) The marsupium of the Unionidae. Biol Bull 19:31–34

Lefevre G, Curtis WC (1910b) Reproduction and parasitism in the Unionidae. J Exp Zool 9:79–115

Lefevre G, Curtis WC (1911) Metamorphosis without Parasitism in the Unionidae. Science 33:863–865

Lefevre G, Curtis WC (1912) Studies on the reproduction and artificial propagation of fresh-water mussels. Bull US Bureau Fish 30:105–201

Lellis WA, White BSJ, Cole JC, Johnson CS, Devers JL, Gray ES, Galbraith HS (2013) Newly documented host fishes for the eastern *Elliptio* mussel *Elliptio complanata*. J Fish Wildl Manag 4:75–85

Leung TL (2017) Fossils of parasites: what can the fossil record tell us about the evolution of parasitism? Biol Rev Cambridge Philos Soc 92:410–430

Lewandowski KB (1976) Unionidae as a substratum for *Dreissena polymorpha* Pall. Polskie Archiwum Hydrobiologii 23:409–420

Li J, O'Foighil D (2012) Host-specific morphologies but no host races in the commensal bivalve *Neaeromya rugifera*. Invertebr Biol 131:197–203

Li J, O'Foighil D, Middelfart P (2012) The evolutionary ecology of biotic association in a mega-diverse bivalve superfamily: sponsorship required for permanent residency in sediment. PLoS One 7:e42121. https://doi.org/10.1371/journal.pone.0042121

Li J, O'Foighil D, Strong EE (2016) Commensal associations and benthic habitats shape macro-evolution of the bivalve clade Galeommatoidea. Proc R Soc B Biol Sci 283:20161006. https://doi.org/10.1098/rspb.2016.1006

Liu GB, Yin CZ, Wang XH, Luo YM, Wang SY (2003) New discovered fishes from *Keichousaurus*-bearing horizon of Late Triassic in Xingyi of Guizhou. Acta Palaeontol Sin 42:346–366

Lopatin AV (1996) The stratigraphy and small mammals from the Aral Formation, the Altynshokysu Locality, Northern Aral Area. Stratigr Geol Correl 4:65–79

Lopes-Lima M, Froufe E, Do VT, Ghamizi M, Mock KE, Kebapci U, Klishko O, Kovitvadhi S, Kovitvadhi U, Paulo OS, Pfeiffer JM 3rd, Raley M, Riccardi N, Sereflisan H, Sousa R, Teixeira A, Varandas S, Wu X, Zanatta DT, Zieritz A, Bogan AE (2017) Phylogeny of the most species-rich freshwater bivalve family (Bivalvia: Unionida: Unionidae): defining modern subfamilies and tribes. Mol Phylogenet Evol 106:174–191

López-Arbarello A (2012) Phylogenetic interrelationships of ginglymodian fishes (Actinopterygii: Neopterygii). PLoS One 7:e39370. https://doi.org/10.1371/journal.pone.0039370

Lucas SG, Tanner LH (2007) Tetrapod biostratigraphy and biochronology of the Triassic–Jurassic transition on the southern Colorado Plateau, USA. Palaeogeogr Palaeoclimatol Palaeoecol 244:242–256

Lützen J, Hong JS, Yamashita H (2009) *Koreamya arculata* (A. Adams, 1856) gen. nov. (Galeommatoidea: Montacutidae), a commensal bivalve associated with the inarticulate brachiopod *Lingula anatina*. J Conchol 39:669–679

Ma B, Jiang H, Sun P, Chen J, Li L, Zhang X, Yuan L (2016) Phylogeny and dating of divergences within the genus *Thymallus* (Salmonidae: Thymallinae) using complete mitochondrial genomes. Mitochondrial DNA 27:3602–3611

Machado J, Lopes-Lima M (2011) Calcification mechanism in freshwater mussels: potential targets for cadmium. Toxicol Environ Chem 93:1778–1787

Mackie GL (1984) Reproduction in Bivalvia. In: Tompa AS, Verdonk NH, van de Biggelaar JAM (eds) The Mollusca. Reproduction. Academic, New York

Malchus N, Sartori AF (2013) Part N, Revised, Volume 1, Chapter 4: The early shell: Ontogeny, features, and evolution. Treatise Online 61:1–114, 44 fig.

Maldanis L, Carvalho M, Almeida MR, Freitas FI, de Andrade JA, Nunes RS, Rochitte CE, Poppi RJ, Freitas RO, Rodrigues F, Siljestrom S, Lima FA, Galante D, Carvalho IS, Perez CA, de Carvalho MR, Bettini J, Fernandez V, Xavier-Neto J (2016) Heart fossilization is possible and informs the evolution of cardiac outflow tract in vertebrates. elife 5:e14698. https://doi.org/10.7554/eLife.14698.001

Mansur MCD, Pereira D, Bergonci PEA, Pimpão DM, Barradas JRdS, Sabaj MH (2019) Morphological assessment of *Rheodreissena* (Bivalvia: Veneroida: Dreissenidae) with an updated diagnosis of the genus, descriptions of two new species, redescription of *R. lopesi*, and the first account of larval brooding in New World dreissenids. Proc Acad Natl Sci Phila 166:1–45

Martill DM (1988) Preservation of fish in the Cretaceous Santana Formation of Brazil. Palaeontology 31:1–18

Martill DM, Harper EM (1990) An application of CPD to the comparison of modern and fossilized soft tissues of fishes. Palaeontology 33:423–428

Marwaha J, Jensen KH, Jakobsen PJ, Geist J (2017) Duration of the parasitic phase determines subsequent performance in juvenile freshwater pearl mussels (Margaritifera margaritifera). Ecol Evol 7:1375–1383

Mauna C, Casadío S, Parras A, Pascual M (2005) Abundance and distribution of *Lithophaga* (Mytilidae) in extant and fossil oysters: taphonomic and paleobiological implications. Ameghiniana 42:395–405

McElwain A, Bullard SA (2014) Histological atlas of freshwater mussels (Bivalvia, Unionidae): *Villosa nebulosa* (Ambleminae: Lampsilini), *Fusconaia cerina* (Ambleminae: Pleurobemini) and *Strophitus connasaugaensis* (Unioninae: Anodontini). Malacologia 57:99–239

McInnes SJ, Pugh PJA (1998) Biogeography of limno-terrestrial Tardigrada, with particular reference to the Antarctic fauna. J Biogeogr 25:31–36

McLain DC, Ross MR (2005) Reproduction based on local patch size of *Alasmidonta heterodon* and dispersal by its darter host in the Mill River, Massachusetts, USA. J N Am Benthol Soc 24:139–147

McLeod JM, Jelks HL, Pursifull S, Johnson NA (2017) Characterizing the early life history of an imperiled freshwater mussel (*Ptychobranchus jonesi*) with host-fish determination and fecundity estimation. Freshwater Sci 36:338–350

McMichael DF (1957) A rewiew of the fossil freshwater mussels (Mollusca, Pelecypoda) of Australasia. Proc Linnean Soc NSW 81:222–244

McMichael DF, Hiscock ID (1958) A monograph of the freshwater mussels (Mollusca: Pelecypoda) of the Australian region. Aust J Mar Freshwat Res 9:372–508

Merrill AS (1960) Living inclusions in the shell of the sea scallop *Placopecten magellanicus*. Ecology 41:385–386

Meyers TR, Millemann RE (1977) Glochidiosis of salmonid fishes. I. Comparitive susceptibility to experimental infection with *Margaritifera margaritifera* (L.) (Pelecypoda: Margaritanidae). J Parasitol 63:728–733

Mikkelsen PM, Bieler R (1989) Biology and comparitive anatomy of *Divariscintilla yoyo* and *D. troglodytes*, two new species of Galeommatidae (Bivalvia) from stomatopod burrows in eastern Florida. Malacologia 31:175–195

Modesto V, Ilarri M, Souza Allan T, Lopes-Lima M, Douda K, Clavero M, Sousa R (2018) Fish and mussels: importance of fish for freshwater mussel conservation. Fish Fish 19:244–259
Mokady O, Arazi G, Bonar DB, Loya Y (1992) Settlement and metamorphosis specificity of *Lithophaga simplex* Iredale (Bivalvia: Mytilidae) on Red Sea corals. J Exp Mar Biol Ecol 162:243–251
Mommsen TP, Vijayan MM, Moon TW (1999) Cortisol in teleosts: dynamics, mechanisms of action, and metabolic regulation. Rev Fish Biol Fish 9:211–268
Morton B (1972) Some aspects of the functional morphology and biology of *Pseudopythina subsinuata* (Bivalvia: Leptonacea) commensal on stomatopod crustaceans. J Zool 166:79–96
Morton B (1978) The biology and functional morphology of *Philobrya munita* (Bivalvia: Philobryidae). J Zool 185:173–196
Morton B (1987) The functional morphology of *Neotrigonia margaritacea* (Bivalvia: Trigoniacea), with a discussion of phylogenetic affinities. Rec Aust Mus 39:339–354
Morton B (1988) Partnerships in the sea: Hong Kong's marine symbioses. Hong Kong University Press, Hong Kong
Morton B (2017) The biology and functional morphology of *Curvemysella paula* (A. Adams, 1856) (Bivalvia: Galeommatoidea: Montacutidae) commensal with *Diogenes goniochirus* Forest, 1956 (Crustacea: Paguridea) on China Beach, Danang, Vietnam. Am Malacol Bull 35:15–30
Morton B, Puljas S (2012) Life-history strategy, with ctenidial and pallial larval brooding, of the troglodytic 'living fossil' *Congeria kusceri* (Bivalvia: Dreissenidae) from the subterranean Dinaric Alpine karst of Croatia. Biol J Linn Soc 108:294–314
Morton B, Scott PH (1989) The Hong Kong Galeommatacea (Mollusca: Bivalvia) and their hosts, with description of new species. Asian Marine Biol 6:129–160
Morton B, Scott PJB (1980) Morphological and functional specializations of the shell, musculature and pallial glands in the Lithophaginae (Mollusca: Bivalvia). J Zool 192:179–203
Morton JE (1957) The habits of *Scintillona zelandica* (Odhner) 1924 (Lamellibranchia: Galeommatidae). J Molluscan Stud 32:185–188
Murphy G (1942) Relationship of the fresh-water mussel to trout in the Truckee River. Cal Fish Game 28:89–102
Mutvei H, Westermark T (2001) How environmental information can be obtained from naiad shells. In: Bauer G, Wächtler K (eds) Ecology and evolution of the freshwater mussels Unionoida. Springer, Berlin, pp. 367-379
Nagler C, Haug JT (2015) From fossil parasitoids to vectors: insects as parasites and hosts. Topics in Parasitology 90:137-200
Narchi W (1969) On *Pseudopythina rugifera* (Carpenter, 1864) (Bivalvia). Veliger 12:43–52
Near TJ, Bolnick DI, Wainwright PC (2007) Fossil calibrations and molecular divergence time estimates in centrarchid fishes (Teleostei: Centrarchidae). Evolution 59:1768–1782
Near TJ, Bossu CM, Bradburd GS, Carlson RL, Harrington RC, Hollingsworth PR Jr, Keck BP, Etnier DA (2011) Phylogeny and temporal diversification of darters (Percidae: Etheostomatinae). Syst Biol 60:565–595
Neves RJ, Weaver LR, Zale AV (1985) An evaluation of host fish suitability for glochidia of *Villosa vanuxemi* and *V. nebulosa* (Pelecypoda: Unionidae). Am Midl Nat 113:13–19
Newell ND, Boyd DW (1975) Parallel evolution in early trigoniacean bivalves. Bull Am Mus Nat Hist 154:53–162
Nezlin LP, Cunjak RA, Zotin AA, Ziuganov VV (1994) Glochidium morphology of the freshwater pearl mussel (*Margaritifera margaritifera*) and glochidiosis of Atlantic salmon (*Salmo salar*): a study by scanning electron microscopy. Can J Zool 72:15–21
Nielsen E (1955) Notes on Triassic fishes from Madagascar. Medd fra Dansk Geol Forening 12:563–578
O'Brien CA, Box JB (1999) Reproductive biology and juvenile recruitment of the Shinyrayed Pocketbook, *Lampsilis subangulata* (Bivalvia: Unionidae) in the Gulf Coastal Plain. Am Midl Nat 142:129–140

O'Dee SH, Watters GT (2000) New or confirmed host identifications for ten freshwater mussels. In: Proceedings of the conservation, captive care, and propagation of freshwater mussels symposium, 1998, p 77–82

O'Foighil D (1986) Prodissoconch morphology is environmentally modified in the brooding bivalve Lasaea subviridis. Mar Biol 92:517–524

O'Foighil D, Graf DL (2000) Prodissoconch morphology of the relict marine paleoheterodont *Neotrigonia margaritacea* (Mollusca: Bivalvia) indicates a non-planktotrophic prejuvenile ontogeny. J Mar Biol Assoc UK 80:173–174

Ogle J (1976) The occurrence of the date mussel, *Lithophaga bisulcata* (Mytilidae) in living oysters off Galveston, Texas. Nautilus 90:74–76

Olsen MM, Kania PW, Heinecke RD, Skjoedt K, Rasmussen KJ, Buchmann K (2011) Cellular and humoral factors involved in the response of rainbow trout gills to *Ichthyophthirius multifiliis* infections: molecular and immunohistochemical studies. Fish Shellfish Immunol 30:859–869

Osés GL, Petri S, Voltani CG, Prado GMEM, Galante D, Rizzutto MA, Rudnitzki ID, da Silva EP, Rodrigues F, Rangel EC, Sucerquia PA, Pacheco MLAF (2017) Deciphering pyritization-kerogenization gradient for fish soft-tissue preservation. Sci Rep 7:1468. https://doi.org/10.7554/eLife.14698.001

Owada M (2009) Organic sheets in the shells of endolithic mytilids (Bivalvia: Mytilidae). Paleontol Res 13:159–166

Owen G (1955) Observations on the stomach and digestive diverticula of the lamellibranchia. Journal of Cell Science 1955(s3-96): 517-537

Paling JE (1968) A method of estimating the relative volumes of water flowing over the different gills of a freshwater fish. J Exp Biol 48:533–544

Parker RS, Hackney CT, Vidrine MF (1984) Ecology and reproductive strategy of a South Louisiana Freshwater Mussel, *Glebula rotundata* (Lamarck) (Unionidae: Lampsilini). Freshw Invertebr Biol 3:53–58

Parkhaev PY (2008) The Early Cambrian radiation of mollusca. In: Ponder WF, Lindberg DR (eds) Phylogeny and evolution of the Mollusca. University of California Press, Berkeley, pp. 33-70

Parodiz JJ, Bonetto AA (1963) Taxonomy and zoogeographic relationships of the South American naiades (Pelecypoda: Unionacea and Mutelacea). Malacologia 1:179–214

Patterson MA (2018) Biology of freshwater mussels. In: Patterson MA, Mair RA, Eckert NL, Gatenby CM, Brady T, Jones JW, Simmons BR, Devers JL (eds) Freshwater mussel propagation for restoration. Cambridge University Press, Cambridge, pp. 25–57

Patterson MA, Jones JW, Gatenby CM (2018) Why propagate freshwater mussels? In: Patterson MA, Mair RA, Eckert NL, Gatenby CM, Brady T, Jones JW, Simmons BR, Devers JL (eds) Freshwater mussel propagation for restoration. Cambridge University Press, Cambridge, pp. 1–24

Peecook BR, Steyer JS, Tabor NJ, Smith RMH (2017) Updated geology and vertebrate paleontology of the Triassic Ntawere formation of northeastern Zambia, with special emphasis on the archosauromorphs. J Vertebr Paleontol 37:8–38

Pennak WR (1971) Toward a classification of lotic habitats. Hydrobiologia 38:321–334

Petit G, Khalloufi B (2012) Paleopathology of a fossil fish from the Solnhofen Lagerstätte (Upper Jurassic, southern Germany). Int J Paleopathol 2:42–44

Petit H, Davis WL, Jones RG (1978) Morphological studies on the mantle of the fresh-water mussel *Amblema* (Unionidae): scanning electron microscopy. Tissue Cell 10:619–627

Pfeiffer JM, Breinholt JW, Page LM (2019) Unioverse: a phylogenomic resource for reconstructing the evolution of freshwater mussels (Bivalvia, Unionoida). Mol Phylogenet Evol 137:114–126

Pfeiffer JM, Graf DL, Cummings KS, Page LM (2018) Molecular phylogeny and taxonomic revision of two enigmatic freshwater mussel genera (Bivalvia: Unionidae incertae sedis: Harmandia and Unionetta) reveals a diverse clade of Southeast Asian Parreysiinae. Journal of Molluscan Studies 84:404–416

Pfeiffer JM, Graf DL (2015) Evolution of bilaterally asymmetrical larvae in freshwater mussels (Bivalvia: Unionoida: Unionidae). Zool J Linnean Soc 175:307–318

Pflugfelder O (1951) Uber das Parasit-Wirtsverhaltnis von *Anodonta*-Glochidien und Spiegelkarpfen (*Cyprinus carpio* L.). Z Parasitenkd 15:119–133
Pinna G (1985) Exceptional preservation in the Jurassic of Osteno. Philos Trans R Soc B 311:171–180
Pimpão DM, Mansur MCD, Bergonci PEA, Beasley CR (2012) Comparative morphometry and morphology of glochidial shells of Amazonian Hyriidae (Mollusca: Bivalvia: Unionida). Am Malacol Bull 30:73–84
Poinar GO (2015) The geological record of parasitic nematode evolution. Advances in parasitology 90:53–92
Pojeta JJ (2000) Cambrian pelecypoda (Mollusca). Am Malacol Bull 15:157–166
Ponder WF (1965) The biology of the genus *Arthritica*. Transact Royal Soc New Zealand Zool 6:75–86
Poulter R, Oliver PG, Hauton C, Sanders T, Ciotti BJ (2018) Infestation of shore crab gills by a free-living mussel species. Mar Biodivers 48:1241–1246
Prezant RS, Chalermwat K (1984) Flotation of the Bivalve *Corbicula fluminea* as a means of dispersal. Science 225:1491–1493
Pugh PJA, McInnes SJ (1998) The origin of Arctic terrestrial and freshwater tardigrades. Polar Biol 19:177–182
Pugh PJA, Scott B (2002) Biodiversity and biogeography of non-marine Mollusca on the islands of the Southern Ocean. J Nat Hist 36:927–952
Pulquerio MJ, Nichols RA (2007) Dates from the molecular clock: how wrong can we be? Trends Ecol Evol 22:180–184
Purnell MA, Donoghue PJC, Gabbott SE, McNamara ME, Murdock DJE, Sansom RS, Smith A (2018) Experimental analysis of soft-tissue fossilization: opening the black box. Palaeontology 61:317–323
Quilhac A, Sire JY (1999) Spreading, proliferation, and differentiation of the epidermis after wounding a cichlid fish, *Hemichromis bimaculatus*. Anat Rec 254:435–451
Qvarnström M, Niedźwiedzki G, Tafforeau P, Žigaitė Ž, Ahlberg PE (2017) Synchrotron phase-contrast microtomography of coprolites generates novel palaeobiological data. Sci Rep 7:2723. https://doi.org/10.1038/s41598-017-02893-9
Reeside JBJ (1927) Two new unionid pelecypods from the Upper Triassic. J Wash Acad Sci 17:476–478
Reis J, Araujo R (2016) Life history of the freshwater mussel *Unio tumidiformis* (Bivalvia: Unionidae) in a temporary Mediterranean-type stream. Invertebr Biol 135:31–45
Reis J, Collares-Pereira JM, Araujo R (2014) Host specificity and metamorphosis of the glochidium of the freshwater mussel *Unio tumidiformis* (Bivalvia: Unionidae). Folia Parasitol 61:81–89
Reznick DN, Ghalambor CK (2001) The population ecology of contemporary adaptations: what empirical studies reveal about the conditions that promote adaptive evolution. Genetica 112:183–198
Ricciardi A, Neves JR, Rasmussen JB (1998) Impending extinctions of North American freshwater mussels (Unionoida) following the zebra mussel (*Dreissena polymorpha*) invasion. J Anim Ecol 67:613–619
Richard PE, Dietz TH, Silverman H (1991) Structure of the gill during reproduction in the unionids *Anodonta grandis, Ligumia subrostrata, and Carunculina parva taxasensis*. Can J Zool 69:1744–1754
Ridewood WG (1903) On the structure of the gills of the lamellibranchia. Philos Trans R Soc B 195:147–284
Roberts D, Breen JP (1985) The biology and functional anatomy of *Modiolarca tumida* (*Musculus marmoratus*) Bivalvia: Mytilidae. J Zool 205:435–443
Robin N, Velasquez M, Boura A, Garcia G, Jauvion C, Boiteau J-M, Gomez B, Daviero-Gomez V, Valentin X (2018) The oldest shipworms (Bivalvia, Pholadoidea, Teredinidae) preserved with soft parts (western France): insights into the fossil record and evolution of Pholadoidea. Palaeontology 61:905–918

Rogers CL, Dimock RV (2003) Acquired resistance of bluegill sunfish *Lepomis macrochirus* to glochidia larvae of the freshwater mussel *Utterbackia imbecillis* (Bivalvia: Unionidae) after multiple infections. J Parasitol 89:51–56

Rogers-Lowery CL, Dimock RV Jr (2006) Encapsulation of attached ectoparasitic glochidia larvae of freshwater mussels by epithelial tissue on fins of naive and resistant host fish. Biol Bull 210:51–63

Rogers-Lowery CL, Dimock RV Jr, Kuhn RE (2007) Antibody response of bluegill sunfish during development of acquired resistance against the larvae of the freshwater mussel *Utterbackia imbecillis*. Dev Comp Immunol 31:143–155

Sahni M, Tewari A (1958) New Unionids from the Triassic (Gondwana) rocks of Tihki, Vindhya Pradesh and Maleri, Hyderabad, Deccan. Records Geol Survey India 87:406–417

Santos-Neto GDC, Beasley CR, Schneider H, Pimpao DM, Hoeh WR, Simone LRL, Tagliaro CH (2016) Genetic relationships among freshwater mussel species from fifteen Amazonian rivers and inferences on the evolution of the Hyriidae (Mollusca: Bivalvia: Unionida). Mol Phylogenet Evol 100:148–159

Sato S, Owada M, Haga T, Hong J-S, Lützen J, Yamashita H (2011) Genus-specific commensalism of the galeommatoid bivalve *Koreamya arcuata* (A. Adams, 1856) associated with lingulid brachiopods. Molluscan Res 31:95–105

Scaps P, Denis V, Haapkylä J (2008) Organisms associated with live scleractinian corals as indicators of coral reefs status in the Wakatobi Marine National Park (SE Sulawesi, Indonesia). In: Proceedings of the 11th International Coral Reef Symposium, Ft. Lauderdale, Florida 2:816–820

Schartum E, Mortensen S, Pittman K, Jakobsen PJ (2017) From pedal to filter feeding: ctenidial organogenesis and implications for feeding in the postlarval freshwater pearl mussel *Margaritifera margaritifera* (Linnaeus, 1758). J Molluscan Stud 83:36–42

Schwalb A, Push MT (2007) Horizontal and vertical movements of unionid mussels in a lowland river. J N Am Benthol Soc 26:261–272

Schwalb AN, Cottenie K, Poos Mark S, Ackerman Josef D (2011a) Dispersal limitation of unionid mussels and implications for their conservation. Freshw Biol 56:1509–1518

Schwalb AN, Poos MS, Ackerman JD (2011b) Movement of logperch—the obligate host fish for endangered snuffbox mussels: implications for mussel dispersal. Aquat Sci 73:223–231

Scott PH (1987) A preliminary review of *Mysella* (Bivalvia, Montacutidae) from the northwestern Pacific. Western Soc Malacol 19:13–14

Scott PJB (1986) A new species of *Lithophaga* (Bivalvia: Lithophaginae) boring corals in the Caribbean. J Molluscan Stud 52:55–61

Scott PJB (1988) Distribution, habitat and morphology of the Caribbean coral—and rock-boring bivalve, *Lithophaga bisulcata* (d'Orbigny) (Mytilidae: Lithophaginae). J Molluscan Stud 54:83–95

Scott PJB, Risk MJ (1988) The effect of *Lithophaga* (Bivalvia: Mytilidae) boreholes on the strength of the coral *Porites lobata*. Coral Reefs 7:145–151

Segers H (2008) Global diversity of rotifers (Rotifera) in freshwater. In: Balian EV, Lévêque C, Segers H, Martens K (eds) Freshwater animal diversity assessment. Springer, Dordrecht, pp. 49–59

Sepkoski JJ Jr, Rex MA (1974) Distribution of freshwater mussels: coastal rivers as biogeographic islands. Syst Zool 23:165–188

Seton M, Müller RD, Zahirovic S, Gaina C, Torsvik T, Shephard G, Talsma A, Gurnis M, Turner M, Maus S, Chandler M (2012) Global continental and ocean basin reconstructions since 200 Ma. Earth Sci Rev 113:212–270

Shedko SV, Miroshnichenko IL, Nemkova GA (2013) Phylogeny of salmonids (Salmoniformes: Salmonidae) and its molecular dating: analysis of mtDNA data. Russ J Genet 49:623–637

Silva-Souza AT, Guardia-Felipi P, Arrebola NR (2011) Embryonic development of *Anodontites trapesialis* (Lamarck, 1819) (Bivalvia: Mycetopodidae). Braz J Biol 71:139–144

Silverman H, Steffens WL, Dietz TH (1985) Calcium from extracellular concretions in the gills of freshwater unionid mussels is mobilized during reproduction. J Exp Zool 236:137–147

Silverman H, Richard PE, Goddard RH, Dietz TH (1989) Intracellular formation of calcium concretions by phagocytic cells in freshwater mussels. Can J Zool 67:198–207
Silverman H, McNeil JW, Dietz TH (1987a) Interaction of trace metals Zn, Cd, and Mn, with Ca concretions in the gills of freshwater unionid mussels. Can J Zool 65:828–832
Silverman H, Todd Kays W, Dietz TH (1987b) Maternal calcium contribution to glochidial shells in freshwater mussels (Eulamellibranchia: Unionidae). J Exp Zool 242:137–146
Simone L, Mezzalira S (1993) Vestígios de partes moles em um Bivalve fóssil (Unionoida, Mycetopodidae) do Grupo Bauru (Cretáceo Superioi), São Paulo, Brasil. An Acad Bras Cienc 65:155–159
Skawina A (2010) Experimental decay of gills in freshwater bivalves as a key to understand their preservation in Upper Triassic lacustrine deposits. PALAIOS 25:215–220
Skawina A, Dzik J (2011) Umbonal musculature and relationships of the Late Triassic filibranch unionoid bivalves. Zool J Linnean Soc 163:863–883
Skrzycki P (2015) New species of lungfish (Sarcopterygii, Dipnoi) from the Late Triassic Krasiejów site in Poland, with remarks on the ontogeny of Triassic dipnoan tooth plates. J Vertebr Paleontol 35:e964357. https://doi.org/10.1080/02724634.2015.964357
Slavík O, Horký P, Douda K, Velisek J, Kolářová J, Lepic P (2017) Parasite-induced increases in the energy costs of movement of host freshwater fish. Physiol Behav 171:127–134
Soler J, Wantzen KM, Jugé P, Araujo R (2018) Brooding and glochidia release in *Margaritifera auricularia* (Spengler, 1793) (Unionoida: Margaritiferidae). J Molluscan Stud 84:182–189
St. John White B, Paola Ferreri C, Lellis WA, Wicklow BJ, Cole JC (2017) Geographic variation in host fish use and larval metamorphosis for the endangered dwarf wedgemussel. Aquat Conserv 27:909–918
Strayer D (1983) The effects of surface geology and stream size on freshwater mussel (Bivalvia, Unionidae) distribution in southeastern Michigan, USA. Freshw Biol 13:253–264
Strayer DL (2008) Freshwater mussel ecology. A multifactor approach to distribution and abundance. University of California Press, Berkeley
Strayer DL, Smith LC (1996) Relationships between zebra mussels (*Dreissena polymorpha*) and unionid clams during the early stages of the zebra mussel invasion of the Hudson River. Freshw Biol 36:771–779
Taeubert J-E, Gum B, Geist J (2013) Variable development and excystment of freshwater pearl mussel (*Margaritifera margaritifera* L.) at constant temperature. Limnologica 43:319–322
Taeubert JE, El-Nobi G, Geist J (2014) Effects of water temperature on the larval parasitic stage of the thick-shelled river mussel (*Unio crassus*). Aquat Conserv 24:231–237
Tankersley RA (1996) Effect of larval brooding on the feeding physiology of freshwater unionid mussels. Invertebr Biol 115:243–255
Tankersley RA, Dimock RV Jr (1992) Quantitative analysis of the structure and function of the marsupial gills of the freshwater mussel *Anodonta cataracta*. Biol Bull 182:145–154
Tankersley RA, Dimock RV Jr (1993a) The effect of larval brooding on the respiratory physiology of the freshwater unionid mussel *Pyganodon cataracta*. Am Midl Nat 130:146–163
Tankersley RV, Dimock R Jr (1993b) The effect of larval brooding on the filtration rate and particle-retention efficiency of *Pyganodon cataracta* (Bivalvia: Unionidae). Can J Zool 71:1934–1944
Tedla S, Fernando CH (1969) Observations on the glochidia of *Lampsilis radiata* (Gmelin) infesting yellow perch, *Perca flavescens* (Mitchill) in the Bay of Quinte, Lake Ontario. Can J Zool 47:705–712
Terui A, Miyazaki Y, Yoshioka A, Kadoya T, Jopp F, Washitani I (2014) Dispersal of larvae of *Margaritifera laevis* by its host fish. Freshwater Sci 33:112–123
Tevesz MJS (1975) Structure and habits of the 'living fossil' pelecypod *Neotrigonia*. Lethaia 8:321–327
Thompson DL, Stilwell JD (2010) Early Aptian (Early Cretaceous) freshwater bivalves from the Australian–Antarctic rift, southeast Victoria. Alcheringa 34:345–357
Tsubaki R, Kato M (2014) A novel filtering mutualism between a sponge host and its endosymbiotic bivalves. PLoS One 9:e108885. https://doi.org/10.1371/journal.pone.0108885

Turner RD (1954) The Family Pholadidae in the Western Atlantic and the Eastern Pacific. Part I—Pholadinae. Johnsonia 3:1–63

Turner RD (1955) The Family Pholadidae in the Western Atlantic and the Eastern Pacific. Part II—Martesiinae, Jouannetiinae and Xylophaginae. Johnsonia 3:65–160

Upeniece I (2001) The unique fossil assemblage from the Lode Quarry (Upper Devonian, Latvia). Fossil Record 4:101–119

Valentich-Scott P, Ó Foighil D, Li J (2013) Where's *Waldo*? A new commensal species, *Waldo arthuri* (Mollusca, Bivalvia, Galeommatidae), from the Northeastern Pacific Ocean. Zookeys:67–80

Van Damme D, Bogan AE, Dierick M (2015) A revision of the Mesozoic naiads (Unionoida) of Africa and the biogeographic implications. Earth Sci Rev 147:141–200

Van Snik Gray E, Lellis W, Cole J, Johnson CS (2009) Host identification for *Strophitus undulatus* (Bivalvia: Unionidae), the Creeper, in the Upper Susquehanna River Basin, Pennsylvania. Am Midl Nat 147:153–161

Vaughn CC (2012) Life history traits and abundance can predict local colonisation and extinction rates of freshwater mussels. Freshw Biol 57:982–992

Vicentini H (2005) Unusual spurting behaviour of the freshwater mussel *Unio crassus*. J Molluscan Stud 71:409–410

Wächtler K, Dreher-Mansur MC, Richter T (2001) Larval types and early postlarval biology in naiads (Unionoida). In: Bauer G, Wächtler K (eds) Ecology and evolution of the freshwater mussels Unionoida. Springer, Berlin, pp. 93–125

Walker KF, Byrne M, Hickey CW, Roper DS (2001) Freshwater mussels (Hyriidae) of Australasia. In: Bauer G, Wächtler K (eds) Ecology and evolution of the freshwater mussels Unionoida. Springer, Berlin, pp. 5–31

Waller DL, Mitchell LG (1989) Gill tissue reactions in walleye *Stizostedion vitreum vitreum* and common carp *Cyprinus carpio* to glochidia of the freshwater mussel *Lampsilis radiata siliquoidea*. Dis Aquat Org 6:81–87

Waller TR (1998) Origin of the molluscan class Bivalvia and a phylogeny of major groups. In: Johnston PA, Haggart JW (eds) Bivalves: an eon of evolution—palaeobiological studies honoring Norman D. Newell. University of Calgary Press, Calgary, pp. 1-45

Wang NZ, Jin F (2007) Actinopterygian fishes from the Permian-Triassic boundary beds in Zhejiang and Jiangxi provinces, South China and fish mass extinction, recovery and radiation. Verteb Palasiat 45:307–329

Wang X, Gan X, Li J, Mayden RL, He S (2012) Cyprinid phylogeny based on Bayesian and maximum likelihood analyses of partitioned data: implications for Cyprinidae systematics. Sci China Life Sci 55:761–773

Warnock RCM, Engelstädter J (2021) The molecular clock as tool for understanding host-parasite evolution. In: De Baets K, Huntley JW (eds) The evolution and fossil record of parasitism:Coevolution and paleoparasitological techniques. Topics in Geobiology 50

Watters GT (1992) Unionids, fishes, and the species-area curve. J Biogeogr 19:481–490

Watters GT (2001) The evolution of the Unionacea in North America, and its implications for the worldwide fauna. In: Bauer G, Wächtler K (eds) Ecology and evolution of the freshwater mussels Unionoida. Springer, Berlin, pp. 281–307

Watters GT (2007) A brief look at freshwater mussel (Unionacea) biology. In: Farris JL, Van Hassel JH (eds) Freshwater bivalve ecotoxicology. Society of Environmental Toxicology and Chemistry (SETAC). CRC Press, New York, pp. 51–64

Watters GT, O'Dee SH (1996) Shedding of untransformed glochidia by fishes parasitized by *Lampsilis fasciola* Rafinesque, 1820 (Mollusca: Bivalvia: Unionidae): evidence of acquired immunity in the field? J Freshw Ecol 11:383–389

Watters GT, O'Dee SH (1998) Metamorphosis of freshwater mussel glochidia (Bivalvia: Unionidae) on amphibians and exotic fishes. Am Midl Nat 139:49–57

Wear RG (1966) Physiological and ecological studies on the bivalve mollusk *Arthritica bifurca* (Webster, 1908) living commensally with the tubicolous polychaete *Pectinaria australis* Ehlers, 1905. Biol Bull 130:141–149

Weinstein SB, Kuris AM (2016) Independent origins of parasitism in Animalia. Biol Lett 12:20160324

Whelan NV, Geneva AJ, Graf DL (2011) Molecular phylogenetic analysis of tropical freshwater mussels (Mollusca: Bivalvia: Unionoida) resolves the position of *Coelatura* and supports a monophyletic Unionidae. Mol Phylogenet Evol 61:504–514

Whyte MA (1991) Phosphate gill supports in living and fossil bivalves. In: Suga S, Nakahara H (eds) Mechanisms and phylogeny of mineralization in biological systems. Springer, Tokyo, pp. 427–431

Wilby PR, Briggs DEG (1997) Taxonomic trends in the resolution of detail preserved in fossil phosphatized soft tissues. Geobios 30:493–502

Wilby PR, Martill DM (1992) Fossil fish stomachs: a microenvironment for exceptional preservation. Hist Biol 6:25–36

Wilby PR, Whyte MA (1995) Phosphatized soft tissues in bivalves from the Portland Roach of Dorset (Upper Jurassic). Geol Mag 132:117–120

Wood EM (1974a) Development and morphology of the glochidium larva of *Anodonta cygnea* (Mollusca: Bivalvia). J Zool 173:1–13

Wood EM (1974b) Some mechanisms involved in host recognition and attachment of the glochidium larva of *Anodonta cygnea* (Mollusca: Bivalvia). J Zool 173:15–30

Yates A, Neumann F, Hancox J (2012) The earliest post-Paleozoic freshwater bivalves preserved in coprolites from the Karoo Basin, South Africa. PLoS One 7:e30228. https://doi.org/10.1371/journal.pone.0030228

Yeager BL, Saylor CF (1995) Fish hosts for four species of freshwater mussels (Pelecypoda: Unionidae) in the Upper Tennessee River Drainage. Am Midl Nat 133:1–6

Yeager MM, Cherry DS, Neves RJ (1994) Feeding and burrowing behaviors of juvenile rainbow mussels, *Villosa iris* (Bivalvia:Unionidae). J N Am Benthol Soc 13:217–222

Yokoyama R, Goto A (2005) Evolutionary history of freshwater sculpins, genus *Cottus* (Teleostei; Cottidae) and related taxa, as inferred from mitochondrial DNA phylogeny. Mol Phylogenet Evol 36:654–668

Zanatta DT, Murphy RW (2006) Evolution of active host-attraction strategies in the freshwater mussel tribe Lampsilini (Bivalvia: Unionidae). Mol Phylogenet Evol 41:195–208

Zieritz A, Sartori AF, Bogan AE, Aldridge DC (2015) Reconstructing the evolution of umbonal sculptures in the Unionida. J Zool Syst Evol Res 53:76–86

Chapter 6
Gastropods as Parasites and Carnivorous Grazers: A Major Guild in Marine Ecosystems

Alexander Nützel

Abstract Parasitism and similar life styles such as carnivorous grazing or mucus feeding without killing the prey are important in marine gastropods. Some of the most diverse living gastropod families have this feeding behavior. Taxonomic uniformitarianism is the most important tool to infer parasitism or similar life styles in fossil gastropods. The extant family groups in question (Eulimidae, Epitoniidae, Pyramidellidae, Architectonicidae, Coralliophilinae, Ovulidae, Cerithiopsidae and Triphoridae) originate mostly in the Late Cretaceous (Cerithiopsidae in the Middle Jurassic) and Paleocene. They are performing an ongoing adaptive radiation and some of the mentioned families belong to the most diverse gastropod groups forming a considerable part of marine ecosystems regarding species richness and relative abundance. At the same time, origination and radiation of the carnivorous, commonly predatory Neogastropoda took place. This points to a trophic revolution in Gastropoda that forms an important aspect of the Mesozoic Marine Revolution. Most modern parasitic gastropods are small, high-spired, show high diversity and low disparity within families and belong to Apogastropoda. By analogy, some extinct gastropod families which show the same properties might have lived parasitic too (e.g., Pseudozygopleuridae, Zygopleuridae, Meekospiridae, Donaldinidae). However, this will remain speculative to a large degree until direct host associations are found. Direct evidence for parasitism is exceptional with the Palaeozoic platyceratid/crinoid interaction being one of the best studied examples. In Gastropoda, functional shell morphology may help to identify parasitism in the fossil record but this field is scarcely studied.

Keywords Parasitism · Diversity · Feeding · Ecology · Adaptive radiation

A. Nützel (✉)
SNSB-Bayerische Staatssammlung für Paläontologie und Geologie, Munich, Germany

Department of Earth and Environmental Sciences, Paleontology and Geobiology, Ludwig-Maximilians-Universität München, Munich, Germany

GeoBio-Center, Ludwig-Maximilians-Universität München, Munich, Germany
e-mail: nuetzel@snsb.de

© The Author(s) 2021

K. De Baets, J. W. Huntley (eds.), *The Evolution and Fossil Record of Parasitism*,
Topics in Geobiology 49, https://doi.org/10.1007/978-3-030-42484-8_6

6.1 Introduction

Among mollusks, most parasites are gastropods. Weinstein and Kuris (2016) estimated that parasitism has evolved convergently at least nine times within Gastropoda out of more than 200 times it evolved within Metazoa. Parasitic or carnivorous grazers more or less permanently attached to their hosts form some of the most diverse living groups of gastropods. An origin of a parasitic lifestyle from carnivorous relatives seems to be a common prerequisite in metazoans, particularly in gastropods (Takano and Kano 2014). Modern groups of parasitic gastropods belong to higher gastropods (Apogastropoda) i.e., Caenogastropoda and Heterobranchia. However, I am not aware of any living parasitic patellogastropods, vetigastropods or neritimorphs (Ponder and Lindberg 1997). The living gastropod family groups Pyramidellidae, Eulimidae, Epitoniidae, Architectonicidae, Mathildidae, Coralliophilinae, Aclididae, Pedicularidae, Caledoniellidae, Ovulidae, Triviidae, Colubrariidae, Cypraeidae, Cancellariidae, Marginellidae and Pseudosacculidae are considered to represent parasites in a strict sense or contain at least one or more parasitic species (e.g., Warén 1983, p. 27, Table 6; Lorenz 2005, Table 5.5). Several of the mentioned families contain few parasitic species and/or have a scarce fossil record and are not considered here in depth.

Eulimidae feed on echinoderms and Pyramidellidae are ectoparasites feeding on the body fluids of various invertebrates, especially polychaete worms and mollusks (e.g., Ankel 1949; Robertson and Mau-Lastovicka 1979). Epitoniidae (Janthinoidea), Architectonicidae, Mathildidae and Ovulidae feed on Cnidaria (anthozoans) either as parasites or as carnivorous grazers (e.g., Kohn 1983). Coralliophilinae (Muricidae, Neogastropoda) live in and feed on corals. Robertson (1970) concluded that adaptations for feeding on Cnidaria in unrelated gastropod families include cuticularized esophaguses and proboscises, ptenoglossate-like radular teeth and total loss of radulae.

The majority of parasitic gastropods are small and high-spired with a generally low shell morphological disparity within each of the families so that alpha taxonomy (species identification and discrimination) is usually very difficult. This also limits the fossil record of parasitic gastropods: Small tiny, mostly aragonitic gastropod shells are rarely preserved at all or in sufficient quality for an identification, especially in formations of high geological age (Paleozoic and Mesozoic) (cf. Roden et al. 2019) (Table 6.1). However, there are also numerous examples in each family where species and genera deviate strongly from the standard morphology so that low disparity within families is not without exceptions. The high diversity of some of these families can be interpreted as result of a pronounced host specificity and co-evolution—but although plausible, this has yet to be demonstrated in detail. Most species of the mentioned families are ectoparasites and are attached by mucous threads or by the proboscis/snout to their hosts. Endoparasitism does occur but is seemingly rare. However, direct observations on feeding and behavior are scarce in relation to the vast diversity of these snails. Parasitic gastropods may have reduced their radula or have adapted their radula for feeding on tissue of their host.

Table 6.1 Most important gastropod families having a parasitic or similar life style and several diverse fossil putative parasitic families based on analogy

	Host	Occurrence	Living species	Subclass
Pyramidellidae	Polychaeta, Mollusca	mJurassic (Bathonian)–Recent	>6000	Heterobranchia
Eulimidae	Echinodermata	lCretaceous (Campanian) –Recent	>1000	Caenogastropoda
Epitoniidae	Anthozoa	lCretaceous (Campanian)–Recent	>700	Caenogastropoda
Architectonicidae	Anthozoa	lTriassic (Carnian)–Recent	140	Heterobranchia
Mathildidae	Anthozoa	mTriassic (Anisian)–Recent	80	Heterobranchia
Coralliophilidae	Anthozoa	lCretaceous (Campanian)–Recent	200–250	Caenogastropoda
Triphoridae	Porifera (Demospongia)	Paleogene (Danian)–Recent	>1000	Caenogastropoda
Cerithiopsidae *ss*	Porifera (Demospongia)	Paleogene (Danian)–Recent	>1000	Caenogastropoda
Ovulidae	Anthozoa	lCretaceous (Cenomanian)–Recent	400	Caenogastropoda
Platyceratidae/Orthonychiidae	Echinodermata (Crinoida)	Ordovician (Arenigian)–Permian (Capitanian)	—	?
By analogy				
Pseudozygopleuridae	—	Carboniferous (Tournaisian)–Early Triassic (Griesbachian)	—	Caenogastropoda
Zygopleuridae	—	Early Triassic (Smithian)– Cretaceous (Albian)	—	Caenogastropoda
Protorculidae	—	Early Triassic (Smithian)–Middle Jurassic	—	Caenogastropoda
Donaldinidae	—	Middle Devonian (Givetian)–Triassic (Carnian) (Recent?)	—	Heterobranchia
Streptacididae	—	Carboniferous (Tournaisian)–Cretaceous (Campanian)	—	Heterobranchia
Meekospiridae	—	Carboniferous (Tournaisian)– Permian (Changhsingian)	—	Caenogastropoda

The highly diverse generally small-sized Cerithiopsidae and sinistrally coiled Triphoridae (Triphoroidea) feed on sponges obviously without necessarily killing their prey. These spongivorous gastropods are not considered to represent parasites in a strict sense but carnivorous grazers and their relationship to sponges resembles parasitism. It is also likely that they feed on microbial biofilms and mucous associated with the sponges. They probably encompass several thousand living species. Cerithiopsidae and Triphoridae were placed in the heterogeneous, probably not-monophyletic group Ptenoglossa together with Epitoniidae and Eulimidae (e.g., Nützel 1998). Takano and Kano (2014) showed based on a molecular phylogeny that Ptenoglossa in their traditional composition are polyphyletic and that Eulimidae are the sister group of carnivorous Vanikoridae.

The census of a marine area at Koumac, an intensively studied coastal site in New Caledonia, undertaken by Bouchet et al. (2002) shows how important parasitic gastropods are in modern tropical environments. These authors counted a total of 2187 mollusk species, 270 (ca. 12%) of which belong to the parasitic Eulimidae (the third most diverse family) and Pyramidellidae (the fourth most species-rich family). The spongivorous Triphoridae and Cerithiopsidae represent the second and fifth most diverse mollusk families. Similarly, a census from the coral triangle (Panglao, Philippines) resulted in the finding of at least 715 species of Pyramidellidae from an area of 15,000 ha (150 km^2) and the real number there may exceed 1100 species (Bouchet 2009). Most of these species are small (41% have adult sizes equal or below 2 mm) (Bouchet 2009). Besides Pyramidellidae, the parasitic Eulimidae and sponge-associated Cerithiopsidae and Triphoridae belong the most diverse groups.

6.2 How to Infer Parasitism in Fossil Gastropods

There are several lines of evidence to infer parasitism (see also Boucot 1990; Nagler and Haug 2015): (1) direct evidence i.e., observation of parasite and prey/host associations or traces of parasitism on fossil prey, (2) taxonomic uniformitarianism i.e., the fossil record of extant parasitic gastropod groups, usually on the family or genus level, (3) functional shell morphology i.e., a shell morphology that is suggestive of parasitism, and connected with this (4) analogy of associated phenomena e.g., the observation that modern parasitic gastropod families generally consist of small, high-spired and highly diverse groups with low disparity can be transferred to extinct groups that show similar properties.

6.2.1 Direct Observations

Direct observations of parasitism in fossil gastropods are rare. Baumiller and Gahn (2002) reported about 30 cases of parasitism from the fossil record, seven of which involved gastropods as parasites. One of the most famous examples is the interaction of platyceratid (respectively orthonychiid) gastropods and crinoids (e.g., Bowsher 1955; Boucot 1990; Webster and Donovan 2012; Baumiller and Gahn 2018) (Fig. 6.1). The cap-shaped or coiled gastropods are found on the crowns of crinoids from the Ordovician to the Permian. These gastropods are attached firmly over or near the crinoid periproct (Baumiller and Gahn 2018). Traditionally, this interaction has been interpreted as coprophagy, not damaging to the crinoid host. However, this interpretation has shifted to kleptoparasitism that hampered growth of the crinoid hosts i.e., infested crinoids are generally smaller than non-infested ones (Baumiller and Gahn 2002, 2018; Baumiller 2003; Gahn and Baumiller 2006). Baumiller and Gahn (2018) observed that "in the rare instances in which snails are found on tubed crinoids, they are never attached over the periproct but rather at the base of the tube, near the crinoid foregut. In order to access nutrients, infesting platyceratids breached the tube by drilling a circular hole through the plates of the tegmen. The development of long tubes in crinoids, which occurred multiple times, and the drilling abilities of platyceratids may represent a case of evolutionary escalation." Sutton et al. (2006) reported a Silurian gastropod with concretionary soft body preservation and assigned it to Platyceratidae. They concluded that the anatomy suggests a sedentary life style. Accordingly, the absence of a proboscis suggests a coprophagous rather than a cleptoparasitic mode of life.

The Ordovician genus *Cyclonema* that can be found attached to tegmens of crinoids is usually assigned to Platyceratidae but its shell morphology is quite distinct from *Platyceras* and according to Knight et al. (1960) *Cyclonema* has a nacreous shell which is unknown for both, *Platyceras* and *Orthonychia*. Therefore it is not unlikely that *Cyclonema* is not a platyceratid and that its possibly parasitic life style attached to crinoids has evolved convergently.

Another example of direct observation of parasitism is certain drill holes or attachment scars (Boucot 1990; Boucot and Poinar 2010). Boucot (1990, p. 72) summarized several holes and malformations on Cretaceous echinid tests likely produced by eulimid gastropods. Neumann and Wisshak (2009) reported the trace

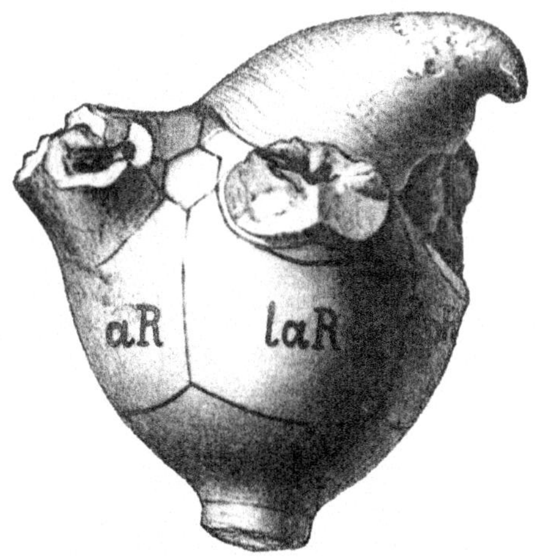

Fig. 6.1 Cap-shaped platyceratid gastropod *Orthonychia varians* attached to the crown of the crinoid *Platycrinites wachsmuthi*, one of the few direct evidences of fossil parasitism; Permian, Timor (from Wanner 1922)

fossil (drill hole) *Oichnus halo* on the tests of Late Cretaceous (Campanian) to Early Paleocene holasteroid echinoids. In this case the drill holes look exactly the same as those produced by the recent cap-shaped eulimid genus *Thyca*. The drill holes are sufficiently complex (concentric scars surmount the hole) and together with an echinoderm being the prey it is reasonable to assume that an eulimid gastropod was the producer. In other cases, traces are not sufficiently complex, e.g. simple drill holes or galls. There are several modern examples of galls produced by parasitic gastropods in skeletons of their host for instance eulimid galls in an echinoderm spine (Warén 1983) and in principal, this has a fossilization potential but according to my knowledge there are few reports of such cases from the fossil record (Boucot 1990; Baumiller and Gahn 2002; Boucot and Poinar 2010). Breton et al. (2017) interpreted a bioerosion trace fossil on Cenomanian oysters as product of parasitic gastropods.

Hayami and Kanie (1980) reported large Late Cretaceous (Campanian) limpets (*Gigantocapulus*) in association with inoceramid bivalves. In analogy with some recent gastropod limpets of the family Capulidae, they interpreted this as parasitism (see also Baumiller and Gahn 2002). In the absence of protoconch and shell microstructure the assignment to Capulidae of this limpet is doubtful. Beu (2007) noted that the systematic placement of *Gigantocapulus* is doubtful and placed it tentatively in the caenogastropod superfamily Vanikoroidea but stated it could also represent Monoplacophora. This author stated an epiparasitic mode of life was possible but a filter feeding sedentarily on bivalves was more likely.

Petit et al. (2014) reported gastropods attached to fossil fishes, but their attribution to scavengers or parasites is not straightforward. They putatively assigned the gastropod specimens to *Aclis* which is normally not associated with fishes.

6.2.2 *Taxonomic Uniformitarianism*

The most commonly used tool to infer parasitism in fossil gastropods is taxonomic uniformitarianism. If for instance all extant pyramidellids are ectoparasites, it is plausible and the most parsimonious assumption that fossil pyramidellid species were parasites too and that this feeding type is a synapomorphy of this clade. Difficulties arise when assignment of fossil species is uncertain and it may be impossible to establish the life style of fossil sister-groups of parasitic families or genera and hence the timing of the acquirement of the parasitic life style in evolutionary lineages. It is also generally true that taxonomic uniformitarianism becomes increasingly problematic the higher the geological age is. As stated above, modern groups of parasitic gastropods belong to 'higher' gastropods (Apogastropoda) i.e., Caenogastropoda and Heterobranchia and the mentioned families are usually present since the Late Cretaceous or Paleocene. Hence, it is evident that parasitism has been pervasive at least since the Paleocene.

In the following the most important parasitic families (or those having a similar lifestyle) are mentioned including their host and fossil record as currently known. This account is not meant to be complete but covers the most diverse groups.

6.2.2.1 Eulimidae

Eulimidae (Caenogastropoda) are mostly ectoparasites that feed on echinoderms (holothurians, sea urchins, sea stars). They are attached to their host with their proboscis or snout more or less permanently (Warén 1983) and can be embedded in the tissue of their host (endoparasitic). Many eulimids have reduced the radula (Aglossa) as result of their parasitic life style. Those eulimids that possess a radula have needle-shaped teeth (Warén 1983) and hence resemble the ptenoglossate radula of Janthinoidea including Epitoniidae. The majority of eulimid species are small, high-spired and have a glossy, extremely smooth shell (Fig. 6.2a, d). However, there is considerable variation regarding shell morphology within this family including cap-shaped or turbiniform morphologies and strongly ornamented forms (e.g., Warén 1983). Some eulimids have abandoned the shell entirely and became

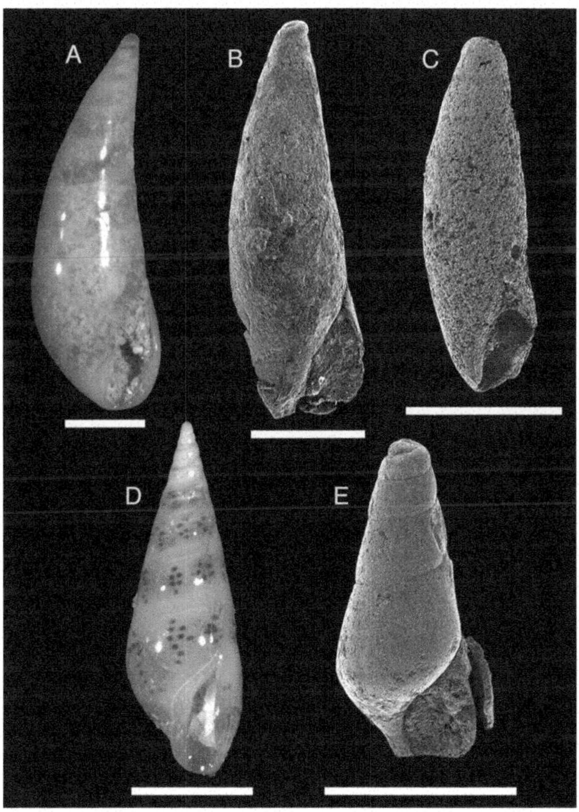

Fig. 6.2 Modern eulimids (**a**, **d**) and similar Late Palaeozoic forms (**b**, **c**, **e**) representing Meekospiridae. Similar shell morphology, smooth shells and small size could suggest a similar mode of life i.e., parasitism. (**a**) Recent eulimid with twisted outline, Lizard Island, Queensland, Australia. (**b**) *Ceraunocochlis fulminula*, with likewise twisted shell axis, Pennsylvanian, Texas (Nützel et al. 2000). (**c**) *Ceraunocochlis* sp. Mississippian, Indiana (Nützel et al. 2000). (**d**) Recent eulimid, Lizard Island, Queensland, Australia. (**e**) Pennsylvanian *Meekospira* sp., Pennsylvanian, Texas (Nützel et al. 2000). Scale bars 1 mm

worm-like - one of the very few cases in Caenogastropoda (Ponder et al. 2008). Eulimids are very diverse but their alpha taxonomy is notoriously difficult because of their small size and the limited disparity of shell morphology among many species of this group. The smooth glossy shell can be interpreted as an adaption to the parasitic life style.

The number of living eulimid species is unknown but certainly >1000. The fossil record shows that the family Eulimidae has a minimum age of Campanian (Dockery 1993)—older reports are doubtful. Eulimids have become increasingly diverse and abundant throughout the Cenozoic (Warén 1983). Some eulimids display a pronounced convergence with Late Palaeozoic subulitoid shells representing the genera *Meekospira* and *Ceraunocochlis* (Fig. 6.2) but the Carboniferous species have a blunt apex/protoconch instead of a pointed larval shell as found in Eulimidae.

6.2.2.2 Epitoniidae

Epitoniidae (wenteltraps) is a highly diverse caenogastropod family that is associated with or feeds on cnidarians. They feed and live in association with anthozoans, such as species of Actiniaria, Zoantharia or Scleractinia (e.g., Gittenberger and Hoeksema 2013 and references therein). Most are considered to be ectoparasitic, others as carnivorous grazers respectively foraging predators or as symbionts (e.g., Robertson 1980; Gittenberger and Hoeksema 2013). Robertson (1980) observed that an *Epitonium* species lives in symbiosis with the Coelenterates feeding, together with another gastropod species representing *Coralliophila*, on mucous, nematocysts and zooxanthellae of its host bot not on its tissue of polyps. Bandel (1991) reported that larger epitoniids feed with their proboscis on anthozoans until their prey dies. Gittenberger and Gittenberger (2005) observed that epitoniids are attached to the underside of mushroom corals by mucous threads. They also found that epitoniid species have specialized on one or few host species.

More than 700 recent epitoniid species have been described validly (Brown and Neville 2015). The diversity may be much greater because unexplored cryptic radiations are likely (cf. Gittenberger and Gittenberger 2005). Epitoniids have small to medium sized shells commonly with lamellar axial ribs and a holostomatous aperture (Fig. 6.3a). Epitoniids are unusual for caenogastropods in having a calcitic shell. They possess the ptenoglossate radula with needle-shaped teeth. Epitoniidae have a smooth larval shell (Fig. 6.3a). By contrast, Nystiellidae, the likewise parasitic sister-group of Epitoniidae, have a larval shell with axial ribs.

Previous reports of a Middle Jurassic first occurrence of Epitoniidae (Cossmann 1912; Tracey et al. 1993) are doubtful because this species, *Proacirsa inornata*, has smooth whorls and its protoconch is unknown. Gründel (2012) placed *Proacirsa* in the heterobranch family Gordenellidae and hence outside Epitoniidae. The oldest known members of Epitoniida with smooth larval shell have a Campanian age (Dockery 1993). Guzhov (2006) reported a nystiellid from the Upper Jurassic of Russia. It has an axially ribbed protoconch and teleoconch and thus resembles the

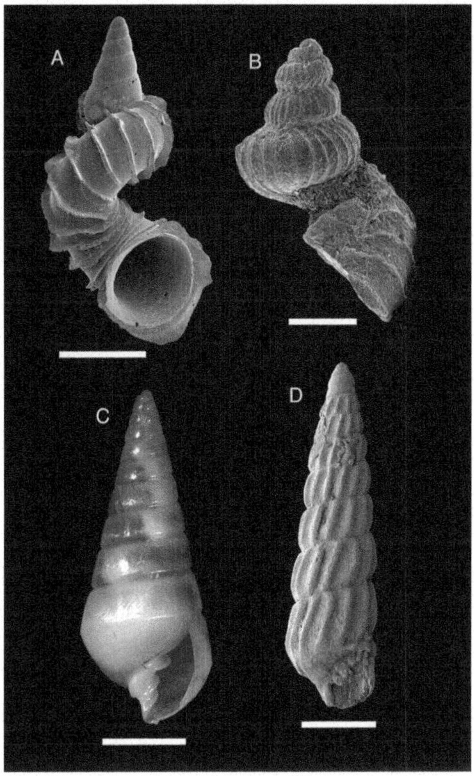

Fig. 6.3 (**a**) Recent epitoniid *Cycloscala hyaline* with smooth larval shell and uncoiled teleoconch with lamellar axial ribs (from Sasaki 2008, fig. 10b). (**b**) Pennsylvanian pseudozygopleurid *Helminthozyga vermicula* with axially ribbed larval shell and uncoiled teleoconch convergent to epitoniid in **a**—this could point to parasitism of this Palaeozoic species (from Nützel 1998). (**c**) Recent pyramidellid, with pronounced folds on inner lip of aperture, Sulawesi, Indonesia. (**d**) Late Triassic zygopleurid, possibly parasitic; Cassian Formation, Italy. Scale bars (**a**): 0.5 mm, (**b**): 0.3 mm, (**c**, **d**): 1 mm

Triassic zygopleurid subfamily Ampezzopleurinae. Nützel (1998) suggested that Epitoniidae arose from Ampezzopleurinae (Zygopleuridae) via Nystiellidae.

6.2.2.3 Coralliophilinae (Muricidae)

Coralliophilinae, a subfamily of Muricidae (Bouchet et al. 2017), is a moderately diverse, small to medium-sized caenogastropod group with approximately 200–250 living species (Marshall and Olivero 2009). The shell morphology including ornamentation is quite variable. Coralliophilinae have reduced the radula and live parasitic on and in coelenterates (Robertson 1970). The genus *Magilus* lives within stony corals and builds a tube within it (Fig. 6.4). Gittenberger and Gittenberger (2011) reported a cryptic radiation of the coralliophilin endoparasitic genus *Leptoconchus*. The species live in parasitic association with mushroom corals.

Coralliophilinae are known since the Late Cretaceous (Sohl 1964). Lozouet and Renard (1998) reported one of the oldest members boring in corals from the Oligocene of the Paris Basin.

Fig. 6.4 Coralliophylin *Magilus antiquus*-tube in a stony coral; Holocene, Sumatra, Indonesia, width 18 cm, photo: G. Janssen

6.2.2.4 Pyramidellidae

Pyramidellidae are ectoparasites that feed on body fluids of their hosts and have reduced the radula (e.g. Robertson and Mau-Lastovicka 1979; Healey 1998a; Ponder and De Kayser 1998). Most of the verified hosts are polychaetes, gastropods and bivalves, but there also appear to be various minor host groups such as polyplacophorans and echinoderms (Robertson and Mau-Lastovicka 1979). Pyramidellidae are one of the most diverse gastropod families with more than 6000 living species (Lygre and Schander 2010). Bouchet (2009) reported that 715 species are present in the Philippines in an area of 15,000 ha (c. 150 km^2) of coastal habitats to a depth of 150 m.

Most pyramidellids are small (<10 mm) and have a high-spired shell with one or more columellar plaits in the aperture (Fig. 6.3c). Shell shape and ornament are quite variable in this family (see Bouchet 2009 for some examples). As heterobranchs they have heterostrophic, sinistral protoconchs which help to separate them from convergent caenogastropods.

The fossil record of Pyramidellidae is hard to track; the earliest finds in the PBDB dating from the Palaeozoic and Triassic refer to the modern genus *Chemnitzia* and these assignments are all outdated—*Chemnitzia* was used as a waste basket taxon, especially in the nineteenth century. Based on Sohl (1964), Tracey et al. (1993) gave a Maastrichtian FAD. Kaim (2004) placed the Middle Jurassic *Chrysallida minuera* (Late Bathonian, Poland) in Pyramidellidae and the similarity to modern *Chrysallida* is indeed stunning although a distinct columellar fold has not been shown; but is has an incipient bulge similar to that of some Recent pyramidellid species. Nevertheless, a Late Bathonian age can be accepted as FAD for Pyramidellidae since folds are also absent or weak in some Recent Pyramidellidae. This relatively old fossil record is somewhat surprising given molecular studies that suggest that they are deeply rooted in Pulmonata (Dinapoli et al. 2011; Schrödl et al. 2011) which would be more congruent with a younger age. Amathinidae Ponder

(1987) (Pyramidelloidea), that are considered to be parasites on bivalves today, were reported by Kaim (2004) from the Early Cretaceous.

6.2.2.5 Architectonicidae and Mathildidae

Architectonicidae are considered to be ectoparasites of colonial cnidarians such as zoantharian, scleractinian and antiparian corals (Healey 1998b and references therein). They show buccal and radular specializations for feeding on hexacorallian coelenterates such as stony corals, sea anemones and zoanthids (Bieler and Petit 2005). A large number of species names have been proposed but according to Bieler and Petit (2005) only about 140 extant species are valid. Their attractive shells are conical and comparatively low-spired (some even planispiral) with an umbilicated base—thus they are unlike most other parasitic gastropod according to shell morphology. Their shell morphology resembles that of Trochidae but they can be separated from them by having a heterostrophic larval shell and lacking nacre.

Bandel (1988) placed two Late Triassic (Carnian) genera (*Amphitomaria* and *Rinaldoconchus*) from the Cassian Formation in Architectonicidae which represents their first appearance date according to the current state of knowledge (Tracey et al. 1993). Attributions of Palaeozoic shells to Architectonicidae are erroneous (Bieler and Petit 2005).

Architectonicidae and the high-spired Mathildoidea are considered to be closely related and are sometimes placed in the same superfamily or as separate superfamilies. Mariottini et al. (2009) stated that "these mollusks are associated to cnidarians for feeding habit". However, the feeding habit of this group is poorly known. If Mathildidae are parasitic or have a similar life style, it would be an important group representing this guild, especially in the Mesozoic. Mathildidae are richly diversified during the Mesozoic at least from the Late Triassic onward and are still present today but more in deeper water environments than in shallow water as it used to be earlier (Bandel 1995; Gründel and Nützel 2013; Bieler 1995). In the Late Triassic Cassian Formation, Mathildidae were much more diverse and abundant than in any site today.

6.2.2.6 Triphoridae and Cerithiopsidae: Carnivorous Grazers (Sponge Feeders)

Triphoridae and Cerithiopsidae are super-diverse gastropod groups that live on sponges (Fig. 6.5). They actually feed on sponge tissue of demosponges—stomach and fecal pellets may contain sponge spicules (Nützel 1998; Fig. 6.6). They also feed on mucous on the sponges. There are observations for some living species that individuals are attached to sponges with mucous threads and that they deposit spawn on or within sponges (Nützel 1998; Fernandes and Pimenta 2019; see Fig. 6.6). Lorenz (2005) reported cypraeid egg clusters in sponges. So in a wider sense, Triphoridae and Cerithiopsidae can be considered ectoparasites, living on their host

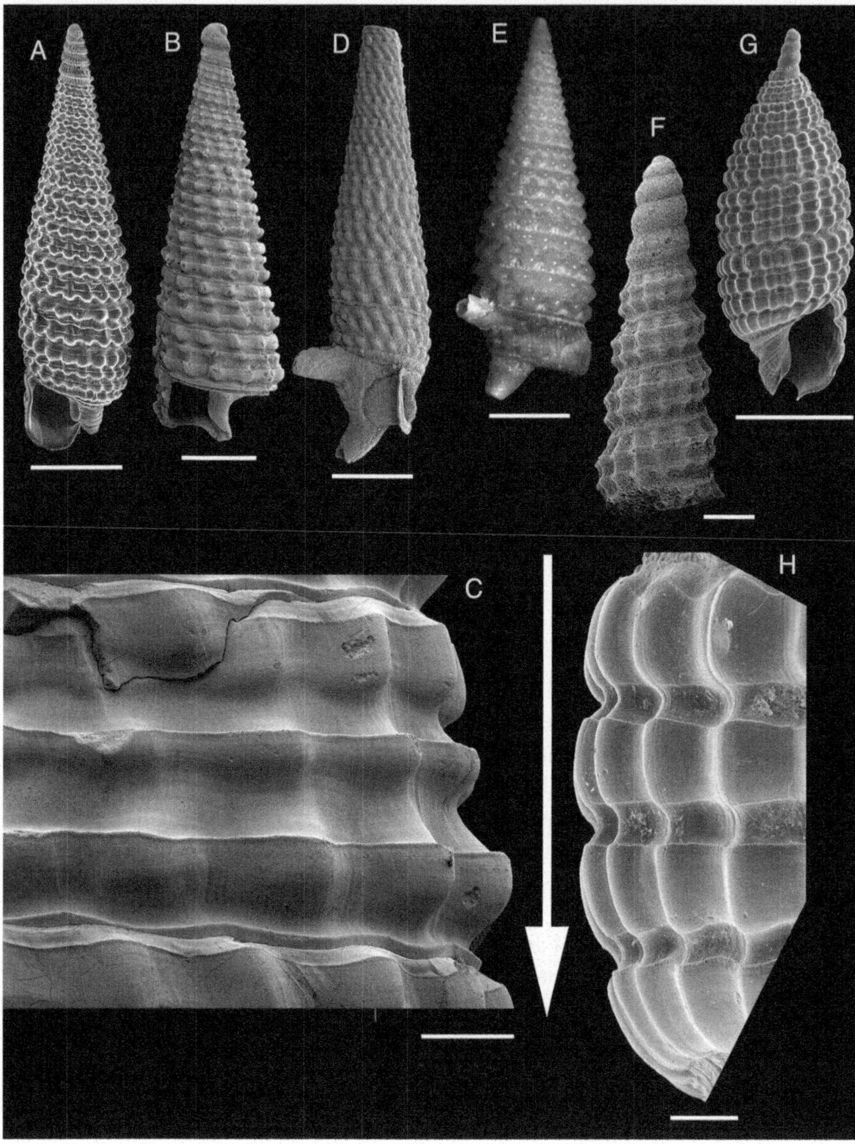

Fig. 6.5 (**a**) Recent sinistral triphorid from Australia, Queensland, Lizard Island (from Nützel 1998). (**b**, **c**) Eocene sinistral triphorid *Ogivia singularis* (J13084) from the Paris Basin. (**c**) Detail of teleoconch ratchet sculpture of spirals cords, anteriorly rounded, gently sloping and posteriorly steep, arrow points in anterior direction . (**d**) Danian sinistral triphorid from Denmark with tube-like anterior and posterior siphonal canals. (**e**) Recent sinistral triphorid from Australia, Queensland, Lizard Island, with tube-like anterior and posterior siphonal canals (from Nützel 1998). (**f**) Danian cerithiopsid with preseved larval shell (from Nützel 1998). (**g**, **h**) Recent barrel-shaped cerithiopsid with slender, smooth larval shell from Australia, Queensland, Lizard Island. (**h**) Detail of teleoconch ratchet sculpture of spirals cords, anteriorly rounded, gently sloping and posteriorly steep, arrow points in anterior direction. Scale bars (**a**), (**b**), (**d**), (**e**), (**g**): 1 mm; (**c**), (**h**): 0.2 mm, (**f**): 0.3 mm

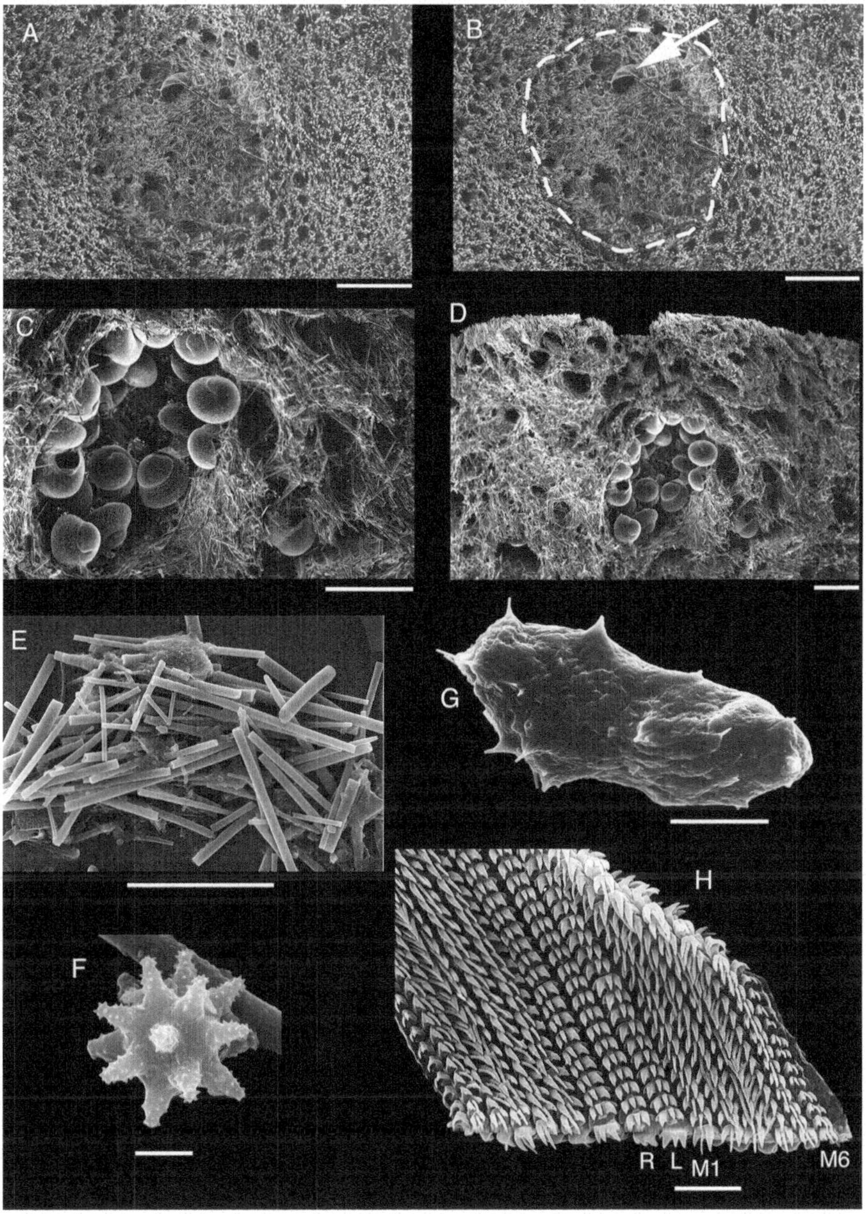

Fig. 6.6 Spawn, stomach contents, fecal pellet and radula of Recent Triphoroidea (from Nützel 1998). (**a**–**d**): Egg chamber of *Cerithiopsis* sp. in the sponge *Suberites domunculus*, Recent, Mediterranean (**a**, **b**). Same view, surface of sponge showing sealed opening (dashed line) with an cerithiopsid embryonic shell emerging from the chamber (arrow). (**c**, **d**). Egg chamber in transverse section with several cerithiopsid embryonic shells. (**e**, **f**) Sponge spicules from a stomach of a cerithiopsid. (**g**) Shrinked fecal pellet of a triphorid containing sponge spicules. (**h**) Rhinioglossate radula of a Triphorinae species, R rachidian, L lateral, M1–M6 marginal teeth. Scale bars (**a**–**d**): 0.3 mm, (**e**): 50μm, (**f**): 3μm, (**g**): 30μm, (**h**): 10μm

without necessarily killing it. In any case, all available evidence suggests a close relationship of these gastropods to their host sponges. There are hints that they are not obligate sponge eaters but it seems that sponges are the preferred source of food. At least they belong, as Turridae and other gastropods, "in guilds of commensals, associates and parasites" (Bouchet et al. 2009). It must be emphasized that feeding observations are scarce covering few species so that generalizations must be considered with caution.

Triphoridae and Cerithiopsidae are small (usually <10 mm), high-spired or barrel shaped and commonly have a constricted body whorl in fully-grown specimens (Fig. 6.5). The great majority of species in both families share a reticulate teleoconch ornament, commonly with asymmetrical knobby spiral cords that are rounded and shallowly sloping anteriorly and steeply sloping posteriorly (Fig. 6.5c, h). This represents probably a ratchet sculpture which facilitates penetration of the sponge tissue and prevents uncontrolled back-slipping. Signor (1982) reported similar ornaments for burrowing gastropods. Commonly cerithiopsids and triphorids have a constricted body whorl and/or a rounded base resulting in an overall elongated pear-shape which may have facilitated intrusion into the host sponge tissue (Fig. 6.5a, d, e, g). Triphorids and Cerithiopsidae sensu strictu have a minimum Paleogene age (see below).

Triphoridae

Triphoridae encompass at least 1000 living species, many of them undescribed. Fernandes and Pimenta (2017) counted more than 600 recent species based on the WORMS database (http://www.marinespecies.org/). Most species belong to the sinistrally coiled subfamily Triphorinae. They have a complex aperture with an anterior and a posterior siphonal canal, sometimes both are tube-like (Fig. 6.5d, e). The posterior siphonal canal forms the outlet for special tentacles emerging form the mantle cavity that may help to ventilate it (Nützel 1998). Triphoridae possess the rhinioglossate radula—the number of teeth per row varies from 5 to 63 (Marshall 1983) (Fig. 6.6h).

Undoubted sinistrally coiled Triphoridae (Triphorinae) are present since the Danian (e.g., Ravn 1933; Nützel 1998; Wesenberg and Schnetler 2014) (Fig. 6.5d) and few Late Cretaceous forms have been reported but those need confirmation (e.g., Kaunhowen 1897; Nützel 1998; Kaim 2004).

Cerithiopsidae

Cerithiopsidae is another highly diverse family of spongivorous small, high-spired caenogastropods. Cerithiopsidae may closely resemble non-parasitic small Cerithioidea for instance the genus *Bittium* and protoconch morphology is needed to separate both groups. However, protoconch preservation is rare, especially in fossil gastropods (e.g., Nützel 2014). Cerithiopsidae are taenioglossate having seven teeth per row. Within the group, the tooth morphology is extremely variable, some having strongly elongated marginal teeth (e.g. Marshall 1978; Nützel 1998). It is obvious that this represents an adaptation to the feeding on sponges but feeding mechanisms have not been studied.

Cerithiopsidae sensu strictu are at least as old as Danian and are richly diversified in the Danian Naesekalk from Denmark (Ravn 1933; Nützel 1998; Wesenberg and Schnetler 2014) (Fig. 6.5f). The higher category Triphoroidea (respectively Cerithiopsoidea) is older with a good fossil record from the Cretaceous and some Jurassic representatives with the earliest record from the Bathonian (Nützel 1998 and references therein). The Triassic Protorculidae (earliest record Early Triassic, Smithian) probably represent its fossil sister group (Nützel 1998; Nützel and Erwin 2002).

6.2.2.7 Ovulidae

According to Wilson (1998) Ovulidae feed on the polyps of cnidarians and belong thus to the guild of carnivorous grazers and they are considered to be parasites sometimes. They belong to Cypraeoidea and have the convolute shell with genera like *Cypraea*. They are of moderate diversity (400 species according to Lorenz 2005) and most species occur in the tropical Indo-Pacific.

Ovulidae originate at the latest in the Late Cretaceous according to Tracey et al. (1993).

6.2.3 Functional Shell Morphology

Functional shell morphology is a potential tool to infer parasitism in fossil gastropods. However, this has not been explored and functional morphology of the gastropod shell is generally not well understood. As outlined above, the ratchet sculpture of triphorids and cerithiopsids as well as their anteriorly constricted and rounded shell may facilitate intrusion into the tissue of the host (Fig. 6.5). The extremely smooth, polished shell with very shallow sutures of many eulimids (Fig. 6.2a, d) forms certainly an adaptation regarding the parasitic life style. This shell morphology is also present in Palaeozoic subulitoid gastropods, such as *Ceraunocochlis* (Fig. 6.2b, c); it is a reasonable speculation that such fossil gastropods were parasitic too (see below).

Gahn and Baumiller (2006) argued that the elongation of the anal tubes of Palaeozoic crinoids had the function to hinder infestations by parasitic platyceratids.

6.2.4 Analogy Based on Associated Phenomena

Taxonomic uniformitarianism fails if gastropod families are extinct, have a relatively high geological age and if the closest living relatives are unknown. Most extant parasitic and carnivorous grazer gastropods are ectosymbiotic and do not leave fossil traces. If, for instance, Pyramidellidae were an extinct Mesozoic or Paleozoic group, their parasitic life style would have remained unrecognized because there is no fossil evidence for it and their preferred host polychaetes have a relatively poor fossil record. The same is true for Triphoridae and Cerithiopsidae. On the other hand, it seems to be unlikely (though not impossible) that a parasitic or similar life style was only present in the mentioned extant highly diverse gastropod families whereas this life style was largely absent or rare in the Palaeozoic and Early Mesozoic. As mentioned, extant families exhibit several traits and if these traits can be found in fossil gastropods, a parasitic life style can be hypothesized by analogy. These traits are as follows:

- belonging to Caenogastropoda or Heterobranchia (Apogastropoda)
- high within family diversity
- low shell disparity within families
- mostly small
- mostly high-spired

Needless to say that the above-mentioned traits are also present in some non-carnivorous or non-parasitic gastropod families (for instance Rissoiidae) and therefore do not represent unequivocal evidence but they do offer the possibility that certain extinct families which display them were parasitic. Potentially, discoveries of rare host associations or effects of parasitic on growth could potentially corroborate such hypotheses. The mentioned traits are present in the following fossil gastropod families which therefore could have had a parasitic life style:

Pseudozygopleuridae	Carboniferous (Tournaisian)–Early Triassic (Griesbachian)
Zygopleuridae	Early Triassic (Smithian)–Cretaceous (Albian)
Protorculidae	Triassic (Smithian)–Middle Jurassic
Donaldinidae	Middle Devonian (Givetian)–Triassic (Carnian)[a]
Streptacididae	Carboniferous (Tournaisian)–Cretaceous (Campanian)
Meekospiridae	Carboniferous (Tournaisian)–Permian (Changhsingian)

[a]Donaldinidae may be still alive as Murchisoniellidae (Bandel 2005; Warén 2013)

Pseudozygopleuridae are small caenogastropods with a worldwide distribution ranging from the Carboniferous to the Early Triassic (Nützel 1998; Foster et al. 2017). Most are high-spired and have a teleoconch ornament of axial ribs, and generally a low disparity with many species differing only in small details. The genus *Helminthozyga* shows uncoiling of teleoconch whorls and resembles extant members of Epitoniidae in this respect and could point to a parasitic lifestyle (Bandel 1991; Sasaki 2008) (Fig. 6.3a, b).

Zygopleuridae range from the Early Triassic (Nützel and Schulbert 2005; Foster et al. 2017) to the Cretaceous (Albian) (Kiel 2006). They closely resemble Palaeozoic Pseudozygopleuridae but differ in larval ornaments (Bandel 1991; Nützel 1998). Undoubtedly, both families are closely related to each other and are included in Pseudozygopleuroidea that may represent the "stem-group" of modern Ptenoglossa (Bandel 1991; Nützel 1998).

Protorculidae are known from the Early Triassic to the Middle Jurassic, especially from the Carnian Cassian Formation (Bandel 1991; Nützel 1998; Nützel and Erwin 2002; Kaim and Conti 2010). They have an anterior siphonal channel and may have a complex teleoconch ornamentation. They share an axially ribbed larval shell with other Zygopleuroidea as well as with living Nystiellidae and basal Cerithiopsidae and were identified as fossil sister group of cerithiopsids (Nützel 1998). It is possible that they had acquired a spongivorous lifestyle already.

Donaldinidae and Streptacididae are small, high-spired, sometimes very slender gastropods with a heterostrophic larval shell. They range from the Devonian to the Cretaceous but their shell morphology is very close to living genera such as *Murchisoniella* or *Ebala* and it thus might be that they persist until today (Bandel 2005; Warén 2013). Their small size, high diversity but low disparity may point to a parasitic life style similar to that of living Pyramidellidae.

Amongst the heterogeneous, largely Palaeozoic group Subulitoidea, there are some diverse genera and families which closely resemble extant parasitic eulimids. Amongst Meekospiridae (Carboniferous–Permian) the genera *Ceraunocochlis* and *Meekospira* closely resemble recent eulimid species (Fig. 6.2). They share the smooth surface, shallow suture and high-spired shape and in the case of *Ceraunocochlis* the bent spire producing an asymmetrical shape is also present in several eulimid genera.

6.3 Conclusions

Direct evidence for parasitism of gastropods such as preserved host/parasite associations or characteristic trace fossils in the host is anecdotic at best in relation with the vast diversity of this class. One of the few exceptions is the Palaeozoic platyceratid/crinoid interaction that has a fairly good record of examples where host and parasite are still attached to each other and additional borings. Most modern highly diverse parasitic (or carnivorous grazers) gastropod families have a Late Mesozoic/Paleocene origin and seem to undergo a major adaptive radiation. At present, taxonomic uniformitarism is the most important tool to infer the fossil record of gastropod parasites. This works for the Cenozoic and Late Mesozoic. Few lines can be traced into the Jurassic and Triassic with some confidence. Several diverse older, extinct families of tiny, high-spired Caenogastropoda and Heterobranchia might have been parasitic but this remains speculative to some degree at this point.

Parasitism and similar feeding types are major trophic modes in modern gastropods and is present in as many as 10,000 extant species with Pyramidellidae being most diverse. This underlines the trophic diversity of gastropods which is most likely one of the major reasons for their evolutionary success.

Acknowledgments I would like to thank Anders Warén and Philippe Bouchet for helpful discussions and providing images. I thank Kenneth de Baets and Andrzej Kaim who provided critical, helpful reviews.

References

Ankel WE (1949) Die Nahrungsaufnahme der Pyramidelliden. Verhandlungen der Deutschen Zoologen, Kiel 1948:478–484

Bandel K (1988) Repräsentieren die Euomphaloidea eine natürliche Einheit der Gastropoden? Geol-Paläont Inst Univ Hamburg 67:1–33

Bandel K (1991) Über triassische "Loxonematoidea" und ihre Beziehungen zu rezenten und paläozoischen Schnecken. Paläontol Z 65:239–268

Bandel K (1995) Mathildoidea (Heterostropha, Gastropoda) from the Upper Triassic St. Cassian formation. Scr Geol 111:1–83

Bandel K (2005) Living fossils among tiny allogastropoda with high and slender shell from the reef environment of the Gulf of Aquba with remarks on fossil and recent relatives. Geol-Paläont Inst Univ Hamburg 89:1–24

Baumiller TK, Gahn, FJ (2002) Fossil record of parasitism on marine invertebrates with special emphasis on the platyceratid-crinoid interaction. In: Kowalewski M, Kelley PH (eds) The fossil record of predation. The Paleontological Society Papers, 8. p 195–209

Baumiller TK, Gahn F (2018) Biotic interactions between platyceratid gastropods and crinoids: a review and new insights. Abstract. In: 5th international paleontological congress, Paris. p. 456

Baumiller TK (2003) Evaluating the interaction between platyceratid gastropods and crinoids: a cost-benefit approach. Palaeogeography, Palaeoclimatology, Palaeoecology 201:199–209

Beu AG (2007) The "*Inoceramus* limpet" *Gigantocapulus problematicus* (Nagao & Otatume, 1938) in New Zealand (Late Cretaceous Gastropoda or Monoplacophora, Gigantocapulidae n. fam). Paläontol Z 81:267–282

Bieler R (1995) Mathildidae from New Caledonia and the Loyalti Islands (Gastropoda: Heterobranchia). Mém Mus Natl Hist Nat 167:595–641

Bieler R, Petit RE (2005) Cataloque of recent and fossil taxa of the family Architectonicidae Gray, 1850 (Mollusca: Gastropoda). Zootaxa 1101:1–119

Bouchet P (2009) From specimens to data, and from seashells to molluscs: the Panglao Marine Biodiversity Project. Vita Malacol 8:1–8

Bouchet P, Lozouet P, Maestrati P, Heros V (2002) Assessing the magnitude of species richness in tropical marine environments: exceptionally high numbers of molluscs at a New Caledonia site. Biol J Linn Soc 75:421–436

Bouchet P, Lozouet P, Sysoev A (2009) An inordinate fondness for turrids. Deep-Sea Res 56:1724–1731

Bouchet P, Rocroi J-P, Hausdorf B, Kaim A, Kano Y, Nützel A, Parkhaev P, Schrödl M, Strong EE (2017) Revised classification, nomenclature and typification of gastropod and monoplacophoran families. Malacologia 61:1–526

Boucot AJ (1990) Evolutionary paleobiology of behavior and coevolution. Elsevier, Amsterdam, 725 p

Boucot AJ, Poinar GO Jr (2010) Fossil behavior compendium. CRC Press, Boca Raton, 391 p

Bowsher AL (1955) Origin and adaptation of platyceratid gastropods. KU Paleontol Contrib Mollusca 5:1–11
Breton G, Wisshak M, Néraudeau D, Morel N (2017) Parasitic gastropod bioerosion trace fossil on Cenomanian oysters from Le Mans, France and its ichnologic and taphonomic context. Acta Palaeontol Pol 62:45–57
Brown LG, Neville BD (2015) Catalog of the recent taxa of the families Epitoniidae and Nystiellidae (Mollusca: Gastropoda) with a bibliography of the descriptive and systematic literature. Zootaxa 3907:1–188
Cossmann M (1912) Essais de paléoconchologie comparée, vol 9. Published by the author and J. Lamarre and Cie, Paris, 215 p
Dinapoli A, Zinssmeister C, Klussmann-Kolb A (2011) New insights into the phylogeny of the Pyramidellidae (Gastropoda). J Molluscan Stud 77:1–7
Dockery DT (1993) The streptoneuran gastropods, exclusive of the Stenoglossa, of the Coffe Sand (Campanian) of northeastern Mississippi. Mississippi Department of Environmental Quality, Office of Geology, Bulletin 129:1–191
Fernandes MR, Pimenta AD (2019) Taxonomic review of *Inella* and *Strobiligera* (Gastropoda: Triphoridae) from Brazil. Zootaxa 4613:1–52
Foster WJ, Danise S, Twitchett RJ (2017) A silicified Early Triassic marine assemblage from Svalbard. J Syst Palaeontol 15:851–877
Fernandes MR, Pimenta AD (2017) Taxonomic overview of Triphoridae in the world: describing triphorids. Informativo SBMa, Ano 48, nº200 (jun/2017):4–10
Gahn JG, Baumiller TK (2006) Using platyceratid gastropod behaviour to test functional morphology. Historical Biology 18:397–404
Gittenberger A, Gittenberger E (2005) A hitherto unnoticed adaptive radiation: epitoniid species (Gastropoda: Epitoniidae) associated with corals (Scleractinia). Contrib Zool 74:125–203
Gittenberger A, Gittenberger E (2011) Cryptic, adaptive radiation of endoparasitic snails: sibling species of *Leptoconchus* (Gastropoda: Coralliophilidae) in corals. Org Divers Evol 11:21–41
Gittenberger A, Hoeksema BW (2013) Habitat preferences of coral-associated wentletrap snails (Gastropoda: Epitoniidae). Contrib Zool 82:1–25
Gründel J (2012) Neubearbeitung der von Laube 1867 beschriebenen Gastropodenfauna aus dem mittleren Jura von Balin/Polen. Ann Naturhist Mus Wien 114:193–288
Gründel J, Nützel A (2013) Evolution and classification of Mesozoic mathildoid gastropods. Acta Palaeontol Pol 58:803–826
Guzhov AV (2006) *Liapinella*, a new epitonioidean genus (Gastropoda) from the Upper Jurassic of European Russia. Ruthenica 16:1–4
Hayami I, Kanie Y (1980) Mode of life of a giant capulid gastropod from the Upper Cretaceous of Saghalien and Japan. Palaeontology 23:689–698
Healey JM (1998a) Superfamily Pyramidelloidea. In: Beesley PL, GJB R, Wells A (eds) Mollusca: the southern synthesis, part B. Fauna of Australia. CSIRO Publishing, Melbourne, pp 865–869
Healey JM (1998b) Superfamily Architectonicoidea. In: Beesley PL, GJB R, Wells A (eds) Mollusca: the southern synthesis, part B. Fauna of Australia. CSIRO Publishing, Melbourne, pp 858–862
Kaim A (2004) The evolution of conch ontogeny in Mesozoic open sea gastropods. Palaeontol Pol 62:1–182
Kaim A, Conti MA (2010) A problematic zygopleuroid gastropod *Acanthostrophia* revisited. Zitteliana A50:21–24
Kaunhowen F (1897) Die Gastropoden der Maestricher Kreide. Palaeontologische Abhandlungen, Neue Folge 4(der ganzen Reihe Band 8):1–132
Kiel S (2006) New and little-known gastropods from the Albian of the Mahajanga Basin, Northwestern Madagascar. J Paleontol 80:455–476
Knight JB, Cox LR, Keen AM, Smith AG, Batten RL, Yochelson EL, Ludbrook NH, Robertson R, Yonge CM, Moore RC (1960) Mollusca - General Features, Scaphopoda, Amphineura, Monoplacophora, Gastropoda. General Features, Archaeogastropoda and some (mainly

Paleozoic) Caenogastropoda and Opisthobranchia. Geological Society of America & University of Kansas Press, Lawrence, xxiii +351 pp.
Kohn AJ (1983) Feeding biology of Gastropoda. In: Wilbur KM (ed) The Mollusca, vol. 5, physiology. Academic, New York, pp 1–64
Lorenz F (2005) Mollusca (molluscs). In: Rohde K (ed) Marine parasitology. CABI Publishing, Oxon, pp 240–245
Lozouet P, Renard P (1998) The Coralliophilidae, gastropoda from the oligocene and lower miocene of the aquitaine (southwestern france): systematics and details of coral hosts. Geobios 31:171–184
Lygre F, Schander C (2010) Six new species of pyramidellids (Mollusca, Gastropoda, Pyramidelloidea) fromWest Africa, introducing the new genus Kongsrudia. Zootaxa 2657:1–17
Mariottini P, Smriglio C, Di Giulio A (2009) Two new mathildids from the south-eastern coast of Africa (Gastropoda, Heterobranchia, Mathildidae). Basteria 73:69–76
Marshall BA (1978) Cerithiopsidae (Mollusca: Gastropoda) of New Zealand, and a provisional classification of the family. New Zealand J Zool 5:47–120
Marshall BA (1983) A revision of the recent Triphoridae of Southern Australia (Mollusca: Gastropoda). Rec Aust Mus Suppl 2:1–119
Marshall BA, Olivero M (2009) The recent Coralliophilinae of the New Zealand region, with descriptions of two new species (Gastropoda: Neogastropoda: Muricidae). Mollusc Res 29:155–173
Nagler C, Haug JT (2015) From fossil parasitoids to vectors: insects as parasites and hosts. In: De Baets K, Littlewood T (eds) Fossil parasites, advances in parasitology 90. Elsevier, Amsterdam, pp 137–200
Neumann C, Wisshak M (2009) Gastropod parasitism on Late Cretaceous to Early Paleocene holasteroid echinoids—evidence from *Oichnus halo* isp. n. Palaeogeogr Palaeoclimatol Palaeoecol 284:115–119
Nützel A (1998) Über die Stammesgeschichte der Ptenoglossa (Gastropoda). Berliner Geowissenschaftliche Abhandlungen. Reihe E 26:1–229
Nützel A (2014) Larval ecology and morphology in fossil gastropods. Palaeontology 57:479–503
Nützel A, Erwin DH, Mapes RH (2000) Identity and phylogeny of the late Paleozoic Subulitoidea (Gastropoda). J Paleontol 74:575–598
Nützel A, Schulbert C (2005) Gastropod lagerstätten in the aftermath of the end-Permian mass extinction—diversity and facies of two major Early Triassic occurrences. Facies 51:495–515
Nützel A, Erwin DH (2002) *Battenizyga*, a new early Triassic gastropod genus with a discussion on the gastropod evolution at the Permian/Triassic boundary. Paläontologische Zeitschrift 76:21–26
Petit G, Robin N, Zorzin R, Merle D (2014) Fossil gastropods (?*Aclis aenigmaticus* n. sp.) on a fish from the Pesciara of Bolca Lagerstätte (Eocene, Northern Italy): an enigmatic association. Studi e ricerche sui giacimenti terziari di Bolca, XV - Miscellanea paleontologica 12:129–136
Ponder WF (1987) The anatomy and relationships of the pyramidellacean limpet, *Amathina tricarinata* (Mollusca: Gastropoda). Asian Marine Biol 4:1–34
Ponder WF, Colgan DJ, Healy JM, Nützel A, Simone LRL, Strong EE (2008) Caenogastropoda. In: Ponder WF, Lindberg DL (eds) Phylogeny and evolution of the Mollusca. University of California Press, Berkeley, pp 331–383
Ponder WF, De Kayzer RG (1998) Superfamily Pyramidelloidea. In: Beesley PL, Ross GJB, Wells A (eds) Mollusca: the southern synthesis, part B. Fauna of Australia. CSIRO Publishing, Melbourne, pp 865–869
Ponder WF, Lindberg DR (1997) Towards a phylogeny of gastropod molluscs: an analysis using morphological characters. Zool J Linnean Soc 119:83–265
Ravn JPJ (1933) Études sur les Pélécypodes et Gasteropodes Daniens. K dansk Vidensk Selsk Skr 9(2):1–71
Robertson R (1970) Review of the predators and parasites of stony corals with special reference to symbiotic prosobranch gastropods. Pac Sci 24:43–54

Robertson R (1980) *Epitonium millecostatum* and *Coralliophila clathrata*: two prosobranch gastropods symbiotic with Indo-Pacific *Palythoa* (Coelenterata: Zoanthidae). Pac Sci 34:1–17

Robertson R, Mau-Lastovicka T (1979) The ectoparasitism of *Boonea* and *Fargoa* (Gastropoda: Pyramidellidae). Biol Bull 157:320–333

Roden VJ, Hausmann IM, Nützel A, Seuss B, Reich M, Urlichs M, Hagdorn H, Kiessling W (2019) Towards an assessment of true biodiversity in fossil ecosystems—the role of liberation lagerstätten. Palaeontology 1–18. https://doi.org/10.1111/pala.12441

Sasaki T (2008) Micromolluscs in Japan: taxonomic composition, habitats, and future topics. Zoosymposia 1:147–232

Schrödl M, Jörger KM, Klussmann-Kolb A, Wilson NG (2011) Bye bye "Opisthobranchia"! A review on the contribution of mesopsammic sea slugs to euthyneuran systematics. Thalassas 27:101–112

Signor PW (1982) Constructional morphology of gastropod ratchet sculpture. Neues Jahrb Geol Palaontol Abh 163:349–368

Sohl NF (1964) Neogastropoda, Opisthobranchia, and Basommatophora of the Ripley, Owl Creek and Prairie Bluff formations. U S Geol Surv Prof Pap 331-B:153–344

Sutton MD, Briggs DEG, Siveter DJ, Siveter DJ (2006) Fossilized soft tissues in a Silurian platyceratid gastropod. Proc R Soc B Biol Sci 273(1590):1039–1044

Takano T, Kano Y (2014) Molecular phylogenetic investigations of the relationships of the echinoderm-parasite family Eulimidae within Hypsogastropoda (Mollusca). Mol Phylogenet Evol 79:258–269

Tracey S, Todd JA, Erwin DH (1993) Mollusca: Gastropoda. In: Benton MJ (ed) The fossil record, vol 2. Chapman & Hall, New York, pp 131–167

Wanner C (1922) XVIII. Die Gastropoden und Lamellibranchiaten der Dyas von Timor. In: Wanner J (ed) Paläontologie von Timor XI. Lieferung. Schweizerbart, Stuttgart, pp 1–82

Warén A (1983) A generic revision of the family Eulimidae. J Molluscan Stud Suppl 13:1–96

Warén A (2013) Murchisonellidae: who are they, where are they and what are they doing? (Gastropoda, lowermost Heterobranchia). Vita Malacol 11:1–14

Webster GD, Donovan SK (2012) Before the extinction—Permian platyceratid gastropods attached to platycrinitid crinoids and an abnormal four-rayed *Platycrinites* s.s. *wachsmuthi* (Wanner) from West Timor. Palaeoworld 21:153–159

Wesenberg B, Schnetler KI (2014) A catalogue of Danian gastropods from the Baunekule facies, Faxe Formation, Denmark. Geological Survey of Denmark and Greenland Bulletin 32:1–117

Weinstein SB, Kuris AM (2016) Independent origins of parasitism in Animalia. Biol Lett 12: 20160324. http://dx.doi.org/10.1098/rsbl.2016.0324

Wilson BW (1998) Superfamily Cypraeoidea. In: Beesley PL, GJB R, Wells A (eds) Mollusca: the southern synthesis, part B. Fauna of Australia. CSIRO Publishing, Melbourne, pp 780–786

Chapter 7
Fossil Constraints on the Timescale of Parasitic Helminth Evolution

Kenneth De Baets, Paula Dentzien-Dias, G. William M. Harrison, D. Timothy J. Littlewood, and Luke A. Parry

Abstract The fossil record of parasitic helminths is often stated to be severely limited. Many studies have therefore used host constraints to constrain molecular divergence time estimates of helminths. Here we review direct fossil evidence for several of these parasitic lineages belong to various phyla (Acanthocephala, Annelida, Arthropoda, Nematoda, Nematomorpha, Platyhelminthes). Our compilation shows that the fossil record of soft-bodied helminths is patchy, but more diverse than commonly assumed. The fossil record provides evidence that ectoparasitic helminths (e.g., worm-like pentastomid arthropods) have been around since the early Paleozoic, while endoparasitic helminths (cestodes, nematodes) arose at least during, or possibly even before the late Paleozoic. Nematode lineages parasitizing terrestrial plant and animal hosts have been in existence at least since the Devonian and Triassic, respectively. All major phyla (Acanthocephala, Annelida, Platyhelminthes, Nematoda, Nematomorpha) had evolved endoparasitic lineages at least since the Mesozoic. Interestingly, although parasitism is considered derived within Metazoa, the oldest evidence for Nematoda and Platyhelminthes includes body fossils of parasitic representatives. Furthermore, the oldest fossil evidence of these parasitic lin-

K. De Baets (✉)
Geozentrum Nordbayern, Friedrich-Alexander University Erlangen-Nürnberg, Erlangen, Germany
e-mail: kenneth.debaets@fau.de

P. Dentzien-Dias
Instituto de Oceanografia, Universidade Federal do Rio Grande, Rio Grande, Brazil
e-mail: pauladentzien@furg.br

G. W. M. Harrison
Naturalis Biodiversity Center, Leiden, Netherlands
e-mail: william.harrison@naturalis.nl

D. T. J. Littlewood
The Natural History Museum, London, UK
e-mail: t.littlewood@nhm.ac.uk

L. A. Parry
Department of Earth Sciences, University of Oxford, Oxford, UK
e-mail: luke.parry@seh.ox.ac.uk

© The Author(s) 2021

K. De Baets, J. W. Huntley (eds.), *The Evolution and Fossil Record of Parasitism*, Topics in Geobiology 49, https://doi.org/10.1007/978-3-030-42484-8_7

eages often falls within molecular divergence times based on host co-evolution suggesting the fossil record of helminths themselves might be just as good or at least complementary (and less circular in justification) to calibration based on host associations. Data also provide evidence for obvious host switches or extinctions, which cautions against models of pure co-divergence where use of host calibrations to constrain divergence time estimates may be considered.

Keywords Parasitic worms · Fossil record · Molecular divergence estimates · Host constraints

7.1 Introduction

Helminths are multicellular parasitic worms varied in size (Caira and Littlewood 2013) from the microscopic (200 μm) to the enormously long (80 m). This polyphyletic grouping includes parasitic flatworms (flukes, tapeworms), nematodes and acanthocephalans (Rohde 2005). They can cause diseases in humans, livestock and wildlife (e.g., helminthiases). Some authors also include tongue worms (pentastomids: e.g., McAllister et al. 2010) or even parasitic annelids (e.g., Öktener 2005) when describing or listing helminths from particular regions. Although an artificial grouping, which contains a number of phylogenetically distantly related clades, the term 'helminths' is still used to describe a group of parasitic organisms sharing a similar form and lifestyle.

Despite their biomedical and economic importance, the deep time evolutionary history of parasitic worms remains poorly understood, which has been largely attributed to their meagre, or rather poorly explored, fossil record (Littlewood and Donovan 2003). As many helminths use vertebrate animal hosts or other hosts with a demonstrably better fossil record, it is helminth hosts that have been used to constrain helminth origins (by phylum) where a certain degree of host specificity might be assumed. With such approaches, some groups of nematodes and parasitic flatworms have been suggested as existing during the Cambrian–Ordovician (Littlewood 2006; Poinar 2011, 2015). In other cases, particularly where host-specificity is common, current biogeographic distribution patterns of host-parasite associations have been used to infer the temporal origin of the interaction (Borda et al. 2008; Verneau et al. 2009; Badets et al. 2011; Cuthill et al. 2016). Some of the most commonly used examples for host or biogeographic constraints include those made for the monogenean group Polystomatidae (Verneau et al. 2002, 2009; Badets et al. 2011; Héritier et al. 2015; Tinsley and Tinsley 2016). However, both approaches might involve circular reasoning (Kodandaramaiah 2011; De Baets et al. 2015, 2016; Warnock and Engelstädter 2021). Fossil associations are among the only direct sources of evidence for the existence of particular lineages in the geological past (De Baets and Littlewood 2015). In addition, the fossil record can provide additional information on morphology, host association and/or life history.

The oldest known unambiguous fossil helminths are isolated remains of pentastomids which are known since the Cambrian-Ordovician (Klompmaker and Boxshall 2015). Older finds have been reported (Cong et al. 2017) although their phylogenetic assignment is poorly resolved and they are interpreted to be likely ectosymbiotic rather than ectoparasitic. There is also fossil evidence that nematodes already parasitized early land plants in the Early Devonian (Poinar et al. 2008) and monogeneans parasitized early gnathostomes in the Middle Devonian (Upeniece 2001, 2011). The first confidently identified evidence for intestinal helminths derives from the Permian (Dentzien-Dias et al. 2013)—although older Carboniferous finds have been reported (Zangerl and Case 1976). These finds suggests that the fossil record of helminths is more ancient than commonly perceived.

Here we review the fossil record of helminths with the aim of providing suitable minimum and maximum constraints for dating their origins and key host transitions during their evolution. The fossil record of helminths is reviewed at greater length in other publications (Poinar 2011, 2015; Huntley and De Baets 2015; Parry et al. 2014, 2019b). We focus particularly on similarities and differences in the fossil record of helminths and in a phylogenetic context. For this purpose, we first briefly review the progress made in disentangling their phylogenetic position within the Metazoa since a major review on the topic (Zrzavý 2001).

7.2 Phylogenetic Affinities and Distribution of Parasitic "Helminths"

Since the early 1990s, the rapid accumulation of molecular sequence data has revolutionized our understanding of animal phylogeny and evolutionary history (Edgecombe et al. 2011). Bilaterian animals are now usually subdivided into three principal clades (Laumer et al. 2019), Deuterostomia, Ecdysozoa and Spiralia (including Lophotrochozoa), with the latter two groups forming the Protostomia. Protostomia contains the most animal phyla and body plans and likewise contains the highest diversity of groups that have evolved a parasitic lifestyle (Weinstein and Kuris 2016). A possible exception are Myxozoa—some of which are worm-like—now considered to be derived cnidarians (Okamura et al. 2015; Okamura and Gruhl 2021 for reviews). Various helminths are so divergent from their close relatives that they have previously been considered separate phyla. Examples include Acanthocephala and Pentastomida, which are now recognized as highly modified members of Rotifera or Syndermata (Struck et al. 2014; Sielaff et al. 2016; Herlyn 2021) and Crustacea in Euarthropoda (Sanders and Lee 2010; Oakley et al. 2012), respectively. In general, parasitic lineages have elevated rates of molecular sequence evolution, which often makes them hard to place robustly in phylogenies, where they often form long branches. Myzostomida are a key example, having been placed as close relatives of flatworms (Eeckhaut et al. 2000), rotifers (Zrzavý et al. 2001) and annelids (Bleidorn et al. 2009) based on different molecular datasets. More recent phylogenomic analyses (Weigert et al. 2014) have confirmed that myzostomids are modified annelids, but their precise phylogenetic position within the phylum remains uncertain.

Nematoda are closely related (sister) to Nematomorpha and are grouped together in the clade Nematoida. Most studies place Nematoida and other Cycloneuralia (Scalidophora) together with Panarthropoda (Onychophora, Tardigrada and Arthropoda) in a clade named Ecdysozoa (characterized by a growth strategy involving periodic moulting of an external cuticle; but see Giribet and Edgecombe 2017). Acoelomorpha are long since removed from the phylum Platyhelminthes which now consists of the paraphyletic "Turbellaria" and the parasitic Neodermata (Egger et al. 2015). The relative position of Platyhelminthes and Rotifera within the Metazoa is less well resolved (Laumer et al. 2019), but they are usually assigned to Spiralia (Struck et al. 2014; Laumer et al. 2015a; Marlétaz et al. 2019). The Spiralia also contain Lophotrochozoa including phyla as diverse as Cycliophora, Entoprocta, Mollusca, Nemertea, Annelida, Brachiopoda, Phoronida and Bryozoa. Most recent studies recover the mostly macrofaunal clade Lophotrochozoa as a sistergroup to a platyhelminth—gastrotrich clade (Rouphozoa sensu Struck et al. 2014), both of which are sister to a clade populated by Gnathifera (Gnathostomulida, Micrognathozoa, Syndermata). The assignment of Chaetognatha as sistergroup to, or in a nested position within Gnathifera, is still debated (Marlétaz et al. 2019).

Even within these phyla, parasitism may have evolved multiple times independently. Weinstein and Kuris (2016) reviewed incidences of parasitism in light of modern phylogenetics and suggested that parasitism might have evolved as many as 143 times in Arthropoda, 18 times in Nematoda, 13 times in Platyhelminthes, nine times in Annelida, seven times in Rotifera (= Syndermata), six times within Acoelomorpha, but only once in the Nematomorpha. These analyses, clearly suggest that parasitic helminths have evolved convergently. This might not only have resulted in similarities in morphology and ecology, but also in their preservation potential, which is largely restricted to sites of exceptional (soft-body) preservation (Konservation-Lagerstätten sensu Seilacher 1970; see De Baets et al. (2021) for a review).

7.3 Preservation Potential

The fossil record of various groups of helminths, or the phyla they belong to, was recently reviewed by various authors (Maas 2013; Poinar 2011, 2014, 2015; De Baets et al. 2015; Klompmaker and Boxshall 2015; Parry et al. 2014, 2019b, Herlyn 2021). We discuss the bearing of these fossils and alternatives for constraining date estimates of the origins and diversification of parasitic helminths.

As many helminths are small and soft-bodied their fossil record is quite patchy and largely restricted to rare finds in conservation traps (e.g., amber) or other sites of exceptional preservation (Littlewood and Donovan 2003). Ectoparasitic forms often use attachment organs (suckers, hooks) and during fossilization might be transported away from their host making it hard to recognize their mode of life or association. Endoparasitic forms might not be discovered at all as they reside within their hosts; few have structures that are not likely decomposed along with the soft-tissues of their hosts. When body fossil evidence is found the often few remaining characters can make it difficult to assign them to particular lineages (Schram 1973; Muir et al. 2014). Additionally, due to convergent losses and gains of traits in the diversification of helminths, they can look quite different from their free-living ancestors and/or very similar to other extant clades of helminths (De Baets and Littlewood 2015).

Soft-body imprints or remains have become particularly known from nematodes, likely related to their resilient external cuticle—characteristic amongst both free-living and parasitic forms. Sclerotized or otherwise more resistant structures such as attachment organs, cocoons and eggs (Figs. 7.1 and 7.2) are considered to have a better preservation potential than body fossils (Littlewood and Donovan 2003). In particular, eggs of particular groups seem to have a high preservation potential—surviving various strong acid and base treatments (Dufour and Le Bailly 2013; Camacho et al. 2018). In the last decades new and impressive findings of parasite

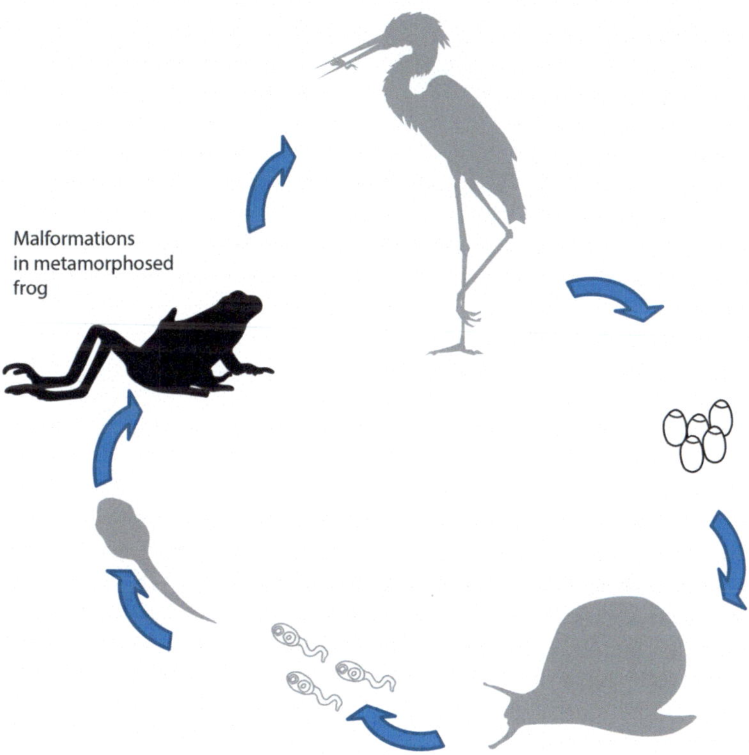

Fig. 7.1 Life-cycle of the trematode *Ribeiroia ondatrae* marking stages which could be preferentially preserved in the fossil record (malformations in metamorphosed frog, eggs in coprolites). Some of the silhouettes were obtained from PhyloPic (http://phylopic.org/)

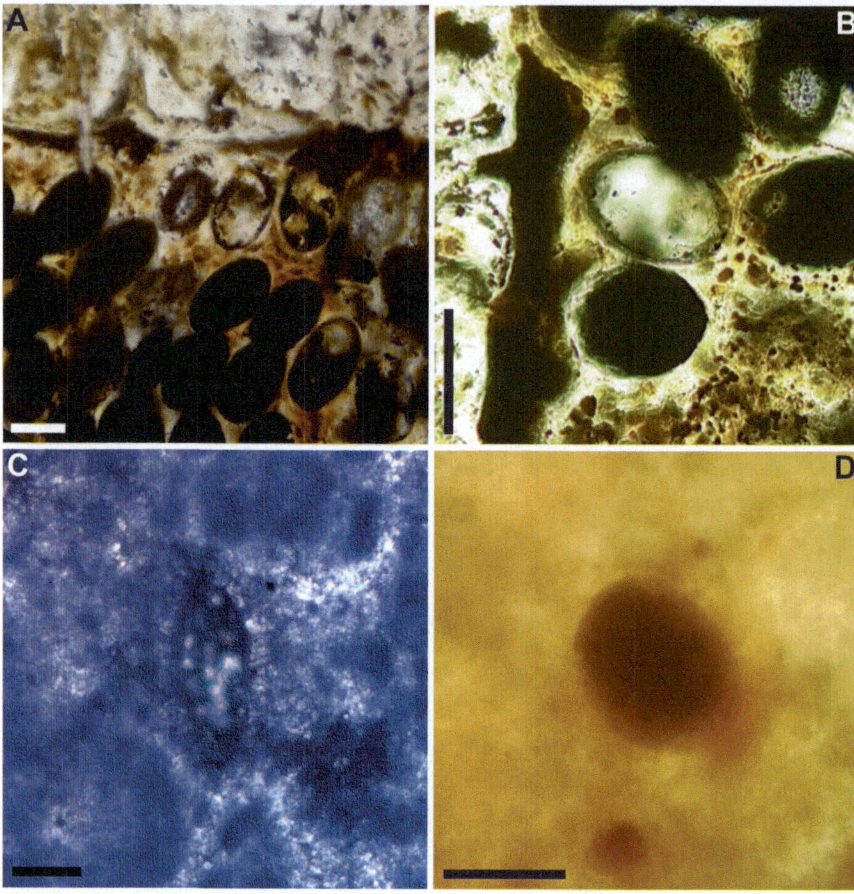

Fig. 7.2 Parasite eggs found inside coprolites thin sections. (**a**) and (**b**) Tapeworm eggs from a Permian shark coprolite (Dentzien-Dias et al. 2013) (**c**) pinworm *Paleoxyuris cockburni* egg from a Triassic coprolite (Francischini et al. 2018); (**d**) nematode egg with 65 μm long and 50 μm wide from a Miocene crocodile coprolite (Dentzien-Dias et al. 2018). Scale bars **a**: 75 μm, **b** and **c**:100 μm, **d**: 50 μm

eggs have been particularly found with different techniques in subfossil (dessicated) to fossil (permineralized) coprolites. Coprolites are known to be a *Konservat-Lagerstätte* (Seilacher et al. 2001; Qvarnström et al. 2016), the bacteria content of feces induce early mineralization that allow the preservation of soft tissues in exquisite detail (e.g., muscles, parasite eggs, hair and feathers: Hollocher and Hollocher 2012; Chin 2021). As most fossil coprolites are permineralized, they are commonly treated with HCl. Using such approaches nematode eggs have been recovered from material dating back to the Triassic (Da Silva et al. 2014; Hugot et al. 2014) with evidence of acanthocephalans and trematodes from the Cretaceous (Cardia et al. 2019a; Poinar and Boucot 2006). However, using such an approach might destroy particular remains. Additional, and including older fossils have been discovered in coprolites using microscopic (Zangerl and Case 1976) or thin section approaches (Dentzien-Dias et al. 2013; Francischini et al. 2018; see fig. 1; Dentzien-Dias et al. 2018)—a common approach to study fossil coprolites (Chin 2002). The assignment of coprolites to their hosts is less straightforward than commonly assumed. It is often done based on composition, fossil content (providing clues to environment or diet), morphology and size. However, these methods are not flawless and identification is best achieved when they are still associated with their owner. Even if the coprolite owner can be identified, there is still a need to establish that putative parasitic remains are not simply present in the coprolite because the microorganism (or its host for that matter) is microbophagous (Poinar 2015) or coprophagous (Bajdek et al. 2016).

Sclerotized hooks from helminth attachment organs have also been identified from fossil remains (Upeniece 2001; De Baets et al. 2015). The best evidence for this is a collection of 69 circlets of microscopic hooks, mostly still attached to their acanthodian and placoderm fish hosts in the Devonian Lode Formation (Upeniece 2001). Similar remains from acanthocephalan or cestode hooks have yet to be found, although the presence of their attachment organs have been interpreted to be present in remains of developing larvae within putative parasite eggs (Cardia et al. 2019a; Dentzien-Dias et al. 2013).

Other types of evidence involve characteristic trace fossils and pathologies left in their hosts (e.g., Huntley et al. 2021), although they are hard to tie to particular parasite taxa with much phylogenetic precision. In some cases, various phylogenetically distantly related modern or now extinct groups with similar behavior could have led to similar structures. More confidence is gained when characteristic parasitic remains or at least signs of in-situ infestation are associated with these structures. Such evidence is however comparatively rare and in some cases, the host structure might be unsuitable for preservation (e.g., early decay of isopod parasite within decapod pathologies: Klompmaker et al. 2017).

7.4 Time Constraints on the Origin of Eumetazoan Helminths

The fossil record can only provide meaningful hard minimum time constraints on the timing of divergences of a clade. Various authors have used records of near-relatives to place soft maxima on the existence of a certain clade at a certain time. This is typically based "on the understanding that they could have been there but are not based on the existence of facies preserving their near relatives, among which are at least some ecological analogues" (Benton et al. 2015). In the case of Eumetazoa a meaningful maximum constraint would be the maximum age of the Lantian biota (Benton et al. 2015). This together with Doushantuo Formation represent the oldest Konservät-Lagerstätten which have yielded putative eumetazoans in similar preservations which have yielded eumetazoans in younger formations. Consequently, a soft-maximum of 636.1 Ma (Condon et al. 2005) time constraint on the origin of free-living eumetazoan clades may be inferred. As parasitism is a derived condition within the various eumetazoan lineages discussed herein, so the origins of parasitic helminths could only arise after the first appearance of Eumetazoa.

Eumetazoan origins considerably predate putative nematoid remains (Muir et al. 2014), with possible traces from the Ediacaran to Ordovician (Balinski et al. 2013) or borings found in Cambrian-Ordovician trilobites (Babcock 2007) reminiscent of possible nematode borings in foraminifers (Sliter 1971; Hope and Tchesunov 1999). These traces even predate the Ediacaran fossils where *Dickinsonia* has occasionally been superficially compared with the annelid *Spinther* (Wade 1972; Conway Morris 1981) associated as ectoparasites or commensals with modern sponges (Rouse and Pleijel 2003; Rouse 2005). However, *Dickinsonia* shows no annelid characters and is now mostly considered a more early diverging metazoan if indeed a metazoan at all (Brasier and Antcliffe 2008; Sperling and Vinther 2010; Bobrovskiy et al. 2018). Considering its comparatively large size it is also difficult to consider it parasitic as no larger-sized suitable host has so far been found (De Baets and Littlewood 2015). More importantly, *Dickinsonia* has been found associated with trace fossils attesting to its free-living mode of life (e.g., Sperling and Vinther 2010).

In the following, we discuss possible time constraints on the origin of particular helminths in their phylogenetic and host context (see Fig. 7.3). We provide details

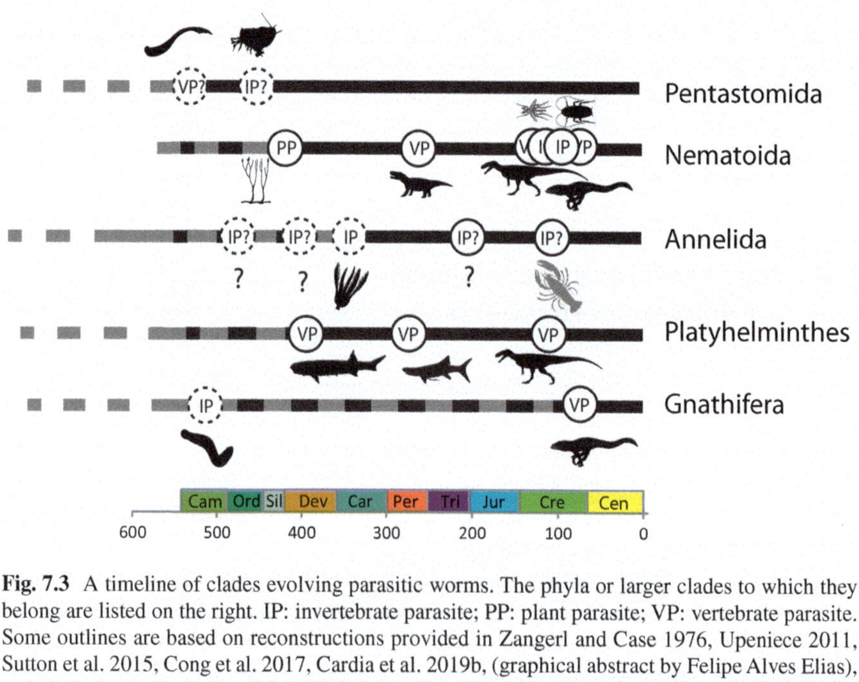

Fig. 7.3 A timeline of clades evolving parasitic worms. The phyla or larger clades to which they belong are listed on the right. IP: invertebrate parasite; PP: plant parasite; VP: vertebrate parasite. Some outlines are based on reconstructions provided in Zangerl and Case 1976, Upeniece 2011, Sutton et al. 2015, Cong et al. 2017, Cardia et al. 2019b, (graphical abstract by Felipe Alves Elias), and phylopic.org (including CC-BY 3.0: Jaime Headden, Mike Keesey; others public domain)

for some possible calibrations, largely following the guidelines by Parham et al. (2012). However, most helminth fossils have not been included in formal phylogenetic analyses. We therefore list the characters which have been used, or could be used, to justify their inclusion within a particular clade.

7.4.1 Fossil Record of Parasitic Annelids

The fossil record of Annelida was recently reviewed by Parry et al. (2014); Parry et al. (2019b). The fossil record of free-living annelids can be traced back to the early Cambrian at least (Morris and Peel 2008; Han et al. 2019), but are typically resolved as members of the annelid stem lineage. Previously, the oldest fossils assigned to the annelid crown group were scolecodonts that are at least Tremadocian in age (Hints and Nõlvak 2006), but a newly discovered magelonid extends the record of crown annelids into the early Cambrian (Chen et al. 2020).

Node. Total group Annelida
Fossil taxon. *Ipoliknus avitus, Adelochaeta sinensis* and at least two unnamed species
Fossil evidence. 6 carbonaceous compression fossils (Han et al. 2019).
Phylogenetic assignment. Both *Ipoliknus* and *Adelochaeta* possess biramous parapodia with capillary chaetae, constraining them to the annelid total group. *Ipoliknus avitus* is resolved as the deepest lineage in the annelid stem group in analysis only considering extinct taxa (Han et al. 2019) as well as a more comprehensive analysis of most annelid lineages and their close relatives (Parry et al. 2019a).
Min age. 515.5 Ma. The fossils are derived from the early Cambrian Chengjiang Biota from the Maotianshan Shales of southwest China. The maximum age of the Chengjiang Biota has been constrained to 518.03 ± 0.69/0.71 using CA-ID-TIMS U-Pb geochronology, whereas the mimimum age is constrained by correlation with the *Antatlasia gutta-pluviae* Zone to an age of 517.0 ± 1.5 (Yang et al. 2018). This minimum age is slightly older than minimum age based on the *Nevadella* trilobite zone for the Sirius Passet Lagerstätte (Harper et al. 2019) which has previously been considered to yield the oldest annelids (Morris and Peel 2008).

Node. Crown group Annelida
Fossil taxon. *Dannychaeta tucolus*
Fossil evidence. 21 carbonaceous compression fossils (Chen et al. 2020)
Phylogenetic assignment. *Dannychaeta tucolus* shares morphological features with extant Magelonidae, including a flattened prostomium, ventrolateral palps that attach near the mouth opening and a sessile, tubicolous lifestyle (Chen et al. 2020). Phylogenetic analysis recovers this genus as a member of stem group Magelonidae, and it is therefore a member of crown Annelida.

Min age. 514 Ma. The fossils derive from the Hongjingshao Member of the Canglangpu Formation which is Cambrian Stage 3 in age, and younger than the Chengjiang Biota (Zeng et al. 2014; Chen et al. 2020).

Annelid fossils of a similar age to those found in Chengjiang have been found in the early Cambrian Sirius Passet Lagerstätte (Conway Morris and Peel 2008). Two species are currently described from this deposit, namely *Phragmochaeta canicularis* (Conway Morris and Peel 2008) and *Pygocirrus butyricampum* (Vinther et al. 2011). Although annelid fossils are more abundant at this site than Chengjiang the age of Sirius Passet is less tightly constrained. This deposit is constrained to the middle to late Cambrian Stage 3 based on its correlation with the *Nevadella* trilobite Zone (Harper et al. 2019). According to GTS 2016 (Ogg et al. 2016), this would yield a minimum age of 515 Ma but the site may be as old as 518 Ma and so this uncertainty encompasses the age of the Chengjiang Lagerstätten (Yang et al. 2018).

Previously, the annelid crown node has been calibrated based on the age of the oldest scolecodonts (Benton et al. 2015), the jaws of eunicidan polychaetes, which first appear unambiguously in the Tremadocian. Eunicida is deeply nested within annelids however (Weigert et al. 2014; Parry et al. 2019a) and so this calibration was always a conservative estimate for the age of the annelid crown group. The above age for the annelid crown node may also be slightly too conservative if the identification of sipunculans from the Chengjiang Biota is correct (Huang et al. 2004) as Sipuncula is recovered as an ingroup annelid taxon in phylogenomic analyses (Weigert et al. 2014). However, no additional sipunculan specimens from the Cambrian have been positively identified (nor are unambiguous sipunculans known from younger deposits) and so the occurrence of this taxon in the Cambrian requires confirmation. *Pygocirrus butyricampum* from Sirius Passet could also conceivably be the oldest crown group annelid, but the phylogenetic position of this taxon is poorly constrained and it is typically resolved in a polytomy with the annelid crown node (Parry et al. 2019a).

All of the fossil annelids known from the Cambrian either have morphological features indicating an epibenthic lifestyle or live in dwelling tubes and so are demonstrably free living and not parasitic. So far, fossil evidence for only four lineages of parasitic annelids has been identified: Branchiobdellida, Myzostomida, Oenonidae and Spionidae. Other groups of annelid parasites have no known fossil record.

Two eunicidan polychaete families that share a prionognath type jaw apparatus are parasitic (Paxton 2009). The family Oenonidae contains species that spend part of their lifecycle as endoparasites of polychaetes, echiurans and bivalves (Hernández-Alcántara and Solís-Weiss 1998). The majority of oenonids parasitize other polychaetes, although some species are thought to be exclusively free-living. This may be a consequence of the parasitic phase of these species not having been recognized however, and all oenonids may in fact share a parasitic lifestyle (Poulin 2001). The Histriobdellidae share a similar jaw architecture and live as commensals on crustaceans, but their close relationship to Oenonidae and other prionognath eunicidans is not universally accepted (Paxton 2009); some authors consider them as close relatives of the Dorvilleidae (Tzetlin 1980). Currently there is no phylogenetic analysis that integrates the diversity of fossil polychaete jaws with extant taxa, although phylogenetic hypotheses of relationships are shown in Paxton (2009) and Parry et al. (2019b).

The prionognath jaw apparatus first appears in the Middle Ordovician (Paxton 2009; Hints and Eriksson 2007), indicating that the oenonid total group had diverged from Lumbrineridae, their sister group (Struck et al. 2015). Jaws that appear more similar to extant oenonids occur in the Silurian (Mierzejewski 1984), with species that are Mesozoic (Jurassic, upper Oxfordian) being assigned to the extant genus *Arabella* (Szaniawski and Gazdzicki 1978), suggesting that the crown group of Oenonidae is at least Oxfordian in age.

Node. Total group Oenonidae

Fossil evidence. Isolated scolecodont finds finds (Paxton 2009 and references therein)

Phylogenetic justification. No explicit phylogenetic analysis has been undertaken for polychaete groups known from isolated scolecodonts. Atraktoprionidae fossils have jaw apparatuses that are of the prionognath type, a type of jaw apparatus that is characteristic of oenonids (Paxton 2009).

Host. *unknown*

Min age. 442.3 Ma. The earliest finds of these taxa are reported from the Ordovician of Estonia and surrounding regions. Several finds derive from loose glacial erratic pebbles which are hard to date as their original provenance is not precisely known (Hints 1998). Some finds like *Atraktoprion* sp. A might at least range into the local Uhaku stage (Hints 2000) which would correlate with the top of the middle Ordovician yielding an age of 443.8–1.5 Ma according to GTS 2016 (Ogg et al. 2016).

Node. Crown group Oenonidae

Phylogenetic justification. Mesozoic fossils have been assigned to the extant genus *Arabella* based on the morphology of jaw elements and apparatus architecture (Szaniawski and Gazdzicki 1978)

Host. unknown

Min age. The specimens derive from a borehole which has has been attributed to the upper Oxfordian (Szaniawski and Gazdzicki 1978). This would correspond with a minimum age of 157.3 Ma according to GTS 2016 (Ogg et al. 2016).

The body plan of Myzostomida is unique among annelids due to their mode of life, but several traits place them in the Annelida, including the presence of parapodia with chaetae (Lanterbecq et al. 2008). Recent myzostomids are associated with echinoderm hosts—chiefly crinoids and to a lesser degree asteroids and ophiuroids (Summers and Rouse 2014). Myzostomid lifestyles range from forms stealing food from their host's food grooves (kleptoparasitic) to consuming host tissue directly (true parasites). Most described species live freely on the exterior of their hosts as adults (although starting life on the host within cysts), while some taxa permanently reside in galls (hard), cysts (soft), or within their host's mouth, digestive system, coelom, or gonads (Summers and Rouse 2014). Typically, galls are expected to have good preservation potential in the fossil record (Welch 1976). A variety of galls in fossil echinoderms—particularly in crinoids—have been attributed to myzostomids (Cameron 1969; Radwańska and Radwańska 2005; Warn 1974). However, only pathologies observed in Carboniferous and Jurassic crinoid arms (Hess 2010; Welch 1976) represent galls which are similar to those produced by extant myzostomids (Jangoux 1987; but see Eeckhaut 1998). So far, no identifiable remains of myzostomids have been identified from such galls which is not surprising as myzostomes have no hard parts other than five pairs of small chitinous hooks (Welch 1976). Other swellings are of uncertain affinity (Ausich and Simms 1999). Eeckhaut (1998) described the first galls induced by myzostomids on extant crinoid stalks, but these look quite different from fossil galls on Paleozoic crinoid stems attributed to myzostomids. In some cases, other types of swellings are associated with phosphatic tubes or rings attributed to *Phosphannulus*—an organism of uncertain affinity. The composition, microstructure and morphology of these tubes and rings rules out an affinity with myzostomids (Welch 1976). Additionally, myzostomids build true galls (Summers and Rouse 2014), penetrating from the outside with the host crinoid reacting with the production of a swelling. *Phosphannulus* are found attached to the outside of crinoid stems and may be subsequently overgrown, a feature not found in association with myzostomids. Although their affinities are unclear, some *Phosphannulus* might have at least been facultative parasites as they were seemingly able to tap in the axial canal of the crinoid when attached (Welch 1976).

Node. Myzostomida
Fossil evidence. Figured gall on the arm of crinoid (Welch 1976)
Host. unidentified crinoid
Min age. 303.4 Ma. The specimen derives from the Seminole Formation of Tulsa County (Okhlahoma). It is attributed to the Missourian (Welch 1976) which is correlated with the Kasimovian—the upper limit of this stage yields a minimum age of 303.4 Ma according to GTS 2016 (Ogg et al. 2016).
Phylogenetic assignment. The galls target the arms and are similar in form and morphology to those caused by extant myzostomids.

Spionid annelids have also been implicated in a variety of pathologies (Cameron 1969) and are usually treated together with parasites by the damage they cause in shellfish and other invertebrates (Martin and Britayev 1998). Spionidae characteristically bore hard substrates, including bivalves shells (Martin and Britayev 1998) and are facultative parasites at best (Huntley 2007). Mud blisters caused by some taxa (e.g., *Polydora*) indicate that they can also affect living bivalves. The oldest confident occurrences of spionid trace fossils derive from the Miocene although candidate trace fossils might be as old as the Silurian (Cameron 1969). Cameron (1967) attributed soft-tissues remains found within a burrow within a bivalve to Spionidae as far back as the Devonian, although the understanding of the preservation of this specimen, and therefore also its phylogenetic affinity, are in need of revision (Blake and Evans 1973; Parry et al. 2014). The bivalve did not show evidence for an in vivo response.

Node. Spionidae

Phylogenetic justification. Cameron (1967) interpreted external structures such as a prostomium, anterior tentacle-like palps, peristomial cirri, parapodia, and dorsal cirri of the parapodia. Its habit together with its residence within a bivalve shell would speak for its assignment to the Spionidae. Nevertheless, the worm is not associated with a U-tube or mud blister attributed to be characteristic for in situ damage of most of this group (Blake and Evans 1973; Huntley 2007).

Host. Bivalve *Cornellites flabellum*

Min age. The fossil assemblage derive from a small quarry north of Morrisville where the middle part of the Cardiff shale member of the marine Marcellus Formation is exposed (Cameron 1967). This member correlates with the upper part of the Marcellus Formation (Oatna Creek Member) at other localities which are considered to represent the upper part of the hemisiatus conodont zone of the Givetian (Brett et al. 2011). The upper boundary of the zone would yield a minimum age of 387.3 Ma according to GTS 2016 (Ogg et al. 2016).

Clitellata can be confidently traced back to the Triassic based on characteristic cocoons (Manum et al.1991, 1994; Bomfleur et al. 2012, 2015; McLoughlin et al. 2016). The most likely producers of these cocoons are considered true leeches and branchiobdellids (Manum et al. 1991; Bomfleur et al. 2015). Clitellate cocoons sometimes function as conservation traps for other remains including nematodes or ciliates (Bomfleur et al. 2015; McLoughlin et al. 2016). In one case, they contain spermatozoa, which suggest branchiobdellids could have produced these coccoons at least since the Eocene and that this group had a significantly different geographic range in the past. This group is usually considered as ectoparasitic of, or at least ectosymbiotic with, crayfish—although they show a variety of relationships ranging from symbiotic, commensal to parasitic (Skelton et al. 2013). More precise evidence is necessary to tie older Triassic cocoons to Branchiobdellids as they could also belong to Euhirudinae (Hirudinida: compare Tessler et al. 2018). True leeches might occur in the Upper Jurassic (Kozur 1970), but this specimen is rather considered a carnivorous then a parasitic species.

Node. Total group Hirudinae

Phylogenetic justification. Characteristic cocoons are only produced by true leeches (Euhirudinae) and branchiobdellids. They may also provide the oldest evidence for clitellata as a whole (Parry et al. 2014), but a putative oligochaete has recently been described from the Lower Triassic (Shcherbakov et al. 2020).

Min age. 201.4 Ma. The earliest finds are attributed to the Rhaetian (Manum et al. 1991; Bomfleur et al. 2015) which would yield a minimum age for this group of 201.4 Ma according to GTS 2016 (Ogg et al. 2016).

Node. Total group Branchiobdellida

Host. Unknown, but modern representatives are mostly leech-like ectoparasites of crayfish (Govedich et al. 2010).

Phylogenetic justification. Annelid sperm described from inclusions in cocoons from the Eocene of Antarctica possess characteristic elongate acrosomes and helical marginal fibres of branchiobdellidan spermatozoans (Bomfleur et al. 2015).

Min age. 47.8 Ma. The Cucullaea I Allomember of the La Meseta Formation has been considered Ypresian (Montes et al. 2013), early Eocene, based on Strontium isotope chemostratigraphy (Ivany et al. 2008) as well as mammal biostratigraphy (Reguero et al. 2013). The age of the Ypresian-Lutetian boundary tabulated in GTS 2016 (Ogg et al. 2016) can therefore be used as a conservative minimum constraint.

7.4.2 Fossil Record of Parasitic Pentastomids

Pentastomids, or tongue worms, have traditionally been considered to be a separate phylum, but they are now considered crustaceans—most likely the sister group of Branchiura (Regier et al. 2010; Sanders and Lee 2010; Oakley et al. 2012; but see Haug et al. 2021). So far no crown-group Branchiura have been reported from the fossil record. Dzik (2008) assigned the enigmatic Cyclida (Carboniferous—Triassic) to Branchiura, but these are free-living (Schram et al. 1997) and their affinities with modern Branchiura is unlikely (Haug et al. 2021).

The fossil record of Pentastomida has been reviewed by various authors (Castellani et al. 2011; Klompmaker and Boxshall 2015). Various isolated fossils from the Cambrian-Ordovician have been identified as larval pentastomids (Walossek and Müller 1994; Walossek et al. 1994; Waloszek et al. 2005; Castellani et al. 2011), and due to their morphology these early forms have been interpreted as being parasitic. They represent the oldest direct evidence for ectoparasitic "worms" in the fossil record. Their hosts are still debated (De Baets et al. 2015). Most authors suspect they were parasitizing early vertebrates (Walossek and Müller 1994; Sanders and Lee 2010), but no direct associations are known. Their hosts must have been different from current hosts as present day lineages were not around at the time of the early fossil pentastomids (De Baets et al. 2015). A recent discovery interpreted as a pentastomid makes interpretations of early life histories even more problematic as these fossils are interpreted as adult and attached to ostracods—a host relationship so far not observed in present day (Siveter et al. 2015; but see Haug et al. 2021 for an alternative interpretation).

Node. Total Group Pentastomida

Fossil Taxon. *Boeckelericambria pelturae*

Fossil evidence. Isolated phosphatic remains of larvae in calcareous nodules

Phylogenetic justification. The diagnostic synapomorphy of a head of *Boeckelericambria pelturae* with two pairs of grasping hooks (Walossek and Müller 1994) as well as a cladistic analysis (Waltécio de Oliveira and Christoffersen 1999) place this species as a stem-group pentastomid. If we accept Pentastomida to be the sister group of branchiura, *B. penturae* must therefore also constrain the first appearance of Ichthyostraca (Wolfe et al. 2016; but see Haug et al. 2021).

Hard minimum. 497 Ma. The Orsten fossils derive from the late Cambrian Alum Shale derive from the *Agnostus pisiformis* zone, which is now considered the upper boundary of the Cambrian series 3 (Nielsen et al. 2014). The minimum age of the uppermost stage of the Cambrian 3 series therefore yields 497 Ma according to GTS 2016 (Ogg et al. 2016).

7.4.3 Fossil Record of Parasitic Platyhelminths

Transitions from free-living to parasitism amongst the Platyhelminthes may have evolved at least nine times (Weinstein and Kuris 2016), but only the divergence of the Neodermata can be reasonably constrained by fossil evidence (De Baets et al. 2015). Neodermata are a monophyletic group comprising exclusively obligate parasites (Fig. 7.4). Present day associations include the use of vertebrates as definitive hosts. The internal phylogeny is still debated, but most authors consider ectoparasitic monogeneans as the earliest diverging lineage while cestode and trematodes are considered sister groups in most molecular analyses (Lockyer et al. 2003; Park et al. 2007; Egger et al. 2015). The oldest fossils of Neodermata include circlets of hooks found on early armored jawed fishes (Leung 2017; Upeniece 2001, 2011). Their appearance, particularly the arrangement of hooks and association with putative sucker discs, are consistent with their identification as monogeneans (Leung 2017; De Baets et al. 2015). Their location near gills and abdominal near fin spines, would also be consistent with an ectoparasitic mode of life. The larger size of hooks when compared to extant taxa is consistent with a trend towards smaller size within monogeneans over time (Poulin 2005).

Node. Neodermata

Fossil evidence. Over 69 circlets of hooks, at least 29 still associated with their fish hosts in position suggestive of ectoparasitism (Upeniece 2001, 2011).

Phylogenetic justification. No formal phylogenetic analysis on the attachment organs was performed. However, the circular arrangement of the bilateral symmetric hooks, traces of cuticular disc as well the maximum number of 16 hooks strongly suggests their assignment as monogenean platyhelminths (Upeniece 2011; De Baets et al. 2015; Leung 2017, 2021).

Host. mostly attached to the acanthodian *Lodeacanthus gajicus* (24) and some (5) also associated with juveniles of the placoderm *Asterolepis ornata* (Upeniece 2001, 2011).

Minimum age. 382.7 Ma. The formation containing the finds was initially assigned to the Early Frasnian (Upeniece 2001), but currently the Lode Formation is assigned to the Upper Givetian (Lukševičs et al. 2012, 2014). This corresponds with 382.7 Ma corresponding to the minimum age assigned to the Givetian-Frasnian boundary according to GTS 2016 (Ogg et al. 2016).

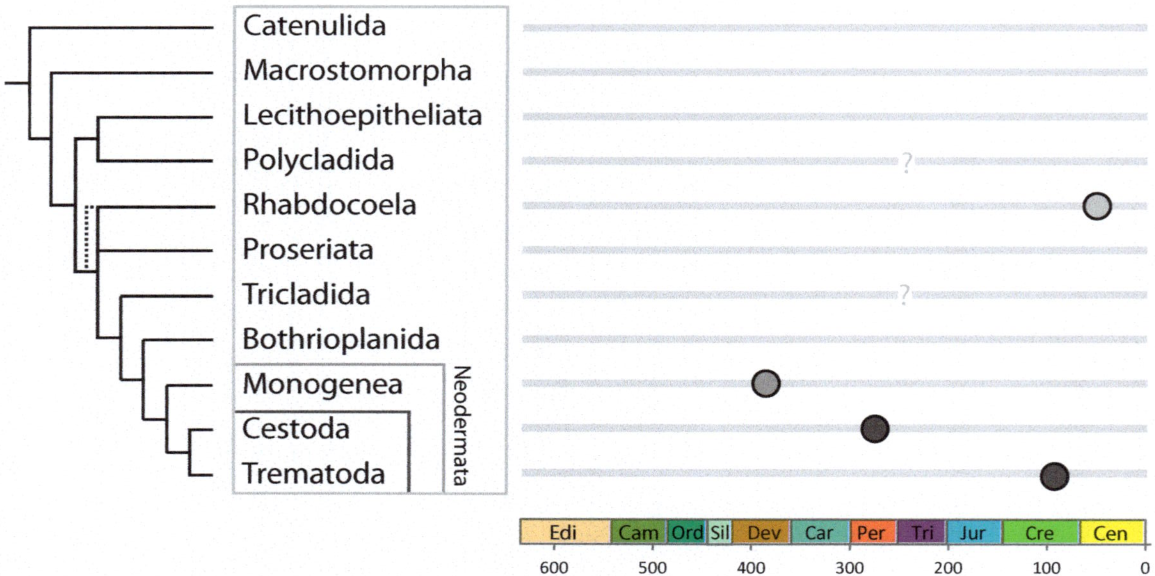

Fig. 7.4 Flatworm body fossils plotted against phylogenetic structure of Platyhelminthes (Egger et al. 2015). Flatworm fossil assignment follow (De Baets et al. 2015) unless otherwise discussed

Most phylogenetic analyses resolve a sister relationship between Trematoda and Cestoda (Park et al. 2007; Lockyer et al. 2003; Egger et al. 2015), both groups with complex life cycles, although some recent phylogenomic analyses suggest otherwise (Laumer and Giribet 2014; Laumer et al. 2015b). Cestodes can be confidently traced back to the Permian (Dentzien-Dias et al. 2013) with fossil remains yielding cestode-like eggs based on morphology and size, the presence of a developing larva with familiar cestode-like hooks and the arrangement of eggs in proglottids. Older finds of eggs within shark coprolites have been attributed to cestodes (Zangerl and Case 1976), but this identification is less convincing and in need of reinvestigation (De Baets et al. 2015). Either way, these finds represent the oldest convincing evidence for endoparasitic helminths in the fossil record (but see Upeniece 2011 for putative earlier finds) and also overlap with divergence time estimates for trypanorhynch tapeworms (Olson et al. 2010).

Node. Eucestoda

Fossil evidence. Eggs in Permian coprolites (Dentzien-Dias et al. 2013)

Phylogenetic justification. No formal phylogenetic analysis was performed. However, the size and morphology of the eggs (ovoid, smooth shelled, with operculum, among others), their interpretation of hooks within one of them as well as their association in protoglottidia speak for the assignment to Eucestoda.

Host. Shark based on spiral coprolite morphology (reflecting the spiral valve morphology of the shark intestine).

Minimum age. 259.8 Ma. The coprolite containing the fossil evidence derives from the upper member of the Rio do Rasto Formation. The vertebrate fauna suggests a Guadalupian (Late Wordian—Capitanian) age (Cisneros et al. 2012; Dentzien-Dias et al. 2012; Dias-Da-Silva 2012) which yields a minimum age of 259.8 Ma defined by the Guadalupian-Lopingian boundary (Ogg et al. 2016).

Although eggs of some lineages of Trematoda such as Dicrocoeliidae (Dufour and Le Bailly 2013) are considered quite resistant, trematodes have so far only been found in a Cretaceous coprolite (Poinar and Boucot 2006) and Middle Pleistocene to younger coprolites (e.g., Jouy-Avantin et al. 1999). Most evidence in the fossil record derives from characteristic pathologies these helminths cause in their hosts (De Baets et al. 2015; Huntley and De Baets 2015; Huntley et al. 2021). Research has focused particularly on characteristic pits and igloo-shaped concretions caused today by gymnophallid trematodes in their bivalve hosts (Ruiz and Lindberg 1989; Ituarte et al. 2001, 2005; Huntley and De Baets 2015). These structures might allow tracing back the origin of (gymnophallid) trematodes to the Eocene (Todd and Harper 2011) and the Cretaceous (Rogers et al. 2018), respectively. However, other trematodes or even distantly related taxa could potentially cause similar pathologies. Superficially similar igloo-shaped pathologies in bivalves (Liljedahl 1985) have been observed as far back as the Silurian. The only fossil vertebrate pathology attributed to a trematode was found in Burmese amber (Poinar et al. 2017). Remains of a lizard trapped in resin contained a cyst which could be convincingly compared to similar metacercarian cysts found in extant lizards. Due to the sister-group relationship between Cestoda and Trematoda, the first appearance of trematodes can be at least constrained to the Permian. If cestodes and monogeneans would be sister taxa, as in some morphological (Brooks 1989) and less commonly molecular analyses (Laumer et al. 2015b), trematodes and cestodes could be constrained to the Devonian as inferred from the first appearance of monogeneans in the fossil record.

The ancient record of parasitic flatworms is quite remarkable since the body fossil record of confidently identifiable free-living flatworms (the paraphyletic 'Turbellaria') start considerably later. Trace fossils associated with soft-body impressions of Permian-to-Triassic age have been attributed to polyclad and triclad flatworms (Knaust 2010). Earlier Ordovician trace fossils associated with putative worm remains could potentially be attributable to free-living 'Turbellaria', but this assignment awaits further scrutiny (Knaust and Desrochers 2019). The oldest confidently identified body fossil of a free-living flatworm is that of a rhabdocoel flatworm from Baltic amber (Poinar 2003a). Rhabdocoela separated even earlier from the lineage leading to the obligate parasitic flatworms (Neodermata) than triclad flatworms (Egger et al. 2015). This establishes the record of more derived Neodermata to be more ancient and relatively more common than that of their free-living relatives.

Acoelomorpha are now assigned to their own clade together with Xenoturbellida and no longer considered platyhelminths (Ruiz-Trillo and Paps 2016; Philippe et al. 2011). Xenacoelomorpha are now considered simplified relatives of the Ambulacraria (Philippe et al. 2019). To our knowledge, there is no fossil record of parasitic acoelomorphs. Some drop-shaped structure in X-ray analyses of the Early Devonian Hunsrück Slate (middle Kaub Formation), considered of Early Emsian age (De Baets et al. 2013), have been interpreted as acoelomorph remains (Fauchald et al. 1988)—but this interpretation (Knaust and Desrochers 2019) and several other of Stürmer's interpretations based on such X-ray analyses have been proven to be erroneous (as reviewed in Otto 1999, 2000). Based on trace fossils associated with body fossil remains—acoelomorphs have potentially been traced back the Late Ordovician, although these comparisons are hard to verify and remain unconfirmed (Knaust and Desrochers 2019).

7.4.4 Fossil Record of Parasitic Nematodes

Molecular phylogenetic analyses have shown that parasitism of plants and animals has arisen at least 15 times independently in nematodes (Blaxter and Koutsovoulos 2015). The sampling coverage of enigmatic nematode lineages is slowly improving (Leduc et al. 2018; Leduc and Zhao 2019), but as long as not all extant families have been molecularly sampled, the independent origins of parasitism might be more numerous (e.g., Sudhaus 2008, 2010; Weinstein and Kuris 2016). The fossil record can so far not constrain all of these transitions, but it can place multiple constraints on the origin of parasitism throughout the nematode tree. The fossil record of parasitic nematodes has recently been reviewed by Poinar (2015), who also reviewed the fossil record of nematodes in general with a focus on amber deposits (Poinar 2011,

2014). In the meantime, various new discoveries of parasitic nematodes have been made in Triassic to Cretaceous coprolites (Da Silva et al. 2014; Hugot et al. 2014; Francischini et al. 2018; Cardia et al. 2018) and in Miocene coprolites (Dentzien-Dias et al. 2018).

We herein map fossil finds in the phylogeny (Fig. 7.5) to demonstrate that several constraints on nematode evolution may be derived from the fossil record.

Based on trace fossils suggestive of free-living nematodes, and therefore also their sister-group Nematomorpha, Nematoida must have been around at least since the Late Ordovician (Knaust and Desrochers 2019; Balinski et al. 2013). Older sinusoidal trace fossils have been reported from the Ordovician (Balinski et al. 2013) and even the Cambrian-Ediacaran (e.g., Parry et al. 2017). Such sinusoidal trace fossils are caused by the specific movement of vermiform animals (possibly nematodes) in waves as their body wall comprises only longitudinal muscles but no ring muscles—an apomorphy for Nematoida (Maas 2013). Confident assignment to Nematoda remains difficult because distantly related forms with similar behavior could potentially also have created such burrows (Collins et al. 2000; Seilacher 2007; Metz 1987). One can be more confident when such traces are associated with worm-like body fossil remains (Knaust 2010; Knaust and Desrochers 2019), but even then it is quite difficult to identify the particular lineages of nematodes that created these traces. Many nematodes lack rich distinguishing external features, which makes them even more difficult to identify in a fossil state, particularly when poorly preserved (Schram 1973; Muir et al. 2014). This is ably illustrated by putative carboniferous nematode remains (Størmer 1963) that have now been re-identified as scorpion spines (Poinar 2011). Well-preserved compression or impression fossils have been attributed to free-living nematodes since the Carboniferous. Schram (1973, 1979) described various specimens of a putative nematode *Nevamermis mackeei*, considering it to be the remains of a free-living form with affinities to modern species of Chromadorida which have comparable cuticular ornamentation and structures possibly identifiable as papillae or cirri (an opinion shared by Poinar 2011). However, these characters are plesiomorphic for Nematoda and are not necessarily helpful for their taxonomic assignment (Maas 2013). According to Poinar (2014), no other fossil nematodes could so far be attributed to Chromadorida.

Node. Nematoida

Fossil evidence. Trace fossils associated with worm-like fossils (Knaust and Desrochers 2019)

Phylogenetic justification. The sinusoidal trace associated with worm-like fossils suggest a nematode or at least a nematoid caused these traces

Minimum age. The rocks of the Vaureal Formation have been assigned to the Late Katian (Knaust and Desrochers 2019), which yield a minimum age of 445.2 Ma according to GTS 2016 (Ogg et al. 2016).

Additional free-living nematode remains include the Jurassic *Eophasma jurasicum* (Arduini et al. 1983), the Early Cretaceous *Captivonema cretacea* (Manum

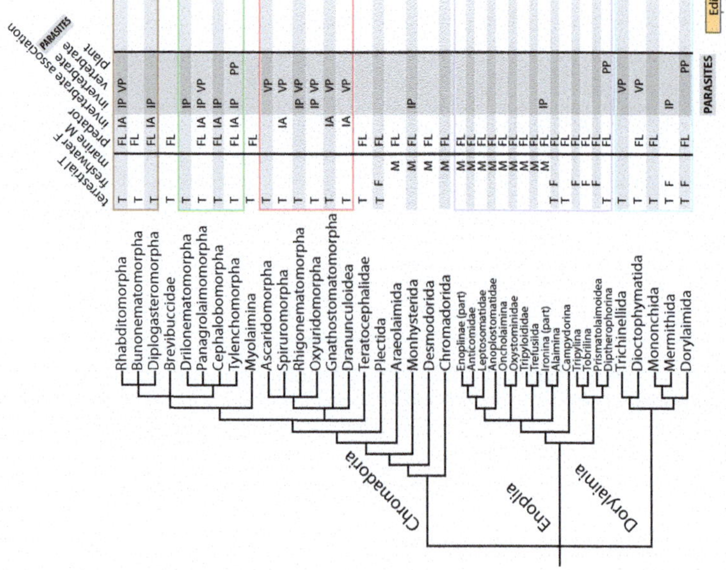

Fig. 7.5 Nematode fossils plotted against phylogenetic structure of Nematoda based on nuclear small subunit ribosomal RNA analyses and interpretation of taxon relationships derived from morphology and origins of particular types of parasitism (Blaxter and Koutsovoulos 2015). Nematode fossil—assignments follow Poinar (2011, 2014, 2015) unless otherwise discussed

et al. 1994) and the Late Cretaceous *Palaeoenoploides alessandrellianus* (Mattavelli and Bracchi 2008). These have been assigned to their own families (Eophasmidae, Captivonematidae, Palaeoenoploididae) which in turn have been attributed to modern orders (Desmoscolecida, Plectida and Enoplida, respectively; Poinar 2011). However, their assignment to modern lineages (Poinar 2011) is based on limited and possibly plesiomorphic traits (e.g., small size, finely annulated cuticle and dense coating of prominent irregularly distributed setae in *Captivonema*; coarsely annulated cuticle in *Eophasma*). The second oldest fossil attributed to the order Plectida is *Oligoplectus succini* derived from Baltic amber (Poinar 2014).

The oldest confidently identified nematode body fossils derives from the Early Devonian (Poinar et al. 2008; Maas 2013). These specimens are associated with the early land plant, *Aglaophyton major*, and are considered the earliest evidence for plant parasitic nematodes (Poinar 2011, 2015). The fossils placed in their own family (Paleonematidae) have also been assigned tentatively to Enoplida.

Node. Crown-group Nematoda
Fossil Taxon. *Palaeonema phyticum*
Fossil Evidence. Up to 500 nematodes and eggs from substomatal chambers and intercellular spaces of early land plants (Poinar et al. 2008).
Host. *Aglaophyton major*
Phylogenetic justification. The body fossils are well-preserved revealing fine morphological details such as buccal cavity, surface ornamentation, pharynx, ovaries, tail and various other features (Poinar et al. 2008) which allows them to be placed unequivocally within Nematoda (Maas 2013). They were assigned to their own family based on the presence of transverse cuticular striations and structure of pharynx and buccal cavity, but could be assigned tentatively to Enoplia (Enoplida according to Poinar 2011) due to inconspicuous lips, a buccal cavity divided in two sections, presence of teeth (two visible, probably three) in the buccal cavity, didelphic and reflexed ovaries, spiral amphidial apertures and a cylindrical pharynx it shares with members of Tripyloididae (Poinar et al. 2008).
Minimum age. 407.6 Ma. The relationship of the radiometric dating of the Milton of Noth Andesite (Parry et al. 2011) with the Rhynie chert is still disputed (Mark et al. 2013; Parry et al. 2013). Therefore, it is more conservative to use the age of the Pragian-Emsian boundary (407.6 Ma according to GTS 2016: Ogg et al. 2016) based on the constraints of spore assemblages of Windyfield and stratigraphically underlying Rhynie chert (Parry et al. 2011).

Burmese amber provides the first evidence for mycetophagous nematodes in the form of fungal-feeding *Cretaciaphelenchoides burmensis* which Poinar (2011) assigned to the Aphelenchoididae; this is also considered the oldest evidence for the order Tylenchomorpha in the fossil record (Poinar 2015).

Node. Tylenchomorpha
Fossil Taxon. *Cretaciaphelenchoides burmensis*
Fossil evidence. Worm body associated with host in Burmese amber (Poinar 2011)
Host. Crane fly (*Gonomyia*)

Phylogenetic justification. Poinar (2011) assigned it to Thelasmotidae based on its morphology, dimensions and host association. He suspects it is closely related to *Cephalobellus*.

Minimum age. Biostratigraphic evidence suggest a late Albian to early Cenomanian age for Burmese amber. Recently, dated zircons suggest a minimum age of 98.17 Ma (Shi et al. 2012).

The oldest record for nematodes exploiting other animals can be traced back to the Mesozoic with evidence of eggs found in a herbivore coprolite from the Triassic (Da Silva et al. 2014). The coprolite is usually assigned to a cynodont (Hugot et al. 2014)—although identifying its owner is not straightforward (Francischini et al. 2018). This coprolite contains eggs attributed to Ascarididae (Ascaridoidea, Ascaridomorpha, Ascaridida) (Da Silva et al. 2014) and overlap in time with the molecular clock divergence time estimates of Ascaridoidea (Li et al. 2018) based on the host fossil records. This suggests, in some cases at least, the fossil record may have value in constraining the initial radiation of terrestrial parasites.

Eggs attributed to Ascaridomorpha are also known from a Lower Cretaceous coprolite attributed to a theropod dinosaur from Belgium (Poinar and Boucot 2006) and an Upper Cretaceous coprolite attributed to crocodylians from the Adamantina Formation of Brazil (Cardia et al. 2018, 2019b). Younger eggs from the Upper Cretaceous were attributed to Heterocheilidae—they fall just outside the divergence times estimated by Li et al. (2018) for the crown-group of this lineage.

Node. Ascaridomorpha

Fossil Taxon. *Ascarites rufferi*

Host. Inferred to a cynodont based on circumstantial evidence.

Fossil Evidence. Egg derived from an Upper Triassic coprolite (Da Silva et al. 2014).

Phylogenetic Justification. No formal phylogenetic analysis was performed but possession of a round, mammillated surface suggests the assignment of eggs to Ascarididae (Da Silva et al. 2014).

Minimum age. 236 Ma. Coprolites derive from the Santa Maria Formation at the Sítio Cortado locality (Pinheiros-Chiniqua Sequence) which is assignable to the Dinodontosaurus Assemblage Zone (Da Silva et al. 2014). This sequence can be assigned a minimum age of 236 Ma. derived from dated zircons from the overlying formation and consistent with the biostratigraphic correlation of this sequence with the Chañares Formation (Francischini et al. 2018).

These Triassic coprolites attributed to cynodonts also contained eggs assigned to Heteroxynematidae (Hugot et al. 2014) which represent the oldest eggs assigned to Oxyuridomorpha (= Oxyurida).

Node. Oxyuridomorpha

Fossil Taxon. *Paleoxyuris cockburni*

Fossil Evidence. Egg derived from an Upper Triassic coprolite (Hugot et al. 2014).

Host. Cynodont? (see Ascaridomorpha).

Phylogenetic Justification. The size, shape and presence of a polar cap speak for the assignment of fossil eggs to the family Heteroxynematidae (Hugot et al. 2014)

Minimum age. 236 Ma. See Ascaridomorpha.

Based on modern host associations and fossil record, oxyurids might have parasitized invertebrates before vertebrates (Poinar 2015). The oldest direct evidence for oxyurids parasitizing invertebrates, however, is *Paleothelastoma tipulae* found associated with the crane fly *Gonomyia* found in Burmese Amber (Poinar 2011). This species was compared with Thelastomatidae parasitizing modern crane flies.

Various other lineages of parasitic nematodes or at least containing parasitic representatives (Mermithidae, Heterorhabditidae, Cosmocercidae) are now known from Cretaceous to Cenozoic amber (Poinar 2015); in various cases these are still associated with their arthropod hosts (Poinar 2011). This does not necessarily mean parasitism of invertebrates started later, but probably rather reflects a preservation and/or collection bias associated with these deposits.

Mermithidae (Mermithida) have been reported from various amber deposits. The oldest fossils which have been assigned to this group (Poinar et al. 1994) derive from Lebanese amber now attributed to the Barremian (Maksoud et al. 2017). Mermithids are reminiscent in size and shape to nematomorphs, but have a smooth outer cuticle as opposed to prominences observed in the cuticle of nematomorphs (Poinar and Buckley 2006).

Node. Mermithida
Fossil Taxon. *Cretacimermis libani*
Phylogenetic justification. The characteristic of the final molt occurring in the host, the absence of cross fibers in the cuticle, the lack of a tail projection on the postparasitic cuticle as well as the host family speak for their similarity to the extant *Heleidomermis* and their assignment to Mermithidae (Poinar et al. 1994). The specimen was interpreted to be moulting in post-parasitic late ontogenetic stage which is rare, but also observed in some extant Mermithidae parasitizing chironomid and biting midges (Poinar 2011).
Fossil Evidence. Body fossil coiled in abdominal cavity of host in Lebanese Amber (Poinar et al. 1994; Poinar 2011).
Host. Midge (Chironomidae)
Minimum age. 126.3 Ma. The age of the Lebanese amber has been pushed back to the Barremian (e.g., Maksoud et al. 2017). These authors argued that amber from middle to late Barremian was reworked from older Early Barremian strata. We conservatively use the age of the Barremian-Aptian boundary as the minimum age until the reworking hypothesis is more generally accepted and it is better established that all amber with inclusions can be confidently attributed to the early Barremian.

The earliest evidence for parasitic Rhabditida (Infraorder Rhabditomorpha) are the Cretaceous *Proheterorhabditis burmanicus* parasitizing a rove beetle found in Burmese Amber (Poinar 2011). Free-living forms (*Vetus libani*) initially unattributed on the family level (Poinar 2011) but later attributed to a closely related family Rhabditidae prove the earliest evidence for the order Rhabditida (Poinar 2014). The latter are slightly older as they can be traced to the Barremian as they were reported from Lebanese amber.

Node. Heterorhabditidae
Taxon. *Proheterorhabditis burmanicus*
Host. Rove beetle (Staphylinidae)
Phylogenetic justification. Their size, ensheathing cuticle with longitudinal striations and small terminal tooth align these fossils interpreted as third-stage juvenile with heterorhabditids (Poinar 2011). The presence of bacterial cells in the body cavity similar in size to that of modern representatives also strengthens this interpretation (Poinar 2015).
Minimum age. 98.17 Ma. See Tylenchomorpha for justification.

Juveniles of the parasitic *Palaeocosmocerca burmanicum* adjacent to a snail in Burmese amber are considered the oldest fossil evidence of family Cosmocercidae (Poinar 2011).

Node. Cosmocercidae
Taxon. *Palaeocosmocerca burmanicum*
Fossil Evidence. Worm bodies associated with putative host in Burmese Amber (Poinar 2011).
Host. Unidentified snail
Phylogenetic Justification. The morphological characteristics of the second-stage juveniles (offset lips, flat head, constricted anterior portion of the pharynx) and dimensions of pre-infective stages are similar to extant Cosmocercidae—particularly *Cosmocercoides dukae* (Poinar 2011)
Minimum age. 98.17 Ma. See Tylenchomorpha for justification

Howardula helenoschini from Baltic amber, considered to belong to Allonematidae (Poinar 2003b), provides the oldest evidence for the presence of animal parasitism in Tylenchomorpha (Poinar 2011, 2015). Other evidence for the presence of this order in the fossil record is *Paleoiotonchium dominicanum* parasitizing a fungus gnat in Dominican amber and a population of *Oligaphelenchoides atrebora* from Mexican amber (Poinar 2011). The former might represent the oldest fossil evidence for the presence of separate life cycles between fungal gnats and fungi in Iotonchidae (Poinar 2011). The latter consist of eggs, juvenile and adults, together with fungal hyphae, providing further evidence for mycophagy in the family Aphelenchoididae (Poinar 2015).

Direct evidence for filariid nematodes goes back to Baltic amber and Dominican amber (Poinar 2015). Their size, shape and ornamentation speak for their identification as filariid nematodes. Horizontal transfers of retroposons suggest that at some time filariids were likely parasitizing tropical birds rather than mammals, thus constraining them to a time interval which at least partially overlaps with the age estimates of Dominican amber (Suh et al. 2016; Suh 2021). In addition to filarial remains, also tropical bird remains (Poinar et al. 2007) have been reported from Dominican amber, suggesting they at least were present at the same time in the same region.

Node. Filarioidea
Fossil Taxon. *Cascofilaria baltica*

Fossil evidence. Microfilaria associated with its female blackfly intermediate host (see Poinar 2011, 2012).

Host. Female blackfly (Simuliidae)

Phylogenetic Justification. Their dimension, morphology and number speak for their identification as filarial nematodes. Due to its body size and morphology with rounded heads and pointed tails, they have been compared with modern *Onchocerca* and *Spendidofilaria* infecting modern mammals and birds, respectively (Poinar 2011).

Minimum age. 34.7 Ma. The age of Baltic amber is still debated. The age of the richest amber-bearing deposits deriving from the Prussian Formation have been attributed to the Late Eocene microfaunistically by Kaplan et al. (1977). However, Ritzkowski (1997) inferred a middle Eocene for the lower layer of the Prussian Formation based on glauconitic dating of its host sediment. However, this reassignment is not generally accepted (Perkovsky et al. 2007). We hereby conservatively use the 34.7 Ma based on Kaplan et al. (1977) which is younger than the minimum age estimate of 43.1 Ma by Ritzkowski (1997).

A summary of our exercise (see Fig. 7.5) illustrates that constraints are available for parasitic lineages or at least free-living relatives assigned to all major nematode clades (I-IV) proposed by Blaxter et al. (1998). Clade II can possible be traced back to the Early Devonian, Clade III to the Upper Triassic, Clades I, IV and V at least to the Cretaceous. Pending reinvestigations and new discoveries, fossil evidence for additional orders can potentially be traced back to the Mesozoic or even Paleozoic.

7.4.5 *Fossil Record of Nematomorphs*

Adult nematomorphs are free-living in marine or freshwater, while larvae are parasitic on terrestrial or marine arthropods (Hanelt et al. 2005). The oldest confidently identified crown-group hairworm derives from Cretaceous amber (Poinar and Buckley 2006), which is considerably later than the inferred origin based on the fossil record of their sister-group. The host of the oldest specimen is unknown, but other finds have been found associated with cockroaches (Poinar 1999). Younger finds derive from Eocene brown coal deposits (Voigt 1938) and Dominican amber (Poinar 1999) which is now usually dated to the Miocene (Iturralde-Vinent 2001; Iturralde-Vinent and MacPhee 1996). All forms so far reported have been attributed to the Gordioidea—parasites of terrestrial arthropods. They are reminiscent of mermithid nematodes, which differ in having a smooth outer cuticular layer. No fossil forms have so far been attributed to the nectonematids, a group which exclusively parasitizes marine invertebrates.

The placement of exclusively parasitic Nematomorpha as sister to Nematoda in molecular analyses raises the interesting question whether the common ancestor of nematodes and nematomorph was also parasitic. However, it is clear that although parasitism arose multiple times in nematodes, the earliest divergent lineages are free-living and lineages parasitizing arthropods similar to nematomorphs like mermithidae are derived (Blaxter and Koutsovoulos 2015). This suggests stem-nematomorphs were likely free-living and may have looked quite different from crown-group nematomorphs.

The diversity of Cycloneuralia in the Cambrian fossil record might suggest an even earlier origin (Maas 2013). Some free-living forms of Palaeoscolecidae even closely resemble some Nematomorpha (Hou and Bergström 1994), but they are usually interpreted as being early offshoots of the lineage towards Priapulida (Harvey et al. 2010). *Shergoldana australensis* from the Cambrian of Queensland resembles nematomorph larvae to some degree (Maas et al. 2007), and probably also lived in the meiofauna, similar to kinorhynch cycloneuralians (Nagler and Haug 2015).

The oldest constraints on the origin of Ecdysozoa are arthropod trace fossils from the Cambrian (Benton et al. 2015).

Node. Gordioida
Fossil Taxon. *Cretachordodes burmitis*
Fossil evidence. Isolated body fossil (Poinar and Buckley 2006)
Fossil Host. Unknown—possibly a cockroach based on circumstantial evidence (Poinar and Buckley 2006).
Phylogenetic Justification. No formal phylogenetic analysis has been performed, but surface ornamentation, dimensions, characteristic coiling after death speak for the assignment of these body fossils to Chordodidae (Gordioidea, Gordioida) (Poinar and Buckley 2006).
Minimum age. 98.17 Ma. See Tylenchomorpha (Nematoda) for justification.

7.4.6 Fossil Record of Syndermata

Weinstein and Kuris (2016) considered that parasitism might have evolved at least nine times independently in Rotifera (= Syndermata). However, earlier reports (May 1989) considered the ectoparasitic *Seison* and Acanthocephala to be independent transitions which is probably not the case (Sielaff et al. 2016). The fossil record of Pararotatoria (Acanthocephala + Seisonidae), which are now considered derived rotifers (syndermatans), is rare. Until recently, the oldest fossil constraint for their presence was that of their close relatives—bdelloid rotifers in Dominican amber (Waggoner and Poinar 1993; Poinar and Ricci 1992). Other more distant relatives have become known from Eocene non-marine deposits in South Australia (Southcott and Lange 1971). Dominican amber has traditionally been dated from the Eocene to Oligocene, but is currently attributed to the Miocene (Iturralde-Vinent and MacPhee 1996; Iturralde-Vinent 2001). New discoveries of acanthocephalan eggs from Upper Cretaceous crocodile coprolites push back this record considerably (Cardia et al. 2019a). However, estimates based on host relationships might place the origin of this group considerably earlier as far back as the Cambrian (Sielaff et al. 2016; Herlyn 2021). Enigmatic hook-like fossils (Cambroclavida) from the Cambrian have been attributed to Acanthocephala (Qian and Yin 1984; Amin 2013), but their size and morphology make this assignment unlikely (Herlyn 2021).

The age of the gnathiferan crown group is complicated by the phylogenetic position of chaetognaths, which have historically been difficult to place in animal phylogeny. Chaetognaths have recently been argued to belong either within, or as the sister taxon of Gnathifera (Laumer et al. 2019; Marlétaz et al. 2019) based on phylogenomic analyses. Furthermore, macrofossils with morphological similarities with both chaetognaths and other gnathiferans have been identified from the middle Cambrian Burgess Shale (Vinther and Parry 2019; Caron and Cheung 2019) further bolstering the hypothesis that chaetognaths and gnathiferans are closely related.

The oldest fossil evidence for chaetognaths are protoconodonts, identified as the remains of grasping spines (Bengtson 1983; Szaniawski 2002; Vannier et al. 2007), which can at least be traced back to the Fortunian (Yang et al. 2016; Kouchinsky et al. 2017). The grasping spines of chaetognaths have been argued to be homologous with the jaw apparatus of gnathiferans (Vinther and Parry 2019). These fossils therefore provide a minimum age for the split between chaetognaths and their sister taxon, which may be all other gnathiferan phyla (e.g., some analyses in Laumer et al. 2019), a clade of rotifers and gnathostomulids (Marlétaz et al. 2019) or Rotifera (Vinther and Parry 2019).

Ectoparasitic, or at least ectosymbiotic, relationships were likely present since the Cambrian (Cong et al. 2017) in forms which are by some considered as possible gnathiferans (Vinther and Parry 2019).

Node. Total Group Gnathifera

Fossil taxon. *Protohertzina anabarica*

Phylogenetic justification. No formal analysis incorporating protoconodonts has been performed, but based on microstructural and gross morphological similarities, protoconodonts have been assigned to the total group of chaetognaths (Bengtson 1983; Szaniawski 2002; Vannier et al. 2007) and would therefore be crown group gnathiferans.

Hard minimum. The oldest protoconodonts assigned to *Protohertzina anabarica* (Bengtson 1983) derive from the middle Fortunian *Anabarites trisulcatus—Protohertzina anabarica* assemblage zone or its correlatives (Kouchinsky et al. 2017; Yang et al. 2016). This would yield a minimum age of 532.7 Ma according to GTS 2016 (Ogg et al. 2016).

Node. Archiacanthocephala

Fossil evidence. Unnamed remains of acanthor larva (Cardia et al. 2019a).

Host taxon. Probably a crocodilian based on the characteristics of the coprolite (Cardia et al. 2018, 2019a, 2019b), but this is probably only a false or facultative host-parasite based on modern host-parasite relationships.

Phylogenetic justification. No formal phylogenetic analysis was performed, but egg size, their elliptical shape and structures wrapped by three concentric, thick shells suggests their identification as acanthocephalan eggs containing remains of acanthor larvae (Cardia et al. 2018). These thick-shelled eggs are most similar in size and morphology to other described Archiacanthocephala eggs from younger subfossil remains and lack a fourth layer present in extant forms (Herlyn 2021). The latter might however be a preservational artefact as this layer is usually poorly present or absent in ancient remains (compare Herlyn 2021).

Hard minimum. 69.9 Ma. The coprolites derive from the Adamantina Formation. The age of the Adamantina Formation is still debated (Castro et al. 2018). The youngest analyses place it formation between 87.7 in the Coniacian (Castro et al. 2018) to early Maastrichtian (Gobbo-Rodrigues et al. 1999) yielding a minimum age of 69.9 Ma according to GTS 2016 (Ogg et al. 2016).

7.5 Discussion

In some cases, confidently identified fossil remains of free-living relatives are more ancient (e.g., annelids) than those of parasitic representatives. In other cases, remains of parasitic helminths are considerably older than those confidently attributable to their closest free-living relatives. Putative free-living Ordovician nematodes have been reported (Muir et al. 2014; Knaust and Desrochers 2019), but confident assignment to Nematoda is unclear. Helminth eggs seems to be more readily preserved in the coprolites of sharks (Dentzien-Dias et al. 2013; De Baets et al. 2015), archosaurs (Poinar and Boucot 2006) or synapsids (Da Silva et al. 2014; Hugot et al. 2014) with age depending on the clade investigated.

At least the rare cases where divergence time estimates based on host constraints can be compared with those of fossil helminths recovered from these hosts (Fig. 7.6) suggest that additional evidence from Paleozoic Lagerstätten or coprolites might be useful to constrain the macroecolutionary and macroecological history of worm parasites (Dentzien-Dias et al. 2013; Da Silva et al. 2014; Cardia et al.2018, 2019b). Such examples concern estimates of crown-group eucestode flatworms (Olson et al. 2010) and stem-group ascaridoid nematodes (Li et al. 2018). The minimum age of eggs attributed to Heterocheilidae (Cardia et al. 2018, 2019b) are slightly younger

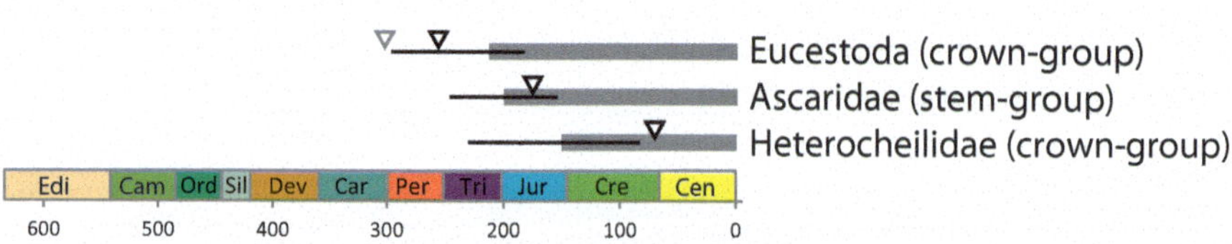

Fig. 7.6 Minimum age of fossils attributable to particular parasitic helminth lineages (Cardia et al. 2019b; Da Silva et al. 2014; Dentzien-Dias et al. 2013) plotted in relationship with available molecular clock divergence estimates based on their hosts (Li et al. 2018; Olson et al. 2010). Older fossils (Zangerl and Case 1976) not confidently attributable to Eucestoda are depicted in gray for reference

then the molecular divergence estimate (Li et al. 2018). However, the age of Lagerstätte is still uncertain (Castro et al. 2018) and it might also be just a matter of time before similar eggs are recovered from older vertebrate coprolites.

In other cases, remains have been recovered from extinct hosts which are phylogenetically placed in between extant hosts suggesting their record might still be consistent (De Baets et al. 2015) with a deep evolutionary origin of these parasites in their common ancestor (e.g., oxyurid nematode in cynodont coprolites, monogenean flatworms attached to fossil acanthodians). In yet other cases, fossil helminth remains are recovered from extinct hosts which are only distantly related to extant hosts suggesting extinct host relationships which are no longer recovered nor predicted from extant host relationships (e.g., putative pentastomid attached to ostracod, gnathiferan ectosymbionts attached to scalidophorans). This suggests caution in considering modern evidence alone when constraining the evolution of past parasites and inferring ancient parasite-host relationships (see also De Baets and Littlewood 2015; De Baets et al. 2015). However, novel methods are currently being developed which might be able to deal with changes in host repertoire (Braga et al. 2020; Warnock and Engelstädter 2021). Fossil finds and their known host associations should be considered when available and further finds might provide crucial information in the context of unravelling the evolutionary history of helminth parasitism.

7.6 Conclusions

The fossil record of soft-bodied helminths is patchy, but more diverse than commonly assumed. This summary provides evidence that ectoparasitic helminths in various lineages have existed since the early Paleozoic, while endoparasitic helminths arose during, or before the late Paleozoic. Lineages parasitizing terrestrial plant and animal hosts have been in existence at least since the Devonian and Triassic, respectively. All major helminth phyla had evolved endoparasitic lineages at least since the Mesozoic. Data also provide evidence for obvious host switches or extinctions, which cautions against models relying solely on co-divergence where use of host calibrations to constrain divergence time estimates may be considered. In some cases, the fossil record of parasitic helminths is better than that of their free-living relatives, yielding constraints on the age of their free-living sister-groups. In at least some cases (Eucestoda, stem-Ascaridoidea), the earliest fossil evidence for particular lineages falls within the divergence time estimates based on host constraints. The fossil record also provides direct evidence for certain modern or now extinct host associations (e.g., putative pentastomid on ostracods).

Acknowledgements We thank Holger Herlyn and John Huntley for valuable discussions and comments which helped to improve the text. This research which led to this work was funded by Swiss National Science Foundation (2012e141438) and FAU Emerging Talents Initative (SS16_NAT_11) to K.D.B. P.D.-D. is funded by Conselho Nacional de Desenvolvimento Científico e Tecnológico (CNPq - 302936/2018-1).

References

Amin OM (2013) Classification of the Acanthocephala. Folia Parasitol 60(4):273–305

Arduini P, Pinna G, Teruzzi G (1983) *Eophasma jurassicum* n.g.n.sp., a new fossil nematode of the Sinemurian of Osteno in Lombardy. Atti Soc ital sci nat Mus civ stor nat Milano 124(1–2):61–64

Ausich WI, Simms MJ (1999) Ecology and ecological interactions. In: Brett CE, Hess H, Simms MJ, Ausich WI (eds) Fossil crinoids. Cambridge University Press, Cambridge, pp 55–60. https://doi.org/10.1017/CBO9780511626159.007

Babcock LE (2007) Role of malformations in elucidating trilobite paleobiology: a historical synthesis. In: Mikulic DG, Landing E, Kluessendorf J (eds) Fabulous fossils–300 years of worldwide research on trilobites, New York State Museum Bulletin, vol 507. The State Education Department, New York, pp 3–19

Badets M, Whittington I, Lalubin F, Allienne JF, Maspimby JL, Bentz S, Du Preez LH, Barton D, Hasegawa H, Tandon V, Imkongwapang R, Ohler A, Combes C, Verneau O (2011) Correlating early evolution of parasitic platyhelminths to Gondwana breakup. Syst Biol 60(6):762–781. https://doi.org/10.1093/sysbio/syr078

Bajdek P, Qvarnström M, Owocki K, Sulej T, Sennikov AG, Golubev VK, Niedźwiedzki G (2016) Microbiota and food residues including possible evidence of pre-mammalian hair in Upper Permian coprolites from Russia. Lethaia 49(4):455–477. https://doi.org/10.1111/let.12156

Balinski A, Sun Y, Dzik J (2013) Traces of marine nematodes from 470 million years old Early Ordovician rocks in China. Nematology 15:567–574. https://doi.org/10.1163/15685411-00002702

Bengtson S (1983) The early history of the Conodonta. Fossils Strata 15(5):5–19

Benton MJ, Donoghue PC, Asher RJ, Friedman M, Near TJ, Vinther J (2015) Constraints on the timescale of animal evolutionary history. Palaeontol Electron 18(1):1–106

Blake J, Evans J (1973) *Polydora* and related genera (Polychaeta: Spionidae) as borers in mollusk shells and other calcareous substrates. Veliger 15:235–249

Blaxter M, Koutsovoulos G (2015) The evolution of parasitism in Nematoda. Parasitology 142(Suppl 1):S26–S39. https://doi.org/10.1017/S0031182014000791

Blaxter ML, Ley P, Garey JR, Liu LX, Scheldeman P, Vierstraete A, Vanfleteren JR, Mackey LY, Dorris M, Frisse LM, Vida JT, Thomas K (1998) A molecular evolutionary framework for the phylum Nematoda. Nature 392. https://doi.org/10.1038/32160

Bleidorn C, Podsiadlowski L, Zhong M, Eeckhaut I, Hartmann S, Halanych KM, Tiedemann R (2009) On the phylogenetic position of Myzostomida: can 77 genes get it wrong? BMC Evol Biol 9(1):150

Bobrovskiy I, Hope JM, Ivantsov A, Nettersheim BJ, Hallmann C, Brocks JJ (2018) Ancient steroids establish the Ediacaran fossil *Dickinsonia* as one of the earliest animals. Science 361(6408):1246–1249. https://doi.org/10.1126/science.aat7228

Bomfleur B, Kerp H, Taylor TN, Moestrup Ø, Taylor EL (2012) Triassic leech cocoon from Antarctica contains fossil bell animal. Proc Natl Acad Sci 109(51):20971–20974. https://doi.org/10.1073/pnas.1218879109

Bomfleur B, Mörs T, Ferraguti M, Reguero MA, McLoughlin S (2015) Fossilized spermatozoa preserved in a 50-Myr-old annelid cocoon from Antarctica. Biol Lett 11(7):20150431. https://doi.org/10.1098/rsbl.2015.0431

Borda E, Oceguera-Figueroa A, Siddall ME (2008) On the classification, evolution and biogeography of terrestrial haemadipsoid leeches (Hirudinida: Arhynchobdellida: Hirudiniformes). Mol Phylogenet Evol 46(1):142–154. https://doi.org/10.1016/j.ympev.2007.09.006

Braga MP, Landis M, Nylin S, Janz N, Ronquist F (2020) Bayesian inference of ancestral host-parasite interactions under a phylogenetic model of host repertoire evolution. Systematic Biology: syaa019. https://doi.org/10.1093/sysbio/syaa019

Brasier MD, Antcliffe JB (2008) *Dickinsonia* from Ediacara: a new look at morphology and body construction. Palaeogeogr Palaeoclimatol Palaeoecol 270(3–4):311–323. https://doi.org/10.1016/j.palaeo.2008.07.018

Brett CE, Baird GC, Bartholomew AJ, DeSantis MK, Ver Straeten CA (2011) Sequence stratigraphy and a revised sea-level curve for the Middle Devonian of eastern North America. Palaeogeogr Palaeoclimatol Palaeoecol 304(1):21–53. https://doi.org/10.1016/j.palaeo.2010.10.009
Brooks DR (1989) The phylogeny of the Cercomeria (Platyhelminthes: Rhabdocoela) and general evolutionary principles. J Parasitol 75(4):606–616
Caira JN, Littlewood DTJ (2013) Worms, Platyhelminthes. In: Levin SA (ed) Encyclopedia of biodiversity, vol 7, 2nd edn. Academic, Waltham, MA, pp 437–469
Camacho M, Araújo A, Morrow J, Buikstra J, Reinhard K (2018) Recovering parasites from mummies and coprolites: an epidemiological approach. Parasit Vectors 11(1):248. https://doi.org/10.1186/s13071-018-2729-4
Cameron B (1967) Fossilization of an ancient (Devonian) soft-bodied worm. Science 155(3767):1246–1248. https://doi.org/10.1126/science.155.3767.1246
Cameron B (1969) Paleozoic shell-boring annelids and their trace fossils. Am Zool 9(3):689–703. https://doi.org/10.1093/icb/9.3.689
Cardia DFF, Bertini RJ, Camossi LG, Letizio LA (2018) The first record of Ascaridoidea eggs discovered in Crocodyliformes hosts from the Upper Cretaceous of Brazil. Rev Bras Paleontol 21(3):238–244. https://doi.org/10.4072/rbp.2018.3.04
Cardia DFF, Bertini RJ, Camossi LG, Letizio LA (2019a) First record of Acanthocephala parasites eggs in coprolites preliminary assigned to Crocodyliformes from the Adamantina Formation (Bauru Group, Upper Cretaceous), São Paulo, Brazil. An Acad Bras Cienc 91(Suppl. 2):e20170848. https://doi.org/10.1590/0001-3765201920170848
Cardia DFF, Bertini RJ, Camossi LG, Letizio LA (2019b) Two new species of ascaridoid nematodes in Brazilian Crocodylomorpha from the Upper Cretaceous. Parasitol Int 72:101947. https://doi.org/10.1016/j.parint.2019.101947
Caron J-B, Cheung B (2019) *Amiskwia* is a large Cambrian gnathiferan with complex gnathostomulid-like jaws. Commun Biol 2(1):164. https://doi.org/10.1038/s42003-019-0388-4
Castellani C, Maas A, Waloszek D, Haug JT (2011) New pentastomids from the Late Cambrian of Sweden—deeper insight of the ontogeny of fossil tongue worms. Palaeontogr Abt A 293(4–6):95–145. https://doi.org/10.1127/pala/293/2011/95
Castro MC, Goin FJ, Ortiz-Jaureguizar E, Vieytes EC, Tsukui K, Ramezani J, Batezelli A, Marsola JCA, Langer MCA (2018) Late Cretaceous mammal from Brazil and the first radioisotopic age for the Bauru Group. R Soc Open Sci 5(5):180482. https://doi.org/10.1098/rsos.180482
Chen H, Parry LA, Vinther J, Zhai D, Xianguang H, Ma XA (2020) A Cambrian crown annelid reconciles phylogenomics and the fossil record. Nature 583:249–252. https://doi.org/10.1038/s41586-020-2384-8
Chin K (2002) Analyses of coprolites produced by carnivorous vertebrates. Paleontol Soc Pap 8:43–50. https://doi.org/10.1017/S1089332600001042
Chin K (2021) Gastrointestinal parasites of ancient non-human vertebrates: evidence from coprolites and other materials. In: De Baets, K., Huntley, J. W. (eds). The evolution and fossil record of parasitism: Coevolution and paleoparasitological techniques. Topics in Geobiology 50
Cisneros JC, Abdala F, Atayman-Güven S, Rubidge BS, Şengör AMC, Schultz CL (2012) Carnivorous dinocephalian from the Middle Permian of Brazil and tetrapod dispersal in Pangaea. Proc Natl Acad Sci 109(5):1584–1588. https://doi.org/10.1073/pnas.1115975109
Collins AG, Lipps JH, Valentine JW (2000) Modern mucociliary creeping trails and the bodyplans of neoproterozoic trace-makers. Paleobiology 26(1):47–55. https://doi.org/10.2307/2666060
Condon D, Zhu M, Bowring S, Wang W, Yang A, Jin Y (2005) U-Pb ages from the Neoproterozoic Doushantuo formation, China. Science 308(5718):95–98. https://doi.org/10.1126/science.1107765
Cong P, Ma X, Williams M, Siveter DJ, Siveter DJ, Gabbott SE, Zhai D, Goral T, Edgecombe GD, Hou X (2017) Host-specific infestation in early Cambrian worms. Nat Ecol Evol 1(10):1465–1469. https://doi.org/10.1038/s41559-017-0278-4
Conway Morris S (1981) Parasites and the fossil record. Parasitology 83(3):489–509

Conway Morris S, Peel JS (2008) The earliest annelids: Lower Cambrian polychaetes from the Sirius Passet Lagerstätte, Peary Land, North Greenland. Acta Palaeontol Pol 53(1):137–149

Da Silva PA, Borba VH, Dutra JMF, Leles D, Da-Rosa AAS, Ferreira LF, Araujo A (2014) A new ascarid species in cynodont coprolite dated of 240 million years. An Acad Bras Cienc 86:265–270. https://doi.org/10.1590/0001-3765201320130036

De Baets K, Klug C, Korn D, Bartels C, Poschmann M (2013) Emsian Ammonoidea and the age of the Hunsrück Slate (Rhenish Mountains, Western Germany). *Palaeontographica A 299*(1–6): 1–113. https://doi.org/10.1127/pala/299/2013/

De Baets K, Dentzien-Dias P, Upeniece I, Verneau O, Donoghue PCJ (2015) Constraining the deep origin of parasitic flatworms and host-interactions with fossil evidence. Adv Parasitol 90:93–135. https://doi.org/10.1016/bs.apar.2015.06.002

De Baets K, Littlewood DTJ (2015) The importance of fossils in understanding the evolution of parasites and their vectors. Adv Parasitol 90:1–51. https://doi.org/10.1016/bs.apar.2015.07.001

De Baets K, Antonelli A, Donoghue PCJ (2016) Tectonic blocks and molecular clocks. Philos Transact R Soc Lond B Biol Sci 371(1699). https://doi.org/10.1098/rstb.2016.0098

De Baets K, Huntley JW, Klompmaker AA, Schiffbauer JD, Muscente AD (2021) The fossil record of parasitism: Its extent and taphonomic constraints. In: De Baets K, Huntley JW(eds) The evolution and fossil record of parasitism: Coevolution and paleoparasitological techniques. Topics in Geobiology 50

Dentzien-Dias PC, de Figueiredo AEQ, Horn B, Cisneros JC, Schultz CL (2012) Paleobiology of a unique vertebrate coprolites concentration from Rio do Rasto Formation (Middle/Upper Permian), Paraná Basin, Brazil. J S Am Earth Sci 40:53–62. https://doi.org/10.1016/j.jsames.2012.09.008

Dentzien-Dias PC, Poinar G Jr, de Figueiredo AEQ, Pacheco ACL, Horn BLD, Schultz CL (2013) Tapeworm eggs in a 270 million-year-old shark coprolite. PLoS One 8(1):e55007. https://doi.org/10.1371/journal.pone.0055007

Dentzien-Dias P, Carrillo-Briceño JD, Francischini H, Sánchez R (2018) Paleoecological and taphonomical aspects of the Late Miocene vertebrate coprolites (Urumaco Formation) of Venezuela. Palaeogeogr Palaeoclimatol Palaeoecol 490:590–603. https://doi.org/10.1016/j.palaeo.2017.11.048

Dias-Da-Silva S (2012) Middle–Late Permian tetrapods from the Rio do Rasto formation, Southern Brazil: a biostratigraphic reassessment. Lethaia 45(1):109–120. https://doi.org/10.1111/j.1502-3931.2011.00263.x

Dufour B, Le Bailly M (2013) Testing new parasite egg extraction methods in paleoparasitology and an attempt at quantification. Int J Paleopathol 3(3):199–203. https://doi.org/10.1016/j.ijpp.2013.03.008

Dzik J (2008) Gill structure and relationships of the Triassic cycloid crustaceans. J Morphol 269(12):1501–1519

Edgecombe G, Giribet G, Dunn C, Hejnol A, Kristensen R, Neves R, Rouse G, Worsaae K, Sørensen M (2011) Higher-level metazoan relationships: recent progress and remaining questions. Org Divers Evol 11(2):151–172. https://doi.org/10.1007/s13127-011-0044-4

Eeckhaut I (1998) *Mycomyzostoma calcidicola* gen. et sp. nov., the first extant parasitic myzostome infesting crinoid stalks, with a nomenclatural appendix by M. J. Grygier. Species Divers 3(1):89–103. https://doi.org/10.12782/specdiv.3.89

Eeckhaut I, McHugh D, Mardulyn P, Tiedemann R, Monteyne D, Jangoux M, Milinkovitch M (2000) Myzostomida: a link between trochozoans and flatworms? Proc R Soc B Biol Sci 267(1451):1383–1392. https://doi.org/10.1098/rspb.2000.1154

Egger B, Lapraz F, Tomiczek B, Müller S, Dessimoz C, Girstmair J, Škunca N, Rawlinson Kate KA, Cameron CB, Beli E, Todaro MA, Gammoudi M, Noreña C, Telford MJ (2015) A transcriptomic-phylogenomic analysis of the evolutionary relationships of flatworms. Curr Biol 25(10):1347–1353. https://doi.org/10.1016/j.cub.2015.03.034

Fauchald K, Stürmer W, Yochelson EL (1988) Two worm-like organisms from the Hunsrück Slate (Lower Devonian), Southern Germany. Palaeontol Z 62(3):205–215. https://doi.org/10.1007/BF02989493

Francischini H, Dentzien-Dias P, Schultz CL (2018) A fresh look at ancient dungs: the Brazilian Triassic coprolites revisited. Lethaia 51(3):389–405. https://doi.org/10.1111/let.12251

Giribet G, Edgecombe GD (2017) Current understanding of Ecdysozoa and its internal phylogenetic relationships. Integr Comp Biol 57(3):455–466

Gobbo-Rodrigues SR, Petri S, Bertini RJ (1999) Ocorrências de ostrácodes na formação Adamantina do grupo Bauru, cretáceo superior da Bacia do Paraná, e possibilidades de correlação com depósitos isócronos argentinos. Parte II- Família Limnocytheridae. An Acad Bras Cienc 71(1):147–148

Govedich FR, Bain BA, Moser WE, Gelder SR, Davies RW, Brinkhurst, RO (2010) Chapter 12 – Annelida (Clitellata): Oligochaeta, Branchiobdellida, Hirudinida, and Acanthobdellida. In: Thorp JH, Covich AP (eds) Ecology and classification of North American freshwater invertebrates (Third Edition). Academic Press, San Diego, pp. 385–436.

Han J, Conway Morris S, Hoyal Cuthill JF, Shu D (2019) Sclerite-bearing annelids from the lower Cambrian of South China. Sci Rep 9(1):4955. https://doi.org/10.1038/s41598-019-40841-x

Hanelt B, Thomas F, Schmidt-Rhaesa A (2005) Biology of the Phylum Nematomorpha. In: Baker JR, Muller R, Rollinson D (eds) Advances in Parasitology, vol 59. Academic, New York, pp 243–305. https://doi.org/10.1016/S0065-308X(05)59004-3

Harper DAT, Hammarlund EU, Topper TP, Nielsen AT, Rasmussen JA, Park T-YS, Smith MP (2019) The Sirius Passet Lagerstätte of North Greenland: a remote window on the Cambrian explosion. J Geol Soc 176(6):1023–1037. https://doi.org/10.1144/jgs2019-043

Harvey THP, Dong X, Donoghue PCJ (2010) Are palaeoscolecids ancestral ecdysozoans? Evol Dev 12(2):177–200. https://doi.org/10.1111/j.1525-142X.2010.00403.x

Haug JT, Haug C, Nagler C (2021) Evolutionary history of crustaceans as parasites. In: De Baets K, Huntley JW (eds) The evolution and fossil record of parasitism: Identification and macroevolution of parasites. Topics in Geobiology 49

Herlyn H (2021) Thorny-headed Worms (Acanthocephala): Jaw-less Members of Jaw-bearing Worms that Parasitize Jawed Arthropods and Jaw-bearing Vertebrates. In: De Baets K, Huntley JW (eds) The Evolution and Fossil Record Of Parasitism: Identification and Macroevolution of Parasites. Topics in Geobiology 49

Héritier L, Badets M, Du Preez LH, Aisien MSO, Lixian F, Combes C, Verneau O (2015) Evolutionary processes involved in the diversification of chelonian and mammal polystomatid parasites (Platyhelminthes, Monogenea, Polystomatidae) revealed by palaeoecology of their hosts. Mol Phylogenet Evol 92:1–10. https://doi.org/10.1016/j.ympev.2015.05.026

Hernández-Alcántara P, Solís-Weiss V (1998) Parasitism among polychaetes: a rare case illustrated by a new species: *Labrorostratus zaragozensis*, n. sp.(Oenonidae) found in the Gulf of California, Mexico. J Parasitol 84(5):978–982

Hess H (2010) Myzostome deformation on arms of the early jurassic crinoid *Balanocrinus gracilis* (Charlesworth). J Paleontol 84(6):1031–1034. https://doi.org/10.1666/10-036.1

Hints O (1998) Late Viruan [Caradoc] polychaete jaws from North Estonia and the St. Petersburg region. Acta Palaeontol Pol 43(3):471–516

Hints O (2000) Ordovician eunicid polychaetes of Estonia and surrounding areas: review of their distribution and diversification. Rev Palaeobot Palynol 113(1–3):41–55

Hints O, Eriksson M (2007) Diversification and biogeography of scolecodont-bearing polychaetes in the Ordovician. Palaeogeogr Palaeoclimatol Palaeoecol 245:95–114. https://doi.org/10.1016/j.palaeo.2006.02.029

Hints O, Nõlvak J (2006) Early Ordovician scolecodonts and chitinozoans from Tallinn, North Estonia. Rev Palaeobot Palynol 139(1):189–209. https://doi.org/10.1016/j.revpalbo.2006.01.004

Hollocher K, Hollocher TC (2012) Early processes in the fossilization of terrestrial feces to coprolites, and microstructure preservation. N M Mus Nat Hist Sci Bull 57:79–92

Hope WD, Tchesunov AV (1999) *Smithsoninema inaequale* n.g. and n.sp. (Nematoda, Leptolaimidae) inhabiting the test of a foraminiferan. Invertebr Biol 118(2):95–108. https://doi.org/10.2307/3227052

Hoyal Cuthill JF, Sewell KB, Cannon LRG, Charleston MA, Lawler S, Littlewood DTJ, Olson PD, Blair D (2016) Australian spiny mountain crayfish and their temnocephalan ectosymbionts: an ancient association on the edge of coextinction? Proc R Soc Lond B Biol Sci 283(1831). https://doi.org/10.1098/rspb.2016.0585

Hou X, Bergstrøm J (1994) Palaeoscolecid worms may be nematomorphs rather than annelids. Lethaia 27(1):11–17. https://doi.org/10.1111/j.1502-3931.1994.tb01548.x

Huang D-Y, Chen J-Y, Vannier J, Saiz Salinas JI (2004) Early Cambrian sipunculan worms from southwest China. Proc R Soc Lond Ser B Biol Sci 271(1549):1671–1676. https://doi.org/10.1098/rspb.2004.2774

Hugot J-P, Gardner S, Borba V, Araujo P, Leles D, Stock Da-Rosa A, Dutra J, Ferreira L, Araujo A (2014) Discovery of a 240 million year old nematode parasite egg in a cynodont coprolite sheds light on the early origin of nematode parasites in vertebrates. Parasit Vectors 7(1):486. https://doi.org/10.1186/s13071-014-0486-6

Huntley JW (2007) Towards establishing a modern baseline for paleopathology: trace-producing parasites in a bivalve host. J Shellfish Res 26(1):253–259. https://doi.org/10.2983/0730-8000(2007)26[253:TEAMBF]2.0.CO;2

Huntley JW, De Baets K (2015) Trace fossil evidence of trematode—bivalve parasite—host interactions in deep time. Adv Parasitol 90:201–231. https://doi.org/10.1016/bs.apar.2015.05.004

Huntley JW, De Baets K, Scarponi D, Linehan LC, Epa YR, Jacobs GS, Todd JA (2021) Bivalve Mollusks as Hosts in the Fossil Record. In: De Baets K, Huntley JW (eds) The evolution and fossil record of parasitism: Coevolution and paleoparasitological techniques. Topics in Geobiology 50

Ituarte CF, Cremonte F, Deferrari G (2001) Mantle-shell complex reactions elicited by digenean metacercariae in *Gaimardia trapesina* (Bivalvia : Gaimardiidae) from the Southwestern Atlantic Ocean and Magellan Strait. Dis Aquat Org 48(1):47–56. https://doi.org/10.3354/dao048047

Ituarte C, Cremonte F, Zelaya DG (2005) Parasite-mediated shell alterations in recent and holocene sub-Antarctic bivalves: the parasite as modeler of host reaction. Invertebr Biol 124(3):220–229. https://doi.org/10.1111/j.1744-7410.2005.00021.x

Iturralde-Vinent MA (2001) Geology of the amber-bearing deposits of the greater Antilles. Caribb J Sci 37(3–4):141–166

Iturralde-Vinent MA, MacPhee RDE (1996) Age and paleogeographical origin of Dominican Amber. Science 273(5283):1850–1852. https://doi.org/10.1126/science.273.5283.1850

Ivany LC, Lohmann KC, Hasiuk F, Blake DB, Glass A, Aronson RB, Moody RM (2008) Eocene climate record of a high southern latitude continental shelf: Seymour Island, Antarctica. Geol Soc Am Bull 120(5–6):659–678. https://doi.org/10.1130/B26269.1

Jangoux M (1987) Diseases of Echinodermata. 111. Agents metazoans (Annelida to Pisces). Dis Aquat Org 3:59–83

Jouy-Avantin F, Combes C, Lumley H, Miskovsky JC, Moné H (1999) Helminth eggs in animal coprolites from a Middle Pleistocene site in Europe. J Parasitol 85(2):376–379. https://doi.org/10.2307/3285652

Kaplan A, Grigelis A, Strelnikova N, Glikman L (1977) Stratigraphy and correlation of palaeogene deposits of South-Western cis-Baltic region. Sovetskaâ Geologiâ 4:30–43

Klompmaker AA, Boxshall GA (2015) Fossil crustaceans as parasites and hosts. Adv Parasitol 90:233–289. https://doi.org/10.1016/bs.apar.2015.06.001

Klompmaker AA, Portell RW, Frick MG (2017) Comparative experimental taphonomy of eight marine arthropods indicates distinct differences in preservation potential. Palaeontology 60(6):773–794. https://doi.org/10.1111/pala.12314

Knaust D (2010) Remarkably preserved benthic organisms and their traces from a Middle Triassic (Muschelkalk) mud flat. Lethaia 43(3):344–356. https://doi.org/10.1111/j.1502-3931.2009.00196.x

Knaust D, Desrochers A (2019) Exceptionally preserved soft-bodied assemblage in Ordovician carbonates of Anticosti Island, eastern Canada. Gondwana Res 71:117–128. https://doi.org/10.1016/j.gr.2019.01.016

Kodandaramaiah U (2011) Tectonic calibrations in molecular dating. Curr Zool 57(1):116–124. https://doi.org/10.1093/czoolo/57.1.116
Kouchinsky A, Bengtson S, Landing E, Steiner M, Vendrasco M, Ziegler K (2017) Terreneuvian stratigraphy and faunas from the Anabar Uplift, Siberia. Acta Palaeontol Pol 62(2):311–440. https://doi.org/10.4202/app.00289.2016
Kozur H (1970) Fossile Hirudinea aus dem Oberjura von Bayern. Lethaia 3(3): 225–232. https://doi.org/10.1111/j.1502-3931.1970.tb01267.x
Lanterbecq D, Bleidorn C, Michel S, Eeckhaut I (2008) Locomotion and fine structure of parapodia in Myzostoma cirriferum (Myzostomida). Zoomorphology 127. https://doi.org/10.1007/s00435-007-0052-2
Laumer CE, Giribet G (2014) Inclusive taxon sampling suggests a single, stepwise origin of ectolecithality in Platyhelminthes. Biol J Linn Soc 111(3):570–588. https://doi.org/10.1111/bij.12236
Laumer CE, Bekkouche N, Kerbl A, Goetz F, Neves RC, Sørensen MV (2015a) Spiralian phylogeny informs the evolution of microscopic lineages. Curr Biol 25. https://doi.org/10.1016/j.cub.2015.06.068
Laumer CE, Hejnol A, Giribet G (2015b) Nuclear genomic signals of the 'microturbellarian' roots of platyhelminth evolutionary innovation. https://doi.org/10.7554/eLife.05503
Laumer CE, Fernández R, Lemer S, Combosch D, Kocot KM, Riesgo A, Andrade SCS, Sterrer W, Sørensen MV, Giribet G (2019) Revisiting metazoan phylogeny with genomic sampling of all phyla. Proc R Soc B Biol Sci 286(1906):20190831. https://doi.org/10.1098/rspb.2019.0831
Leduc D, Zhao ZQ, Verdon V, Xu Y (2018) Phylogenetic position of the enigmatic deep-sea nematode order Rhaptothyreida: a molecular analysis. Mol Phylogenet Evol 122:29–36. https://doi.org/10.1016/j.ympev.2018.01.018
Leduc D, Zhao ZQ (2019) Phylogenetic position of the parasitic nematode Trophomera (Nematoda, Benthimermithidae): a molecular analysis. Mol Phylogenet Evol 132:177–182
Leung TL (2017) Fossils of parasites: what can the fossil record tell us about the evolution of parasitism? Biol Rev Camb Philos Soc 92(1):410–430. https://doi.org/10.1111/brv.12238
Leung TLF (2021) Parasites of fossil vertebrates: what we know and what can we expect from the fossil record? In: De Baets K, Huntley JW (eds) The evolution and fossil Record of Parasitism: Identification and macroevolution of parasites. Topics in Geobiology 49
Li L, Lu L, Nadler SA, Gibson DI, Zhang LP, Chen HX, Zhao WT, Guo YN (2018) Molecular phylogeny and dating reveal a terrestrial origin in the early carboniferous for ascaridoid nematodes. Syst Biol 67(5):888–900. https://doi.org/10.1093/sysbio/syy018
Liljedahl L (1985) Ecological aspects of a silicified bivalve fauna from the Silurian of Gotland. Lethaia 18(1): 53–66. https://doi.org/10.1111/j.1502-3931.1985.tb00684.x
Littlewood DTJ (2006) The evolution of parasitism in flatworms. In: Maule AG, Marks NJ (eds) Parasitic flatworms: molecular biology, biochemistry, immunology and physiology. CABI, Wallingford, pp 1–36
Littlewood DTJ, Donovan SK (2003) Fossil parasites: a case of identity. Geol Today 19(4):136–142
Lockyer AE, Olson PD, Littlewood DTJ (2003) Utility of complete large and small subunit rRNA genes in resolving the phylogeny of the Neodermata (Platyhelminthes): implications and a review of the cercomer theory. Biol J Linn Soc 78(2):155–171
Lukševičs E, Stinkulis Ģ, Mūrnieks A, Popovs K (2012) Geological evolution of the Baltic Artesian Basin. In: Delina A, Kalvans A, Saks T, Bethers U, Vircavs V (eds) Highlights of groundwater research in the Baltic Artesian Basin. University of Latvia, Riga, pp 7–52
Lukševičs E, Stinkulis Ģ, Saks T, Popovs K, Jātnieks J (2014) The Devonian stratigraphic succession and evolution of the Baltic Sedimentary Basin. In: Rocha R, Pais J, Kullberg JC, Finney S (eds) STRATI 2013. Springer International Publishing, Heidelberg, pp 539–541. https://doi.org/10.1007/978-3-319-04364-7_103
Maas A (2013) Gastrotricha, Cycloneuralia and Gnathifera: the fossil record. In: Schmidt-Rhaesa A (ed) Handbook of zoology, vol 1. De Gruyter, Berlin, pp 11–28

Maas A, Waloszek D, Haug JT, Muller KJ (2007) A possible larval roundworm from the Cambrian 'Orsten' and its bearing on the phylogeny of Cycloneuralia. Memoirs Assoc Aust Palaeontol 34:499

Maksoud S, Azar D, Granier B, Gèze R (2017) New data on the age of the Lower Cretaceous amber outcrops of Lebanon. Palaeoworld 26(2):331–338. https://doi.org/10.1016/j.palwor.2016.03.003

Manum SB, Bose MN, Sawyer RT (1991) Clitellate cocoons in freshwater deposits since the Triassic. Zool Scr 20(4):347–366. https://doi.org/10.1111/j.1463-6409.1991.tb00300.x

Manum SB, Bose MN, Sayer RT, Boström S (1994) A nematode (*Captivonema cretacea* gen. et sp. n.) preserved in a clitellate cocoon wall from the Early Cretaceous. Zool Scr 23(1):27–31. https://doi.org/10.1111/j.1463-6409.1994.tb00370.x

Mark DF, Rice CM, Trewin NH (2013) Discussion on 'A high-precision U–Pb age constraint on the Rhynie Chert Konservat-Lagerstätte: time scale and other implications': Journal, Vol. 168, 863–872. J Geol Soc 170(4):701–703. https://doi.org/10.1144/jgs2011-110

Marlétaz F, Peijnenburg KTCA, Goto T, Satoh N, Rokhsar DS (2019) A new spiralian phylogeny places the enigmatic arrow worms among gnathiferans. Curr Biol 29(2):312–318.e313. https://doi.org/10.1016/j.cub.2018.11.042

Martin D, Britayev TA (1998) Symbiotic polychaetes: review of known species. Oceanogr Mar Biol Annu Rev 36:217–340

Mattavelli A, Bracchi G (2008) Palaeoenoploides alessandrellianus n. gen. n. sp: a Late Cenomanian 'roundworm' from the Upper Cretaceous of Lebanon (Nematoda: Enoplidae). Atti Soc ital sci nat Mus civ stor nat Milano 149(2):179–194

May L (1989) Epizoic and parasitic rotifers. In: Rotifer symposium V. Springer, pp 59–67

McAllister C, Bursey C, Freed P (2010) Helminth parasites (Cestoidea, Nematoda, Pentastomida) of selected herpetofauna from Cameroon, West Africa. Acta Parasitol 55(1):90–93

McLoughlin S, Bomfleur B, Mörs T, Reguero M (2016) Fossil clitellate annelid cocoons and their microbiological inclusions from the Eocene of Seymour Island, Antarctica. Palaeontol Electron 19.1(11A):1–27

Metz R (1987) Sinusoidal trail formed by a recent biting midge (Family Ceratopogonidae): trace fossil implications. J Paleontol 61(2):312–314

Mierzejewski P (1984) Synclinophora synclinalis Eisenack—the oldest Arabellid polychaete. Rev Palaeobot Palynol 43(1–3):285–292

Montes M, Nozal F, Santillana S, Marenssi S, Olivero E (2013) Mapa Geológico de la isla Marambio (Seymour) Escala 1: 20.000. Serie Cartográfica Geocientífica Antártica. Instituto Antártico Argentino, Buenos Aires

Muir LA, Ng T-W, Li X-F, Zhang Y-D, Lin J-P (2014) Palaeoscolecidan worms and a possible nematode from the Early Ordovician of South China. Palaeoworld 23(1):15–24. https://doi.org/10.1016/j.palwor.2013.06.003

Nagler C, Haug JT (2015) From fossil parasitoids to vectors: insects as parasites and hosts. Adv Parasitol 90:137–200. https://doi.org/10.1016/bs.apar.2015.09.003

Nielsen AT, Weidner T, Terfelt F, Høyberget M (2014) Upper Cambrian (Furongian) biostratigraphy in Scandinavia revisited: definition of superzones. GFF 136(1):193–197. https://doi.org/10.1080/11035897.2013.878748

Oakley TH, Wolfe JM, Lindgren AR, Zaharoff AK (2012) Phylotranscriptomics to bring the understudied into the fold: monophyletic Ostracoda, fossil placement and pancrustacean phylogeny. Mol Biol Evol 30(1):215–233. https://doi.org/10.1093/molbev/mss216

Ogg JG, Ogg G, Gradstein FM (2016) A concise geologic time scale: 2016. Elsevier, Amsterdam

Okamura B, Gruhl A, Bartholomew JL (2015) An introduction to myxozoan evolution, ecology and development. In: Okamura B, Gruhl A, Bartholomew J (eds) Myxozoan evolution, ecology and development. Springer, Cham, pp 1–20

Okamura B, Gruhl A (2021) Evolution, origins and diversification of parasitic cnidarians. In: De Baets K, Huntley JW (eds) The evolution and fossil record of parasitism: Identification and macroevolution of parasites. Topics in Geobiology 49

Öktener A (2005) A checklist of parasitic helminths reported from sixty-five species of marine fish from Turkey including two new records of monogeneans. Zootaxa 1063(1):33–52. https://doi.org/10.11646/zootaxa.1063.1.2

Olson PD, Caira JN, Jensen K, Overstreet RM, Palm HW, Beveridge I (2010) Evolution of the trypanorhynch tapeworms: parasite phylogeny supports independent lineages of sharks and rays. Int J Parasitol 40(2):223–242. https://doi.org/10.1016/j.ijpara.2009.07.012

Otto M (1999) The problem of fossil interpretation. Lethaia 32(2):140–142. https://doi.org/10.1111/j.1502-3931.1999.tb00532.x

Otto M (2000) Supposed soft tissue preservation in the Hunsrückschiefer (Lower Devonian, Rheinisches Schiefergebirge): the example of brachiopods. Palaeontol Z 74(1–2):79–89. https://doi.org/10.1007/BF02987954

Parham JF, Donoghue PCJ, Bell CJ, Calway TD, Head JJ, Holroyd PA, Inoue JG, Irmis RB, Joyce WG, Ksepka DT, Patané JSL, Smith ND, Tarver JE, van Tuinen M, Yang Z, Angielczyk KD, Greenwood JM, Hipsley CA, Jacobs L, Makovicky PJ, Müller J, Smith KT, Theodor JM, Warnock RCM, Benton MJ (2012) Best practices for justifying fossil calibrations. Syst Biol 61(2):346–359. https://doi.org/10.1093/sysbio/syr107

Park JK, Kim KH, Kang S, Kim W, Eom KS, Littlewood DTJ (2007) A common origin of complex life cycles in parasitic flatworms: evidence from the complete mitochondrial genome of Microcotyle sebastis (Monogenea : Platyhelminthes). BMC Evol Biol 7:11. https://doi.org/10.1186/1471-2148-7-11

Parry LA, Boggiani PC, Condon DJ, Garwood RJ, Leme JDM, McIlroy D, Brasier MD, Trindade R, Campanha GAC, Pacheco MLAF, Diniz CQC, Liu AG (2017) Ichnological evidence for meiofaunal bilaterians from the terminal Ediacaran and earliest Cambrian of Brazil. Nature Ecology & Evolution 1(10):1455–1464. https://doi.org/10.1038/s41559-017-0301-9`

Parry L, Tanner A, Vinther J (2014) The origin of Annelids. Palaeontology 57:1091–1103. https://doi.org/10.1111/pala.12129

Parry LA, Edgecombe GD, Sykes D, Vinther J (2019a) Jaw elements in Plumulites bengtsoni confirm that machaeridians are extinct armoured scaleworms. Proc R Soc B Biol Sci 286(1907):20191247. https://doi.org/10.1098/rspb.2019.1247

Parry LA, Eriksson M, Vinther J (2019b) The Annelid fossil record. In: Purschke G, Böggemann M, Westheide W (eds) Handbook of zoology. Annelida. Volume 1: Annelida Basal Groups and Pleistoannelida, Sedentaria I. DeGruyter, Berlin, pp 69–88. https://doi.org/10.1515/9783110291582-003

Parry SF, Noble SR, Crowley QG, Wellman CH (2011) A high-precision U–Pb age constraint on the Rhynie Chert Konservat-Lagerstätte: time scale and other implications. J Geol Soc 168(4):863–872. https://doi.org/10.1144/0016-76492010-043

Parry SF, Noble SR, Crowley QG, Wellman CH (2013) Reply to discussion on 'A high-precision U–Pb age constraint on the Rhynie Chert Konservat-Lagerstätte: time scale and other implications': Journal, 168, 863–872. J Geol Soc 170(4):703–706. https://doi.org/10.1144/jgs2012-089

Paxton H (2009) Phylogeny of Eunicida (Annelida) based on morphology of jaws. 2:241–264. https://doi.org/10.11646/zoosymposia.2.1.18

Perkovsky E, Rasnitsyn A, Vlaskin A, Taraschuk M (2007) A comparative analysis of the Baltic and Rovno amber arthropod faunas: representative samples. Afr Invertebr 48(1):229–245

Philippe H, Brinkmann H, Copley RR, Moroz LL, Nakano H, Poustka AJ, Wallberg A, Peterson KJ, Telford MJ (2011) Acoelomorph flatworms are deuterostomes related to Xenoturbella. Nature 470(7333):255–258. https://doi.org/10.1038/nature09676

Philippe H, Poustka, AJ, Chiodin M, Hoff KJ, Dessimoz C, Tomiczek B, Schiffer PH, Müller S, Domman D, Horn M (2019) Mitigating anticipated effects of systematic errors supports sister-group relationship between Xenacoelomorpha and Ambulacraria. Curr Biol 29(11): 1818–1826. https://doi.org/10.1016/j.cub.2019.04.009

Poinar GO (1999) Paleochordodes protus n.g., n.sp. (Nematomorpha, Chordodidae), parasites of a fossil cockroach, with a critical examination of other fossil hairworms and helminths of extant cockroaches (Insecta: Blattaria). Invertebr Biol 118(2):109–115. https://doi.org/10.2307/3227053

Poinar GO (2003a) A rhabdocoel turbellarian (Platyhelminthes, Typhloplanoida) in Baltic amber with a review of fossil and sub-fossil platyhelminths. Invertebr Biol 122(4):308–312

Poinar GO (2003b) Fossil evidence of phorid parasitism (Diptera: Phoridae) by allantonematid nematodes (Tylenchida: Allantonematidae). Parasitology 127(6):589–592. https://doi.org/10.1017/s0031182003004141

Poinar GO, Boucot AJ (2006) Evidence of intestinal parasites of dinosaurs. Parasitology 133(02):245–249. https://doi.org/10.1017/S0031182006000138

Poinar GO, Kerp H, Hass H (2008) *Palaeonema phyticum* gen. n., sp. n. (Nematoda: Palaeonematidae fam. n.), a Devonian nematode associated with early land plants. Nematology 10(1):9–14. https://doi.org/10.1163/156854108783360159

Poinar GO, Philbrick KA, Cohn MJ, Turner RT, Iwaniec UT, Wunderlich J (2017) X-ray microcomputed tomography reveals putative trematode metacercaria in a 100 million year-old lizard (Squamata: Agamidae). Cretac Res 80:27–30. https://doi.org/10.1016/j.cretres.2017.07.017

Poinar GO (2011) The evolutionary history of nematodes: as revealed in stone, amber and mummies, vol 9. Nematology monographs and perspectives. Brill, Leiden

Poinar GO (2014) Palaeontology of nematodes. In: Schmidt-Rhaesa A (ed) Handbook of zoology: Gastroricha, Cycloneuralia and Gnathifera, vol 2: Nematoda. Berlin De Gruyter, Berlin, pp 173–178

Poinar GO (2015) The geological record of parasitic nematode evolution. Adv Parasitol 90:53–92. https://doi.org/10.1016/bs.apar.2015.03.002

Poinar GO, Ricci C (1992) Bdelloid rotifers in Dominican amber: evidence for parthenogenetic continuity. Experientia 48(4):408–410. https://doi.org/10.1007/bf01923444

Poinar GO, Acra A, Acra F (1994) Earliest fossil nematode (Mermithidae) in Cretaceous Lebanese amber. Fundam Appl Nematol 17(5):475–477

Poinar GO, Buckley R (2006) Nematode (Nematoda: Mermithidae) and hairworm (Nematomorpha: Chordodidae) parasites in Early Cretaceous amber. J Invertebr Pathol 93(1):36–41. https://doi.org/10.1016/j.jip.2006.04.006

Poinar GO, Voisin C, Voisin JF (2007) Bird eggshell in Dominican amber. Palaeontology 50(6):1381–1383

Poinar GO (2012) New fossil nematodes in Dominican and Baltic amber. Nematology 14(4):483–488. https://doi.org/10.1163/156854111X612199

Poulin R (2001) Body size and segmentation patterns in free-living and parasitic polychaetes. Can J Zool 79(5):741–745. https://doi.org/10.1139/z01-033

Poulin R (2005) Evolutionary trends in body size of parasitic flatworms. Biol J Linn Soc 85(2):181–189. https://doi.org/10.1111/j.1095-8312.2005.00481.x

Qian Y, Yin G (1984) Zhijinitae and its stratigraphical significance. Acta Palaeontol Sin 23:216–223

Qvarnström M, Niedźwiedzki G, Žigaitė Ž (2016) Vertebrate coprolites (fossil faeces): an underexplored Konservat-Lagerstätte. Earth-Science Reviews 162:44–57

Radwańska U, Radwańska A (2005) Myzostomid and copepod infestation of Jurassic echinoderms: a general approach, some new occurrences, and/or re-interpretation of previous reports. Acta Geol Pol 55(2):109–130

Regier JC, Shultz JW, Zwick A, Hussey A, Ball B, Wetzer R, Martin JW, Cunningham CW (2010) Arthropod relationships revealed by phylogenomic analysis of nuclear protein-coding sequences. Nature 463(7284):1079–1083. https://doi.org/10.1038/nature08742

Reguero M, Goin F, Hospitaleche CA, Marenssi S, Dutra T (2013) Late Cretaceous/Paleogene stratigraphy in the James Ross Basin. In: Reguero M et al (eds) Late Cretaceous/Paleogene West Antarctica terrestrial biota and its intercontinental affinities. Springer, London, pp 19–25

Ritzkowski S (1997) K-Ar-Altersbestimmungen der Bernsteinführenden Sedimente des Samlandes (Paläogen, Bezirk Kaliningrad). Metalla (Sonderheft) 66:19–23

Rogers RR, Curry Rogers KA, Bagley BC, Goodin JJ, Hartman JH, Thole JT, Zatoń M (2018) Pushing the record of trematode parasitism of bivalves upstream and back to the Cretaceous. Geology 46(5):431–434. https://doi.org/10.1130/G40035.1

Rohde K (2005) Definitions, and adaptations to a parasitic way of life. In: Rohde K (ed) Marine parasitology. CSIRO, Oxon, pp 1–6
Rouse GW (2005) Polychaeta (Bristle Worms). In: Rohde K (ed) Marine parasitology. CABI Publishing, Oxon, pp 193–196
Rouse GW, Pleijel F (2003) Problems in polychaete systematics. Hydrobiologia 496(1):175–189. https://doi.org/10.1023/A:1026188630116
Ruiz-Trillo I, Paps J (2016) Acoelomorpha: earliest branching bilaterians or deuterostomes? Org Divers Evol 16(2):391–399. https://doi.org/10.1007/s13127-015-0239-1
Ruiz GM, Lindberg DR (1989) A fossil record for trematodes: extent and potential uses. Lethaia 22(4):431–438
Sanders KL, Lee MSY (2010) Arthropod molecular divergence times and the Cambrian origin of pentastomids. Syst Biodivers 8(1):63–74. https://doi.org/10.1080/14772000903562012
Schram FR (1973) Pseudocoelomates and a nemertine from the Illinois Pennsylvanian. J Paleontol 47(5):985–989. https://doi.org/10.2307/1303083
Schram FR (1979) Worms of the Mississippian Bear Gulch Limestone of central Montana, USA. Transactions of the San Diego Society of Natural History 19:107–120
Schram FR, Vonk R, Cees HJH (1997) Mazon Creek Cycloidea. J Paleontol 71(2):261–284
Seilacher A (1970) Begriff und bedeutung der Fossil-Lagerstätten, Neues Jahrbuch für Geologie und Paläontologie, Monatshefte 1970(1):34–39
Seilacher A (2007) Trace fossil analysis. Springer, New York
Seilacher A, Marshall C, Skinner HCW, Tsuihiji T (2001) A fresh look at sideritic "coprolites". Paleobiology 27(1):7–13
Shcherbakov DE, Timm T, Tzetlin AB, Vinn O, Zhuravlev AY (2020) A probable oligochaete (Clitellata, Annelida) from an Early Triassic Lagerstätte of the southern Cis-Urals and its evolutionary implications. Acta Palaeontologica Polonica. https://doi.org/10.4202/app.00704.2019
Shi G, Grimaldi DA, Harlow GE, Wang J, Wang J, Yang M, Lei W, Li Q, Li X (2012) Age constraint on Burmese amber based on U–Pb dating of zircons. Cretac Res 37:155–163. https://doi.org/10.1016/j.cretres.2012.03.014
Sielaff M, Schmidt H, Struck TH, Rosenkranz D, Mark Welch DB, Hankeln T, Herlyn H (2016) Phylogeny of Syndermata (syn. Rotifera): mitochondrial gene order verifies epizoic Seisonidea as sister to endoparasitic Acanthocephala within monophyletic Hemirotifera. Mol Phylogenet Evol 96:79–92. https://doi.org/10.1016/j.ympev.2015.11.017
Siveter DJ, Briggs DE, Siveter DJ, Sutton MD (2015) A 425-million-year-old silurian pentastomid parasitic on ostracods. Curr Biol 25(12):1632–1637. https://doi.org/10.1016/j.cub.2015.04.035
Sliter WV (1971) Predation on benthic foraminifers. J Foraminif Res 1(1):20–28. https://doi.org/10.2113/gsjfr.1.1.20
Skelton J, Farrell KJ, Creed RP, Williams BW, Ames C, Helms BS, Stoekel J, Brown BL (2013) Servants, scoundrels, and hitchhikers: current understanding of the complex interactions between crayfish and their ectosymbiotic worms (Branchiobdellida). Freshwater Science 32(4):1345–1357. https://doi.org/10.1899/12-198.1
Southcott RV, Lange RT (1971) Acarine and other microfossils from the Maslin Eocene, South Australia. Rec South Australian Mus 16:1–21
Sperling EA, Vinther J (2010) A placozoan affinity for Dickinsonia and the evolution of late Proterozoic metazoan feeding modes. Evol Dev 12(2):201–209. https://doi.org/10.1111/j.1525-142X.2010.00404.x
Størmer L (1963) Gigantoscorpio willsi, a new scorpion from the Lower Carboniferous of Scotland and its associated preying microorganisms. Skrifter utgitt av Det Norske VidenskapsAkademi i Oslo I. Mat.-Naturv. Klasse Ny Serie 8:1–171
Struck TH, Golombek A, Weigert A, Franke FA, Westheide W, Purschke G, Bleidorn C, Halanych KM (2015) The evolution of annelids reveals two adaptive routes to the interstitial realm. Curr Biol 25:1993–1999
Struck TH, Wey-Fabrizius AR, Golombek A, Hering L, Weigert A, Bleidorn C, Klebow S, Iakovenko N, Hausdorf B, Petersen M, Kück P, Herlyn H, Hankeln T (2014) Platyzoan paraphyly based on phylogenomic data supports a non-coelomate ancestry of Spiralia. Mol Biol Evol. https://doi.org/10.1093/molbev/msu143

Sudhaus W (2008) Evolution of insect parasitism in rhabditid and diplogastrid nematodes. In: Makarov SE, Dimitrijevic RN (eds) Advances in arachnology and developmental biology: papers dedicated to Professor Bozidar P.M. Curcic, Monographs, vol 12. Institute of Zoology, Belgrade, pp 143–161

Sudhaus W (2010) Preadaptive plateau in Rhabditida (Nematoda) allow the repeated evolution of zooparasites, with an outlook on evolution of life cycles with Spiroascarida. Palaeodiversity 3(supplement):117–130

Suh A, Witt CC, Menger J, Sadanandan KR, Podsiadlowski L, Gerth M, Weigert A, McGuire JA, Mudge J, Edwards SV, Rheindt FE (2016) Ancient horizontal transfers of retrotransposons between birds and ancestors of human pathogenic nematodes. Nat Commun 7(1):11396. https://doi.org/10.1038/ncomms11396

Suh A (2021) Horizontal transfer of transposons as genomic fossils of host-parasite interactions. In: De Baets K, Huntley JW (eds) The evolution and fossil record of parasitism: Coevolution and paleoparasitological techniques. Topics in Geobiology 50

Summers MM, Rouse GW (2014) Phylogeny of Myzostomida (Annelida) and their relationships with echinoderm hosts. BMC Evol Biol 14(1):170

Szaniawski H (2002) New evidence for the protoconodont origin of chaetognaths. Acta Palaeontol Pol 47(3)

Szaniawski H, Gazdzicki A (1978) A reconstruction of three Jurassic polychaete jaw apparatuses. Acta Palaeontol Pol 23(1)

Tessler M, de Carle D, Voiklis ML, Gresham OA, Neumann JS, Cios S, Siddall ME (2018) Worms that suck: phylogenetic analysis of Hirudinea solidifies the position of Acanthobdellida and necessitates the dissolution of Rhynchobdellida. Molecular Phylogenetics and Evolution 127:129-134. https://doi.org/10.1016/j.ympev.2018.05.001.

Tinsley RC, Tinsley MC (2016) Tracing ancient evolutionary divergence in parasites. Parasitology 143(14):1902–1916

Todd JA, Harper EM (2011) Stereotypic boring behaviour inferred from the earliest known octopod feeding traces: early Eocene, southern England. Lethaia 44(2):214–222. https://doi.org/10.1111/j.1502-3931.2010.00237.x

Tzetlin A (1980) Ophryotrocha schubravyi sp. n. and the problem of evolution of the mouth parts in the Eunicemorpha (Polychaeta). Zoologichesky Zhurnal 59(5):666–676

Upeniece I (2001) The unique fossil assemblage from the Lode Quarry (Upper Devonian, Latvia). Fossil Rec 4(1):101–119

Upeniece I (2011) Palaeoecology and juvenile individuals of the Devonian Placoderm and Acanthodian fishes from Lode Site, Latvia. University of Latvia, Riga

Vannier J, Steiner M, Renvoisé E, Hu SX, Casanova JP (2007) Early Cambrian origin of modern food webs: evidence from predator arrow worms. Proc R Soc B Biol Sci 274(1610):627–633. https://doi.org/10.1098/rspb.2006.3761

Verneau O, Bentz S, Sinnappah ND, Ld P, Whittington I, Combes C (2002) A view of early vertebrate evolution inferred from the phylogeny of polystome parasites (Monogenea: Polystomatidae). Proc R Soc B Biol Sci 269(1490):535–543

Verneau O, Du Preez L, Badets M (2009) Lessons from parasitic flatworms about evolution and historical biogeography of their vertebrate hosts. C R Biol 332(2–3):149–158

Vinther J, Eibye-Jacobsen D, Harper DA (2011) An Early Cambrian stem polychaete with pygidial cirri. Biol Lett 7(6):929–932. https://doi.org/10.1098/rsbl.2011.0592

Vinther J, Parry LA (2019) Bilateral jaw elements in *Amiskwia sagittiformis* bridge the morphological gap between Gnathiferans and Chaetognaths. Curr Biol 29(5):881–888.e881. https://doi.org/10.1016/j.cub.2019.01.052

Voigt E (1938) Ein fossiler Saitenwurm (Gordius tenuifibrosus n. sp.) aus der eozänen Braunkohle des Geiseltales. Nova acta Leopoldina 5:351–360

Wade M (1972) Dickinsonia: polychaete worms from the late Precambrian Ediacara fauna, South Australia. Mem Queensl Mus 16(2):171–190

Waggoner BM, Poinar GO Jr (1993) Fossil habrotrochid rotifers in Dominican amber. Experientia 49(4):354–357. https://doi.org/10.1007/BF01923421

Walossek D, Müller KJ (1994) Pentastomid parasites from the lower Paleozoic of Sweden. Trans R Soc Edinburgh-Earth Sci 85:1–37

Walossek D, Repetski JE, Muller KJ (1994) An exceptionally preserved parasitic arthropod, Heymonsicambria taylori n.sp. (Arthropoda incertae sedis: Pentastomida), from Cambrian-Ordovician boundary beds of Newfoundland, Canada. Can J Earth Sci 31(11):1664–1671

Waloszek D, Repetski JE, Maas A (2005) A new Late Cambrian pentastomid and a review of the relationships of this parasitic group. Trans R Soc Edinburgh-Earth Sci 96:163–176. https://doi.org/10.1017/S0263593300001280

Waltécio de Oliveira A, Christoffersen ML (1999) A cladistic approach to relationships in Pentastomida. J Parasitol 85(4):695–704. https://doi.org/10.2307/3285745

Warn JM (1974) Presumed myzostomid infestation of an ordovician crinoid. J Paleontol 48(3):506–513

Warnock RCM, Engelstädter J (2021) The molecular clock as tool for understanding host-parasite evolution. In: De Baets K, Huntley JW (eds) The evolution and fossil record of parasitism: Coevolution and paleoparasitological techniques. Topics in Geobiology 50

Weigert A, Helm C, Meyer M, Nickel B, Arendt D, Hausdorf B, Santos SR, Halanych KM, Purschke G, Bleidorn C (2014) Illuminating the base of the annelid tree using transcriptomics. Mol Biol Evol 31(6):1391–1401

Weinstein SB, Kuris AM (2016) Independent origins of parasitism in Animalia. Biol Lett 12(7). https://doi.org/10.1098/rsbl.2016.0324

Welch JR (1976) *Phosphannulus* on Paleozoic crinoid stems. J Paleontol 50(2):218–225

Wolfe JM, Daley AC, Legg DA, Edgecombe GD (2016) Fossil calibrations for the arthropod Tree of Life. Earth-Sci Rev 160:43–110. https://doi.org/10.1016/j.earscirev.2016.06.008

Yang B, Steiner M, Zhu M, Li G, Liu J, Liu P (2016) Transitional Ediacaran–Cambrian small skeletal fossil assemblages from South China and Kazakhstan: implications for chronostratigraphy and metazoan evolution. Precambrian Res 285:202–215. https://doi.org/10.1016/j.precamres.2016.09.016

Yang C, Li X-H, Zhu M, Condon DJ, Chen J (2018) Geochronological constraint on the Cambrian Chengjiang biota, South China. J Geol Soc 175(4):659–666. https://doi.org/10.1144/jgs2017-103

Zangerl R, Case GR (1976) *Cobelodus aculeatus* (Cope), an anacanthous shark from Pennsylvanian black shales of North America. Palaeontogr Abt A 154(4–6):107–157

Zrzavý J (2001) The interrelationships of metazoan parasites: a review of phylum-and higher-level hypotheses from recent morphological and molecular phylogenetic analyses. Folia Parasitol 48(2):81–103

Zrzavý J, Hypša V, Tietz DF (2001) Myzostomida are not annelids: molecular and morphological support for a clade of animals with anterior sperm flagella. Cladistics 17(2):170–198

Chapter 8
Thorny-Headed Worms (Acanthocephala): Jaw-Less Members of Jaw-Bearing Worms That Parasitize Jawed Arthropods and Jawed Vertebrates

Holger Herlyn

Abstract Stem-acanthocephalans in the millimeter range might already have parasitized mandibulates in the Cambrian, while larger body sizes presumably evolved along with the upward-inclusion of gnathostome hosts. The characteristic morphology of modern acanthocephalans including the mostly hooked attachment organ (proboscis) should have emerged in the same context. Due to their rigidity, acanthocephalan hooks and copulatory caps are candidates for fossilization, but soft-tissue preservation might also have occurred under exceptional circumstances. Nonetheless, eggs represent the only ancient remains assigned to acanthocephalans to date. These were mostly retrieved from dried mammalian coprolites of up to ca. 12,000 years old. However, the recent discovery of eggs in a coprolite from the Upper Cretaceous illustrates that acanthocephalan eggs can also occur in fossilized remains. These and other aspects of acanthocephalan preservation, morphology, phylogeny, evolution, and pathogenicity are discussed in the present chapter that additionally includes a reflection of why Cambroclavida unlikely have an acanthocephalan origin.

Keywords Spiny-headed worms · Wheel animals · Phylogeny · Parasite evolution · Host usage

8.1 Introduction

Extant spiny- or thorny-worms (Acanthocephala) are endoparasites that occur worldwide where suitable hosts enable the establishment of their life cycle (Fig. 8.1). The intermediate host is recruited from Crustacea, Insecta, and Myriapoda and thus

H. Herlyn (✉)
Institute of Organismic and Molecular Evolution, Anthropology, University of Mainz, Mainz, Germany
e-mail: herlyn@uni-mainz.de

© The Author(s) 2021

K. De Baets, J. W. Huntley (eds.), *The Evolution and Fossil Record of Parasitism*, Topics in Geobiology 49, https://doi.org/10.1007/978-3-030-42484-8_8

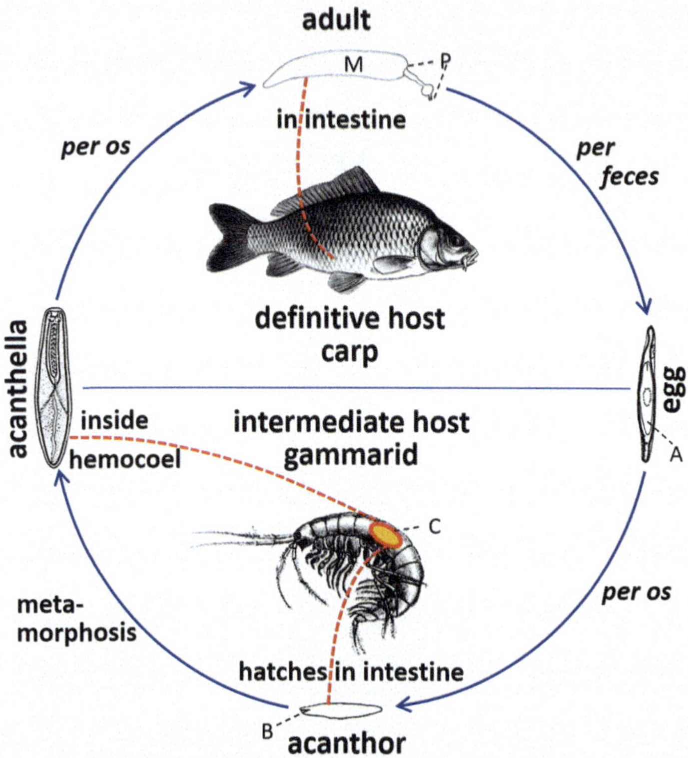

Fig. 8.1 Aquatic two-host life cycle of extant acanthocephalans. The palaeacanthocephalan genus *Pomphorhynchus* serves as an example. Presumably, the last common ancestor of crown-Acanthocephala already had an aquatic life cycle. Sketches of animals modified after Sielaff et al. (2016). *A* Acanthor inside egg, *B* blade of acanthor, *C* cystacanth, *M* metasoma, *P* presoma

from Mandibulata (Arthropoda) (Schmidt 1971; Rota-Stabelli et al. 2013). Cartilaginous fishes (Chondrichthyes), teleost fishes (Teleostei), amphibians (Amphibia), sauropsids (Sauropsida), and mammals (Mammalia) and thus jawed vertebrates (Gnathostomata) serve as definitive hosts (see, e.g., Petrochenko 1956, 1958; Gibson et al. 2014). Thereby, the specificity of host usage seems generally to be weaker for definitive than intermediate hosts (Conway Morris and Crompton 1982; Parker et al. 2015).

Following mating in a definitive host, female acanthocephalans shed eggs into the intestine or cloaca from where they are released along with the excrements into the environment (Figs. 8.1 and 8.2). The 'lucky' egg is orally taken up by a suitable mandibulate whereupon the larva (acanthor) hatches inside the alimentary tract and then penetrates the intestinal wall, thus entering the host's hemocoel. Thereafter, the larva experiences a drastic reorganization (catastrophic metamorphosis) to a growing juvenile (acanthella) which finally is encysted (cystacanth) (Fig. 8.1; Meyer 1932). The life cycle is closed when an intermediate host carrying one or more juveniles or cystacanths is ingested by a suitable definitive host (e.g., Conway Morris and Crompton 1982; Goater et al. 2014). Such an inclusion of a second host

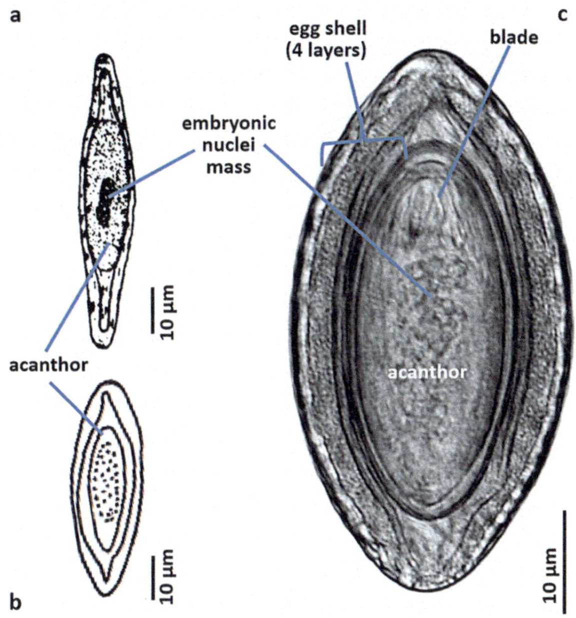

Fig. 8.2 Acanthocephalan eggs. (**a**) Drawing of an egg of *Echinorhynchus coregoni* (Palaeacanthocephala) (modified after Van Cleave 1921). (**b**) Sketch of an egg of *Neoechinorhynchus dighaensis* (Eoacanthocephala) (modified after Gautam et al. 2018). (**c**) Light micrograph of an egg of *Macracanthorhynchus hirudinaceus* (Archiacanthocephala: Oligacanthorhynchida) (modified after Kamimura et al. 2018). (**a–c**) The acanthella develops from the embryonic nuclei mass inside the acanthor

should increase the chance that a single male or female eventually meets a worm of the opposite sex so that they can mate and reproduce (Parker et al. 2015). However, the requirement for subsequent infection of two hosts lowers the chances of maintaining the life cycle. Acanthocephalans defy this adversity with two strategies. In particular, there is evidence that acanthocephalans manipulate the behavior of the intermediate host in such a way that the probability of host-transfer is increased (Bakker et al. 2017). Yet, while only some acanthocephalan species might be able to do so, the adults of all of them have increased fecundity (see Parker et al. 2015). In fact, males and females appear to be adapted to the production of high quantities of gametes and offspring (see below). In both respects, acanthocephalans resemble other endoparasites with complex life cycles such as tapeworms (Platyhelminthes, Cestoda) or trematodes (Platyhelminthes, Trematoda) (e.g., Goater et al. 2014).

In extension of the obligatory two-host cycle, at least some of the extant acanthocephalan species can continue their life in a second definitive host. This can happen when a higher-level predator feeds on a first definitive host (post-cyclic transmission). It also occurs that acanthocephalans survive but do not mature in a first gnathostome host (paratenic host), and complete their development in a definitive host that had preyed on the paratenic host (e.g., Hernández-Orts et al. 2019). In addition, acanthocephalans can end up in an accidental host inside which the worms survive without having the ability to reproduce, but also without a (realistic) chance of transfer into a definitive host (e.g., Kennedy 2006). Taken together, every jawed vertebrate that at least occasionally feeds on mandibulates appears to be a potential host in the one way or another. Correspondingly, acanthocephalan diversity might be much higher than suggested by the figure of about 1,300 species currently known (Gibson et al. 2014). In support of this expectation, ongoing descriptions of species from previously under-investigated gnathostome vertebrates let the count of acanthocephalan species continuously grow (e.g., Gomes et al. 2015; Gautam et al. 2019).

Inside the alimentary tract of their definitive hosts, thorny-headed worms grow to body sizes of mostly several millimeters up to centimeters (Fig. 8.3), while only few of the extant species exceed body sizes of 10 cm or more. Whether a tiny creature or a considerable worm, males generally remain smaller than females (see, e.g., Petrochenko 1956, 1958). However, except for this inverse sexual dimorphism in body size, male and female acanthocephalans have pretty much the same outer appearance: The anterior-most section is occupied by the holdfast organ (proboscis) followed by the neck. Proboscis and neck together constitute a retractable

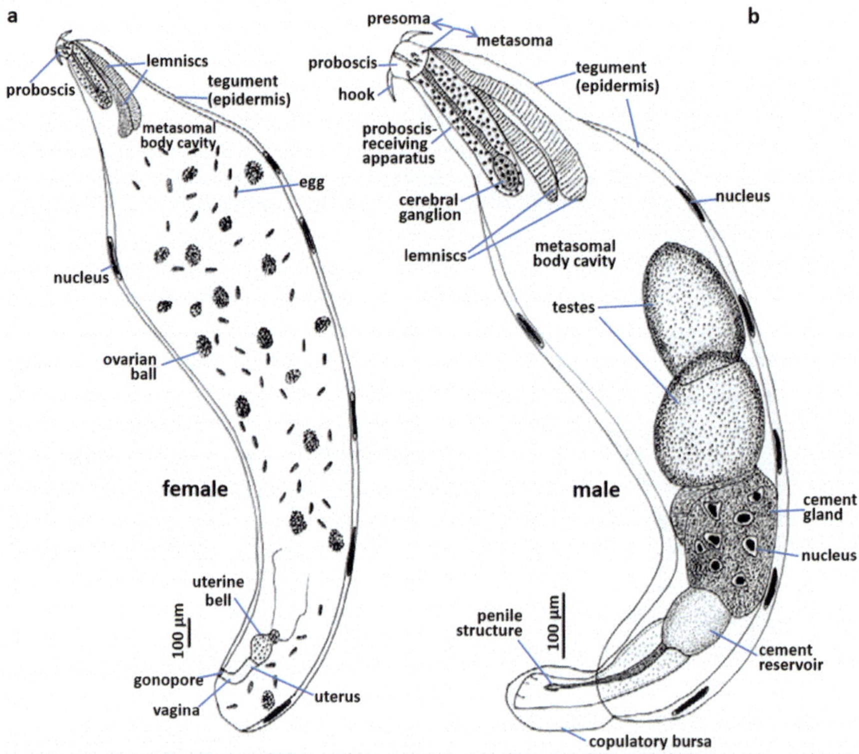

Fig. 8.3 Drawings of the fish-parasite *Neoechinorhynchus mexicoensis* (Eoacanthocephala). (**a**) Adult female with ovarian balls and eggs floating inside the metasomal body cavity. (**b**) Adult male with syncytial cement gland and two tandem-arranged testes. Except for the proboscis-receiving apparatus, which in eoacanthocephalans is comprised of receptacle and receptacle protrusor, the musculature is not shown (Modified after Pinacho-Pinacho et al. 2014)

functional unit named presoma (also praesoma). Farther posterior follows the trunk or metasoma, which mostly is more voluminous than the presoma in adult thorny-headed worms (Figs. 8.3 and 8.4a; Herlyn and Taraschewski 2017). This basic

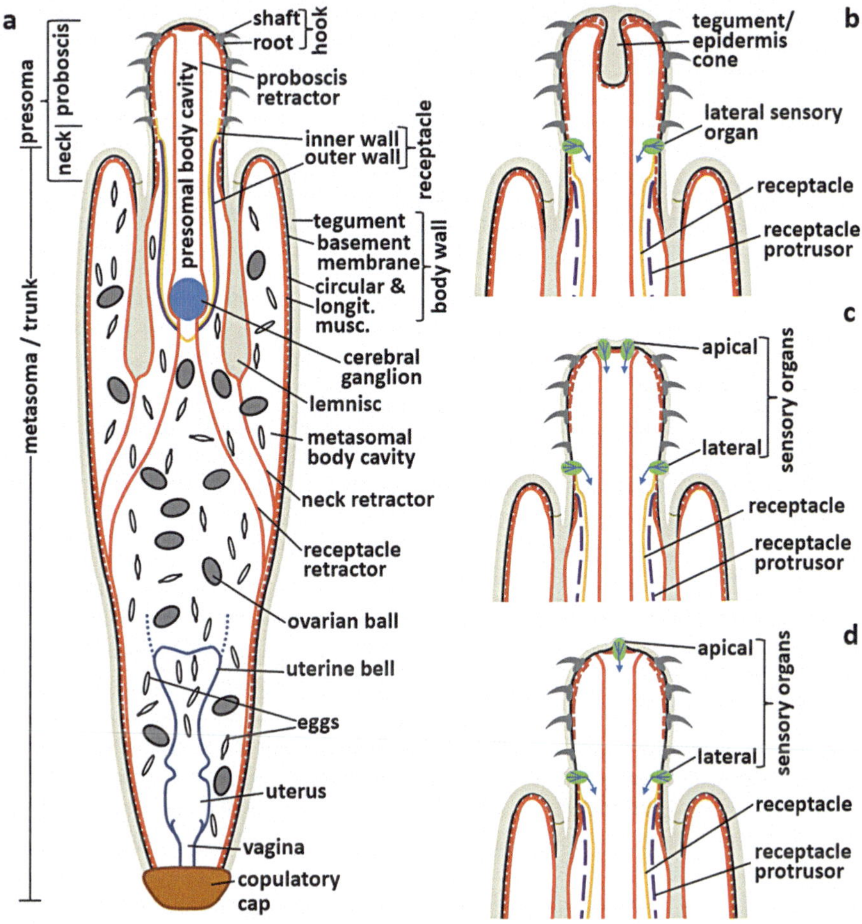

Fig. 8.4 Schematic depiction of the morphology of extant acanthocephalans (horizontal section plane). Not drawn to scale. (**a**) Palaeacanthocephala. (**b**) Eoacanthocephala. (**c**) Archiacanthocephala: Gigantorhynchida. (**d**) Archiacanthocephala: Oligacanthorhynchida. In acanthocephalans, hooks, copulatory cap, and eggs represent prime candidates for preservation but also the tegument is rather rigid and might occasionally have undergone fossilization. Internally, early crown-acanthocephalans should have possessed a two-layered muscular apparatus (light orange and lilac lines in (**a–d**)) suspending the cerebral ganglion and receiving the inverted proboscis (receptacle; see also Fig. 8.5a). Most likely, presomal sensory organs and tegument/epidermis cone evolved later on (**b–d**). Each presomal sensory organ consists of a bulbous swelling of the support cell (green), into which dendritic endings are imbedded (blue) which unite to nerves that extend to the cerebral ganglion (blue arrows). Successions of red lines and white dots underneath the basement membrane (also basal lamina) symbolize anastomosing strands of the circular musculature of the body wall. Light red solid lines highlight musculature with mainly longitudinally orientated anastomosing muscle strands. See main text for Apororhynchida (Archiacanthocephala), Moniliformida (Archiacanthocephala), and Polyacanthocephala. *longit.* longitudinal, *musc.* musculature

organization is widely conserved in three of the up to four acanthocephalan taxa with the traditional rank of a class, i.e. Palaeacanthocephala, Eoacanthocephala, and Polyacanthocephala. The same is true for Gigantorhynchida, Moniliformida, and Oligacanthorhynchida, which are ranked as orders within the fourth acanthocephalan class, Archiacanthocephala. In contrast, archiacanthocephalans belonging to the order Apororhynchida have an inflated proboscis and a comparably small metasoma (e.g., Petrochenko 1956, 1958; Herlyn 2017).

8.2 Acanthocephalans in Hominoids and Potential Reservoirs for Human Infections

The probably first case report of an infection of a human with an acanthocephalan was reported from Bohemia in the middle of the nineteenth century (Lambl 1859). The single female of *Macracanthorhynchus hirudinaceus* was excised from the intestine of a 9-year-old boy who had died of leukemia. How exactly the boy acquired the worm has to remain unanswered but one possible route of infection would be the intake of an infected grub of a scarabeid beetle (Coleoptera, Scarabeidae) (Pavlović et al. 2010). Indeed, people can generally acquire infections with acanthocephalans when they eat raw or insufficiently cooked intermediate hosts (e.g., Grassi and Calandruccio 1888; Zhong et al. 1983; Sahar et al. 2006; Mehlhorn 2016). Obviously, this happens quite regularly as demonstrated by Mathison et al. (2016). The study lists cases of infections with *M. hirudinaceus* and *Macracanthorhynchus ingens* from 11 countries in Asia, Australia, Europe, North America, and South America, and infections with *Moniliformis moniliformis* from 16 countries in the same continents plus Africa. In addition, humans may get infected when eating raw crayfish and fish, as discussed for acanthocephalans of the genera *Bolbosoma* and *Corynosoma* (e.g., Kaito et al. 2019; Sasaki et al. 2019).

The risk of human infections should be higher when mandibulates such as scarabeid grubs are infected at increased rate, and this should also be more likely to be the case when reservoir hosts live in vicinity of people. These can be omnivorous farm animals kept in the open. Outdoor management is still common in many parts of the world, and previously has been widespread in countries where it is now the exception. Thus, in the nineteenth century a veterinarian reported high morbidity in a herd of grazing piglets (*Sus scrofa domestica*, Suinae) in Hungary as a result of infection with *M. hirudinaceus* (Kocourek 1877). According to a note by the editorial office scarabeid beetles of the genus *Melolontha* represented the probable source of infection (see also Pavlović et al. 2010). Another report from the second half of the nineteenth century states that 40% of the swine slaughtered in Catania (Italy) were infected with *M. hirudinaceus* (Grassi and Calandruccio 1888). However, archiacanthocephalans have also found ways to exploit farm animals until more recently. Thus, *M. hirudinaceus* specimens could be collected for a study published in 2000 in the abattoir of La Paz (Bolivia), from butchered swine that were kept under outdoor management before. Yet macracanthorhynchiasis is not restricted to domestic swine as illustrated by an infected herd of white-lipped peccaries (*Tayassu pecari*, Suinae) kept in pasture in Brazil (de Almeida et al. 2006). Last but not least, archiacanthocephalans also occur in farm animals other than mammals. For example, chicken (*Gallus gallus domesticus*) can be parasitized by *Mediorhynchus gallinorum*, as has been observed in Indonesia (Amin et al. 2013). Accordingly, eggs of Archiacanthocephala might also be contained in old manure of human farm animals, although the preservation of pig and chicken manure might be a rarity per se.

Infections of farm animals and humans probably take place again and again somewhere in the world. However, it is also clear that the ongoing industrialization of animal production, in particular the separation of livestock from intermediate hosts, impairs the establishment of acanthocephalan life cycles with humans as hosts (compare Dunagan and Miller 1980). On the other hand, acanthocephalans do not depend on farm animals for getting into the vicinity of people since other mammals with a more or less close connection to humans can also be infected. In Tunisia, for instance, the occurence of *M. hirudinaceus* was shown for stray dogs (*Canis familiaris*, Carnivora) (Lahmar et al. 2017). This illustrates that the human companion can act as a reservoir for infections of people. But also mice and rats (*Mus* or *Rattus*, both Rodentia), raccoons (*Procyon lotor*, Carnivora), wild boar etc. can carry acanthocephalans, and due to their occurrence in or near settlements can be a reservoir for infections of humans and farm animals (Grassi and Calandruccio 1888; Dingley and Beaver 1985; Gassó et al. 2016; Kogi et al. 2016).

Raw animal food, including insects, should have contributed more to human nutrition in prehistoric times than today (Reinhard 1990; Gonçalves et al. 2003 Reinhard 2017). The mandibulates might have been collected in the surrounding landscape but also in human accomodations such as caves. In fact, the chance of ingesting an acanthella or cystacanth is not to be underestimated when eating a protein source like an insect in a raw or insufficiently cooked state. For example, the archiacanthocephalan *Moniliformis dubius* was found in up to 35% of the cockroaches (*Periplaneta americana*, Blattodea) collected from the lecture halls of a Nigerian university (Kogi et al. 2016). Thereby, the intensity of infection can reach considerable levels of more than 100 juvenile stages per individual intermediate host, as reported for *M. moniliformis* in beetles of the species *Blaps mucronata* (Coleoptera, Tenebrionidae) (Grassi and Calandruccio 1888). Accordingly, the uptake of a single intermediate host can be sufficient to enable acanthocephalan reproduction. In fact, Grassi and Calandruccio (1888) demonstrated that acanthocephalan eggs are contained in human stool upon infection with several cystacanths. Not least, eggs in human coprolites demonstrate that our species can serve as definitive host for archiacanthocephalans (see Table 8.1).

The anatomically modern *Homo sapiens* is not the only species within hominoids (Primates, Hominoidea) that can be parasitized by acanthocephalans. Also small apes (Hylobatidae), orangutans (*Pongo*), gorillas (*Gorilla*) and chimpanzees (*Pan*) sporadically take in insects, which can carry developmental stages of archiacanthocephalans (e.g., Van Thiel and Bruss 1945; Myers and Kuntz 1972). This phenomenon is actually not restricted to hominoids and rather affects other primates as well (Richart and Benirschke 1963; Solórzano-García and Pérez-Ponce

Table 8.1 Summary of acanthocephalan eggs found at archeological sites

Archiacanthocephalan taxon	Probable origin of coprolites	Archeological site	Age or comment	Reference
Echinopardalis 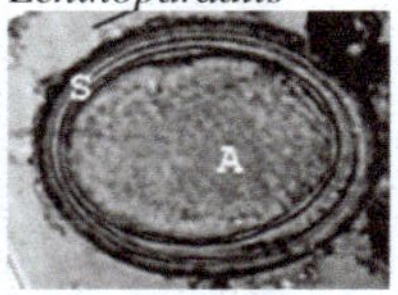	C; small felid, probably *Felis tigrina* or *Felis pardalis*	São Raimundo Nonato, Piaui, NE Brazil	9,000 BP	Noronha et al. (1994)
Gigantorhynchus echinodiscus 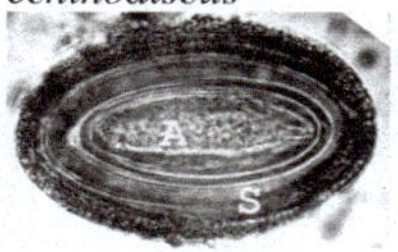	C; anteater, *Tamandua tetradactyla* or *Myrmecophaga tridactyla*	Baixão da Vaca/São Raimundo Nonato, Piauí State, Brazil	2,955 BC to 625 AD	Ferreira et al. (1989)
Gigantorhynchus echinodiscus	C; anteater, *T. tetradactyla* or *M. tridactyla*	Boqueirão Soberbo/Varzelândia, Minas Gerais State, Brazil	2,955 BC to 625 AD	Ferreira et al. (1989)
Macracanthorhynchus hirududinaceus 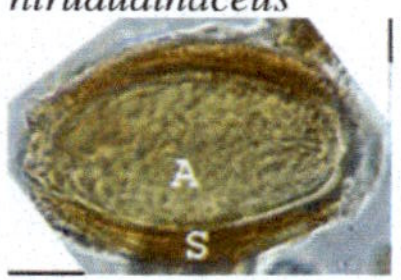	C; canid, dog?	Chehrabad salt mine/NW Iran	Fourth/fifth century CE	Mowlavi et al. (2015)
Macracanthorhynchus[a]	C; human	Antelope Cave/NW Arizona, USA	1,230 ± 40 BP	Fugassa et al. (2011)
Moniliformis clarki	C; human	Danger cave/Utah, USA	7,839 ± 630 BC to 1,869 ± 160 BC	Moore et al. (1969)

Archiacanthocephalan taxon	Probable origin of coprolites	Archeological site	Age or comment	Reference
Macracanthorhynchus 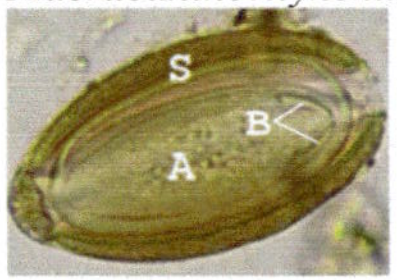	R; probably barn owl, *Tyto alba*	Epullán Chica cave/NW Patagonia, Argentina	1,980 ± 80 BP	Beltrame et al. (2015)
Moniliformis clarki	C; human	Danger Cave/Utah; Dirty Shame Shelter/Oregon; Hogup Cave/Utah; all USA	10,000 BC to 20 AD	Table 2 in Reinhard (1990)
Probably *Prosthenorchis* 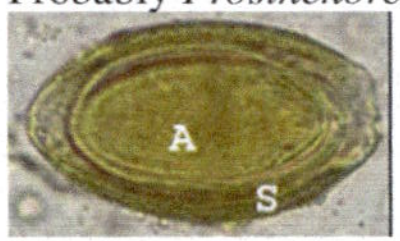	C; potentially skunk, *Conepatus chinga*	Epullán Chica cave/NW Patagonia, Argentina	1,980 ± 80 BP to 1,740 ± 60 BP	Beltrame et al. (2018)
Unspecified 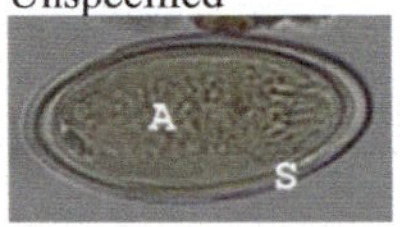	C; carnivore/human/unidentified	Hierakonpolis/Egypt	3,900 BC to 3,300 BC	Nezamabadi (2014)
Unspecified 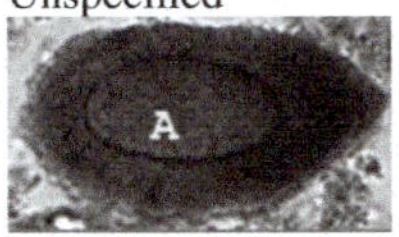	Sq; human artifact	Guapi/Rio de Janeiro, Brazil	"The outer eggshell is poorly preserved with an amorphous contour"	Camacho et al. (2013)
Unspecified[a]	C; human	Antelope Cave/NW Arizona, USA	ca. 1,100 BP	Hunt et al. (2012)
Unspecified	CC, human (spontaneously mummified female)	Dakhleh Oasis, Egypt	AD 30–395 BC	Horne (2002)

(continued)

Table 8.1 (continued)

Archiacanthocephalan taxon	Probable origin of coprolites	Archeological site	Age or comment	Reference
Unspecified	C; human	Clyde's Cavern/Utah; Black Mesa/ Arizona; Glen Canyon/Utah; all USA	460 AD to 1,500 AD	Table 2 in Reinhard (1990)
Unspecified	C; human?	Boqueirão Soberbo/Minas Gerais, Brazil	4,905 ± 85 BP to 1,325 ± 60 BP	Gonçalves et al. (2003)
Unspecified	C; human?	Gentio Cave/Minas Gerais, Brazil	3,490 ± 120 BP to 430 ± 70 BP	Gonçalves et al. (2003)
Unspecified 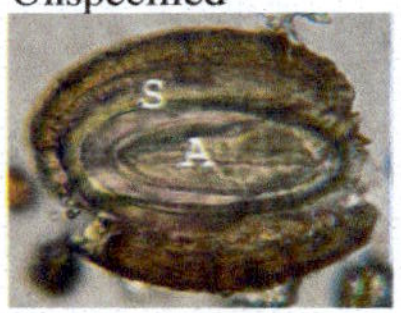	C, Crocodiloformes	Santo Anastácio Municipality, Southwestern São Paulo State, Brazil	Bauru Group, Upper Cretaceous from the Paraná Basin	Cardia et al. (2019)

Micrographs in the left column are modified after the references in the right column. For age ranges relating to sites reported in Table 2 of Reinhard (1990), see also Table XIII in Gonçalves et al. (2003), Table 3 of Sianto et al. (2009), and Table I in Fugassa et al. (2011).
A acanthor, *B* blades, *C* coprolite, *CC* colon content, *R* regurgitation pellet, *S* shell, *Sq* Sambaqui (shell mound).
[a]The corresponding egg depicted in Fugassa et al. (2011) and Hunt et al. (2012) appears to be identical.

de León 2018). Consequently, one should assume for the Last Common Ancestor (LCA) of diverse primate taxa that they were insectivorous to at least some extent. In the present context though, it is of particular interest that the LCAs of the crown-group taxa on the lineage to humans (Hominoidea, Hominidae, Homininae, *Pan* + *Homo*, and Hominini) should also have been insectivorous to some extent (for other character states of these LCAs, see Herlyn 2016). In expansion of this argument, insects are considered part of the diet of the Hominini genera *Ardipithecus* and *Australopithecus*. In the case of early members of *Homo* and prehistoric anatomical modern humans, there is even direct evidence for insectivory—and last but not least, in some human societies, insects continue to be part of the diet (Reinhard and Bryant 1992; Sayers and Lovejoy 2014; van Huis 2017). To put it the other way: There appears to be no principal reason why *Ardipithecus ramidus*, *Australopithecus afarensis*, *Homo erectus* etc. should not have been infected with (archi)acanthocephalans at least sporadically. Now *only* their eggs have to be discovered in coprolites of the corresponding taxa (compare Sistiaga et al. 2014; Chin 2021).

8.3 Solid-Parts and Their Preservation Potential

8.3.1 Acanthocephalan Propagules: Eggs in Space and Time

In acanthocephalans, the egg-producing units are not ovaries in the traditional sense. Instead, numerous so-called ovarian balls of unclear ontogenetic origin freely float in the fluid-filled metasoma, and release immature eggs into the body cavity wherein they mature (Figs. 8.3a and 8.4a; e.g., Dunagan and Miller 1991). The fluid remains in motion through movements of the body so that the eggs sporadically pass by an elaborated egg sorting apparatus. This so-called uterine bell is situated in the hind trunk and allows only mature eggs to enter the terminal genital tract from where the eggs are discharged through the genital pore into the alimentary tract or cloaca of the definitive host (Fig. 8.4a; Herlyn and Röhrig 2003 and references therein). The eggs are eventually released with the excrements into the environment (Fig. 8.1), a process that can take place at high rates upon mating. Gravid females of the larger species might even produce 82,000 eggs per day on average, and this for a patent period of 10 months (see Dunagan and Miller 1991). Consequently, acanthocephalan eggs should be quite common in the droppings of infected gnathostomes—and the hosts might not only have been taxa with extant members. As acanthocephalan life cycles can be extended by post-cyclic transmission in at least some extant species (Kennedy 1999, 2006), acanthocephalan eggs may also be contained in fossilized coprolites of higher-level predators belonging to Ichthyosauria and Pterosauria (both Sauropsida), to name just a few (compare Leung 2021). In addition, acanthocephalan eggs may be preserved in petrified sediments under favorable conditions.

The contour of acanthocephalan eggs enables a rough taxonomic diagnosis. Thus, eggs of archiacanthocephalans and polyacanthocephalans usually have a rounded-oval appearance. The same applies to at least some eoacanthocephalans and palaeacanthocephalans, while eggs of other palaeacanthocephalans are stronger elongated or spindle-shaped (Fig. 8.2; Uglem 1972; Schmidt 1985; Amin 1987; Taraschewski and Peters 1992; Golvan 1994; Amin and Heckmann 2017). Since also the next phylogenetic relatives of Acanthocephala as a whole, i.e., the different taxa traditionally regarded as wheel animals (Rotifera) (e.g., Verweyen et al. 2011; Weber et al. 2013), have rounded to oval eggs (see e.g., de Beauchamp 1909; Murray 1910; Edmondson 1965; Dartnall 1995; Habdija et al. 2011), this should be the plesiomorphic character state. Consequently, ancient acanthocephalans should have possessed rounded to oval eggs, too. These eggs presumably had thin shells because life cycles were most likely aquatic in early acanthocephalans (see below). Indeed, the shell is thinner in extant acanthocephalans with aquatic life cycles than in species exploiting terrestrial hosts. Such thicker eggshells are particularly well known from archiacanthocephalans (Fig. 8.2; e.g., Van Cleave 1921, 1947; Uglem 1972), and probably protect the acanthors from dehydration.

As far as the shells of eggs were investigated in more detail, most authors distinguished four envelopes or layers, of which the second-outermost one was found to contain keratin or a keratin-like substance. The same may be true for the second-innermost layer, as demonstrated for an archiacanthocephalan species. Lastly, a chitin-like constituent was reported for the innermost eggshell layer of palae- and archiacanthocephalans (compare Fig. 8.2a–c; Whitfield 1973; Peters et al. 1991; Taraschewski and Peters 1992; Taraschewski et al. 1992). The incorporation of keratin- and chitin-like compounds into the eggshell might account for its rigid nature and, in any case, should be advantageous for preservation. In fact, leaving aside a fossil from the Triassic of India with uncertain phylogenetic affiliation (see Kumar and Kumar 2001), eggs are the only known remains of higher age that clearly refer to acanthocephalans (Table 8.1 and references therein). With a single exception (next paragraph), the age of the finds ranges from at least several hundred up to about 12,000 years before present. The said eggs are from sites in North and South America, Asia and Africa, but there is no apparent reason why eggs that would testify to ancient infections with archiacanthocephalans should not also be found in Europe. Either way, most of the corresponding remains were obtained from dried feces, probably left by anteaters (Xenarthra), a dog (Carnivora, Canidae), a small felid (Carnivora, Felidae), a skunk (Carnivora, Mephtitidae), an unspecified carnivore, and humans (Table 8.1). Six of the eggs found in colon content of an Egyptian mummy from Roman times complete the list of ancient mammalian infections with acanthocephalans (Table 8.1: Horne 2002). In addition, acanthocephalan eggs retrieved from a preserved hairball could reflect an infection of one or more smaller mammals (Rodentia) which an owl (Aves, Strigiformes) preyed upon (Table 8.1: Beltrame et al. 2015).

According to the shape of the eggs as well as the thickness and structuring of the eggshell layers, these ancient acanthocephalan eggs belong to Archiacanthocephala. They were especially assigned to the genera *Echinopardalis*, *Gigantorhynchus*, *Macracanthorhynchus*, *Moniliformis* and *Prosthenorchis* (Table 8.1: Moore et al. 1969; Ferreira et al. 1989; Reinhard 1990; Noronha et al. 1994; Fugassa et al.

2011; Beltrame et al. 2015, 2018; Mowlavi et al. 2015), although the mostly poor preservation of the outer shell layer impairs comparisons with modern eggs (compare Table III in Fugassa et al. 2011; Beltrame et al. 2015). In other cases, the archiacanthocephalan genus remained unspecified (Table 8.1: Reinhard 1990; Horne 2002; Gonçalves et al. 2003 Hunt et al. 2012; Camacho et al. 2013; Nezamabadi 2014). A very recent paper additionally reports eggs of *Oligacanthorhynchus*, in addition to *Macracanthorhynchus* and *Gigantorhynchus*, from *Tamandua tetradactyla* coprolites which were collected from 3,190 to 8,870 year old layers at archeological sites in Piauí state, Brazil (de Souza et al. 2020). However, keratin- and chitin-like substances are not restricted to archiacanthocephalan eggs (see above). So the question arises why archiacanthocephalan eggs predominate and why are there so many human and carnivore coprolites among the egg-carrying samples. This could partly reflect sampling bias because terrestrial caves usually attract high research attention when they were human shelters in pre-historic times (e.g., Araújo et al. 2015). But the skew might have a natural background in addition: First, higher thickness of their shells should render archiacanthocephalan eggs particularly resistant to damaging effects (compare Camacho et al. 2013, 2018). Second, preservation through dehydration and undisturbed storage should be more likely to happen to droppings in terrestrial caves than to feces released in other environments. Yet, the gnathostomes that have left larger droppings in land caves in the relevant time were mostly members of Carnivora and humans (e.g., Camacho et al. 2018). In any case, these finds suggest that archiacanthocephalan eggs could be contained not only in preserved human stool from mummies and latrines but also in mummified cats etc., as has already been shown for eggs of endoparasitic nematodes (Nematoda) and flatworms (Platyhelminthes) (Ferreira et al. 1983; Novo and Ferreira 2016; Yeh and Mitchell 2016).

A recent report of ancient acanthocephalan eggs refers for the first time to fossilized remains. The corresponding four eggs were included in a phosphatized coprolite from Upper Cretaceous sediments in São Paulo State, Brazil (Cardia et al. 2019). The assignment of the four eggs to Acanthocephala and especially to Archiacanthocephala is supported by the thickness and multi-layered appearance of their eggshells, and also by the structures enclosed which in at least three cases are reminiscent of acanthors (see Fig. 6b–d in Cardia et al. 2019). The fact that these egg remains maximally show three eggshell layers does not preclude an archiacanthocephalan origin since the outermost egg shell layer generally tends to be poorly preserved (see above). However, the extant species of Archiacanthocephala are primarily known to use mammals and birds as definitive hosts (e.g., Near 2002), while the Cretaceous coprolite presumably has a crocodyliform origin (Cardia et al. 2019). Although a wider spectrum of taxa from Tetrapoda may serve as paratenic host (see e.g., Petrochenko 1958), there appears to be no indication that crocodiles belong to the usual hosts. It would therefore be conceivable that—comparable to the above example of a hairball—the Cretaceous animal, which left the dropping, devoured at least one paratenic or definitive host of an archiacanthocephalan. In support of such possibility, Cardia et al. (2019) state that the fluvial deposits of the respective Brazilian Lagerstätte contain fossilized bones of fishes, lizards (Lacertilia), turtles

(Testudines), stem-group representatives of birds commonly referred to as "dinosaurs", mammals and other taxa, besides remains of Crocodyliformes. Nonetheless, it may also be that the Cretaceous creature ingested one or more acanthocephalan developmental stages when preying on a crustacean or other intermediate host. Either way, the finding demonstrates that acanthocephalan eggs can fossilize, and that details of eggshell composition and acanthor morphology remain discernible under advantageous conditions. With such excellent preservation, the find reminds of putative hooklets inside the egg of a tapeworm (Platyhelminthes, Cestoda) discovered in a 270 million year old shark coprolite (Dentzien-Dias et al. 2013). Likewise, eggs of presumably cestode origin have been reported from rectum content of a Carboniferous shark fossil (Zangerl and Case 1976). Perhaps, acanthocephalan eggs will once be discovered in coprolites of similar or even higher age.

8.3.2 Hooks

While trunk and neck bear spines in some of the extant acanthocephalan species and not in others, the proboscis is almost always armed with recurved hooks (Figs. 8.4 and 8.5a, b). These hooks are extracellular differentiations consisting of protein (see Miller and Dunagan 1985), with roots resting in the basement membrane (basal lamina) underlying the tegument (also integument, epidermis, cutis) (Fig. 8.6a, b). However, this anchoring might actually be somewhat flexible due to the probable discontinuity of the fibers constituting hook roots and basement membrane (Fig. 8.6b; Taraschewski et al. 1989). In histological preparations, hook roots and shafts show a peripheral rind of condensed material and a core of less densely woven fibers (see labelled hook in Fig. 8.6a). This structuring seems to be reflected in a radial gradient of element abundances as revealed by X-ray analysis. Presumably, the incorporation of substances like calcium, sulfur, and phosphorus contributes to the stiffening of the hooks, which reportedly proceeds in anterio-posterior direction along the proboscis (Taraschewski 1989a, b; Amin and Heckmann 2017). In any case, the sclerotization should increase the chances that acanthocephalan hooks are contained in the fossil record (compare Littlewood and Donovan 2003).

Although exceptional preservation conditions are required, fossilization of "invertebrate" hooks is possible. This is exemplified by fossil hooks in abdominal and gill regions of Devonian remains from Latvia assigned to Placodermi and Acanthodii (Upeniece 2001, 2011; De Baets et al. 2015). Their probable location on the surface of the fish fossils, their appearance and partially also their circular arrangement suggest that the hooks once belonged to ectoparasitic monogeneans (Platyhelminthes, Monogenea) (Upeniece 2011; Leung 2017; De Baets et al. 2021), and not acanthocephalans. Still, the size of the fossil hook-like remains (0.02–0.40 mm; Upeniece 2011) is in a range known from extant acanthocephalans (Figs. 8.5a and 8.6a). However, when likewise small monogenean hooks can fossilize, why have no fossil hooks of acanthocephalans been discovered so far? The most likely reason is that acanthocephalans live inside their gnathostome hosts—and

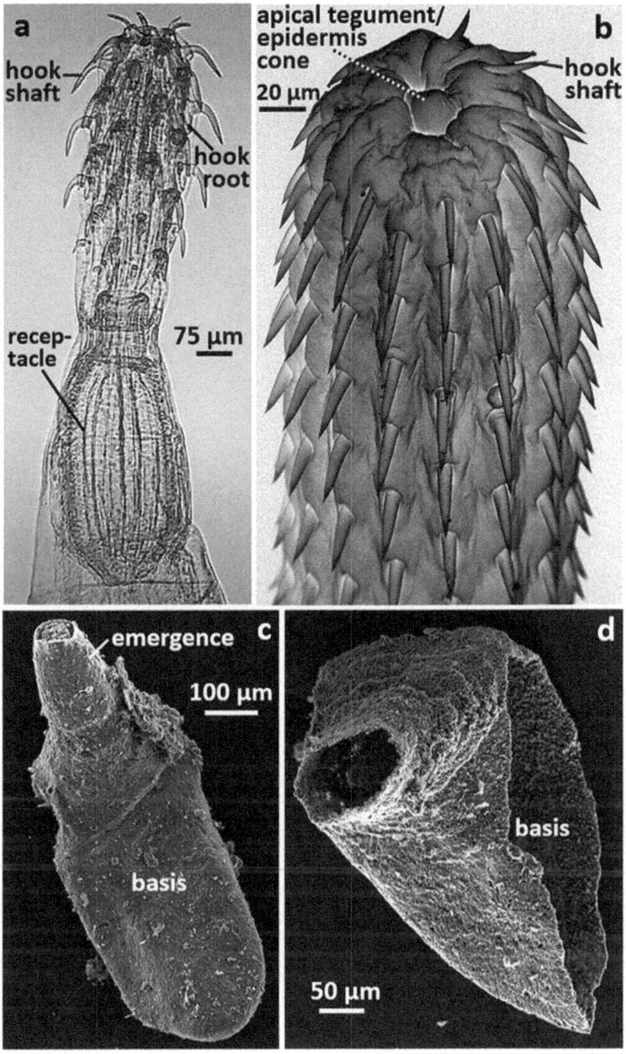

Fig. 8.5 Comparison of hooks in the extant acanthocephalan species (**a**, **b**) and cambroclavid microfossils from the Cambrian (**c**, **d**). (**a**) Light micrograph of a mounted specimen of *Acanthocephalus anguillae* (Palaeacanthocephala): The proboscis bears recurved hooks. (**b**) *Paratenuisentis ambiguus* (Eoacanthocephala): Scanning electron micrograph of about the anterior two thirds of the proboscis. The tegument or epidermis cone is visible at the proboscis apex. (**c**, **d**) Scanning electron micrographs of *Cambroclaves* microfossils (Courtesy of Thomas Wotte, Geological Institute at TU Bergakademie Freiberg, Germany)

even "worse", inside the alimentary canal. Probably, acanthocephalans do so since they exploit gnathostomes as hosts (see evolutionary scenario below). For this reason, fossilization would require that acanthocephalan hooks first withstand autolysis of the bowels upon the death of a host, and then encounter excellent preservation conditions.

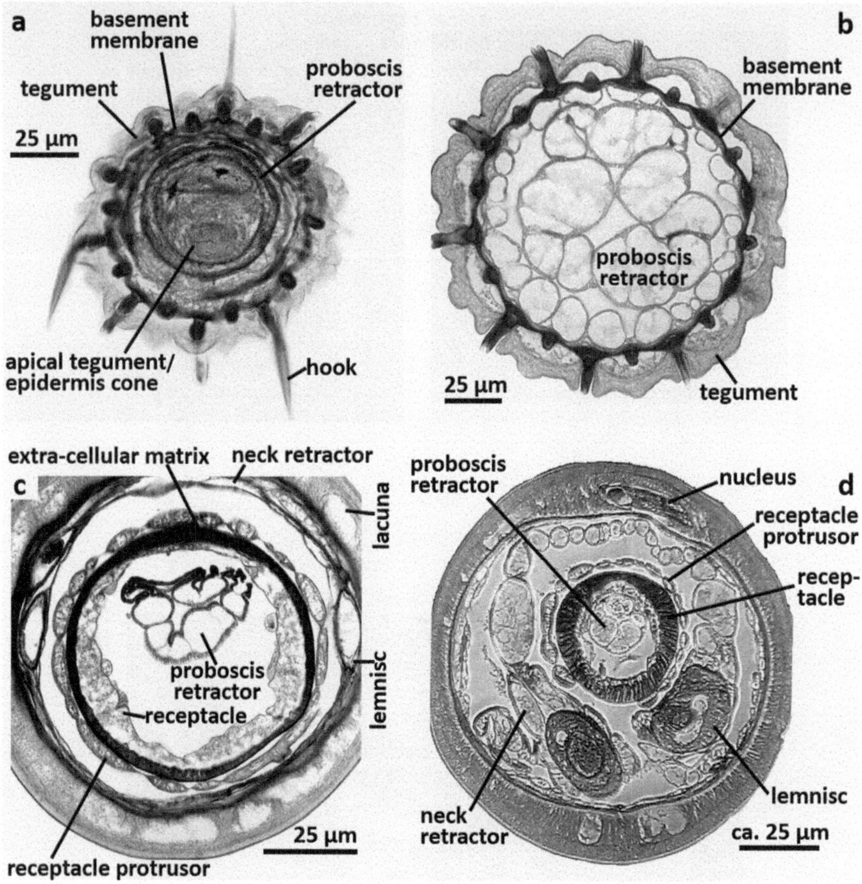

Fig. 8.6 *Paratenuisentis ambiguus* (Eoacanthocephala): Light micrographs of consecutive transverse sections. (**a**) Proboscis at the level of the apical tegument or epidermis cone. (**b**) About halfway along the proboscis. (**c**, **d**) Foretrunk with the proboscis-receiving apparatus (receptacle protrusor plus receptacle) at its center. The receptacle has an anterior medullary (**c**) and a posterior contractile portion (**d**)

8.3.3 Copulatory Cap

Besides hooks and eggs, there is another structure, for which preservation seems to be in the range of the possible, and this structure has to do with acanthocephalan reproduction: In the course of mating, the male worm encloses the female hind end with the everted copulatory bursa (Fig. 8.3b), and introduces spermatozoa through a penile structure into the female genital tract. Before the male detaches from the female, it seals the female genital pore with the proteinaceous secretion from one or more so-called cement glands (Fig. 8.3b) (Dezfuli et al. 2001). Obviously, the female is hindered from additional copulations as long as the copulatory cap is present (Fig. 8.4a). Thus, cement and cement apparatus might reflect increased competition between males, as does the enlargement of the usually two testes (Fig. 8.3b; Poulin and Morand 2000). In any case, a capping structure at the posterior ending of a fossil intestinal parasite would be an indication of a female acanthocephalan.

8.4 Soft Tissue, Functional Morphology and the Ideal Fossil

Fossils from the lower Cambrian of China and Canada show that the preservation of worms can be so excellent that besides contour and appendages also internal structures like the digestive tract appear to be visible (Hu et al. 2008; Briggs and Caron 2017; Shu et al. 2017). The mentioned fossils are reminiscent of priapulids and might have dwelled the sediment, which buried them later on. Still, soft tissue preservation is not restricted to Cambrian sediment dwellers but also occurred in epibiotic or ectoparasitic helminths. A first example is *Inquicus fellatus*, a parasite from the Cambrian of China, which lived on priapulids or priapulid-like worms (Cong et al. 2017). Another example is a special find amongst the already-mentioned remains of fish ectoparasites from the Upper Devonian of Latvia: The corresponding fossil shows not only a circlet of six hooks reminiscent of a monogenean opisthaptor, but also the contour of the hook-bearing soft tissue (Upeniece 2001, 2011; De Baets et al. 2015). But when soft-tissue preservation is rare in ectoparasites, it should be even rarer in intestinal parasites.

8.4.1 Outer Contour and Tegument

Under the premise that preservation of intestinal parasites is unlikely, the tegument or epidermis should not be less suitable for fossilization in acanthocephalans than other helminths (compare Littlewood and Donovan 2003). Whether the tegument experiences sclerotization (Taraschewski et al. 1989) or not, it is very resistant to mechanical destruction and enzymatic decomposition (personal observation; but see Reinhard 1990). This toughness is probably due to the syncytial organization of the tegument and a presumably proteinaceous lamina underneath its distal plasma membrane, a character complex shared by acanthocephalans and their closer phylogenetic relatives (Ahlrichs 1997; Herlyn and Ehlers 2001; Near 2002). Provided that fossilization of the acanthocephalan tegument occurred, the outer contour of the hypothetical remains might show an increase in body diameter from neck to metasoma as it is common in the extant species (Figs. 8.3 and 8.4; e.g., Petrochenko 1956, 1958). Furthermore, if a larger bulbous differentiation shines through at the proboscis apex, this could be an intrusion of the tegument ((apical) tegument or epidermis cone), as it is characteristic of the monophylum including polyacanthocephalans and eoacanthocephalans (Figs. 8.4b and 8.5b; Herlyn 2001; Amin 2013;

Gazi et al. 2016). However, if the contour of the hypothetical fossil is more reminiscent of a balloon with a shorter worm-like appendix, it could originate from an archiacanthocephalan, particularly a member of Apororhynchida (Herlyn 2017).

8.4.2 Presomal Musculature and Anchoring

Internally, the transition from presoma to metasoma is marked by a muscular apparatus which suspends the cerebral ganglion or brain and probably consists of two muscular layers in all extant species. The only complication in this general scheme is that each layer can be either a continuous muscular sheath or a muscular mesh of anastomosing strands. In Palaeacanthocephala, both muscular layers are continuous and thereby have such intimate contact that the entire apparatus has a double-walled appearance (Fig. 8.4a). Archiacanthocephalans belonging to Moniliformida show the same pattern. In archiacanthocephalans belonging to Gigantorhynchida and Oligacanthorhynchida as well as in all members of Eoacanthocephala, only the inner layer is a sheath-like muscle, whereas the outer layer forms a loose muscular mesh (Figs. 8.4b–d). Accordingly, the muscular apparatus suspending the cerebral ganglion looks single-walled in these species (Herlyn 2002; Herlyn and Taraschewski 2017). The same pattern seems to be realized in Polyacanthocephala (Amin 1987). Yet another way in which both muscle layers can be arranged is shown by the archiacanthocephalan taxon Apororhynchida: here both muscle layers consist of delicate vela-like muscular strands (Herlyn 2017).

Unfortunately, the established terminology does not reflect the homology of the two internal muscular layers at the presoma-metasoma transition. Thus, it is common practice to refer to a double-walled receptacle, when both muscular layers are sheath-like and have intimate contact (Palaeacanthocephala, Moniliformida). The layers themselves are then considered as inner and outer wall of the receptacle (Fig. 8.4a). Yet, the receptacle is commonly regarded as single-walled, when only the inner layer is sheath-like (Eoacanthocephala inclusively Polyacanthocephala, Gigantorhynchida, Oligacanthocephala). Then, this inner layer is termed just receptacle, while the established name for the strands of the receptacle-surrounding muscle is receptacle protrusor (Figs. 8.4b–d and 8.6c, d; Herlyn and Taraschewski 2017; also Amin 1987). The use of alternative names for the same muscles further complicates matters (for a survey, see Herlyn and Taraschewski 2017).

The alternative organization of the muscular apparatus carrying the cerebral ganglion has functional consequences: When both muscular layers are mesh-like (Apororhynchida), their contraction should not increase the pressure inside the presomal part of the body cavity relative to its metasomal portion (Herlyn 2017). However, as long as at least one of the muscular layers has a sheath-like organization (all other taxa of extant Acanthocephala; Figs. 8.4, 8.5a and 8.6c, d), presomal and metasomal body cavities are separated. Then, contraction of the either single- or double-walled muscular apparatus increases the hydrostatic pressure inside the presomal body cavity. This again leads to the eversion of neck and proboscis, which can be accompanied by partial rotation alongside the body axis and a bending relative to the trunk (Hammond 1966; Herlyn and Ehlers 2001). Reversal of the process results from relaxation of the "eversion muscles" and contraction of an retractor muscle. The presomal portion of the latter is termed proboscis retractor whereas the metasomal portion is commonly referred to as receptacle retractor (Fig. 8.4a). The proboscis retractor consists of circularly arranged longitudinal muscular strands (Figs. 8.4 and 8.6a, b) that anastomose and thus form a tube-like contractile mesh. The mesh can display a complicated folding in cross section depending on the taxon investigated (e.g., Dunagan and Miller 1991; Herlyn 2002). However, in all extant acanthocephalans the muscle inserts anteriorly at the inner site of the body wall just beneath the proboscis apex. After having extended through most of the presomal (part of the) body cavity, the muscle splits into two or three portions that separately pass through the bottom of the muscular apparatus suspending the cerebral ganglion (Fig. 8.4a). The separate portions continue through the metasomal (part of the) body cavity before they attach to the inner surface of the metasomal body wall (Fig. 8.4a) (Herlyn 2017; Herlyn and Taraschewski 2017 and references therein).

Eversion and inversion of the presoma are usually repeated until the worm is attached to the intestinal mucosa or deeper (Hammond 1966; Aguiar et al. 2018). Characteristic for the resting position is a subsequent slight withdrawal of the neck by contraction of the neck retractor, which also has a mesh-like organization (Figs. 8.4a and 8.6c, d). As some of its strands enclose the lemniscs, usually paired processes of the presomal tegument extending into the metasomal (part of the) body cavity (Figs. 8.4 and 8.6d), contraction of the neck retractor presumably presses fluid from the lemniscs into the presomal tegument, thus stiffening the everted proboscis. Thereby, the transportation of fluid takes place via a so-called lacunar system extending through tegument and lemniscs (e.g., Fig. 8.6c; Hammond 1966; Herlyn 2002, 2017; Herlyn and Taraschewski 2017). This may be the case or not, but it is almost certain that there is no connection between the lacunar system inside the tegument and lemniscs and any sub-tegumental structures including the body wall musculature (compare Nielsen 2012). In addition, the musculature is not hollow, although it may occasionally appear hollow due to preparation artifacts (for a discussion, see Herlyn and Taraschewski 2017; see also Nikishin 2004).

From a paleoparasitological point of view, the practical value of the above details on the functional morphology of acanthocephalan anchoring is currently limited. In principle, however, musculature and other decay-prone structures can fossilize under certain conditions (Parry et al. 2018). In addition, most of the aforementioned muscles are visible in total preparations of acanthocephalans (e.g., Fig. 5a, b in Herlyn and Taraschewski 2017). Accordingly, the one or other of the aforementioned muscles might shine through the body wall of the yet to be discovered fossilized thorny-headed worm, comparable to the presumed alimentary tract in *I. fellatus* (see Fig. 2a in Cong et al. 2017). Thus, if the fossil of a suspected endoparasite ever discloses details of its internal organization, then a broader longitudinal strand extending through the anterior body section (presoma) could correspond to the proboscis retractor of an acanthocephalan. Furthermore, strands with smaller diameter that extend through the foretrunk (anterior portion of metasoma) could represent the receptacle retractor and/or neck retractor of an acanthocephalan. In addition, a bag-like structure between a smaller (presomal) and a wider (metasomal) body cavity could testify to the muscular apparatus that once suspended the cerebral ganglion of a palaeacanthocephalan, eoacanthocephalan, gigantorhynchid, moniliformid, or oligacanthorhynchid. But if such a septum is not visible in the hypothetical fossil and the outer contour approximates a balloon with a worm-like appendix, the remain could originate from an apororhynchid.

8.4.3 Presomal Sensory Organs

Whether involved in the eversion of the proboscis or not, the muscular apparatus always carries the cerebral ganglion (Fig. 8.4a). Obviously, the acanthocephalan cerebral ganglion should not only steer muscular activity but also integrate incoming signals. In some acanthocephalans, such information will come from comparably prominent sense or sensory organs at the base of the neck (lateral sensory organs) and the proboscis apex (apical sensory organs). These organs reside just beneath the bottom of circular pits shaped by tapering tegument, which resemble pores (Gee 1987). Depending on the taxon, these pits can reside on the tips of conical elevations of the presomal tegument, comparable to the crater of a volcano (Fig. 8.4c, d). While this may sound quite complicated, things are getting easier with respect to the sensory organs themselves. In fact, lateral and apical sensory organs are identical in their general organization: They are essentially terminal bulbous swellings of the processes of a so-called support cell (which is a syncytium) into which dendritic endings are imbedded (Fig. 8.4b–d). Proximally, the dendritic differentiations leave the bulbs and unite to nerves that extend to the cerebral ganglion (e.g., Harada 1931; Gee 1987; Herlyn et al. 2001).

Although their occurrence is widespread, these sensory organs are not present in all extant acanthocephalan species. According to the available data, extant acanthocephalans possess either two apical sensory organs in addition to a pair of lateral sensory organs (Archiacanthocephala: Gigantorhynchida, Moniliformida) or one apical organ plus two lateral sensory organs (Archiacanthocephala: Oligacanthorhynchida) or two lateral sensory organs only (probably all eoacanthocephalans and some palaeacanthocephalans). The last alternative is the absence of any sensory organs of the described type (other palaeacanthocephalans) (e.g., Gee 1987; survey in Herlyn et al. 2001). How apororhynchids fit into the picture remains to be elucidated. Nonetheless, the available data suggest that a duplication of the presomal sensory apparatus occurred in archiacanthocephalans, namely from a state of only two lateral sensory organs (Fig. 8.4b) to the formation of two lateral plus two apical sensory organs (Fig. 8.4c). Once established, both apical sensory organs seem to have fused to an unpaired structure in the stem line of crown-Oligacanthorhynchida (Fig. 8.4d; Weber et al. 2013).

Here, too, the practical value of this morphological description can only arise if a fossil endoparasite with corresponding soft tissue preservation will ever be discovered. Nonetheless, the fossilization of decay-prone structures such as the nervous system is possible (Parry et al. 2018). Thus, if a fossil endoparasite would reveal subtegumental bulbous differentiations at the transition from a smaller body section (neck) to a broader body section (metasoma) and additionally at the presumed anterior tip, the structures could represent presomal sensory organs of an archiacanthocephalan (Moniliformida, Gigantorhynchida, Oligacanthorhynchida). However, if such bulbs would only be discernible at the base of the neck, the worm could rather have been an eoacanthocephalan or paleacanthocephalan.

8.4.4 Lack of an Intestinal Tract

In addition to the typical outer contour, a muscular septum between the presomal and metasomal body cavities (or between the corresponding parts of a singular body cavity) and sensory organs of the type described in the previous section, a fossilized crown-acanthocephalan should lack an intestinal tract. Indeed, the alimentary tract got reduced in the stem line of the taxon and acanthocephalans of all developmental stages take up nutrients exclusively via the surface (e.g., Near 2002), just as it is the case in tapeworms (compare Figs. 8.3 and 8.4; e.g., Goater et al. 2014). Obviously, the non-identification of a structure in a fossil does not necessarily imply that it was absent in the living animal. But if there should be evidence for a digestive tract in a fossil remain of a presumed endoparasitic worm, then the animal behind cannot have been a crown-acanthocephalan.

8.5 Pathological Manifestations of Infections with Acanthocephalans

The symptoms of humans suffering from infections with acanthocephalans have been studied in a self-experiment: Three weeks after swallowing *M. moniliformis* cystacanths, which before were excised from beetles (*B. mucronata*), increasing amounts of eggs appeared in the stool of S. Calandruccio. He presented with "severe abdominal pain, sometimes perceived as tearing and increased by pressure on the aching spot, and here and there some diarrhea, strong buzzing in the ears (later in the whole head), as well as great fatigue and flaccidity" (translated from German after Grassi and Calandruccio 1888). Additional symptoms reported for macracanthorhynchiasis and moniliformiasis are poor appetite, perianal itching, besides the appearance of worms in the stool. Consequently, the mostly young patients, who usually become infected when they take in an intermediate host, present to a doctor quite quickly (e.g., Sahar et al. 2006; Mathison et al. 2016).

Intensities of hundreds or even >1,500 thorny-headed worms per individual host, as reported for teleost fishes and birds, can cause life-threatening constipation (Wurmbach 1937; Perry 1942; Sanford 1978). The lives of gnathostome hosts may be additionally threatened if acanthocephalans penetrate the intestinal wall, enter the body cavity, other organs or mesenteries and thus elicit peritonitis (Choi et al.

2010). But even if death is not brought about, such high intensities can weaken gnathostome hosts, as primarily studied in fishes. In particular, the damaging of the intestinal wall by the activity of the proboscis (Hammond 1966; Herlyn and Taraschewski 2017 and references therein; also Aguiar et al. 2018) can cause inflammations, lesions and necroses and thus a decrease of the intact absorptive surface (Taraschewski et al. 1989; de Matos et al. 2017; Jerônimo et al. 2017). Additionally, acanthocephalans absorb components from disintegrating host tissue and infiltrating blood as well as nutrients from the intestinal contents, which are no longer available to the host (Taraschewski and Mackenstedt 1991a, b; Sures 2002; see also Sures et al. 2000). Depending on the acanthocephalan species, it also belongs to the normal attachment behavior that an acanthocephalan pierces the intestinal wall with the proboscis, whereupon a nodule forms toward the body cavity which encloses the anterior-most body section of the worm (Wurmbach 1937; Taraschewski 2000; Dezfuli et al. 2015).

Considering the different ways of how acanthocephalans can damage vertebrates it may surprise that fishes can tolerate high intensities of infection (e.g., Đikanović et al. 2010). Nonetheless, the pathological manifestations outlined above can negatively affect their growth rate, general condition, and survival rate (Martins et al. 2001; Malta et al. 2001; Jerônimo et al. 2017). In addition, growing fishes can develop spinal deformations due to the reduced mineral availability resulting from intense infections with acanthocephalans (Silva-Gomes et al. 2017). Consequently, acanthocephalans could have contributed to spinal deformations such as shown by some remains of Miocene killifish (Teleostei, Cyprinodontidae) from Kenya (Altner and Reichenbacher 2015). Another indirect indication of infection could be the presence of the aforementioned nodules on the outer surface of intestinal remains of fossil Teleostei, Elasmobranchii, Ichthyosauria etc.

8.6 Phylogenetic Relationships of Acanthocephala and Taxonomic Implications

Penis worms (Priapulida) long belonged to the circle of candidate taxa for the acanthocephalan sister-group due to similarities in body organization and other morphological features (Conway Morris and Crompton 1982). For priapulids share with acanthocephalans the presence of hooks or scalids, as they are called in priapulids, in the anterior body section (Habdija et al. 2011). However, it is meanwhile quite certain that Priapulida do not belong to the closer relatives of Acanthocephala, which rather have a nested position inside Gnathifera (Ahlrichs 1997; Witek et al. 2009; Fröbius and Funch 2016). The taxon name refers to the evolutionary novelty of jaw-like solid-parts inside the pharynx (Ahlrichs 1997) as they are present in Gnathostomulida, Micrognathozoa, and already-mentioned wheel animals or Rotifera. The jaw-like elements serve in food uptake, whereby rod-like elements with a characteristic substructure seem to provide flexibility to the "jaws" (Rieger and Tyler 1995; Herlyn and Ehlers 1997; Kristensen and Funch 2000). The rods again might be homologous to the grasping spines in arrow worms (Chaetognatha; Shu et al. 2017), and, indeed, arrow worms currently appear to be close relatives of Gnathifera (Fig. 8.7; Fröbius and Funch 2016). Alternatively, arrow worms may branch off within Gnathifera, then as a sister to a clade comprised of micrognathozoans, rotifers and acanthocephalans (Marlétaz et al. 2019). In any case, the naming after jaw-like elements (ἡ γνάθος) is a recurrent theme in the kinship circle of Gnathifera. Yet, the eponymous differentiations obviously got lost in the stem line of crown-Acanthocephala, along with the reduction of the alimentary tract (e.g., Conway Morris and Crompton 1982)—which has the noteworthy consequence that acanthocephalans are jaw-less members of a clade, which is named after jaw-like solid-parts (Gnathifera). To top it all, these jaw-less members of Gnathifera parasitize jawed arthropods (Mandibulata) and jawed vertebrates (Gnathostomata).

Inside Gnathifera, the already-mentioned rotifers or wheel animals represent the next phylogenetic relatives of acanthocephalans amongst the extant species (Fig. 8.7). Such a relationship was already expected by von Haffner (1950) and others (Rieger and Tyler 1995; see also Conway Morris and Crompton 1982), and in the meantime gained support from molecular analyses (e.g., Mark Welch 2000; Herlyn et al. 2003; Struck et al. 2014). Inspired by the presumed evolutionary novelty of a syncytial organization of the tegument, the name Syndermata was introduced for the Rotifera-Acanthocephala clade (Ahlrichs 1997). However, while some authors use the newly introduced name, others preferentially regard acanthocephalans as highly derived rotifers (e.g., Mark Welch 2000; García-Varela and Nadler 2006). In fact, the naming of the taxon heated some authors in a surprising way although there is wide agreement on the decisive point, i.e., the monophyletic origin of Acanthocephala and the three traditional rotifer taxa Monogononta, Bdelloidea, and Seisonidea (also Seisonacea, Seisonidae) (e.g., Nielsen 2012). Apart from its syncytial organization, the tegument is special in all members of the Acanthocephala-Rotifera group, by having an intra-syncytial lamina—a fact mentioned above with respect to the preservation potential of the acanthocephalan tegument. Another evolutionary novelty of the Acanthocephala-Rotifera group seems to be that the distal plasma membrane of the tegument shapes crypt-like infoldings (Ahlrichs 1997; Near 2002). These infoldings increase the surface of the tegument, which should be of special relevance for nutrient uptake via surface in gut-less acanthocephalans (Graeber and Storch 1978). Members of the Rotifera-Acanthocephala clade are also specific with respect to sperm ultrastructure, as far as spermatozoa are produced at all (only females are known for bdelloid rotifers). In particular, the flagellum inserts at the anterior pole of the sperm head, instead of at its rear end (Ahlrichs 1997; Ferraguti and Melone 1999).

Molecular studies suggest that Rotifera in the traditional understanding (Monogononta, Bdelloidea, Seisonidea) represents a paraphyletic assemblage. In particular, bdelloids appear to be closer related to acanthocephalans than to monogononts (e.g., Near et al. 1998; Near 2002; García-Varela and Nadler 2006; Witek et al. 2008). Even closer related to acanthocephalans could be seisonids (Fig. 8.7). A seisonid-acanthocephalan sister-group relationship gains support from part of the molecular studies, whereby the choice of the substitution model seems to be

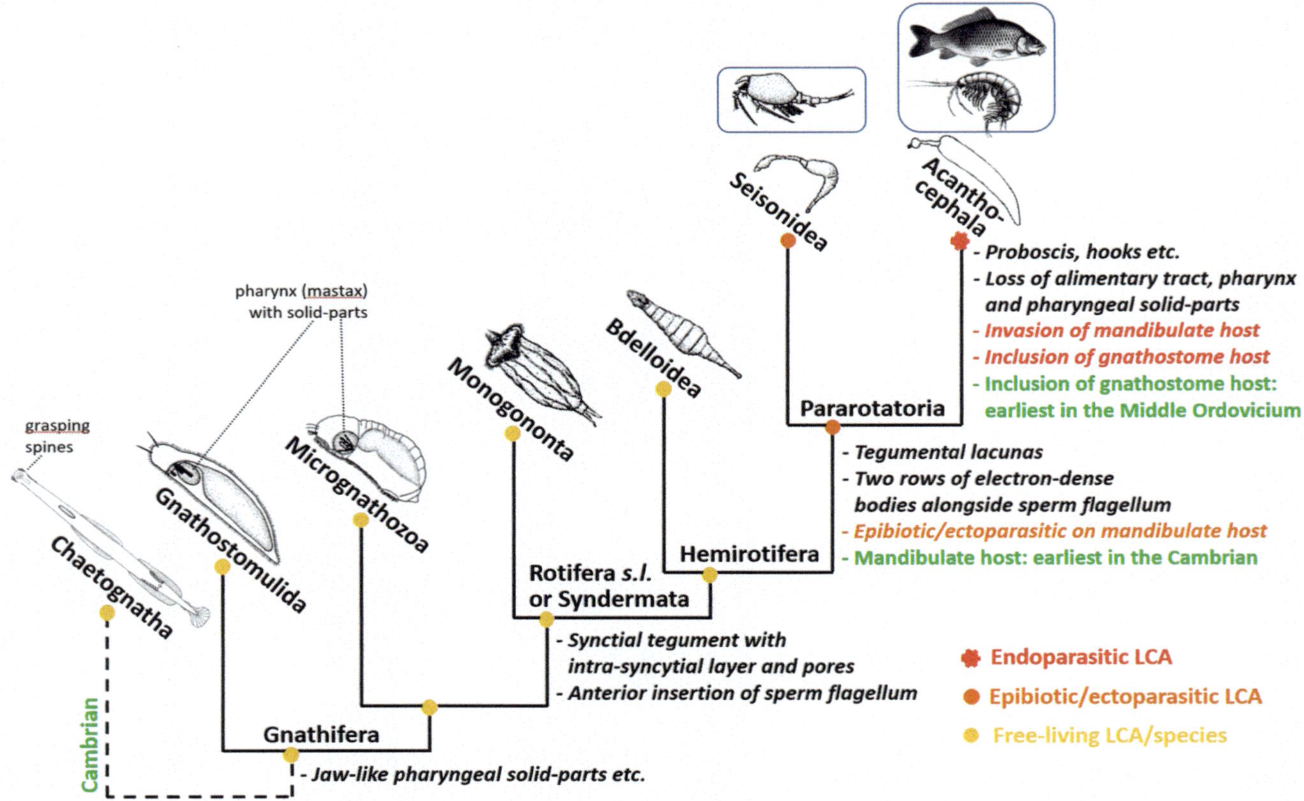

Fig. 8.7 Evolution of the endoparasitic two-host cycle of crown-Acanthocephala via an epibiotic/ectoparasitic stage. Italics give evolutionary novelties. Changes in lifestyle and host usage are highlighted by orange and red labelling, respectively. Colored symbols at nodes and leaves of the tree give presumed lifestyles of the last common ancestor (LCA) of the corresponding taxa each. Time estimates in green refer to fossil members of arrow worms or Chaetognatha (according to Shu et al. 2017; Briggs and Caron 2017), and to the onset of the fossil record of potential hosts, i.e., mandibulate arthropods and gnathostome vertebrates (according to Sansom et al. 2015; Daley et al. 2018; see also Janvier 2003). Extant host species are boxed. The phylogenetic tree shown combines results from Ahlrichs (1997, 1998), Herlyn and Ehlers (1997), Ferraguti and Melone (1999), Herlyn et al. (2003), Witek et al. (2008, 2009), Wey-Fabrizius et al. (2014), Sielaff et al. (2016), and Fröbius and Funch (2016). Chaetognatha may alternatively be sister to a monophylum comprised of Micrognathozoa and Syndermata/Rotifera (Marlétaz et al. 2019) or to Syndermata/Rotifera alone (Vinther and Parry 2019). However, both alternatives would have no influence on the proposed stepwise establishment of an endoparasitic two-host cycle on the lineage to crown-acanthocephalans via an epibiotic or ectoparasitic stage (as retained in seisonids), invasion of mandibulates, and upward inclusion of gnathostome hosts. Sketches of animals after Ahlrichs (1997), Sielaff et al. (2016), and Kovatkina (2016).

crucial for its recognition (Wey-Fabrizius et al. 2014; Supplementary Figure S1 to Laumer et al. 2015; also Herlyn et al. 2003). Mitochondrial gene order additionally accords with a monophyletic origin of Seisonidea and Acanthocephala (Sielaff et al. 2016). A grouping of Seisonidea and Acanthocephala (Pararotatoria) is further in line with morphological data, whereby some evolutionary novelties refer to the tegument, again. In particular, the already-mentioned infoldings of the distal plasma membrane widen inside the tegument to larger caverns in seisonids and acanthocephalans. This characteristic has obviously undergone expansion in the stem line of crown-Acanthocephala, towards the already mentioned lacunar system (Fig. 8.6c). The tegument of seisonids and acanthocephalans is further distinguished by containing larger filament bundles (Ahlrichs 1997). The third evolutionary novelty in support of monophyletic Seisonidea-Acanthocephala relates to ultrastructural details of the spermatozoa again, which in both taxa have two rows of electron-dark bodies that accompany the anterior portion of the sperm flagellum (Ahlrichs 1997, 1998). Furthermore, there is considerable similarity in the cytomorphology of spermatogenesis states between seisonids and acanthocephalans (Marchand and Mattei 1976; Ferraguti and Melone 1999). In contrast, there seems to be no morphological feature that could represent an evolutionary novelty of a clade comprising acanthocephalans and bdelloids only (Ricci 1998), although such grouping re-occurs in part of the molecular analyses (e.g., García-Varela and Nadler 2006). Summing up all evidence, a plausible tree topology for Syndermata or Rotifera (inclusively Acanthocephala) appears to be: (Monogononta, (Bdelloidea, (Seisonidea, Acanthocephala))) (Fig. 8.7). This phylogenetic hypothesis provides the backbone for the inference of a scenario for the evolution of the endoparasitic two-host cycle in the stem line of crown-Acanthocephala in the next section.

8.7 Evolution of Acanthocephalan Endoparasitism: A Conditional Hypothesis

Given the life styles of extant species, the last common ancestors (LCAs) of crown-Gnathostomulida, crown-Micrognathozoa, crown-Monogononta, and crown-Bdelloidea were most probably free-living (e.g., Near et al. 1998; Kristensen and Funch 2000; Sterrer and Sørensen 2015). The same should apply to the LCAs of crown-Gnathifera and crown-Rotifera/Syndermata. The LCA of crown-Hemirotifera, from which Bdelloidea, Seisonidea, and Acanthocephala evolved, should also have been free-living. Lifestyles of monogonont, bdelloid and seisonid LCAs may be regarded as semi-sessile (compare Ahlrichs and Riemann 2019) but is herein referred to as free-living because the animals can easily detach from their substrates. However, while the substrate of monogononts and bdelloids is variable, the LCA of seisonids probably lived on mandibulates, possibly also from such hosts, just as the extant species of the group are doing or are assumed to do (e.g., Sørensen et al. 2005; see also Fontaneto and de Smet 2015). Lastly, the endoparasitic two-host cycle of all extant Acanthocephala should be a heritage of their LCA.

When matching the character states presumed for single LCAs with the phylogenetic tree in Fig. 8.7, it is the most parsimonious to postulate a shift from free-living to an epibiotic (epizoic) or ectoparasitic lifestyle on mandibulate arthropods for the branch uniting Seisonidea and Acanthocephala. Presumably, a second shift in the way of living followed on the acanthocephalan stem line, namely towards an endoparasitic lifestyle with mandibulates as hosts. In a third step, the acanthocephalan life cycle was expanded by upward-inclusion of gnathostomes as hosts (Conway Morris and Crompton 1982; Herlyn et al. 2003; Wey-Fabrizius et al. 2014; Sielaff et al. 2016). If the steps towards the two-host cycle of crown-Acanthocephala involved several host species or only a single one, each, has to remain unanswered. Nonetheless, several up to many individuals of at least one host species should have been the ground on which each of the evolutionary steps took place. In this sense, at least, the use of plural seems appropriate in respect to acanthocephalan hosts.

The stepwise establishment of a two-host cycle should have taken place in an aquatic environment. This is indeed very likely considering that extant chaetognaths, gnathostomulids, micrognathozoans, monogononts, bdelloids, and seisonids live in aquatic environments and many acanthocephalans use aquatic hosts. In turn, terrestrial life cycles as in extant archiacanthocephalans and some palaeacanthocephalan species should reflect secondary changes (Near et al. 1998). Yet, it is less clear whether the LCA of crown-acanthocephalans used marine or freshwater species as hosts since both alternatives occur in extant thorny-headed worms and their closer phylogenetic relatives (Ax 2001; Petrochenko 1956, 1958; Kristensen and Funch 2000; Fontaneto and de Smet 2015; Sterrer and Sørensen 2015). However, if seisonids retained not only an epibiotic/ectoparasitic lifestyle from their LCA with acanthocephalans but also the exploitation of marine crustaceans, acanthocephalan evolution should also have begun in a marine environment.

The establishment of the acanthocephalan two-host cycle was evidently accompanied by the evolution of several morphological novelties such as an invertible proboscis, a particular muscular apparatus suspending the cerebral ganglion, absence of an alimentary tract, a re-organized tegument and specialties in the reproductive systems of both sexes (Dezfuli et al. 2001; Near 2002; Herlyn and Röhrig 2003; Herlyn and Taraschewski 2017, etc.). Early stem-acanthocephalans, however, should have displayed the plesiomorphic alternatives. In particular, they presumably did not reach the body sizes known from extant species (see, e.g., compilation in Petrochenko 1956, 1958). According to body sizes in extant gnathostomulids, micrognathozoans, monogononts, bdelloids and seisonids, adults of ancient thorny-headed worms have presumably measured in the range of less than one to a few limiting their fossilisation potential. Subsequent increases of body size along with the evolution of gigantism in "invertebrate" host lineages can not be ruled out (Klug et al. 2015). However, if evolution took such a path it unlikely happened in the stem line of crown-acanthocephalans. With respect to this lineage, the presumed upward-inclusion of gnathostomes into the life cycle more likely paved the way for larger body sizes. In extension of this argument, it was found for extant acanthocephalans that size is positively correlated with body mass of their vertebrate hosts (Poulin et al. 2003). Another positive correlate was the temperature regime imposed by the host. In particular, adult acanthocephalans were found to grow to larger body sizes in endothermic than ectothermic hosts. It is to be said, however, that both effects largely disappeared when correlation analyses included a correction against possible phylogenetic bias.

8.8 Acanthocephala and Gnathifera: Fossil Report and Time Line

The following estimates of the time line of gnathiferan and especially acanthocephalan evolution have as a condition that the phylogenetic and temporal assignment of fossils is at least approximately correct. This especially applies to the dating of the earliest appearance of potential hosts (for a discussion, see De Baets and Littlewood 2015; Warnock and Engelstädter 2021). In Gnathifera, another obstacle is that fossil evidence is comparably sparse. In fact, there seems to be only one report on fossilized eggs attributed to Acanthocephala, from the Upper Cretaceous (see above). All other ancient acanthocephalan eggs are remains of several hundred to about 12,000 years (Table 8.1). In addition, only few fossil wheel animals have been found so far, i.e., a monogonont from Eocene North Maslin Sands in South Australia (Southcott and Lange 1971) and bdelloids in Dominican amber from the Miocene (Poinar and Ricci 1992; Waggoner and Poinar 1993; Iturralde-Vinent and MacPhee 1996). However, Eocene and Miocene fossils almost certainly do not shed light on the emergence of Rotifera-Acanthocephala. In fact, if the already-mentioned Cambrian species *I. fellatus* really belongs to Gnathifera, the stem line of Rotifera-Acanthocephala may go back to the Early Palaeozoic (Cong et al. 2017). An Early Paleozoic origin of Gnathifera and Rotifera-Acanthocephala would receive additional confirmation if arrow worms (Chaetognatha) really are sister to or occupy a nested position within Gnathifera (Fig. 8.7; Fröbius and Funch 2016; Marlétaz et al. 2019; Vinther and Parry 2019). Thus, well preserved fossils of arrow worms are documented from the Cambrian Chengjiang Lagerstätte in China and from the Burgess Shale in Canada (Shu et al. 2017; Briggs and Caron 2017), and a potential stem chaetognath, *Amiskwia sagittiformis*, also lived in the Cambrian (Vinther and Parry 2019).

Due to the lack of direct evidence, time estimates regarding acanthocephalan evolution have to rely on the appearance of mandibulate and gnathostome hosts in the fossil record (compare De Baets et al. 2015). In particular, the presumed ancestors of Seisonidea and Acanthocephala should not have lived on mandibulates prior to the emergence of such hosts in the Cambrian (Daley et al. 2018). Probably, these hosts had a crustacean-like appearance or were crustaceans (see Zhang and Pratt 2012; Harvey et al. 2012). It might even be possible to narrow down the spectrum of first hosts to single taxa within Crustacea. Thus, extant seisonids live on Phyllocarida (Crustacea), especially on Leptostraca (Fontaneto and de Smet 2015; see also Sørensen et al. 2005), which emerged in the Permian according to the present knowledge. However, leptostracans very much resemble Cambrian-Carboniferous phyllocarids collectively called Archaeostraca (Collette and

Hagadorn 2010). Consequently, ancient seisonids may already have lived on phyllocarids in the Paleozoic, and they might have retained this host usage from their LCA with acanthocephalans, just as they presumably kept the epibiotic/ectoparasitic lifestyle. If so, first stem-acanthocephalans should have exploited phyllocarids as well. But parasitization of other Paleozoic "invertebrates" by early acanthocephalans can also not be ruled out. Among them might have been species of Trilobita, the potential sister group of crown-Mandibulata (Scholtz and Edgecombe 2006).

Whether only phyllocarids or (also) other mandibulates or even trilobites were exploited in the beginning of acanthocephalan evolution will probably remain elusive. It will also be difficult if not impossible to assess more precisely when the presumed shift from an ecto- to an endoparasitic lifestyle might have taken place in acanthocephalan evolution. However, the establishment of a two-host cycle should post-date the emergence of fish-like gnathostomes in the Middle Ordovician (Sansom et al. 2015; also Janvier 2003) or later (Friedman and Sallan 2012; Brazeau and Friedman 2015; Klug et al. 2017). In case that early acanthocephalans also used fish-like vertebrates without jaws as definitive hosts, the two-host cycle could have been established even earlier. Such possibility can not be ruled out since extant acanthocephalans were occasionally reported from lampreys (Petromyzontida) (Petrochenko 1956; Conway Morris and Crompton 1982). In any case, acanthocephalan diversity might subsequently have increased along with the diversification of crown-Gnathostomata upon extinction of placoderms in the Upper Devonian (Trinajstic et al. 2007; Sansom et al. 2015). When conquering new hosts, acanthocephalans may have benefited from generally less tight bonds to definitive than intermediate hosts (Conway Morris and Crompton 1982; Parker et al. 2015).

The emergence of the individual gnathostome taxa used as hosts provides an approximate orientation for the earliest possible origin of individual lineages within Acanthocephala. Thereby, life cycles involving tetrapods should have evolved from cycles with fish-like gnathostomes (Near 2002). Although not necessarily representing a suitable model for such transitions, it is worth noting that some of the extant acanthocephalan species exploit an aquatic intermediate host and a terrestrial definitive host (e.g., Dezfuli and Giari 1999). Either way, most of the extant species retained the ancestral condition of an aquatic cycle. Thus, life cycles in extant eoacanthocephalans involve sharks (Elasmobranchii, Selachii) and ray-finned fishes (Actinopterygii), in particular bowfin (Amiiformes) and teleost fishes, besides turtles (Petrochenko 1956; Near et al. 1998). Polyacanthocephalans also have aquatic life cycles, with teleost fishes and caimans (Crocodilia) serving as gnathostome hosts (Amin 1987; Echi et al. 2015). Since Polyacanthocephala either has a nested position inside Eoacanthocephala or is sister to Eoacanthocephala (Verweyen et al. 2011; Echi et al. 2015; Gazi et al. 2016), the LCAs of both taxa might already have used fish-like gnathostomes in the Middle Ordovician or later (see above, for references). Under the renewed assumption that the exploitation of fish-like gnathostomes represents the ancestral state, the origin of the palaecanthocephalan stem line could also go back to the Paleozoic. In fact, extant palaeacanthocephalans infect diverse fish-like gnathostomes, especially sharks and rays (Elasmobranchii,

Batoidea), sturgeons (Acipenseriformes), and bowfin and teleost fishes in addition to Amphibia, Sauropsida and Mammalia (Petrochenko 1956, 1958; Near et al. 1998). As mentioned above, the taxon Archiacanthocephala is specific because the extant species have life cycles with terrestrial mammals and birds as definitive hosts (e.g., Near et al. 1998). If this definitive host spectrum reflects an association of early archiacanthocephalans with stem-Amniota, the transition to the usage of terrestrial definitive hosts should not have occurred much earlier than about 346–358 million years ago, which approximately marks the divergence of Amphibia and Amniota. However, if early archiacanthocephalans had originally exploited solely mammals and later on additionally conquered birds, the transition to a terrestrial cycle should postdate the divergence of Sauropsida and Mammalia 297–326 million years ago (split estimates according to timetree.org). The age of fossil archiacanthocephalan eggs obtained from an Upper Cretaceous coprolite (Cardia et al. 2019) is in line with both possibilities.

Above estimates on the emergence of individual taxa contrast to some degree with phylogenetic reconstructions suggesting a closer relationship of Eoacanthocephala (inclusively Polyacanthocephala) to Palaeacanthocephala than Archiacanthocephala (e.g., García-Varela and Nadler 2005; Verweyen et al. 2011), thus giving the following tree: (Archiacanthocephala, (Palaeacanthocephala, Eoacanthocephala)). In fact, such a phylogeny implicates that the stem line of archiacanthocephalans goes back further in time than the stem lines of palaeacanthocephalans and eoacanthocephalans. However, archiacanthocephalans could have obtained an aquatic life cycle long before a terrestrial cycle was established. In any case, some inconsistency between the distribution of traits and phylogenetic trees is not uncommon in acanthocephalan research: The presence or absence of lateral sensory organs (see above), for example, can only be aligned with the aforementioned phylogeny if one assumes their secondary loss within Palaeacanthocephala (Weber et al. 2013). The complex situation is also reflected in the naming of the three major acanthocephalan taxa: The prefixes palae- (old, ancient or primitive), eo- (earliest), and archi- (primary) all express the view that the respective taxon combines to a remarkable extent ancient characters (compare Meyer 1932; Van Cleave 1936).

8.9 Cambroclavida: Microfossils of Questionable Acanthocephalan Affiliation

Certain Cambrian sclerites were repeatedly regarded as acanthocephalan remains (Qian and Yin 1984; Amin 2013). These fossils are collectively referred to as Cambroclavida and actually are of unclear phylogenetic affiliation (Clausen and Álvaro 2006; Kouchinsky et al. 2012). The microfossils were discovered in peri-Gondwanan deposits of China and Europe, amongst others (e.g., Elicki and Wotte 2003). Part of them is remotely reminiscent of the hooks of extant acanthocephalans (Fig. 8.5): They show hook-like recesses that emerge from a basis that might appear similar to the root of an acanthocephalan hook. On the other hand, the basis of the cambroclavid sclerites is hollow (Fig. 8.5d), which is not the case in the roots of acanthocephalan hooks (Fig. 8.6a, b). Moreover, the hooks of extant acanthocephalans are imbedded into the basement membrane underlying the tegument (Fig. 8.6a, b). In addition, even if the fibers of basement membrane and hook roots should be discontinuous (c.f. Taraschewski et al. 1989), it remains uncertain whether fossil remains of acanthocephalan hooks would present as isolated units as it is usually the case in cambroclavid microfossils. Furthermore, hooks of extant acanthocephalan species are more or less recurved, while this is not the case in at least part of the cambroclavid microfossils (Figs. 8.5 and 8.6a). Most of all, the size of cambroclavid fossils does not accord with the expectation for early acanthocephalans, which should have remained much smaller than the extant species (see above). For example, the maximum extension of *Cambroclaves* fossils can easily reach >500 μm (Elicki and Wotte 2003). Recalling that early stem-acanthocephalans should have measured in the range of one or few millimeters, hooks of cambroclavid dimension would appear huge. Not least, the question arises why acanthocephalan hooks—if cambroclavid microfossils are such—occur in peri-Gondwanan deposits, whereas they have not been found in any other context so far. Thus, a closer affinity of cambroclavids to acanthocephalans seems unlikely at present.

8.10 Conclusions

Analyses of molecular and morphological data have shown that the taxon Acanthocephala (thorny-headed worms) has a nested position inside Gnathifera, a clade that also includes Gnathostomulida, Micrognathozoa and Rotifera. Especially, Rotifera appears to be a paraphyletic assemblage as long as Acanthocephala is excluded (Fig. 8.7). In addition, arrow worms (Chaetognatha) seem to belong to the kinship of the Gnathifera. In support of this possibility, recent studies suggest that arrow worms, for which Cambrian fossils are known (Shu et al. 2017; Briggs and Caron 2017), are either sister to Gnathifera or occupy a nested position within the gnathiferan clade (Fröbius and Funch 2016; Marlétaz et al. 2019; Vinther and Parry 2019). Beyond that, Cambrian fossils have been attributed to Gnathifera (Caron and Cheung 2019; Vinther and Parry 2019), some of which might even have been epibionts or ectoparasites (Cong et al. 2017). Accordingly, a Cambrian origin of Gnathifera is likely.

With regard to the Rotifera-Acanthocephala group, the temporal origin is less clear. In fact, the oldest known fossils of Rotifera (inclusively Acanthocephala) or Syndermata, as the group is also called, are acanthocephalan eggs from an Upper Cretaceous coprolite (Cardia et al. 2019). Before this recently published finding, only few remains of monogonont and bdelloid rotifers from the Eocene and Miocene were known (Southcott and Lange 1971; Poinar and Ricci 1992; Waggoner and Poinar 1993; Iturralde-Vinent and MacPhee 1996). On the other hand, estimates for the appearance of mandibulates and gnathostomes enable rough time constraints for the earliest possible associations with members of these taxa. In particular, the postulated first epibiotic or ectoparasitic association with mandibulates in the common stem line of seisonids and acanthocephalans can not have occurred prior to the emergence of jawed arthropods in the Cambrian (Daley et al. 2018). Such a one-host cycle was probably passed on to the seisonid and acanthocephalan lineages, followed by a change from living *on* to living *in* mandibulates in acanthocephalan evolution (Herlyn et al. 2003; Wey-Fabrizius et al. 2014; Sielaff et al. 2016). Likewise, the presumed upward-inclusion of gnathostomes into the acanthocephalan life cycle should not have occurred prior to the emergence of corresponding hosts in the Middle Ordovician (Sansom et al. 2015; also Janvier 2003) or later (Brazeau and Friedman 2015; Klug et al. 2017). Although we cannot be sure whether evolution has taken the path outlined, the following appears to be more certain: The LCA of crown-acanthocephalans probably showed an obligate two-host cycle involving mandibulates and gnathostomes as intermediate and definitive hosts, respectively (Fig. 8.1). Extensions of this two-host cycle by paratenic and second definitive hosts could have occurred subsequently.

The presumed one-host-cycle in early acanthocephalan evolution implicates that adult worms should have differed considerably with respect to morphology, when compared to the adults in extant species. In particular, early acanthocephalans should not have grown to body sizes as known from extant species. A marked increase in body size rather followed the upward-inclusion of gnathostomes as hosts. Several other evolutionary novelties should also have evolved along with the two-host cycle. Especially, metamorphosis of the larval stage inside the mandibulate intermediate host (acanthor) to a young adult (acanthella) is obviously a developmental correlate of the two-host cycle (compare Meyer 1932). A hooked proboscis and a muscular apparatus suspending the cerebral ganglion (receptacle and receptacle-surrounding muscle) likely evolved in the same context. Likewise, traits that are related to an increase in fecundity (large testes, fragmented ovaries, uterine bell, etc.) should have emerged in the stem line of crown-acanthocephalans, along with the establishment of a two-host cycle (Herlyn and Röhrig 2003; Poulin and Morand 2000; Parker et al. 2015). However, there might also be characters in extant acanthocephalans that already existed in the supposed one-host stage (Sielaff et al. 2016). In particular, a digestive tract might then already have been lacking as suggested by its absence in all developmental stages of the extant species (compare Near et al. 1998; Wey-Fabrizius et al. 2014). Correspondingly, morphological and physiological changes that enable nutrient uptake via the tegument at least in part occurred prior to the establishment of a two-host cycle (Mauer et al. 2020).

Eggs are the only free propagules in the life cycles of the extant acanthocephalan species (Figs. 8.1 and 8.2). They are also the sole ancient remains of acanthocephalans known to date. This probably reflects their enhanced preservability due to the incorporation of keratin and, depending on the taxon, chitin (Whitfield 1973; Peters et al. 1991; Taraschewski and Peters 1992; Taraschewski et al. 1992). The ancient eggs discovered so far have most likely an archiacanthocephalan origin, as suggested by their size and the increased thickness and structure of their shells (Table 8.1 and references therein). In most of the cases, the eggs were retrieved from human, carnivoran and xenarthran coprolites of several hundred to about 12,000 years. However, there seems to be no reason why gnathostome vertebrates feeding on intermediate, paratenic, or definitive hosts should not have been infected tens or hundreds of thousands or millions of years ago. In line with this, a coprolite from the Upper Cretaceous of Brazil was recently found to contain remains reminiscent of archiacanthocephalan eggs (Cardia et al. 2019). The defecating animal might have been a member of Crocodyliformes but various extinct predators such as ichthyosaurs as well as taxa with extant species like sharks might also have been infected by ancient acanthocephalans.

Though nothing corresponding has been found or recognized so far, acanthocephalan hooks should be the prime candidates for preservation, besides eggs (Figs. 8.4, 8.5a, b and 8.6a, b). They are rather rigid and undergo sclerotization, as reported for some of the extant species (Taraschewski 1989a, b). The hooks might still be in position in the hypothetical ideal fossil of an acanthocephalan, such that their arrangement gives the contour of the proboscis. However, disaggregated acanthocephalan hooks might also be contained in the fossil record. If so, it will be anything but trivial to distinguish them from fossil hooks of other endoparasites (see, e.g., Plate IV in Lambl 1859). Either way, acanthocephalan hooks might once be detected in the abdomen and especially in the intestine of fossilized gnathostomes. In addition, presence of a structure covering the hind end of a fossil endoparasite could be an indication of a female acanthocephalan (Fig. 8.4a): Such a structure would correspond to the copulatory cap, which males of extant acanthocephalans produce from the secretion of their cement gland(s) (Dezfuli et al. 2001). Obviously, such excellent preservation of an acanthocephalan is not too likely.

The chance for preservation of acanthocephalan soft-tissue is certainly smaller than for comparably rigid structures such as eggs, hooks and copulatory cap. Still, it can not be ruled out, not even for the tegument. Actually, the tegument (also integument or epidermis) of extant acanthocephalans is rather resistant to mechanical damage and enzymatic digestion, which probably is due to its syncytial organization and a presumably proteinaceous lamina inside (e.g., Díaz Cosín 1972; Graeber and Storch 1978; Ahlrichs 1997; Herlyn and Ehlers 2001). Provided that a corresponding remain will ever be found, an increase in body diameter behind the hooked attachment organ could indicate the presoma-metasoma transition. The ideal acanthocephalan fossil might also show a tegument cone (apical epidermis cone) in the anterior proboscis section as known from extant members of Eoacanthocephala and Polyacanthocephala (Figs. 8.4b, 8.5b and 8.6a). Subtegumental sensory structures of the presoma or the muscular apparatus suspending the cerebral ganglion (receptacle plus receptacle protrusor) might also be discernable in the ideal, though still hypothetical, fossil of an acanthocephalan worm (Figs. 8.4, 8.5a and 8.6c, d). All this may seem unlikely, but the preservation of fragile structures such as muscles and the nervous system is possible per se, even in worm-like organisms (Parry et al. 2018).

Acknowledgements My thanks go to Dr. Kenneth De Baets and Dr. John Warren Huntley for giving the opportunity to contribute this chapter and for useful comments on earlier drafts. I am also grateful to Prof. Thomas Wotte (Freiberg, Germany) who made available the scanning electron micrographs of *Cambroclaves* fossils shown in Fig. 8.5. Not least, I thank the holders of rights who have agreed to the usage of various previously published illustrations.

References

Aguiar LS, de Oliveira MIB, de Matos LV, Gomes ALS, da Costa JI, da Silva GS (2018) Distribution of the acanthocephalan *Neoechinorhynchus buttnerae* and semiquantitative analysis of histopathological damage in the intestine of tambaqui (*Colossoma macropomum*). Parasitol Res 117:1689–1698

Ahlrichs WH (1997) Epidermal ultrastructure of *Seison nebaliae* and *Seison annulatus*, and a comparison of epidermal structures within the Gnathifera. Zoomorphology 117:41–48

Ahlrichs WH (1998) Spermatogenesis and ultrastructure of the spermatozoa of *Seison nebaliae* (Syndermata). Zoomorphology 118:255–261

Ahlrichs WH, Riemann O (2019) Seisonidae. In: Schmidt-Rhaesa A (ed) Handbook of zoology, Gastrotricha and Gnathifera. De Gruyter, Berlin

Altner M, Reichenbacher B (2015) †*Kenyaichthyidae* fam. nov. and †*Kenyaichthys* gen. nov. - first record of a fossil aplocheiloid killifish (Teleostei, Cyprinodontiformes). PLoS One 29:e0123056

Amin OM (1987) Key to the families and subfamilies of Acanthocephala, with the erection of a new class (Polyacanthocephala) and a new order (Polyacanthorhynchida). J Parasitol 73:1216–1219

Amin OM (2013) Classification of the Acanthocephala. Folia Parasitol 60:273–305

Amin OM, Heckmann R (2017) *Rhadinorhynchus oligospinosus* n. sp. (Acanthocephala, Rhadinorhynchidae) from mackerels in the Pacific Ocean off Peru and related rhadinorhynchids in the Pacific, with notes on metal analysis. Parasite 24:19

Amin OM, Heckman RA, Yudhanto S (2013) The finding of *Mediorhynchus gallinorum* (Acanthocephala: Gigantorhynchidae) in chickens from Indonesia, with expanded description using SEM. Comp Parasitol 80:39–46

Araújo A, Reinhard K, Ferreira LF (2015) Palaeoparasitology – human parasites in ancient material. Adv Parasitol 90:349–387

Ax P (2001) Das System der Metazoa III. Spektrum Akademischer Verlag, Heidelberg

Bakker TCM, Frommen JG, Thünken T (2017) Adaptive parasitic manipulation as exemplified by acanthocephalans. Ethology 2017:1–6

Beltrame MO, Fernández FJ, Sardella NH (2015) First paleoparasitological record of acanthocephalan eggs from Northwestern Patagonia (Late Holocene, Argentina). Acta Trop 146:33–35

Beltrame MO, Bellusci A, Fernández FJ, Sardella NH (2018) Carnivores as zoonotic parasite reservoirs in ancient times: the case of the Epullán Chica archaeological cave (Late Holocene, Northwestern Patagonia, Argentina). Archaeol Anthropol Sci 10:795–804

Brazeau MD, Friedman M (2015) The origin and early phylogenetic history of jawed vertebrates. Nature 520:490–497

Briggs DEG, Caron J-B (2017) A large Cambrian chaetognath with supernumerary grasping spines. Curr Biol 27:2536–2543

Camacho M, Pessanha T, Leles D, Dutra JMF, Silva R, de Souza SC, Araújo A (2013) Lutz's spontaneous sedimentation technique and the paleoparasitological analysis of sambaqui (shell mound) sediments. Mem Inst Oswaldo Cruz 108:155–159

Camacho M, Araújo A, Morrow J, Buikstra J, Reinhard K (2018) Recovering parasites from mummies and coprolites: an epidemiological approach. Parasit Vectors 11:248

Cardia DFF, Bertini RJ, Camossi LG, Letizio LA (2019) First record of Acanthocephala parasites eggs in coprolites preliminary assigned to Crocodyliformes from the Adamantina Formation (Bauru Group, Upper Cretaceous), São Paulo, Brazil. An Acad Bras Cienc 91(Suppl 2):e20170848

Caron J-B, Cheung B (2019) *Amiskwia* is a large Cambrian gnathiferan with complex gnathostomulid-like jaws. Commun Biol 2:164

Chin K (2021) Gastrointestinal parasites of ancient non-human vertebrates: evidence from coprolites and other materials. In: De Baets, K., Huntley, J. W. (eds). The evolution and fossil recordof Parasitism: Coevolution and paleoparasitological techniques. Topics in Geobiology 50

Choi C-J, Lee H-J, Go J-H, Park Y-K, Chai J-Y, Seo M (2010) Extraintestinal migration of *Centrorhynchus* sp. (Acanthocephala: Centrorhynchidae) in experimentally infected rats. Korean J Parasitol 48:139–143

Clausen S, Álvaro JJ (2006) Skeletonized microfossils from the Lower–Middle Cambrian transition of the Cantabrian Mountains, northern Spain. Acta Palaeontol Pol 51:223–238

Collette JH, Hagadorn JW (2010) Early evolution of phyllocarid arthropods: phylogeny and systematics of Cambrian–Devonian archaeostracans. J Paleontol 84:795–820

Cong P, Ma X, Williams M, Siveter DJ, Siveter DJ, Gabbot SE, Zhai D, Goral T, Edgecombe GD, Hou X (2017) Host-specific infestation in early Cambrian worms. Nat Ecol Evol 1:1465–1469

Conway Morris S, Crompton DWT (1982) The origins and evolution of the Acanthocephala. Biol Rev 57:85–115

Daley AC, Antcliffe JB, Drage HB, Pates S (2018) Early fossil record of Euarthropoda and the Cambrian explosion. Proc Natl Acad Sci U S A 115:5323–5331

Dartnall H (1995) The rotifers of Heard island: preliminary survey with notes on other freshwater groups. Pap Proc R Soc Tasmania 129:7–15

de Almeida Souza A, Sobestiansky J, Linhares GFC, de Oliveira WP, de Barros Araújo JL (2006) Surto de Macracantorrincose em queixada (*Tayassu pecari*) criado extensivamente nos arredores de Goiânia – Estado de Goiás. Patologia 34:213–220

De Baets K, Littlewood TJ (2015) The importance of fossils in understanding the evolution of parasites and their vectors. In: De Baets K, Littlewood TJ (eds) Advances in parasitology 90: fossil parasites. Academic, Heidelberg

De Baets K, Dentzien-Dias PC, Upeniece I, Verneau O, Donoghue PCJ (2015) Constraining the deep origin of parasitic flatworms and host-interactions with fossil evidence. Advances in Parasitology 90:93–135.

De Baets K, Dentzien-Dias P, Harrison GWM, Littlewood DTJ, Parry LA (2021) Fossil Constraints on the Timescale of Parasitic Helminth Evolution. In: De Baets K, Huntley JW (eds) The evolution and fossil Record of Parasitism: Identification and macroevolution of parasites. Topics in Geobiology 49

de Beauchamp P (1909) Recherches sur les Rotifères: les formations tégumentaires et l'appareil digestif. Arch Zool Exp Gén 10(4):1–410

de Matos LV, de Oliveira MIB, Gomes ALS, da Silva GS (2017) Morphological and histochemical changes associated with massive infection by *Neoechinorhynchus buttnerae* (Acanthocephala: Neoechinorhynchidae) in the farmed freshwater fish *Colossoma macropomum* Cuvier, 1818 from the Amazon State, Brazil. Parasitol Res 116:1029–1037

de Souza MV, Chame M, Mendonça de Souza SMF, Daltrini Felice G, Guidon N, Sianto L (2020) New parasite occurences in *Tamandua tetradactyla* (Pilosa: Myrmecophagidae) in the Northeast of Brazil: a paleoparasitological study. Oecologia autralis 24:141–153

Dentzien-Dias PC, Poinar G, de Figueiredo AE, Pacheco AC, Horn BL, Schultz CL (2013) Tapeworm eggs in a 270 million-year-old shark coprolite. PLoS One 8:e55007

Dezfuli BS, Giari L (1999) Amphipod intermediate host of *Polymorphus minutus* (Acanthocephala), parasite of water birds, with notes on ultrastructure of host-parasite interface. Folia Parasitol 46:117–122

Dezfuli BS, Simoni E, Mischiati C (2001) The cement apparatus of larval and adult *Acanthocephalus anguillae* (Acanthocephala), with notes on the copulatory cap and origin of gland secretion. Parasitol Res 87:299–305

Dezfuli BS, Bo T, Lorenzoni M, Shinn AP, Giari L (2015) Fine structure and cellular responses at the host–parasite interface in a range of fish–helminth systems. Vet Parasitol 208:272–279

Díaz Cosín DJ (1972) La pared del cuerpo de *Macracanthorhynchus hirudinaceus*. Bol R Soc Española Hist Nat 70:239–270

Đikanović V, Nikolić V, Simić V, Jakovčev-Todorović D, Cakić P (2010) The intestinal parasite *Pomphorhynchus laevis* Müller, 1776 (Acanthocephala) from Barbel *Barbus Barbus* L. from the Danube river in the area of Belgrade. BALWOIS 2010

Dingley D, Beaver PC (1985) *Macracanthorhynchus ingens* from a child in Texas. Am J Trop Med Hyg 34:918–920

Dunagan TT, Miller DM (1980) *Macracanthorhynchus hirudinaceus* from swine: an eighteen-year record of Acanthocephala from Southern Illinois. Proc Helminthol Soc Wash 47:33–36
Dunagan TT, Miller DM (1991) Acanthocephala. In: Harrison FW, Ruppert EE (eds) Microscopic anatomy of invertebrates 4: Aschelminthes. Wiley-Liss, New York
Echi PC, George S, Kumar S (2015) Differential size-biased parasitism between *Polyacanthorhynchus nigerianus* (*nomen nudum*) and *Polyacanthorhynchus echiyensis* (*nomen nudum*) (Acanthocephala: Polyacanthocephala). Braz J Biol Sci 3:113–119
Edmondson WT (1965) Reproductive rate of planktonic rotifers as related to food and temperature in nature. Ecol Monogr 35:61–111
Elicki O, Wotte T (2003) Cambroclaves from the Cambrian of Sardinia (Italy) and Germany: constraints for the architecture of western Gondwana and the palaeogeographical and palaeoecological potential of cambroclaves. Palaeogeogr Palaeoclimatol Palaeoecol 195:55–71
Ferraguti M, Melone G (1999) Spermiogenesis in *Seison nebaliae* (Rotifera, Seisonidea): further evidence of a rotifer-acanthocephalan relationship. Tiss Cell 31:428–440
Ferreira LF, de Araújo AJ, Confalonieri UE (1983) The finding of helminth eggs in a Brazilian mummy. Trans R Soc Trop Med Hyg 77:65–67
Ferreira LF, Araújo A, Confalonieri U, Chame M (1989) Acanthocefalan eggs in animal coprolites from archeological sites from Brazil. Mem Inst Oswaldo Cruz 84:201–203
Fontaneto D, de Smet W (2015) Rotifera. In: Schmidt-Rhaesa A (ed) Handbook of zoology, Gastrotricha and Gnathifera. De Gruyter, Berlin. (electronic resource)
Friedman M, Sallan LC (2012) Five hundred million years of extinction and recovery: a phanerozoic survey of large-scale diversity patterns in fishes. Palaeontology 55:707–742
Fröbius AC, Funch P (2016) Rotiferan *Hox* genes give new insights into the evolution of metazoan bodyplans. Nat Commun 8:9
Fugassa MH, Reinhard KJ, Johnson KL, Gardner SL, Vieira M, Araújo A (2011) Parasitism of prehistoric humans and companion animals from antelope cave, Mojave County, Northwest Arizona. J Parasitol 97:862–867
García-Varela M, Nadler SA (2005) Phylogenetic relationships of Palaeacanthocephala (Acanthocephala) inferred from SSU and LSU rDNA gene sequences. J Parasitol 91:1401–1409
García-Varela M, Nadler SA (2006) Phylogenetic relationships among Syndermata inferred from nuclear and mitochondrial gene sequences. Mol Phylogenet Evol 40:61–72
Gassó D, Serrano E, Castillo-Contreras R, Aguilar XF, Cadena AC, Velarde R, Mentaberre G, López-Olvera JR, Risco D, Gonçalves P, Lavín S, Fernandez-Llário P, Segalés J, Ferrer D (2016) Coprological tests underestimate *Macracanthorhynchus hirudinaceus* burden in wild boar. Parasitol Res 115:2103–2105
Gautam NK, Mohapatra A, Saxena AM (2018) On a new species of *Neoechinorhynchus* Hamann, 1892 (Neoechinorhynchinae Travassos, 1926) from *Eleutheronema tetradactylum* (Shaw, 1804) from Digha coast, West Bengal, India. J Parasit Dis 42:462–466
Gautam NK, Misra PK, Saxena AM (2019) Four new species of the genus *Pallisentis* (Quadrigyridae, Van Cleave, 1920) from freshwater fish in Uttar Pradesh, India. Acta Parasitol 15.
Gazi M, García-Varela M, Park C, Littlewood DTJ, Park JK (2016) Mitogenomic phylogeny of Acanthocephala reveals novel class relationships. Zool Scripta 45:437–454
Gee RJ (1987) A comparative morphological study of the stutzzelle (support cell) in the phylum Acanthocephala. Can J Zool 65:660–668
Gibson DI, Bray RA, Hunt D, Georgiev BB, Scholz T, Harris PD, Bakke TA, Pojmanska T, Niewiadomska K, Kostadinova A, Tkach V, Bain O, Durette-Desset M-C, Gibbons L, Moravec F, Petter A, Dimitrova ZM, Buchmann K, Valtonen ET, de Jong Y (2014) Fauna europaea: helminths (animal parasitic). Biodivers Data J 2:e1060
Goater TG, Goater CP, Esch GW (2014) Parasitism – the diversity and ecology of animal parasites, 2nd edn. Cambridge University Press, Cambridge
Golvan YJ (1994) Nomenclature of the Acanthocephala. Res Rev Parasitol 54:135–205
Gomes AP, Olifiers N, Souza JG, Barbosa HS, D'Andrea PS, Maldonado A (2015) A new acanthocephalan species (Archiacanthocephala: Oligacanthorhynchidae) from the crab-eating fox (*Cerdocyon thous*) in the Brazilian pantanal wetlands. J Parasitol 101:74–79

Gonçalves ML, Araújo A, Ferreira LF (2003) Human intestinal parasites in the past: new findings and a review. Mem Inst Oswaldo Cruz 98(Suppl 1):103–118

Graeber K, Storch V (1978) Elektronenmikroskopische und morphometrische Untersuchungen am Integument der Acanthocephala (Aschelminthes). Z Parasitenkd 57:121–135

Grassi B, Calandruccio S (1888) Ueber einen *Echinorhynchus*, welcher auch in Menschen parasitirt und dessen Zwischenwirth ein Blaps ist. Centralbl Bakt u Parasitenk 3:521–525

Habdija I, Primc Habdija B, Radanović I, Špoljar M, Matoničkin Kepčija R, Vujčić Karlo S, Miliša M, Ostojić A, Sertić Perić M (2011) Protista - Protozoa; Metazoa – Invertebrata. Alfa d.d, Zagreb

Hammond RA (1966) The proboscis mechanism of *Acanthocephalus ranae*. J Exp Biol 45:203–213

Harada I (1931) Das Nervensystem von *Bolbosoma turbinella* (Dies.). Jpn J Zool 3:161–199

Harvey THP, Vélez MI, Butterfield NJ (2012) Exceptionally preserved crustaceans from western Canada reveal a cryptic Cambrian radiation. Proc Natl Acad Sci U S A 109:1589–1594

Herlyn H (2001) First description of an apical epidermis cone in *Paratenuisentis ambiguus* (Acanthocephala: Eoacanthocephala) and its phylogenetic implications. Parasitol Res 87:306–310

Herlyn H (2002) The musculature of the praesoma in *Macracanthorhynchus hirudinaceus* (Acanthocephala, Archiacanthocephala) – re-examination and phylogenetic significance. Zoomorphology 121:173–182

Herlyn H (2016) The phylogenetic system of primates – character evolution in the light of a consolidated tree. Org Divers Evol 16:689–713

Herlyn H (2017) Organization and evolution of the proboscis musculature in avian parasites of the genus *Apororhynchus* (Acanthocephala: Apororhynchida). Parasitol Res 116:1801–1810

Herlyn H, Ehlers U (1997) Ultrastructure and function of the pharynx of *Gnathostomula paradoxa* (Gnathostomulida). Zoomorphology 117:135–145

Herlyn H, Ehlers U (2001) Organisation of the praesoma in *Acanthocephalus anguillae* (Acanthocephala, Palaeacanthocephala) with special reference to the muscular system. Zoomorphology 121:13–18

Herlyn H, Röhrig H (2003) Ultrastructure and overall organization of ligament sac, uterine bell, uterus and vagina in *Paratenuisentis ambiguus* (Acanthocephala, Eoacanthocephala) – the character evolution within the Acanthocephala. Acta Zool 84:239–247

Herlyn H, Taraschewski H (2017) Evolutionary anatomy of the muscular apparatus involved in the anchoring of Acanthocephala to the intestinal wall of their vertebrate hosts. Parasitol Res 116:1207–1225

Herlyn H, Martini N, Ehlers U (2001) Organisation of the praesoma of *Paratenuisentis ambiguus* (Van Cleave, 1921) (Acanthocephala: Eoacanthocephala), with special reference to the lateral sense organs and musculature. Syst Parasitol 50:105–116

Herlyn H, Piskurek O, Schmitz J, Ehlers U, Zischler H (2003) The Syndermatan phylogeny and the evolution of acanthocephalan endoparasitism as inferred from 18S rDNA sequences. Mol Phylogenet Evol 26:155–164

Hernández-Orts JS, Montero FE, García NA, Crespo EA, Raga JA, García-Varela M, Aznar FJ (2019) Transmission of *Corynosoma australe* (Acanthocephala: Polymorphidae) from fishes to South American sea lions *Otaria flavescens* in Patagonia, Argentina. Parasitol Res 118:433–440

Horne PD (2002) First evidence of enterobiasis in ancient Egypt. J Parasitol 88:1019–1021

Hu S, Li Y, Luo H, Fu X, You T, Pang J, Liu Q, Steiner M (2008) New record of palaeoscolecids from the Early Cambrian of Yunnan, China. Acta Geol Sin 82:244–248

Hunt AP, Lucas SG, Milàn J, Spielmann JA (2012) Vertebrate coprolites. New Mex Mus Nat Hist Sci Bull 57:5–23

Iturralde-Vinent MA, MacPhee RDE (1996) Age and paleogeographical origin of Dominican amber. Science 273:1850–1852

Janvier P (2003) Vertebrate characters and the Cambrian vertebrates. C R Palevol 2:523–531

Jerônimo GT, de Pádua SB, de Andrade Belo MA et al (2017) *Neoechinorhynchus buttnerae* (Acanthocephala) infection in farmed *Colossoma macropomum*: a pathological approach. Aquaculture 469:124–127

Kaito S, Sasaki M, Goto K, Matsusue R, Koyama H, Nakao M, Hasegawa H (2019) A case of small bowel obstruction due to infection with *Bolbosoma* sp. (Acanthocephala: Polymorphidae). Parasitol Int 68:14–16

Kamimura K, Yonemitsu K, Maeda K, Sakaguchi S, Setsuda A, Varcasia A, Sato H (2018) An unexpected case of a Japanese wild boar (*Sus scrofa leucomystax*) infected with the giant thorny-headed worm (*Macracanthorhynchus hirudinaceus*) on the mainland of Japan (Honshu). Parasitol Res 117:2315–2322

Kassatkina P (2016) Arrow worms (Chaetognatha) from the Arctic Seas of Russia: five new species of the family Sagittidae from the Laptev Sea. Biol Bull 43:1184–1194

Kennedy CR (1999) Post-cyclic transmission in *Pomphorhynchus laevis* (Acanthocephala). Folia Parasitol 46:111–116

Kennedy CR (2006) Ecology of Acanthocephala. Cambridge University Press, Cambridge

Klug C, De Baets K, Kröger B, Bell MA, Korn D, Payne JL (2015) Normal giants? Temporal and latitudinal shifts of Palaeozoic marine invertebrate gigantism and global change. Lethaia 48:267–288

Klug C, Frey L, Pohle A, De Baets K, Korn D (2017) Palaeozoic evolution of animal mouthparts. Bull Geosci 92:511–524

Kocourek F (1877) Der Riesenkratzer (*Echinorhynchus gigas*) als Ursache einer seuchenartigen Sterblichkeit in einer Schweineherde. Oesterreichische Monatsschrift fuer Thierheilkunde mit Beruecksichtigung der Viehzucht und Landwirtschaft 2:89–93

Kogi E, Yaro CA, Mbah CE (2016) Occurrence of acanthocephalan, *Moniliformis dubius* in cockroach, *Periplaneta americana* and *Blattella germanica* in Zaria, Nigeria. World J Biol Med Sci 3:157–164

Kouchinsky A, Bengtson S, Runnegar B, Skovsted C, Steiner M, Vendrasco MJ (2012) Chronology of early Cambrian biomineralisation. Geol Mag 149:221–251

Kristensen RM, Funch P (2000) Micrognathozoa: a new class with complicated jaws like those of Rotifera and Gnathostomulida. J Morphol 246:1–49

Kumar P, Kumar P (2001) Phthirapteran insect and larval Acanthocephala from the late Triassic sediments of the Satpura basin, India. J Palaeontol Soc India 46:141–146

Lahmar S, Arfa I, Ben Othmen S, Jguirim W, Saïd Y, Dhibi A, Boufana B (2017) Intestinal helminths of stray dogs from Tunisia with special reference to zoonotic infections. Parasitol Open 3:1–9

Lambl VDF (1859) Mikroskopische Untersuchungen der Darm-Excrete. Beitrag zur Pathologie des Darms und Diagnostik am Krankenbette. Vrtljschr Prakt Heilk 16:1–58

Laumer CE, Bekkouche N, Kerbl A, Goetz F, Neves RC, Sørensen MV, Kristensen RM, Hejnol A, Dunn CW, Giribet G, Worsaae K (2015) Spiralian phylogeny informs the evolution of microscopic lineages. Curr Biol 25:2000–2006

Leung TLF (2017) Fossils of parasites: what can the fossil record tell us about the evolution of parasitism? Biol Rev 92:410–430

Leung TLF (2021) Parasites of fossil vertebrates: what we know and what can we expect from the fossil record? In: De Baets K, Huntley JW (eds) The evolution and fossil Record of Parasitism: Identification and macroevolution of parasites. Topics in Geobiology 49

Littlewood DTJ, Donovan SK (2003) Fossil parasites: a case of identity. Geol Today 19:136–142

Malta JDO, Gomes ALS, de Andrade SMS et al (2001) Massive infestation by *Neoechinorhynchus buttnerae* Golvan, 1956 (Eoacanthocephala: Neochinorhynchidae) in young "tambaquis" *Colossoma macropomum* (Cuvier, 1818) cultured in the Central Amazon. Acta Amazon 31:133–143

Marchand B, Mattei X (1976) La Spermatogenèse des Acanthocéphales. I. L'Appareil Centriolaire et Flagellaire au Cours de la Spermiogenèse d'*Illiosentis furcatus* var *africana* Golvan, 1956 (Palaeacanthocephala, Rhadinorhynchidae). J Ultrastruct Res 54:347–358

Mark Welch DB (2000) Evidence from a protein-coding gene that acanthocephalans are rotifers. Invert Biol 119:17–26

Marlétaz F, Peijnenburg KTCA, Goto T, Satoh N, Rokhsar DS (2019) A new spiralian phylogeny places the enigmatic arrow worms among gnathiferans. Curr Biol 29:312–318.e3

Martins ML, de Moraes FR, Fujimoto RY et al (2001) Prevalence and histopathology of *Neoechinorhynchus curemai* Noronha, 1973 (Acanthocephala: Neoechinorhynchidae) in *Prochilodus lineatus* Valenciennes, 1836 from Volta Grande Reservoir, MG, Brazil. Braz J Biol 61:517–522

Mathison BA, Bishop HS, Sanborn CR, dos Santos Souza S, Bradbury R (2016) *Macracanthorhynchus ingens* infection in an 18-month-old child in Florida: a case report and review of Acanthocephaliasis in humans. Clin Infect Dis 63:1357–1359

Mauer K, Hellmann SL, Groth M et al (2020) The genome, transcriptome, and proteome of the fish parasite *Pomphorhynchus laevis* (Acanthocephala). PLoS One 15:e0232973

Mehlhorn H (2016) Human parasites – diagnosis, treatment, prevention. Springer, Heidelberg

Meyer A (1932) Acanthocephala. In: Bronn's Klassen und Ordnungen des Tierreichs, vol 4. Akademische Verlagsgesellschaft, Leipzig

Miller DM, Dunagan TT (1985) Functional morphology. In: Crompton DWT, Nickol BB (eds) Biology of Acanthocephala. Cambridge University Press, Cambridge

Moore JG, Fry GF, Englert E (1969) Thorny-headed worm infection in North American prehistoric man. Science 163:1324–1325

Mowlavi G, Makki M, Heidari Z, Rezaeian M, Mohebali M, Araújo A, Boenke N, Abolfazl A, Stollner T, Mobedi I (2015) *Macracanthorhynchus hirudinaceus* eggs in canine coprolite from the Sasanian Era in Iran (4th/5th century CE). Iran J Parasitol 10:245–249

Murray J (1910) British Antarctic Expedition 1907–9, under the command of Sir E.H. Shackleton, c.v.o. Reports on the scientific investigations 1

Myers BJ, Kuntz RE (1972) A checklist of parasites and commensals reported for the chimpanzee (*Pan*). Primates 13:433–471

Near TJ (2002) Acanthocephalan phylogeny and the evolution of parasitism. Integr Comp Biol 41:668–677

Near TJ, Garey JR, Nadler SA (1998) Phylogenetic relationships of the Acanthocephala inferred from 18S ribosomal DNA sequences. Mol Phylogenet Evol 10:287–298

Nezamabadi M (2014) New contribution of paleoparasitology in the middle east and first data on the Iranian plateau adjacent area. Dissertation thesis, Université de Franche-Comté

Nielsen C (2012) Animal evolution – interrelationships of living phyla, 3rd edn. Oxford University Press, Oxford

Nikishin VP (2004) Subsurface musculature of spiny-headed worms (Acanthocephala) and its role in formation of intercellular matrix. Biol Bull 31:598–612

Noronha D, Ferreira LF, Rangel A, Araújo A, Gomes DC (1994) *Echinopardalis* (Acanthocephala, Oligacanthorhynchidae) eggs in felid coprolites dated from 9,000 years before present found in the Brazilian Northeast. Mem Inst Oswaldo Cruz 89:119–120

Novo SPC, Ferreira LF (2016) The paleoparasitology in Brazil and findings in human remains from South America: a review. Korean J Parasitol 54:573–583

Parker GA, Ball MA, Chubb JS (2015) Evolution of complex life cycles in trophically transmitted helminths. Host incorporation and trophic ascent. J Evol Biol 28:267–291

Parry LA, Smithwick F, Nordén KK, Saitta ET, Lozano-Fernandez J, Tanner AR, Caron JB, Edgecombe GD, Briggs DEG, Vinther J (2018) Soft-bodied fossils are not simply rotten carcasses – toward a holistic understanding of exceptional fossil preservation: exceptional fossil preservation is complex and involves the interplay of numerous biological and geological processes. BioEssays 40:1700167

Pavlović IN, Kulišić ZB, Tambur ZŽ, ProtiŽ NM (2010) Scarabidae – intermediate host for *Macracanthorhynchus hirudinaceus*. Zb Mat Srp Prir Nauke 119:89–95

Perry ML (1942) A new species of the acanthocephalan genus *Filicollis*. J Parasitol 28:385–388

Peters W, Taraschewski H, Latka I (1991) Comparative investigations of the morphology and chemical composition of the eggshells of Acanthocephala I. *Macracanthorhynchus hirudinaceus* (Archiacanthocephala). Parasitol Res 77:542–549

Petrochenko VI (1956) Acanthocephala of domestic and wild animals I. English translation from Russian published in 1971. Keter Press, Jerusalem

Petrochenko VI (1958) Acanthocephala of domestic and wild animals II. English translation from Russian published in 1971. Keter Press, Jerusalem

Pinacho-Pinacho CD, Sereno-Uribe AL, García-Varela M (2014) Morphological and molecular data reveal a new species of *Neoechinorhynchus* (Acanthocephala: Neoechinorhynchidae) from *Dormitator maculatus* in the Gulf of Mexico. Parasitol Int 63:763–771
Poinar GO, Ricci C (1992) Bdelloid rotifers in Dominican amber: evidence for parthenogenetic continuity. Experientia 48:408–410
Poulin R, Morand S (2000) Testes size, body size and male-male competition in acanthocephalan parasites. J Zool 250:551–558
Poulin R, Wise M, Moore J (2003) A comparative analysis of adult body size and its correlates in acanthocephalan parasites. Int J Parasitol 33:799–805
Qian Y, Yin G (1984) Zhijinitae and its stratigraphical significance. Acta Palaeontol Sin 23:216–223
Reinhard KJ (1990) Archaeoparasitology in North America. Am J Phys Anthropol 82:145–163
Reinhard KJ (2017) Reestablishing rigor in archaeological parasitology. Int J Paleopathol 19:124–134
Reinhard KJ, Bryant VM (1992) Coprolite analysis: a biological perspective on archaeology. Pap Nat Resour 46:245–288
Ricci C (1998) Are lemnisci and proboscis present in the Bdelloidea? Hydrobiologia 387/388:93–96
Richart R, Benirschke K (1963) Causes of death in a colony of marmoset monkeys. J Pathol Bacteriol 86:221–223
Rieger RM, Tyler S (1995) Sister-group relationship of Gnathostomulida and Rotifera-Acanthocephala. Invertebr Biol 114:186–188
Rota-Stabelli O, Daley AC, Pisani D (2013) Molecular timetrees reveal a Cambrian colonization of land and a new scenario for ecdysozoan evolution. Curr Biol 23:292–398
Sahar MA, Madani TA, Al Mohsen IZ, Almodovar EL (2006) A child with an acanthocephalan infection. Ann Saudi Med 26:321–324
Sanford SE (1978) Mortality in mute swans in Southern Ontario associated with infestation with the thorny-headed worm, *Polymorphus boschadis*. Can Vet J 19:234–236
Sansom RS, Randle E, Donoghue PCJ (2015) Discriminating signal from noise in the fossil record of early vertebrates reveals cryptic evolutionary history. Proc R Soc B 282:20142245
Sasaki M, Katahira H, Kobayashi M, Kuramochi T, Matsubara H, Nakao M (2019) Infection status of commercial fish with cystacanth larvae of the genus *Corynosoma* (Acanthocephala: Polymorphidae) in Hokkaido, Japan. Int J Food Microbiol 305:108256
Sayers K, Lovejoy CO (2014) Blood, bulbs, and bunodonts: on evolutionary ecology and the diets of *Ardipithecus*, *Australopithecus*, and early *Homo*. Q Rev Biol 89:319–357
Schmidt GD (1971) Acanthocephalan infections of man, with two new records. J Parasitol 57:582–584
Schmidt GD (1985) Development and life cycles. In: Crompton DWT, Nickol BB (eds) Biology of Acanthocephala. Cambridge University Press, Cambridge
Scholtz G, Edgecombe GD (2006) The evolution of arthropod heads: reconciling morphological, developmental and palaeontological evidence. Dev Genes Evol 216:395–415
Shu D, Conway Morris S, Han J, Hoyal Cuthill JF, Zhang Z, Cheng M, Huang H (2017) Multi-jawed chaetognaths from the Chengjiang Lagerstätte (Cambrian, Series 2, Stage 3) of Yunnan, China. Palaeontology 60:763–772
Sianto L, Chame M, Silva CSP, Gonçalves MLC, Reinhard K, Fugassa M, Aráujo A (2009) Animal helminths in human archaeological remains: a review of zoonoses in the past. Rev Inst Med trop S Paulo 51:119–130
Sielaff M, Schmidt H, Struck TH, Rosenkranz D, Mark Welch DB, Hankeln T, Herlyn H (2016) Phylogeny of Syndermata (syn. Rotifera): mitochondrial gene order verifies epizoic Seisonidea as sister to endoparasitic Acanthocephala within monophyletic Hemirotifera. Mol Phylogenet Evol 96:79–92
Silva-Gomes AL, Gomes Coelho-Filho J, Viana-Silva W, Braga-Oliveira MI, Bernardino G, Costa JI (2017) The impact of *Neoechinorhynchus buttnerae* (Golvan, 1956) (Eoacanthocephala: Neochinorhynchidae) outbreaks on productive and economic performance of the tambaqui *Colossoma macropomum* (Cuvier, 1818), reared in ponds. Lat Am J Aquat Res 45:496–500
Sistiaga A, Mallol C, Galván B, Summons RE (2014) The Neanderthal meal: a new perspective using faecal biomarkers. PLoS One 9:e101045

Solórzano-García B, Pérez-Ponce de León G (2018) Parasites of Neotropical primates: a review. Int J Primatol 39:155–182
Sørensen MV, Segers H, Funch P (2005) On a new *Seison* Grube, 1861 from coastal waters of Kenya, with a reappraisal of the classification of the Seisonida (Rotifera). Zool Stud 44:34–43
Southcott RV, Lange RT (1971) Acarine and other microfossils from the Maslin Eocene, South Australia. Rec South Austr Mus 16:1–21
Sterrer W, Sørensen MV (2015) Gnathostomulida. In: Schmidt-Rhaesa A (ed) Handbook of zoology, Gastrotricha and Gnathifera. De Gruyter, Berlin
Struck TH, Wey-Fabrizius AR, Golombek A, Hering L, Weigert A, Bleidorn C, Klebow S, Iakovenko N, Hausdorf B, Petersen M, Kück P, Herlyn H, Hankeln T (2014) Platyzoan paraphyly based on phylogenomic data supports a noncoelomate ancestry of Spiralia. Mol Biol Evol 31:1833–1849
Sures B (2002) Competition for minerals between *Acanthocephalus lucii* and its definitive host perch (*Perca fluviatilis*). Int J Parasitol 32:1117–1122
Sures B, Franken M, Taraschewski H (2000) Element concentrations in the archiacanthocephalan *Macracanthorhynchus hirudinaceus* compared with those in the porcine definitive host from a slaughterhouse in La Paz, Bolivia. Int J Parasitol 30:1071–1076
Taraschewski H (1989a) Host-parasite interface of *Paratenuisentis ambiguus* (Eoacanthocephala) in naturally infected eel and in laboratory-infected sticklebacks and juvenile carp and rainbow trout. J Parasitol 75:911–919
Taraschewski H (1989b) Host-parasite interface of *Neoechinorhynchus rutili* (Eoacanthocephala) in naturally infected salmonids. J Fish Dis 12:39–48
Taraschewski H, Mackenstedt U (1991a) Autoradiographic and morphological studies on the uptake of the triglyceride [3H]-glyceroltrioleate by acanthocephalans. Parasitol Res 77:247–254
Taraschewski H, Mackenstedt U (1991b) Autoradiographic and morphological investigations on the uptake and incorporation of tritiated lysin by acanthocephalans. Parasitol Res 77:536–541
Taraschewski H (2000) Host-parasite interactions in Acanthocephala: a morphological approach. Adv Parasitol 46:1–179
Taraschewski H, Peters W (1992) Comparative investigations of the morphology and chemical composition of the eggshells of Acanthocephala II. Palaeacanthocephala. Parasitol Res 78:376–381
Taraschewski H, Sagani C, Mehlhorn H (1989) Ultrastructural study of the host-parasite interface of *Moniliformis moniliformis* (Archiacanthocephala) in laboratory-infected rats. J Parasitol 75:288–296
Taraschewski H, Peters W, Latka I (1992) Comparative investigations of the morphology and chemical composition of the eggshells of Acanthocephala. III. Eoacanthocephala. Parasitol Res 78:382–387
Trinajstic K, Marshall C, Long J, Bifield K (2007) Exceptional preservation of nerve and muscle tissues in Late Devonian placoderm fish and their evolutionary implications. Biol Lett 3:197–200
Uglem GL (1972) The life cycle of *Neoechinorhynchus cristatus* Lynch, 1936 (Acanthocephala) with notes on the hatching of eggs. J Parasitol 58:1071–1074
Upeniece I (2001) The unique fossil assemblage from the Lode Quarry (Upper Devonian, Latvia). Mitt Mus Natkd Berl Geowiss 4:101–119
Upeniece I (2011) Palaeoecology and juvenile individuals of the Devonian placoderm and acanthodian fishes from Lode site, Latvia. Dissertation thesis, Riga
Van Cleave HJ (1921) Acanthocephala from the eel. Trans Amer Microsc Soc 40:1–13
Van Cleave HJ (1936) The recognition of a new order in the Acanthocephala. J Parasitol 22:202–206
Van Cleave HJ (1947) A critical review of terminology for immature stages in acanthocephalan life histories. J Parasitol 33:118–125
van Huis A (2017) Did early humans consume insects. J Insects Food Feed 3:161–163
Van Thiel PH, Bruss W (1945) Presence of *Prosthenorchis spirula* chez les chimpanzes, son role pathogene et son developpement dans *Blattella germanica*. Ann Parasitol 20:304–320

Verweyen L, Klimpel S, Palm HW (2011) Molecular phylogeny of the Acanthocephala (class Palaeacanthocephala) with a paraphyletic assemblage of the orders Polymorphida and Echinorhynchida. PLoS One 6:e28285

Vinther J, Parry LA (2019) Bilateral jaw elements in *Amiskwia sagittiformis* bridge the morphological gap between gnathiferans and chaetognaths. Curr Biol 29(5):881–888.e1

von Haffner K (1950) Organisation und systematische Stellung der Acanthocephalen. Zool Anz 145:243–274

Waggoner BM, Poinar GO (1993) Fossil habrotrochid rotifers in Dominican amber. Experientia 49:354–357

Warnock RCM, Engelstädter J (2021) The molecular clock as tool for understanding host-parasite evolution. In: de Baets K, Huntley JW (eds) The Evolution and Fossil Record Of Parasitism: Coevolution and Paleoparasitological Techniques. Topics in Geobiology 50

Weber M, Wey-Fabrizius AR, Podsiadlowski L, Witek A, Schill RO, Sugár L, Herlyn H, Hankeln H (2013) Phylogenetic analysis of endoparasitic Acanthocephala based on mitochondrial genomes suggests secondary loss of sensory organs. Mol Phylogenet Evol 66:182–189

Wey-Fabrizius AR, Herlyn H, Rieger B, Rosenkranz D, Witek A, Mark Welch DB, Ebersberger I, Hankeln T (2014) Transcriptome data reveal syndermatan relationships and suggest the evolution of endoparasitism in Acanthocephala via an epizoic stage. PLoS One 9:e88618

Whitfield PJ (1973) The egg envelopes of *Polymorphus minutus* (Acanthocephala). Parasitology 66:387–403

Witek A, Herlyn H, Meyer A, Boell L, Bucher G, Hankeln T (2008) EST based phylogenomics of Syndermata questions monophyly of Eurotatoria. BMC Evol Biol 8:345

Witek A, Herlyn H, Ebersberger I, Mark Welch DB, Hankeln T (2009) Support for the monophyletic origin of Gnathifera from phylogenomics. Mol Phylogenet Evol 53:1037–1041

Wurmbach H (1937) Zur krankheitserregenden Wirkung der Acanthocephalen. Z f Fischerei u Hilfswiss 35:217–232

Yeh H-Y, Mitchell PD (2016) Ancient human parasites in ethnic Chinese populations. Korean J Parasitol 54:565–572

Zangerl R, Case GR (1976) *Cobelodus aculeatus* (Cope), an anacanthous shark from Pennsylvanian black shales of North America. Palaeontogr Abt A 154:107–157

Zhang X-g, Pratt BR (2012) The first stalk-eyed phosphatocopine crustacean from the Lower Cambrian of China. Curr Biol 22:2149–2154

Zhong HL, Feng LB, Wang CX, Kang B, Wang ZZ, Zhou GH, Zhao Y, Zhang YZ (1983) Human infection with *Macracanthorhychus hirudinaceus* causing serious complications in China. Chin Med J 96:661–668

Chapter 9
Chelicerates as Parasites

Jason A. Dunlop

Abstract Among Chelicerata, larval instars of sea spiders (Pycnogonida) can be parasitic. The oldest putative sea spider from the Cambrian 'Orsten' is immature and resembles comparable instars of modern species with a parasitic phase to their life cycle. All other parasitic chelicerates are mites, with several examples in both the Acariformes and Parasitiformes clades. Fossils revealing parasitic behaviour, or belonging to purely parasitic clades, come from various amber sources from the mid-Cretaceous onwards. From Acariformes there are records of Parasitengona, Myobiidae, Pterygosomatoidea, Resinacaridae, Acarophenacidae, Pyemotidae and Apotomelidae. Parasitiformes is represented by several ticks (Ixodida) and potentially Laelapidae from the Mesostigmata. Parasitism appears to have evolved independently within mites on several occasions. Possible transitions to this lifestyle via nest associations and/or phoresy are discussed. Arachnids as victims of parasites include amber records of nematode worms (Mermithidae), erythraeid mites (Erythraeidae), mantid flies (Neuroptera: Mantispidae), ichneumon wasps (Hymenoptera: Ichneumonidae) and spider flies (Diptera: Acroceridae).

Keywords Chelicerata · Pycnogonida · Arachnida · Acariformes · Parasitiformes · Parasitism · Fossil · Amber

9.1 Introduction

Chelicerata are one of the major arthropod clades and encompass the largely terrestrial Arachnida (arachnids), together with their marine relatives Pycnogonida (sea spiders), Xiphosura (horseshoe crabs), the extinct Eurypterida (sea spiders) and the little-known extinct group Chasmataspidida. Most living chelicerates are predators and this lifestyle was probably true as well for most of the fossil taxa. The name Chelicerata refers to the first pair of head appendages, the chelicerae or chelifores, which are modified into mouthparts. These may be chelate (= pincer-like), as in

J. A. Dunlop (✉)
Museum für Naturkunde, Leibniz Institute for Evolution and Biodiversity Science, Berlin, Germany
e-mail: Jason.dunlop@mfn.berlin

© The Author(s) 2021

K. De Baets, J. W. Huntley (eds.), *The Evolution and Fossil Record of Parasitism*, Topics in Geobiology 49, https://doi.org/10.1007/978-3-030-42484-8_9

scorpions for example, or shaped like a pocket-knife as in spiders, in which a movable fang articulates against a basal element. The chelicerae are typically used to bite and/or masticate food items, and at least in spiders they bear the openings of the venom glands.

Compared to predation, parasitism among chelicerates is fairly rare. In terms of definition (e.g., Esch and Fernández 1993) a parasite is taken here to mean an animal that lives in intimate association with another organism, and benefits from this relationship while harming the host. There are some problems with this definition in that the amount of 'harm' caused is relative, and it is important to differentiate parasitism from a one-off predatory attack which kills (and consumes) the host. Spiders can autotomise a leg grabbed by a predator and escape their attacker alive, but this non-lethal attack does not make the predator a parasite. Parasitism may alternatively be envisaged as a relationship characterised by the parasite generally being smaller than the host and to a certain degree host-specific and reliant on the host for nourishment. In general, parasites only feed on part of the host's tissues. This is usually non-lethal, although large-scale infestation may reduce the host's fitness or even become fatal. In this context, we should also reiterate the distinction between a parasite and a parasitoid. A parasitoid spends much of its life cycle on (or in) the body of the host and infection is invariably fatal, such as a wasp larvae consuming a caterpillar and eventually erupting from its body. A handful of mites can be considered parasitoids (see Sects. 9.4.1 and 9.4.2).

Parasitism in chelicerates is restricted to the larvae of several sea spiders (Sect. 9.2), as well as a number of subgroups of mites (Sects. 9.4.1 and 9.4.2) within the arachnids. Ticks are perhaps the most obvious and familiar examples of arachnid parasites, and are of considerable economic importance as disease vectors. Other mite groups are also parasitic, at least at some stage of their life cycle. Chiggers, feather mites, scabies mites and the *Varroa* mite affecting honeybees are further examples of arachnid parasites which are significant for humans (Diaz 2010: Table 1) or domestic animal husbandry.

9.1.1 The Chelicerate Fossil Record

The significance of the fossil record for understanding parasitism was reviewed by De Baets and Littlewood (2015); De Baets et al. (2021a) and Leung (2017), who discussed, among other subjects, the value and limitations of fossil data for calibrating evolutionary frameworks and investigating co-evolutionary relationships between hosts and their parasites: see also Warnock and Engelstädter (2021). Although focussing in insects, chelicerates were mentioned as parasitic terrestrial arthropods by Nagler and Haug (2015). The fossil record of Chelicerata in general was reviewed by Dunlop (2010), Dunlop and Penney (2012), and more recently by Wolfe et al. (2016) as part of a wider survey of the arthropods. For a review of recent developments in chelicerate phylogeny see Giribet (2018).

Table 9.1 Parasitic clades of acariform and parasitiform mites, summarizing the type of parasitism and the typical host group, as well as the fossil record (if any) of the lineage in question. See also Fig. 9.3

Taxon	Parasitism	Fossil record
Acariformes		
Prostigmata: Eupodides: Halacaroidea	Some species parasitic on aquatic invertebrates	–
Prostigmata: Eupodides: Tydeoidea	Some species haematophagous on gastropods, or in the nasal cavities of amphibians and birds	Devonian—Recent
Prostigmata: Anystides: Parasitengona	Larval instar (which differs from adult) parasitizes invertebrates or vertebrates	Cretaceous—Recent
Prostigmata: Raphignathina: Myobioidea	Ectoparasites in the fur of various mammals	Eocene—Recent
Prostigmata: Raphignathina: Pterygosomatoidea	Ectoparasites of lizards, tortoises and occasionally arthropods	Cretaceous—Recent
Prostigmata: Raphignathina: Cheyletoidea	Ectoparasites of arthropods, reptiles birds and mammals	Cretaceous—Recent
Prostigmata: Heterostigmata: Scutacaroidea	A few species are parasites or parasitoids of cockroaches or ants	–
Prostigmata: Heterostigmata: Pyemotoidea	Females of most species are parasitoids of insects	Cretaceous—Recent
Prostigmata: Heterostigmata: Tarsonemoidea	Several internal or external parasites of insects	Subfossils
Oribatida: Astigmata: Hemisarcoptoidea	A few species are haematophagous on beetles and scale insects	Neogene—Recent
Oribatida: Astigmata: Glycyphagoidea	Deutonymphs of some species infest hair follicles of rodents	–
Oribatida: Astigmata: Hypoderatoidea	Deutonymphs infest subcutaneous tissue of birds (rarely rodents)	–
Oribatida: Astigmata: Psoroptida	Inhabit the fur, feathers, respiratory tract or upper skin layers of vertebrates	Neogene—Recent
Parasitiformes		
Ixodida	Haematophagous on vertebrates	Cretaceous—Recent
Mesostigmata: Antennophorina: Antennophoroidea	One genus induces ants to regurgitate food (kleptoparasitism?)	–
Mesostigmata: Uropodina: Uropodoidea	At least one species is a parasitoid of ants	Eocene—Recent
Mesostigmata: Heterozerconia: Heterozerconoidea	At least one species is haematophagous on reptiles	–
Mesostigmata: Gamasina: Ascoidea	A few species are haematophagous on insects	Eocene—Recent
Mesostigmata: Gamasina: Phytoseiodea	One family (Otopheidomenidae) parasitizes several groups of insects	–
Mesostigmata: Gamasina: Dermanyssoidea	A wide range of ecto- and endoparasites affecting both arthropods and vertebrates	Eocene—Recent

In brief, the oldest chelicerates are Cambrian in age (Fig. 9.1); either middle if one accepts the Canadian Burgess Shale fossil *Sanctacaris* as a chelicerate (e.g. Legg 2014), or late Cambrian if dated on a putative larval sea spider from the Orsten of Sweden (see below). The oldest horseshoe crabs and eurypterids are Ordovician in age, while the oldest arachnids (e.g. Dunlop 1996; Waddington et al. 2015) are Silurian (Fig. 9.1). These comprise scorpions (Scorpiones) from the mid-Silurian and the extinct spider-like Trigonotarbida from the late Silurian. Harvestmen (Opiliones), pseudoscorpions (Pseudoscorpiones) and some mites (Acariformes) are known from the Devonian. Spiders (Araneae), whip spiders (Amblypygi), whip scorpions (Thelyphonida) and camel spiders (Solifugae) were present by the time of the Late Carboniferous Coal Measures. The remaining mites (Parasitiformes), palpigrades (Palpigradi) and schizomids (Schizomida) are first recorded from Cretaceous (ca. 100 Ma) Burmese amber. All three are typically small and rather cryptic groups, and this late occurrence in the fossil record is probably an artefact of their low chances of being preserved. Thus a Palaeozoic origin for all the arachnid orders seems likely (Fig. 9.1) and is also implied by several fossil-calibrated molecular studies (e.g. Rota-Stabelli et al. 2013; Sharma and Giribet 2014: Fig. 3). Warnock et al. (2012) also inferred that parasitiform mites should have been present by at least the Carboniferous.

9.2 Sea Spiders

Sea spiders are an enigmatic group whose position within the arthropods has long proved controversial (reviewed by Dunlop and Arango 2005). Most workers now accept them as chelicerates (see e.g. Giribet 2018), with which they share the chelate first pair of appendages—termed chelifores in sea spiders—although these are reduced or absent in some derived groups. All sea spiders are characterised by a unique feeding apparatus, the proboscis, which they use to ingest food, typically sucking up nutrients from sessile organisms. While adult sea spiders are not strictly parasitic, the larval stages of numerous species have been reported as ecto- or endoparasites of other marine organisms; for a review see Brenneis et al. (2017), and references therein. These authors recognised five basic patterns of development, three of which involve a hatching (protonymphon) larva which begins its life cycle as a parasite. It should be noted here that (a) the entire life cycle is not known for all pycnogonid species and (b) that the observed patterns are not necessarily taxon-specific, such that different developmental pathways may occur even within the same genus:

1. In developmental pattern 1, the protonymphon lives as an ectoparasite on cnidarians and occasionally molluscs.
2. In pattern 2, there is no parasitic phase; the animals hatch at the later postlarval stage and are lecithotrophic—i.e., they receive nourishment from a repository of yolk held in the midgut—before eventually becoming free-living.

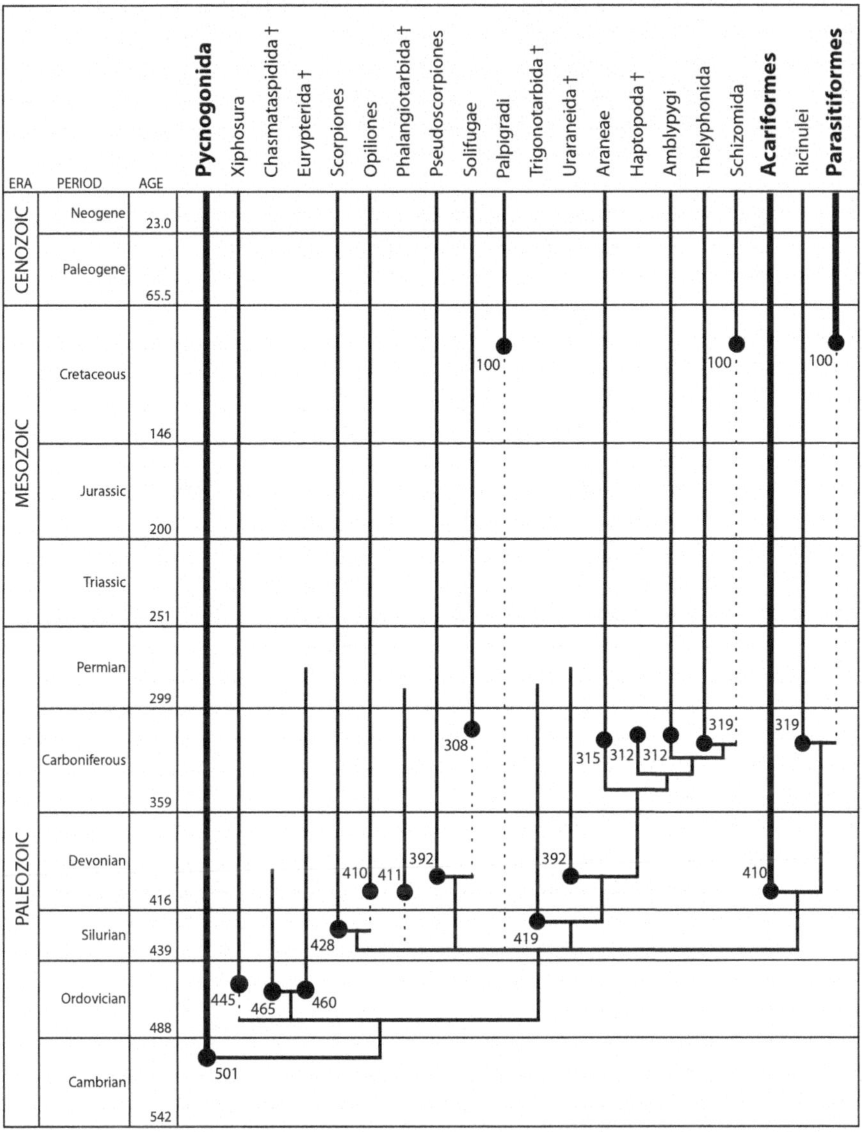

Fig. 9.1 Chelicerate fossil record modified from Dunlop (2010), in turn based on the phylogeny of Shultz (2007). Dated circles indicate oldest body fossil records; dashed lines ghost ranges for which a record would be expected; extinct orders marked with a †. Other hypotheses (reviewed by Giribet 2018) place, for example, scorpions closer to spiders and other lung-bearing arachnids, or camel spiders close to the acariform mites. Parasitism, indicated here in bold and thicker lines, occurs in Pycnogonida (among juveniles only) and the two mite lineages Acariformes and Parasitiformes

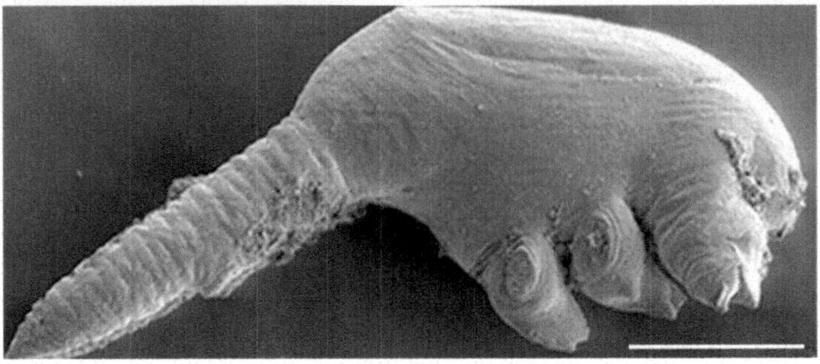

Fig. 9.2 The putative fossil sea spider (Pycnogonida) *Cambropycnogon klausmuelleri* (Waloszek and Dunlop 2002) from the early Cambrian 'Orsten' of Sweden. Probably at a post-protonymphon stage, it shows similarities to juvenile instars of some modern sea spiders which have a parasitic mode of life while still immature. Scale bar equals 100 μm. (Image courtesy of Dieter Waloszek)

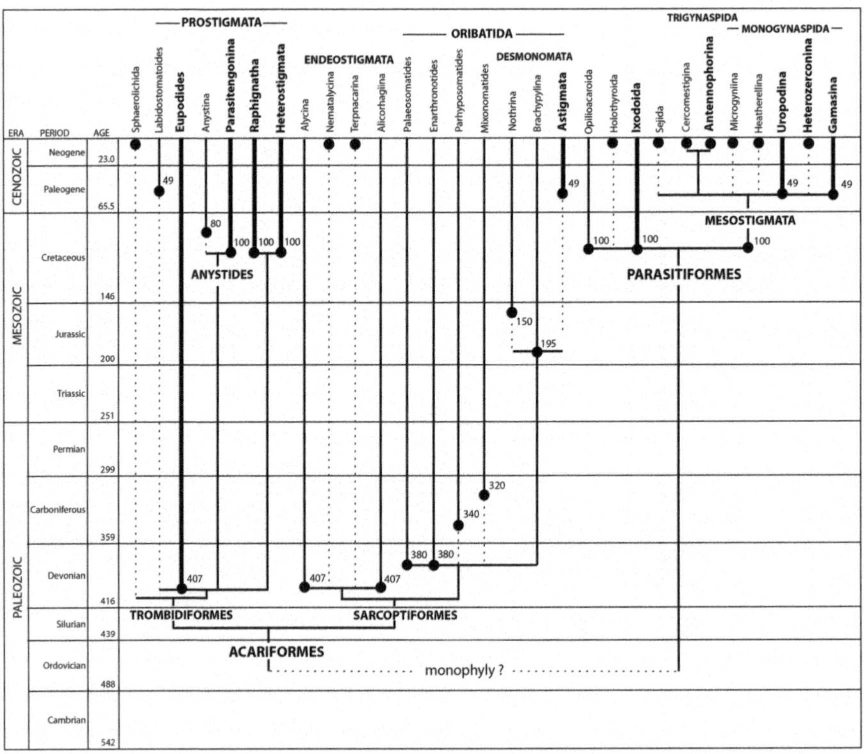

Fig. 9.3 Mite fossil record based on the systematics of Lindquist et al. (2009). Dated circles indicate oldest body fossil records; dashed lines ghost ranges for which a record would be expected. Parasitism, indicated here in bold and thicker lines, can be observed in at least ten clades. Due to the large number of families and superfamilies, lineages are primarily shown here at the suborder/cohort level. The figure thus masks several cases of convergent acquisition of parasitic behaviours, within a given clade, and at least 21 (independent?) shifts to parasitism may have taken place across mites in general (Table 9.1); see also text for details

3. In pattern 3, the protonymphon is an ectoparasite of nudibranch or bivalve molluscs, or on sedentary polychaete worms.
4. In pattern 4, the protonymphon is an endoparasite of cnidarians or other hydrozoans. Note that the postlarval instars in modes 1, 3 and 4 are all then free-living.
5. In pattern 5, the animals hatch at a later stage (an advanced postlarva) which is again lecithotrophic and thereafter free-living.

9.2.1 *Cambropycnogon*

Fossil Pycnogonida are rare (summarised by Bamber 2007), with a handful of records from the Ordovician, Silurian, Devonian and Jurassic. All of these are based on adult material. The oldest putative sea spider is a tiny, ca. 270 μm long, phosphatised juvenile instar from the late Cambrian (ca. 501 Ma) 'Orsten' of Sweden (Fig. 9.2) described as *Cambropycnogon klausmuelleri* (Waloszek and Dunlop 2002). Its pycnogonid affinities were subsequently questioned by Bamber (2007), although it does resemble the post-protonymphon stage of modern sea spiders, and other authors (Wolfe et al. 2016; Brenneis et al. 2017) have been more sympathetic towards its inclusion in Pycnogonida. If *Cambropycnogon* is an early sea spider larva, it would indicate that these animals also underwent anamorphic development back in the Cambrian; in other words early instars do not have the full complement of adult appendages and add limb pairs with successive postembryonic stages. It is thus tempting to speculate whether Cambrian sea spiders also had a parasitic hatching stage. Brenneis et al. (2017) noted that *Cambropycnogon* showed similarities to living species such as *Pycnogonum littorale*, which use the (parasitic) type 1 development pattern in their scheme. However, it should be reiterated that the *Cambropycnogon* fossils are not at the protonymphon stage—the next set of limb buds (or anlage) was already developing—and its mode of life remains equivocal. Further data is needed to assess the extent to which *Cambropycnogon* preserves characters from the chelicerate ground pattern, as opposed to adaptions for a parasitic mode of life.

9.3 Horseshoe Crabs and Eurypterids

Excluding sea spiders, the remaining chelicerates are usually grouped together as the Euchelicerata (reviewed by Giribet 2018). This clade traditionally comprises Xiphosura, Eurypterida, Chasmataspidida and Arachnida although, as elucidated by Lamsdell (2013), several fossils traditionally considered to be basal horseshoe crabs may be better interpreted as basal euchelicerates instead. It should also be noted that the textbook 'Merostomata' grouping of Xiphosura and Eurypterida has not been supported in more recent phylogenies, which tend to place eurypterids closer to arachnids. Irrespective of phylogenetic details, the feeding ecology of modern horseshoe crabs (Botton et al. 2003) is probably a fair guide to the mode of life in early euchelicerates. Living horseshoe crabs generally use their prosomal legs to dig worms and molluscs out of the substrate, masticate them with the spiny gnathobases at the base of these legs and then transfer the resulting sediment with food particles forwards to the mouth via a current of water. In other words they are active predators, with no evidence for parasitism, and it seems reasonable to assume that early fossil euchelicerates from the mid-Palaeozoic—which had a similar general body plan—were not parasites either. Eurypterids also had gnathobases (Selden 1981), although some achieved quite large body sizes of more than 2 m (Lamsdell and Braddy 2010) and had correspondingly huge chelicerae (mouthparts), which suggests they may have been able to capture larger prey items such as fish. Again, a parasitic mode of life seems highly unlikely and none of the known eurypterids known show any morphological adaptations that would support this hypothesis. The feeding ecology of the extinct chasmataspidids is equivocal. Given their general morphological similarities to both horseshoe crabs and eurypterids a similar mode of life would be expected.

9.4 Arachnids

Like horseshoe crabs and eurypterids, arachnids are predominantly predatory arthropods (Beccaloni 2009). Prey capture is normal for groups like spiders, scorpions and pseudoscorpions, and several evolutionary novelties such as modified limbs (claws, spines) or venom have evolved to help them entrap and subdue their victims. A predatory mode of life seems likely for the Palaeozoic ancestors of the arachnids and spider-like biting mouthparts have been documented in, for example, exceptionally preserved Early Devonian fossils belonging to the extinct arachnid order Trigonotarbida (Haug 2017). Note that some spiders practice kleptoparasitism, stealing food from the webs of larger spider species (e.g. van Helsdingen 2011), but this behaviour should not be counted as parasitism *sensu stricto*. Harvestmen primarily capture live prey, but can have a more varied diet (Acosta and Machado 2007) supplemented by material such as decaying animals, lichen, fungi and even bird droppings. However, as these authors note, one should not make generalisations about *all* harvestmen since individual taxa have individual food-preferences. It may also be worth noting in this context that while most arachnids are liquid-feeders, harvestmen are still able to ingest solid particles.

The other exception to the general predacious rule among the arachnids are the mites. These arachnids tend to have smaller body sizes and have evolved to present a wide range of ecologies, which include both predatory and non-predatory lifestyles. They are also the only arachnids to have evolved parasitic behaviour and will thus be the focus of the remainder of this chapter. Their fossil record was recently summarised by Sidorchuk (2018) who argued that they have been small throughout their geological history. It is important to point out that while all mites were traditionally grouped as a single order, Acari, it has become apparent that there are two major clades which are now usually treated as separate orders: Acariformes (or Actinotrichida) and Parasitiformes (or Anactinotrichida). The two mite clades differ from each other in several important characters (summarised by Dunlop and Alberti 2008) thus Acari in its traditional sense may not be monophyletic; see e.g. Pepato et al. (2010) and Pepato and Klimov (2015), but see also Lozano-Fernandez et al. (2019) for a recent defence of mite monophyly. A time-calibrated summary of the major mite groups, also indicating lineages in which parasitism has been recorded is presented in Fig. 9.3. At least ten major lineages (five acariforms, five parasitiforms) include taxa with parasitic behaviour, although some of these clades appear to show multiple acquisitions of this lifestyle and there may be at least 21 independent evolutionary shifts to these ecologies (Table 9.1).

9.4.1 Acariform Mites

Acariform mites are the more diverse of the two main clades with more than 42,000 species described so far and presumably many more undiscovered. They can be broadly divided into two main lineages: Trombidiformes and Sarcoptiformes. The following survey follows the higher systematics and superfamily classification recognised in the *Manual of Acarology* (Lindquist et al. 2009); see also Fig. 9.3 and Table 9.1.

9.4.1.1 Trombidiform Mites

Trombidiform mites are the most diverse assemblage (ca. 25,800 species) and exhibit a correspondingly diverse range of ecologies; summarised by Krantz (2009). Two main divisions can be recognised. Sphaerolichida is a small clade of poorly-known mites which may be fungal feeders and/or predators. None are considered parasites. Most trombidiforms belong to the Prostigmata, a group which as a whole can potentially be traced back to the Early Devonian (ca. 407 Ma) based on material first described by Hirst (1923), although Sidorchuk (2018) cautioned that the precise affinities of these fossils are unclear. Prostigmatids include free-living predators, such as snout mites or rake-legged mites, as well as plant pests such as spider mites. A number of prostigmatan superfamily lineages are parasitic, or at least contain parasitic species, as summarised below:

1. Halacaroidea within the cohort Eupodides are a group of mites inhabiting marine or freshwater environments (Bartsch 1989); they should not to be confused with water mites (see below). While most halacaroid mites are believed to be predators, or to feed on algae, several taxa are known or suspected parasites (e.g. Bartsch 1987) of other aquatic invertebrates. There is no fossil record.
2. Tydeoidea, also in Eupodides, is a diverse assemblage of mites which includes predatory and omnivorous species. Some members of the tydeoid family

Erynetidae feed on the blood of gastropods (snail mites) while others are haematophagous in the nasal cavities of amphibians and birds (André and Fain 2000). The entire superfamily potentially goes back to the Early Devonian (Dunlop and Garwood 2018) with fossils assigned to Tydeidae, but the Erynetidae (with parasitic species) lacks a fossil record.

3. Parasitengona is a cohort accommodating a substantial number of species, including velvet mites, chiggers and water mites. They all share a particular developmental strategy in which inactive and active juvenile stages alternate with one another, and in which the larva—which is morphologically quite different from the adult—is active and parasitises invertebrate or vertebrate hosts, see Wohltmann (2000) for a review. The larvae of water mites (Hydrachnidiae) parasitise aquatic insects and occasionally sponges and molluscs. The larvae of terrestrial parasitengonids usually parasitise other arthropods and are quite commonly found in amber attached to a host. Larvae of Trombiculidae (chiggers) typically parasitise vertebrates and can induce lesions in humans with intense itching. Chiggers are also vectors of diseases such as scrub typhus. Parasitengona can be traced back to the Cretaceous—specifically the families Erythraeidae and Tanupodidae. For example the ca. 105 Ma San Just (Spain) amber includes a larval mites parasitizing a fly (Arillo et al. 2018). The San Just fossils were assigned to *Leptus* sp., but may be misidentified and not congeneric (Sidorchuk and Khaustov 2018). The ca. 78 Ma Canadian amber includes an erythraeid feeding on a midge (Poinar Jr et al. 1997). Numerous examples of larval parasitengonids attached to insect hosts (Fig. 9.4a, b) have been recorded in the younger Baltic and Dominican ambers (e.g. Poinar Jr 1985a; Poinar Jr et al. 1991; Eichmann 2002; Dunlop and Penney 2012).
4. Myobioidea, in the cohort Raphignathina, contains a single family, Myobiidae, consisting of ectoparasites found in the fur of a variety of marsupial and placental mammals. Their origins may be South American and potentially predate the marsupial/placental split, which may have occurred in the Jurassic. Fossil of these mites were recently recognised in Eocene (ca. 44–49 Ma) Baltic amber (Sidorchuk et al. 2019). The authors found the mites associated with fossil hairs (Fig. 9.4c) from which they inferred that the host may have belonged to the extinct Amphilemuridae, the probable sister-family to hedgehogs. There is also a subfossil record of these mites on a Pleistocene vole (Dubinina and Bochkov 1996).
5. Pterygosomatoidea, also in Raphignathina, contains a single family, Pterygosomatidae, which is typically ectoparasitic on lizards and tortoises, and occasionally also arthropods such as cockroaches. A fossil pterygosomatid was described from Cretaceous (ca. 100 Ma) Archingeay amber from France by Sidorchuk and Khaustov (2018). The authors could not match it to a definitive host, but speculated that it may have lived on the cockroaches that are also found quite commonly in this amber.
6. Cheyletoidea, also in Raphignathina, is an important assemblage of families, of which Cheyletidae includes free-living predators, nest-associates and parasites of arthropods, birds and small mammals. The remaining six cheyletoid families

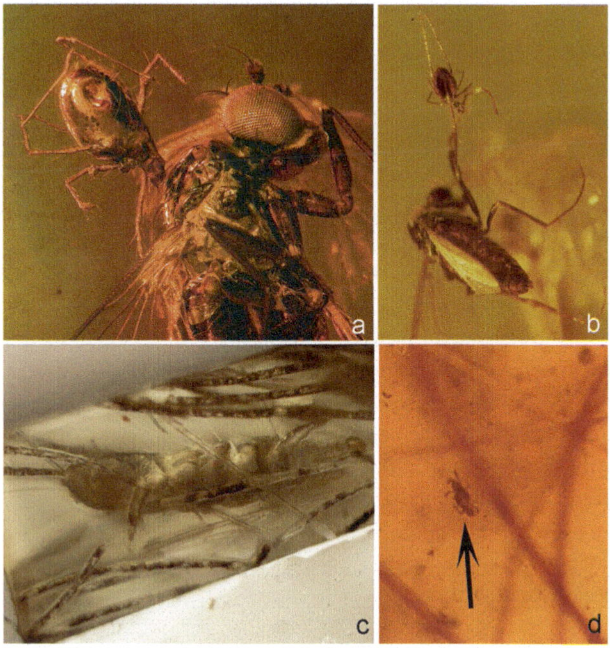

Fig. 9.4 Fossil acariform mites exhibiting parasitic behaviour, or potentially belonging to parasitic clades. (**a**, **b**) Juveniles of *Leptus* sp. (Erythraeidae) attached to flies (Diptera) in Eocene Baltic amber. (**a** Reproduced from Dunlop and Penney (2012) with permission from Siri Scientific Press, **b** courtesy of Michael Zwanzig). (**c**) *Protohylomysobia erinaceophilus* (Sidorchuk and Bochkov 2019) (Myobiidae) associated with hairs of its mammalian host's hairs; also in Baltic amber. (Image courtesy of Ekaterina Sidorchuk). (**d**) Fossil sarcoptiform fur mite (?Apotomelidae) in Miocene Dominican amber. (Image courtesy of George Poinar Jr.)

are all parasitic on reptiles, birds and mammals. They can infect areas such as feather quills, the upper skin layers or hair follicles, and induce diseases such as 'sheep itch'. The human face mite *Demodex folliculorum* belongs to this group and may be a causative factor in acne. Other members of this superfamily are found in the cloaca of turtles and closely related forms in the respiratory tract of birds; these may indicate an origin in deep time back to early reptiles. Free-living Cheyletidae are known from the Cretaceous (Cockerell 1917), but the six purely parasitic families lack a fossil record.

7. Scutacaroidea, in the cohort Heterostigmata includes two families, and they are often associated with other arthropods. Members of one family (Microdispidae) are thought to be parasites or parasitoids on cockroaches and ants. There is no fossil record.
8. Pyemotoidea, also in Heterostigmata, includes four families, the adult females of which are usually parasitoids of the eggs and larvae of a range of insects. Some species paralyse their prey with a toxin and these mites are sometimes found in stored food products where they have been implicated in human dermatitis, or 'grain itch'. The family Resinacaridae is known from Cretaceous

Burmese amber (Khaustov and Poinar Jr 2010), Acarophenacidae from Cretaceous (ca. 100 Ma) Taimyr amber (Russia) (Magowski 1994, 1995), and Pyemotidae is known from Ukrainian Rovno amber (Khaustov and Perkovsky 2010), which is probably about the same age as Baltic amber.

9. The extinct superorder Nasutiacaroidea with a single family Nasutiacaridae from Cretaceous (ca. 85–97 Ma) Vendean amber (France) is probably a heterostigmatan and has stylet like mouthparts which could indicate an association with another group of animals, but explicit evidence for a parasitic lifestyle is lacking (Sidorchuk et al. 2016).
10. Tarsonemoidea, also in Heterostigmata, includes two families. Some members of the Tarsonemidae are parasites or parasitoids of insects, for example *Acarapis woodi* the honeybee tracheal mite (Gary and Page Jr 1989). All members of the second family, Podapolipidae, are external or internal parasites of arthropods, again sometimes in the trachea or the air sacs of bees, or in the reproductive tissues of beetles. Tarsonemidae are known as subfossils (probably less than a million years old) from Japanese copal (Aoki 1974).

9.4.1.2 Sarcoptiform Mites

Sarcoptiforms are sometimes referred to as ‘chewing mites’, although it would be more correct to state that they ingest particulate matter with chewing restricted to the oribatids. They can be divided into two main groups (Fig. 9.3): Endeostigmata and Oribatida. The endeostigmatids are traditionally a small but ancient lineage, with putative records going back to the early Devonian (Dunlop and Garwood 2018). Recent work suggested that gall mites may be endeostigmatids too (Bolton et al. 2017), in which case they would massively increase the number of known species. A subsequent molecular study (Klimov et al. 2018) found support for gall mites being either endeostigmatid or having the more traditional trombidiform affinities, depending on the genes used. Living species of endostigmatids (excluding gall mites) have been variously reported feeding on nematodes, other mites, fungi or algae. Parasitic lifestyles have not been recorded. Oribatid mites are a highly diverse group of more than 9000 species, with a fairly good fossil record from the mid-Devonian onwards thanks to their often quite heavily-sclerotised bodies. Most oribatids are saprophages or mycophages with occasional records of predatory behaviour or scavenging on dead arthropods. Again, there are no records of parasitic oribatids, although a few species are known to act as intermediate hosts for cestode worms (e.g. Akrami et al. 2007).

The cohort Astigmata is now widely regarded as having evolved from within the Desmonomata clade of the oribatids (Fig. 9.3). Astigmatans are typically minute, soft bodied mites which (as the name implies) lack tracheae and simply respire over the cuticle. The ca. 5000 species of astigmatans are ecologically diverse and it is difficult to make generalisations about their biology, but in many cases they are specialised for exploiting ephemeral habitats. To this end they have evolved a specialised dispersing deutonymph stage, the hypopus, which has an attachment organ and is often transported by a larger animal in an example of phoresy (see also Sect. 9.5). Thus intimate associations with larger arthropods or even vertebrates have evolved in several groups, and in at least four astigmatan lineages these lifestyles have gone further and can be considered parasitic:

1. Hemisarcoptoidea contains a few species in the Hemisarcoptidae which have been recorded as haematophagous parasite of beetles or scale insects. These mites have even been employed in pest control against scale insects and one hemisarcoptid has been used as a model organism for how parasitism may have evolved from phoresy (Holte et al. 2001; see also Sect. 9.5). Some Winterschmidtiidae have been recorded feeding on the paralysed prey of a wasp, or on the larva or the pupa of the wasp itself. The entire superfamily can be dated back to a record of Winterschmidtiidae in Miocene (ca. 16 Ma) Chiapas amber (Mexico) (Türk 1963).
2. Members of the Glycyphagoidea are thought to have been originally associated with vertebrate nests (OConnor 1994). The families Pedetopodidae, Chortoglyphidae, Echimyopodidae and some Glycyphagidae have deutonymphs which parasitize the hair follicles of rodents causing dermatitis. Some glycyphagids are again thought to induce dermatitis in people handling food products (see also above). There is no fossil record.
3. All members of the family Hypoderatidae, the only family of the Hypoderatoidea, have deutonymphs which parasitise the subcutaneous tissues of birds (e.g. Pence et al. 1997) and (rarely) desert rodents. The deutonymph is the only stage of the lifecycle which acquires food—i.e. adults do not feed—and can engorge themselves to increase their body volume by up to 1000 times, for a mechanism see Alberti et al. (2016). There is no fossil record and many extant species are only known from the deutonymph.
4. Psoroptida are a putatively monophyletic assemblage of three superfamilies: Pterolichoidea and Analgoidea (feather mites), and Sarcoptoidea (fur mites). An important characteristic of this group is that they have lost the deutonymph stage of the lifecycle and are transferred from host to host by direct contact. Psoroptid mites typically feed on oils or sebaceous secretions and can be considered parasites in that their presence can damage the plumage or fur. Some psoroptids infect the respiratory tract of the host and others burrow into the upper layer of the skin. Classic examples here would be mange mites, causing hair loss in dogs, and scabies mites inducing itching in humans. Possible feather mite eggs described from the Cretaceous (ca. 115 Ma) Crato Formation of Brazil (Martill and Davis 1998) were rejected as unconvincing by Proctor (2003). The oldest psoroptid is thus a putative fur mite (Fig. 9.4d)—now thought to belong to the family Apotomelidae—from Miocene (ca. 16 Ma) Dominican amber (Poinar Jr 1988).

9.4.2 *Parasitiform Mites*

As the name implies, the order Parasitiformes includes several parasitic lineages of mites, although it should be stressed that this is not the lifestyle for the majority of the species, including the genus *Parasitus* (see e.g. Rapp 1959) which gives the group its name. Four main parasitiform clades can be recognised (Fig. 9.3): Opilioacarida, Holothyrida, Ixodida and Mesostigmata.

9.4.2.1 Opilioacarids

Opilioacarids are rare mites (ca. 40 species) which resemble small harvestmen. They have been observed feeding on pollen, fungal spores and small arthropods (Walter and Proctor 1998), and like harvestmen they are capable of ingesting particulate matter. Opilioacarids are known since the Cretaceous (Dunlop and de Oliveira Bernardi 2014).

9.4.2.2 Holothyrids

Holothyrids are large, dome-shaped mites which appear to be scavengers and have been recorded sucking haemolymph from dead invertebrates (Walter and Proctor 1998). In this context several authors regard Holothyrida as the sister-group of Ixodida (e.g. Klompen 2010), and if this phylogeny is correct holothyrid ecology could offer valuable insights into the blood-feeding origins of ticks. Holothyrids lack a fossil record.

9.4.2.3 Ticks (Ixodida)

Ticks are probably the most familiar parasitic arachnids, being quite commonly encountered and of medium to large size. Engorged females of some tropical species can be a couple of centimetres long. About 900 tick species have been described, all of which are haematophagous ectoparasites of terrestrial vertebrates. Ticks have attracted a considerable body of research as they are important disease vectors both for humans and domestic animals. For a general summary of tick biology see Sonenshine and Roe (2013). More detailed accounts of their phylogeny and evolution can be found in, for example, Estrada-Peña and de la Fuente (2018), Klompen et al. (1996), Barker and Murrell (2002, 2004), de la Fuente (2003), Nava et al. (2009) and Mans et al. (2012, 2016). The ecology and epidemiological significance of ticks was reviewed by Estrada-Peña and de la Fuente (2014), and references therein.

In brief, ticks have a complex life cycle in which the larva and subsequently the nymph ingest a blood meal before dropping off and moulting. Adult ticks also feed on blood, may mate on the host, and the engorged female usually produces thousands of eggs. Different instars often use different hosts, such that three-host lifecycles are commonplace while two- or one-host lifecycles are rarer (Estrada-Peña and de la Fuente 2014). Ticks usually live dormant in the undergrowth, but specific environmental conditions (e.g. weather or photoperiod) induce 'questing' behaviour in which the parasites typically climb up grass stems, or similar vegetation, and wait for a host to brush by. Once on the host they seek a suitable feeding site, 'cement' their serrated mouthparts into the host's skin and then subsequently pass compounds from the salivary glands into the host which help maintain a constant flow of blood during feeding (Kazimírova and Štibrániová 2013). Any pathogens in the host's blood enter the gut of the tick. In order to be transferred to new hosts these pathogens must migrate into the salivary glands, where they will enter a new host during the next round of feeding behaviour. This is referred to as trans-stadial (or horizontal) transmission. In some cases the pathogen migrates into the ovaries and eggs where it is passed onto a new generation of (already infected) ticks. This is referred to as trans-ovarial (or vertical) transmission. Pathogens transmitted by ticks include viruses, bacteria and protozoans, and they are significant vectors of conditions such as typhus or Rocky Mountain spotted fever (both caused by *Rickettsia* bacteria) and Lyme disease (borreliosis caused by spirochetes) among others; see e.g. de la Fuente et al. 2008; Poinar 2021 for a review.

Extant ticks can be divided into three families. Nuttalliellidae is restricted to Africa and has a single species that is probably sister to the remaining extant ticks. Argasidae are often referred to as soft ticks and occasionally also as bird ticks, although this is somewhat misleading as several use mammals as hosts. Ixodidae are the hard ticks and have a shield-like dorsal surface, or scutum. Nuttalliellidae lacks a fossil record, but both Argasidae and Ixodidae can be traced back to the Cretaceous. The fossil record of the ticks can be summarised (form oldest to youngest) as follows:

1. One of the oldest ticks is *Deinocroton draculi* (Fig. 9.5a) from Cretaceous (ca. 100 Ma) Burmese amber (Peñalver et al. 2017). These authors placed the inclusions in a new, extinct family Deinocrotonidae which appears to be more closely related to Nuttalliellidae. *Deinocroton* is associated with specialised setae (hastisetae) belonging to the larvae of dermestid beetles, modern examples of which are typically inhabitants of bird nests. This may imply that these extinct ticks also inhabited the nests of feathered dinosaurs and used them as hosts.
2. The oldest hard ticks (Ixodidae) also come from Burmese amber, from which four genera have now been recognised. *Cornupalpatum* (Fig. 9.5d) and *Compluriscutula* are extinct taxa, described by Poinar Jr and Brown (2003) and Poinar Jr and Buckley (2008), respectively. Poinar Jr (2015a) subsequently claimed to have observed rickettsial-like cells in *Cornupalpatum* Poinar (2021). A nymph of the same genus was recorded grasping a pennaceous feather in Burmese amber (Peñalver et al. 2017). This strongly suggests it used either non-avian feathered dinosaurs and/or early birds as hosts.

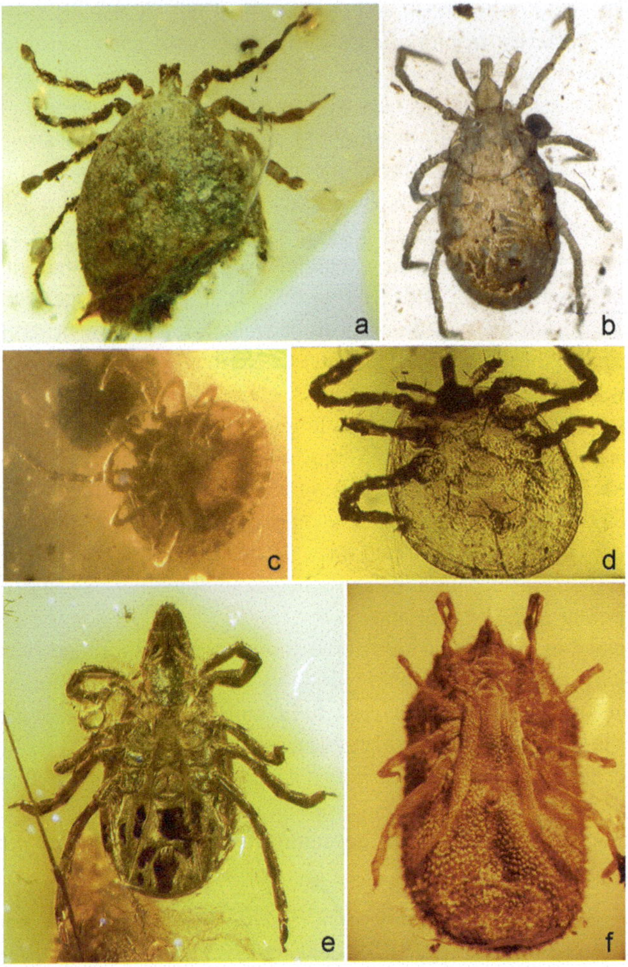

Fig. 9.5 Fossil ticks (obligate ectoparasites of vertebrates) preserved in amber. (**a**) *Deinocroton draculi* (Peñalver et al. 2017) (Deinocrotonidae) from mid-Cretaceous Burmese amber. (**b**) *Haemaphysalis* (*Alloceraea*) *cretacea* (Chitimia-Dobler et al. 2017) (Ixodidae) from Burmese amber. (**c**) *Amblyomma birmitum* (Chitimia-Dobler et al. 2017) (Ixodidae) from Burmese amber. (**d**) *Cornupalpatum burmanicum* (Poinar Jr and Brown 2003) from Burmese amber. (**e**) *Ixodes succineus* (Weidner 1964) (Ixodidae) from Eocene Baltic amber. (**f**) *Ornithodoros antiquus* (Poinar Jr 1995) from Miocene Dominican amber. (Image **a** courtesy of Enrique Peñalver, **b** and **c** courtesy of Lidia Chitimia-Dobler, **d** and **f** courtesy of George Poinar Jr.)

3. The two extinct hard tick genera may be related to the extant genus *Amblyomma*; first reported in Burmese amber by Klompen *in* Grimaldi et al. (2002) with a formally described *Amblyomma* species (Fig. 9.5c) added by Chitimia-Dobler et al. (2017). *Amblyomma* usually feed on reptiles today. Another extant genus, *Haemaphysalis* (Fig. 9.5b) has also been described from Burmese amber (Chitimia-Dobler et al. 2018). Modern *Haemaphysalis* species typically

parasitise mammals, such that this fossil nymph may be among the oldest examples of a mammalian ectoparasite.

4. The oldest soft tick (Argasidae) comes from the Cretaceous (ca. 90–94 Ma) New Jersey amber (Klompen and Grimaldi 2001). This larval specimen was described in the extant genus *Carios*, although this is usually subsumed under the genus *Ornithodoros* now. Klompen and Grimaldi (2001) noted that most ticks originally assigned to *Carios* were bat parasites, although bats had probably not evolved by that time. Birds, or even pterosaurs, as possible alternative hosts were discussed. An undescribed, bloated tick from Burmese amber figured as an argasid was figured by Shi et al. (2012), although it likely belongs to a different tick group.
5. Eocene (ca. 49 Ma) Baltic amber has yielded the oldest record of the extant genus *Ixodes* (Fig. 9.5f). First described by Weidner (1964), it was redescribed with the help of computed tomography by Dunlop et al. (2016), who determined that it is not closely related to the modern sheep tick *Ixodes ricinus* as originally thought, but is instead closer to the Asian species *Ixodes ovatus*. These authors also discussed molecular estimates for when tick-borne pathogens are thought to have appeared compared to the fossil record of the ticks that may have carried them. A putative Baltic amber record of another extant hard tick genus, *Hyalomma*, figured in de la Fuente (2003) is probably a misidentified, and non-parasitic, caeculid mite (Chitimia-Dobler et al. 2017), but see Estrada Peña and de la Fuente (2018) for a rebuttal. A contemporary non-amber record from the Eocene Green River Formation of the USA (Scudder 1885) is a *nomen dubium*, and probably not even a tick (Dunlop 2011).
6. Miocene (ca. 16 Ma) Dominican Republic amber hosts records of the soft tick *Ornithodoros* Fig. 9.5g) described by Poinar Jr (1995) as well as several records of the hard tick *Amblyomma* which appear to be very close to living Neotropical species (Lane and Poinar Jr 1986; Keirans et al. 2002). Spirochete-like cells were reported by Poinar Jr (2015b) from the body cavity of some of these *Amblyomma* inclusions.
7. There are two subfossil records of living hard tick species, namely a *Dermacentor* from the ear of a fossil rhinoceros (Kulczyński in Schille 1916) and an *Ixodes* discovered in an owl pellet (Sanchez et al. 2010).

9.4.2.4 Mesostigmatids

Mesostigmatids—sometimes referred to as gamasid mites—are a diverse assemblage with around 11,400 extant species. They are sometimes referred to as predatory mites, but this conceals a range of lifestyles from free-living predators of nematode worms or microarthropods, to animals which feed on fungi, pollen or nectar and, of course, several parasites (Table 9.1). It should be noted that the feeding ecology remains equivocal for a number of groups. Many mesostigmatids are found in association with insects or vertebrates, often with phoretic relationships (see also Sect. 9.5). Note that the family Parasitidae was created for non-parasitic species (e.g. Elbardy 1972) which were initially assumed to be parasites since they were found associated with other arthropods. Three main clades within Mesostigmata can be recognised: Sejida, Trigynaspida and Monogynaspida. For a recent overview of the mesostigmatid fossil record see Dunlop et al. (2018: Table 1). There are reports of Cretaceous mesostigmatids in amber (Poinar Jr 1998; Peñalver et al. 2017), but none of them have been formally assigned to a family. Of the three main clades, sejids are predators, the deutonymphs are often phoretic on beetles, but parasitism has not been recorded. Parasitic species are found within the two remaining subgroups, albeit somewhat sporadically in five of the superfamilies listed below:

1. The superfamily Antennophoroidea in the cohort Antennophorina contains the only parasites within the Trigynaspida clade. Members of the genus *Antennophorus* (Antennophoridae) are associated with ants and induce the host to regurgitate food for them (Franks et al. 1991). This may actually be an example of kleptoparasitism rather than parasitism *per se* as the host is not attacked directly. There is no fossil record.
2. Uropodoidea in the cohort Uropodina are a monogynaspid group well known for their phoretic behaviours (e.g. Faasch and Schaller 1966; see also Sect. 9.5.2) and their association with other arthropods. At least one species of the genus *Macrodinychus* (Macrodinychidae) is a parasitoid of ants which feeds on the ants' pupal stage. The wider Uropodoidea lineage dates back to at least Palaeogene (ca. 44–49 Ma) Baltic amber, but these are phoretic fossils rather than parasites.
3. Heterozerconidae, the only family in Heterozerconoidea and the cohort Heterozerconina, are unusual mites associated with millipedes and occasionally reptiles. They have a pair of sucker-like discs for attaching themselves to their host (Gerdeman and Alberti 2007). Relationships are thought to be mostly commensal, but at least one species is a haematophagous parasite of worm-lizards (Amphisbaena) (Flechtmann and Johnston 1990). There is no fossil record.
4. Ascoidea in the cohort Gamasina is a diverse superfamily which includes a number of taxa that feed on fungi or pollen (the family Ameroseiidae) as well as species with a more typical predatory lifestyle (e.g. Walter 1987). The genus *Proctolaelaps* (Ascidae) includes a handful of species known (or suspected) to be parasites of their insect hosts (Egan and Hunter 1975). Ascoidea as a group are known from Eocene Baltic amber, but again the record here appears to be phoretic rather than parasitic.
5. Phytoseiodea is a gamasine superfamily which includes several predatory taxa, some of which have potential as biological control agents against, e.g., spider mites. One family (Otopheidomenidae) is exclusively parasitic on insects (e.g. Zhang 1995) with different subgroups feeding on moths, true bugs, crickets or termites. There is no fossil record.

The majority of the parasitic mesostigmatids occur in the final gamasine superfamily, Dermanyssoidea, which is thus considered in more detail below. The original lifestyle was probably free-living, as is still the case in some Laelapidae. The remaining dermanyssoid mites are, or are thought to be, either blood-feeding ectoparasites, or endoparasites of the respiratory tract and parasitism probably arose on multiple occasions (Dowling and OConnor 2010). Some families are known from only one—or a handful—of species and can have quite specific host relationships (see below). Data on host taxa was primarily taken from Valiente et al. (2005) and Lindquist et al. (2009). The entire Dermanyssoidea group can be dated to an Eocene (ca. 44–49 Ma) Baltic amber record of Laelapidae, but the other dermanyssoid families lack as fossil record:

1. Laelapidae include both free living and parasitic forms. Some species parasitise other arthropods and a remarkable fossil discovery was an example of the extant genus *Myrmozercon* attached to the head of an ant (Dunlop et al. 2014) in Eocene Baltic amber (Fig. 9.6a, b). Living members of this genus are thought to be parasitic on ants—as opposed to merely commensal—but this has yet to be demonstrated experimentally (Joharchi and Moradi 2013). Other laelapids are typically associated with the nests of small mammals or birds and several species have shifted to a parasitic lifestyle and feed on blood or lymph. At least one species, the spiny rat mite *Laelaps echidnina*, can induce dermatitis in humans (Diaz 2010).
2. Varroidae are an economically important group which parasitise bees. These pests attack both adults and larvae—which in itself may kill the host—and also transmit pathogens (e.g. Rosenkrantz et al. 2010). Varroid mites are thus thought to be one of the factors causing a decline in honeybee numbers (e.g. Le Conte et al. 2010).
3. Iphiopsididae are associated with myriapods, arachnids and terrestrial crustaceans and are sometimes treated as a subgroup of Laelapidae (Nemati et al. 2015). Their exact mode of life remains equivocal, but they are thought to be either commensal or parasitic on their hosts.
4. Ixodorhynchidae are ectoparasites of snakes (Dowling 2009).
5. Omentolaelapidae is a monotypic family, whose sole species is an ectoparasite of colubrid snakes (Fajfer 2012).
6. Dasyponyssidae are a rare group with two genera which parasitise armadillos (Radovsky and Yunker 1971).
7. Manitherionyssidae is a monotypic family whose sole species parasitises pangolins (Valiente et al. 2005).
8. Hystrichonyssidae is a monotypic family whose sole species parasitises Asian porcupines (Keegen et al. 1960) and (questionably) snakes.
9. Spinturnicidae are ectoparasites of the wing membranes of microbats (= Microchiroptera) (Gettinger and Gribel 1989).
10. Spelaeorhychidae are a rare group of quite large mesostigmatids which were originally believed to be ticks. These mites parasitise bats (Martyn 1988). Their demanyssoid affinities have been questioned in the literature and they may deserve their own superfamily.
11. Macronyssidae are a fairly large group of haematophagous ectoparasites which can be found on reptiles, birds and mammals. Some species such as the tropical rat mite (*Ornithonyssus bacoti*) can also affect humans causing dermatitis

Fig. 9.6 Fossil mesostigmatid mites exhibiting possible parasitic or phoretic behaviour. (**a**) An ant, *Ctenobethylus goepperti* (Mayr 1868), from Eocene Baltic amber carrying a putative parasitic mesostigmatid mite assigned to *Myrmozercon* sp. (**b**) Detail of the mite. (**c**) A longhorn beetle (Coleoptera: Cerambycidae) from Baltic amber carrying several phoretic mesostigmatid mites (arrowed). Inset shows details of a deutonymph of a uropodine mite (Uropodina,?*Uroobovella*). (Image **c** courtesy of Michael Zwanzig)

(Beck 2008) and are also of note as potential vectors for pathogens such as *Rickettsia* (Reeves et al. 2007)

12. Dermanyssidae are another substantial group that are ectoparasites of birds (e.g. the red poultry mite *Dermanyssus gallinae*), or rodents. Again humans can be affected (Diaz 2010) through, e.g., dermatitis in poultry workers or via *Rickettsia* from mites associated with rodents.
13. Halarachnidae, sometimes referred to as nasal mites, generally infect the respiratory tract of a range of terrestrial (Furman 1954) and marine (Konishi and Shimazaki 1998) mammals. Their presence can induce excessive mucus production, rhinitis and sinusitis. One subfamily parasitises the ear instead.

14. Rhinonyssidae are another group of nasal mites, this time found in the respiratory tract of various birds (de Rojas et al. 2001).
15. Entonyssidae are endoparasites found in the lungs of snakes (Stiller et al. 1977).

9.5 The Origins of Parasitic Behaviour

Early derivative lineages of both acariform and parasitiform mites are not parasitic and as noted above this lifestyle has clearly evolved several times independently (Fig. 9.3, Table 9.1). This begs the question of how, and when, parasitism evolved. The origins of lineages which infest specific groups of arthropods or vertebrates may be constrained by the time when the host first appeared; see also De Baets and Littlewood (2015); Warnock and Engelstädter (2021). Thus there has been considerable debate about which vertebrates were the original hosts of ticks, and correspondingly diverse estimates for tick origins from as young as the Devonian to as late as the Cretaceous; see Chitimia-Dobler et al. (2018) for a review. The fossil record of mites and ticks is largely unhelpful here as the few relevant fossils of parasitic lineages are restricted to amber deposits from the mid-Cretaceous onwards. Earlier origins and/or radiations remain obscure.

Alternatively, the fossil record of the host, or molecular clock data for their likely origins, could be used as a proxy here. For example, the monotypic mesostigmatid family Manitherionyssidae lacks a fossil record and today is only found on pangolins. These unusual mammals have a fossil record going back to the Eocene and molecular data suggests divergence from their sister group a little earlier in the Late Cretaceous (Meredith et al. 2009). This data could be used to constrain the origins of the mites as well. The problem with this line of reasoning is the implicit assumption that early members of parasitic clades used the same hosts as today. As noted for flatworms (De Baets et al. 2015), biogeography and/or fossils calibrations if available may represent more reliable guides to the origination of a given parasite—as opposed to the fossil record of its modern host(s). Furthermore, we cannot exclude the possibility of host switching (reviewed by Poulin 2011) in the past or the utilisation of now extinct groups. A case in point would be bird ticks (Argasidae: Argasinae), which are estimated to have split off from the other argasid subfamily (the mammal parasites, Ornithodorinae) during the Triassic (234 Ma: Mans et al. 2012). This obviously predates the oldest definitive bird (*Archaeopteryx*: ca. 150 Ma), but perhaps earlier argasid ticks lived on [feathered?] dinosaurs; see also Chitimia-Dobler et al. (2018).

Within hard ticks (Ixodidae) there is an estimated split between the so-called prostriates (the genus *Ixodes*) and all other hard ticks (the Metastriata) which corresponds with the end-Permian mass extinction event (reviewed by Chitimia-Dobler et al. 2018). This event in turn led to a major shift in vertebrate faunas. *Ixodes* ticks are usually found on mammals today and we might speculate (Chitimia-Dobler et al. 2018) that during the Triassic the prostriate (*Ixodes*) ticks became associated with cynodonts, from which mammals eventually evolved, while the metastriates became associated first with reptiles (e.g. the tick genus *Amblyomma*) and later—perhaps in the mid Cretaceous?—other genera underwent a further shift to mammalian hosts; as evidenced by the first example of the typically mammal-associated genus *Haemaphysalis* in Burmese amber.

9.5.1 Nest Associates

Irrespective of the timing of this behavioural shift, the modality by which a free-living species becomes a parasite is also of interest. A free-living predator may, of course, switch to parasitism directly; see e.g. Weinstein and Kuris (2016) for a discussion. It is notable that among the acariform mites and the mesostigmatid mites in general there are many species associated with the nests of birds and small mammals. It may not require a huge shift from feeding on nest microarthropods or organic material within the nest, to feeding opportunistically and eventually exclusively on the vertebrates within the nest environment. Intimate associations with the nest builder also facilitate transfer to new habitats by riding on the larger animal's body. For a review of this specifically within the Eleutherengona group and their shift to parasitism on mammals see Bochkov (2009, and references therein).

9.5.2 From Phoresy to Parasitism

In this context it is also interesting to note how widespread phoresy is among both acariform and parasitiform mites. To reiterate, phoresy is defined as a commensal relationship, of benefit to the mite but at little or no cost to the host, in which the mite effectively acts as a 'stowaway' attaching itself to a larger animal (reviewed by Binns 1982). Phoresy is typically used for dispersal and/or to exploit ephemeral sources of food. It is particularly well-developed in the astigmatans (Houck and OConnor 1991), several mesostigmatan lineages (Hunter and Rosario 1988) and in early-derivative heterostigmatans (Lindquist 1986). Given that phoresy introduces an intimate relationship between the mite and its host it is easy to envisage a process by which the mite acquired nourishment from the host during transport, and eventually came to rely on the host as the only source of food. A potential model species is the astigmatan *Hemisarcoptes cooremani*, which is carried by a beetle, but which also feeds on it during this transport (Holte et al. 2001). The situation here is complicated by the fact that the beetle can obtain water from the mite, which suggests a mutualistic relationship rather than parasitism *per se*.

9.5.2.1 Phoresy in the Fossil Record

The oldest putative example of phoresy is an oribatid from the Carboniferous (ca. 320 Ma) of Xiaheyan in China (Robin 2021; Robin et al. 2016) found on the thorax of a grasshopper-like insect in the extinct order Archaeorthoptera. The mite was assigned to the Mixonomata clade, the authors noting there was no indication of parasitism and that this was probably an example of phoresy instead. Phoresy is uncommon among oribatid mites today, but mixonomatids have been reported phoretic on, e.g., harvestmen (Townsend Jr et al. 2008) in modern ecosystems. The phoretic deutonymph of an astigmatan mite (possibly Histiostomatidae) was documented from Eocene Baltic amber attached to a woodlouse spider (Dysderidae) by Dunlop et al. (2012). Again, spiders are not the usual host of these mites, but such associations are known today (see discussion in Dunlop et al. 2012). It should be added that computer tomography (μCT) was particularly helpful in resolving morphological details such as the underside of the astigmatid's body with its characteristic sucker plate. This methodology may be applicable to other phoretic mites both in sediments and amber; see also comments in De Baets and Littlewood (2015). Other apparently phoretic acariform mites from Baltic amber were figured by Eichmann (2002), although as the author noted himself some of them may actually be parasites as its sometimes difficult to determine whether the mite is using its mouthparts.

Among the parasitiform mites there are phoretic examples of mesostigmatids in Cretaceous Burmese amber (unpublished observations). Deutonymphs of tortoise mites (Urodinychidae) were documented from Rovno and Baltic ambers (Lyubarsky and Perkovsky 2012; Dunlop et al. 2013). In both cases the juvenile mites were hosted by beetles (Coleoptera: Cryptophagidae and Cerambycidae), and the characteristic stalk-like anal pedicel which they use for attachment was clearly visible (Fig. 9.6c). The Baltic amber beetles also carried tentative records of Microgynioidea and Ascidae, leading Dunlop et al. (2013) to suggest that the rarity of fossil mesostigmatid mites may be an artefact and that further taxa may be found by searching for parasitic or phoretic mites attached to their more mobile hosts, which are more likely to have come into contact with sticky tree resin.

9.6 Chelicerates as Victims

Finally, chelicerates, are also affected by parasites. Horseshoe crabs can host a diverse fauna of epibionts (Botton et al. 2003; Leibovitz and Lewbart 2003) including cnidarians, bryozoans, bivalves and barnacles, as well as algal infections (Braverman et al. 2012). While many of these organisms are commensal animals, rather than parasites, heavy infestation of attached or encrusting animals probably hinders the crab to some extent. True parasites of horseshoe crabs can be found attacking the haemolymph-rich gills, and include certain protozoans and turbellarid worms, while a nematode is known to burrow into the cuticle. Of particular note is the trematode *Microphallus limuli*, which first infects a marine snail, before being consumed by a horseshoe crab as an intermediate host (Stunkard 1951). The presence of the parasite may impair the crab's righting behaviour during excursions onto land, causing it to be more likely to be consumed by herring gulls, which are the definitive host. Fossil examples of epibionts affecting horseshoe crabs are not known to the author, although this may yet be discovered as it has been observed as possible bacterial damage in fossil decapods (Klompmaker et al. 2016) and also in trilobites (De Baets et al. 2021b; Kenneth de Baets, pers. comm. 2020).

9.6.1 Arachnid Parasites and Parasitoids

Spiders, and other arachnids, can fall victim to a range of parasites or parasitoids (see e.g. van Helsdingen 2011) and occasionally these associations can be preserved in the fossil record too (e.g. Nagler and Haug 2015). Several can potentially be traced back to the Mesozoic, others are first recorded in Eocene ambers:

1. Parasitic nematode worms in the family Mermithidae have been reported from extant spiders and harvestmen (Poinar Jr 1985b). An extinct representative of this worm family, *Heydenius araneus*, has also been documented by Poinar Jr (2000) attacking a crab spider (Thomisidae) in Eocene Baltic amber (Fig. 9.7b).
2. The larvae of parasitengonid mites can be regularly found as ectoparasites on spiders (e.g. Mąkol and Felska 2011) and harvestmen (Townsend Jr et al. 2008;

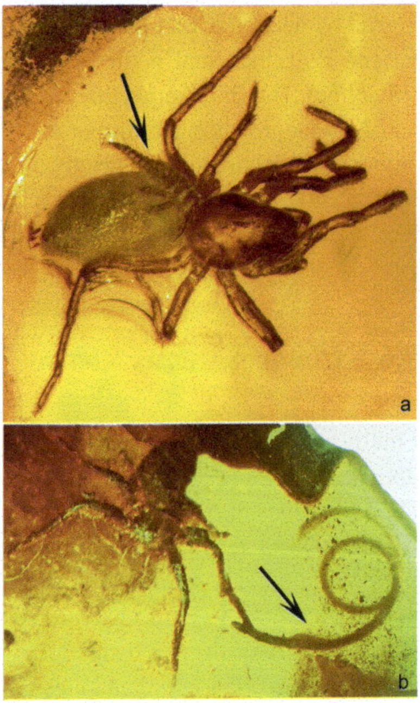

Fig. 9.7 Fossil arachnids as victims of parasites. (**a**) Larval mantispid (Neuroptera: Mantispidae) (arrowed) attached to a clubionid spider; also in Baltic amber. (**b**) The nematode worm *Heydenius araneus* (Poinar Jr 2000) (Mermethidae) (arrowed), attacking a crab spider in Eocene Baltic amber. (Images courtesy of Michael Ohl and George Poinar Jr. respectively)

Gabryś et al. 2011). Several examples of putative parasitengonid mites attached to spiders were documented from Baltic amber by Wohltmann in Wunderlich (2004); see also above.

3. Mantid flies (Neuroptera: Mantispidae) have larvae that typically feed on spider eggs. The larva either seeks an already available egg sac or attaches itself to a spider and waits for the eggs to be laid. Whether this constitutes parasitism *sensu stricto* is debatable—it represents egg parasitoidism if the egg is presumably killed—but Ohl (2011) reported a remarkable case from Baltic amber of a mantispid larva attached to a clubionid spider (Fig. 9.7a). Adult mantid flies are known as far back as the Jurassic (e.g. Jepson et al. 2013).
4. At least three families of wasps (Hymenoptera) are spider parasitoids, or at least include species which attack spiders. Pompilidae (spider wasps) are dedicated spider hunters, depositing an ectoparasitoid egg on the paralysed victim. The fossil record of Pompilidae is subject to debate. A Cretaceous example was described from Burmese amber by Engel and Grimaldi (2006), but Rodriguez et al. (2016) questioned this interpretation and restricted the group to Eocene and younger records.
5. Ichneumon wasps (Ichneumonidae) are parasitoids which usually attack the larvae of butterflies, beetles or other hymenopterans, but occasionally also spiders. Again, the egg is lain on (or in) the host, which may or may not be paralysed, and the host is eventually consumed alive. Fossil ichneumonids (Labandeira and Li 2021) can be traced back to the Cretaceous (e.g. McKellar et al. 2013). An apparent case of egg parasitism of a clubionid spider by an ichneumonid was documented in Eocene Baltic amber by Poinar Jr (2004). An ichneumonid larva, again on a clubionid spider, was described from Miocene Dominican amber by Poinar Jr (1987).
6. Some species of digger wasp (Crabronidae) attack spiders. The family as a whole is known from the Cretaceous, but the spider-hunting Trypoxylini lineage is first recorded from Eocene French amber (Nel 2005).
7. Spider flies (Diptera: Acroceridae) are parasitoids whose larva enters the host's body and often lodges near the book lung, before consuming the hosts tissues. The oldest fossil examples come from the Jurassic of central Asia (Gillung and Winterton 2017; Labandeira and Li 2021) and they have also been recorded from Burmese, Baltic and Dominican ambers. Spiders are not the only victims, and Kerr and Winterton (2008) described a remarkable case of a larval acrocerid attached to a whirligig mite (Anystidae) in Baltic amber.

Acknowledgements I thank Kenneth de Baets for inviting this contribution and Lidia Chitimia-Dobler, Enrique Peñalver, David Penney, Michael Ohl, George Poinar Jr., Ekaterina Sidochuk, Dieter Waloszek and Michael Zwanzig for providing images of specimens. Kenneth de Baets, an anonymous reviewer and Ekaterina Sidochuk provided valuable comments on the typescript. Ekaterina died in January 2019, and I would like to dedicate this work to her memory for her extensive contributions to our understanding of fossil mites.

References

Acosta LE, Machado G (2007) Diet and foraging. In: Pinto-da-Richa R, Machado G, Giribet G (eds) Harvestmen. The biology of Opiliones. Harvard University Press, Cambridge, MA, pp 309–338

Akrami MA, Saboori A, Eslami A (2007) Observations on oribatid mites (Acari: Oribatida) serving as intermediate hosts of *Moniezia expansa* (Cestoda: Anoplocephalidae) in Iran. Int J Acarol 33:365–369

Alberti G, Kanarek G, Dabert J (2016) Unusual way of feeding by the deutonymph of *Neottialges evansi* (Actinotrichida, Astigmata, Hypoderatidae), a subcutaneous parasite of cormorants, revealed by fine structural analyses. J Morph 277:1368–1389

André HM, Fain A (2000) Phylogeny, ontogeny and adaptive radiation in the superfamily Tydeoidea (Acari: Actinedida), with a reappraisal of morphological characters. Zool J Linnean Soc 130:405–448

Aoki J (1974) [On the fossil mites in Mizunami amber from Gifu Prefecture, Central Japan.] Bull Mizunami Foss Mus 1:397–399 [in Japanese with English summary]

Arillo A, Blagoderov V, Peñalver E (2018) Early Cretaceous parasitism in amber: a new species of *Burmazelmira* fly (Diptera: Archizelmiridae) parasitized by a *Leptus* sp. mite (Acari, Erythraeidae). Cretac Res 86:24–32

Bamber RN (2007) A holistic re-interpretation of the phylogeny of the Pycnogonida Latreille, 1810 (Arthropoda). Zootaxa 1668:295–312

Barker SC, Murrell A (2002) Phylogeny, evolution and historical zoogeography of ticks: a review of recent progress. Exp Appl Acarol 28:55–68

Barker SC, Murrell A (2004) Systematics and evolution of ticks with a list of valid genus and species names. Parasitology 129:S15–S36

Bartsch I (1987) *Australacarus inexpectatus* gen. et spec. nov. (Halacaroidea, Acari), mit einer Ubersicht über parasitisch lebende Halacariden. Zool Anz 218:17–24

Bartsch I (1989) Marine mites (Halacaroidea: Acari): a geographical and ecological survey. Hydrobiologia 178:21–42

Beccaloni J (2009) Arachnids. The Natural History Museum, London. 320 pp

Beck W (2008) Occurrence of a house-infesting Tropical rat mite (*Ornithonyssus bacoti*) on murides and human beings. Trav Med Infect Dis 6:245–249

Binns ES (1982) Phoresy as migration – some functional aspects of phoresy in mites. Biol Rev 57:571–620

Bochkov AV (2009) A review of mites of the parvorder Eleutherengona (Acariformes: Prostigmata)—permanent parasites of mammals. Acarina 1(Suppl):1–149

Bolton SJ, Philipp E, Chetverikov PE, Klompen H (2017) Morphological support for a clade comprising two vermiform mite lineages: Eriophyoidea (Acariformes) and Nematalycidae (Acariformes). Syst Appl Acarol 22:1096–1131

Botton ML, Shuster CN Jr, Keinath JA (2003) Horseshoe crabs in a food web: who eats whom? In: Shuster CN Jr, Barlow HB, Brockmann HJ (eds) The American horseshoe crab. Harvard University Press, Cambridge, MA, pp 133–153

Braverman H, Leibovitz L, Lewbart GA (2012) Green algal infection of American horseshoe crab (*Limulus polyphemus*) exoskeletal structures. J Invert Pathol 111:90–93

Brenneis G, Bogomolova EV, Arango CP, Krapp F (2017) From egg to "no-body": an overview and revision of developmental pathways in the ancient arthropod lineage Pycnogonida. Front Zool 16:6. https://doi.org/10.1186/s12983-017-0192-2

Chitimia-Dobler L, De Araujo BC, Ruthensteiner B, Pfeffer T, Dunlop JA (2017) *Amblyomma birmitum* a new species of hard tick in Burmese amber. Parasitology 144:1441–1448

Chitimia-Dobler L, Pfeffer T, Dunlop JA (2018) *Haemaphysalis cretacea* a nymph of a new species of hard tick in Burmese amber. Parasitology 145:1440–1451

Cockerell TDA (1917) Arthropods in Burmese amber. Psyche 24:40–45

De Baets K, Littlewood DTJ (2015) The importance of fossils in understanding the evolution of parasites and their vectors. Adv Parasitol 90:1–51

De Baets K, Dentzian-Dias PC, Upeniece I, Verneau O, Donoghue PCJ (2015) Constraining the deep origin of parasitic flatworms and host-interactions with fossil evidence. Adv Parasitol 90:93–135
De Baets K, Huntley JW, Klompmaker AA, Schiffbauer JD, Muscente AD (2021a) The fossil record of parasitism: Its extent and taphonomic constraints. In: De Baets K, Huntley JW (eds) The evolution and fossil record of parasitism: Coevolution and paleoparasitological techniques. Topics in Geobiology 50
De Baets K, Budil P, Fatka O, Geyer G (2021b) Trilobites as hosts for parasites: From paleopathologies to ethiologies. In: De Baets K, Huntley JW (eds). The evolution and fossil record of parasitism: Coevolution and paleoparasitological techniques. Topics in Geobiology 50
de la Fuente J (2003) The fossil record and the origin of ticks (Acari: Parasitiformes: Ixodida). Exp Appl Acarol 29:331–344
de la Fuente J, Estrada-Peña A, Venzal JM, Kocan KM, Sonenshine DE (2008) Overview: ticks as vectors of pathogens that cause disease in humans and animals. Front Biosci 13:6938–6946
de Rojas M, Mora MD, Ubeda JM, Cutillas C, Navajas M, Guevara DC (2001) Phylogenetic relationships in rhinonyssid mites (Acari: Rhinonyssidae) based on mitochondrial 16S rDNA sequences. Exp Appl Acarol 25:957–967
Diaz JH (2010) Mite-transmitted dermatoses and infectious diseases in returning travelers. J Travel Med 17:21–31
Dowling APG (2009) *Ixobioides truncatus* (Johnston) comb. nov. (Acari: Mesostigmata: Ixodorhynchidae): a synonymy and redescription. Syst Appl Acarol 14:216–224
Dowling APG, OConnor BM (2010) Phylogeny of Dermanyssoidea (Acari: Parasitiformes) suggests multiple origins of parasitism. Acarologia 50:113–129
Dubinina HV, Bochkov AV (1996) Pleistocene mites (Acariformes) from the narrow-skulled vole *Microtus gregalis egorovi* Feigin (Rodentia: Arvicolinae). Acarina 4:35–38
Dunlop JA (1996) A trigonotarbid arachnid from the Upper Silurian of Shropshire. Palaeontology 39:605–614
Dunlop JA (2010) Geological history and phylogeny of Chelicerata. Arthropod Struct Dev 39:124–142
Dunlop JA (2011) The fate and status of the supposed fossil tick *Ixodes tertiarius* Scudder, 1885. Acarologia 51:399–404
Dunlop JA, Alberti G (2008) The affinities of mites and ticks: a review. J Zool Syst Evol Res 46:1–18
Dunlop JA, Arango CP (2005) Pycnogonid affinities: a review. J Zool Syst Evol Res 43:8–21
Dunlop JA, de Oliveira Bernardi LF (2014) An opilioacarid mite in Cretaceous Burmese amber. Naturwissenschaften 101:759–763
Dunlop JA, Garwood RJ (2018) Terrestrial invertebrates in the Rhynie chert ecosystem. Philos Trans R Soc B 373:20160493
Dunlop JA, Penney D (2012) Fossil arachnids. Siri Scientific Press, Manchester
Dunlop JA, Wirth S, Penney D, McNeil A, Bradley RS, Withers PJ, Preziosi RF (2012) A minute fossil phoretic mite recovered by phase-contrast X-ray computed tomography. Biol Lett 8:475–460
Dunlop JA, Kontschán J, Zwanzig M (2013) Fossil mesostigmatid mites (Mesostigmata: Gamasina, Microgyniina, Uropodina), associated with longhorn beetles (Coleoptera: Cerambycidae) in Baltic amber. Naturwissenschaften 100:337–344
Dunlop JA, Kontschán J, Walter DE, Perrichot V (2014) An ant-associated mesostigmatid mite in Baltic amber. Biol Lett 10:20140531. https://doi.org/10.1098/rsbl.2014.0531
Dunlop JA, Apanaskevich DA, Lehmann J, Hoffmann R, Fusseis F, Ehlke M, Zachow S, Xiao X (2016) Microtomography of the Baltic amber tick *Ixodes succineus* reveals affinities with the modern Asian disease vector *Ixodes ovatus*. BMC Evol Biol 16:203. https://doi.org/10.1186/s12862-016-0777-y
Dunlop JA, Walter DE, Kontschán J (2018) The status of the putative fossil sejid mite in Baltic amber. Acarologia 58:665–672
Egan MK, Hunter PE (1975) Redescription of a cockroach mite, *Proctolaelaps nauphoetae*, with notes on its biology. Ann Entomol Soc Am 68:361–364
Eichmann F (2002) Paläosymbiose im Bernstein. Arbeitskreis Päl Hannover 30:1–28
Elbardy EA (1972) Observations on the biology of *Pergamasus crassipes* (L.), a predaceous gamasid mite inhabiting forest soils in Bavaria (Acarina: Mesostigmata: Parasitidae). J Appl Entomol 71:296–303

Engel MS, Grimaldi DA (2006) The first Cretaceous spider wasp (Hymenoptera: Pompilidae). J Kansas Entomol Soc 79:359–368
Esch GW, Fernández JC (1993) A functional biology of parasitism. Chapman and Hall, London
Estrada-Peña A, de la Fuente J (2014) The ecology of ticks and epidemiology of tick-borne viral diseases. Antivir Res 108:104–128
Estrada-Peña A, de la Fuente J (2018) The fossil record and the origin of ticks revisited. Exp Appl Acarol 75:255–261
Faasch H, Schaller F (1966) Das Phoresie-Verhalten der Milben *Uroobovella marginata* Koch 1839 und *Uropoda orbicularis* Müller (Acari, Uropodina). Zool Anz 176:176–182
Fajfer M (2012) Acari (Chelicerata) – parasites of reptiles. Acarina 20:108–129
Flechtmann CHW, Johnston DE (1990) *Zeterohercon*, a new genus of Heterozerconidae (Acari: Mesostigmata) and the description of *Zeterohercon amphisbaenae* n. sp. from Brazil. Int J Acarol 16:143–148
Franks NR, Healy KJ, Byrom L (1991) Studies on the relationship between the ant ectoparasite *Antennophorus grandis* (Acarina: Antennophoridae) and its host *Lasius flavus* (Hymenoptera: Formicidae). J Zool 225:59–70
Furman DP (1954) A revision of the genus *Pneumonyssus* (Acarina: Halarachnidae). J Parasitol 40:31–42
Gabryś G, Felska M, Kłosińska A, Staręga W, Mąkol J (2011) Harvestmen (Opiliones) as hosts of Parasitengona (Acari: Actinotrichida, Prostigmata) larvae. J Arachnol 39:349–351
Gary NE, Page RE Jr (1989) Tracheal mite (Acari: Tarsonemidae) infestation effects on foraging and survivorship of honey bees (Hymenoptera: Apidae). J Econ Entomol 82:734–739
Gerdeman B, Alberti G (2007) First ultrastructural observations on the paired suckers of a heterozerconid mite (Heterozerconidae; Gamasida). In: Morales-Malacara JB et al (eds) Acarology XI: Proceedings of the International Congress, pp 581–584
Gettinger D, Gribel R (1989) Spinturnicid mites (Gamasida: Spinturnicidae) associated with bats in Central Brazil. J Med Entomol 26:491–493
Gillung JP, Winterton SL (2017) A review of fossil spider flies (Diptera: Acroceridae) with descriptions of new genera and species from Baltic Amber. J Syst Palaeontol 16:325–350
Giribet G (2018) Current views on chelicerate phylogeny—a tribute to Peter Weygoldt. Zool Anz 273:7–13
Grimaldi DA, Engel MS, Nascimbene PC (2002) Fossiliferous Cretaceous amber from Myanmar (Burma): its rediscovery, biotic diversity, and paleontological significance. Am Mus Novit 3361:1–71
Haug C (2017) Feeding strategies in arthropods from the Rhynie and Windyfield cherts: ecological diversification in an early non-marine biota. Philos Trans R Soc B 373:20160492
Hirst S (1923) On some arachnid remains from the Old Red Sandstone (Rhynie Chert bed, Aberdeenshire). Ann Mag Nat Hist 12(9):455–474
Holte AE, Houck MA, Collie NL (2001) Potential role of parasitism in the evolution of mutualism in astigmatid mites: *Hemisarcoptes cooremani* as a model. Exp Appl Acarol 25:97–107
Houck MA, OConnor BM (1991) Ecological and evolutionary significance of phoresy in the Astigmata. Annu Rev Entomol 36:611–636
Hunter PE, Rosario RMT (1988) Associations of Mesostigmata with other arthropods. Annu Rev Entomol 33:393–417
Jepson JE, Heads SW, Makarkin VN, Ren D (2013) New fossil mantidflies (Insecta: Neuroptera: Mantispidae) from the Mesozoic of North-Eastern China. Palaeontology 56:603–613
Joharchi O, Moradi M (2013) Review of the genus *Myrmozercon* Berlese (Acari: Laelapidae), with description of two new species from Iran. Zootaxa 3686:244–254
Kazimírova M, Štibrániová I (2013) Tick salivary compounds: their role in modulation of host defences and pathogen transmission. Front Cell Infect Microbiol 3(43):1–19
Keegen HK, Yunker CR, Baker EW (1960) *Hystrichonyssus turneri*, n. sp., n. g., representing a new subfamily of Dermanyssidae (Acarina) from a Malayan porcupine. Malaya 29:205–208
Keirans JE, Lane RS, Cauble R (2002) A series of larval *Amblyomma* species (Acari: Ixodidae) from amber deposits in the Dominican Republic. Int J Acarol 28:61–66

Kerr PH, Winterton SL (2008) Do parasitic flies attack mites? Evidence in Baltic amber. Biol J Linn Soc 93:9–13
Khaustov AA, Perkovsky EE (2010) The first fossil record of mites of the family Pyemotidae (Acari: Heterostigmata), with description of a new species from Rovno Amber. Palaeontol J 44:418–421
Khaustov AA, Poinar GO Jr (2010) *Protoresinacarus brevipedis* gen. n., sp. n. from Early Cretaceous Burmese amber: the first fossil record of mites of the family Resinacaridae (Acari: Heterostigmata: Pyemotoidea). Hist Biol 23:219–222
Klimov PB, OConnor BM, Chetverikov PE, Bolton SJ, Pepato AR, Mortazavi AL, Tolstikov AV, Bauchan GR, Ochoa R (2018) Comprehensive phylogeny of acariform mites (Acariformes) provides insights on the origin of the four-legged mites (Eriophyoidea), a long branch. Mol Phylogenet Evol 119:105–117
Klompen H (2010) Holothyrids and ticks: new insights from larval morphology and DNA sequencing, with the description of a new species of *Diplothyrus* (Parasitiformes: Neothyridae). Acarologia 50:269–285
Klompen H, Grimaldi D (2001) First Mesozoic record of a parasitiform mite: a larval argasid tick in Cretaceous amber (Acari: Ixodida: Argasidae). Ann Entomol Soc Am 94:10–15
Klompen JSH, Black WC, Keirans JE, Oliver JH Jr (1996) Evolution of ticks. Annu Rev Entomol 41:141–161
Klompmaker AA, Chistoserdov AY, Felder DL (2016) Possible shell disease in 100 million-year-old crabs. Dis Aquat Org 119:91–99
Konishi K, Shimazaki K (1998) Halarachnid mites infesting the respiratory tract of Stellar sea lions. Biosphere Cons 1:45–48
Krantz GW (2009) Habits and habitats. In: Krantz GW, Walter DE (eds) A manual of acarology, 3rd edn. Texas Tech University Press, Lubbock, pp 64–82
Labandeira CC, Li L (2021). The history of insect parasitism and the mid-Mesozoic parasitoid revolution. In: De Baets K, Huntley JW (eds) The Evolution and Fossil Record Of Parasitism: Identification and Macroevolution of Parasites. Topics in Geobiology 49
Lamsdell JC (2013) Revised systematics of Palaeozoic 'horseshoe crabs' and the myth of monophyletic Xiphosura. Zool J Linnean Soc 167:1–27
Lamsdell JC, Braddy SJ (2010) Cope's Rule and Romer's theory: patterns of diversity and gigantism in eurypterids and Palaeozoic vertebrates. Biol Lett 6:265–269
Lane RS, Poinar GO Jr (1986) First fossil tick (Acari: Ixodidae) in New World amber. Int J Acarol 12:75–78
Le Conte Y, Ellis M, Ritter W (2010) *Varroa* mites and honey bee health: can *Varroa* explain part of the colony losses. Apidologie 41:353–363
Legg DA (2014) *Sanctacaris uncata*: the oldest chelicerate (Arthropoda). Naturwissenschaften 101:1065–1073
Leibovitz L, Lwebart GA (2003) Diseases and symbionts: vulnerability despite tough shells. In: Shuster CN Jr, Barlow HB, Brockmann HJ (eds) The American horseshoe crab. Harvard University Press, Cambridge, MA, pp 245–267
Leung TLF (2017) Fossils of parasites: what can the fossil record tell us about the evolution of parasitism? Biol Rev 92:410–430
Lindquist EE (1986) The world genera of Tarsonemidae (Acari: Heterostigmata): a morphological, phylogenetic, and systematic revision, with a reclassification of family-group taxa in the Heterostigmata. Mem Entomol Soc Canada 136:1–517
Lindquist EE, Krantz GW, Walter DE (2009) Classification. In: Krantz GW, Walter DE (eds) A manual of acarology, 3rd edn. Texas Tech University Press, Lubbock, pp 97–103
Lozano-Fernandez J, Tanner AR, Giacomelli M, Carton R, Vinther J, Edgecombe GD, Pisani D (2019) Increasing species sampling in chelicerate genomic-scale datasets provides support for monophyly of Acari and Arachnida. Nat Commun 10:2295
Lyubarsky GY, Perkovsky EE (2012) The first Eocene species of the genus *Cryptophagus* (Coleoptera, Clavicornia, Cryptophagidae). Vest Zool 46:36–40
Magowski WŁ (1994) Discovery of the first representative of the mite subcohort Heterostigmata (Arachinida: Acari) in the Mesozoic Siberian amber. Acarologia 35:229–241

Magowski WŁ (1995) Fossil heterostigmatid mites in amber – 85 million year-old an arthropod mite relationships. In: Kropczynska D, Boczek J, Tomczyk A (eds) The Acari: physiological and ecological aspects of Acari – host relationships. Dabor, Warsaw, pp 53–58
Mąkol J, Felska M (2011) New records of spiders (Araneae) as hosts of terrestrial Parasitengona mites (Acari: Actinotrichida: Prostigmata). J Arachnol 39:352–354
Mans BJ, de Klerk D, Pienaar R, Castro MH, Latif AA (2012) The mitochondrial genomes of *Nuttalliella namaqua* (Ixodoidea: Nuttalliellidae) and *Argas africolumbae* (Ixodoidea: Argasidae): estimation of divergence dates for the major tick lineages and reconstruction of ancestral blood-feeding characteristics. PLoS One 7(11):e4946
Mans BJ, de Castro MH, Pienaar R, de Klerk D, Gaven P, Genu S, Latif AA (2016) Ancestral reconstruction of tick lineages. Ticks Tick Borne Dis 7:509–535
Martill DM, Davis PG (1998) Did dinosaurs come up to scratch? Nature 396:528–529
Martyn KP (1988) A new species of the mite genus *Spelaeorhynchus* (Acarina: Mesostigmata) parasitic on bats of the family Phyllostomidae. J Nat Hist 22:757–765
Mayr GL (1868) Die Ameisen des baltischen Bernsteins. Koch, Königsberg
McKellar RC, Kopylov DS, Engel MS (2013) Ichneumonidae (Insecta: Hymenoptera) in Canadian Late Cretaceous amber. Foss Rec 16:217–227
Meredith RW, Gatesy J, Murphy WJ, Ryder OA, Springer MS (2009) Molecular decay of the tooth gene enamelin (*ENAM*) mirrors the loss of enamel in the fossil record of placental mammals. PLoS Genet 5(9):e1000634
Nagler C, Haug JT (2015) From fossil parasitoids to vectors: insects as parasites and hosts. Adv Parasitol 90:137–200
Nava S, Guglielmone AA, Mangold AJ (2009) An overview of systematics and evolution of ticks. Front Biosci 14:2857–2877
Nel A (2005) Oldest representatives of the Sphecidae: Trypoxylini in the Early Eocene French amber (Insecta: Hymenoptera). C R Palevol 4:17–24
Nemati A, Riahi E, Gwiazdowicz DJ (2015) Description of a new species of *Julolaelaps* (Acari, Mesostigmata, Laelapidae) from Iran. ZooKeys 526:105–116
OConnor BM (1994) Life-history modifications in astigmatid mites. In: Houck MA (ed) Mites. Chapman and Hall, New York, pp 136–159
Ohl M (2011) Aboard a spider—a complex developmental strategy fossilized in amber. Naturwissenschaften 98:453–456
Peñalver E, Arillo A, Delclòs X, Peris D, Grimaldi DA, Anderson SR, Nascimbene PC, Pérez-de la Fuente R (2017) Ticks parasitized feathered dinosaurs as revealed by Cretaceous amber assemblages. Nat Commun 8:1924. https://doi.org/10.1038/s41467-017-01550-z
Pence DB, Spalding MG, Bergan JF, Cole RA, Newman S, Gray PN (1997) New records of subcutaneous mites (Acari: Hypoderatidae) in birds, with examples of potential host colonization events. J Med Entomol 34:411–416
Pepato AR, Klimov PB (2015) Origin and higher-level diversification of acariform mites – evidence from nuclear ribosomal genes, extensive taxon sampling, and secondary structure alignment. BMC Evol Biol 15:178. https://doi.org/10.1186/s12862-015-0458-2
Pepato AR, Rocha CEF, Dunlop JA (2010) Phylogenetic position of the actinotrichid mites: sensitivity to homology assessment under total evidence. BMC Evol Biol 10:235. https://doi.org/10.1186/1471-2148-10-235
Poinar GO Jr (1985a) Fossil evidence of insect parasitism by mites. Int J Acarol 11:37–38
Poinar GO Jr (1985b) Mermithid (Nematoda) parasites of spiders and harvestmen. J Arachnol 13:121–128
Poinar GO Jr (1987) Fossil evidence of spider parasitism by Ichneumonidae. J Arachnol 14:399–400
Poinar GO Jr (1988) Hair in Dominican amber: evidence for Tertiary land mammals in the Antilles. Experientia 44:88–89
Poinar GO Jr (1995) First fossil soft tick, *Ornithodoros antiquus* n. sp. (Acari: Argasidae) in Dominican amber with evidence of their mammalian host. Exp Dermatol 51:584–587
Poinar GO Jr (1998) Fossils explained 22: palaeontology of amber. Geol Today 1998:154–160
Poinar GO Jr (2000) *Heydenius araneus* n.sp. (Nematoda: Mermithidae), a parasite of a fossil spider, with an examination of helminths from extant spiders (Arachnida: Araneae). Invert Biol 119:388–393

Poinar GO Jr (2004) Fossil evidence of spider egg parasitism by ichneumonid wasps. Beitr Araneol 3B:1874–1877
Poinar GO Jr (2015a) Rickettsial-like cells in the Cretaceous tick, *Cornupalpatum burmanicum* (Ixodida: Ixodidae). Cretac Res 52:623–627
Poinar GO Jr (2015b) Spirochete-like cells in a Dominican amber *Amblyomma* tick (Arachnida: Ixodidae). Hist Biol 27:565–570
Poinar G (2021) Fossil record of viruses, parasitic bacteria and parasitic protozoa. In: De Baets K, Huntley JW (eds) The Evolution and Fossil Record Of Parasitism: Identification and Macroevolution of Parasites. Topics in Geobiology 49
Poinar GO Jr, Brown AE (2003) A new genus of hard ticks in Cretaceous Burmese amber (Acari: Ixodida: Ixodidae). Syst Parasitol 54:199–205
Poinar GO Jr, Buckley R (2008) *Compluriscutula vetulum* (Acari: Ixodida: Ixodidae), a new genus and species of hard tick from Lower Cretaceous Burmese amber. Proc Entomol Soc Washington 110:445–450
Poinar GO Jr, Treat AE, Southcott RV (1991) Mite parasitism of moths: examples of paleosymbiosis in Dominican amber. Experientia 47:210–212
Poinar GO Jr, Krantz GW, Boucot AJ, Pike TM (1997) A unique Mesozoic parasitic association. Naturwissenschaften 84:321–322
Poulin R (2011) Evolutionary ecology of parasites. Princeton University Press, Princeton
Proctor H (2003) Feather mites (Acari: Astigmata): ecology behaviour, and evolution. Annu Rev Entomol 48:185–209
Radovsky FJ, Yunker CR (1971) *Xenarthronyssus furmani*, n. gen., n. sp. (Acarina: Dasponyssidae), parasites of armadillos, with two subspecies. J Med Entomol 8:135–142
Rapp A (1959) Zur Biologie und Ethologie der Käfermilbe *Parasitus coleoptratorum* L. 1758 (Ein Beitrag zum Phoresie-Problem). Zool Jahrb Syst 86:303–366
Reeves WK, Loftis AD, Szumlas DE, Abbassy MM, Helmy IM, Hanafi HA, Dasch GA (2007) Rickettsial pathogens in the tropical rat mite *Ornithonyssus bacoti* (Acari: Macronyssidae) from Egyptian rats (*Rattus* spp.). Exp Appl Acarol 41:101–107
Robin N, Béthoux O, Sidorchuk E, Cui Y, Li Y, Germain D, King A, Berenguer F, Ren D (2016) A Carboniferous mite on an insect reveals the antiquity of an inconspicuous interaction. Curr Biol 26:1–7
Robin N (2021) Importance of data on fossil symbioses for parasite-host evolution. In: De Baets K, Huntley JW (eds) The evolution and fossil record of parasitism: Coevolution and paleoparasitological techniques. Topics in Geobiology 50
Rodriguez J, Waichert C, von Dohlen CD, Poinar GO Jr, Pitts JP (2016) Eocene and not Cretaceous origin of spider wasps: fossil evidence from amber. Acta Palaeotol Polon 61:89–96
Rosenkrantz P, Aumeier P, Zigelmann B (2010) Biology and control of *Varroa destructor*. J Invertebr Pathol 103:S96–S119
Rota-Stabelli O, Daley AC, Pisani D (2013) Molecular timetrees reveal a Cambrian colonization of land and a new scenario for ecdysozoan evolution. Curr Biol 23:1–7
Sanchez JP, Nava S, Lareschi M, Ortiz PE, Guglielmone AA (2010) Finding of an ixodid tick inside a Late Holocene owl pellet from northwestern Argentina. J Parasitol 96:820–822
Schille F (1916) Entomologie aus der Mammut- und Rhinoceros-Zeit Galiziens. Entomol Z 30:42–43
Scudder SH (1885) 3. Classe. Arachnoidea. Spinnen. Skorpione. In: Zittel KA (ed) Handbuch der Palaeontologie. I. Abtheilung. Palaeozoologie 2. R. Oldenbourg, München, pp 732–746
Selden PA (1981) Functional morphology of the prosoma of *Baltoeurypterus tetragonophthalmus* (Fischer) (Chelicerata: Eurypterida). Trans R Soc Edinburgh Earth Sci 72:9–48
Sharma PP, Giribet G (2014) A revised dated phylogeny of the arachnid order Opiliones. Front Genet 5:255
Shi G, Grimaldi DA, Harlow GE, Wang J, Wang J, Yang M, Lei W, Li Q, Li X (2012) Age constraint on Burmese amber based on U–Pb dating of zircons. Cretac Res 37:155–163
Shultz JW (2007) A phylogenetic analysis of the arachnid orders based on morphological characters. Zool J Linnean Soc 150:221–265
Sidorchuk EA (2018) Mites as fossils: forever small? Int J Acarol 44:349–359

Sidorchuk EA, Khaustov AA (2018) A parasite without host: the first fossil pterygosomatid mite (Acari: Prostigmata: Pterygosomatidae) from French Lower Cretaceous amber. Cretac Res 91:131–139

Sidorchuk EA, Perrichot V, Lindquist EE (2016) A new fossil mite from French Cretaceous amber (Acari: Heterostigmata: Nasutiacaroidea superfam. nov.), testing evolutionary concepts within the Eleutherengona (Acariformes). J Syst Palaeontol 14:297–317

Sidorchuk EA, Bochkov AV, Weiterschan T, Chernova OF (2019) A case of mite-on-mammal ectoparasitism from Eocene Baltic amber (Acari: Prostigmata: Myobiidae and Mammalia: Erinaceomorpha). J Syst Palaeontol 17:331–347

Sonenshine D, Roe RM (eds) (2013) Biology of ticks, 2nd Edn. 1–2. Oxford University Press, Oxford

Stiller D, Lim BL, Nadchatram M (1977) *Entonyssus asiaticus* Fain, 1960 (Acari: Entonyssidae), a lung parasite of Malaysian snakes with notes on the immature stages. Southeast Asian J Trop Med Public Health 8:129–130

Stunkard HW (1951) Observation on the morphology and life history of *Microphallus* n. sp. (Trematoda: Microphallidae). Biol Bull 191:307–318

Townsend VR Jr, Proud DN, Moore MK, Tibbetts JA, Burns JA, Hunter RK, Lazarowitz SR, Felgenhauer BE (2008) Parasitic and phoretic mites associated with Neotropical harvestmen from Trinidad, West Indies. Ann Entomol Soc Am 101:1026–1032

Türk E (1963) A new tyroglyphid deutonymph in amber from Chiapas, Mexico. Uni Cal Pub Entomol 31:49–51

Valiente Moro C, Chauve C, Zenner L (2005) Vectorial role of some dermanyssoid mites (Acari, Mesostigmata, Dermanyssoidea). Parasite 12:99–109

van Helsdingen PJ (2011) Spiders in a hostile world (Arachnoidea, Araneae). Arachnol Mitt 40:55–64

Waddington J, Rudkin DM, Dunlop JA (2015) A new mid-Silurian aquatic scorpion—one step closer to land? Biol Lett 11:20140815

Waloszek D, Dunlop JA (2002) A larval sea spider (Arthropoda: Pycnogonida) from the Upper Cambrian Orsten of Sweden, and the phylogenetic position of pycnogonids. Palaeontology 45:421–446

Walter DE (1987) Life history, trophic behavior, and description of *Gamasellodes vermivorax* n. sp. (Mesostigmata: Ascidae), a predator of nematodes and arthropods in semiarid grassland soils. Can J Zool 65:1689–1685

Walter DE, Proctor HC (1998) Feeding behavior and phylogeny: observations on early derivative mites. Exp Appl Acarol 22:39–50

Warnock RCM, Yang Z, Donoghue PCJ (2012) Exploring uncertainty in the calibration of the molecular clock. Biol Lett 8:156–159

Warnock RCM, Engelstädter J (2021) The molecular clock as tool for understanding host-parasite evolution. In: De Baets K, Huntley JW (eds) The evolution and fossil record of parasitism: Coevolution and paleoparasitological techniques. Topics in Geobiology 50

Weidner H (1964) Eine Zecke, *Ixodes succineus* sp. n. im Batischen Bernstein. Veöffent Überseemus Bremen 3:143–151

Weinstein SB, Kuris AM (2016) Independent origins of parasitism in Animalia. Biol Lett 12:20160324

Wohltmann A (2000) The evolution of life histories in Parasitengona (Acari: Prostigmata). Acarologia 41:145–204

Wolfe J, Daley AC, Legg DA, Edgecombe GD (2016) Fossil calibrations for the arthropod Tree of Life. Earth Sci Rev 160:43–110

Wunderlich J (2004) Parasites, parasitoids and other enemies of fossil spiders and their egg sacs. Beitr Araneol 3A:115–126

Zhang Z-Q (1995) Review of the systematics and biology of Otopheidomenidae (Acari: Mesostigmata) with a description of *Eickwortius* gen. nov. from a termite (Isoptera: Termitidae). Syst Entomol 20:239–246

Chapter 10
Evolutionary History of Crustaceans as Parasites

Joachim T. Haug, Carolin Haug, and Christina Nagler

Abstract Modern crustaceans are extremely diverse, not only in their morphologies, but also in their life styles. It is therefore not surprising that parasitism evolved in various lineages of Eucrustacea independently, in groups such as amphipodan, isopodan and copepodan crustaceans, but also barnacles and fish lice. Parasitic crustaceans have become specialized to many different host species and show a wide variety of attachment and feeding specializations. Among the parasitic crustaceans, different groups are especially interesting to study for reconstructing the evolution of parasitism within this group. This chapter summarizes the modern aspects, evolutionary history and fossil record of parasitic crustacean groups. By reviewing the parasitic crustaceans with emphasis on their fossil record, this chapter aims to improve our understanding of parasitism in general.

Keywords Palaeoparasitology · Amphipoda · Isopoda · Cymothoida · Copepoda · Thecostraca · Branchiura · Pentastomida · Evolutionary history

J. T. Haug (✉)
Department of Biology II, University of Munich (LMU), Planegg-Martinsried, Germany

GeoBio-Center, University of Munich (LMU), Planegg-Martinsried, Germany
e-mail: joachim.haug@palaeo-evo-devo.info

C. Haug
GeoBio-Center, University of Munich (LMU), Planegg-Martinsried, Germany
e-mail: carolin.haug@palaeo-evo-devo.info

C. Nagler
Department of Biology II, University of Munich (LMU), Planegg-Martinsried, Germany
e-mail: christina.nagler@palaeo-evo-devo.info

© The Author(s) 2021

K. De Baets, J. W. Huntley (eds.), *The Evolution and Fossil Record of Parasitism*, Topics in Geobiology 49, https://doi.org/10.1007/978-3-030-42484-8_10

10.1 Introduction

Parasites are ubiquitous and abundant among most metazoan groups (Poulin and Morand 2000), in other words: in every larger recognised group of animals we find parasitic forms. Due to the influence of parasites on their hosts, hence on food webs, co-evolution, population dynamics and with all this also on the economy, there is a specific research interest in the evolution of parasitism (Klompmaker et al. 2014; De Baets and Littlewood 2015; De Baets et al. 2021; Jephcott et al. 2016; Leung 2021; Robin 2021; van Dijk and De Baets 2021).

Arthropoda, or better Euarthropoda is a group including forms such as spiders and scorpions (chelicerates), shrimps and crabs (eucrustaceans), centipedes and millipedes (myriapods) or wasps and flies (insects), to name just a few examples. In terms of species richness, biomass, individual richness and morphological and ecological diversity, the group Euarthropoda is often considered to be the most successful group of metazoans. Yet this statement is a mere logical mistake, every larger group which includes Euarthropoda would be more successful (Haug et al. 2016); still representatives of Euarthropoda are indeed almost everywhere and dominate modern ecosystems. Naturally within this group, numerous ingroups evolved independently a parasitic lifestyle leading to various adaptive radiations (Poulin and Morand 2000; Baeza 2015; Klompmaker and Boxshall 2015; Morand 2015; Tavares-Dias et al. 2015; Briggs et al. 2016; Weinstein and Kuris 2016).

Here we address the evolution of parasitism towards modern day parasites among crustaceans. Crustaceans is used here in a typological way excluding a large ingroup of Crustacea sensu lato with numerous parasites, namely insects. As entomology has not yet generally recognised that insects are, simply put, 'flying crustaceans' (e.g., Moura and Christoffersen 1996; Schwentner et al. 2017), these are treated in a separate chapter Labandeira and Li (2021). While among insects there are several prominent examples of parasitism, also many lineages of (non-insect) aquatic crustaceans have independently evolved a variety of different strategies generally interpreted as parasitic, such as haematophagy, kleptoparasitism, parasitoidism, and parasitic castration (Kuris 1974; Torchin et al. 2003; Lafferty and Kuris 2009; Baeza 2015; Poulin and Randhawa 2015). These strategies are usually linked to specialised morphologies, leading to a large morphological diversity.

Investigating crustaceans as parasites is not a mere intellectual exercise. Parasitic crustaceans are ecologically and economically significant, due to the fact that infested hosts such as fishes or crabs have reduced fitness and survival rates ultimately causing significant economic losses in fisheries and aquaculture (Rohde 2005).

Yet, due to the framework of the book we are taking a less well-established view on parasitism in crustaceans. We address the evolutionary history of the parasitic crustacean lineages by looking at fossils. Why should we address the evolution of parasitism in all these different lineages from a deep-time perspective?

Fossils are an important source of information, for example, for understanding the evolution of different lifestyles in specific groups as they contribute unique character combinations, namely morphological intermediates, to the character evolution towards modern representatives. They may also provide important additional data for character polarisation and character transformation order (Donoghue et al. 1989; Rust 2007; Edgecombe and Legg 2013; Legg et al. 2013; Leung 2017). Even more modern day methods such as molecular clocks have to rely on calibration by fossils in order to provide accurate and reliable divergence data for specific groups (Warnock and Engelstädter 2021). In short, fossils provide additional and direct information that can, for logical reasons, never be gathered by looking at modern forms exclusively.

Besides this strong advantage, considering fossils when it comes to lifestyle questions comes with a difficulty: We cannot directly observe the life habits of a once living organism, but we have to infer it. Inferring the life strategy of fossil parasites can be based on four different types of observations (Nagler and Haug 2015): association reflecting a direct interaction (of host and parasite, both fossilised in contact with each other), indirect pathological changes (of the host), indirect functional morphological inference (of structures of the putative parasite), indirect phylogenetic inference (concerning the position of the putative parasite). The latter can include forms which have particular parasitic life stages which have so far only been inferred from presence of non-parasitic adults.

Crustaceans are quite good candidates for a 'fossil parasites' approach due to the fact that representatives are relatively well preserved in rocks and amber due to their relatively large size and the specifics of the arthropod cuticle with a comparably high fossilization potential and thus often exceptional preservation (e.g., Briggs 1999; Gupta et al. 2006; Pohl et al. 2010; Edgecombe and Legg 2013). As we will see in some groups it may prove to be more difficult to identify a fossil parasite as such as some forms have a rather reduced morphology, as for example in parasitic copepods with a sack-like body form. Also some modern day parasitic crustaceans display a low abundance in natural populations (Rohde 2005). Still many parasitic crustaceans are comparably large in relation to their hosts and to other parasitic organisms (for example, parasitic isopods; Brusca 1981; Bunkley-Williams and Williams 1998) and thus, also their fossils have a comparably high preservation potential and can be integrated into an evolutionary reconstruction.

In the following, we discuss examples of parasites in various crustacean groups, namely Amphipoda (beach hoppers, scuds, whale lice), Isopoda (pill bugs, wood lice, sea slaters), Copepoda (copepods), Thecostraca (barnacles and their relatives), and Branchiura (fish lice). All these should have the potential to actually yield fossil representatives and we discuss which of them indeed do and which do not and speculate why. Furthermore, we will also discuss Pentastomida (tongue worms), a group which has by some authors been interpreted as an ingroup of Eucrustacea (see discussion below). We discuss the forms of parasitism in the widest sense, including parasitism sensu stricto ("traditional" type of parasitism) as well as different more exotic-appearing types such as parasitic castrators or "micropredators". We cannot provide a strict categorisation as there can be no strict threshold criteria, hence all categories of parasitism should be seen as flexible concepts. Developing a less ambiguous categorisation of parasitism types will be an enterprise for future research. Such an endeavour might benefit from also studying functional morphology of putative parasitic representatives or their close relatives (Nagler and Haug 2016).

10.2 Amphipoda

10.2.1 General Aspects

Amphipoda is a group of malacostracan crustaceans, more precisely an ingroup of Peracarida, a group characterised by the evolution of extensive brood care with the aid of a brood pouch (marsupium) found among all the different (adult female) representatives of the group (e.g., Richter and Scholtz 2001). Other examples of peracaridan crustaceans besides amphipods include mysid shrimps, wood lice, and hooded shrimps. The more widespread known forms of amphipods are gammarideans (scuds). Most amphipod species are marine, but several also occur in freshwater, and some species live on the beach (sandhoppers) or are fully terrestrial (landhoppers) (e.g., Bousfield 2007). In two ingroups of Amphipoda lifestyles that have been interpreted as parasitic have evolved, in Cyamida and in Hyperiida.

Cyamidans, whale lice, exhibit a dorso-ventrally flattened body shape unusual for amphipodans, and possess prominent grasping appendages (e.g., Leung 1967). The overall arrangement with these sub-chelate appendages is distantly reminding of the arrangement in some parasitic isopods (see below). Cyamidans have been proposed to be ectoparasites; they cling to the skin of whales and feed on the outer layer of the whale skin (Seger and Rowntree 2018 and references therein). The diet might also be supplemented with algae growing on the surface of the whale (Leung 1976). It is unclear if the whale loses a significant amount of energy by the cyamidan. They might therefore also be categorised as commensalistic organisms. In contrast to other organisms that attach to a larger animal, cyamidans lack a mobile phase of life. The transmission from one host to another, also across species, occurs via direct contact of the hosts (e.g., Wardle et al. 2000; Seger and Rowntree 2018 and references therein), i.e. their entire life cycle is directly bound to their hosts.

Hyperiidans are pelagic marine amphipodans usually occurring at greater depths. They are associated with cnidarian medusae, ctenophorans or tunicates, in some species also with colonial radiolarians (e.g. Laval 1980; Anderson 1983), or more general, with gelatinous planktic organisms. Females of some hyperiidan species bite small holes into the gelatinous body of the host and transfer their offspring into the holes for further development (e.g., Crossley et al. 2009). Others crawl into the tunice of salps and consume the entire animal besides the tunic, the latter then being used as a floating aid and shelter for the offspring. Additionally, the adults use the planktic organisms as floating aid as their swimming abilities are rather poorly developed (Laval 1980). Hyperiidans in captivity are known to cling to any larger planktic organism. Both adults and immatures feed on the host or on the host's prey, with the different species differing significantly in their feeding mode. Hence, hyperiidans have an ecological role straddling in the range of phoresy, parasitism, parasitoidism and predation.

10.2.2 Phylogenetic Inference of Appearance and Molecular Estimations of Early Evolution

Cyamida has been interpreted as the sistergroup to the group of raptorial skeleton shrimps, Caprellida, ambush predators with large sub-chelate grasping appendages; alternatively, Cyamida could represent an ingroup of Caprellida (e.g., Ito et al. 2011 and references therein). The latter scenario would indicate that cyamidans evolved from a caprellidan-like ancestor. Cyamidans might have evolved rather recently, in dependence on the relatively young geological age of their exclusive hosts (c. 50 mya; e.g., Thewissen and Williams 2002). There are currently no molecular estimations available on the evolution of cyamidans or hyperiidans, but the origin of Amphipoda lies presumably at least in the Triassic (Copilaş-Ciocianu et al. 2019).

10.2.3 Fossil Representatives

The oldest unequivocal amphipodan fossils are known from Baltic amber (c. 50 mya; e.g., Jażdżewski and Kupryjanowicz 2010; a putative older fossil amphipodan from the Triassic described by McMenamin et al. 2013 has been re-interpreted as a remain of a decapodan crustacean by Starr et al. 2016). Concerning the oldest occurrence of other peracaridan ingroups, e.g., Tanaidacea, Amphipoda are likely to be much older, probably late Palaeozoic (see Vonk and Schram 2007, their Fig. 5). However, all fossil amphipodans found to date are gammarideans. No fossil hyperiidans or cyamidans have been identified yet, despite the relatively large size of cyamidans compared to many other parasites, or supposed parasites, reaching into the centimetre range.

10.3 Isopoda

10.3.1 General Aspects

Isopoda is, like Amphipoda, a group of peracaridan crustaceans, best known by their terrestrial representatives, the wood lice. Yet, isopods have evolved very different lifestyles and morphologies and show a very wide geographical distribution (Wägele 1989; Brandt 1992; Dreyer and Wägele 2001; Poore and Bruce 2012).

Cymothoida sensu Wägele (1989) is the only isopodan ingroup with parasitic representatives and includes various different groups that evolved different strategies. These strategies range from scavengers, predators, "micropredators" (in fact temporary parasites; see also Robin 2021; Robin et al. 2019 and references therein) to obligate, permanently attached parasites completely relying on their hosts (e.g., Nagler et al. 2017a and references therein). These different strategies seem to be quite strictly coupled to distinct ingroups:

- Representatives of Cirolanidae typically show a scavenging lifestyle (Bruce 1986).
- Representatives of Corallanidae are scavengers, predators, but some are also temporary ectoparasites on fishes and zooplankton (Monod 1969; Bruce et al. 1982; Gentil-Vasconcelos and Tavares-Dias 2015).
- Representatives of Aegidae are temporary ectoparasites on fishes: they attach just shortly for feeding and leave their hosts afterwards (Brusca 1983). Representatives of Aegidae can be interpreted as "micropredators" sensu Lafferty and Kuris (2002), but in the framework of understanding the evolution of parasitism, they should be better considered as 'temporary parasites', i.e. organisms that feed on a host without killing it, but interacting with it only for the time of feeding.
- Immature representatives of Cymothoidae have a lifestyle similar to representatives of Aegidae, attaching just shortly for feeding to their host fishes and thus can be considered as temporary ectoparasites; adults are obligatory, permanent ectoparasites on or in their host fishes (Brusca 1981; Bunkley-Williams and Williams 1998; Smit et al. 2014). One of the best studied cases of the parasitic strategies of Cymothoidae, and the most popularised one, are probably isopods that are replacing the tongue of their host fishes and feed in a kleptoparasitic manner (Brusca and Gilligan 1983; Thatcher 2000; Costa et al. 2010; Hadfield et al. 2011; Parker and Booth 2013).
- Gnathiidae is not considered as an ingroup of Cymothoida by many authors, yet their lifestyle and the associated morphology strongly argues for such an ingroup position (see e.g. Nagler et al. 2017a). During a specific larval phase (praniza larva), gnathiids feed similar to representatives of Aegidae and juvenile Cymothoidae on host fishes, representing temporary ectoparasites; their adults do not consume food (Smit and Davies 2004; Hispano et al. 2014).
- Representatives of Epicaridea have a specific lifecycle involving also a host change coupled to their ontogenetic phase. After hatching as epicaridium larva, they infest intermediate hosts, small crustaceans such as copepods, as microniscus stage. As cryptoniscus larva, they look for and infest their final hosts, decapodan crustacean (Williams and Boyko 2012). Depending on their infection site and their adult morphology, adult epicarideans can be considered as permanent ecto-parasites or endo-parasites, with some additionally acting as parasitic castrators (Beck 1980).

The evolution of parasitism within Cymothoida has been proposed to have occurred stepwise by a shift from non-parasitic to parasitic lifestyles (Fig. 10.1; Nagler et al. 2017a). Due to their enormous diversity in morphology and ecology, Cymothoida is an interesting group for studying the evolution of parasitism, especially together with deep-time aspects based on fossils and their diversity and abundance in marine habitats.

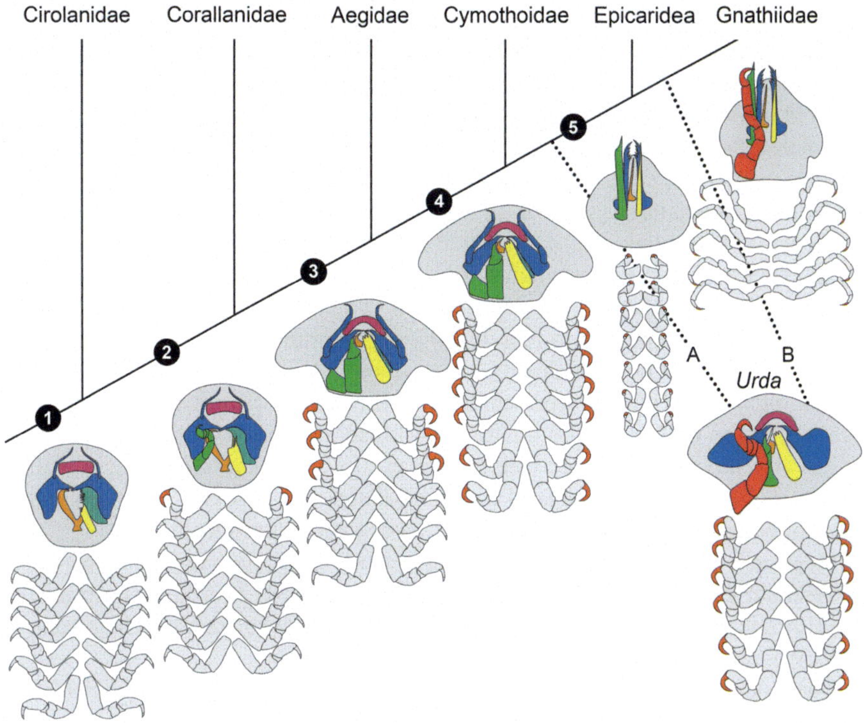

Fig. 10.1 Schematic character evolution of the mouthparts and thoracopods within Cymothoida, from Nagler et al. (2017a). Color-marks: labrum = purple, mandibles = blue, paragnaths = orange, maxillula = cyan, unclear mouthpart = white, maxilla = yellow, maxilliped = green, first free thoracopod = red, claw-like dactyli on thoracopods = dark orange. Important steps are (with oldest known occurrences): (1) cirolanid-like ancestor with carnivorous mouthparts, swimming thoracopods (possibly since Permian, at least since Jurassic); (2) modified mouthparts for piercing, maxilliped and first free thoracopod modified for attachment; (3) mouthparts forming a sucking mouth cone, three thoracopods modified for attachment (at least since Late Miocene); (4) sucking mouth cone, all seven thoracopods modified for attachment (at least since Upper Jurassic); (5) mouthparts elongated still forming a mouth cone for piercing, reduced mandibular palp and maxillula, reduced seventh thoracopod (possibly since Early Jurassic, at least since Late Cretaceous). For detailed discussion see Nagler et al. (2017a)

10.3.2 Phylogenetic Inference of Appearance and Molecular Estimations of Early Evolution

The first fossil isopodans date back to the Carboniferous (355 mya). Therefore, Schram (1977, 1982) suggested an Early to Middle Devonian age (390 mya) for the origin of Peracarida. This must have been followed by a quick radiation leading to a high diversity and wide distribution in the late Palaeozoic (419–252 mya; Brusca and Wilson 1991; Lins et al. 2012, 2017).

The first fossil representatives of Cymothoida are generally attributed to Cirolanidae from the Permian (290 mya) in Brazil (Martins-Neto 2001). Given this ancestral type of life style and no well formulated apomorphies, the group might turn out as paraphyletic. Fossil cirolanid-type fossils have been found scavenging on whales, fishes and water bugs (Wilson et al. 2011; Pasini and Garassino 2012; Hyžný et al. 2013).

Within the remaining group of Cymothoida (excluding the cirolanid-like forms) molecular and morphological data support a sistergroup relationship of Corallanidae to all other parasitic representatives of Cymothoida (Wägele 1989; Dreyer and Wägele 2001; Brandt and Poore 2003; Wetzer et al. 2013; Hata et al. 2017; Nagler et al. 2017a). Some (modern) representatives of Corallanidae are scavengers and predators. Yet, they possess already longer, thinner and more pointed mouthparts typical for parasites than do representatives of Cirolanidae (Bruce 1986; Bruce et al. 1982; Bunkley-Williams and Williams 1998). Also, some representatives are known to act as temporary parasites (see Nagler et al. 2017a and references therein). However, the group might as well be paraphyletic, or otherwise some representatives of the group have evolved temporary parasitism independently.

Aegidae is usually interpreted as the sistergroup to all other remaining parasitic representatives of Cymothoida (Brusca and Wilson 1991; Wilson 2009; Hadfield 2012; Nagler et al. 2017a). Yet, also this group has been suggested to be paraphyletic (Wägele 1989).

The sister group to Aegidae, with all the permanently attached isopodan parasites has been suggested to have arisen in the Early Jurassic (180 mya) in the deep sea (Schram 1977; Brusca 1981; Brandt and Poore 2003; Smit et al. 2014; Nagler et al. 2016; Hata et al. 2017). Within this group, Cymothoidae is sister group to all the remaining ones. Cymothoiids seem to have expanded their habitats to shallower seas and also brackish or freshwater by shifting their fish host species at the latest in the Late Jurassic (150 mya; Nagler et al. 2016; Hata et al. 2017).

The sister group to Cymothoidae includes three groups with unresolved relationships among them. Epicaridea has been considered to represent the direct sister group to Cymothoidae by many authors (see discussion in Nagler et al. 2017a), but Gnathiidae also shares many characters with Epicaridea, hence very likely representing also an ingroup of Cymothoida, closer related to Epicaridea. This relationship has been disputed by many authors (see discussion in Nagler et al. 2017a) as only the larval forms exhibit the shared characters of Gnathiidae and Epicaridea, while adults have a derived biology and morphology. Yet, as the group Thecostraca (sensu lato) demonstrates larval characters can be of prime importance to deduce evolutionary relationships especially in lineages with parasitic forms (see further below). The third group in the sister group to Cymothoidae is an exclusively fossil one, *Urda*. More precisely, at least one representative of the group could be identified as a parasite based on functional morphology (Fig. 10.2b). It represents the oldest indication for parasitism within Cymothoida, to the best of our knowledge, back in the Jurassic (168 mya; Nagler et al. 2017a). Sharing morphological characters with modern representatives of Cymothoidae (plesiomorphic features) and Gnathiidae, it seems that *Urda* is either closely related to Gnathiidae + Epicaridea

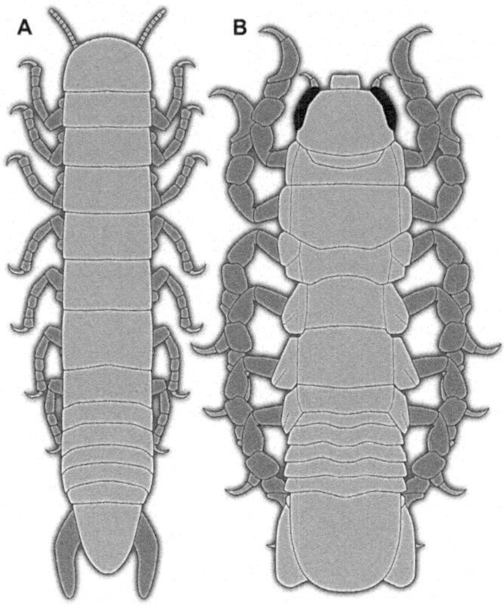

Fig. 10.2 Reconstructions of fossil parasitic isopodan crustaceans in dorsal view (part 1). (**a**) Cymothoiid-like fossil found on fishes; lithographic limestones of Southern Germany, ca. 150 million years old, ca. 30 mm body length. Based on Nagler et al. (2016). (**b**) *Urda rostrata*, Bethel-Bielefeld, Germany, ca. 168 million years old, ca. 37 mm body length. Based on Nagler et al. (2017a). Note that in both reconstructions the thoracic legs are artificially "forced" outside to main body to provide access to the morphology of these appendages

or directly to Gnathiidae (Nagler et al. 2017a). The exact relationship between Gnathiidae, Epicaridea and *Urda* cannot be resolved with the currently available data, especially as the mouthparts of *Urda* are not fully understood yet (Nagler et al. 2017a). Furthermore, it will be necessary to investigate all different fossil forms currently attributed to *Urda* to evaluate the monophyly of this group.

Epicarideans parasitizing other crustaceans evolved from fish-parasitizing ancestors (Dreyer and Wägele 2001, 2002). Due to fossil findings of cryptoniscus larvae in amber, this host change must have occurred at the latest in the Cretaceous (Fig. 10.3; Serrano-Sánchez et al. 2016; Néraudeau et al. 2017). The evolutionary origin of Epicaridea has been suggested to be in the Jurassic (180 mya) parasitizing squat lobsters (though the oldest possible occurrence is on an erymidan lobster; Soergel 1913 after Klompmaker et al. 2014) because of the large abundance of their decapod hosts (e.g., Klompmaker et al. 2013), fossil findings of pathologically transformed hosts (Wienberg Rasmussen et al. 2008; Klompmaker et al. 2014, 2021) and based on a molecular phylogeny (Markham 1986; Boyko et al. 2012, 2013). Finally, it has been suggested that the highly specialized parasitic lifestyle, mainly on fishes, and the diverse morphology of modern representatives within the parasitic ingroup of Cymothoida originated stepwise from non-parasitic cirolaniid-like ancestors with a scavenging life style on fishes (Nagler et al. 2017a).

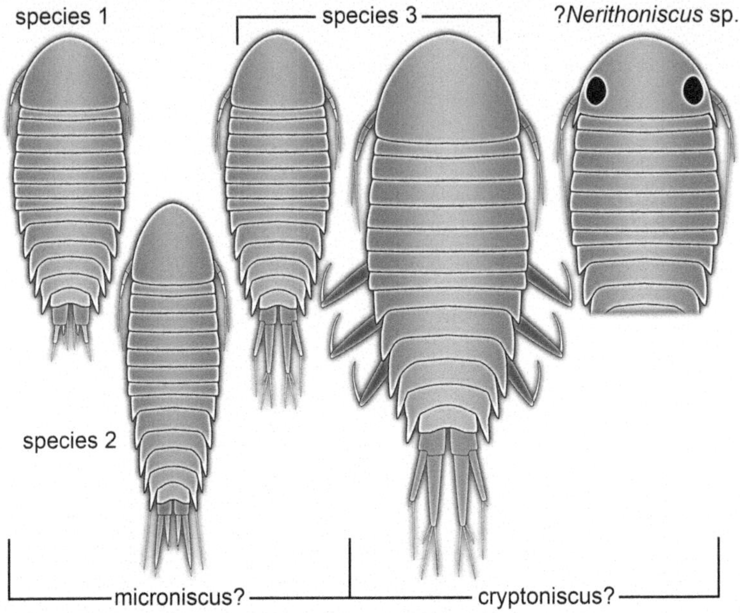

Fig. 10.3 Reconstructions of fossil parasitic isopodan crustaceans in dorsal view (part 2). Epicaridean larvae preserved in Mexican amber, about 25 million years old, between 450 and 695 μm body length. Based on Serrano-Sánchez et al. (2016). At least four different species seem to be present in the material based on differences of the uropods and the presence of compound eyes in one specimen (only in one extant group, *Nerithoniscus*). Two distinct size classes can be differentiated. The smaller specimens may represent microniscus stages, i.e. those parasitic on copepodan crustaceans. The larger specimens most likely represent cryptoniscus larvae, i.e. infective stages searching for a final host. One specimen of the smaller and one of the larger size class may be conspecific based on the uropod morphology

10.3.3 Fossil Representatives

The fossil record for isopods in general is sparse and even poorer for Cymothoida (Smit et al. 2014; partly probably due to the limited preservation potential, see Klompmaker et al. 2017) except for the group Cirolanidae (at least numerous fossils have been interpreted as representatives of Cirolanidae; Wieder and Feldmann 1992; Hyžný et al. 2013; Etter 2014; Smit et al. 2014; Robin et al. 2019). This might be partly caused by often ambiguous identification of fossils related with their preservation (Maguire et al. 2018; Vega et al. 2019) and their interpretation as representatives of a specific group (Bowman 1971; Maguire et al. 2018).

The oldest fossil representatives of Cirolanidae suggesting directly (associated with their hosts) their scavenging lifestyle have been reported from the Jurassic (150 mya), associated with a water bug (Polz 2004) and from the Cretaceous (100 mya), associated with a fish carcass (Wilson et al. 2011). Feldmann (2009;

indirect functional morphology indication) reported a supposed cirolaniid in pre-molting condition from the Cretaceous. Many cirolaniids have been reported from the Miocene from South America and Japan, especially from the group of giant isopods, *Bathynomus* (Kato et al. 2016; Maguire et al. 2018). Two ambiguous cases of direct associations may question the scavenging lifestyle of cirolaniids (Klompmaker and Boxshall 2015): These are (1) a single cirolaniid isopodan associated with a shark (Bowman 1971; Boucot and Poinar Jr 2010) and (2) an individual cirolaniid isopodan attached to the ventral side of a squid (Polz et al. 2006).

So far, no fossil representative of Corallanidae has been reported.

Only one fossil representative of Aegidae has been reported from the Late Miocene (20 mya; indirect phylogenetic/functional morphology indication; Hansen and Hansen 2010). The oldest fossil of representatives of Cymothoidae indicating directly a parasitic lifestyle have been reported from the Jurassic (150 mya) from the Solnhofen limestones (Fig. 10.2a; Nagler et al. 2016), although within the above discussed phylogenetic framework it could also be possible that these forms are offshoots of the lineage towards Cymothoidae + (Epicaridea + *Urda* + Gnathiidae) or even of the lineage towards (Epicaridea + *Urda* + Gnathiidae). The oldest fossil representatives of Cymothoida with a parasitic lifestyle (indicated indirectly by functional morphology) have been reported from the Early Jurassic (168 mya) from the group *Urda* (Nagler et al. 2017a). First body fossils of representatives of Epicaridea (infective larva) have been reported from Cretaceous amber (100 mya) from France (Néraudeau et al. 2017). An older and more indirect support for fossil representatives of Epicaridea are pathological swellings in the branchial chamber of decapods (Fig. 10.4; the swellings are usually referred to as *Kanthyloma crusta*) the oldest swellings have been reported from the Early Jurassic (180 mya, though this is a bit doubtful, see discussion in Klompmaker et al. 2014, the next oldest ones would be Late Jurassic; Hessler 1969; Conway Morris 1990; Feldmann et al. 1993; Wienberg Rasmussen et al. 2008; Klompmaker et al. 2014, 2016, 2021; Klompmaker and Boxshall 2015; Robins and Klompmaker 2019). No fossil representative of Gnathiidae has been reported yet.

10.4 Copepoda

10.4.1 General Aspects

Copepoda is a species-rich group of small and morphologically very diverse crustaceans, ubiquitous in all aquatic habitats, even occurring in some terrestrial habitats (e.g., Reid 2001). Copepodans are extremely abundant, dominating the zooplankton (Humes 1994). Yet, similar to isopodans, the current abundance of copepodans is not reflected in the fossil record due to their small size and fragility (Selden et al. 2010), but they may have appeared already during the early Palaeozoic (Andres 1989, see below). Their lifestyle varies from planktic to benthic, with very different

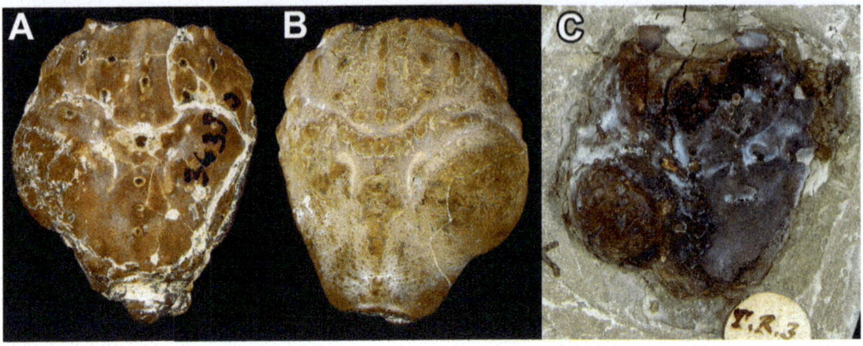

Fig. 10.4 Indications of fossil parasitic isopodan crustaceans. Fossil brachyuran hosts, likely representatives of Raninoidea show distinct deformations (usually referred to as *Kanthyloma crusta*) that are similar to bopyrid isopodan infestations in modern forms. All specimens from NHM London, UK. (**a**) Specimen 34380, ca. 22 mm length. (**b**) Specimen 29796, ca. 23 mm length. (**c**) Specimen I3702, ca. 18 mm length

feeding habits including, among others, predation and parasitism; probably about one third of all copepodan species are parasitic (Humes 1994).

Parasitism evolved several times independently within Copepoda (Boxshall and Halsey 2004). This might explain the vast diversity of hosts that range from sponges and cnidarians to molluscs, other crustaceans, echinoderms, fishes and also mammals. Relatively few representatives of Copepoda are parasitic on Nemertea, Platyhelminthes, Bryozoa, Phoronida, Echiura, Brachiopoda, Enteropneusta, Hemichordata, Vestimentifera and Sipuncula (Boxshall and Halsey 2004). Representatives of some copepodan groups, such as Mytilicolidae or Siphonostomatoida, are parasitic on molluscs and fishes respectively and are major pests in commercial bivalve cultures and fisheries (Pike and Wadsworth 1999; Boxshall and Defaye 2006). Most parasitic representatives are ectoparasites, but there are also some endoparasites, e.g., in the mesogloea of corals (Stock 1975; 1981), in the coelomic cavities of representatives of Sipunculida (Ho et al. 1981; Schwabe and Maiorova 2015) or under the integument of gastropods (Anton and Schrödl 2013 and references therein).

Parasitic copepods usually feed by rasping the surface of the host using their mandibles, piercing the host tissues by an oral tube formed by the mandibles and, subsequently, sucking body fluids out of the host (Boxshall 1990a; Boxshall et al. 2005). All parasitic copepods show a body organisation in which the body is divided into a functional head (head + 1 segment) anterior trunk (5 segments with swimming appendages + genital segment) and posterior trunk (4 segments + telson) (Boxshall et al. 2005). However, most representatives of parasitic groups within Copepoda appear to lack some body subdivisions (conjoined segments, reduction and loss of appendage elements and setation) up to an extreme "simplification" with an almost amorphous body that lacks any segmentation and appendages (Kabata 1979; Huys and Boxshall 1991; Boxshall and Halsey 2004). Representatives of some groups, such as Herpyllobiidae, Nicothoidae and Chitonophilidae, show an amorphous body comparable to rhizocephalan barnacles (see below) with a nutrient-absorbing network embedded inside the host and an external egg sac (already known for quite a while, see Lankester 1880).

After hatching as nauplius larva, they metamorphose after up to six stages into a copepodid (Boxshall 1990b). Most parasitic copepodids attach to their host and develop through up to five copepodid stages. If the adult parasitic copepodan is an obligate parasite, the female is permanently anchored into the host by claw-like head appendages, a modified ventral body surface or appendages, or a combination of it, whereas the smaller males hold onto the females (Boxshall 1990b).

10.4.2 Phylogenetic Inference of Appearance and Molecular Estimations of Early Evolution

The molecular data still provide no clear phylogenetic position of Copepoda within Eucrustacea, and no convergence with morphological data has been achieved yet (e.g., Huys and Boxshall 1991; Regier et al. 2005; Eyun 2017; Khodami et al. 2017). Apparently, the parasitic groups within Copepoda evolved parasitism independently (Huys and Boxshall 1991; Boxshall and Halsey 2004; Boxshall et al. 2005). Within these parasitic groups, also changes to a non-parasitic life style have been inferred, e.g. in Siphonostomatoida or Cyclopoida (Huys and Boxshall 1991).

10.4.3 Fossil Representatives

Probably due to their small size, fragility and cryptic nature, parasitic copepods have a scarce fossil record; only few body fossils of parasitic copepods have been found. The oldest fossil remains (supposed fragments of antennae and maxillulae) of representatives of Copepoda (though without hints on their life habits) have been reported from the Carboniferous (303 mya; Selden et al. 2010). However, Andres (1989) depicted several presumed parasitic crustaceans from the Ordovician of Sweden (c. 470 mya), which possess a sucking mouth cone and biramous swimming appendages closely reminding of those of copepods (Fig. 10.5b).

Kabatarina pattersoni from the Cretaceous Santana Formation of Serra do Araripe, Brazil (120–110 mya) is known from two complete female specimens from two fish skulls (direct host-parasite association) (Figs. 10.5a and 10.6; Cressey and Patterson 1973) and 13 more fragmented specimens, apparently including parts of males (Cressey and Boxshall 1989). Although the latter record displays only fragments, these were attached to the skull of a fish, hence we consider the record as direct indication for parasitism. Another possible parasitic copepodan is one specimen attached to a shrimp-like tanaidacean crustacean in amber from the Lower Cretaceous of Spain (Vonk and Schram 2007).

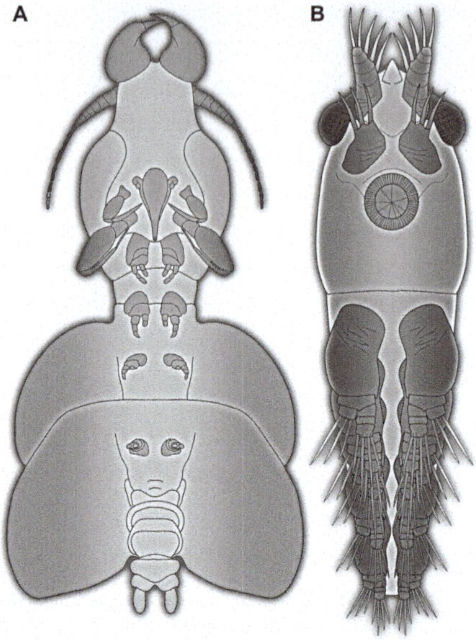

Fig. 10.5 Reconstructions of (possible) fossil parasitic copepodan crustaceans. (**a**) *Kabatarina pattersoni*, female in ventral view, ca. 1 mm length, modified from Cressey and Boxshall (1989). (**b**) Reconstruction (ventral view) of small crustacean depicted in Andres (1989) from Ordovician of Sweden, about 470 million years old, ca. 600–650 μm length; the overall habitus would be well compatible with the interpretation as a parasitic copepodan crustacean

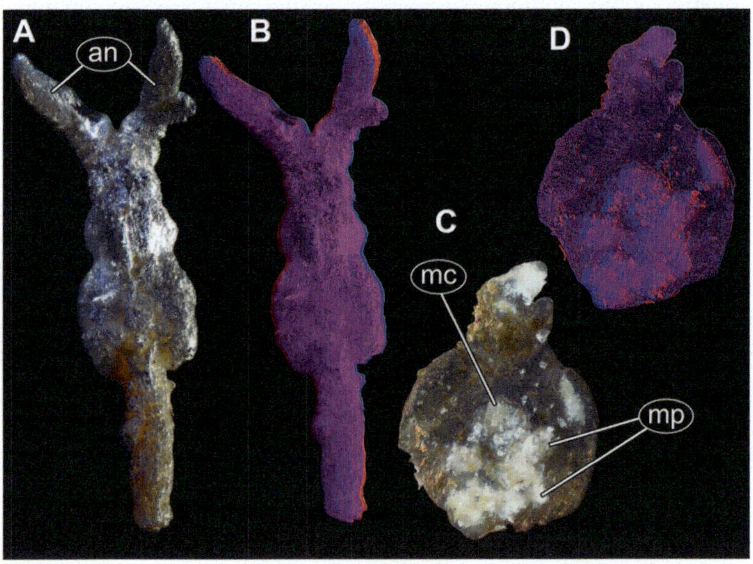

Fig. 10.6 Actual specimens of fossil parasitic copepodan crustaceans, *Kabatarina pattersoni*, Brazil, about 110 million years old. (**a**, **b**) Specimen In63469, ca. 1 mm length; male in ventral view mainly preserving the (2nd) antennae (an). (**c**, **d**) Specimen In63625, ca. 0.7 mm length; ventral view on anterior body, preserving the mouth cone (mc) and the maxillipeds (mp). All images composite-super-macrographs under cross-polarised light; (**a**, **c**) Colour images. (**b**, **d**) Red-blue stereo-anaglyphs of virtual surface reconstructions

Due to the pathological changes in the host morphology caused by modern copepods, several ambiguous fossils have been reported possessing comparable modifications: Punctured fish scales from the Middle Devonian from Estonia (390 mya) (Lukševics et al. 2009), external galls (termed *Castexia douvillei*) in echinoderms were inferred to be caused by copepods from the Jurassic (Radwańska and Radwański 2005) and external cysts in echinoderms from the Jurassic, the so-called Halloween pumpkin-mask cysts (Weinfurtner 1989; Radwanska and Poirot 2010). Although pathological changes of the host morphology have to be seen critical, the reported cases of external galls in echinoderms are very similar to modern copepodan infections (e.g., Stock 1968) and thus represent a relatively reliable indirect indication of parasitic copepods or at least particular strategies.

10.5 Thecostraca

10.5.1 General Aspects

Thecostraca sensu lato sensu Haug (2011) (see also Haug and Haug 2015) is a group of eucrustaceans with numerous lineages that have evolved highly specialised forms of parasitism (see also Klompmaker and Boxshall 2015). Thecostraca s.l. shares specific developmental patterns with Copepoda (Haug and Haug 2015) and therefore most likely forms a monophyletic group with Copepodoida as a sister group (Haug et al. 2011). Although this larger group has been termed "Hexanauplia" (Oakley et al. 2013, in principle referring to Dahms 2004) this term is misleading as Copepodoida and Thecostraca sensu lato have evolved six naupliar stages independently from each other (Haug and Haug 2015). It is therefore more consistent to refer to this group as 'Maxillopoda' in the sense of Haug and Haug (2015), i.e. the group characterised by a specialised delayed developmental pattern of the trunk appendages (see Haug and Haug 2015 for details).

Thecostraca s.l. has representatives as early as the middle Cambrian, about 500 million years ago (Müller and Walossek 1988; Walossek and Müller 1998). The first representative *Bredocaris admirabilis* (sister species to Thecostraca sensu stricto sensu Haug 2011) does not show any clear specialisation for a parasitic life style.

The next possible offshoot of the lineage is the group Tantulocarida, sister group to the remaining thecostracans (= Euthecostraca sensu Haug 2011, Thecostraca sensu Høeg et al. 2009, see also Haug and Haug 2015; yet the ingroup position of Tantulocarida within Thecostraca s.l. has been doubted; see Petrunina et al. 2014 and references therein). These tiny crustaceans are parasites on different crustaceans, including copepods, hence are even smaller than these (Boxshall and Lincoln 1983) and have therefore a limited fossilization potential.

Facetotecta is the sister group to the remaining euthecostracans. All representatives known so far are larval forms, so-called Y-nauplii and Y-cypris larvae (Høeg et al. 2004). A forced metamorphosis of the latter provided a highly dissolved and dedifferentiated organism (ypsigon) (Glenner et al. 2008). This morphology indicates that later stage facetotectans are parasitic, most likely endoparasitic. Yet, it still remains totally unclear which organisms could be the host for facetotectans.

Ascothoracida is the next group branching off in Euthecostraca. Adults live on echinoderms or cnidarians, possess a distinct piercing mouth cone and have therefore been interpreted as parasites, but also as commensalistic forms (e.g., Grygier and Salvat 1984; Grygier and Høeg 2005).

Cirripedia is the sister group to Ascothoracida. Within Cirripedia, Acrothoracica is sistergroup to the remaining forms. Acrothoracicans interact with other organisms as they bore into different calcareous substrates, including corals, molluscan shells, and crinoidan skeletons (e.g., Petriconi 1971; Tomlinson 1987). Hence, they are at least commensalistic, possibly even parasitic.

The remaining cirripedians form two distinct groups, Thoracica and Rhizocephala.

Rhizocephalans are obligate parasites as adults and heavily affect their host. Their adult body is a mycel-like root-system inside another crustacean (most famous in crabs, but also in various other groups of crustaceans including many types of decapodans, but also barnacles). Only their gonads, called externae, are still differentiated true organs (e.g., Glenner and Høeg 2002). Rhizocephalans often castrate their host and also seem to influence its behaviour. Interestingly, they are often associated with their non-parasitic relatives (O'Brien and Van Wyk 1985). Longer-living rhizocephalans influence the moulting frequency of their hosts, i.e. they slow it down (e.g., O'Brien and Van Wyk 1985). In this way, barnacles have longer time to attach and prosper on the back of an infested crab. Barnacle-bearing crabs are therefore interesting candidates for being checked for rhizocephalan infestations.

Also, the exact role of barnacles, or better non-parasitic lineages of Cirripedia, should be evaluated carefully. There are modern forms of lepadomorphan cirripedians that are fully parasitic on sharks, apparently having evolved many traits similar to rhizocephalans convergently (Rees et al. 2014). Furthermore, also "normal" barnacles sitting on other organisms need to be seen in a differentiated view. In most cases, the barnacles may affect the organisms not recognisably. Yet, if there are too many attached to one organism they will indeed have a negative effect on their host by causing additional energy costs, while they benefit from a stable substrate that is even able to move to more beneficial environments. In such a case, we can employ an ecological point of view: the barnacle benefits from this interaction, while the host has costs. Based on this calculation the barnacles could indeed be considered as facultative parasites.

10.5.2 Phylogenetic Inference of Appearance and Molecular Estimations of Early Evolution

As pointed out above, the position of Thecostraca within Eucrustacea can still not easily be resolved.

10.5.3 Fossil Representatives

The oldest representative of Thecostraca s.l., *Bredocaris admirabilis* from the Cambrian, shows no indications of having been parasitic. However, several forms of thecostracans have become fossilised, which may be considered as having been parasitic.

So far, no fossil tantulocaridans or facetotectans have been reported. Due to the small size of tantulocaridans and the fact that we already have a very scarce fossil record of copepodans, it is unlikely to find them as fossils. The fossil record of Ascothoracida is scarce so far. Body fossils are unknown, only structures interpreted as deformations on the host induced by ascothoracidans have been found (e.g., Voigt 1959, 1967; Madsen and Wolff 1965, see summary in Klompmaker and Boxshall 2015). Trace fossils of supposed acrothoracican borings are known from Devonian and younger findings (see recent review by Klompmaker and Boxshall 2015). So far, the possible fossil record of rhizocephalans is very scarce. A possible larval specimen has been reported from the Upper Jurassic (c. 150 mya) limestones of Southern Germany (Fig. 10.7a; Nagler et al. 2017b). Indirect changes of the host, such as partial feminization of male individuals, have been interpreted as indications for a rhizocephalan parasite in fossil crabs (Fig. 10.7b; Bishop 1974, 1983; Feldmann 1998). Furthermore, numerous epibiontic barnacles on different organisms have been described; some of these may represent facultative parasites (see discussion above; Figs. 10.7c and 10.8; Nagler et al. 2017c).

10.6 Branchiura

10.6.1 General Aspects

Branchiurans (fish lice) are ectoparasites on fishes. They possess attachment structures and a mouth cone specialised for sucking blood from the host. Their body is dorso-ventrally flattened to reduce drag on the parasite as well as the risk to get removed by the host (e.g., Møller 2009 and references therein).

Naturally, the group is well characterised by numerous autapomorphies. Among these is the specialised maxillula. In younger stages the maxillula forms a hook (Fig. 10.9a, b), but becomes transformed into a suction disk during post-embryonic development in most species (Møller et al. 2008).

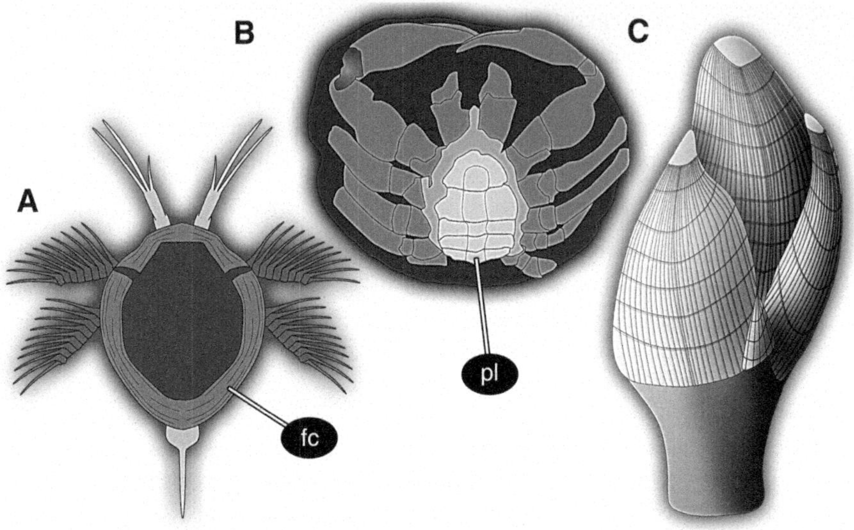

Fig. 10.7 Reconstructions of possible fossil parasitic euthecostracan crustaceans. (**a**) Super-sized possible nauplius larva, lithographic limestones of Southern Germany, about 150 million years old, ca. 4.7 mm length. Based on Nagler et al. (2017b). The specimen seems to have fronto-lateral horns (clear indication for an ingroup position within Cirripedia) and a flotation collar (fc); in modern forms such a structure is restricted to parasitic barnacles. Lighter grey structures absent in the fossil, amended based on comparison to other nauplii. (**b**) Fossil male crab with femininised pleon (pl); in modern forms this morphology is indicative for an infection by a parasitic barnacle (Rhizocephala). Interpretive drawing based on Feldmann (1998). (**c**) Fossil barnacle found on a sponge, lithographic limestones of southern Germany, about 150 million years old, ca. 5 mm length. Modified after Nagler et al. (2017c). Based on the fact that the barnacle benefits from sitting on the sponge and the sponge is most likely negatively affected, this may be considered a case of facultative parasitism

10.6.2 Phylogenetic Inference of Appearance and Molecular Estimations of Early Evolution

Branchiura is a group difficult to interpret in a phylogenetic framework (e.g., Møller 2009), i.e. we do not really know the relationship to other groups (see also below 10.7.2).

10.6.3 Fossil Representatives

So far, no fossil fish louse has been found. It is possible that such fossils have been overlooked on fossil fishes, as even larger parasites such as cymothoid isopods seem to have been largely overlooked (see discussion in Nagler et al. 2016).

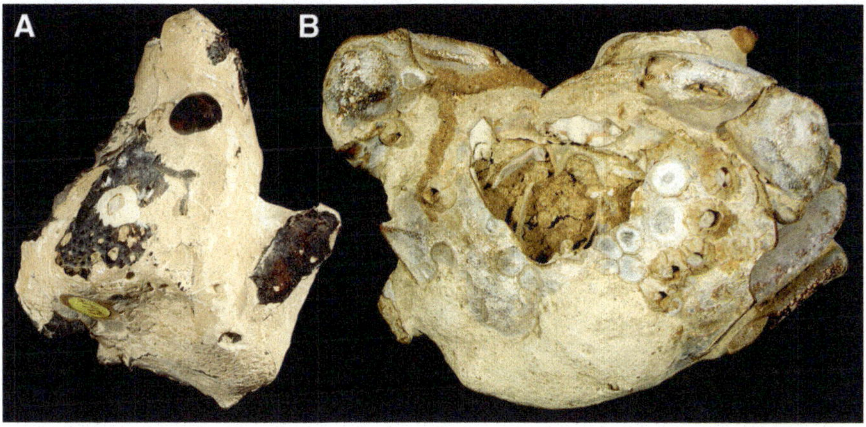

Fig. 10.8 Fossil barnacles as epibionts on decapod crustaceans. Both specimens from NHM London, UK. (**a**) Specimen 43617, long side of stone ca. 55 mm. (**b**) Specimen I3754, long side of stone ca. 80 mm. In case of very few barnacles as in **a**, little or no negative effect for the carrying animal is expected. In cases of a heavy infestation, the carrying animal will be negatively affected, while the barnacles are benefitting; here barnacles may be considered as facultative parasites

Interestingly it has been suggested that Branchiura is closely related to (or nested within?) Cycloida, a group of still enigmatic, exclusively fossil crustaceans (Fig. 10.9c, d; Dzik 2008; see also Schram et al. 1997; Schweigert et al. 2009). Yet, Cycloida itself is highly problematic to understand, in fact we cannot even be sure whether all of these are crustaceans at all. The group included forms that have in the meantime been interpreted as larval forms either of meiuran decapod crustaceans or mantis shrimps (Hyžný et al. 2016). Other forms have been functionally compared to crabs (see Dzik 2008). These possess a rounded shield in dorsal view, that is distantly resembling the shield of branchiurans. A further aspect in common is that both groups have been supposed to represent an ingroup of Maxillopoda, yet neither branchiurans nor cycloidans are known to show the developmental patterns characterising Maxillopoda. Although it has been indicated that cycloidans could be parasitic (Müller 1955), there are no real indications (of whatever type) that would support such an interpretation (see also discussion in Schram et al. 1997).

10.7 Pentastomida

10.7.1 General Aspects

Pentastomida (tongue worms) is a group of exclusively parasitic forms living in the respiratory tracts of tetrapods (e.g., Böckeler 2005). Pentastomidans possess two pairs of prominent appendages; together with the mouth this makes five attachment structures in the modern forms, hence the latinised name, "five mouths". Unlike

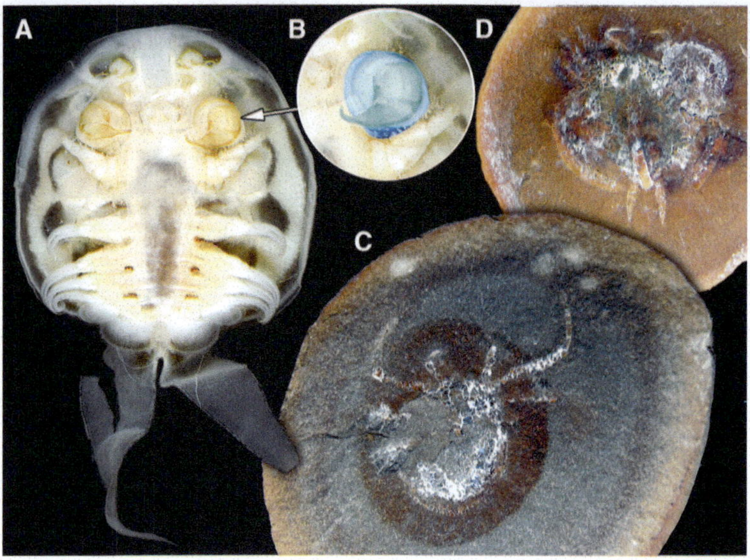

Fig. 10.9 Extant parasitic branchiuran crustacean and supposed fossil relatives (Cycloida). (**a**, **b**). *Dolops dorades*; specimen K14900, CeNak, Hamburg, ca. 13 mm length. (**a**) Ventral view. (**b**) Close-up on colour-marked image of maxillula transformed into a hook. (**c**, **d**) Representatives of Cycloida, which have been considered to be parasitic; both specimens from Mazon Creek, about 300 million years old, FMNH Chicago, USA. (**c**) Specimen 25662, ca. 14 mm main body length without appendages. (**d**) Specimen pe22421, ca. 13 mm main body length without appendages. All images composite macrographs under cross-polarised light

many other groups discussed here, the ecological role of extant pentastomidans as parasites of tetrapodans is unquestioned (e.g. Haugerud 1989; Christoffersen and De Assis 2015). However, the relationship of Pentastomida to other arthropod groups is more problematic, and will be discussed in detail below. Nevertheless, pentastomidans are discussed here as some readers may expect them among the crustaceans.

10.7.2 Phylogenetic Inference of Appearance and Molecular Estimations of Early Evolution

Basically, there are two strongly opposing hypotheses concerning the phylogenetic position of Pentastomida. Most morphological characters point to a position of Pentastomida within Arthropoda sensu lato, but outside Arthropoda sensu stricto, the group of sclerotised arthropodans (Waloszek et al. 2006; Castellani et al. 2011). More precisely, Pentastomida is in this case interpreted as being closer related to sclerotised arthropods than other unsclerotised forms, such as onychophorans (velvet worms) or tardigradans (water bears), as they already possess pivot joints which are absent in onychophorans and tardigradans, but still lack the strict sclerotisation of the body surface.

On the contrary, molecular data suggest a position of Pentastomida deep among sclerotised arthropodans, inside Euarthropoda, inside Eucrustacea as sistergroup to Branchiura (e.g., Regier et al. 2010; Oakley et al. 2013; Giribet and Edgecombe 2013). Interestingly also a morphological view has supported this interpretation (Wingstrand 1972; Storch and Jamieson 1992). Unfortunately, this phylogenetic interpretation was based on two problematic aspects: a simplified sperm morphology shared by both groups and the fact that both are parasites (see Walossek and Müller 1994; Chesunov 2002).

If Pentastomida would represent indeed an ingroup of Eucrustacea we would face a severe problem: They would represent the only case in which all crustacean characters would have been lost in all life stages. Even highly modified parasitic crustaceans that "dissolve" their entire morphology as parasites (see above, Thecostraca) retain clear crustacean type characters in the one or other life phase (the softness of representatives of Pentastomida could, of course, be explained also with such a reduction interpretation). We should therefore still be very cautious with accepting pentastomidans as crustaceans. The branch leading to the modern forms and also that of Branchiura is extremely long, the effect of the occurrence of artefacts based on such branches is well known (e.g., De Baets and Littlewood 2015).

10.7.3 Fossil Representatives

Larval forms of pentastomidans have been reported from Cambrian and Ordovician deposits in exceptional Orsten-type preservation (Andres 1989; Walossek and Müller 1994; Waloszek et al. 2006; Castellani et al. 2011). Although some of these forms appear very unusual (e.g., *Heymonsicambria taylori*; Walossek et al. 1994) and thus their interpretation as pentastomidans has been questioned, others ("hammer-type" larvae) share an astonishing amount of characters with modern pentastomidans, making alternative interpretations difficult.

A younger supposed occurrence is less easy to be interpreted. A small Silurian fossil attached to an ostracodan crustacean has been interpreted as a pentastomidan (Siveter et al. 2015). Although also this type of preservation ("Herefordshire-type") must be called exceptional, it is to a certain degree limited. The factual spatial resolution is about 100 μm, as the voxel size is about 20 μm. Hence not too many characters can be resolved that would support interpretation as a pentastomidan besides five structures in the front. The fossil could potentially also represent a derived parasitic copepodan crustacean or even a non-arthropodan structure as no distinct arthropodan features appear to be present.

10.8 Conclusion and Outlook

The fossil record of parasitic crustaceans is still fragmentary, but has been growing in recent years triggered by active search. Numerous new studies have illustrated that it is not just a preservational issue, but remains have also been overlooked. We will therefore need to continue active and directed search for more cases of fossil crustacean parasites or traces in the hosts left by them. Yet, it will also be necessary to further improve our concept of parasitism and how to identify cases of a parasitic life style. As pointed out above this is less simple in some cases as generally accepted. Also, the currently available categories for different types of parasitism are not well conceptualised. It may likely turn out that we will be unable to provide clear-cut thresholds for identifying specific types of parasitism or parasitism in general, but also relational categories demand to be properly outlined.

Acknowledgements We thank Kenneth de Baets and John Warren Huntley for the invitation to contribute to this volume. The authors have been supported by different funding agencies, mostly the German Research Foundation (DFG) under Grants HA-6300/3-2, the German Academic Exchange Service (DAAD), the Volkswagen Foundation, the European Commission's (FP6) Integrated Infrastructure Initiative Programme SYNTHESYS, and Lehre@LMU Munich. Curators from different museums have provided access to the material, namely Claire Mellish (NHM London), Paul Mayer (FMNH Chicago) and Martin Schwentner (formerly CeNak Hamburg). During an excursion to the CeNak Hamburg we kindly received support with documentation and image processing from several students of the LMU Munich. The authors also thank Henrik Glenner, University of Bergen, and J. Matthias Starck, LMU Munich, for their support.

References

Anderson OR (1983) Radiolaria. Springer, New York

Andres D (1989) Phosphatisierte Fossilien aus dem unteren Ordoviz von Südschweden. Berl Geowiss Abh A 106:9–19

Anton RF, Schrödl M (2013) The gastropod–crustacean connection: towards the phylogeny and evolution of the parasitic copepod family Splanchnotrophidae. Zool J Linn Soc 167(4):501–530

Baeza JA (2015) Crustaceans as symbionts: an overview of their diversity, host use and life styles. In: Thiel M, Watling L (eds) Lifestyles and feeding biology. The Natural History of the Crustacea. Oxford University Press, Oxford, pp 163–189

Beck JT (1980) The effects of an isopod castrator, *Probopyrus pandalicola*, on the sex characters of one of its caridean shrimp hosts, *Palaemonetes paludosus*. Biol Bull 158:1–15

Bishop GA (1974) A sexually aberrant crab (*Dakoticancer overanus* Rathbun, 1917) from the Upper Cretaceous Pierre Shale of South Dakota. Crustaceana 26:212–218

Bishop GA (1983) A second sexually aberrant specimen of *Dakoticancer overanus* Rathbun, 1917, from the Upper Cretaceous *Dakoticancer* assemblage, Pierre Shale, South Dakota (Decapoda, Brachyura). Crustaceana 44:23–26

Böckeler W (2005) Pentastomida (tongue worms). In: Rohde K (ed) Marine parasitology. CABI Publishing, Wallingford, pp 235–240

Boucot AJ, Poinar GO Jr (2010) Fossil behavior compendium. CRC Press, Boca Raton

Bousfield EL (2007) Talitridae. In: Carlton JT (ed) The Light and Smith manual: intertidal invertebrates from Central California to Oregon. University of California Press, Berkeley, pp 611–613

Bowman TE (1971) *Palaega lamnae*, new species (Crustacea; Isopoda) from the upper Cretaceous of Texas. J Paleontol 45:540–541
Boxshall GA (1990a) The skeletomusculature of siphonostomatoid copepods, with an analysis of adaptive radiation in structure of the oral cone. Philos Trans R Soc Lond B 328:167–212
Boxshall GA (1990b) Precopulatory mate guarding in copepods. Bijdr Tot Dierkd 60:209–213
Boxshall GA, Defaye D (2006) Pathogens of wild and farmed fish: sea lice. CRC Press, Boca Raton
Boxshall GA, Halsey SH (2004) An introduction to copepod diversity. Ray Society, Andover
Boxshall GA, Lincoln RJ (1983) Tantulocarida, a new class of Crustacea ectoparasitic on other crustaceans. J Crustac Biol 3(1):1–16
Boxshall GA, Lester R, Grygier M, Høeg J, Glenner H, Shields J, Lützen J (2005) Crustacean parasites. In: Rohde K (ed) Marine parasitology. CSIRO Publishing, Collingwood, Australia, pp 123–169
Boyko CB, Williams JD, Markham JC (2012) Recent and fossil Isopoda Bopyridae parasitic on squat lobsters and porcelain crabs (Crustacea: Anomura: Chirostyloidea and Galatheoidea), with notes on nomenclature and biogeography. Zootaxa 3150:1–35
Boyko CB, Moss J, Williams JD, Shields JD (2013) A molecular phylogeny of Bopyroidea and Cryptoniscoidea (Crustacea: Isopoda). Syst Biodivers 11:495–506
Brandt A (1992) Origin of Antarctic Isopoda (Crustacea, Malacostraca). Mar Biol 113:415–423
Brandt A, Poore GC (2003) Higher classification of the flabelliferan and related Isopoda based on a reappraisal of relationships. Invertebr Syst 17:893–923
Briggs DE (1999) Molecular taphonomy of animal and plant cuticles: selective preservation and diagenesis. Philos Trans R Soc Lond B 354(1379):7–17
Briggs DE, Siveter DJ, Siveter DJ, Sutton MD, Legg D (2016) Tiny individuals attached to a new Silurian arthropod suggest a unique mode of brood care. Proc Natl Acad Sci 113:4410–4415
Bruce NL (1986) Cirolanidae (Crustacea: Isopoda) of Australia. Rec Aust Mus Suppl 6:1–239
Bruce NL, Brusca RC, Delaney PM (1982) The status of the isopod families Corallanidae Hansen, 1890, and Excorallanidae Stebbing, 1904 (Flabellifera). J Crustac Biol 2:464–468
Brusca RC (1981) A monograph on the Isopoda Cymothoidae (Crustacea) of the eastern Pacific. Zool J Linnean Soc 73:117–199
Brusca RC (1983) A monograph on the isopod family Aegidae in the tropical Eastern Pacific: I. The genus *Aega*. Allan Hancock Monogr Mar Biol 12:1–39
Brusca RC, Gilligan MR (1983) Tongue replacement in a marine fish (*Lutjanus guttatus*) by a parasitic isopod (Crustacea: Isopoda). Copeia 1983:813–816
Brusca R, Wilson G (1991) A phylogenetic analysis of the Isopoda with some classificatory recommendations. Mem Qld Mus 31:143–204
Bunkley-Williams L, Williams EH (1998) Isopods associated with fishes: a synopsis and corrections. J Parasitol 84:893–896
Castellani C, Maas A, Waloszek D, Haug JT (2011) New pentastomids from the Late Cambrian of Sweden–deeper insight of the ontogeny of fossil tongue worms. Palaeontogr Abt A 293:95–145
Chesunov AV (2002) A case of tongueworms (Pentastomida): a specific problem in context of the modern phylogenetics. Zh Obshch Biol 63:209–226
Christoffersen ML, De Assis JE (2015) Class Eupentastomida Waloszek, Repetski & Maas, 2006. In: von Vaupel Klein JC (ed) Treatise on zoology-anatomy, taxonomy, biology. The Crustacea, vol 5. Brill, Leiden, pp 5–75
Conway Morris S (1990) Parasitism. In: Briggs DE, Crowther P (eds) Palaeobiology: a synthesis. Blackwell, Oxford, pp 376–380
Copilaş-Ciocianu D, Sidorov D, Gontcharov A (2019) Adrift across tectonic plates: molecular phylogenetics supports the ancient Laurasian origin of old limnic crangonyctid amphipods. Org Div Evol 19:191–207
Costa EFDS, de Oliveira MR, Chellappa S (2010) First record of *Cymothoa spinipalpa* (Isopoda: Cymothoidae) parasitizing the marine fish Atlantic bumper, *Chloroscombrus chrysurus* (Osteichthyes: Carangidae) from Brazil. Mar Biodivers Rec 3:e1

Cressey R, Boxshall G (1989) *Kabatarina pattersoni*, a fossil parasitic copepod (Dichelesthiidae) from a Lower Cretaceous fish. Micropaleontology 35:150–167

Cressey R, Patterson C (1973) Fossil parasitic copepods from a Lower Cretaceous fish. Science 180:1283–1285

Crossley SMG, George AL, Keller CJ (2009) A method for eradicating amphipod parasites (Hyperiidae) from host jellyfish, *Chrysaora fuscescens* (Brandt, 1835), in a closed recirculating system. J Zoo Wildl Med 40:174–180

Dahms H-U (2004) Postembryonic apomorphies proving the monophyletic status of the Copepoda. Zool Stud 43:446–453

De Baets K, Littlewood DTJ (2015) The importance of fossils in understanding the evolution of parasites and their vectors. Adv Parasitol 90:1–51

De Baets K, Huntley JW, Klompmaker AA, Schiffbauer JD, Muscente AD (2021) The fossil record of parasitism: Its extent and taphonomic constraints. In: De Baets K, Huntley JW (eds) The evolution and fossil record of parasitism: Coevolution and paleoparasitological techniques. Topics in Geobiology 50

Donoghue MJ, Doyle JA, Gauthier J, Kluge AG, Rowe T (1989) The importance of fossils in phylogeny reconstruction. Annu Rev Ecol Syst 20:431–460

Dreyer H, Wägele J (2001) Parasites of crustaceans (Isopoda:Bopyridae) evolved from fish parasites: molecular and morphological evidence. Zoology 103:157–178

Dreyer H, Wägele J-W (2002) The Scutocoxifera tax. nov. and the information content of nuclear SSU rDNA sequences for reconstruction of isopod phylogeny (Peracarida: Isopoda). J Crustac Biol 22:217–234

Dzik J (2008) Gill structure and relationships of the Triassic cycloid crustaceans. J Morphol 269(12):1501–1519

Edgecombe GD, Legg DA (2013) The arthropod fossil record. In: Arthropod biology and evolution. Springer, Berlin, pp 393–415

Etter W (2014) A well-preserved isopod from the Middle Jurassic of southern Germany and implications for the isopod fossil record. Palaeontology 57:931–949

Eyun S (2017) Phylogenomic analysis of Copepoda (Arthropoda, Crustacea) reveals unexpected similarities with earlier proposed morphological phylogenies. BMC Evol Biol 17:23. https://doi.org/10.1186/s12862-017-0883-5

Feldmann RM (1998) Parasitic castration of the crab, *Tumidocarcinus giganteus* Glaessner, from the Miocene of New Zealand: coevolution within the Crustacea. J Paleontol 72:493–498

Feldmann RM (2009) A new cirolanid isopod (Crustacea) from the Cretaceous of Lebanon: dermoliths document the pre-molt condition. J Crustac Biol 29:373–378

Feldmann RM, Tshudy DM, Thomson MR (1993) Late Cretaceous and Paleocene decapod crustaceans from James Ross Basin, Antarctic Peninsula. J Paleontol 67(S28):1–41

Gentil-Vasconcelos HC, Tavares-Dias M (2015) First study on infestation of *Excorallana berbicensis* (Isopoda: Corallanidae) on six fishes in a reservoir in Brazilian Amazon during dry and rainy seasons. Lat Am J Aquat Res 43:936–943

Giribet G, Edgecombe GD (2013) The Arthropoda: a phylogenetic framework. In: Minelli A, Boxshall GA, Fusco G (eds) Arthropod biology and evolution. Springer, Berlin, pp 17–40

Glenner H, Høeg JT (2002) A scenario for the evolution of the Rhizocephala. In: Modern approaches to the study of Crustacea. Springer, Boston, MA, pp 301–310

Glenner H, Høeg JT, Grygier MJ, Fujita Y (2008) Induced metamorphosis in crustacean y-larvae: towards a solution to a 100-year-old riddle. BMC Biol 6:21

Grygier MJ, Høeg JT (2005) Ascothoracida (ascothoracids). In: Rohde K (ed) Marine parasitology. CABI Publishing, Wallingford, pp 149–154

Grygier MJ, Salvat MB (1984) *Dendrogaster argentinensis*, new species, a South American seastar parasite (Crustacea: Ascothoracida). Proc Biol Soc Wash 97:43–48

Gupta NS, Michels R, Briggs DE, Evershed RP, Pancost RD (2006) The organic preservation of fossil arthropods: an experimental study. Proc R Soc Lond B 273(1602):2777–2783

Hadfield K (2012) The biodiversity and systematics of marine fish parasitic isopods of the Family Cymothoidae from Southern Africa. PhD thesis, University of Johannesburg, Johannesburg, South Africa

Hadfield KA, Bruce N, Smit NJ (2011) *Cymothoa hermani* sp. nov. (Isopoda, Cymothoidae, Crustacea), a parasitic isopod, collected off the Zanzibar coast, Tanzania from the mouth of a parrotfish (Scaridae). Zootaxa 2876:57–68

Hansen T, Hansen J (2010) First fossils of the isopod genus *Aega* Leach, 1815. J Paleontol 84:141–147

Hata H, Sogabe A, Tada S, Nishimoto R, Nakano R, Kohya N, Takeshima H, Kawanishi R (2017) Molecular phylogeny of obligate fish parasites of the family Cymothoidae (Isopoda, Crustacea): evolution of the attachment mode to host fish and the habitat shift from saline water to freshwater. Mar Biol 164:105. https://doi.org/10.1007/s00227-017-3138-5

Haug C (2011) Head shield evolution and development in the Arthropoda. PhD thesis, University of Ulm, Germany

Haug JT, Haug C (2015) "Crustacea": comparative aspects of larval development. In: Wanninger A (ed) Evolutionary developmental biology of invertebrates 4: Ecdysozoa II: Crustacea. Springer, Wien, pp 1–37

Haug JT, Olesen J, Maas A, Waloszek D (2011) External morphology and post-embryonic development of *Derocheilocaris remanei* (Crustacea: Mystacocarida) revisited, with a comparison to the Cambrian taxon *Skara*. J Crustac Biol 31:668–692

Haug JT, Haug C, Garwood R (2016) Evolution of insect wings and development – new details from Palaeozoic nymphs. Biol Rev 91:53–69

Haugerud RE (1989) Evolution in the pentastomids. Parasitol Today 5:126–132

Hessler RR (1969) Peracarida. In: Moore R (ed) Treatise on invertebrate paleontology, Arthropoda. University of Kansas and Geological Society of America, Lawrence

Hispano C, Bultó P, Blanch AR (2014) Life cycle of the fish parasite *Gnathia maxillaris* (Crustacea: Isopoda: Gnathiidae). Folia Parasitol 61:277–284

Ho JS, Katsumi F, Honma Y (1981) *Coelotrophus nudus* gen. et sp. nov., an endoparasitic copepod causing sterility in a sipunculan *Phascolosoma scolops* (Selenka and De Man) from Sado Island, Japan. Parasitology 82(3):481–488

Høeg JT, Lagersson NC, Glenner H (2004) The complete cypris larva and its significance in thecostracan phylogeny. In: Scholtz G (ed) Evolutionary developmental biology of Crustacea, Crustacean issues, vol 15. Balkema, Lisse, pp 197–215

Høeg JT, Pérez-Losada M, Glenner H, Kolbasov GA, Crandall KA (2009) Evolution of morphology, ontogeny and life cycles within the Crustacea Thecostraca. Arthropod Syst Phylogeny 67(2):199–217

Humes AG (1994) How many copepods? Hydrobiologia 292(293):1–7

Huys R, Boxshall GA (1991) Copepod evolution. Ray Society, London

Hyžný M, Bruce NL, Schloegl J (2013) An appraisal of the fossil record for the Cirolanidae (Malacostraca: Peracarida: Isopoda: Cymothoida), with a description of a new cirolanid isopod crustacean from the Early Miocene of the Vienna Basin (Western Carpathians). Palaeontology 56:615–630

Hyžný M, Haug C, Haug JT (2016) *Mesoprosopon triasinum* from the Triassic of Austria revisited: the oldest eumalacostracan larva known to date and its significance for interpreting fossil cycloids. Gondwana Res 37:86–97

Il-Hoi Kim, Hideyuki Yamashiro (2007) Two species of poecilostomatoid copepods inhabiting galls on scleractinian corals in Okinawa, Japan. J Crustacean Biol 27(2): 319–326

Ito A, Aoki MN, Yahata K, Wada H (2011) Complicated evolution of the caprellid (Crustacea: Malacostraca: Peracarida: Amphipoda) body plan, reacquisition or multiple losses of the thoracic limbs and pleons. Dev Genes Evol 221(3):133–140

Jażdżewski K, Kupryjanowicz J (2010) One more fossil niphargid (Malacostraca: Amphipoda) from Baltic amber. J Crustac Biol 30(3):413–416

Jephcott TG, Sime-Ngando T, Gleason FH, Macarthur DJ (2016) Host–parasite interactions in food webs: diversity, stability, and coevolution. Food Webs 6:1–8

Kabata Z (1979) Parasitic Copepoda of British fishes, vol 152. Ray Society, London, pp 1–468

Kato H, Kurihara Y, Tokita T (2016) New fossil record of the genus *Bathynomus* (Crustacea: Isopoda: Cirolanidae) from the middle and upper Miocene of Central Japan, with description of a new supergiant species. Paleontol Res 20:145–156

Khodami S, McArthur JV, Blanco-Bercial L, Arbizu PM (2017) Molecular phylogeny and revision of copepod orders (Crustacea: Copepoda). Sci Rep 7:9164

Klompmaker A, Boxshall G (2015) Fossil crustaceans as parasites and hosts. Adv Parasitol 90:233–289

Klompmaker AA, Schweitzer CE, Feldmann RM, Kowalewski M (2013) The influence of reefs on the rise of Mesozoic marine crustaceans. Geology 41(11):1179–1182

Klompmaker AA, Artal P, van Bakel BWM, Fraaije RHB, Jagt JWM (2014) Parasites in the fossil record: a Cretaceous fauna with isopod-infested decapod crustaceans, infestation patterns through time, and a new ichnotaxon. PLoS One 9:e92551. https://doi.org/10.1371/journal.pone.0092551

Klompmaker AA, Hyžný M, Portell RW, Kowalewski M (2016) Growth, inter- and intraspecific variation, palaeobiogeography, taphonomy and systematics of the Cenozoic ghost shrimp *Glypturus*. J Syst Palaeontol 14:99–126

Klompmaker AA, Portell RW, Frick MG (2017) Comparative experimental taphonomy of eight marine arthropods indicates distinct differences in preservation potential. Palaeontology 60:773–794

Klompmaker AA, Robins CM, Portell RW, De Angeli A (2021) Crustaceans as hosts of parasites throughout the Phanerozoic. In: De Baets K, Huntley JW (eds) The evolution and fossil record of parasitism: Coevolution and paleoparasitological techniques. Topics in Geobiology 50

Kuris AM (1974) Trophic interactions: similarity of parasitic castrators to parasitoids. Q Rev Biol 49:129–148

Labandeira CC, Li L (2021) The History of Insect Parasitism and the Mid-Mesozoic Parasitoid Revolution. In: De Baets K, Huntley JW (eds) The evolution and fossil record of parasitism: Identification and macroevolution of parasites. Topics in Geobiology 49

Lafferty KD, Kuris AM (2002) Trophic strategies, animal diversity and body size. Trends Ecol Evol 17:507–513

Lafferty KD, Kuris AM (2009) Parasitic castration: the evolution and ecology of body snatchers. Trends Parasitol 25:564–572

Lankester ER (1880) Degeneration: a chapter in Darwinism, vol 12. Macmillan and Company, London

Laval P (1980) Hyperiid amphipods as crustacean parasitoids associated with gelatinous zooplankton. Oceanogr Mar Biol Annu Rev 18:11–56

Legg DA, Sutton MD, Edgecombe GD (2013) Arthropod fossil data increase congruence of morphological and molecular phylogenies. Nat Commun 4:3485. https://doi.org/10.1038/ncomms3485

Leung YM (1967) An illustrated key to the species of whale-lice (Amphipoda, Cyamidae), ectoparasites of Cetacea, with a guide to the literature. Crustaceana 12(3):279–291

Leung YM (1976) Life cycle of *Cyamus scammoni* (Amphipoda: Cyamidae), ectoparasite of gray whale, with a remark on the associated species. Sci Rep Whales Res Inst 28:153–160

Leung TL (2017) Fossils of parasites: what can the fossil record tell us about the evolution of parasitism? Biol Rev 92:410–430

Leung TLF (2021) Parasites of fossil vertebrates: what we know and what can we expect from the fossil record? In: De Baets, K., Huntley, J. W. (eds). The evolution and fossil record of parasitism: Identification and macroevolution of parasites. Topics in Geobiology 49

Lins LS, Ho SY, Wilson GD, Lo N (2012) Evidence for Permo-Triassic colonization of the deep sea by isopods. Biol Lett 8(6):979–982

Lins LS, Ho SY, Lo N (2017) An evolutionary timescale for terrestrial isopods and a lack of molecular support for the monophyly of Oniscidea (Crustacea: Isopoda). Org Div Evol 17(4):813–820
Lukševics E, Lebedev O, Mark-Kurik E, Karatajūtė-Talimaa V (2009) The earliest evidence of host–parasite interactions in vertebrates. Acta Zool 90:335–343
Madsen FJ, Wolff T (1965) Evidence of the occurrence of Ascothoracica (parasitic cirripeds) in Upper Cretaceous. Meddel Dansk Geolog For 15:556–558
Maguire EP, Feldmann RM, Jones WT, Schweitzer CE, Casadío S (2018) The first fossil isopod from Argentina: a new species of Cirolanidae (Crustacea: Peracarida) from the Miocene of Patagonia. J Crustac Biol 38:34–44
Markham JC (1986) Evolution and zoogeography of the Isopoda Bopyridae, parasites of Crustacea Decapoda. Crustaceana 4:143–164
Martins-Neto RG (2001) Review of some Crustacea (Isopoda and Decapoda) from Brazilian deposits (Paleozoic, Mesozoic and Cenozoic) with descriptions of new taxa. Acta Geol Leopoldensia 24:237–254
McMenamin MA, Zapata LP, Hussey MC (2013) A Triassic giant amphipod from Nevada, USA. J Crustac Biol 33(6):751–759
Møller OS (2009) Branchiura (Crustacea)-survey of historical literature and taxonomy. Arthropod Syst Phylogeny 67(1):41–55
Møller OS, Olesen J, Avenant-Oldewage A, Thomsen PF, Glenner H (2008) First maxillae suction discs in Branchiura (Crustacea): development and evolution in light of the first molecular phylogeny of Branchiura, Pentastomida, and other "Maxillopoda". Arthropod Struct Dev 37(4):333–346
Monod T (1969) Sur trois crustacés isopodes marins de la région Guyane-Amazone. Cah Orstom Oceanogr 7:47–68
Morand S (2015) (macro-)Evolutionary ecology of parasite diversity: from determinants of parasite species richness to host diversification. Int J Parasitol Parasites Wildl 4:80–87
Moura G, Christoffersen ML (1996) The system of the mandibulate arthropods: Tracheata and Remipedia as sister groups, "Crustacea" nonmonophyletic. J Comp Biol 1(3/4):95–113
Müller AH (1955) Über einen Neufund von *Halicyne plana* und die systematische Stellung von *Halicyne* (Crustacea?). Paläontol Z 29:131–135
Müller KJ, Walossek D (1988) External morphology and larval development of the Upper Cambrian maxillopod *Bredocaris admirabilis*. Fossils Strata 23:1–70
Nagler C, Haug JT (2015) From fossil parasitoids to vectors: insects as parasites and hosts. Adv Parasitol 90:137–200
Nagler C, Haug JT (2016) Functional morphology of parasitic isopods: understanding morphological adaptations of attachment and feeding structures in *Nerocila* as a pre-requisite for reconstructing the evolution of Cymothoidae. PeerJ 4:e2188. https://doi.org/10.7717/peerj.2188
Nagler C, Haug C, Resch U, Kriwet J, Haug JT (2016) 150 million years old isopods on fishes: a possible case of palaeo-parasitism. Bull Geosci 91:1–12
Nagler C, Hyžný M, Haug JT (2017a) 168 million years old "marine lice" and the evolution of parasitism within isopods. BMC Evol Biol 17:76. https://doi.org/10.1186/s12862-017-0915-1
Nagler C, Høeg JT, Haug C, Haug JT (2017b) A possible 150 million years old cirripede crustacean nauplius and the phenomenon of giant larvae. Contr Zool 86:213–227
Nagler C, Haug JT, Glenner H, Buckeridge J (2017c) *Litholepas klausreschi* gen. et sp. nov., a new neolepadine barnacle (Cirripedia, Thoracica) on a sponge from the Upper Jurassic lithographic limestones of southern Germany. Neues Jahrb Geol Paläontol Abh 284:29–42
Néraudeau D, Perrichot V, Batten DJ, Boura A, Girard V, Jeanneau L, Nohra YA, Polette F, Saint Martin S, Saint Martin J-P (2017) Upper Cretaceous amber from Vendée, North-Western France: age dating and geological, chemical, and palaeontological characteristics. Cretac Res 70:77–95
O'Brien J, Van Wyk P (1985) Effects of crustacean parasitic castrators (epicaridean isopods and rhizocephalan barnacles) on growth of crustacean hosts. In: Wenner AM (ed) Crustacean issues 3. Factors in adult growth. A. A. Balkema, Rotterdam, pp 191–218

Oakley TH, Wolfe JM, Lindgren AR, Zaharoff AK (2013) Phylotranscriptomics to bring the under-studied into the fold: monophyletic Ostracoda, fossil placement, and pancrustacean phylogeny. Mol Biol Evol 30(1):215–233

Parker D, Booth A (2013) The tongue-replacing isopod *Cymothoa borbonica* reduces the growth of largespot pompano *Trachinotus botla*. Mar Biol 160:2943–2950

Pasini G, Garassino A (2012) *Palaega pisana* n. sp. (Crustacea, Isopoda, Cirolanidae) from the Pliocene of Orcíano Pisano, Pisa (Toscana, Central Italy). Nat Hist Sci 153:3–11

Petriconi V (1971) Zur Schwimmrichtung der Belemniten und Ökologie bohrender Cirripedier. Palaeogeogr Palaeoclimat Palaeoecol 9(2):133–147

Petrunina AS, Neretina TV, Mugue NS, Kolbasov GA (2014) Tantulocarida versus Thecostraca: inside or outside? First attempts to resolve phylogenetic position of Tantulocarida using gene sequences. J Zool Syst Evol Res 52(2):100–108

Pike A, Wadsworth S (1999) Sealice on salmonids: their biology and control. In: Advances in parasitology. Elsevier, Amsterdam, pp 233–337

Pohl H, Wipfler B, Grimaldi D, Beckmann F, Beutel RG (2010) Reconstructing the anatomy of the 42-million-year-old fossil *Mengea tertiaria* (Insecta, Strepsiptera). Naturwissenschaften 97(9):855–859

Polz H (2004) Asselansammlung auf einer Wasserwanze aus den Solnhofener Plattenkalken. Archaeopteryx 22:51–60

Polz H, Schweigert G, Maisch MW (2006) Two new species of *Palaega* (Isopoda: Cymothoida: Cirolanidae) from the Upper Jurassic of the Swabian Alb, South Germany. Stuttg Beitr Naturk B 362:1–17

Poore GCB, Bruce NL (2012) Global diversity of marine isopods (except Asellota and crustacean symbionts). PLoS One 7:e43529. https://doi.org/10.1371/journal.pone.0043529

Poulin R, Morand S (2000) The diversity of parasites. Q Rev Biol 75:277–293

Poulin R, Randhawa HS (2015) Evolution of parasitism along convergent lines: from ecology to genomics. Parasitology 142:S6–S15

Radwanska U, Poirot E (2010) Copepod-infested Bathonian (Middle Jurassic) echinoids from northern France. Acta Geol Pol 60:549–555

Radwańska U, Radwański A (2005) Myzostomid and copepod infestation of Jurassic echinoderms: a general approach, some new occurrences, and/or re-interpretation of previous reports. Acta Geol Pol 55:109–130

Rees DJ, Noever C, Høeg JT, Ommundsen A, Glenner H (2014) On the origin of a novel parasitic-feeding mode within suspension-feeding barnacles. Curr Biol 24(12):1429–1434

Regier JC, Shultz JW, Kambic RE (2005) Pancrustacean phylogeny: hexapods are terrestrial crustaceans and maxillopods are not monophyletic. Proc R Soc Lond B Biol Sci 272:395–401

Regier JC, Shultz JW, Zwick A, Hussey A, Ball B, Wetzer R, Martin JW, Cunningham CW (2010) Arthropod relationships revealed by phylogenomic analysis of nuclear protein-coding sequences. Nature 463:1079–1083

Reid JW (2001) A human challenge: discovering and understanding continental copepod habitats. Hydrobiologia 453(454):201–226

Richter S, Scholtz G (2001) Phylogenetic analysis of the Malacostraca (Crustacea). J Zool Syst Evol Res 39:113–136

Robin N, Marramà G, Vonk R, Kriwet J, Carnevale G (2019) Eocene isopods on electric rays: tracking ancient biological interactions from a complex fossil record. Palaeontology 62:287–303

Robins CM, Klompmaker AA (2019) Extreme diversity and parasitism of Late Jurassic squat lobsters (Decapoda: Galatheoidea) and the oldest records of porcellanids and galatheids. Zool J Linn Soc 187(4):1131–1154

Robin N (2021) Importance of data on fossil symbioses for parasite-host evolution. In: De Baets K, Huntley JW (eds) The evolution and fossil record of parasitism: Coevolution and paleoparasitological techniques. Topics in Geobiology 50

Rohde K (2005) Marine parasitology. CSIRO Publishing, Collingwood

Rust J (2007) Die Bedeutung von Fossilien für phylogenetische Rekonstruktionen. Species Phylogeny Evol 1:75–87

Schram FR (1977) Paleozoogeography of late paleozoic and Triassic Malacostraca. Syst Biol 26:367–379

Schram FR (1982) The fossil record and evolution of Crustacea. In: Abele LG (ed) The biology of Crustacea, Systematics, the fossil record and biogeography, vol 1. Academic, New York, pp 93–147

Schram FR, Vonk R, Cees HJH (1997) Mazon Creek Cycloidea. J Paleontol 71:261–284

Schwabe E, Maiorova A (2015) *Golfingicola abyssalis* gen. et sp. nov., a new endoparasitic copepod (Crustacea) in a sipunculan from abyssal depths of the Northwest Pacific Ocean. Deep Sea Res II Top Stud Oceanogr 111:135–146

Schweigert G, Fraaije R, Wulf M (2009) Die Cycliden – eine rätselhafte Krebsgruppe. Fossilien 1:17–22

Schwentner M, Combosch DJ, Nelson JP, Giribet G (2017) A phylogenomic solution to the origin of insects by resolving crustacean-hexapod relationships. Curr Biol 27(12):1818–1824

Seger J, Rowntree VJ (2018) Whale lice. In: Würsig B, Thewissen JGM, Kovacs KM (eds) Encyclopedia of marine mammals, 3rd edn. Academic, London, pp 1051–1054

Selden PA, Huys R, Stephenson MH, Heward AP, Taylor PN (2010) Crustaceans from bitumen clast in Carboniferous glacial diamictite extend fossil record of copepods. Nat Commun 1:50. https://doi.org/10.1038/ncomms1049

Serrano-Sánchez L, De M, Nagler C, Haug C, Haug JT, Centeno-García E, Vega FJ (2016) The first fossil record of larval stages of parasitic isopods: cryptoniscus larvae preserved in Miocene amber. Neues Jahrb für Geol Paläontol Abh 279:97–106

Siveter DJ, Briggs DE, Siveter DJ, Sutton MD (2015) A 425-million-year-old Silurian pentastomid parasitic on ostracods. Curr Biol 25(12):1632–1637

Smit N, Davies A (2004) The curious life-style of the parasitic stages of gnathiid isopods. Adv Parasitol 58:289–391

Smit NJ, Bruce NL, Hadfield KA (2014) Global diversity of fish parasitic isopod crustaceans of the family Cymothoidae. Int J Parasitol Parasites Wildl 3:188–197

Starr HW, Hegna TA, McMenamin MA (2016) Epilogue to the tale of the Triassic amphipod: *Rosagammarus* McMenamin, Zapata and Hussey, 2013 is a decapod tail (Luning formation, Nevada, USA). J Crustac Biol. 36(4):525–529

Stock JH (1968) The Calvocheridae, a family of copepods inducing galls in sea-urchin spines. Bijdr Tot Dierk 38(1):85–90

Stock JH (1975) Corallovexiidae, a new family of transformed copepods endoparasitic in reef corals with two new genera and ten new species from Curaçao. Stud Fauna Curaçao Other Caribbean Islands 47(1):1–45

Stock JH (1981) Associations of hydrocorallia stylasterina with gall-inhabiting copepoda siphonostomatoidea from the south-west Pacific. Part II. On six species belonging to four new genera of the copepod family Asterocheridae. Bijdragen tot de Dierkunde, 51(2):287–312

Storch V, Jamieson BGM (1992) Further spermatological evidence for including the Pentastomida (tongue worms) in the Crustacea. Int J Parasitol 22:95–108

Tavares-Dias M, Dias-Júnior MBF, Florentino AC, Silva LMA, da Cunha AC (2015) Distribution pattern of crustacean ectoparasites of freshwater fish from Brazil. Rev Bras Parasitol Veterinária 24:136–147

Thatcher VE (2000) The isopod parasites of South American fishes. In: Salgado-Maldonado G, García Aldrete AN, Vidal-Martínez VM (eds) Metazoan parasites in the Neotropics: a systematic and ecological perspective. Instuto de Biología, Universidad Nacional Autónoma de México, Ciudad de México, pp 193–226

Thewissen JG, Williams EM (2002) The early radiations of Cetacea (Mammalia): evolutionary pattern and developmental correlations. Annu Rev Ecol Syst 33(1):73–90

Tomlinson JT (1987) The burrowing barnacles (Acrothoracica). In: Southward AJ (ed) Barnacle biology, Crustacea Issues 5. Routledge, London, pp 63–71

Torchin ME, Lafferty KD, Dobson AP, McKenzie VJ, Kuris AM (2003) Introduced species and their missing parasites. Nature 421:628–630

van Dijk J, De Baets K (2021) Biodiversity and host-parasite (co)extinction. In: De Baets K, Huntley JW(eds). The evolution and fossil record of parasitism: Coevolution and paleoparasitological techniques. Topics in Geobiology 50

Vega FJ, Bruce NL, González-León O, Jeremiah J, Serrano-Sánchez MDL, Alvarado-Ortega J, Moreno-Bedmar JA (2019) Lower Cretaceous marine isopods (Isopoda: Cirolanidae, Sphaeromatidae) from the San Juan Raya and Tlayúa formations, Puebla, Mexico. J Crustac Biol 39:121–135

Voigt E (1959) *Endosacculus moltkiae* n. g. n. sp., ein vermutlicher fossiler Ascothoracide (Entomostr.) als Cystenbildner bei der Oktokoralle *Moltkia minuta*. Paläontol Z 33:211–223

Voigt E (1967) Ein vermutlicher Ascothoracide (*Endosacculus* (?) *najdini* n. sp.) als Bewohner einer kretazischen *Isis* aus der UdSSR. Paläontol Z 41:86–90

Vonk R, Schram FR (2007) Three new tanaid species (Crustacea, Peracarida, Tanaidacea) from the Lower Cretaceous Álava amber in northern Spain. J Paleontol 81:1502–1509

Wägele J-W (1989) Evolution und phylogenetisches System der Isopoda. Zoologica 140:1–262

Walossek D, Müller KJ (1994) Pentastomid parasites from the Lower Palaeozoic of Sweden. Earth Environ Sci Trans R Soc Edinb 85:1–37

Walossek D, Müller KJ (1998) Early arthropod phylogeny in light of the Cambrian "Orsten" fossils. In: Edgecombe GD (ed) Arthropod fossils and phylogeny. Columbia University Press, New York, pp 185–231

Walossek D, Repetski JE, Müller KJ (1994) An exceptionally preserved parasitic arthropod, *Heymonsicambria taylori* n. sp.(Arthropoda incertae sedis: Pentastomida), from Cambrian–Ordovician boundary beds of Newfoundland, Canada. Can J Earth Sci 31(11):1664–1671

Waloszek D, Repetski JE, Maas A (2006) A new Late Cambrian pentastomid and a review of the relationships of this parasitic group. Trans R Soc Edinburgh Earth Sci 96:163–176

Wardle WJ, Haney TA, Worthy GA (2000) New host record for the whale louse *Isocyamus delphinii* (Amphipoda, Cyamidae). Crustaceana 73(5):639–641

Warnock RCM, Engelstädter J (2021) The molecular clock as tool for understanding host-parasite evolution. In: De Baets K, Huntley JW (eds) The evolution and fossil record of parasitism: Coevolution and paleoparasitological techniques. Topics in Geobiology 50

Weinfurtner G (1989) Seelilienstielglieder mit Parasiten. Fossilien 6:64–65

Weinstein SB, Kuris AM (2016) Independent origins of parasitism in Animalia. Biol Lett 12(7):20160324

Wetzer R, Perez-Losada M, Bruce NL (2013) Phylogenetic relationships of the family Sphaeromatidae Latreille, 1825 (Crustacea: Peracarida: Isopoda) within Sphaeromatidea based on 18S-rDNA molecular data. Zootaxa 3599:161–177

Wieder RW, Feldmann RM (1992) Mesozoic and Cenozoic fossil isopods of North America. J Paleontol 66:958–972

Wienberg Rasmussen H, Jakobsen SL, Collins JSH (2008) Raninidae infested by parasitic Isopoda (Epicaridea). Bull Mizunami Foss Mus 34:31–49

Williams JD, Boyko CB (2012) The global diversity of parasitic isopods associated with crustacean hosts (Isopoda: Bopyroidea and Cryptoniscoidea). PLoS One 7:e35350. https://doi.org/10.1371/journal.pone.0035350

Wilson GD (2009) The phylogenetic position of the Isopoda in the Peracarida (Crustacea: Malacostraca). Arthropod Syst Phylogeny 67:159–198

Wilson GD, Paterson JR, Kear BP (2011) Fossil isopods associated with a fish skeleton from the Lower Cretaceous of Queensland, Australia–direct evidence of a scavenging lifestyle in Mesozoic Cymothoida. Palaeontology 54:1053–1068

Wingstrand KG (1972) Comparative spermatology of a pentastomid, *Raillietiella hemidactyli*, and a branchiuran crustacean, *Argulus foliaceus*, with a discussion of pentastomid relationships. Kong Dansk Videnskab Selskab Biol Skrifter 19:1–72

Chapter 11
The History of Insect Parasitism and the Mid-Mesozoic Parasitoid Revolution

Conrad C. Labandeira and Longfeng Li

Abstract Insect parasites and parasitoids are a major component of terrestrial food webs. For parasitoids, categorization is whether feeding activity is located inside or outside its host, if the host is immobilized or allowed to grow, and if the feeding is done by one or many conspecific or heterospecific individuals, and other features. Fossil evidence for parasitism and parasitoidism consists of taxonomic affiliation, morphology, gut contents, coprolites, tissue damage and trace fossils. Ten hemimetabolous and holometabolous orders of insects developed the parasite condition whereas seven orders of holometabolous insects evolved the parasitoid life habit. Modern terrestrial food webs are important for understanding the Mid Mesozoic Parasitoid Revolution. The MMPR began in late Early Jurassic (Phase 1), in which bottom-to-top regulation of terrestrial food webs dominated by inefficient clades of predators were replaced by top-to-bottom control by trophically more efficient parasitoid clades. The MMPR became consolidated in Phase 2 by the end of the Early Cretaceous. These clades later expanded (phases 3 and 4) as parasitoids became significant ecological elements in terrestrial food webs. Bottom-to-top food webs explained by the resource concentration hypothesis characterize pre-MMPR time. During phases 1 and 2 of MMPR (Middle Jurassic to Early Cretaceous), a shift

This is a U.S. government work and its text is not subject to copyright protection in the United States; however, it may be subject to foreign copyright protection.

C. C. Labandeira (✉)
Department of Paleobiology, National Museum of Natural History, Smithsonian Institution, Washington, DC, USA

Department of Entomology and BEES Program, University of Maryland, College Park, MD, USA

College of Life Sciences, Capital Normal University, Beijing, People's Republic of China
e-mail: labandec@si.edu

L. Li
College of Life Sciences, Capital Normal University, Beijing, People's Republic of China

Institute of Vertebrate Paleontology, College of Life Science and Technology, Gansu Agricultural University, Lanzhou City, People's Republic of China

© The Author(s) 2021

K. De Baets, J. W. Huntley (eds.), *The Evolution and Fossil Record of Parasitism*, Topics in Geobiology 49, https://doi.org/10.1007/978-3-030-42484-8_11

ensued toward top-to-down food webs, explained by the trophic cascade hypothesis, exemplified by hymenopteran parasitoid clades Stephanoidea and Evanioidea. Clade-specific innovations spurring the MMPR included long, flexible ovipositors (wasps), host seeking, triungulin and planidium larvae (mantispids, beetles, twisted-wing parasites, flies), and extrudable, telescoped ovipositors (flies). After the MMPR, in phases 3 and 4 (Late Cretaceous to Recent), parasitoids increased in taxonomic diversity, becoming integrated into food webs that continue to the present day.

Keywords Food webs · Idiobiont · Koinobiont · Ovipositor drill · Telescoped ovipositor · Triungulin larvae

> It seems clear that the ancestral hymenopteran and dipteran parasitoids found themselves in a relatively unexploited adaptive zone. The resultant adaptive radiation onto different host species, possibly occurring simultaneously with the adaptive radiation of the modern insect orders, is responsible for the huge number of species we observe today. (H.C.J. Godfray 1994)

> … parasites [and parasitoids] strongly affect food web structure. Indeed, they disproportionately dominate food web links. … yet some parasites [parasitoids] have population dynamic impacts that are hugely disproportionate to their small size … recognition of parasite [parasitoid] links may have important consequences for ecosystem stability because they can increase connectance and nestedness. (Lafferty et al. 2006, insertions ours to conform with modern terminology)

11.1 Introduction

One of the enduring features of continental ecosystems during and since the mid Mesozoic has been the ecological expansion of insect parasitism in general and the emergence of the insect parasitoid guild in particular. This underappreciated fact only recently has been recognized (Labandeira 2002, 2015; Li et al. 2018a), given the increasingly important role that insect (Freeland and Boulton 1992; Mills 1994; Lafferty et al. 2006; Dunne et al. 2013) and other (Kuris et al. 2008; Hughes et al. 2011a) parasitoids play in the trophic structure of modern terrestrial ecosystems. For example, one of the best-studied systems has been the leaf miner–parasitoid community in Central America, which spotlights the importance of top-to-bottom parasitoid regulation of leaf-mining herbivores in local food webs (Memmott et al. 1994, 2000). According to the fossil record, parasitism—in the broad sense that includes parasitoidism (Lafferty et al. 2006)—began in the marine realm, evolving at least 136 times across 15 of the conventionally recognized 43 phyla ranging from Cnidaria to Arthropoda (Weinstein and Kuris 2016). Of these phyla, it is Arthropoda, both marine and continental, that had the most numerous independent originations of parasitism, constituting about 64% of total originations. It is in the continental realm that encompasses terrestrial and freshwater ecosystems where known cases of parasitism expand dramatically. The parasitoid guild in continental ecosystems is represented by at least 84 separate originations among 10 orders of insects, amounting to 10% of all described insect species (Gaston 1991; Weinstein and Kuris 2016). Considering only parasitoids, rather than the broader category that would include inquilines, parasites and cleptoparasitoids, there are 68,000 out of 850,000 described insect species, or about 8% of the total, that are obligately parasitoid species (Gaston 1991; Godfray 1994). The majority of insect parasitoids are typically small, inconspicuous wasps of Hymenoptera (sawflies, wasps, ants and bees), constituting about 75% of the total, and Diptera (true flies) account for another 20% (Eggleton and Belshaw 1992). The remaining 5% is scattered among lineages of Neuroptera (lacewings and antlions), Coleoptera (beetles), Strepsiptera (twisted-wing parasites), Trichoptera (caddisflies) and Lepidoptera (moths and butterflies) (Askew 1971; Eggleton and Belshaw 1992; Godfray 1994). However, because of their inconspicuousness, rarity and frequent fidelity to a single host, there is significant underreporting of parasitoid taxa in local, community-level assessments. This under-representation especially is true for apocritan wasps (LaSalle and Gauld 1991) and tachinid Diptera (Crosskey 1980; also see Stireman III 2005). Given these and other recent updates and trends in species descriptions, it is probable that up to 25% of insect species have parasite or parasitoid life habits (Godfray 1994), and that 20% of insect species are solely parasitoids (Hochberg and Hawkins 1992).

Historically, a major issue regarding macroevolutionary patterns in parasite and especially parasitoid insects has been whether diet specialization, as opposed to consumption of an eclectic spectrum of food, has been a pathway for increased diversification (Rainford and Mayhew 2015). One early test of these two contending hypotheses was sister clade comparisons of lineages in which one clade possessed a non-parasitoid diet and its sister clade engaged in an exclusively parasitoid diet (Wiegmann et al. 1993). The question posed by the authors of that study was whether the parasitoid life habit (see below for a definition) was an evolutionary dead end involving specialization, or rather a key innovation that led to expansive speciation (Wiegmann et al. 1993). The authors found, of 15 clades examined, that 6 were significantly more diverse and 9 less diverse than their saprophage or predator sister clades, leading to the conclusion that a parasitic or parasitoid life style does not result in increased diversification rates and may very well be an evolutionary dead end. However, the "push of the past" phenomenon, whereby clades that persist for a significant length of deep time typically experience high levels of early diversification but subsequently have substantially decreased levels (Budd and Mann 2018), is a relevant consideration. The push-of-the-past phenomenon may indicate that currently depauperate parasitoid clades have experienced elevated speciation levels in the past. Nevertheless, this seeming paradox of elevated specialization and evolution of a highly restricted life habit yielding a dramatic increase in speciosity (Drake 2003) was retested by Rainford and Mayhew (2015) by employing a different approach, methodology and analyses (details are provided in Rainford and Mayhew 2015). The results of Rainford and Mayhew (2015) indicated that there are well-characterized bouts of diversification of parasitoidism involving parasitoid wasps (Rainford and Mayhew 2015), where the bulk of parasitoidism, approximately 75%, resides (Eggleton and Belshaw 1992). Contributing to this preeminence of hymenopteran parasitoids is that about 85% of insect parasitoids have an adult female as the host-seeking stage (Eggleton and Belshaw 1992). By contrast, the relatively rare life-habit of ectoparasitism (defined below) was identified as an ecological dead end for insect diversification, occurring prominently in the Phthiraptera (lice). Importantly, the majority of these parasitoid and ectoparasite lineages (Rainford and Mayhew 2015), whether or not they led to specialized dead ends or elevated diversification of taxa, originated during the Middle Jurassic to Early Cretaceous (Labandeira 2002). During the mid Mesozoic, the diversification of parasitoid taxa, compared to the slowly evolving ectoparasite taxa, likely was linked to the presence of complete metamorphosis (holometabolous development) as a major key innovation (Rainford et al. 2014), rather than any clade-specific transformation.

This contribution is organized into two sections. The first section is a review of the biology and fossil history of insect parasitism and parasitoidism. The review includes relevant definitions of a parasite and parasitoid and their relationship to predation, the several types of parasitoidism, the kinds of evidence for demonstrating all three feeding behaviors in the fossil record, their evolutionary biology, and an extended exploration of their fossil record. In the second section, the Mid Mesozoic Parasitoid Revolution (MMPR) is proposed as a major biological event in terrestrial food-web history. As the MMPR is the principal emphasis of this report, evidence for the MMPR is provided from the fossil record and modern studies of terrestrial food webs, indicating that ecological communities were transformed during this prolonged event. This transformational change started from resource-driven control of consumers that emphasized bottom–up links in which photosynthesis was regulated by the availability of plant resources for their herbivore consumers. Later, a shift toward parasitoid-driven control of consumers focused on top–down, more efficient regulation. Four fossil biotas from each of the time intervals—before, during and after the MMPR—illustrate this trophic transformation of food webs by the dramatic increase of parasitoid groups in the mid Mesozoic. Last, the diversification events of two lineages of wasp parasitoids, Stephanoidea and Evanioidea, are examined to understand the role of hymenopteran parasitoids during the early phases of the MMPR. The contribution concludes by an assessment of the role that trophic specialization provided in launching the MMPR and the consequences of this major ecological event.

11.2 Defining the Insect Consumption of Animals

The terms predation and parasitism are well established in the paleobiological and entomological literature. The indelicately pronounced term, parasitoidism, has been a more recent term and formalized concept, and has received a considerable amount of scrutiny, particularly in the older literature (Clausen 1940; Askew 1971). In the earlier literature, the term parasitism was used confusingly to mean the traditional parasitism of insects that often feed on a variety of integumental tissues or less commonly in internal tissues of large animals, often vertebrates, whose hosts remain alive after attack, as well as true parasitoids that consume and eventually kill their considerably larger hosts. In this contribution, a distinction is made between these three, very different life habits and consideration is made of parasitism and parasitoidism as trophically distinctive feeding modes (Frank and Gillett-Kaufman 2006; Labandeira 2002).

A recurring issue that informs the definition of parasitism is whether herbivory, in the broadest sense of the interaction, a form of parasitism (Windsor 1998; Eggleton and Gaston 1990). Efforts also have been made to circumscribe a special type of herbivory that occurs in concealed plant tissues, such as leaf mines or galls, as parasitism (Janzen 1975; Price 1980). Whether the broader or the narrower version of herbivory is considered as parasitism, it would render moot much of the specialized trophic biology of animal–animal interaction inherent in true parasitism. Moreover, if parasites are defined as species that exploit other free-living species (Windsor 1998), then minimally about half of all species are parasites (May 1988; Bush et al. 2001), or parasites probably outnumber free-living species by a factor of 4 (May 1992). Such an inordinate broadening of the definition of parasitism would provide sustenance to the claim that parasitism is a confounding concept (Araújo et al. 2003). Historically, herbivory has been considered as a fundamental trophic interaction limited to plants as the consumed organisms (Ings et al. 2009), whereas parasitism—as well as predation and parasitoidism—have considered animals as the consumed organisms. That distinction is retained herein.

11.2.1 Predation

Predation is a type of trophic interaction of a consumer that requires multiple prey items throughout its lifetime and always results in death of its prey (Morris 1998). Such a definition excludes consumption of multiple organisms that do not result in their individual deaths, nor is predation the prolonged consumption of a single organism (a host) that eventually results in its death. Consequently, distinctions are made between prey and host. Prey is the term that applies to an organism that is killed quickly by a consumer involved in predation. By contrast, a host, unlike prey, is defined as a temporally prolonged food resource used by a parasite or parasitoid. One type of host is an organism that is not killed by its consumer after a brief encounter, which involves minor consumption such as a blood-feeding mosquito on a vertebrate host. Such a relationship is parasitism. A second type of host is an organism that is killed only after a prolonged process of consumption, such as a wasp larva feeding internally on an herbivorous caterpillar. Such an interaction is parasitoidism.

11.2.2 Parasitism

Parasitism is an intimate trophic interaction between a consumer and its' typically much larger host that does not result in death of the host (Morris 1998). Parasitism does not include herbivory, for reasons mentioned above, because one fundamental difference of a food web is based on the trophic interactions of the consumption of plants versus the consumption of animals. Thus, the exclusion of herbivory from parasitism makes good ecological sense. An alternative view is that parasitism is a perplexing concept difficult to define owing to conceptual difficulties in separating parasitism from herbivory and from the insensible trophic gradations among commensalism, mutualism and symbiosis (Araújo et al. 2003)—a view that is not taken here.

11.2.3 Parasitoidism

Parasitoidism is a trophic interaction whereby a free-living adult has a dispersive stage whose parasitic larva gradually consumes and eventually obligately kills its host from within or without upon the parasitoid emerging as an adult (Gauld and Bolton 1988; Morris 1998). The feeding behavior of parasitoids can be considered intermediate between that of a predator and a parasite (Knutson and Berg 1966). Although a parasitoid does kill its host, analogous to a predator killing its prey, it also feeds and has life habits like that of a parasite, existing in intimate association and extracting in incremental fashion nutrition from its host (Askew 1971). The definition of what constitutes a parasitoid has undergone many changes in the century-long history of the concept. These shifts in definition are probably why there are several categories of parasitoids.

11.2.4 Parasitoidism: A History of the Term

Reutter (1913) was the first to coin the term parasitoid, which he defined, insightfully, as a feeding behavior intermediate between predation and parasitism. The term was redefined by Waage and Greathead (1986), who detailed the parasitoid life history as follows.

> Adult female parasitoids are free-living, feed on nectar, pollen or as predators and forage actively for their arthropod hosts on plants and other substrates. Usually, on locating a host, the female lays one or more eggs on or in it, and the ensuing larvae consume the host tissue, killing the host in the process.

By contrast, Eggleton and Gaston (1990) mentioned that Price (1984) offered a more unambiguous definition based on a circumscribed definition of parasitoid life-history patterns.

> A species of insect that requires and eats only one animal in its life span by living parasitically as a larva on a host; but the adult is free-living and may ultimately kill many hosts by leaving eggs or larvae near or on the host that consume the host.

A few years later Gauld and Bolton (1988) provided a more succinct definition with a greater economy of words.

> Parasitoids are insects whose larvae develop by feeding on or within an arthropod host, and this host individual is almost always killed by the developing parasitoid larva.

Finally, the definition by Eggleton and Gaston (1990) is presented, which is a minor elaboration of Gauld and Bolton's (1988) formulation. However, the definition was borrowed mostly from Kuris (1974).

> A parasitoid is an organism which develops on or in another single ("host") organism, extracts nourishment from it, and kills it as a direct or indirect result of that development.

These definitions vary in scope, with emphases placed variously on the larval or adult stages. The more restrictive definition is that of Gauld and Bolton (1988), for which Eggleton and Gaston (1990) references the deep understanding of hymenopteran biology by the authors. The less circumscribed definitions of the first two quotes are not as inclusive and probably exclude clades such as Strepsiptera. Strepsiptera do not kill their hosts, but rather castrate them, but in any event, a genetic death is the result. (Consequently, strepsipterans are considered herein as parasitoids.) More to the point, the first three definitions would exclude organisms other than insects, an unsound restriction that would disallow mermithinid nematodes (Poinar Jr 2003) and ophiocordycipitacean fungi (Hughes et al. 2011a; Evans et al. 2011), which clearly are non-insectan parasitoids, often on insects. However, the last definition of Eggleton and Gaston (1990) would encompass non-insectan taxa, including the myriad of phyla with parasitoid taxa in the marine realm (Weinstein and Kuris 2016). Because the definitions of trophic groups such as herbivore, parasite and predator are functional descriptions, the simple, functional definition immediately above by Eggleton and Gaston (1990) is proposed for a parasitoid, with minor exceptions, based on an earlier meaning by Kuris (1974). Such an ecumenical definition would include mermithid nematodes and ophiocordycipitacean fungi mentioned above and probably hemiepiphytic plants that ultimately kill their host plants (Putz and Holbrook 1989).

11.2.5 Types of Parasitoidism

Parasitoidism represents a complex interplay of ecological and behavioral phenomena. These phenomena can be categorized in many ways, some of which are binary contrasts and others that are terms for singular, distinctive types of parasitoidism. The major types of parasitoid relationships in common usage are detailed below.

11.2.5.1 Ectoparasitoidism Versus Endoparasitoidism

Parasitoids that develop within the body and feed on the internal tissues of their host are endoparasitoids. Ectoparasitoids, by contrast, live on the external surface of their host although often their mouthparts are buried into deeper subcutaneous tissues. Complexities to these two contrasting definitions are parasitoids which include species that spend part of their development as endoparasitoids and the other part as ectoparasitoids (Godfray 1994), or parasitoids that have their bodies partly located deep in host internal tissue and partly exposed to the surface (Cook 2014).

11.2.5.2 Koinobiont Parasitoidism Versus Idiobiont Parasitoidism

Koinobiont parasitoids initially immobilize but continually allow their host to grow. After a parasitoid oviposits an egg on its host, a koinobiont parasitoid allows its host to develop to maturity while simultaneously feeding on it. Koinobiont development of the host frequently is promoted by endoparasitoidism. By contrast, after initial oviposition, an idiobiont parasitoid prevents further development of its host while feeding on it, resulting subsequently in limited host resources (Haeselbarth 1979). Idiobiont development of its host is frequently fostered by ectoparasitoidism.

11.2.5.3 Solitary Versus Gregarious Parasitoidism

Solitary parasitoids are those that feed alone on a host without other parasitoid accompaniment (Mackauer and Chau 2001). Gregarious parasitoids, by contrast, feed on a host in multiple numbers, ranging from two to thousands such that typically one parasitoid individual remains after all conspecifics have died (Mackauer and Chau 2001). Generally, solitary versus gregarious parasitoidism is a property of the particular parasitoid species and is regulated by complex hormonal interactions with the host and other potential colonizing parasitoids.

11.2.5.4 Superparasitoidism Versus Multiparasitoidism

Superparasitoidism is a condition whereby multiple eggs, typically many, are oviposited on or in the same host individual by a female of the same parasitoid species (Fisher 1961; Mackauer and Chau 2001). However, if a second parasitoid species lays eggs in addition to the first parasitoid species, multiparasitoidism is the result (Fisher 1961). In either situation, monospecific or heterospecific competition ensues among the parasitoid larvae for limited resources of the host. Superparasitoidism sometimes is accompanied by the oviposition of a single polyembryonic egg that produces multiple, genetically identical larvae that number from tens to thousands of individuals (Silvestri 1906). Unlike gregarious parasitism, multiple, often many, individuals survive to complete their development on the same host.

11.2.5.5 Hyperparasitoidism

Primary parasitoids have a simple relationship between a host and its parasitoid. Secondary parasitoids, or hyperparasitoids, are facultative or obligate parasitoids of a primary parasitoid on the same source host. Tertiary parasitoids have been documented (Godfray 1994; Frank and Gillett-Kaufman 2006), and up to five levels of hyperparasitoidism can occur in the case of oak galls that involve ichneumon wasps (Askew 1961). Facultative hyperparasitoids are situations where a parasitoid can attack either a host or the parasitoid of that host, as opposed to an obligate hyperparasitoid that can only attack a parasitoid of the host (Godfray 1994). Obligate hyperparasitoids exhibit very high host specificity (Schär and Vorburger 2013). The concept of hyperparasitoidism is inextricably linked to hyperparasitism in the older biological literature, where the two life histories were not distinguished. The French naturalist Maria Sibylla Merian (1647–1717) was probably the first person to recognize hyperparasitoidism (as hyperparasitism), as illustrated in her drawings (Todd 2011). The concept later was recognized in verse in 1733 by the English writer Jonathan Swift, who undoubtedly borrowed the idea from earlier authors.

11.2.5.6 Egg Parasitoidism, Larval Parasitoidism, Pupal Parasitoidism and Adult Parasitoidism

Parasitoids of holometabolous insects can attack any of the four major developmental stages, or any two of adjacent stages of the egg, multiple instars of the larva, pupa and adult. Frequently parasitoids are not only species specific but also target particular developmental instars, such as egg parasitoids (Malyshev 1968; Huber 1986; Whitfield 1998), larval parasitoids (Askew 1971; Weinstein and Austin 1991; Whitfield 2003), pupal parasitoids (Clausen 1940; Quevillon and Hughes 2018) and adult parasitoids (Askew 1971; Whitfield 1998, 2003), although it is the earlier stages that are most often attacked. For hemimetabolous insect hosts, parasitoids attack their eggs (Muldrew 1953; Greathead 1963; Brown 1973); nymphs, if terrestrial (Eggleton and Belshaw 1992; Cook 2014); naiads, if aquatic (Clausen 1940; Askew 1971); or adults (DeBach 1964; Kirkpatrick 1947; Elzinga 1977; Cook 2014), although there is little somatic differentiation between nymphal and their conspecific adult instars. Parasitoids of egg, pupal and adult stages of their hosts typically are idiobionts, as are those parasitoids of larvae whose sting causes permanent paralysis (Gauld and Bolton 1988; Godfray 1994). Koinobiont parasitoids are those that attack the egg–larval and larval–pupal instar couplets or are parasitoids of adjacent larval instars, and do not paralyze their hosts. A special category is an egg parasitoid, which often is a miniscule insect that deposits typically very small eggs on the surface or in the interstitial tissues of a much larger host egg. Other egg parasitoids oviposit in an egg sac containing multiple eggs, in which the hatched parasitoid larva either slowly consumes nutritive tissues of the egg that eventually starves and kills the embryo, or alternatively successively consumes eggs within an egg sac (Askew 1971; Vetter et al. 2012).

11.2.5.7 Cleptoparasitoidism

Cleptoparasitoidism is a rare life style in which a parasitoid absconds a vital resource such as food that results in the death of the host after it is deprived of the supply (Frank and Gillett-Kaufman 2006; Dehon et al. 2017). Food resources can include foliage, galls, insects or other provisions intended for the host (Eggleton and Belshaw 1992). A related concept is social parasitism, which is a special type of parasitism constituting a relationship between two species such that one species is dependent parasitically on the other (Brandt et al. 2005; Smith et al. 2007).

An example is a social parasite that benefits from brood care and access to resources that are managed by and benefits the host colony. Social parasitism occurs among a broad variety of insects in which the hosts and the parasites often are closely related, in some cases forming a sympatric species pair. By contrast, inquilines are animals that exploit the living space, such as a termite or ant nest, resulting in a commensal relationship in which one member benefits and the other member derives a reward or is not harmed (Malyshev 1968). Inquilism often is a precursor to social parasitism, which in turn may be a precursor to cleptoparasitoidism (Askew 1971).

11.3 The Evidence: Distinguishing Predation, Parasitism and Parasitoidism

Several types of evidence are important for detecting the presence of predation, parasitism and parasitoidism in the fossil record (Fig. 11.1). Some of this evidence is associated with particular techniques, such as synchrotron X-ray microtomography (van de Camp et al. 2018), or time-of flight secondary-ion mass spectrometry

Fig. 11.1 (continued) New York State (Shear et al. 1989). (**c**) Vertebrate coprolite containing remains of a cockroach, including wings (w), legs (l) and ovipositor (o) from the Middle Pennsylvanian Mazon Creek locality of Illinois (FMNH PE 54114). (**d**) From the same locality as (**c**) is the insect *Protdiamphipnoa woodwardi* Brongniart (Cnemidolestodea: Cnemidolestidae), with prominent forewing eyespots (Carpenter 1971). (**e**) The enigmatic flea or flea-like insect *Strashila incredibilis* Rasnitsyn (?Siphonaptera: Saurophthiridae), from the Late Jurassic of Transbaikalia, Russia, exhibiting aptery, chelate hind tarsus and piercing-and-sucking mouthparts (Rasnitsyn 1992). (**f**) The parasitoid ensign wasp *Leptephialtites caudatus* Rasnitsyn (Hymenoptera: Ephialtitidae) from the Late Jurassic of Karatau, Kazakhstan, with elongate ovipositor (o) and ovipositor valves (ov) (Rasnitsyn 1975). (**g**) The digger wasp *Angarosphex beiboziensis* Hong (Hymenoptera: Angarosphecidae) from the Jurassic–Cretaceous boundary of China, exhibiting abdominal banding indicating a Batesian model (Hong 1984). (**h**) The peculiar chewing louse *Saurodectes vrsanskyi* Rasnitsyn & Zherikhin (?Phthiraptera: Saurodectidae) from the Early Cretaceous of Transbaikalia, Russia (Rasnitsyn and Zherikhin 1999) Note gut contents. (**i**) Gut contents of the middle Eocene bat *Palaeochiropteryx tupaiodon* Revilliod (Chiroptera: Palaeochiropterygidae) from Messel, Germany, containing butterfly scales and other insect fragments (Richter and Storch 1980). Approximate length of scale at right is 90 µm. (**j**) Robber fly (Diptera: Asilidae) from the middle Eocene of Colorado, USA (USNM501477), displaying raptorial forelegs and mouthparts of a single, dagger-like stylet (arrow). (**k**) The egg (nit) of an undetermined sucking louse (Phthiraptera: Anoplura) on a mammalian hair shaft, from middle Eocene Baltic amber of Germany (Voigt 1952). (**l**) *Palaeopsylla klebsiana* Dampf (Siphonaptera: Hystrichopsyllidae), a mammal-parasitizing flea from the same provenance as (**k**), showing socketed antennae, head comb and maxillary lever–stylet complex typical of fleas (Dampf 1910). (**m**) The seed *Tectocarya rhenana* (Cornaceae), with exit hole of a seed predator, from the early Miocene of Germany (Schmidt et al. 1958). (**n**) A bee cell (Hymenoptera: Stenotritidae) from the Pleistocene of South Australia, with a small exit hole of a probable parasitoid (Houston 1987). All subfigures are redraw from original images or are camera lucida drawings of specimens from the Field Museum of Natural History in Chicago (FMNH) or the National Museum of Natural History in Washington, DC (USNM). Scale bars: solid, 1.0 cm; striped, 0.1 cm. (Reproduced, with permission of the Paleontological Society, from Fig. 2 of Labandeira 2002)

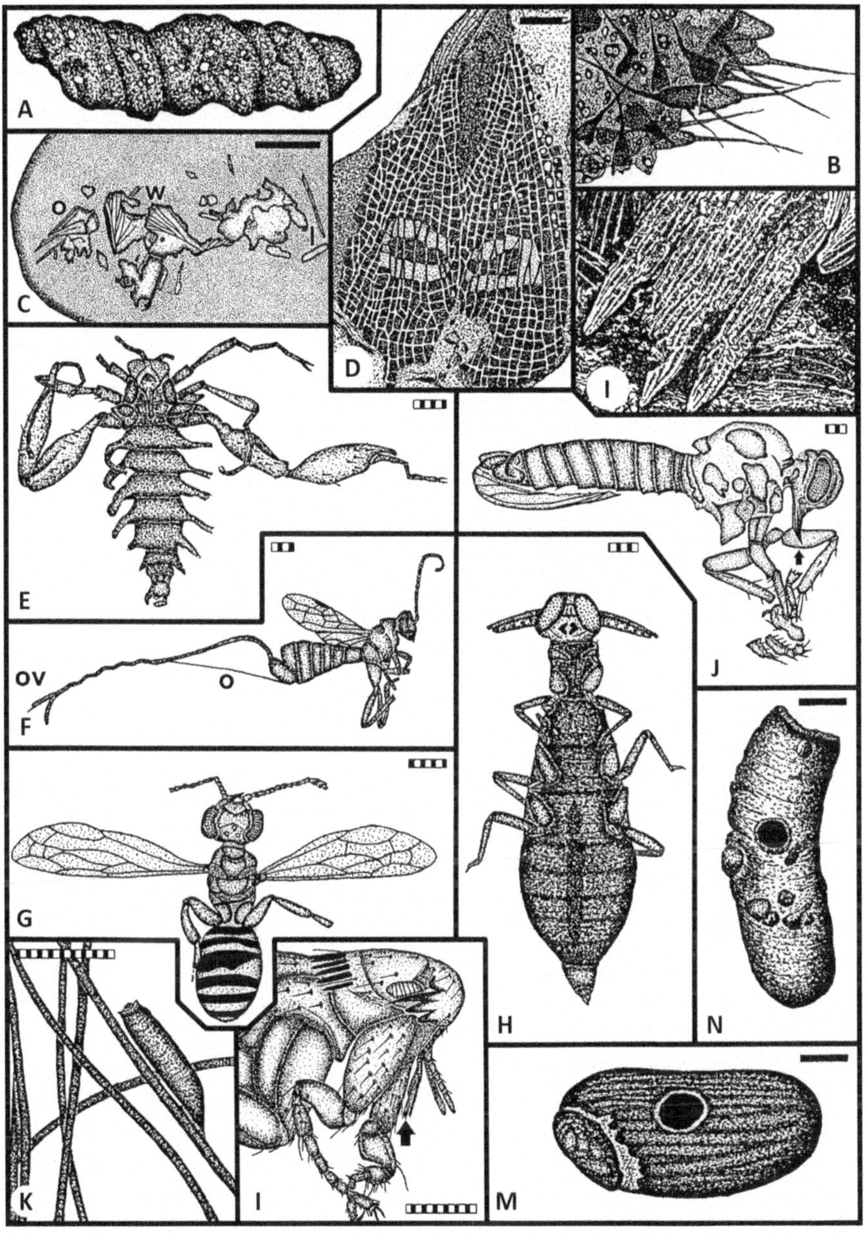

Fig. 11.1 Evidence from taxonomic affiliation (**f**, **h**, **j**–**l**), structural and functional attributes (**b**, **e**, **f**, **h**, **j**, **l**), organismic damage (**m**, **n**), gut contents (**i**), coprolites (**a**, **c**), and predation avoidance (**d**, **g**) for insect predation, parasitism and parasitoidism in the fossil record. Examples are predation (**a**–**d**, **g**, **i**, **j**, **m**), parasitism (**e**, **h**, **k**, **l**) and parasitoidism (**f**, **n**). (**a**) A coprolite containing early land-plant spores from the Early Devonian of Wales (Edwards et al. 1995). (**b**) A fossil spider spinneret, showing a cluster of attached and detached spigots, from the Middle Devonian of Gilboa,

(Greenwalt et al. 2013; Greenwalt 2021), that allows for assignment of these insect fossils, respectively, to parasitoid and parasite life habits. The evidence is highly variable and consists of modern and fossil molecular data, broadly construed (Nagler and Haug 2015), but also includes morphological and interaction-related features of the fossil organisms themselves or their taxonomic affiliations (Fig. 11.1f, h, j–l) (De Baets and Littlewood 2015; Leung 2017). Functional features also are important (Fig. 11.1b, e, f, h, j, l), as are damage to organisms (Fig. 11.1m, n), gut contents (Fig. 11.1h), coprolites (Fig. 11.1a, c), and predator avoidance traits (Fig. 11.1d, g) (Labandeira 2002). Evidence of parasitism is seen in Fig. 11.1e, h, k, l; evidence of parasitoidism is seen in Fig. 11.1f, n.

11.3.1 Biomolecular Data

Within the deep-time context of the insect fossil record, study of biomolecules has two basic applications for understanding the occurrence of predation and especially parasitism and parasitoidism. The first approach involves understanding of the feeding habits and other host relationships of the fossil parasite containing the ingested biomolecule such as blood (Greenwalt 2021; Greenwalt et al. 2013; Yao et al. 2014). In a complementary sense, identification of the ingested biomolecule itself is necessary for specification of the host taxon (De Baets and Littlewood 2015). Identification of both parasite and host is still is in their infancy (De Baets and Littlewood 2015); but characterization of certain biomolecules, such as fossil keratin, shows considerable promise (Wappler et al. 2004; Briggs and Summons 2014).

The other major use of biomolecular data relevant for inferring predation, parasitism and parasitoidism in the fossil record is the production of phylogenies (De Baets and Littlewood 2015; Nagler and Haug 2015; Warnock and Engelstädter 2021). The influence of molecular phylogenies of insects—particularly those of fossil-calibrated cladistic studies of predator, parasite and parasitoid groups—has provided considerable assistance in inferring the predatory, parasitic or parasitoidic life habits of many insect lineages (Pohl and Beutel 2005; Winterton et al. 2007; Bologna et al. 2008; Heraty et al. 2013; Winkler et al. 2015; Peters et al. 2017; Gillung et al. 2018). In addition, such studies are a principal basis for understanding the inter-relationships among predator and especially parasite and parasitoid lineages and their relationships to sister clades (Grimaldi and Engel 2005a).

11.3.2 Taxonomic Affiliation

The insect fossil record consists of evidence for predatory, parasitic and parasitoidic interactions based on taxonomic similarity to their recent relatives (Boucot and Poinar Jr 2010; Leung 2017). For example, bee cleptoparasitism was found in the fossil record of a late Paleocene (60 Ma) site from Menat, France (Martins et al.

2018). The cleptoparasite was identified based on distinctive features of its wing that are affiliated with modern cleptoparasite bees. Similarly, modern Strepsiptera (twisted-wing parasites) are endoparasitoids that have distinctive features, including extreme sexual dimorphism between males and females and high levels of morphological specialization between immatures and adults (Askew 1971; Kathirithamby 2009). Three species of fossil Strepsiptera were found in 21 million-year-old Dominican Amber (Kathirithamby and Grimaldi 1993) that display the same distinctive features as their extant strepsipteran relatives, indicating that the same parasitoid relationships have survived approximately 21 million years.

Amber provides a wealth of morphological, and indirectly behavioral, detail occasionally revealing the consuming predator and its consumed prey item, often *in flagrante delicto* (Labandeira 2014a). Examples include a dance fly with a nonbiting midge clutched by its forelegs and an insect larva consuming a scuttle fly's head (Grimaldi 1996; Janzen 2002). As for parasitism, there are several occurrences of modern-aspect Paleogene fleas attributable to a modern clade (Poinar Jr 2015). Older, Late Jurassic to Early Cretaceous stem-group lineages have been attributed to modern Siphonaptera (Gao et al. 2012, 2014) based on several common, structural features (but see Dittmar et al. 2016; Leung 2021). Mesozoic evidence for parasitoids also occurs in amber, but is qualitatively different by providing more detail of external insect structure (Labandeira 2014a). Poinar (2013) shows several such examples of parasitoids that involve insects.

11.3.3 Structural and Functional Attributes

Structural evidence historically has been the standard mode for inferring the presence of parasitism or parasitoidism in the fossil record. Many examples illustrate body structures, especially mouthparts, attachment devices and ovipositors, of a particular larva or adult that indicate a parasite or parasitoid life habit during the later Mesozoic and Cenozoic (Leung 2017). Body structures strongly implicating parasitoidism include those from Coleoptera (Engel 2005a; Poinar Jr 2009), Diptera (Rocha et al. 2015; Zhang et al. 2016) and Hymenoptera (Barling et al. 2013; Spasojevic et al. 2017). Of special note is synchrotron X-ray microtomography (Labandeira 2014a), that has been used for documenting parasitoid interactions with their hosts caught in the act (van de Kamp et al. 2018). In that study (van de Kamp et al. 2018) 55 parasitoidization events were three-dimensionally imaged that recorded four species of hymenopteran endoparasitoids in a dipteran pupal host.

The life habits of predators, parasites and parasitoids frequently can be deduced from their functional morphology (Leung 2017). The fossil record provides considerable evidence for predation, such as silk-producing spigots in a Devonian insectivorous spider likely used for trapping of prey items (Fig. 11.1b). The fossil record also provides evidence, such as wing eyespots of a cnemidolestid archaeorthopteran (Fig. 11.1d), evidently a potential prey item, to deflect attention and avoid consumption by a potential predator. The presence of a long drilling ovipositor (Fig. 11.1f) is strong evidence for parasitoidism.

Evidence for social parasitism is available from a rove beetle, *Cretotrichopsenius burmiticus*, (Coleoptera: Staphylinidae) from Myanmar amber (99 Ma). These beetles were assigned to the subfamily Aleocharinae and consisted of a horseshoe crab or "limuloid" larval ecomorph with associated structural features that indicated termitophily and nest parasitism (Cai et al. 2017). Many specialized structures of this aleocharine beetle have a mimetic body form of a termite worker caste that allowed accommodation within termite society, probably Mastotermitidae or Kalotermitidae. Notably, evidence for social parasitism was provided by another aleocharine rove beetle from the same deposit, *Mesosymbion compactus* (Yamamoto et al. 2016), which possessed a similar limuloid habitus and other structural features that are associated with termitophily.

11.3.4 Host Tissue Damage

Physical evidence for parasitism can take many forms, including indirect and direct evidence for parasites and parasitoids on a variety of arthropod and vertebrate hosts (Boucot and Poinar Jr 2010; De Baets and Littlewood 2015; Dunlop 2021; Poinar 2021). One previously unrecorded type of paleopathological evidence is punctures in the osteoderms of armadillos from the late Miocene of the Pampas, in Argentina (Tomassini et al. 2016). Osteoderms occur occasionally in mammals and are particularly common in edentates that include Pleistocene glyptodonts and modern armadillos. In these mammals, osteoderms are osseous or keratinous deposits that form plates, scales or other flat structures embedded in the integument. Osteoderms from the extinct armadillo *Ciasicotatus ameghinoi* (Cingulata: Dasypodidae) bear punctures, sometimes clustered, that indicate feeding by the jigger flea *Tunga* (Siphonaptera: Tungidae) when the host was still alive.

One mode of evidence involving tissue damage is the rare emergence of a parasitoid from its host, as revealed in 44 million-year-old middle Eocene Baltic Amber (Leung 2017). From this deposit Poinar Jr and Miller (2002) document a parasitoid wasp larva, identified as an extinct genus of the subfamily Neoneurinae (Hymenoptera: Braconidae), exiting the abdomen of its host, a species of the garden ant *Lasius* (Hymenoptera: Formicidae). Both host and parasitoid were alive during this episode, as the ant displays contorted body features in response to a stimulus while the wasp larva began to secrete silk for its pupal case. Both behavioral features occurred immediately before their demise. This relationship, typical of certain braconid wasps attacking *Lasius* ants today, is an interaction that has been pushed back to the middle Eocene, demonstrating the antiquity of some highly specific parasitoid–ant host interactions. Other types of evidence are more difficult to acquire; for example, host tissues occasionally are altered such as a small puparium of a strepsipteran parasitoid ensconced within the body of a stingless bee (Boucot and Poinar Jr 2010).

11.3.5 Plant–Insect Interactions

Plant–insect interactions also can reveal the presence of parasitoids, recording the identities of the plant host, often the insect herbivore, but rarely the insect parasitoid. The parasitoid often remains unidentifiable because of the lack of distinctive features of the parasitoid entry hole, the induction of anomalous host behavior resulting in atypical plant damage, or exit holes. For example, the galler damage type DT83 (Labandeira et al. 2007b, page 13) has features consistent with a gall midge (Diptera: Cecidomyiidae) gall, occurring on plant morphotype TY46 (Euphorbiaceae) from early Eocene Laguna del Hunco, Argentina. However, this gall individual contains six, undistinctive, circular, parasitoid exit holes of varying diameters that are unidentifiable to a culprit taxon. Similarly, although often considered a plant–insect interaction, an example of predation occurs in a seed with a circular exit hole (Fig. 11.1m), but lacks sufficient characters to identify the seed predator, such as a seed bug (Lattin 1999).

Leaf mines also can record the presence of parasitoid behavior from oviposition scars (Krassilov 2008a). As leaf miners are concealed herbivores, an adult parasitoid must penetrate through or otherwise pierce foliar tissue with their ovipositor to deposit an egg on or in the larval host or mine. Evidence for ovipositor piercings or larval entry holes does occur on fossil leaf mines. These penetrations often are termed “cut-outs” (Krassilov 2008a), or alternatively “predation holes” (Krassilov 2008b), some of which may be linear oviposition scars adjacent the leaf mine from ovipositing parasitoids.

Another type of evidence reveals the presence of a parasitoid, in this case a fungus, in a series of distinctive, successive, bilaterally symmetrical holes adjacent major veins found on the undersides of dicotyledonous leaves in humid, tropical environments (Harper and Krings 2021). Such features are the “death grips” of a zombie ant, such as the carpenter ant *Camponotus*, as it dies following zombification by the parasitoid fungus *Ophiocordyceps* (Hypocreales: Ophiocordycipitaceae), a widespread plant–fungus–ant interaction across the modern tropics (Hughes et al. 2011a). Such distinctive evidence also has been found on a dicot leaf from the 48 million-year-old Messel Formation in central–west Germany (Hughes et al. 2011b). This occurrence indicates that the parasitoid association has been present since the early Eocene, and probably originated in the Cretaceous (Sung et al. 2007), during the initial diversification of the parasitoid fungus clade. As in the case of plant galls caused by mites and insects, the infected zombie ant and its characteristic leaf damage is considered the extended phenotype of the parasitoid fungus (Hughes 2014).

11.3.6 Gut Contents

Gut contents in insects assume a variety of forms, ranging from solid contents such as pollen and indigestible plant-tissue fragments in guts of insects (Krassilov et al. 1997; Rasnitsyn and Krassilov 2000), to shaft fragments of bird feathers and vertebrate blood (Wappler et al. 2004; Greenwalt et al. 2013; Greenwalt 2021) in parasites and parasitoids. The 48 million-year-old Messel Formation of Germany preserves gut contents spectacularly well, an example of which are butterfly scales in the gut of the bat *Palaeochiropteryx tupaiodon* (Fig. 11.1i) described by Richter and Storch (1980). Another instance is the chewing louse *Saurodectes vrsanskyi* (?Phthiraptera: Saurodectidae) that shows fluidized gut contents consisting of likely secreted dermal fluids or blood (Fig. 11.1h). For parasites, the detection of blood in gut contents is especially important (Greenwalt 2021).

Four studies have used the gut contents in a louse, a bug, a flea and a mosquito to indicate the presence of parasitism and reveal the specific diets of these parasites. First, a bird louse (Phthiriaptera: Amblycera: Menoponidae), with mandibulate mouthparts typical for this clade, was discovered among a well-preserved middle Eocene (44 Ma) biota at Eckfeld, Germany (Wappler et al. 2004). The presence of remains of keratinous scales from feather shaft bases in the gut of the louse, detected by light-microscope imaging, securely indicated an ectoparasitic relationship with a bird (Wappler et al. 2004). A second study involved bugs (Hemiptera: Heteroptera) from the Early Cretaceous (125 Ma) Yixian Formation of northeastern China (Yao et al. 2014). This study established *Torirostratus pilosus* (Hemiptera: Torirostratidae) as closely related to Cimicidae (bed bugs) and Polyctenidae (bat bugs). An energy-dispersive X-ray spectroscopic examination of the head, antennae, prothorax, metathoracic leg, central abdomen and adjacent mudstone matrix (the latter used as a standard), revealed elevated iron content, and hence the presence of degraded hemoglobin, from an opaque region in the central abdomen (Yao et al. 2014). In a third study, a Mesozoic-aspect female flea with a substantially distended abdomen was examined from the same deposit as the example immediately above. The flea, *Pseudopulex tanlan* (Siphonaptera: Pseudopulicidae), a member of an extinct, mid-Mesozoic lineage related to modern fleas (Huang 2014), had a ballooned abdomen with stretched intersegmental membranes that contained 15 times the volumetric intake of blood as that of modern fleas (Gao et al. 2014). This attribution of blood as the cause of abdominal distension was based on the taxonomic affinity of fleas rather than a chemical analysis. In the last study, a fossil mosquito (Diptera: Culicidae) was discovered in the 44 million-year-old middle Eocene Kishenehn Formation of southwestern Montana, and subsequently analyzed for a large opaque residue in its moderately distended abdomen. Energy-dispersive X-ray spectroscopy and time-of-flight secondary ion mass spectrometry indicated that hemoglobin was present in the mosquito's abdomen (Greenwalt et al. 2013). While these studies can identify with remarkable accuracy the diet of a parasite's last meal—whether keratin flakes from bird feather shafts or vertebrate blood—the precise species identification of the hosts of the last meal remain mostly unknown.

11.3.7 Coprolites

Arthropod (Fig. 11.1a) and vertebrate (Fig. 11.1c) coprolites often contain identifiable material assignable to particular insect predators and the hosts of parasites and parasitoids, but are rare as fossils (Labandeira 2002; Robin et al. 2016; Chin 2021). Coprolites of hosts are usually difficult to assign to a producer; nevertheless, they potentially are a significant archive of fossilized trophic relationships that include evidence for parasitism (De Baets and Littlewood 2015; Qvarnström et al. 2016). Such coprolites occasionally reveal prey items consumed by predators such as swallowed insect parasites combed from hair and feathers by vertebrate grooming behavior (Qvarnström et al. 2016). More remotely, parasitoid remains can occur in the coprolites of small terrestrial vertebrates. However, for insects during their 412 million-year-long existence, by far the greatest populations of insect coprolites involve plant cuticle, trichomes and vascular tissue remnants from detritivores and herbivores, and pollen from pollinivores (Labandeira and Phillips 1996; Labandeira 1998). Fluid-feeding parasites and parasitoids would rarely produce a detectible coprolite record.

11.3.8 Sedimentary Ichnological Evidence

Sedimentary ichnological evidence provides support for recognizing inquilinism, cleptoparasitism, parasitoidism and scavenging in the fossil record, particularly in the trace fossil record of Pleistocene and Holocene deposits that involve beetle, wasp and bee hosts (Ellis and Ellis-Adam 1993; Bown et al. 1997; Mikulás and Genise 2003; Genise and Cladera 2004). Cleptoparasitism, for example, occurs in the sedimentary ichnological record as fossil traces of *Tombownichnus pepei* that consist of pits in the walls or infillings of *Coprinosphaera*, the brood balls of beetles (Sánchez and Genise 2009). The *Tombownichnus–Coprinosphaera* parasitoid and host relationship has modern equivalents (Halffter and Edmonds 1982; Halffter and Matthews 1999), although the material documented from the middle Eocene–lower Miocene Saramiento Formation of Patagonia, Argentina (Sánchez and Genise 2009) indicates an ancient association. The cleptoparasite culprits remain unknown, although the *Tombownichnus pepei* pit type likely represents pupation chambers excavated by last-instar larvae.

Another example is *T. plenus* from Semnan Province, Iran, that involves curvilinear, narrow burrows up to 8 mm long that invade bee colony cells (Bagheri et al. 2013). These bee nest structures in *Celliforma* nests are attributed to one of the three main groups of bee parasitoids: Meloidae (blister beetles), Bombyliidae (bee flies) or Mutillidae (velvet ants). Similarly, possible parasitoid attack of bee cells are present in ichnofossils of *Celliforma*, attributed to modern Anthophoridae (carpenter bees), from Fuerteventura and Lanzarote of the Canary Islands, Spain. The outer sedimentary linings of individual unopened cells exhibit numerous perforations that suggest considerable pre-imaginal mortality, attributed to predation or fungal attacks (Ellis and Ellis-Adam 1993), but equally could represent parasitoid attack. In another example, direct evidence of a parasitoid entry or an exit hole is provided by a bee cell of Stenotritidae, from the Pleistocene of South Australia (Houston 1987).

Perforations of thickened bee cell walls and small cocoons within the bee cells also have been described from the Paleocene–Eocene Claron Formation of southwestern Utah (Bown et al. 1997). Perforations of these cell walls are rounded, incomplete or complete holes that suggest parasitoid presence, but of unknown taxonomic affiliation. A second instance is the association of *Lazaichnus fistulosus* with beetle pupal chambers, consisting of holes circular to broadly ellipsoidal in outline that penetrate the cell wall and are connected to a single cavity of *Monesichnus ameghinoi* (Mikulás and Genise 2003). This probable parasitoid trace fossil is from the Late Cretaceous to Early Paleogene Ascencio Formation of Uruguay and represents the earliest occurrence of probable parasitoidism in the sedimentary ichnological record (Martin and Varricchio 2011).

11.4 Evolutionary and Ecological Biology

Parasitism, in the broadest sense of the term to include parasitoidism, is a very ancient relationship extending into Paleozoic (Leung 2017) and likely even Precambrian (De Baets and Littlewood 2015) time. The varieties of parasitism exist as a continuum (Robin 2021) that ranges from symbiosis where two interacting organisms mutually benefit, to commensalism where one organism benefits and the other has a neutral interaction, and then to predation, parasitism and eventually parasitoidism where one interacting organism benefits and the other is disadvantaged, including death (Araújo et al. 2003). For parasitism, one conservative estimate of the animal kingdom indicates that this life habit has originated at least 223 times in 15 phyla, with the greatest representation occurring overwhelmingly in Arthropoda (Weinstein and Kuris 2016). Within arthropods, it is not necessarily the most diverse clades that exhibit the greatest incidence of parasitism. Clearly, at the family level, originations of parasitism in insects were overwhelmingly concentrated among the holometabolous insects where development included larval and pupal instars (Poulin and Morand 2000). These originations consisted of 84 separate events for holometabolous insects, compared with five originations for hemimetabolous insects, where larval and pupal instars are absent. Hemimetabolous insect occurrences are Dermaptera (one occurrence), Psocoptera (one occurrence), Phthiraptera (one occurrence, included in Psocodea) and Hemiptera (one occurrence) (Fig. 11.2). At the family level of analysis, the distribution of the incidence of originations of parasitism is unrelated to underlying total species diversity (Fig. 11.2). Coleoptera (beetles) are the most diverse clade, yet exhibit only ten occurrences of parasitism. Diptera displays 60 separate occurrences of parasitism (Feener and Brown 1997; Weinstein and Kuris 2016), whereas Hymenoptera (sawflies, wasps, ants and bees), which have a species diversity of 75% that of the Diptera

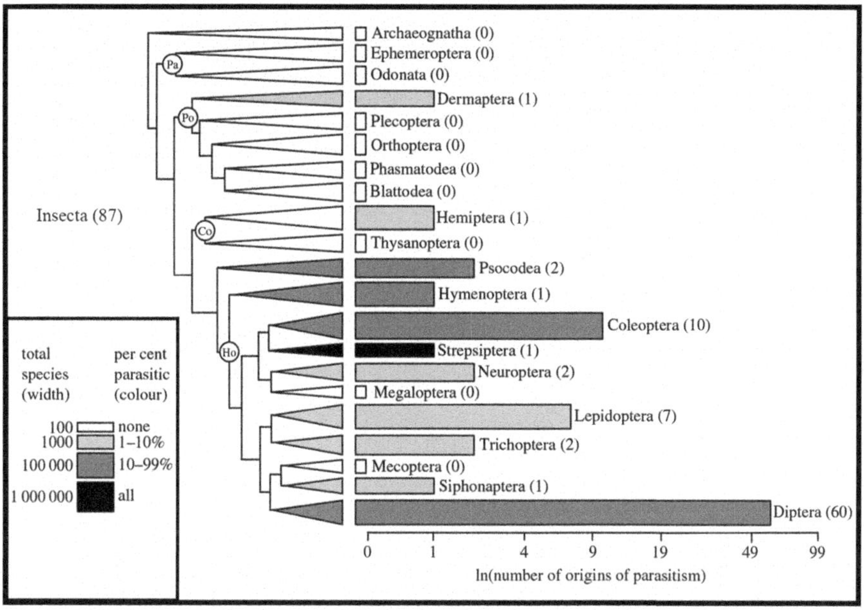

Fig. 11.2 The origins of parasitism across Insecta. The bar plots are the natural log-transformed number of independent acquisitions of parasitism/parasitoidism within the Insecta, arranged based on taxonomic affiliation (Misof et al. 2014). Bar shading denotes the percent of the group that is parasitic or parasitoidic and bar width is proportional to the log transformed number of species per group. Taxonomic group circled abbreviations: *Pa* Paleoptera, *Po* Polyneoptera, *Co* Condylognatha, *Ho* Holometabola. (Reproduced with permission from the bottom panel of Fig. 1, in Weinstein and Kuris (2016)

(flies), display one or two originations of parasitism (Eggleton and Belshaw 1992; Dowton and Austin 1995b). Nevertheless, it appears that parasitic lineages of insects do not differ from their respective free-living confamilial lineages in the potential for speciation.

In a separate study involving maximum likelihood reconstruction of the major diets of hexapod families, it appears that the hypothesis of larval dietary substrates shaping the major pattern of insect clade richness largely is not borne out (Rainford and Mayhew 2015). An accumulation curve based on dietary originations through geologic time indicated that dietary ecologies of fungivory, phytophagy, predation and ectoparasitism appear early in clade history and display a steady, modest rate of origination (Fig. 11.3). The exceptions were the ecologies of detritivory and especially parasitoidism, whose originations reflect an upward and marked trend during the mid to late Mesozoic; the latter trend is attributable to the radiation of parasitoid Hymenoptera during this interval. The origination data also indicate that the ecology of ectoparasitism behaves very different from that of parasitoidism, with the former originating at a very low rate compared to the latter, indicating that ectoparasitism may not be a viable long-term evolutionary strategy. The evidence for solid phylogenetic conservatism regarding a specialized ecology such as parasitoidism is borne out by the 12 types of parasitoid guilds that ecologically characterize

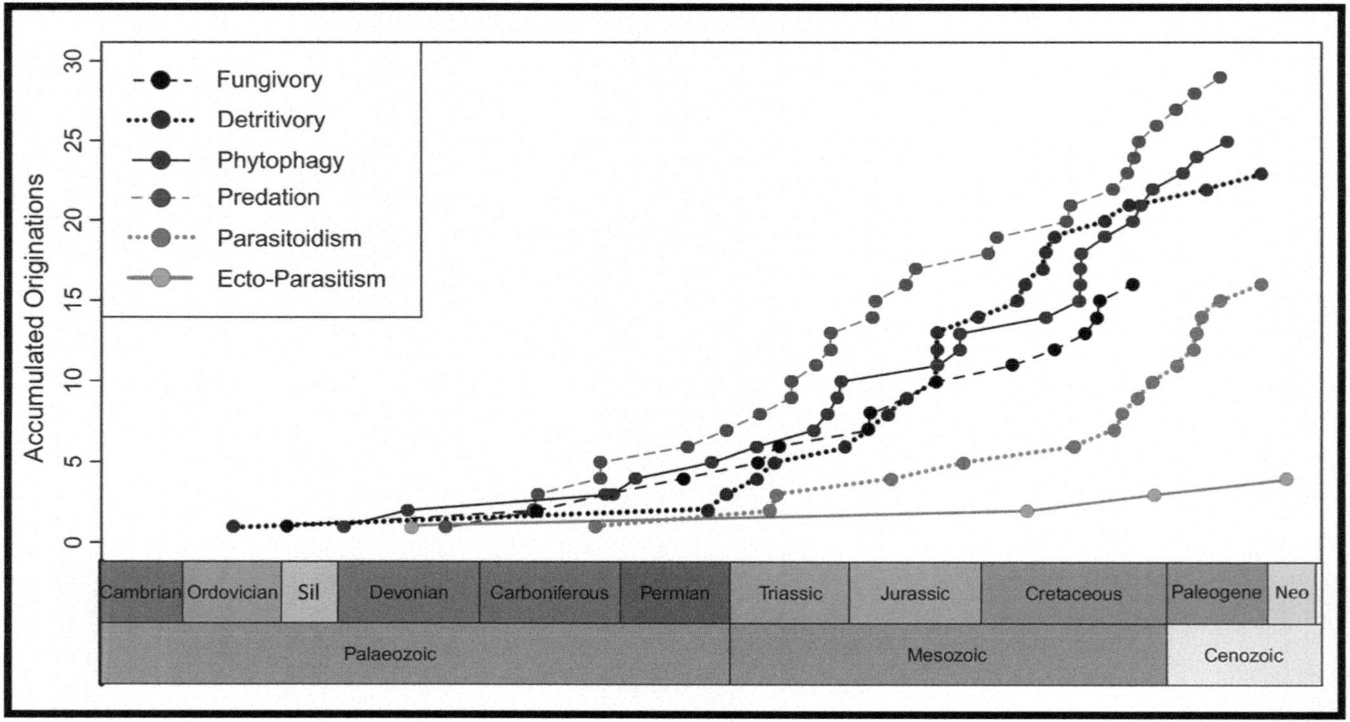

Fig. 11.3 Accumulation of major dietary-guild originations for the immature stages of insect clades. These data are based on maximum likelihood reconstruction; see Rainford and Mayhew (2015) for details. Abbreviations: *Sil* Silurian, *Neog* Neogene. The thin slice of time next to the recent is the (unlabeled) Pleistocene + Holocene. (Modified for greater clarity and reproduced with permission from Fig. 3 of Rainford and Mayhew 2015)

holometabolous insect communities (Mills 1994). Mills (1994) regarded that parasitoid guilds are defined by three basic features. First, what stage of the host is attacked by the parasitoid? Is it the egg; early, middle or late larval instar; prepupae; pupa; or adult? Second, what is the stage of the host in which it is killed? Third, is the mode of parasitoidism internal (endoparasitoidism) or external (ectoparasitoidism)? The 12 documented combinations of these three features that form modern parasitoid guilds delineate ecological diversity across the seven parasitoid-bearing orders of insects, and provide a phenomenological explanation for Rainford and Mayhew's (2015) conclusion of the evolutionarily dynamic nature of the parasitoid life habit.

The three, major parasitoid harboring groups—Hymenoptera, Diptera and Coleoptera—each display trajectories of particular lineages to and from the parasitoid life habit via particular before-and-after life-habit transitions (Poulin 2011), indicated by arrows in Fig. 11.4a–c. For example, in Coleoptera (Fig. 11.4a), although the thicknesses of the arrows reveal a small number of parasitoid species, the principal movement *toward* parasitoidism has been from mycophagy to clepto-parasitoidism by the families Bothrideridae, Meloidae, Rhipiceridae, Ripiphoridae and Strepsiptera (the latter clade considered a separate order in this report). A second source of movement *toward* parasitoidism has been from specialized egg and clepto-provisioning predation to cleptoparasitoidism by the families Carabidae, Cleridae, Passandridae and Staphylinidae. The principal movement *away from* parasitoidism has been from cleptoparasitoidism to special egg predation by the families Cleridae and Meloidae. The patterns for Diptera (Fig. 11.4b) and Hymenoptera (Fig. 11.4c) display a different pattern and have a greater number of trajectories, higher numbers of transiting species, and a greater number of transiting family-level lineages to and from the parasitoid life habit. These evolutionary shifts are explained in more detail below for each of the three discussed insect orders.

11.5 Parasite and Parasitoid Taxa

Insect parasites and parasitoids have evolutionary and ecologically very different life habits. The two life habits have a different set of effects on their hosts and on themselves, based on the particular mode of extracting host resources. These different life habits may be related to the distinctive taxonomic spectra of parasites versus parasitoids. Parasites consist of hemimetabolous lineages—Blattodea, Dermaptera, Psocoptera, Phthiraptera and Hemiptera—that are not represented as parasitoids. Additionally, parasites are represented by a holometabolous lineage, Siphonaptera, which is not represented as parasitoids, and are also represented by another holometabolous lineage, Coleoptera, that is very poorly represented as parasitoids. Only Diptera and Hymenoptera overlap significantly in containing many dominant parasite and parasitoid lineages. In comparison to parasites, the entirely holometabolous character of parasitoids is notable, consisting of Neuroptera, Coleoptera, Strepsiptera, Diptera, Trichoptera, Lepidoptera and Hymenoptera.

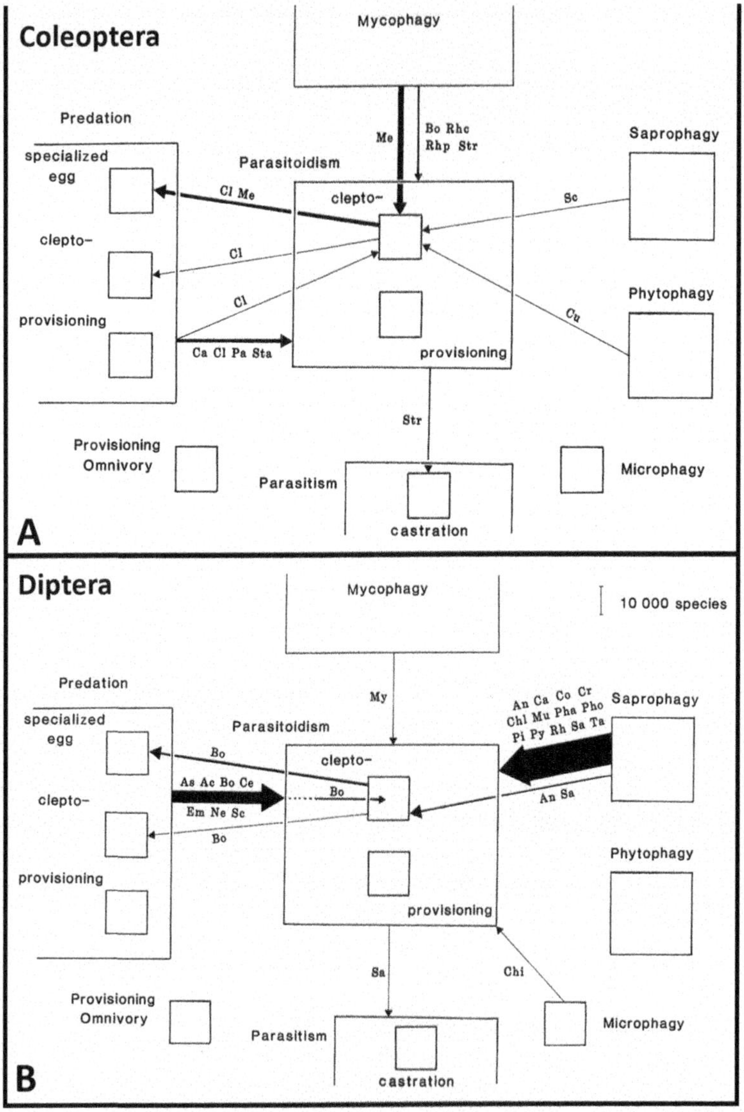

Fig. 11.4 Evolutionary shifts to and from the parasitoid habit in (**a**) Coleoptera, including Strepsiptera, (**b**) Diptera and (**c**) Hymenoptera. The thickness of the arrow is directly proportional to the number of described parasitoid species in families that are derived from the ancestor making the shift. It does not indicate the number of times the shift has occurred; data compiled as of 1991. Abbreviations of Coleoptera Taxa: *Bo* Bothrideridae, *Ca* Carabidae, *Cl* Cleridae, *Cu* Curculionidae, *Me* Meloidae, *Pa* Passandridae, *Rhc* Rhipiceridae, *Rhp* Ripiphoridae, *Sc* Scarabaeidae, *Sta* Staphylinidae, *Str* Strepsiptera. Abbreviations of Diptera taxa: *Ac* Acroceridae, *An* Anthomyiidae, *As* Asilidae, *Bo* Bombyliidae, *Ca* Calliphoridae, *Ce* Cecidomyiidae, *Chi* Chironomidae, *Chl* Chloropidae, *Co* Conopidae, *Cr* Cryptochaetidae, *Em* Empididae, *Mu* Muscidae, *My* Mycetophilidae, *Ne* Nemestrinidae, *Pho* Phoridae, *Pha* Phaeomyiidae, *Pi* Pipunculidae, *Py* Pyrgotidae, *Rh* Rhinophoridae, *Sa* Sarcophagidae, *Sc* Sciomyzidae, *Ta* Tachinidae. Abbreviations Fig. 11.4 (continued) of Hymenoptera taxa: *Ag* Agaonidae, *Ap* Apidae, *Be* Bethylidae, *Ch* Chrysididae, *Cy* Cynipidae, *Eul* Eulophidae, *Eum* Eumenidae, *Eup* Eupelmidae, *Eur* Eurytomidae, *Ev* Evaniidae, *Fo* Formicidae, *Ga* Gasteruptiidae, *Ic* Ichneumonidae, *Ma* Masaridae, *Po* Pompilidae, *Pt* Pteromalidae, *Sa* Sapygidae, *Sp* Sphecidae, *Ta* Tanaostigmatidae, *To* Torymidae, *Ve* Vespidae. (Reproduced with permission from Figs. 2–4 in Eggleton and Belshaw 1992)

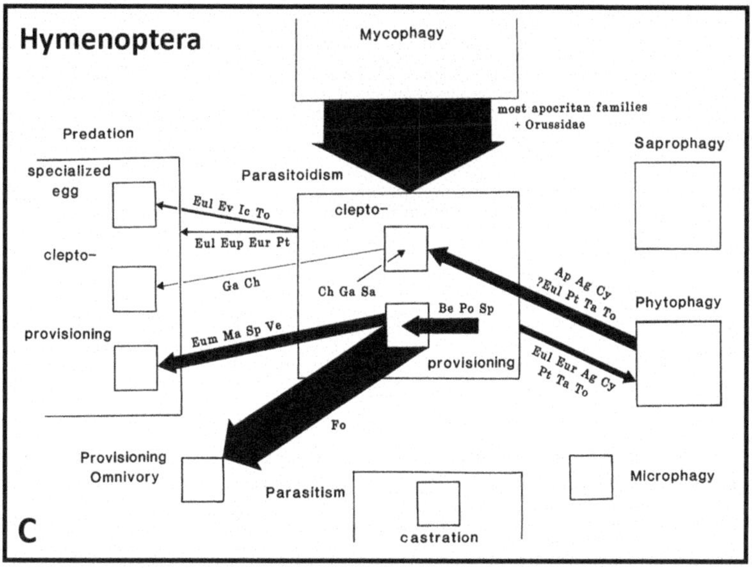

Fig. 11.4 (continued)

11.5.1 Parasite Taxa

The parasite life style and feeding mode is confined to ten insect orders. The Blattodea, Dermaptera, Psocoptera, Phthiraptera and Hemiptera undergo incomplete metamorphosis, or hemimetabolous development, in which an egg develops into multiple immature developmental stages (instars) that are termed nymphs if terrestrial and naiads if aquatic, ending with a reproductively viable adult insect. The other parasite orders, Coleoptera, Siphonaptera, Diptera, Lepidoptera and Hymenoptera undergo complete metamorphosis, or holometabolous development, in which the egg develops into multiple developmental instars of the distinctive larva stage, followed by another distinctive stage, the pupa, in turn followed by emergence of the adult from the pupa. Whereas parasites occur in insect orders that are hemimetabolous and holometabolous in development, parasitoids occur only in orders with holometabolous development.

11.5.1.1 Blattodea (Cockroaches)

Blattodea (Cockroaches) are not considered as having parasitic members; typically, they are considered on the receiving end of any parasitic interaction (Askew 1971). Currently, no cockroach is known to be a parasite or parasitoid of another species (Bell et al. 2007), the probable occurrence of cockroach parasites in the fossil record is noteworthy (Vršanský et al. 2019). From 99 million-year-old Late Cretaceous (Cenomanian) Myanmar Amber, two cockroaches, *Spinka fussa* and *Bimodala ohmkuhnlei* of the Blattidae (American cockroaches) have been described as associated with complex ant nests. These two species occurred in the same, large, amber pieces as the ant nests, and likely were myrmecophiles (Vršanský et al. 2019). (A myrmecophile is a commensal or parasite of the nest of an ant species.) The authors interpret the two cockroach myrmecophile species as engaged in commensalistic and parasitic relationships with their ant hosts, consistent with morphological features in extant ant myrmecophiles. The significance of these extinct, social parasitic cockroaches was their impact on the evolution of complex ant nests in late Mesozoic tropical forests soon after the earliest evidence for ant eusociality in the fossil record.

11.5.1.2 Dermaptera (Earwigs)

Modern Dermaptera (earwigs) consist of 3 suborders, 11 families and 203 genera, and are nocturnal, hemimetabolous insects with distinctively short forewings and terminal abdominal cerci modified into forceps-like pincers (Haas 2018). Two dermapteran lineages, Arixeniidae (bat earwigs) and Hemimeridae (rodent earwigs), often are placed in their own suborders, and live ectoparasitically on mammals (Popham 1984), although recent evidence suggests that the relationships may be more commensalistic than parasitic (Haas 2018). Hemimeridae occur in sub-Saharan Africa on hamster rats and possess a suite of structures, such as dorsoventrally flattened bodies; short, grooved legs to cling to fur; and mouthparts for abrading and feeding on host skin and surface fungi (Rehn and Rehn 1935, 1937; Ashford 1970; Nakata and Maa 1974). Arixeniidae occur with molossid bats in caves in Southeast Asia, and possess a matted pubescent body, long antennae and long legs for rapid movement (Nakata and Maa 1974). A phylogenetic study of Dermaptera (Kocarek et al. 2013) indicates that the Hemimeridae and Arixeniidae each has a sister-group relationship with another dermapteran family, indicating that the parasite lineages are convergently evolved, highly modified earwigs (Fig. 11.5). These relationships indicate the potential for rapid change of parasite external morphology when a new trophic niche becomes available (Kocarek et al. 2013), likely during the Eocene when their rodent and bat hosts initially diversified.

11.5.1.3 Psocoptera (Booklice, Psocids)

The term, Psocoptera (booklice, psocids) currently is not widely used, which is partly attributable to the paraphyletic status that this ancient group of hemimetabolous insects has with respect to the nested clade of Phthiraptera, the parasitic lice (Nagler and Haug 2015; Mockford 2018). Nevertheless, Psocoptera as used in this report is a convenient reference to all intervening, free-living lineages of bark lice that form the broader clade. The Psocoptera includes the variously ranked Trogiomorpha, Amphientometae, Sphaeropsocidae, Psocomorpha (the overwhelming bulk of taxonomic diversity), and the book lice, Liposcelididae (Johnson et al. 2018). Of these, particular attention should be devoted to Liposcelididae. The

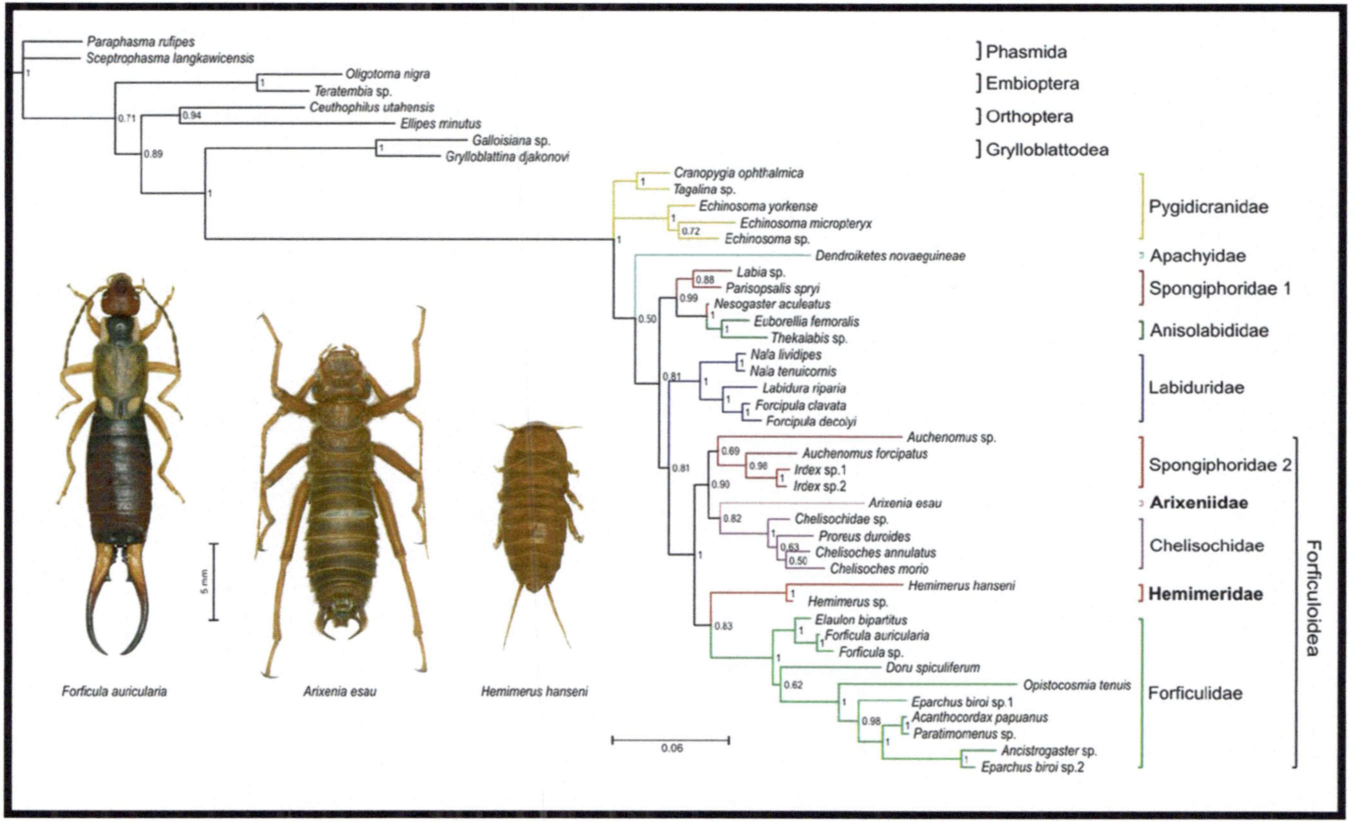

Fig. 11.5 Bayesian phylogram of Dermaptera (earwig) families based on nuclear sequence data of 18S and 28S ribosomal DNA and histone-3. Numbers above branches indicate posterior probabilities. The Arixenidae and Hemimeridae (bold font) are parasitic clades. (Reproduced with permission from Fig. 2 of Kocarek et al. 2013)

Liposcelididae is the sister-group to the parasitic lice clade of Phthiraptera, consisting of the Amblycera + [Anoplura + (Rhyncophthirina + Ischnocera)] clade, equivalent to the common names of chewing lice 1 + [sucking lice + (chewing lice 2 + elephant/warthog lice)], treated separately below. In particular, the Liposcelididae provides significant clues to the origin of parasitism in the Phthiraptera and provides evidence that demonstrates tendencies toward parasitism in certain psocopteran lineages.

Liposcelididae consist of 9 genera and approximately 200 species of small, pale, dorsoventrally compressed, and often wingless barklice that are inquilines of insect nests (Mockford 1971). In particular, Liposcelididae are morphologically and behaviorally intermediate between free-living Pachytroctidae (thick book lice), from which they originated, and the ectoparasitic Phthiraptera (parasitic lice) clade with which they are sister groups (Johnson et al. 2004, 2018). Notably, the diet of probably the best-studied extant booklouse, synanthropic *Liposcelis bostrychophila*, consists of raw cereal grains treated with yeast (Green and Turner 2005). Other species of the genus evidently consume sloughed off integument such as flakes of skin and covering sheaths of feathers (Lin et al. 2004; Grimaldi and Engel 2005b). Fossils of Liposcelididae are rare, but the lineage has its earliest occurrence in Myanmar Amber (99 Ma), providing a minimal date for the divergence of Liposcelididae and Phthiraptera (Grimaldi and Engel 2005b). Nonetheless, a dated molecular phylogenetic analysis indicates an older separation date at the Triassic–Jurassic boundary (Johnson et al. 2018).

11.5.1.4 Phthiraptera (Parasitic Lice)

Phthiraptera (parasitic lice) consist of 4 suborders, 24 families, 304 genera and 5316 species that are obligate ectoparasites of birds and mammals (Galloway 2018). Phthirapterans are wingless, dorsoventrally flattened, have small eyes or are blind, and possess a variety of sense organs located especially on their antennae and mouthparts for detection of and their positioning on their hosts (Marshall 1981; Clayton et al. 2016). As ectoparasites, they complete all life stages of their development on the bodies of their hosts, requiring their host's moist, warm skin and adjacent humid microenvironment to live and reproduce by attaching their elongate eggs to hair shafts with a cementing substance (Fig. 11.1k). Lice are sensitive to minute temperature gradients across their host's body and die if deprived of their thermally constrained integumental microenvironment for more than several hours to a few days (Tompkins and Clayton 1999).

The major clades of the monophyletic Phthiraptera are the Amblycera and the Ish nocera + (Rhyncophthirina + Anoplura) clades (Johnson et al. 2018). In common parlance these clades are equivalent to, respectively, the chewing lice 1 clade and the chewing lice 2 + (elephant/warthog lice + sucking lice) clade. Phthiraptera were derived from the psocopteran lineage Liposcelididae, with which it shares a sister-group relationship (Grimaldi and Engel 2005b; Johnson et al. 2018). Within Phthiraptera, the basal-most lineage is Amblycera, the chewing lice 1 clade. These lice principally infest birds, but one-eighth of the species represent three nonspeciose, minor families that are on mammals (Galloway 2018). The next derived (and most diverse) clade is Ischnocera, the chewing lice 2 clade, representing 54% of phthirapteran species. Ischnocera primarily parasitize birds, although, as in the Amblycera, about one-eighth is on mammalian hosts. The Amblycera and Ischnocera are characterized by mandibulate mouthparts, broad and flattened heads, and typically consume flaked off skin, integumental exudates such as sebum and sweat, other debris such as fungi associated with feathers and occasionally blood and lymph at the skin surface (Lehane 1991). The major structural differences between these two major types of chewing lice involve their heads and mouthparts. The third clade, Rhyncophthirina, has highly modified chewing mouthparts that have been reduced to tiny pincer-like mandibles at the end of a long tubular extension of the head capsule for imbibing of blood and dermal secretions, akin to piercing and sucking. Rhyncophthirina is a highly specialized, blood-feeding clade that occurs only on mammals and consists of one genus and three species that parasitize the African elephant, wart hog and Red River hog as hosts (Galloway 2018).

Sister-group to the Rhyncophthirina is the Anoplura, the sucking lice, which are the only truly piercing-and-sucking clade with stylate mouthparts within Phthiraptera. Anoplura secure their food by targeting blood vessels and attach their eggs, known as nits, to the hair of their hosts by large, grasping and curved tarsal claws (Fig. 11.1k). Anoplura parasitize most orders of placental mammals by feeding on their blood. About 70% are associated with rodents (Light et al. 2010), but are not found on armadillos, pangolins, elephants, aquatic species except for seals, and are notably absent on bats (Durden and Musser 1994). Based on a phylogenetic analysis of Anoplura (Light et al. 2010), the clade originated during the Late Cretaceous (Campanian Stage) around 77 million years ago, but did not diversify until the early Paleogene, after the ecologic crisis marking the Cretaceous–Paleogene boundary, and presumably coincident with the diversification of many warm-blooded mammal lineages. A notable exception to anopluran and other phthirapteran hosts are bats, which are inferred to be present and diverse during the Paleogene diversification of placental mammals (Teeling et al. 2005). Phthirapteran absence from bat hosts may be attributable to competition from their previously acquired, rich fauna of parasites that included mites, bat bugs, bat flies and fleas, or were limited by the significant drops in bat body temperatures during hibernation (Grimaldi and Engel 2015a). Both limitations would eliminate phthirapterans as hosts (Clayton et al. 2016).

Phthiraptera have a poor, but rather remarkable fossil record. The oldest yet most bizarre specimen is a large, 17 mm long *Saurodectes vrsanskyi* (Fig. 11.1h) of the monotypic Saurodectidae from the Early Cretaceous (130 Ma) Baissa locality in Transbaikalia, Russia (Rasnitsyn and Zherikhin 1999). This specimen bears several anomalous features: a large size of 17 mm in length; well-developed, compound eyes; a pair of peculiar, horn-like appendages extending laterally from each side of the head; long and ambulatory legs; widely separated leg bases; a pair of robust spines behind the forecoxae; and exceptionally small claws. These characters are atypical compared to modern species of Phthiraptera. The initial, reasonable, attribution of the specimen's life style was an ectoparasite of a warm-blooded vertebrate possessing hair, such as a pterosaur, or possibly a mammal of large size (Rasnitsyn and Zherikhin 1999). Nevertheless, the specimen does reveal structures similar to chewing lice of the Ishnocera such as a highly flattened body, a single clawed tarsus and robust truck tracheae and spiracles, but its' very large size would be inconsistent with parasitization of any host except a large vertebrate. However, contrary interpretations regarding the affinities of the fossil suggest that this specimen is not a phthirapteran (Wappler et al. 2004; Dalgleish et al. 2006). Major features indicating that the specimen is not a phthirapteran is the absence or lack of preservation of mandibles and the presence of large lateral head processes (Dalgleish et al. 2006).

From the Eckfeld maar crater beds in Germany, of middle Eocene (44 Ma) age, is *Megamenopon rasnitsyni* (Amblycera: Menoponidae), a member of primitive chewing lice (Wappler et al. 2004). This specimen exhibits similarities to amblyceran feather lice that parasitize Anseriformes (ducks, geese and swans) and Charadiformes (shorebirds). The presence of feather chaff in gut contests of the specimen confirms its life habit as an ectoparasite (Wappler et al. 2004). From nearby Baltic Amber, a deposit of approximately the same age as Eckfeld, Voigt (1952) described phthirapteran eggs cemented to mammalian hair, likely belonging to an unknown sucking louse of Anoplura.

11.5.1.5 Hemiptera (Bugs)

Within the Hemiptera, it is only the suborder Heteroptera (true bugs) and its constituent infraorder, Cimicomorpha, which contains parasitic lineages engaged in hematophagy. Four cimicomorph lineages, Cimicidae (bed bugs), Polyctenidae (bat bugs), Reduviidae (assassin bugs), and extinct Torirostratidae, contain the hematophagous genera within Hemiptera. Given established phylogenies within the Cimicomorpha (Schuh et al. 2009; Weirauch and Munro 2009; also see Yao et al. 2014), hematophagous parasitism originated in the Hemiptera three times: once in Reduviidae, once in the Polyctenidae + Cimicidae clade, and once in the extinct Torirostratidae.

Cimicidae consists of 6 subfamilies and 21 genera, including the common bed bug *Cimex lectularis*, which are obligate parasites of warm-blooded vertebrates, including humans. As a group, cimicids are frequently narrow host specialists, and rarely are vectors for disease-causing viruses and other pathogens (Reinhardt and Siva-Jothy 2007). Cimicids are attracted to hosts by several cues, including temperature, carbon dioxide gradients and animal kairomones, and feed once every 3–7 days (Reinhardt and Siva-Jothy 2007).

Polyctenidae are a family of rare subtropical bugs, the probable sister-group of Cimicidae, that consist of two subfamilies and five genera that are ectoparasites of bats and exhibit substantial host specialization. The family derives its name from the presence of prominent combs that cover several regions of the body, including the antennal bases, gena of the head, and pronotum, prosternum, mesonotum, ventral surfaces of the abdominal segments and tarsi (Askew 1971). Members of

Polyctenidae have hairy bodies, lack ocelli and compound eyes, display wings that are featureless flaps, and have abbreviated fore legs (Maa 1964).

Among predatory Reduviidae (assassin bugs), one subfamily, Triatominae (kissing bugs, conenose bugs), consists of 6 tribes and 18 genera that are responsible for feeding on vertebrate blood and vectoring trypanosomes that cause Chagas disease in humans, a debilitating disorder of the Americas. *Rhodnius prolixus* (common kissing bug) and *Triatoma infestans* (winchuka) are major vectors of Chagas disease and frequently are associated with triggering anaphylaxis (Klotz et al. 2010). The Triatominae is a monophyletic subfamily within the Reduviidae (Weirauch and Munro 2009), although the Reduviidae is phylogenetically very distant from the Cimicidae + Polyctenidae clade within Heteroptera (Schuh et al. 2009).

Presently the only extinct, family-level lineage of hematophagous Hemiptera is Torirostratidae, from 125 million-year-old Early Cretaceous Yixian Formation of northeastern China (Yao et al. 2014). Two well-preserved genera are placed in the Torirostratidae but remain unplaced within the Cimicomorpha. The fossil record of other hematophagous Heteroptera is represented by two occurrences. One is a fossil of Cimicidae from mid-Cretaceous Myanmar Amber, *Quasicimex eilapinastes*. *Quasicimex* exhibits many characters of Cimicidae sensu lato by possessing several apomorphies while concurrently retaining some primitive characters not found in the crown-group (Engel 2008a). The Polyctenidae, however, lack a fossil record. The other occurrence is the infamous reduviid *Triatoma* (kissing bugs), a major vector for blood-borne diseases, which has a fossil record, as *T. dominicana*, that extends to Dominican Amber approximately 21 million-years ago (Poinar Jr 2005). A second specimen reveals that a triatomine–trypanosomid vector association existed through the presence of *Trypanosoma antiquus* individuals in a fecal droplet of *T. dominicana* adjacent to mammalian hairs (Poinar Jr 2005, 2021). This association suggests a vertebrate host was a precursor to Chagas Disease that currently affects millions of humans in Central and South America (Lent and Wygodzinsky 1979).

11.5.1.6 Coleoptera (Beetles)

Within Coleoptera, the representation of parasites is scattered across the taxonomically vast order. Often, descriptions of the life habits of taxa are unclear as to whether a particular species is a parasite or a parasitoid, particularly since the term "parasite" occasionally meant "parasitoid" in the older literature. However, parasite habits have been established for Leiodidae and Meloidae and probably Rhipiceridae, Passandridae and Bothrideridae, although the latter three lineages are overwhelmingly represented by parasitoids and are reviewed more appropriately in Sect. 11.5.2.2. It is highly likely other beetle lineages will be determined to have obligate parasitic life habits in the near future, as the life histories of individual species become better known.

Leiodidae (round fungus beetles) consist of 6 subfamilies and about 3800 species that are small to very small and typically are saprophagous or feed on a variety of fungi in punky wood, although a broad spectrum of feeding habits are present, including parasitism (Newton 2005). One subfamily, Platypsyllinae, consists of 4 genera and about 20 species that are ectoparasites of semiaquatic and aquatic mammals such as insectivores (Talpidae), beavers (Castoridae), mountain beavers (Aplodontidae) and river otters (Mustelidae), consuming flakes of decorticated skin, dermal exudates and probably surficial blood (Newton 2005; Peck 2006). The highly specialized parasite *Platypsyllus* (beaver beetle) is the best documented platypsylline, and the adult, flea-like in overall appearance, is a flat, wingless and blind form that has thoracic hooks for attachment to the hair of its host. The larvae and adults are parasites of the North American beaver, *Castor canadensis*, and the North American river otter, *Lontra canadensis*, and leave their host only to pupate (Peck 2006).

Meloidae (blister beetles) are a diverse, cosmopolitan family of beetles occurring primarily in warm temperate to arid regions that consist of four subfamilies and about 120 genera (Bologna et al. 2010). The diets of adult Meloidae are variable, with phytophagy dominant and predation subdominant. Most larvae are predaceous, feeding on grasshopper egg pods and immatures of aculeate wasps and bees, with development occurring in the soil or wood (Bologna et al. 2010). However, parasite and parasitoid forms occur throughout the family. The life cycle of Meloidae is complex, with the larva undergoing hypermetamorphosis and consisting of four distinct larval instars prior to entering the pupal stage. The first larval instar is an actively mobile, well sclerotized triungulin, which is the dispersal stage for seeking parasite and parasitoid hosts. The triungulin instar is followed by a first instar grub, a feeding stage of robust proportions; then a coarctate instar, a resting stage in diapause; and finally a second grub instar that is active but nonfeeding prior to pupation (Selander and Mathieu 1964). Parasitic (and parasitoidic) larvae use grasshoppers, ground beetles, aculeate wasps and bees as hosts (Engel 2005a; Bologna et al. 2010). Although meloid consumers of other animals are over-represented by parasitoids, the family does harbor several lineages of ectoparasites (Engel 2005a; Bologna and Di Giulio 2011).

The fossil record of beetle parasites, as would be expected, is poor, particularly for non-diverse families. There is no fossil record of the Platypsyllinae, although the Leiodidae has a decent fossil record of 16 occurrences that extends to the Late Jurassic (Perkovsky 1999). Rhipiceridae apparently has a fossil record extending to the middle Eocene (Ponomarenko 1995). Bothrideridae has a known first occurrence from the middle Eocene (Ponomarenko 1995). The Passandridae has a sparse fossil record of three Cenozoic occurrences; the earliest occurrence is from an approximately 44 million-year-old middle Eocene (Lutetian) deposit in Argentina (Ramírez et al. 2016). The Meloidae is a largely Cenozoic lineage that has a sparse fossil record when compared to its extant diversity. The meloid fossil record, however, includes records of triungulin larvae (Engel 2005a; Bologna et al. 2008) that provide evidence for phoresy and other features consistent with an ectoparasitic mode of life (Engel 2005a). The meloid fossil record includes 12 fossil occurrences; the oldest is from the 35 million-year-old late Eocene. The Ripiphoridae have a fossil record extending to the early Cenomanian (Cai et al. 2018; Batelka et al. 2016, 2019). It is unclear if any of these occurrences demonstrate a specific parasitic relationship with an actual, extinct host.

11.5.1.7 Siphonaptera (Fleas)

Flea-like insects, fleas and basal lineages of flies are closely related to one or more lineages of mid-Mesozoic Mecoptera (scorpionflies) (Huang et al. 2012; Lin et al. 2019; but see Byers 1996). The long-proboscid pollinating scorpionfly lineage Aneuretopsychidae (Ren et al. 2011) is a prime suspect as an ancestor of flea-like insects, fleas and the earliest Diptera (Lin et al. 2019). For mid-Mesozoic flea-like insects, mouthpart structure, body form, other specialized features, and known relationships between specific external adaptations for living on a host in modern fleas (Huang et al. 2012; Gao et al. 2012, 2014) indicate that these taxa were ectoparasites feeding primarily on blood. These ectoparasites likely had intimate associations with endothermic vertebrates during the mid Mesozoic (Leung 2017; Labandeira 2019), ranging from the Middle Jurassic to Early Cretaceous (Huang et al. 2012; Gao et al. 2014). However, lineages of mid Mesozoic, early flea-like lineages (Huang et al. 2012, Huang et al. 2013a), known only from East Asia and Australia, are morphologically different from all modern flea taxa that occur worldwide (Leung 2017, 2021; Rasnitsyn and Strelnikova 2017). Unlike modern fleas (Fig. 11.11), Middle Jurassic to Early Cretaceous flea lineages (Fig. 11.6) possessed cylindrical, dorsoventrally or laterally compressed bodies of large size (Strelnikova and Rasnitsyn 2016; Rasnitsyn and Strelnikova 2017). They also had mouthpart modifications such as overall robustness, stylets with variously positioned serrations, and stylet envelopment by two sutured half-tubes of the labium (Labandeira 2019). These mid Mesozoic forms typically lacked jumping hind legs and stereotypical ctenidial combs (Huang et al. 2013a).

A variety of vertebrate groups have been suggested as hosts for mid-Mesozoic "fleas", particularly those with vestitures of feathers such as nonavian dinosaurs and possibly birds, and pelages of hair such as pterosaurs and large mammals (Ji et al. 2006; Wellnhofer 2008; Godefroit et al. 2014). Nevertheless, these oldest fleas, even

Fig. 11.6 Habitus of the holotype specimen PIN 3064/189 of *Saurophthirus longipes* (?Siphonaptera: Saurophthiridae) was initially described by Ponomarenko in 1976. This specimen exhibits a distended abdomen interpreted to reflect a blood meal. Body length is 12 mm. (Photograph by D. Grimaldi; reproduced with permission from Fig. 1 in Rasnitsyn and Strelnikova 2017)

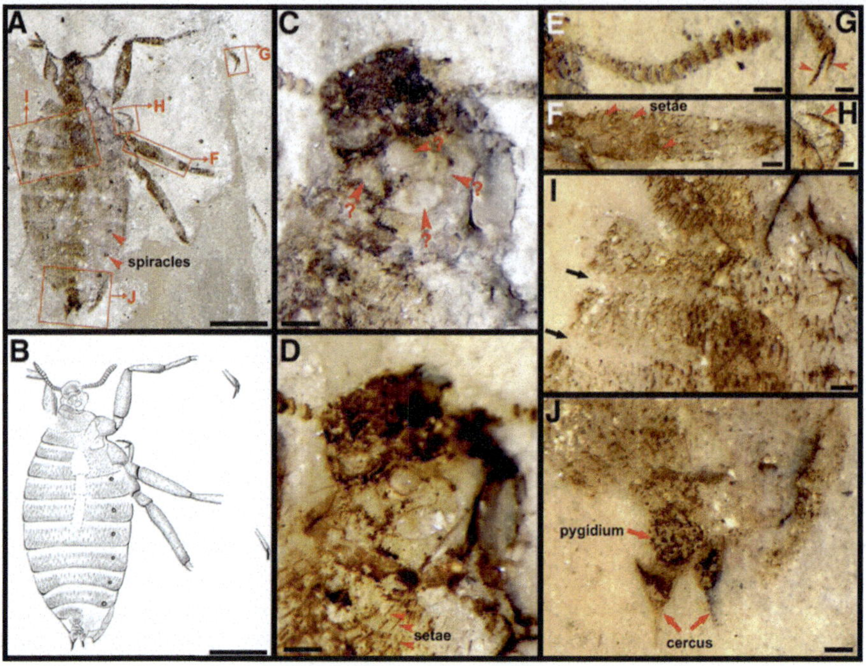

Fig. 11.7 The holotype female of *Pseudopulex tanlan* (Siphonaptera: Pseudopulicidae) (Gao et al. 2014), from the Early Cretaceous Yixian Formation of northeastern China (CNU-SIP-LL2013002). (**a**, **b**) Photograph and line drawing. Photograph of the head region with (**c**) and without (**d**) alcohol. Red arrows point to four, ovoidal, perhaps membranous structures of unknown function. (**e**) Antenna. (**f**) Mid-leg femur. (**g**, **h**) Tarsal claw. (**i**) Segmental boundary (black arrows) of the abdomen. (**j**) Abdominal terminalia. Scale bars: (**a**, **b**) 2 mm; (**c**–**j**) 0.2 mm. (Reproduced with permission from Fig. 1 of Gao et al. 2014)

with their larger size and massive mouthparts may not have had dinosaurs such as theropods for hosts (Dittmar et al. 2016). Many features described for mid-Mesozoic giant "fleas" (Gao et al. 2012; Huang et al. 2013a) are not found on modern fleas that are known blood-feeding parasites of mammals and birds. The relationship that large parasite size is reflective of large host size has never been tested empirically (Dittmar et al. 2016).

There are three principal lineages of mid-Mesozoic fleas: Pseudopulicidae, Tarwiniidae and Saurophthiridae (Ponomarenko 1976; Huang et al. 2013a). The Pseudopulicidae are latest Middle Jurassic in age and from East Asia. Pseudopulicidae are large bodied, robust, flea-like insects that have a cylindrically shaped body, lack pronotal and genal ctenidial combs as well as saltatorial hind legs, and have an abdomen capable of significant distension (Fig. 11.7a, b, f–j) (Gao et al. 2012, 2014). The head has well developed eyes, moniliform antennae, robust mouthparts with prominent stylets of inwardly directed teeth, and four ovoidal, perhaps membranous structures of unknown function on the dorsal part of the head (Fig. 11.7c–e) (Gao et al. 2012, 2014). A ballooned abdomen evidently had a high volumetric capacity for storing blood (Fig. 11.7a, b). A second, related, group of fleas with some modern, siphonapteran morphology are an Early Cretaceous (Aptian) lineage, Tarwiniidae, from Australia, consisting of laterally compressed and small bodies, slender hind limbs for jumping, pseudoctenidial combs, diminutive eyes, and robust, grasping tarsal claws (Huang 2015; Leung 2017). A third lineage is Saurophthiridae, related to Pseudopulicidae and Tarwiniidae (Huang 2014), or alternatively with Tarwiniidae as sister clade to Strashilidae (Rasnitsyn and Quicke 2002; Nagler and Haug 2015). *Saurophthirus* resembles modern fleas by having a smaller size, more abbreviated mouthparts, and very long legs (Fig. 11.6). However, once considered as pterosaur ectoparasites (Ponomarenko 1976), Saurophthiridae lack ctenidia necessary for their ensconcement during flight on a pterosaur's wing membrane, as proposed by Ponomarenko (1976). By comparison, ichnopsyllid fleas on modern bats possess ctenidia that affix them securely to bat wings in flight (Huang 2014). For Saurophthiridae, life habits were more likely as nest inquilines or cleptoparasites feeding on a variety of dermal fluid substances, rather than as ectoparasites on the body directly feeding on pterosaur blood (Huang 2014). Recent evidence based on reconstruction of massive tracheae required for respiration indicates a larval aquatic existence, while adults might have maintained a parasitic presence on pterosaurs (compare Leung 2021; Shcherbakov 2017; Rasnitsyn and Strelnikova 2017).

These three lineages likely had a similar timing of origin during the mid-Mesozoic. One study indicates that modern fleas originated considerably later, during the Paleogene, and possibly earlier in the Late Cretaceous (Zhu et al. 2015). Such origins and phylogenetic relationships would indicate that mid-Mesozoic lineages of fleas, which evidently became extinct during the mid Cretaceous, constituted a separate clade or group of related clades of flea-like organisms, possibly affiliated with Siphonaptera or originating from a separate scorpionfly ancestor (Huang et al. 2012). Accordingly, mid-Mesozoic "giant" fleas and modern, diminutive fleas were the two, distinctive subclades of a more broadly inclusive Siphonaptera clade (Huang et al. 2012; also see Gao et al. 2014). Both subclades might have evolved independently in parallel with the similarly independent evolution of homoeothermic body temperatures on their respective host clades in a manner similar to blood-feeding Diptera and their major vertebrate host lineages (Lukashevich and Mostovski 2003). The three lineages of Pseudopulicidae, Tarwiniidae and Saurophthiridae evidently are closely related, share several features in common typical of ectoparasites, and likely fed on the blood, lymph, sebum and other dermal secretions of haired early mammals, birds or feathered dinosaurs (Huang 2014).

Modern aspect fleas, typical of all modern Siphonaptera lineages, are found in major amber deposits of the Cenozoic (Leung 2017). The flea lineages fed on the blood of birds and monotreme, metatherian and eutherian mammals (Zhu et al. 2015), through stylate mouthparts that were more gracile in almost every structural aspect than those of the mid Mesozoic (Gao et al. 2014). An extensive list of characters, including mouthpart features, differentiate and link the five, major, mid-Mesozoic flea-like lineages that are related by a *Tarwinia* + [(*Pseudopulex* + *Hadropsylla*) + *Tyrannopsylla*] + [*Saurophthirus* + modern fleas] phylogeny (Huang et al. 2012; Gao et al. 2014). Characters defining the six lineages include those of the head, eyes, antennae, pronotum, coxae, tarsi, genitalia and cerci, but mouthpart traits are important. For example, the labial palps form a short, segmented, robust structure in *Saurophthirus* and modern fleas, whereas they form a long, non-segmented, streamlined, more gracile structure enveloping the mouthpart stylets in all other mid-Mesozoic flea-like lineages except *Tarwinia* (Gao et al. 2014). This indicates that the *Saurophthirus* + modern fleas constitute an expanded Siphonaptera clade that is distantly related to all other mid-Mesozoic fleas.

The earliest hosts for these early true flea lineages were mammals (Perrichot et al. 2012), although four, separate, parasite shifts apparently colonized birds most likely through introductions from ground nesting flea taxa (Whiting et al. 2008; Zhu et al. 2015). During the Cenozoic, several lineages of fleas are documented in ambers, although the most commonly encountered are common fleas (Pulicidae), particularly in Baltic and Dominican ambers (Lewis and Grimaldi 1997; Perrichot et al. 2012; Poinar Jr 2015). Rodent fleas of Ctenophthalmidae and Hystrichopsyllidae (Fig. 11.1l) were described from Baltic Amber (Dampf 1910; Pielowska et al. 2018). Marsupial fleas of Rhopalopsyllidae are known from Dominican Amber (Grimaldi and Engel 2005a).

11.5.1.8 Diptera (Flies)

Diptera constitute the greatest variety of parasitism, principally through blood-feeding (hematophagy), of any insect group, and are responsible for the widespread vectoring of parasites that often produce blood diseases principally to vertebrates, including humans. Eight major clades encompass 15 families that collectively are responsible for almost all of the documented parasitic relationships between Diptera and their vertebrate and other hosts (Lehane 1991; Lukashevich and Mostovski 2003). At the family level, hematophagy originated minimally 12 times (Wiegmann et al. 2011), although at lower taxonomic levels of origination are much greater. Dipteran hematophagous clades often are assembled into a nematocerous group (Culicoidea, Chironomoidea, Psychodiodea), a brachycerous group (Tabanoidea, Rhagionoidea), and a cyclorrhaphous group (Muscoidea, Hippoboscoidea, Oestroidea). Notably, these clades collectively represent the widest variety of mouthpart types in insects (Labandeira 1997, 2019), and house a diversity of feeding styles. Most of these families have members that are vectors of a variety of infectious diseases (Lehane 1991), and apparently opportunities were seized in occupation of new ecological niches for transmission of blood parasites sometime during the Early Cretaceous (Leung 2017). The pathogen fossil record indicates that such disease transmission commenced during the Early Cretaceous and has expanded to the Recent (Labandeira 2014a; Poinar Jr 2018, 2021).

The Culicoidea is a major clade of nematocerous flies (Wiegmann et al. 2011) that evolved lineages parasitic on vertebrates through blood feeding. One of the earliest occurrences of Culicoidea is from Late Triassic (late Carnian) strata of the Solite Quarry in southernmost Virginia, U.S.A. (Blagoderov et al. 2007). A particular specimen (Fig. 11.8), unassignable to family, displays the anterior thoracic region, head capsule, eyes, and sufficiently well preserved mouthparts to determine that the proboscis likely has a design consistent with feeding on pollination drops or

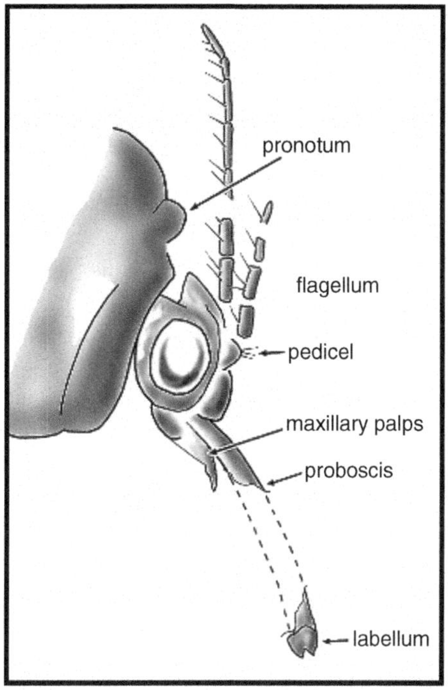

Fig. 11.8 Detail of head and preserved proboscis of a Culicomorpha specimen from the Solite quarries of southern Virginia, USA. This specimen (VMNH951) shows a long proboscis of a type generally associated with blood feeding, indicating that this form may be the earliest instance of blood feeding with specialized mouthparts in the fossil record. Distance from the top-most part of the head to proboscis tip is 0.5 mm. (Reproduced with permission from Fig. 9 in Blagoderov et al. 2007)

blood (Blagoderov et al. 2007). The specimen lacks the midsection of the proboscis but does reveal its base and terminus with a labellum and an intact maxillary palp. One possibility for the feeding style of this proboscis morphology is that it is a long-proboscid siphonate tube for imbibing pollination drops or related fluids from seed plants (Labandeira 2005). Alternatively, it could represent a stiffened stylate condition consistent with piercing and sucking for blood meals. As the adult diet ground plan of the Culicomorpha is blood feeding (Lukashevich and Mostovski 2003; Labandeira 2005), it is thus more likely that this specimen represents blood feeding. If this is the case, then this specimen represents the earliest documented example parasitic blood feeding in the fossil record.

One of the major families of Culicoidea is Culicidae (mosquitoes), which consist of approximately 40 extant genera that typically attack vertebrate hosts by bending their flexible stylet fascicles through host integument to obtain blood from capillaries, a feeding strategy known as solenophagy (Bouchet and Lavaud 1999). The earliest fossil occurrence of Culicidae is early Cenomanian Myanmar amber on the Early Cretaceous–Late Cretaceous boundary interval at 99 million years ago (Borkent and Grimaldi 2004). This particular specimen also is phylogenetically ancient, a sister-group to all other extant and extinct species of Culicidae. Nevertheless, it is probable that substantially older culicids will be found, extending the blood-feeding habit to the Jurassic (Borkent and Grimaldi 2004). The Cenozoic record of Culicidae is sporadic and specimens are found in several compression and amber deposits (e.g., Szadziewski 1998). Perhaps the most stunning discovery was a 46 million-year-old Eocene mosquito from the Kishenehn Formation of northwestern Montana with blood in its distended abdomen (Greenwalt 2021; Greenwalt et al. 2013). Heme-derived porphyrin molecules of hemoglobin were detected and molecularly characterized by advanced mass spectrometry and other techniques, indicating the presence of blood feeding, although the source of the vertebrate blood remains unknown.

The Chironomoidea include the principal blood-feeding groups of Ceratopogonidae (biting midges, no-see-ums) and Simuliidae (black flies, buffalo gnats). While extant Chironomidae (non-biting midges) are considered as predominately feeding on pollen, nectar, honeydew and other carbohydrate-rich food, extinct forms were hematophagous based on mouthpart structure (Azar and Nel 2012). Chironomidae extend to the Late Triassic–Early Jurassic boundary interval (Grimaldi and Engel 2005a), making them among the earliest hematophagous Diptera. Ceratopogonidae are represented by several, prominent, extant genera that are blood feeders, and the family has a good fossil record (Borkent 2001), extending to the Early Cretaceous (Berriasian Stage) at about 140 Ma. There are several instances of ceratopogonid parasitism in the amber fossil record. One example is a ceratopogonid specimen from Myanmar Amber (99 Ma) that hosted in its gut oocysts of the malarial parasite, *Paleohaemoproteus burmacis*, a pathogen similar to modern species found in reptiles and birds (Poinar Jr and Telford Jr 2005), and likely representing an ancient association. From the same deposit, a second ceratopogonid specimen, *Leptoconops nosopheris*, vectored the trypanosome *Paleotrypanosoma burmanicus* that may have produced trypanosomiasis-like symptoms in an unknown vertebrate host (Poinar 2021).

The fossil record of the Simuliidae begins during the Late Jurassic (Oxfordian Stage) and has a fair record of occurrences up to the recent (Currie and Grimaldi 2000). However, black fly–parasite associations are rare. The earliest fossil Simuliidae has been described from Kazakhstan (Kalugina 1991). From these compression-impression deposits, it appears that the blood-feeding habit in Simuliidae originated early within the clade, and may be the ancestral diet (Kalugina 1991).

The Psychodoidea is a more ancient nematocerous lineage than Chironomoidea that extends to the Late Triassic (Fraser et al. 1996). The only blood-feeding lineage of Psychodoidea is Psychodidae (moth flies), of which the subfamily Phlebotominae (sand flies) represent the only origin of hematophagy within the larger clade (Grimaldi and Engel 2005a, b) and are a major cause globally for transmission of several modern diseases in animals (Lehane 1991). One Cretaceous association involves the phlebotomine *Palaeomyia burmitis* that vectored a trypanosome parasite, *Paleoleishmania proterus*, which presumably developed in reptilian blood cells consumed by the sand fly (Poinar Jr 2004a; Poinar Jr and Poinar 2004a, b). This sand fly–trypanosome–reptile parasitic interaction occurs in mid Cretaceous Myanmar amber (Poinar 2021), and suggests a pathological relationship similar to modern leishmaniasis, a debilitating disease creating ulcers of the skin and viscera in vertebrates.

Tabanoidea consist of two families, Tabanidae (horse flies, deer flies) and Athericidae (watersnipe flies) that are involved in blood feeding. Tabanidae constitute about 32 modern genera of brachycerous flies that bite vertebrates using a slashing and cutting strategy to obtain blood and fluids that pool up within lesions, feeding strategy known as telmophagy (Bouchet and Lavaud 1999). Although the earliest Tabanidae originate from the latest Middle Jurassic of northeastern China (Ren 1998; Labandeira 2010), their original diet has been a mystery as to whether they consumed nectar-like pollination drops or blood (Grimaldi and Engel 2005a). This uncertainty arises because tabanids occur in the fossil record about 40 million years before the appearance of angiosperms, although a broad variety of vertebrates were present at this time. One hypothesis is that pangionine tabanids possessed siphonate mouthparts and were consumers of pollination drops (Labandeira 2010), whereas tabanine tabanids instead bore stylate mouthparts and were consumers of vertebrate blood (Martins-Neto 2003; Lukashevich and Mostovski 2003; Labandeira et al. 2007a, b). A less likely, alternative hypothesis is that for species of mid-Mesozoic tabanids, males consumed plant fluids of ovulate organs in a pollination mutualism. By contrast, conspecific females consumed vertebrate blood necessary for oögenesis and indicative of parasitism, a feeding behavior that occurs in some extant, hematophagous flies (Labandeira 2005).

Tabanids are known from younger earliest Cretaceous deposits of Brazil, the United Kingdom and Russia, some with probable hematophagous habits (Martins-Neto 2003; Lukashevich and Mostovski 2003). One particular specimen from the early Cretaceous of Transbaikalia is of large size with a robust body and gracile legs, a head having an inflated clypeus and elongate proboscis encompassing hardened stylets, and antennae and palps that bear numerous, multiple types of sensillae (Mostovski et al. 2003). These features are consistent with modern blood-feeding tabanids (Lukashevich and Mostovski 2003). In Paleogene deposits, such as Baltic Amber, the diversity of blood-feeding tabanids can be substantial, consisting contemporaneously of four genera and seven species of Chrysopinae and one genus and species of Tabaninae (Pielowska et al. 2018).

Athericidae are a modestly speciose family of 13 genera in which adults feed on nectar or blood and the aquatic larvae are predatory (Stuckenberg 1973). The athericid fossil record is unspectacular, with the exception of its earliest occurrence, the unusual *Qiyia jurassica* from the Jiulongshan Formation of northeastern China (Chen et al. 2014). *Qiya jurassica* possesses the most atypical attachment mechanisms for any ectoparasite: a thoracic sucker (Fig. 11.9). The aquatic fly larva *Q. jurassica* is a stem-group athericid from the latest Middle Jurassic of northwestern China (Chen et al. 2014). The larva is interpreted to have inhabited streams with rapid currents. In addition to this broad, ventral, circular attachment device occupying all three thoracic segments, *Q. jurassica* possesses a diminutive head with distinct piercing-and-sucking mouthparts, and stump-like abdominal prolegs for both attachment and movement. The authors (Chen et al. 2014) suggest that feeding occurred on salamander blood, although recent aquatic insect larvae in rapidly flowing streams with suction attachment discs are frequently predatory rather than parasitic in feeding habits (Leung 2017).

Rhagionoidea consist of one hematophagous family, Rhagionidae (snipe flies), of medium- to large sized flies that contain about 16 genera, several of which are hematophagous and do not transmit disease (Krenn and Aspöck 2012). Nagatomi and Soroida (1985) list only the distantly related *Spaniopsis* and *Symphoromyia* as blood feeders, indicating the hematophagy originated multiple times within the family. Although Rhagionidae enter the fossil record during the Early Jurassic (Lukashevich and Mostovski 2003), the earliest hematophagous forms are of Late Jurassic age, from the Glushkovo Formation of Transbaikalia, Russia (Kovalev and Mostovski 1997). One species with particularly evident hematophagous structures is the female of *Palaeoarthroteles mesozoicus* that houses a robust, hypognathous proboscis, downturned palps and an apparently rigid labium enveloping the stylets (Kovalev and Mostovski 1997; Labandeira 2019). An alternative explanation is that *Palaeoarthroteles* was an insect predator (Lukashevich and Mostovski 2003); however, the absence of spinose legs, stiff bristles surrounding the mouthparts and other features of predatory insects strongly suggests the blood-feeding nature of this insect.

Mormotomyiidae (frightful hairy fly) consists of the monotypic, spiderlike and hirsute *Mormotomyia hirsuta* that occurs on a bat guano substrate below a local cliff face in Kenya (Copeland et al. 2011). Once thought to be a bat parasite because of its habitus and association with guano, recent evidence indicates that it is not a bat parasite (Copeland et al. 2011). Mormotomyiidae evidently are a member of the Ephydroidea, an acalyptrate group basal to the Calyptratae Clade (Kirk-Spriggs et al. 2011).

Within Calyptratae (calyptrate flies), the three superfamilies Muscoidea (often treated as a grade), Hippoboscoidea and Oestroidea (Kutty et al. 2001) collectively bear nine families that contain parasitic taxa. Calyptrate flies are monophyletic and the most derived major clade of flies (Grimaldi and Engel 2005a, b; Kutty et al. 2001), responsible for a considerable amount of pestilence on vertebrates, including humans, particularly as feeders on blood and other fluids in their role as parasites and parasitoids (Oldroyd 1964; Marshall 2012). Many of the blood-feeding forms are found in the ubiquitous family Muscidae (house flies, stable flies), which consist of about 110 genera, including the familiar, synanthropic fly *Musca domestica*. Typically, muscids are predatory or feed on fluid exudates from plants and animals, such as sugar, honeydew, sweat, tears and blood. Blood-feeding muscids principally include *Stomoxys*, *Hydrotaea* and *Haematobia* of the tribe Stomoxyini (Krenn and

Fig. 11.9 Reconstruction of the watersnipe fly larva *Qiyia jurassica* (Diptera: Athericidae), from the latest Middle Jurassic of Inner Mongolia, China. The ectoparasitic fly specimen is portrayed in left lateral view displays the ventral thoracic sucker for host attachment and a reduced, partly retractile head. (Reproduced with permission from Fig. 3 of Chen et al. 2014)

Aspöck 2012) that are parasitic on vertebrates, mostly livestock (Cupp et al. 1998; Salem et al. 2012), and represent probably a single origination of hematophagy (Schremmer 1961). Feeding in this tribe is accomplished by sharp, prestomal teeth on the labellum, resulting in surface abrasion of the integument and sponging of blood and other dermal fluids that pool in the lesions. The Tribe Stomoxyini lacks a fossil record, but because the clade is a member of the post-Cretaceous Calyptratae (Grimaldi and Engel 2005a), the lineage likely originated during the Paleogene and acquired parasitic interactions with large mammals at that time.

The Hippoboscoidea, termed Pupipara in the older literature, consists of the four constituent lineages of Glossinidae (tsetse flies), Hippoboscidae (louse flies, keds), Nycteribiidae (bat flies) and Streblidae (bat flies), although the last two lineages are not monophyletic (Griffiths 1972; Kutty et al. 2001). Hippoboscoid flies are united by several, common and distinctive reproductive features. The mature female lays a single mature larva, the puparium, after which the larva immediately pupariates, emerging from the hardened, last-instar larval skin that encloses the pupa (Ferrar 1987). The emergent, hematophagous adult then locates and parasitizes a suitable bird or mammalian host. This unique larval development is probably the most significant K-selection strategy in insects, with females laying only one egg at a time, albeit in succession, and larvae nursed internally in the mother by special uterine "milk" glands (Oldroyd 1964). Hippoboscoidea have considerable mouthpart penetrability and styletal dexterity for puncturing the often-hardened integument of most vertebrate hosts. Perhaps because of thickened and indurated host integument, the adult mouthparts consist of a short, rigid proboscis with prominent, prestomal teeth mounted on a labellum used for abrading and cutting through integumental tissue to feed on upwelling fluids such as lymph and blood (Jobling 1926, 1928, 1929, 1933).

Glossinidae (tsetse flies) are large, hematophagous flies with distinctive mouthparts that are obligately parasitic on vertebrates, including humans (Jobling 1933). From their blood feeding, glossinids are vectors of the trypanosome *Trypanosoma* that cause the debilitating and fatal diseases of rinderpest and nagana in livestock and sleeping sickness in humans throughout modern, sub-Saharan Africa (Lambrecht 2018). Glossinids consist of a single modern and fossil genus, *Glossina*. *Glossina* contains 15 species that is split into three subgenera, each of which occupies a savanna, forest or riverine–lacustrine margin habitat. Two *Glossina* species have been found in the 34 million-year-old, latest Eocene Florissant Formation in Colorado, U.S.A. (Grimaldi 1992), and an additional species is known from a 10 million-year-older occurrence at the Enspel locality in Germany (Wedmann 2000). This biogeographical distribution indicates that *Glossina* had a widespread distribution in the Northern Hemisphere during the Paleogene, and became restricted to sub-Saharan Africa in the Neogene. It remains unknown whether *Glossina* was a vector for diseases similar to sleeping sickness during its earlier, Paleogene existence (Martins-Neto 2003).

Hippoboscidae (louse flies, keds) consist of 3 subfamilies and 21 genera of obligate parasites of birds and mammals. Hippoboscid fossil history is extremely limited and is comprised of two occurrences. One specimen is from Rott, Rhineland,

Germany, of late Oligocene age (Chattian Stage) at approximately 25.5 Ma (Statz 1940; Maa 1966). Maa (1966) concluded that the specimen was an early, basal lineage of Hippoboscidae. A more recent find of a Hippoboscid is from the latest Miocene Messinian Stage, approximately 6 million years in age (Bradley and Landini 1984). However, these occurrences say little about the evolutionary biology of hippoboscids (Dittmar et al. 2006). An alternative method is a molecular phylogenetic approach (Dittmar et al. 2006) that places the Hippoboscidae as the sister-group to the Glossinidae, with an origin approximately at the same time as the separation of the Nycteribiidae and Streblidae during the late Eocene (De Moya 2019).

Nycteribiidae (bat flies) are 12 genera of dorsoventrally flattened, spider-like flies lacking eyes and wings that are obligate, host specific parasites on bats. Nycteribiids have been implicated in the transmission of bat malaria in Africa (Obame-Nkoghe et al. 2016). They lack a fossil record. Related to Nycteribiidae are Streblidae (bat flies), consisting of 5 subfamilies and approximately 33 genera of blood-feeding flies that are ectoparasitic specialists on a variety of bats (Lehane 1991). Some genera are involved in transmission of bat malaria (Obame-Nkoghe et al. 2016). Although the oldest streblid is early Miocene in age (Poinar Jr and Brown 2012), the clade is thought to have originated during the Eocene, based on the fossil record of the closely related Glossinidae (Dittmar et al. 2006). From the same early Miocene deposit, an unaffiliated streblid harbored *Vetufebrus ovatus* (Haemospororida: Plasmodiidae), a bat malarial parasite (Poinar Jr 2011) that apparently is not vectored by any known extant streblid (Leung 2017). This suggests that this particular parasitic relationship of malaria may be a disease that has become extinct for streblid vectors and their bat hosts (compare van Dijk and De Baets 2021 for other putative examples).

Oestroidea contains four families, Oestridae (bot flies), Calliphoridae (blow flies, carrion flies, greenbottles), Mesembrinellidae (blow flies) and Mystacinobiidae (New Zealand bat fly), which contain significant numbers of parasitic, blood-feeding species. Parasitism likely originated five times in the Oestroidea, based on genus-level phylogenies (Pape 2001; Kutty et al. 2001). As cyclorrhaphan calyptrates, Oestroidea are a recent clade that originated during the late Paleocene to early Eocene epochs of the Cenozoic, about 60–40 million years ago (Cerretti et al. 2017). The most iconic of the oestroid parasitic families, Oestridae consist of about 160 species (Wood 2006) and contain the most conspicuous, diverse and noxious of parasites, as they are internal parasites of mammals and frequently inhabit their host's flesh, occasionally using an intermediate vector such as a dipteran to complete their life cycle (Colwell et al. 2006). A typical life cycle is illustrated by the reindeer warble fly, *Hypoderma tarandi*, documented on reindeer, *Rangifer tarandus*. The fly initially lays eggs during Arctic summer at the base of thin hairs adjacent the skin, followed by hatching and then penetration of the first-instar larva into the skin, soon resulting in development of nodules (warbles) under the skin while the host's proteins from subcutaneous tissue are ingested by the larva (Asbakk et al. 2014). During the following Arctic winter, the larva continues to consume proteins, increasing in size through several instar shifts up to 5 cm long, eventually maturing and leaving the host during early Arctic spring and pupating on the ground, where they emerge as adult flies after several weeks to complete the life cycle (Asbakk et al. 2014). The cumulative load of reindeer warble fly infestation is particularly debilitating on reindeer cows, particularly during calving season in late summer (Ballesteros et al. 2011). Thematically, this life cycle of infestation is similar to bot flies attacking a variety of other large mammals, especially bovids (e.g., Oryan et al. 2009).

Calliphoridae are a diverse, worldwide family consisting of 75 genera, some of which have larvae (screwworms) that cause myiasis and other diseases such as dysentery and paratuberculosis that especially affect livestock (Thomas and Mangan 1989) (Table 11.1). Calliphoridae are of vital importance in carrion communities, assisting the degradation of animal tissues. Several occurrences (puparia) of Calliphoridae are late Neogene in age, the oldest is upper Pliocene, from 3.6 to 2.6 million years ago (Kitching 1980). An earlier record of two puparia in an ironstone nodule attributed to the Calliphoridae, from the late Cretaceous of Canada (Campanian Stage) and described by McAlpine (1970), is erroneous and are attributed to Cyclorrhapha (Grimaldi and Engel 2005a). Since no calyptrate group is known from the Mesozoic (Grimaldi and Engel 2005a), the calyptrate Calliphoridae likely originated during the more recent Paleogene and was parasitic on mammals of medium to large size.

A closely related lineage to the Calliphoridae is the Mesembrinellidae, consisting of three subfamilies and nine genera that occur in humid, primary forests of the Neotropics. Some adults feed on blood and are parasites whereas other species are found on decomposing animal material and fermented substances in fruits (Marinho et al. 2017). The only fossil occurrence is *Mesembrinella caenozoica* from early Miocene amber (Fig. 11.10), from the Dominican Republic (21 Ma). Their origin is placed during the Eocene (Cerretti et al. 2017).

Mystacinobiidae is a closely related family to Calliphoridae, consisting of the enigmatic, monotypic *Mystacinobia zelandica* (New Zealand bat fly) that recently has been discovered (Gleeson et al. 2000). This group is a spider-like, wingless, structurally highly modified fly that feeds on bat guano and is phoretic and possibly parasitic on the New Zealand short-tailed bat (Holloway 1976).

Excluded from the list of parasitic Diptera, Strashilidae is a group of mid-Mesozoic, eastern Eurasian insects exemplified by the morphologically bizarre *Strashila incredibilis* (Fig. 11.1e; Rasnitsyn 1992). They possess features that superficially resemble some ectoparasitic insects. These features include dorsoventral flattening; legs with highly swollen metafemora and grasping terminal claws; prominent lateral, gill-like appendages of the abdomen; and a head with sunken antennae and a ventrally directed, short beak housing reduced piercing-and-sucking mouthparts (Huang et al. 2013b). Males were winged whereas females were wingless. Once thought as terrestrial ectoparasites of pterosaurs and possibly affiliated with Siphonaptera (Rasnitsyn 1992), the group together with Tarwiniidae were considered as sister group to Saurophthiridae (Rasnitsyn and Quicke 2002; Nagler and Haug 2015). Others proposed that Strashilidae as a lineage originating deeper

Table 11.1 Larval biologies of modern and fossil parasitoid insect taxa

Taxon[a]	Parasitoid larval life-habit[b]	Dominant larval hosts[c]	Host stage of larval parasitoid attack[d]	Geochronologic range[e]	References[f]
NEUROPTERA					
Hemerobioformia					
Mantispidae	Ectoparasitoid predator	Coleoptera: Scarabaeidae; Lepidoptera: Noctuidae; Hymenoptera: Apidae	Larvae and pupae of soil insects; egg sacs of spiders	Callovian–Recent	Askew (1971), Lambkin (1986), Eggleton and Belshaw (1992), Grimaldi and Engel (2005a), Poinar Jr and Buckley (2010), Haug et al. (2018), Chen et al. (2019)
COLEOPTERA					
Caraboidea					
Carabidae	Ectoparasitoid	Diplopoda; Orthoptera: Stenopelmatidae; Coleoptera: Chrysomelidae; Diptera	Pupae, puparia and eggs of soil arthropods; on plants	Carnian–Recent	Clausen (1940), Erwin (1979), Eggleton and Belshaw (1992), Grimaldi and Engel (2005a)
Staphylinoidea					
Staphylinidae	Ectoparasitoid	Diptera	Puparia in ephemeral habitats	Carnian–Recent	Clausen (1940), Klimaszewski (1984), Eggleton and Belshaw (1992), Fraser et al. (1996), Chatzimanolis et al. (2012)
Scarabaeoidea					

Taxon[a]	Parasitoid larval life-habit[b]	Dominant larval hosts[c]	Host stage of larval parasitoid attack[d]	Geochronologic range[e]	References[f]
Scarabaeidae	Cleptoparasitoid, predator	Coleoptera: Geotrupidae; other Scarabaeidae	Eggs, larvae and dung brood balls	Tithonian–Recent	Hammond (1976), Eggleton and Belshaw (1992), Krell (2006), Bai et al. (2012)
Dascilloidea					
Rhipiceridae	Endoparasitoid	Hemiptera: Cicadidae	Nymphs	Lutetian–Recent	Elzinga (1977), Eggleton and Belshaw (1992), Ponomarenko (1995), Grimaldi and Engel (2005a)
Cleroidea					
Cleridae	Ectoparasitoid, cleptoparasitoid predator	Hymenoptera: Apidae; Lepidoptera; Orthoptera: Acrididae	Larvae in nests and galls; egg pods	Cenomanian–Recent	Clausen (1940), Foster (1976), Eggleton and Belshaw (1992), Rasnitsyn and Ross (2000), Kolibáč and Huang (2016)
Cucujoidea					
Bothrideridae	Ectoparasitoid	Coleoptera; Hymenoptera: Apidae	Larvae, pupae of dead wood	Lutetian–Recent	Clausen (1940), Roberts (1980), Eggleton and Belshaw (1992), Grimaldi and Engel (2005a)

(continued)

Table 11.1 (continued)

Taxon[a]	Parasitoid larval life-habit[b]	Dominant larval hosts[c]	Host stage of larval parasitoid attack[d]	Geochronologic range[e]	References[f]
Passandridae	Ectoparasitoid, hyperparasitoid	Coleoptera: Cerambycidae; Hymenoptera: Braconidae	Pupae in dead wood, larvae of parasitoids	Cenomanian–Recent	Clausen (1940), Askew (1971), Eggleton and Belshaw (1992), Grimaldi and Engel (2005a)
Tenebrionoidea					
Meloidae	Ectoparasitoid, cleptoparasitoid	Hymenoptera: Apidae; Orthoptera: Acrididae	Larvae in soil and nests; egg pods and masses	Cenomanian–Recent	Clausen (1940), Pinto and Selander (1970), Askew (1971), Eggleton and Belshaw (1992), Bologna et al. (2008)
Ripiphoridae	Ectoparasitoid, endoparasitoid	Blattodea; Coleoptera: Cerambycidae; Hymenoptera: Scoliidae, Tiphiidae, Vespidae	Larvae in wood and nests; mobile adults	Callovian–Recent	Clausen (1940), Linsley et al. (1952), Grimaldi et al. (2005), Batelka et al. (2016, 2019), Cai et al. (2018), Hsiao et al. (2017), Hsiao and Huang (2018)
Curculionoidea					
Anthribidae	Ectoparasitoid, predator	Hemiptera: Coccoidea	Nymphs; egg masses	Hauterivian–Recent	Clausen (1940), Askew (1971), Eggleton and Belshaw (1992), Ülgentürk (2001), Gratshev and Zherikhin (2003)
STREPSIPTERA[g]					
"***Protostrepsiptera***"[h]					

Taxon[a]	Parasitoid larval life-habit[b]	Dominant larval hosts[c]	Host stage of larval parasitoid attack[d]	Geochronologic range[e]	References[f]
Cretostylopidae†	Endoparasitoid[i]	—	Nymphs or larvae[i]	Cenomanian	Grimaldi et al. (2005), Kathirithamby and Engel (2014)
Phthanoxenidae†	Endoparasitoid[i]	—	Nymphs or larvae[i]	Cenomanian	Engel and Huang (2016)
Protoxenidae†	Endoparastoid[i]	—	Nymphs or larvae[i]	Lutetian	Pohl et al. (2005)
Eleostrepsiptera[h]					
Mengeidae†	Endoparasitoid[i]	Zygentoma: Lepismatidae[i]	Nymphs[i]	Lutetian	Clausen (1940), Askew (1971), Kulicka (1979), Godfray (1994), Grimaldi and Engel (2005a), Pohl and Beutel (2008), Kathirithamby (2009), Engel et al. (2016a)
Neostrepsiptera[h]					
Bahiaxenidae	Endoparasitoid[i]	—	Nymphs or larvae[i]	Recent	Bravo et al. (2009)
Bohartillidae	Endoparasitoid[i]	—	Nymphs or larvae[i]	Aquitanian–Recent	Pohl and Kinzelbach (2001), Kinzelbach and Pohl (1994), Cook (2014)
Corioxenidae	Endoparasitoid	Hemiptera: Heteroptera	Nymphs	Lutetian–Recent	Henderickx et al. (2013), Cook (2014)

(continued)

Table 11.1 (continued)

Taxon[a]	Parasitoid larval life-habit[b]	Dominant larval hosts[c]	Host stage of larval parasitoid attack[d]	Geochronologic range[e]	References[f]
Elenchidae	Endoparasitoid	Hemiptera: Auchenorrhyncha	Nymphs	Aquitanian–Recent	Askew (1971), Kinzelbach and Pohl (1994), Cook (2014)
Halictophagidae	Endoparasitoid	Blattodea; Orthoptera: Caelifera; Hemiptera: Auchenorrhyncha; Diptera	Nymphs	Cenomanian–Recent	Kathirithamby (2009), Cook (2014), Engel et al. (2016a)
Kinzelbachillidae[†]	Endoparasitoid[i]	—	Nymphs or larvae[i]	Cenomanian	Pohl and Beutel (2016)
Lychnocolacidae	Endoparasitoid[i]	—	Nymphs or larvae[i]	Recent	Engel et al. (2016a)
Mengenillidae	Endoparasitoid	Zygentoma: Lepismatidae	Nymphs	Cenomanian–Recent	Pohl and Beutel (2008), Cook (2014), Engel et al. (2016a), Pohl and Beutel (2019), Pohl et al. (2018)
Myrmecolacidae	Parasite, then (?) endoparasitoid	Females on Hymenoptera: Formicidae; males on Mantodea and Orthoptera	Larvae or nymphs	Thanetian–Recent	Kinzelbach (1983), Kinzelbach and Pohl (1994), Pohl and Kinzelbach (1995), Cook (2014), Wang et al. (2015)
Protelencholacidae[†]	Endoparasitoid[i]	—	Nymphs or larvae[i]	Aquitanian	Pohl and Beutel (2005), Engel et al. (2016a)
Protoxenidae[†]	Endoparasitoid[i]	—	Nymphs or larvae[i]	Priabonian	Pohl et al. (2005)

Taxon[a]	Parasitoid larval life-habit[b]	Dominant larval hosts[c]	Host stage of larval parasitoid attack[d]	Geochronologic range[e]	References[f]
Stylopidae	Endoparasitoid	Hymenoptera: Vespoidea, Apoidea	Larvae	Lutetian–Recent	Kinzelbach and Lutz (1985), Pohl and Kinzelbach (2001), Kogan and Poinar Jr (2010), Cook (2014)
Xenidae	Endoparasitoid	Hymenoptera: Vespoidea, Apoidea	Larvae	Lutetian–Recent	Cook (2014), Engel et al. (2016a)
DIPTERA					
Culicomorpha					
Chironomidae	Ectoparasitoid	Gastropoda: Pulmonata; Ephemeroptera: Rithrogeniidae	Immature; naiad	Norian–Recent	Clausen (1940), Krzeminski and Jarzembowski (1999), Grimaldi and Engel (2005a)
Bibionomorpha					
Cecidomyiidae	Endoparasitoid	Hemiptera: Sternorrhyncha	Nymphs, adults	Albian–Recent	Clausen (1940), Arillo and Nel (2000), Grimaldi and Engel (2005a)
Mycetophilidae	Endoparasitoid	Platyhelminthes	Immatures, adults	Cenomanian–Recent	Hickman (1965), Eggleton and Belshaw (1992), Grimaldi and Engel (2005a)

(continued)

Table 11.1 (continued)

Taxon[a]	Parasitoid larval life-habit[b]	Dominant larval hosts[c]	Host stage of larval parasitoid attack[d]	Geochronologic range[e]	References[f]
Nemestrinimorpha[j]					
Acroceridae	Endoparasitoid	Araneae: Clubionidae, Lycosidae, Salticidae; Acari	Adults	Kimmeridgian–Recent	Clausen (1940), Askew (1971), Grimaldi and Engel (2005a), Winterton et al. (2007), Kerr and Winterton (2008), Gillung et al. (2018)
Nemestrinidae	Endoparasitoid, ectoparasitoid	Orthoptera: Caelifera; Coleoptera: Scarabaeidae, Cerambycidae	Eggs, larvae and nymphs	Toarcian–Recent	Clausen (1940), Greathead (1963), Grimaldi (1995), Mostovski (1998), Ansorge and Mostovski (2002), Grimaldi and Engel (2005a), Wedmann (2007), Zhang et al. (2017)
Archisargoidea					
Eremochaetidae†	Endoparasitoid	—	Probably larvae	Tithonian–Cenomanian	Zhang (2017), Grimaldi and Barden (2016)
Muscomorpha					
Asiloidea					
Asilidae	Ectoparasitoid	Coleoptera: Scarabaeidae, Buprestidae, Cerambycidae, Curculionidae; Hymenoptera: Siricidae	Larvae occurring in dead wood	Kimmeridgian–Recent	Knutson (1972), Eggleton and Belshaw (1992), Grimaldi and Engel (2005a), Haug et al. (2017)

Taxon[a]	Parasitoid larval life-habit[b]	Dominant larval hosts[c]	Host stage of larval parasitoid attack[d]	Geochronologic range[e]	References[f]
Bombyliidae	Ectoparasitoid, endoparasitoid, hyperparasitoid	Orthoptera: Acrididae; Coleoptera: Tenebrionidae; Lepidoptera: Tortricidae, Noctuidae; Diptera: Tachinidae; Hymenoptera: Tenthredinidae, Apidae, Ichneumonidae	Eggs, larvae and pupae variously of free feeders and parasitoids	Callovian–Recent	Clausen (1940), Hull (1973), Rasnitsyn (1985), Eggleton and Belshaw (1992), Yeates and Greathead (1997), Nel and De Ploëg (2004), Grimaldi and Engel (2005a), Wedmann and Yeates (2008)
Empidoidea					
Empididae	Endoparasitoid, ectoparasitoid	Trichoptera: Glossosomatidae, Rhyacophilidae	Pupae	Tithonian–Recent	Knutson and Flint (1979), Vinikour and Anderson (1981), Eggleton and Belshaw (1992), Coram et al. (2000)
Aschiza[j]					
Phoridae	Endoparasitoid, cleptoparasitoid	Oligochaeta; Gastropoda: Pulmonata; Opiliones; Araneae; Myriapoda; Hemiptera; Coleoptera; Diptera; Trichoptera; Hymenoptera: Vespoidea, Apoidea	Eggs, nymphs, larvae, other immatures; adults	Albian–Recent	Robinson (1971), Ferrar (1987), Arillo and Mostovski (1999), Eggleton and Belshaw (1992), Brown (1997, 1998), Coupland and Barker (2004), Grimaldi and Engel (2005a)

(continued)

Table 11.1 (continued)

Taxon[a]	Parasitoid larval life-habit[b]	Dominant larval hosts[c]	Host stage of larval parasitoid attack[d]	Geochronologic range[e]	References[f]
Pipunculidae	Endoparasitoid	Hemiptera: Auchenorrhyncha	Nymphs occurring on plants	Lutetian–Recent	Clausen (1940), Askew (1971), Eggleton and Belshaw (1992), Grimaldi and Engel (2005a)
Acalyptratae[j]					
Chamaemyiidae	Endoparasitoid	Hemiptera: Sternorrhyncha	Nymphs	Lutetian–Recent	Hennig (1965), Eggleton and Belshaw (1992)
Chloropidae	Endoparasitoid, cleptoparasitoid	Hemiptera: Sternorrhyncha	Nymphs	Lutetian–Recent	Hennig (1965), Narchuk (1972), Ferrar (1987), Grimaldi and Engel (2005a)
Conopidae	Endoparasitoid	Blattodea; Orthoptera; Diptera: Calyptratae; Hemiptera: Coccoidea; Hymenoptera: Aculeata	Nymphs, larvae, puparia	Lutetian–Recent	Smith (1966), Hennig (1966), Smith and Cunningham-von Someren (1985), Gibson and Skevington (2013), Rocha et al. (2015)
Cryptochaetidae	Endoparasitoid	Hemiptera: Coccoidea: Margarodidae	Nymphs occurring on plants	Lutetian–Recent	Thorpe (1941), Hennig (1965), Eggleton and Belshaw (1992)
Pyrgotidae	Endoparasitoid	Coleoptera: Scarabaeidae	Adults	Lutetian–Recent	Hennig (1965), Askew (1971), Eggleton and Belshaw (1992), Grimaldi and Engel (2005a)

Taxon[a]	Parasitoid larval life-habit[b]	Dominant larval hosts[c]	Host stage of larval parasitoid attack[d]	Geochronologic range[e]	References[f]
Sciomyzidae	Endoparasitoid	Oligochaeta; Bivalvia; Gastropoda: Pulmonata; Diplopoda	Immatures and adults in freshwater and terrestrial habitats	Lutetian–Recent	Hennig (1965), Bailey (1989), Vala et al. (1990), Eggleton and Belshaw (1992), Grimaldi and Engel (2005a)
Schizophora					
Anthomyiidae	Endoparasitoid	Orthoptera	Late nymph, adult	Lutetian–Recent	Hennig (1965), Ferrar (1987), Eggleton and Belshaw (1992), Michelson (2000), Grimaldi and Engel (2005a)
Calliphoridae	Ectoparasitoid, endoparasitoid, cleptoparasitoid	Oligochaeta; Gastropoda: Pulmonata; Isoptera; Hymenoptera: Formicidae; Mammalia	Adults	Lutetian–Recent	Hennig (1965), Stevens (2003), Stevens et al. (2006), Coupland and Barker (2004), Stevens et al. (2006)
Muscidae	Endoparasitoid	Gastropoda: Pulmonata; Diplopoda; Lepidoptera	Immatures and adults of soft-bodied invertebrates	Lutetian–Recent	Hennig (1965), Ferrar (1987), Bailey (1989), Vala et al. (1990), Eggleton and Belshaw (1992), Coupland and Barker (2004)

(continued)

Table 11.1 (continued)

Taxon[a]	Parasitoid larval life-habit[b]	Dominant larval hosts[c]	Host stage of larval parasitoid attack[d]	Geochronologic range[e]	References[f]
Rhinophoridae	Endoparasitoid	Crustacea: Isopoda	All instars	Lutetian–Recent	Hennig (1965), Bedding (1965), Ferrar (1987), Eggleton and Belshaw (1992), Grimaldi and Engel (2005a)
Sarcophagidae	Endoparasitoid, cleptoparasitoid	Oligochaeta; Gastropoda: Pulmonata; Araneae; Orthoptera: Acrididae; Isoptera; Diptera: Tabanidae; Coleoptera: Curculionidae; Lepidoptera: Bombycoidea; Hymenoptera: Bradybaenidae, Aculeata	Eggs, larvae, other immatures; adults	Lutetian–Recent	Hennig (1965), Ferrar (1987), Coupland and Barker (2004), Grimaldi and Engel (2005a), Stevens et al. (2006)
Tachinidae	Endoparasitoid	Scorpionida; Araneae; Chilopoda; Isopoda; Embioptera; Orthoptera; Dermaptera; Blattodea; Mantodea; Phasmatodea; Hemiptera: Heteroptera; Coleoptera: Scarabaeidae; Lepidoptera: Ditrysia; Hymenoptera: Symphyta	Nymphs, larvae and adults variously of wood borers and phytophages	Lutetian–Recent	Clausen (1940), Askew (1971), Ferrar (1987), Eggleton and Belshaw (1992), Evenhuis (1994), Lehmann (2003), Stireman III et al. (2006), Cerretti et al. (2014, 2017), Winkler et al. (2015)
TRICHOPTERA					
Hydroptilidae	Ectoparasitoid	Trichoptera	Pupae	Santonian–Recent	Labandeira (1994), Wells (1992, 2005), Morris (1998)
LEPIDOPTERA					
Zygaenoidea					

Taxon[a]	Parasitoid larval life-habit[b]	Dominant larval hosts[c]	Host stage of larval parasitoid attack[d]	Geochronologic range[e]	References[f]
Epipyropidae	Ectoparasitoid	Hemiptera: Auchenorrhyncha; Lepidoptera	Nymphs, larvae on plants	Recent	Askew (1971), Eggleton and Belshaw (1992), Grimaldi and Engel (2005a)
Pyraloidea					
Pyralidae	Ectoparasitoid	Lepidoptera: Saturniidae	Larvae of externally feeding herbivore	Lutetian–Recent	Askew (1971), Eggleton and Belshaw (1992), Sohn et al. (2012), Grimaldi and Engel (2005a)
HYMENOPTERA					
Symphyta					
Orussoidea					
Orussidae	Ectoparasitoid	Coleoptera: Buprestidae; Hymenoptera: Siricidae	Larvae of wood borers	Turonian–Recent	Basibuyuk et al. (2000), Whitfield (2003), Grimaldi and Engel (2005a), Vilhelmsen and Zimmermann (2014)
Apocrita					
Stephanoidea					
Aptenoperissidae[†]	Ectoparasitoid[i]	—	Larvae of wood borers[i]	Cenomanian	Rasnitsyn et al. (2017), Rasnitsyn and Öhm-Kühnle (2018), Zhang et al. (2018a)

(continued)

Table 11.1 (continued)

Taxon[a]	Parasitoid larval life-habit[b]	Dominant larval hosts[c]	Host stage of larval parasitoid attack[d]	Geochronologic range[e]	References[f]
Myanmarinidae[†]	Ectoparasitoid[i]	—	Larvae of wood borers[i]	Cenomanian	Li et al. (2018b), Zhang et al. (2018c)
Stephanidae	Ectoparasitoid	Coleoptera: Buprestidae, Cerambycidae; Hymenoptera: Siricidae	Larvae of wood borers	Cenomanian–Recent	Whitfield (2003), Li et al. (2017b), Engel and Huang (2017), Zhang et al. (2018a)
"Ephialitoidea"					
"Ephialtitidae"[†]	Ectoparasitoid[i]	Coleoptera[i]; Hymenoptera[i]	Larvae of wood borers[i]	Toarcian–Aptian	Zhang et al. (2002a), Rasnitsyn et al. (2003), Grimaldi and Engel (2005a), Peñalver and Engel (2006), Li et al. (2013b, 2015b)
Megalyroidea					
Megalyridae	Ectoparasitoid	Coleoptera: Cerambycidae	Larvae of wood borers	Cenomanian–Recent	Rasnitsyn (1975), Whitfield (2003), Grimaldi and Engel 2005a; Perrichot (2009), Zhang et al. (2018a)
Trigonalyroidea					
Trigonalidae	Endoparasitoid, hyperparasitoid	Symphytan or lepidopteran caterpillar; then Ichneumonoidea or Vespidae larva	Larvae of externally feeding folivores	Aptian–Recent	Weinstein and Austin (1991), Whitfield (2003), Nel et al. (2003), Poinar Jr and Shaw (2007), Grimaldi and Engel (2005a)

Taxon[a]	Parasitoid larval life-habit[b]	Dominant larval hosts[c]	Host stage of larval parasitoid attack[d]	Geochronologic range[e]	References[f]
Maimetshidae†	Endoparasitoid[i]	—	Externally feeding folivores[i]	Barremian–Santonian	Rasnitsyn (1975), Rasnitsyn and Brothers (2009), Engel (2016)
Evanioidea					
Andreneliidae†	Ectoparasitoid, predator[b]	—	Larvae of wood borers[i]	Barremian	Rasnitsyn and Martínez-Delclòs (2000), Grimaldi and Engel (2005a), Li et al. (2018a, b)
Anomopterellidae†	Ectoparasitoid[i]	Coleoptera[i]; Hymenoptera[i]	Larvae of wood borers[i]	Callovian–Tithonian	Li et al. (2013a, 2014b, 2018a)
Aulacidae	Ectoparasitoid	Coleoptera: Buprestidae, Cerambycidae; Hymenoptera: Xiphydriidae	Larvae of wood borers	Hauterivian–Recent	Carlson (1979), Rasnitsyn (1980), Jennings et al. (2004), Grimaldi and Engel (2005a), Turrisi and Vilhelmsen (2010), Li et al. (2018a)
Baissidae†	Ectoparasitoid[i]	Coleoptera[i]; Hymenoptera[i]	Larvae of wood borers[i]	Berriasian–Turonian	Rasnitsyn (1975, 1991a), Rasnitsyn et al. (1998), Basibuyuk et al. (2002), Nel et al. (2004), Engel (2013), Li et al. (2018a)

(continued)

Table 11.1 (continued)

Taxon[a]	Parasitoid larval life-habit[b]	Dominant larval hosts[c]	Host stage of larval parasitoid attack[d]	Geochronologic range[e]	References[f]
Evaniidae	Predator, ectoparasitoid	Blattodea	Oothecate eggs	Hauterivian–Recent	Brown (1973), Basibuyuk et al. (2002), Deans et al. (2004), Grimaldi and Engel (2005a), Li et al. (2018a)
Gasteruptiidae	Predator–inquiline	Hymenoptera: Apoidea	Eggs and larvae of wasps and solitary bees	Barremian–Recent	Malyshev (1968), Rasnitsyn (1991b), Jennings et al. (2004), Li et al. (2018a)
Othniodellithidae[†]	Ectoparasitoid[i], predator[i]	—	Larvae of wood borers[i]	Cenomanian	Engel et al. (2016b)
Praeaulacidae[†]	Ectoparasitoid[i]	Coleoptera[i]	Larvae of wood borers[i]	Callovian–Cenomanian	Rasnitsyn (1972), Jell and Duncan (1986), Oberprieler et al. (2012); Li et al. (2014a, c, d, 2015a, 2017c, 2018a)
Vectevaniidae[†]	Ectoparasitoid[i]	Coleoptera[i]; Hymenoptera[i]	Larvae of wood borers[i]	Priabonian	Rasnitsyn (2013)
Proctotrupoidea					
Austroniidae	Endoparasitoid[i]	—	Larvae[i]	Cenomanian–Recent	Grimaldi and Engel (2005a), Zhang et al. (2018a)
Diapriidae	Endoparasitoid	Diptera: Syrphidae, Sarcophagidae, Tephritidae; Hymenoptera: Formicidae	Fly puparia; brood nest associates of ants	Cenomanian–Recent	Clausen (1940), Whitfield (2003), Engel et al. (2013a), Zhang et al. (2018a), Quevillon and Hughes (2018)

Taxon[a]	Parasitoid larval life-habit[b]	Dominant larval hosts[c]	Host stage of larval parasitoid attack[d]	Geochronologic range[e]	References[f]
Heloridae	Endoparasitoid	Neuroptera: Chrysopidae	Larvae of predators	Callovian–Recent	Shih et al. (2011), Shi et al. (2014), Li et al. (2017c)
Maamingidae	Endoparasitoid[i]	—	Larvae[i]	Recent	Early et al. (2001)
Mesoserphidae[†]	Endoparasitoid[i]	—	Larvae[i]	Callovian–Barremian	Kozlov (1968), Shih et al. (2011), Shi et al. (2013), Garrouste et al. (2016)
Monomachidae	Endoparasitoid[i]	—	Larvae[i]	Kimmeridgian–Recent	Rasnitsyn (1988), Whitfield (2003)
Pelecinidae	Endoparasitoid	Coleoptera: Scarabaeidae	Larvae of soil root feeders	Callovian–Recent	Kozlov (1974), Johnson and Musetti (1999), Zhang et al. (2002b), Shih et al. (2009)
Peleserphidae[†]	Endoparasitoid[i]	—	—	Cenomanian	Zhang et al. (2018b)
Peradeniidae	Endoparasitoid[i]	—	—	Lutetian–Recent	Naumann and Masner (1985), Johnson et al. (2001), Grimaldi and Engel 2005a, 2005b
Proctorenyxidae	Endoparasitoid[i]	—	—	Recent	Grimaldi and Engel (2005a)
Proctotrupidae	Endoparasitoid	Coleoptera: Staphylinidae, Carabidae, Elateridae	Young larvae of leaf litter	Kimmeridgian–Recent	Askew (1971), Rasnitsyn (1988), Whitfield (2003)
Roproniidae	Endoparasitoid	Hymenoptera: Tenthredinidae	Larvae of external feeders	Barremian–Recent	Zhang and Zhang (2001), Whitfield (1998), Garrouste et al. (2016)

(continued)

Table 11.1 (continued)

Taxon[a]	Parasitoid larval life-habit[b]	Dominant larval hosts[c]	Host stage of larval parasitoid attack[d]	Geochronologic range[e]	References[f]
Spathiopterygidae[†]	Endoparasitoid[i]	—	Insect eggs[i]	Cenomanian	Engel and Ortega-Blanco (2013), Engel et al. (2015), Zhang et al. (2018a)
Vanhorniidae	Endoparasitoid	Coleoptera: Eucnemidae	Larvae of wood borers	Santonian–Recent	Clausen (1940), Whitfield (2003)
Cynipoidea					
"Archaeocynipidae"[†]	Endoparasitoid[i]	—	Larvae[i]	Hauterivian	Rasnitsyn and Kovalev (1988), Grimaldi and Engel (2005a)
Austrocynipidae	Endoparasitoid to ectoparasitoid	Lepidoptera: Oecophoridae	Larvae of leaf miners	Recent	Whitfield (2003)
Charapidae	Endoparasitoid, hyperparasitoid	Hemiptera: Sternorrhyncha; Hymenoptera: Braconidae	Nymphs, larvae	Recent	Clausen (1940), Askew (1971), Godfray (1994)
Eucoilidae	Endoparasitoid	Diptera: Cyclorrhapha	Larvae	Turonian–Recent	Clausen (1940), Askew (1971), Peñalver et al. (2013), Buffington et al. (2014)
Figitidae	Endoparasitoid to ectoparasitoid, hyperparasitoid	Diptera: Sarcophagidae, Tephritidae; Neuroptera; Hymenoptera: Cynipidae	Larvae in rotting meat	Turonian–Recent	Clausen (1940), Whitfield (2003), Liu et al. (2007), Buffington et al. (2014)
Ibaliidae	Endoparasitoid to ectoparasitoid	Hymenoptera: Siricidae	Eggs and larvae of wood borers	Turonian–Recent	Clausen (1940), Whitfield (2003), Grimaldi and Engel (2005a), Liu and Engel (2010)

Taxon[a]	Parasitoid larval life-habit[b]	Dominant larval hosts[c]	Host stage of larval parasitoid attack[d]	Geochronologic range[e]	References[f]
Liopteridae	Endoparasitoid to ectoparasitoid	Coleoptera: Buprestidae, Cerambycidae, Curculionidae	Larvae of wood borers	Turonian–Recent	Whitfield (2003), Liu et al. (2007)
Protimaspidae[†]	Endoparasitoid[i]	—	Larvae of wood borers[i]	Turonian	Liu et al. (2007)
Stolamissidae[†]	Endoparasitoid[i]	—	Larvae of wood borers[i]	Turonian	Liu et al. (2007)
Platygastroidea					
Platygastridae	Endoparasitoid	Hemiptera: Sternorrhyncha; Diptera: Cecidomyiidae; Lepidoptera; Hymenoptera	Insect eggs, larvae of gallers, scale insects	Hauterivian–Recent	Clausen (1940), Whitfield (1998, 2003), Grimaldi and Engel (2005a)
Scelionidae	Endoparasitoid	Orthoptera; Mantodea; Hemiptera; Neuroptera; Coleoptera; Diptera: Tabanidae; Lepidoptera; Araneae	Insect eggs	Aptian–Recent	Brues (1940), Clausen (1940), Grimaldi and Engel (2005a), Nel and Azar (2005), Talamas et al. (2017), Haas et al. (2018)
Ceraphronoidea					
Ceraphronidae	Endoparasitoid, hyperparasitoid	Thysanoptera; Neuroptera; Trichoptera; Diptera: Cecidomyiidae, Cyclorrhapha; Hymenoptera: Sphecoidea	Eggs, larvae, pupae and puparia	Santonian–Recent	Alekseyev and Rasnitsyn (1981), Whitfield (1998, 2003)

(continued)

Table 11.1 (continued)

Taxon[a]	Parasitoid larval life-habit[b]	Dominant larval hosts[c]	Host stage of larval parasitoid attack[d]	Geochronologic range[e]	References[f]
Megaspilidae	Ectoparasitoid, hyperparasitoid	Hemiptera: Sternorrhyncha; Neuroptera: Hemerobiidae, Coniopterygidae; Mecoptera; Diptera: Cecidomyiidae, Syrphidae, Chloropidae, Chamaemyiidae, Muscidae	Nymphs and larvae of parasitoids and gallers	Cenomanian–Recent	Whitfield (1998, 2003), Grimaldi and Engel (2005a), Zhang et al. (2018a, b)
Radiophronidae[†]	Ectoparasitoid[i]	—	Larvae[i]	Albian–Campanian	Ortega-Blanco et al. (2010), Zhang et al. (2018a)
Stigmaphronidae[†]	Endoparasitoid[i]	—	Insect eggs and larvae[i]	Hauterivian–Campanian	Rasnitsyn (1991b), Whitfield (1998), Grimaldi and Engel (2005a), Ortega-Blanco et al. (2011b)
Mymarommatoidea					
Gallorommatidae	Endoparasitoid[i]	—	Insect eggs[i]	Barremian–Santonian	Gibson et al. (2007)
Mymarommatidae	Endoparasitoid[i]	—	Insect eggs	Cenomanian–Recent	Yoshimoto (1975), Huber (1986), Grimaldi and Engel (2005a), Engel and Grimaldi (2007a), Heraty and Darling (2009), Haas et al. (2018), Zhang et al. (2018c)

Taxon[a]	Parasitoid larval life-habit[b]	Dominant larval hosts[c]	Host stage of larval parasitoid attack[d]	Geochronologic range[e]	References[f]
Serphitoidea					
Serphitidae[†]	Endoparasitoid[i]	—	Insect eggs[i]	Hauterivian–Turonian	Brues (1937), Kozlov and Rasnitsyn (1979), Grimaldi and Engel (2005a), Engel et al. (2011), Haas et al. (2018)
Chalcidoidea[j]					
Agaonidae	Pollinator, inquiline, galler	Moraceae: *Ficus*	Fig synconia	Priabonian–Recent	Weiblen (2002), Compton et al. (2010), Farache et al. (2016)
Aphelinidae	Ectoparasitoid, endoparasitoid, hyperparasitoid	Hemiptera: Sternorrhyncha; Diptera: Cecidomyiidae; Lepidoptera; Hymenoptera: Dryinidae	Nymphs, larvae, eggs	Lutetian–Recent	Clausen (1940), Burks et al. (2015), Peters et al. (2017), Haas et al. (2018)
Chalcididae	Ectoparasitoid, endoparasitoid, seed predator, herbivore	Coleoptera; Diptera: Tachinidae; Neuroptera: Myrmeleontidae; Lepidoptera; Hymenoptera: Apoidea; angiosperms	Larvae, pupae, seeds	Ypresian–Recent	Clausen (1940), Askew (1971), Whitfield (2003), Peters et al. (2017), Quevillon and Hughes (2018)
Diversinitidae[†]	Endoparasitoid[i]	—	Eggs[i]	Cenomanian	Haas et al. (2018)
Encyrtidae	Endoparasitoid, hyperparasitoid predator	Hemiptera: Sternorrhyncha; Coleoptera: Coccinellidae; Lepidoptera; Hymenoptera: Chalcidoidea; Acari	Nymphs, larvae, eggs	Lutetian–Recent	Clausen (1940), Askew (1971), Simutnik (2002), Simutnik and Perkovsky (2006), Heraty and Darling (2009), Peters et al. (2017)

(continued)

Table 11.1 (continued)

Taxon[a]	Parasitoid larval life-habit[b]	Dominant larval hosts[c]	Host stage of larval parasitoid attack[d]	Geochronologic range[e]	References[f]
Eucharitidae	Ectoparasitoid	Hemiptera: Sternorrhyncha; Hymenoptera: Formicidae	Larvae, prepupae and pupae of ant broods	Lutetian–Recent	Clausen (1940), Askew (1971), Heraty and Darling (2009), Quevillon and Hughes (2018)
Eulophidae	Ectoparasitoid, endoparasitoid, hyperparasitoid, predator	Orthoptera: Saltatoria; Hemiptera: Sternorrhyncha; Coleoptera: Chrysomelidae; Lepidoptera; Hymenoptera: Cynipidae; Diptera; Araneae	Larvae, nymphs, pupae and eggs of concealed insects: leaf miners, gallers and case bearers	Lutetian–Recent	Clausen (1940), Askew (1971), Hong (2002), Haas et al. (2018)
Eupelmidae	Endoparasitoid, ectoparasitoid, hyperparasitoid, predator	Orthoptera; Hemiptera; Coleoptera: Chrysomelidae; Neuroptera; Hymenoptera: Chalcidoidea, Cynipoidea	Eggs, larvae of leaf miners, fly puparia, and other insects; spider egg sacs	Lutetian–Recent	Clausen (1940), Thorpe (1941), Trjapitsyn (1963), Askew (1971), Whitfield (2003), Gibson (2009)
Eurytomidae	Ectoparasitoid, endoparasitoid, seed predator, herbivore	Orthoptera: Grylloidea; Coleoptera; Lepidoptera; Diptera: Tephritidae; Hymenoptera: Formicidae, Cynipoidea	Eggs, larvae in galls, stems and nests; ants	Ypresian–Recent	Clausen (1940), Askew (1971), Hong (2002), Heraty and Darling (2009), Quevillon and Hughes (2018)
Leucospidae	Ectoparasitoid	Hymenoptera: Apoidea	Larvae	Aquitanian–Recent	Askew (1971), Engel (2002)

Taxon[a]	Parasitoid larval life-habit[b]	Dominant larval hosts[c]	Host stage of larval parasitoid attack[d]	Geochronologic range[e]	References[f]
Mymaridae	Endoparasitoid	Orthoptera; Coleoptera: Hydrophilidae, Dytiscidae	Eggs	Cenomanian–Recent	Kaddumi (2005), Huber (1986, 2005), Heraty and Darling (2009), Heraty et al. (2013), Engel et al. (2013b), Zhang et al. (2018c)
Ormyidae	Ectoparasitoid	Gall-forming insects	Larvae of gallers		Hanson (1992), Peters et al. (2017)
Perilampidae	Hyperparasitoid, ectoparasitoid, galler	Neuroptera: Chrysopidae; Diptera: Tachinidae; Coleoptera: Anobiidae, Curculionidae; Hymenoptera: Diprionidae, Ichneumonidae, Cynipidae, Halictidae	Larvae, pupae of wood borers, gallers and parasitoids	Ypresian–Recent	Clausen (1940), Askew (1971), Peñalver and Engel (2006), Heraty and Darling (2009)
Pteromalidae	Endoparasitoid, ectoparasitoid, hyperparasitoid	Hemiptera: Sternorrhyncha; Coleoptera: Curculionidae, (including Scolytinae), Anobiidae; Hymenoptera: Xiphydriidae, Cynipidae, Braconidae; Lepidoptera; Diptera: Pupipara; Siphonaptera	Larvae, pupae and adults of wood borers, gallers, seed predators and other parasitoids	Aptian–Recent	Askew (1971), Godfray (1994), Whitfield (2003), Barling et al. (2013), Krogman (2013)
Rotoitidae	Endoparasitoid[i]	—	—	Campanian–Recent	Bouček and Noyes (1987), Gumovsky et al. (2018)
Signiphoridae	Ectoparasitoid, hyperparasitoid	Hemiptera: Sternorrhyncha	Nymphs, “pupae”	Lutetian–Recent	Clausen (1940), Perkovsky et al. (2010)

(continued)

Table 11.1 (continued)

Taxon[a]	Parasitoid larval life-habit[b]	Dominant larval hosts[c]	Host stage of larval parasitoid attack[d]	Geochronologic range[e]	References[f]
Tanaostigmatidae	Galler, parasitoid	Various plant lineages	Larvae of gallers	Lutetian–Recent	Grimaldi and Engel (2005a), Weitschat and Wichard (2010)
Tetracampidae	Parasitoid, herbivore	Coleoptera: Chrysomelidae; Hymenoptera: Diprionidae; Diptera: Agromyzidae	Eggs, larvae	Campanian–Recent	Yoshimoto (1975), Gumovsky and Perkovsky (2005)
Torymidae	Ectoparasitoid, herbivore	Hymenoptera: Cynipidae	Larvae of gallers	?Campanian–Recent	Askew (1971), Grissell (1980), Heraty and Darling (2009), McKellar and Engel (2012)
Trichogrammatidae	Endoparasitoid	Hemiptera; Megaloptera; Neuroptera; Coleoptera; Diptera; Lepidoptera; Hymenoptera	Eggs	Lutetian–Recent	Clausen (1940), Schmidt et al. (2010), Burks et al. (2015), Coty et al. (2016), Peters et al. (2017), Haas et al. (2018)
Ichneumonoidea					
Braconidae	Ectoparasitoid, some endoparasitoids	Psocoptera; Hemiptera: Aphididae; Coleoptera; Neuroptera; Hymenoptera; Orthoptera	Eggs, larvae and adults of wood borers	Hauterivian–Recent	Rasnitsyn (1983), Rasnitsyn and Sharkey (1988), Whitfield (2003), Shaw (2004), Ortega-Blanco et al. (2011c), Perrichot et al. (2008)

Taxon[a]	Parasitoid larval life-habit[b]	Dominant larval hosts[c]	Host stage of larval parasitoid attack[d]	Geochronologic range[e]	References[f]
Ichneumonidae	Ectoparasitoid, some endoparasitoids	Coleoptera; Neuroptera; Diptera; Trichoptera; Lepidoptera; Hymenoptera: Xiphydriidae; Araneae	Aphid nymphs, larvae and pupae of wood borers; ants	Berriasian–Recent	Brues (1937), Rasnitsyn (1983), Whitfield (2003), Perrichot et al. (2008), Li et al. (2017a), Spasojevic et al. (2017), Quevillon and Hughes (2018)
Aculeata					
Chrysidoidea					
Bethylidae	Ectoparasitoid	Coleoptera: Curculionidae, Cerambycidae; Lepidoptera	Larvae	Cenomanian–Recent	Clausen (1940), Evans (1969), Grimaldi and Engel (2005a), Perrichot and Nel (2008), Zhang et al. (2018a)
"Bethylonymidae"[†]	Ectoparasitoid[i]	—	Larvae[i]	Tithonian–Hauterivian	Rasnitsyn (1975, 1988), Grimaldi and Engel (2005a)
Chrysididae	Ectoparasitoid, endoparasitoid, cleptoparasitoid	Phasmatodea; Hymenoptera: Tenthredinidae, Diprionidae, Vespidae, Apoidea; Lepidoptera	Eggs, larvae, pupae	Cenomanian–Recent	Cockerell (1917), Clausen (1940), Kimsey and Bohart (1990), Whitfield (1998), Grimaldi and Engel (2005a), Zhang et al. (2018a)

(continued)

Table 11.1 (continued)

Taxon[a]	Parasitoid larval life-habit[b]	Dominant larval hosts[c]	Host stage of larval parasitoid attack[d]	Geochronologic range[e]	References[f]
Dryinidae	Ectoparasitoid→endoparasitoid	Hemiptera: Auchenorrhyncha	Nymphs, adults	Hauterivian–Recent	Jervis (1980), Whitfield (1998), Engel (2003), Grimaldi and Engel (2005a)
Embolemidae	Ectoparasitoid→endoparasitoid	Phasmatodea; Hemiptera: Fulgoroidea; Hymenoptera: Formicidae	Nymphs, larvae	Hauterivian–Recent	Rasnitsyn (1996), Whitfield (1998), Olmi (1998), Grimaldi and Engel (2005a), Ortega-Blanco et al. (2011a)
Falsiformicidae†	Ectoparasitoid[i]	—	Larvae[i]	Hauterivian–Cenomanian	Rasnitsyn (1975, 2002), Perrichot et al. (2014), Zhang et al. (2018a)
Plumariidae	Ectoparasitoid[i]	—	—	?Turonian–Recent	Roig-Alsina (1994), Grimaldi and Engel (2005a)
Sclerogibbidae	Ectoparasitoid	Embioptera	Nymphs, adults	Hauterivian–Recent	Whitfield (1998), Penteado-Dias and van Achterberg (2002), Engel and Grimaldi (2006)
Scolebythidae	Ectoparasitoid	Coleoptera: Cerambycidae, Anobiidae	Larvae of wood borers	Hauterivian–Recent	Prentice et al. (1996), Melo (2000), Engel and Grimaldi (2007b), Engel et al. (2013c)

Taxon[a]	Parasitoid larval life-habit[b]	Dominant larval hosts[c]	Host stage of larval parasitoid attack[d]	Geochronologic range[e]	References[f]
Vespoidea					
Bradynobaenidae	Ectoparasitoid	Arthropods; Solifugida	Larvae and other immatures	Recent	Brothers (1995), Grimaldi and Engel (2005a)
Mutillidae	Ectoparasitoid, predator	Blattodea; Coleoptera; Lepidoptera: Limacodidae; Diptera; Hymenoptera: Formicidae, Apoidea	Nymphs, larvae	Lutetian–Recent	Clausen (1940), Lelej (1996), O'Neill (2001), Quevillon and Hughes (2018)
Pompilidae	Ectoparasitoid, predator	Araneae	Adults	Cenomanian–Recent	O'Neill (2001), Grimaldi et al. (2002)
Rhopalosomatidae	Ectoparasitoid	Orthoptera: Grylloidea	Nymphs?, adults	Cenomanian–Recent	Gurney (1953), Townes (1977), Darling and Sharkey (1990), Lohrmann and Engel (2017), Zhang et al. (2018a)
Sapygidae	Ectoparasitoid, cleptoparasitoid, predator	Apoidea: Eumenidae, Megachilidae, Apidae	Eggs and larvae of social insects	Cenomanian–Recent	Torchio (1972), Spahr (1987), Grimaldi and Engel (2005a), Bennett and Engel (2005)
Scoliidae	Ectoparasitoid, predator	Coleoptera: Scarabaeidae; Hymenoptera: Sphecidae	Larvae	Barrremian–Recent	Clausen (1940), Lai (1988), Rasnitsyn and Martínez-Delclòs (1999, 2000), Zhang et al. (2002c)

(continued)

Table 11.1 (continued)

Taxon[a]	Parasitoid larval life-habit[b]	Dominant larval hosts[c]	Host stage of larval parasitoid attack[d]	Geochronologic range[e]	References[f]
Sierolomorphidae	Ectoparasitoid[i], predator[i]	—	—	Turonian–Recent	Evans (1961), Rasnitsyn (2000), Grimaldi and Engel (2005a)
Tiphiidae	Ectoparasitoid, predator	Orthoptera: Stenopelmatidae; Coleoptera: Carabidae, Cincidellidae, Scarabaeidae, perhaps Curculionidae	Nymphs, last instar larvae	Aptian–Recent	Clausen (1940), Knisley et al. (1989), Darling and Sharkey (1990), Grimaldi and Engel (2005a)
Apoidea					
Ampulicidae	Ectoparasitoid to endoparasitoid, predator	Blattodea	Nymphs, adults	?Albian–Recent	Clausen (1940), Bohart and Menke (1976), Brandão et al. (1989), Antropov (2000), Rasnitsyn (2002), Engel and Grimaldi 2005a
Crabronidae	Ectoparasitoid to endoparasitoid, predator	Ephemeroptera; Hemiptera: Sternorrhyncha; Coleoptera; Diptera: Syrphidae; Lepidoptera; Hymenoptera; Acari	Larvae, adults, nymphs	Cenomanian–Recent	Clausen (1940), Bohart and Menke (1976), Antropov (2000)
Heterogynaidae	Ectoparasitoid to endoparasitoid, predator[i]	—	—	Recent	Bohart and Menke (1976), Grimaldi and Engel (2005a)
Sphecidae	Ectoparasitoid to endoparasitoid predator	Orthoptera: Acrididae, Gryllidae, Stenopelmatidae; Blattodea; Lepidoptera: Noctuidae	Nymphs, adults, larvae	Cenomanian–Recent	Clausen (1940), Bohart and Menke (1976), Darling and Sharkey (1990), Antropov (2000)

[a]Subclades lacking parasitoid behaviors are not included. A dagger (†) indicates an extinct lineage

[b]Only those family-level taxa with parasitoid or related life habits are listed under their respective, encompassing clades. Obligate or facultative parasitic groups that do not kill their hosts are excluded from this list. Parasitoid types—ectoparasitoid, endoparasitoid, hyperparasitoid, cleptoparasitoid—are listed in approximate order of abundance

[c]Major host clades are listed, occasionally with family-level examples to flesh out taxonomic details. This is not an exhaustive list and new parasitoid clades are discovered each year. The extended dash (—) indicates that the taxonomic identities of the hosts remain unknown or would have been based on inadequate data

[d]The major developmental stage of hosts undergoing parasitoid attack is provided. The North American rather European system of naming larval stages is used. A larva designates a developmental stage intercalated between the egg and pupal stages in holometabolous insects. Accordingly, a naiad is the developmental stage of a nonholometabolous insect in the aquatic realm whereas a nymph is the analogous developmental stage in the terrestrial realm. The extended dash (—) indicates that the life habits of the hosts remain unknown or would have been based on inadequate data

[e]The geologic time scale of Walker et al. (2013) is used, one that represents advances in several fields that have contributed to greater precision of age estimates. Range-through data is assumed. Geochronologic ranges are for the broader clade encompassing the parasitoid taxa, and not for individual parasitoid lineages

[f]References include primary sources of original studies as well as reviews of original sources. The sources are not meant to be complete, but rather indicative of the more important life-habit studies and fossil occurrences

[g]All strepsipteran lineages are considered parasitoids, rather than their attribution to parasites mentioned by Eggleton and Belshaw (1992). The parasitoid life habit of Strepsiptera is based on Kathirithamby's (2009) criterion that after strepsipteran hosts are parasitized, they "…develop and obtain nutrients from a single host, thereby castrating and eventually killing it" (page 234)

[h]Major taxonomic subdivisions of the Strepsiptera are after Engel et al. (2016a)

[i]Attribution to a particular parasitoid larval life habit for these fossil and poorly known groups are inferred from one or more of the following four criteria. First is the phylogenetic position of the parasitoid lineage within a broader phylogeny. Second are the known biologies of closely related lineages. Third is the availability of particular host lineages for parasitoidism within the local, contemporaneous biota. Fourth are the limitations of small size for host choice

[j]A probable paraphyletic group

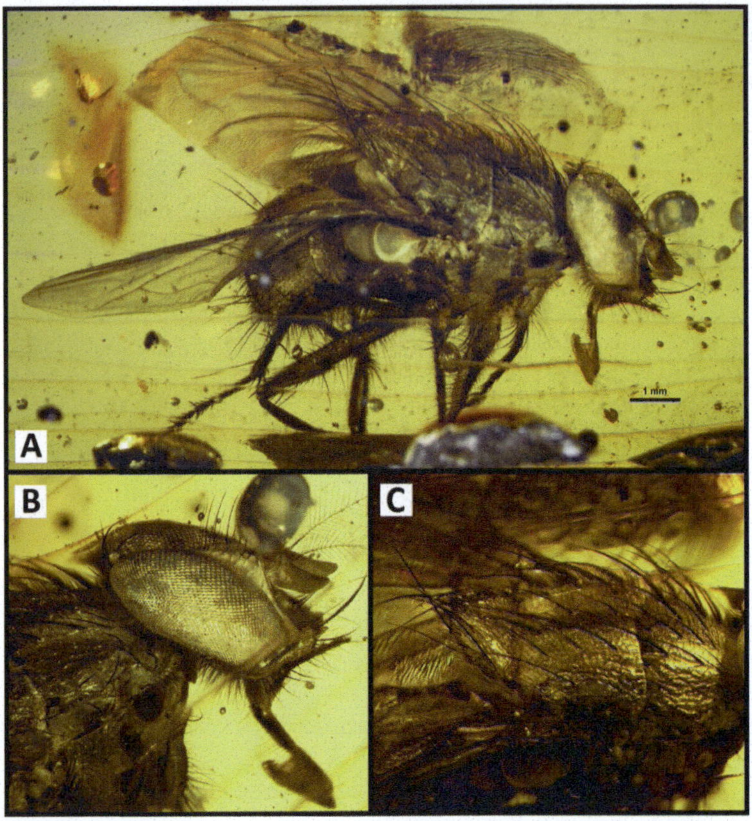

Fig. 11.10 Holotype of the adult bot fly *Mesembrinella caenozoica* (Diptera: Mesembrinellidae), from early Miocene amber of the Dominican Republic. (**a**) Habitus in dorsolateral view. (**b**) Head and the anterior thorax in right dorsolateral view. (**c**) Thorax in right dorsolateral view. (Reproduced with permission from Fig. 1 in Cerretti et al. 2017)

within Mecoptera (Vršanský et al. 2010). Current evidence suggests, however, a nonparasitic existence. Several features, especially genitalia, are structurally linked to nematocerous Diptera (Huang et al. 2013b), particularly Nymphomyiidae (nymphomyiid crane flies), which presently are a specialized, relict group of neotenic amphibious insects with adults lacking functional mouthparts.

11.5.1.9 Lepidoptera (Moths)

In Southeastern Asia a clade of erebid moths recently have developed a unique modification of the lepidopteran siphon that is adapted to piercing the often-thick integument of large mammals for the imbibition of blood (Bänziger 1968, 1971). The clade consists of ten species of *Calyptra*, formerly assigned to Noctuidae (owlet moths) that now is placed in the Erebidae (vampire moths), and affiliated with the

Tribe Calpini of the Subfamily Calpinae (Zaspel et al. 2012). The clade of *Calyptra* consist of ten species of hematophagous vampire moths in which evidently only males imbibe blood from large mammals that include cattle, zebu, tapirs and occasionally elephants and humans, centered in Southeast Asia but with a geographic range extending to India and southeastern Russia (Bänziger 1971; Zaspel et al. 2007). Related species and genera within Calpini frequent the more delicate membranes of the eyes, rectum and genitalia for feeding on tears, sebum, sweat and other dermal secretions (Bänziger 1968, 1996). A more distantly related clade of fruit piercing and surface scarifying moths are known from the tropics of Africa and South America (Zenker et al. 2010). The mouthpart siphons of these fruit feeding, and blood- and lymph feeding species, are rigid, lack internal stylets, and contain external, spine-like structures at the proboscis tip that are hydraulically stiffened for penetration into skin. The siphonal stylet is powered by a cibarial food pump for ingestion of fluid host tissue (Bänziger 1996; Labandeira 2019). The clade that bears these distinctive mouthparts lacks a fossil record.

11.5.1.10 Hymenoptera (Wasps)

Social parasitism frequently is found in Hymenoptera and is particularly prevalent among bees. A recurring feature of social parasitism is embodied in Emery's Rule, which is narrowly defined as social parasites are the closest relatives of their hosts, frequently at the sister-species level (Wilson 1971). A broader version of this rule states that a clade or related group of social parasites is solely related to their hosts (Smith et al. 2007). An illustration of this phenomenon occurs in allodapine carpenter bees (Apidae: Xylocopinae) (Smith et al. 2007), which exhibit a looser version of relationships between a host clade and its sister social parasite clade (Fig. 11.11), providing evidence for sympatric speciation. As seen in the lack of rigorous correlation of sympatry in Fig. 11.11, there may be inherent flaws in the notion of Emery's Rule, at least in some instances. Confounding processes such as extinction or speciation within one or both of the clades, host-switching, and incomplete sampling (Smith et al. 2007) could derail the observed patterns. Notably, older host–social parasite relationships tend to support sympatric speciation whereas younger such relationships do not. This discontinuity likely is because extinctions may result in a pruning of the older clades such that they appear retrospectively to have originated sympatrically (Smith et al. 2007).

A second type of common parasitic relationship among bees is cleptoparasitism. Cleptoparasitism, the theft of resources by one animal from another animal, is rarely found in the fossil record. However, based on a distinctive forewing shape and other morphometrically-gleaned features, it was possible to assign a cleptoparasite identity to a wing of a bee from the middle Paleocene, at the 60.5 million-year-old Menat locality in central France (Dehon et al. 2017). The presence of cleptoparasite taxa is significant, as it establishes that before the midpoint of the geochronologic history of bees, cleptoparasites were present.

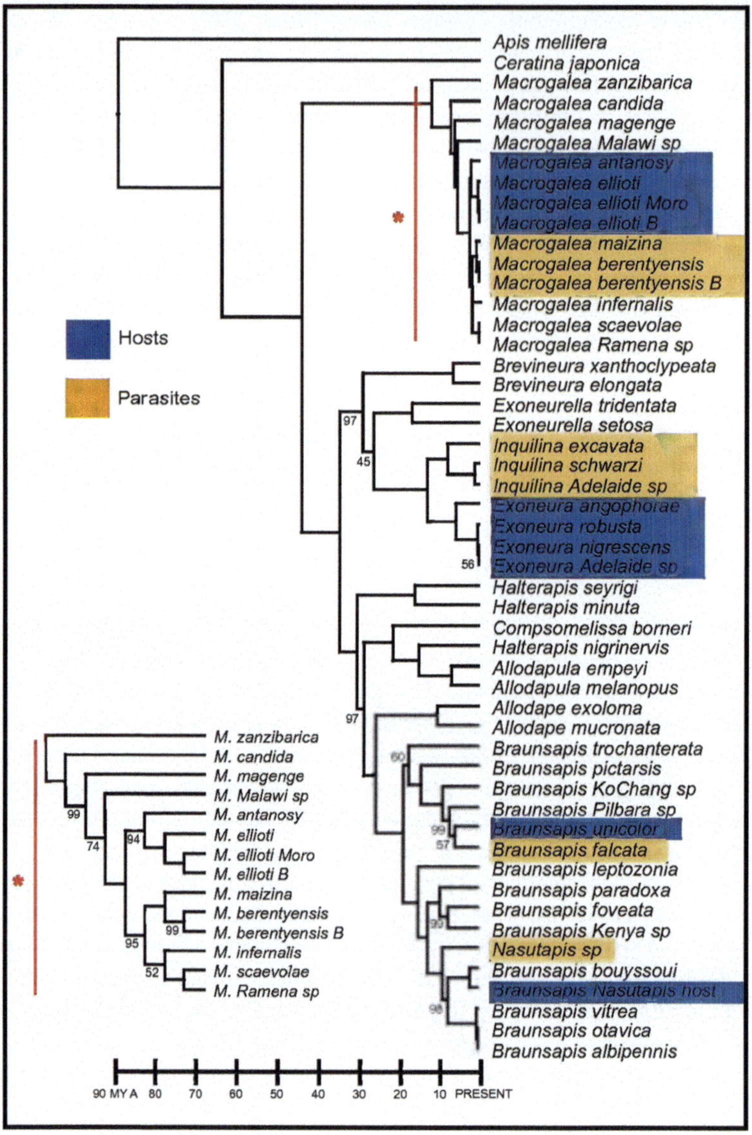

Fig. 11.11 Chronogram of allodapine bee species based on a penalized likelihood transformed Bayesian consensus phylogram (Smith et al. 2007). The insert shows an enlarged view of the chronogram for the *Macrogalea* species, whose divergences are very recent. The divergence ages are indicated by the geologic time scale at bottom, and posterior probability support values less than 100 are shown. These values ae indicated on the enlarged inset of the chronogram for *Macrogalea*. Host species are indicated within a blue rectangle whereas corresponding parasite species are shown in an adjacent orange rectangle. (Reproduced with permission from Fig. 2 of Smith et al. 2007)

11.5.2 Parasitoid Taxa

The parasitoid life style and feeding mode is confined to seven insect orders, all of which have holometabolous development (Table 11.1). Parasitoids occur in Coleoptera (beetles), Strepsiptera (twisted-wing parasites), Diptera (flies) and Hymenoptera (wasps), and with small numbers in Neuroptera (mantidflies), Trichoptera (caddisflies) and Lepidoptera (moths). There are 146 extant and extinct families of insect parasitoids, in decreasing rank order of familial diversity: Hymenoptera, Diptera, Strepsiptera, Coleoptera, Lepidoptera, Trichoptera and Neuroptera (Tables 11.1 and 11.2). Of these families, 116 (79.5%) are extant and 30 (20.5%) are extinct (Table 11.2).

Morris (1998) cites five criteria that characterize insect parasitoids. However, Morris also states that occasional exceptions occur. The general criteria for an arthropod to be deemed a parasitoid are the following:

1. The larval stage assumes the parasitoid strategy.
2. The adult stage is free living and serves as the agent of dispersal.
3. Host death is obligatory and often is coordinated with the emergence of the adult parasitoid.
4. Host pathology, including death, is generally not dependent on the intensity of parasitoid activity.
5. Only one host is killed in an interaction with a parasitoid.

Criteria 1 and 2 apply to all insects and mites, although some exceptions occur in mites. For criterion 3, some insects fail to kill their hosts, such as some anthomyiid and tachinid fly species, in which their hosts survive a bout with parasitoids and subsequently complete their development (Clausen 1940; English-Loeb et al. 1990). Because of the rarity of these examples, such species are included as viable parasitoids. Criterion 4 is a way of expressing the relationship that parasitoids, such as

Table 11.2 The fossil record of insect parasitoid families

Order	Number of families	Percentage representation
Breakdown by order		
Neuroptera	1	0.7
Coleoptera	10	6.8
Strepsiptera	17	11.6
Hymenoptera	92	63.0
Diptera	23	15.8
Trichoptera	1	0.7
Lepidoptera	2	1.4
Totals	***146***	***100.0%***
Breakdown by extant vs extinct		
Extinct	30	20.5
Extant	116	79.5
Totals	***146***	***100.0%***

braconid larvae on a caterpillar, at maturity are often within an order of magnitude the size of their host, whereas parasites (e.g. a mosquito on a human) are typically multiple orders of magnitude smaller than their host (Morris 1998). These relationships admit of rare exceptions where the ability of a parasitoid to kill its host is dependent on an elevated number of conspecific parasitoids on the same host (Blumberg and Luck 1990). Nevertheless, such density dependency is rare and these arrangements would remain as parasitoid interactions. The close relationship between predation and parasitism is illustrated in criterion 5, whereby if more than one host succumbs to the same parasitoid individual, then the relationship becomes predation (Morris 1998). A related but frequently unstated factor is the size of the host relative to the size of the parasitoid. If host size is large to very large, such as a vertebrate, compared to the size of its insect parasitoid, then successful development of the attacker is not contingent upon death of the host. Such a result would invalidate an essential, obligate feature of the host–parasitoid relationship.

11.5.2.1 Neuroptera (Mantidflies)

Although almost all neuropteran larvae are predators on arthropods, Mantispidae (mantidflies) are the sole family that is a parasitoid of arthropods (Table 11.1) (Morris 1998). One subfamily, the diverse Mantispinae, have larvae that are egg predators of spiders. By contrast, Subfamily Symphrasinae, consisting of about 60 species, are ectoparasitoids of scarab beetle larvae, the pupal cocoons of owlet moths (Noctuidae), and larvae and pupae of aculeate Hymenoptera, particularly vespoid and sphecoid wasps and solitary bees (Parfin 1958; Eggleton and Belshaw 1992). Host-seeking parasitoid larvae of Symphrasinae find their host larvae in the soil or in subterranean cells. Similar to many Coleoptera, larval Symphrasinae undergo hypermetamorphosis with a mobile triungulin stage for detection of an appropriate host, and represent a behavior that originated once (Eggleton and Belshaw 1993). (A triungulin larva is the first larval instar of hypermetamorphic insects, such as blister beetles, which is mobile, active, sclerified and host seeking that becomes legless, grub-like and parasitoid in subsequent instars.)

The Mantispinae and Symphrasinae have earliest occurrences in 99 million-year-old mid-Cretaceous Myanmar Amber (Haug et al. 2018; Shi, pers. comm. 2018); earlier occurrences of Mantispidae extending to the Jurassic lack subfamily placement (Jepson 2015). In Neuroptera, parasitoidism arose once in Symphrasinae, likely from a life habit of soil predation (Haug et al. 2018). The parasitoid life habit evolved into egg predation on spider egg sacs associated with a host seeking, first instar larva that is phoretic on the adult female spider (Gilbert and Rayor 1983; Eggleton and Belshaw 1992). This larval life habit has been documented for an undescribed mantispid, attributed to Mantispinae, for a first instar phoretic larva that is an egg parasitoid on a female disc-web spider (Oecobiidae) in Myanmar Amber (Haug et al. 2018). The other modern subfamily, Symphrasinae, has a fossil record (Jepson 2015) and is known from the same deposit (C. Shi, pers. comm. 2018), although nothing is known of its Cretaceous larval life habits.

11.5.2.2 Coleoptera (Beetles)

The parasitoid habit in Coleoptera is expressed in ten families that occur sporadically across the order, variously as ectoparasitoids, endoparasitoids and cleptoparasitoids that extend to the Middle Jurassic and conceivably as early as the Late Triassic (Table 11.1). For Coleoptera, host-searching behavior occurs typically in first-instar larvae.

Carabidae (ground beetles) consist of approximately 1500 genera and 40,000 species that are allocated into approximately 80 tribes (Arndt et al. 2005). Of these tribes, 7 have parasitoid life histories, including termite symbionts, ant symbionts and ectoparasitoids on various invertebrates (Erwin 1979). When other suspected tribes that house parasitoids are included (Erwin 1979), about a fourth of all carabid tribes have one or more parasitoid species. In general, two major ectoparasitoid groups of carabids are present. The first group attacks hosts in soils that include beetle pupae and millipedes, whereas a second group parasitoidizes leaf beetles on above-ground organs of plants (Erwin 1979). The earliest fossil Carabidae is Late Triassic (Grimaldi and Engel 2005a, b), occurring approximately 230 million years ago, although the parasitoid taxa are undoubtedly more recent than that, given their phylogenetic placement (Maddison et al. 1999).

Staphylinidae (rove beetles) include the subfamily Aleocharinae that consist of many ectoparasitoid genera on cyclorrhaphan, typically Pupipara, flies (Maus et al. 1998). Staphylinidae constitute about 48,000 species of which about one quarter are in the Subfamily Aleocharinae (Thayer 2005). Of the Aleocharinae, only the genus *Aleochara* are parasitoids, consisting of 400 described species and constituting about one-third of the species-level diversity of the subfamily (Maus et al. 1998) and about 0.08% of Staphylinidae. Typical microhabitats for parasitoid aleocharines are moist, fleeting habitats such as carrion, dung, polypore fungi, kelp-laden beach wrack, and decaying plant material in the nests of vertebrates and ants (Eggleton and Belshaw 1992). Aleocharine parasitoid eggs, upon hatching, produce a mobile triungulin stage analogous to the planidium of most dipteran parasitoids (Lawrence et al. 1991; Arnett Jr and Thomas 2001). The triungulin first-instar larva actively searches for potential hosts, attacks the appropriate host and either becomes attached or enters the host, initiating the process of parasitoidization, seen in a variety of hypermetamorphic parasitoids (Falin 2002; Engel 2005a; Haug et al. 2018). Subsequent instars are typically sessile, grub-like and continue the process of parasitoidization to its consummation through starvation of social insect colonies (Askew 1971; Godfray 1994). The earliest fossil Staphylinidae are Late Triassic in age, occurring approximately 230 million years ago and, like Carabidae, are significantly older than their descendant parasitoid taxa that originate during the Early Cretaceous (Chatzimanolis et al. 2012).

Scarabaeidae constitute approximately 27,000 species, of which approximately 5% or less in our estimation are cleptoparasitoids associated with deprivation of food and other resources from social insects (Scholtz and Grebennikov 2005). Five subfamilies of Scarabaeidae contain cleptoparasitoid taxa. Aphodiinae adults have specialized associations, evidenced by distinctive morphological adaptations, with termites and ants (Tangelder and Krikken 1982; Howden and Storey 1992; Stebnicka 1999). Scarabaeinae larvae, some known as "kleptocoprids", breed in dung that consume food resources provisioned by other, larger dung beetles (Halffter and Edmonds 1982; Halffter and Matthews 1999). Dynastinae include larvae that feed on brood and stored honey reserves in beehives (Glaser 1976; Evans and Nel 1989). Cetoniinae consist of Cremastocheilini adults that abscond the food stores in termites, ant and bee nests (Deloya 1988; Alpert 1994). Valginae adults and larvae are associated with termite mounds (Krikken 1978). Some Scarabaeidae (scarab beetles) are cleptoparasitoids that consume provisions of other scarab and closely related geotrupid beetles (Hammond 1976; Halffter and Matthews 1999; Scholtz and Grebennikov 2005). In the most common cleptoparasitoid group, the adult lays eggs into dung balls or analogous brood structures that hatch, followed by the first-instar larvae immediately burrowing and eventually consuming the host egg and all provisions in a dung ball or analogous brood structure (Halffter and Edmonds 1982; Paulian 1988). Some taxa of the scarabaeid subfamilies Valginae and Cetoniinae (Cremastocheilini) that are associates of social insect nests probably are cleptoparasitoids (Krikken 1978; Alpert 1994; Scholtz and Chown 1995). The earliest documented Scarabaeidae is latest Middle Jurassic in age (Bai et al. 2012), although parasitoid forms more likely are no older than mid Cretaceous in age based on their position within Scarabaeoidea phylogeny (Krell 2006).

Rhipiceridae (cedar beetles, cicada parasite beetles), constitute 7 genera and 100 species that are globally distributed, all of which are ectoparasitoids on cicadas (Lawrence 2005). Female rhipicerids such as *Sandalus* oviposit eggs in the cracks and interstices of elm tree bark, where cicadas also have oviposited (Elzinga 1977). Later, triungulin-like larvae and nymphs are flushed by rain and redeposited on the ground, where, after some time, a rhipicerid late-instar ectoparasitoid larva becomes attached to the cicada nymph host (Evans and Steury 2012). In eastern North America, there appears to be annual tracking by the rhipicerid parasite of its cicada host based on fluctuating population levels. The earliest occurrence of rhipicerid beetles in the fossil record is considerably recent, the Lutetian Stage of the middle Eocene (Ponomarenko 1995).

Cleridae (checkered beetles) consist of approximately 3400 species, are mostly predaceous (Kolibác 2010), but have two major parasitoid strategies. Their first is a role as cleptoparasitoids of apid bees, depriving larvae of nest provisions in soil cells and decaying plant material. The first instar larva is host searching or phoretic and feeds only a short time on the host's provisions (Eggleton and Belshaw 1992). Clerids have a second role as parasitoids of lepidopteran caterpillars occurring in galls that occur on plant shoots, although the host locating process is poorly understood (Eggleton and Belshaw 1992). The origin of the parasitoid life habit in Cleridae likely was derived from predation of insects in dead wood habitats. By contrast, the parasitoid habit evolved into predation on bees in soil and plant litter habitats and possibly on grasshopper egg pod predation. The oldest fossils of Cleridae are from the latest Middle Jurassic of China (Kolibáč and Huang 2016).

Bothrideridae (dry bark beetles) consists of 38 genera, 400 species and 4 subfamilies, where one subfamily, Bothriderinae, has ectoparasitoid life habits (Philips and Ivie 2002). Bothrideridae are highly elongate beetles whose adults occur under tree bark or in composted soil. Their nonparasitic larvae feed on fungi whereas their hypermetamorphic, host-seeking, first-instar larvae is a spinose triungulin that attacks and becomes an ectoparasitoid on wood-boring larvae of metallic wood-boring beetles, ambrosia beetles, bark beetles, deathwatch beetles, augur beetles, longhorn beetles, and wood wasps (Horion 1961; Ślipiński et al. 2010). Subsequent larval instars are grub-like and complete an ectoparasitoid existence on their host (Philips and Ivie 2002; Lawrence and Ślipiński 2013). The ectoparasite habit likely originated from mycophagy in an ancestral bothriderid lineage (Crowson 1981), and subsequently evolved into ectoparasitoidism. Ectoparasitoidism currently is confined to the diverse Subfamily Bothriderinae, where adults of several genera bear lightly sclerotized, swollen abdomens probably involved in very fecund levels of egg production. The oldest Bothrideridae are from the middle Eocene (Ponomarenko 1995).

Passandridae (flat bark beetles) consist of 9 genera and 109 species that are ectoparasitoids of the larvae or pupae of wood boring insects such as weevils, longhorned beetles, ambrosia beetles, bark beetles or wood wasps that live in dead wood or under bark (Crowson 1981; Thomas 2002; Burkhardt and Ślipiński 2010). Passandrids also assume a second major parasitoid strategy as hyperparasitoids on braconid wasps (Burkhardt and Ślipiński 2010). The larval body is highly modified for ectoparasitoid habits and display changes in larval morphology from instar to instar (Burkhardt and Ślipiński 2003), similar to hypermetamorphic larvae. Passandrid larval morphology externally is variously sclerotized, flattened and spiny, with an enlarged abdomen and reduced mouthparts. Evidently, parasitoid larvae of Passandridae display limited host specificity. Adults are considerably flat dorsoventrally and occupy subcortical bark habitats. The oldest fossils of Passandridae evidently are Cenomanian in age (Ponomarenko 1995).

Meloidae (blister beetles) are mostly parasitoids of grasshoppers and non-domesticated bees (Lawrence et al. 1991; Arnett Jr et al. 2002). Meloid first-instar triungulin larvae frequently are encountered on flowers, are acquired by foraging male bees, and then are transported to female bees during mating (Lawrence et al. 1991; Arnett Jr et al. 2002). At the bee nest site, the triungulin larvae invade bee cells where eggs reside and successively consume either the egg or the developing larva. Meloid larvae undergo hypermetamorphosis and initially are ambulatory, host-seeking triungulins, followed by sessile, grub-like larval instars that emerge from their host insect nest as adult blister beetles.

About 120 genera and 2500 species of Meloidae have larvae that are predaceous or more frequently parasitoids (Bologna et al. 2010). Larval hypermetamorphosis is common and involves host-searching behavior, and after attack of the host, the larva transforms into a fat, fleshy grub that in turn results in an adult beetle (Bologna et al. 2010). The fossil history of Meloidae extends to the earliest Cenomanian at 99 million years ago, in Myanmar Amber (Poinar Jr and Brown 2014), and consists of a distinctive, miniscule, triungulin larva attached to a jumping ground bug (Hemiptera: Schizopteridae) in a seeming parasitoid association (Poinar Jr and Brown 2014). This suggests that blister beetles and their parasitoid life cycle extend to the Early Cretaceous.

Ripiphoridae (wedge-shaped beetles) are closely related to Meloidae and currently are all parasites or parasitoids. Ripiphorids are endoparasitoids of cockroach adults; ectoparasites of the wood-boring larvae of longhorn beetles; and endoparasitoids, rarely ectoparasites, of aculeate hymenoptera larvae, especially bees (Eggleton and Belshaw 1992). The eggs of ripiphorid adults typically are laid in the host environment, with the triungulin larvae becoming phoretic, eventually finding and immediately attacking their hosts (Askew 1971). A relevant, postulated fossil interaction, from the Middle Jurassic Yanliao Biota of northeastern China (Fig. 11.12), involves parasitization of wood-boring beetles (Hsiao et al. 2017). In a younger deposit, Ripiphorids are associated with Early Cretaceous parasitism on insects such as cockroaches, based on plesiomorphic taxa discovered from 99 million-year-old Myanmar Amber (Beutel et al. 2016; Batelka et al. 2016, 2019). Certain features of one of these Myanmar Amber taxa, such as a miniaturized ripiphorid, indicate that parasitism rather than parasitoidism (Batelka et al. 2019), was present, suggesting an initial phase of parasitism (Hsiao et al. 2017; Fig. 11.12) that subsequently became replaced by parasitoidism in descendant clades of the Late Cretaceous and Cenozoic.

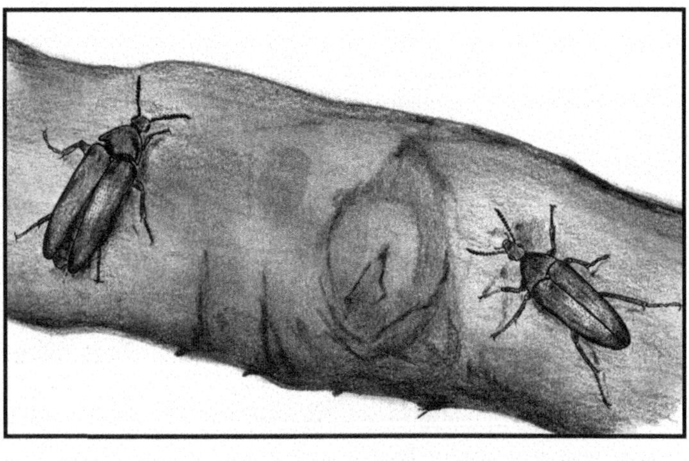

Fig. 11.12 A reconstruction of the parasitoid wedge-shaped beetle *Archaeoripiphorus nuwa* (Coleoptera: Ripiphoridae) from the Middle Jurassic Yanliao Biota of northeastern China. This setting illustrates a hypothesized behavior whereby ovipositing females search for damaged xylem cells that are caused by xylophagous larvae of beetles. (Reproduced with permission from Fig. 6 in Hsiao et al. 2017)

Anthribidae (fungus weevils) consist of about 360 genera and 4000 species, a few of which, particularly species of *Araecerus* and *Anthribus*, are cleptoparasitoids of soft scales such as *Lecanium* (Ülgentürk 2001; Ülgentürk and Toros 1996; Valentine 2002). The female weevil oviposits her eggs under a soft scale insect and the weevil larvae develop at the expense of the host's eggs (Valentine 2002; Mermudes and Leschen 2014). The earliest fossil record of Anthribidae is mid Early Cretaceous, about 132 million years ago (Gratshev and Zherikhin 2003), although it remains unknown when the cleptoparasitoid habit commenced in this lineage.

Coleoptera parasitoids, compared to the two other major parasitoid offenders, Diptera and Hymenoptera, have a relatively narrow range of hosts, attacking eight orders within the single phylum Arthropoda (Eggleton and Belshaw 1992). Host searching is overwhelmingly accomplished by the larva in Coleoptera, as in Diptera, but unlike Hymenoptera in which the host-searching stage is an adult. Among the ten families of Coleoptera, the parasitoid condition originated at least 14 times (Eggleton and Belshaw 1992), and likely several more times once a reliable phylogeny of Scarabaeidae has been established. These originations have arisen from two major ancestral strategies (Eggleton and Belshaw 1992; Morris 1998), although the transitions to parasitoidism involve mycophagy, nest provisioning, saprophagy and phytophagy (Fig. 11.4a). The first strategy is from mycophagy, often associated with nest provisioning, typical of Bothrideridae, Meloidae, Rhipiceridae and Ripiphoridae. In this pathway, the candidate parasitoid larva is transported via phoresy. It then first feeds on fungi in bark-beetle wood borings, followed by parasitism, or parasitoidizing or killing the existing beetle occupant (Morris 1998). Cleptoparasitoidism originated in blister beetles that feed on pollen in bee nests, and among weevils that feed in the wood borings of bark and ambrosia beetles. A modification of this strategy is found in scarabaeid parasitoids that may have originated ancestrally as saprophages or mycophages associated with nest provisioning (Eggleton and Belshaw 1992). The second strategy is exemplified by aleocharine parasitoids that originated from predaceous precursors. Both strategies have resulted in a greater number of cleptoparasitoids than ectoparasitoids plus endoparasitoids, amounting to a majority of all beetle parasitoids. This cleptoparasitoid to ectoparasitoid and endoparasitoid ratios are unlike analogous proportions in all other parasitoid containing insect orders. Shifts away from the parasitoid life habit have been rare among Coleoptera, involving egg predation and cleptoparasitic provisioning.

11.5.2.3 Strepsiptera (Twisted-Wing Parasites)

Strepsiptera are (controversially) internal parasites or parasitoids; they consist of two suborders, 17 extinct and extant families and about 610 species (Table 11.1). Males resemble normal, adult, winged insects, but have vestigial mandibulate mouthparts, distinctive compound eyes and halteres as forewings (Cook 2014). Female strepsipterans are larviform, live in the host wedged within the intersegmental membrane between abdominal sclerites but have a projecting orifice, the brood canal that is used for mating. Strepsipteran triungulin larvae are structurally convergent on larvae of Ripiphoridae and Meloidae (Kathirithamby 2009) and frequently are confused with each other in the fossil record because of the structural similarity (Beutel et al. 2016). As the triungulin larvae emerge from the female's body, they then consume her. The strepsipteran hypermetamorphic larva is a mobile, hexapodous, host-seeking triungulin instar (Pohl and Beutel 2019), which immediately after burrowing through their host's exocuticle molts into a second instar, a featureless, vermiform individual. This vermiform larva induces the host to produce a gall-like structure from which the larva feeds as it undergoes four more molts, during which the host is castrated. Larval hosts of Strepsiptera include silverfish (Thysanura), cockroaches (Blattodea), mole crickets (Orthoptera), planthoppers and leafhoppers (Hemiptera–Auchenorrhyncha), and bees and wasps (Hymenoptera–Aculeata) (Kathirithamby 2009). For Myrmecolacidae the larval host is different from the host to which the female is embedded (Hayward et al. 2011). The unique life cycle of Strepsiptera likely originated once. Recently, another distinctive hypermetamorphic stage of the strepsipteran life cycle was confirmed by discovery of the first free-living, late-instar larva, probably a female, belonging to a stem-group lineage of Strepsiptera from 44 million-year-old middle Eocene Baltic Amber (Pohl et al. 2019).

Considerable discussion has centered on whether strepsipterans, with the exception of the Mengenillidae, are parasites or parasitoids. Features mitigating against a parasitoid designation are (1) the host does not typically die immediately after adult emergence; (2) multiple females and multiple males can live within a single host; and (3) the adults are parasitic and not the larvae (Kathirithamby 2009). Nevertheless, Strepsiptera castrate their hosts, which renders them genetically dead, closely resembling the standard parasitoid life habit. Additionally, the basal-most extant lineage of Strepsiptera are the Mengeidae, which are considered true parasitoids, indicating that the life habits of the other extant, likely derived families evolved some morphological and behavioral features inconsistent with the ancestral parasitoid condition. Because of castration of their hosts and other nominal parasitoid features, Strepsiptera are here considered as parasitoids. However, if a traditional definition of parasitism in the Strepsiptera is accepted, in which host death is not imminent upon emergence of the parasitoid, then the transition from Mengeidae to other strepsipteran lineages could be construed as a shift from the parasitoid to a parasite life habit.

Although mengeid strepsipterans are endoparasitoids of soil-dwelling silverfish, Strepsiptera probably evolved from a Permian lineage of beetle taxa involved in mycophagy and wood boring (Eggleton and Belshaw 1992). This is particularly relevant since Strepsiptera likely are closely related to or the sister clade to Coleoptera (Misof et al. 2014), and early beetles are associated with wood, such as fungi-consuming polyphagan beetles in conifer host trees from the late Permian of northern China (Feng et al. 2017). After an approximate 190 million-year-long interval of a presumed divergence event from Coleoptera during the Early Permian, early Strepsiptera consisted of three extinct clades. First is extinct "Protostrepsiptera", possessing plesiomorphic features and appearing in 99 million-year-old Myanmar Amber (Grimaldi et al. 2005; Pohl et al. 2005; Kathirithamby and Engel 2014;

Engel et al. 2016a). Second is the oldest known strepsipteran attributable to a modern family, the mobile, primary larva of Mengenellidae, also from earliest Late Cretaceous Myanmar amber (Fig. 11.13). This specimen likely was associated with a wood-boring host larva (Pohl et al. 2018). Third is extinct Mengeidae, occurring in Baltic amber, a true parasitoid. Notably, the current host spectrum of Strepsiptera does not reflect this Late Cretaceous to Paleogene pattern (Poinar Jr 2004b;

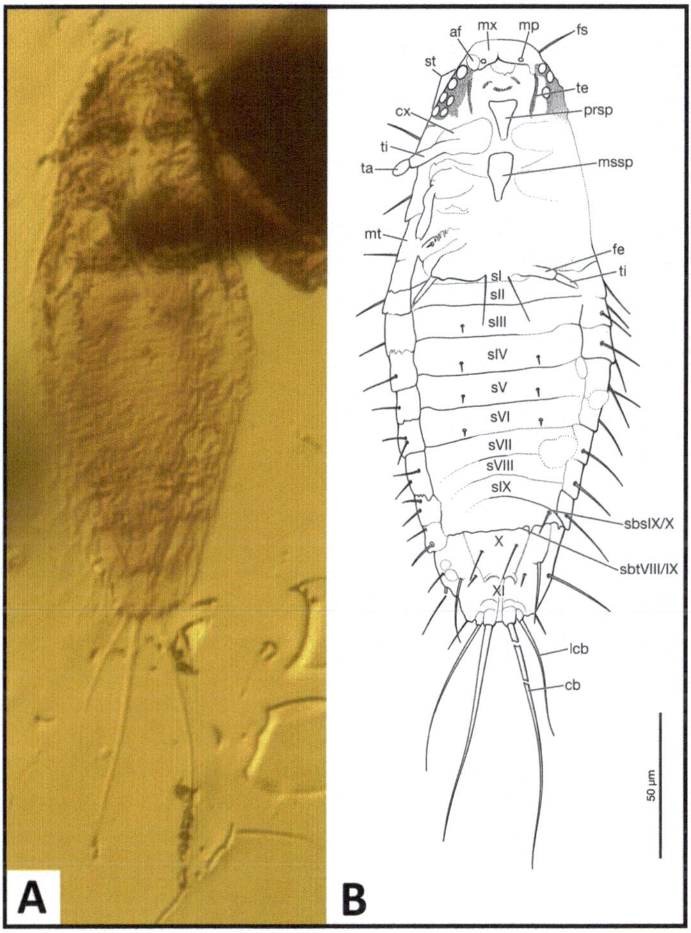

Fig. 11.13 A ventral view of an unnamed primary larva of a twisted-wing parasite (Strepsiptera: Mengenillidae) in Myanmar amber. (**a**) Photomicrograph of larva. (**b**) Drawing base on photomicrographs with a fluorescence microscope. Details of microscopy from Pohl et al. (2018). Abbreviations: *af* antennal field, *cb* caudal seta, *cx* coxa, *fe* femur, *fs* frontal seta, *lcb* lateral caudal seta, *mp* maxillary palp, *mssp* mesosternal plate, *mt* metanotum, *mx* maxilla, *prsp* prosternal plat, *sbsIX/X* segmented border between abdominal sternites IX/S, *sbtVIII/IX* segmental border between abdominal tergites VIII/IX, *sI–IX* abdominal sternites I–XI, *st* stemmata, *Ta* tarsus, *te* tentorium, *ti* tibia, *X* abdominal segment X, *XI* abdominal segment XI. (Reproduced with permission from Fig. 1 of Pohl et al. 2018)

Kathirithamby 2009). It appears that the early hosts of Strepsiptera may have been more peripheral to their current host distribution, such as wood frequenting cockroaches.

11.5.2.4 Diptera (Flies)

Dipteran parasitoids represent a much wider spectrum of hosts than does their other major parasitoid competitor, Hymenoptera. Dipteran parasitoids attack 22 orders of organisms within the five phyla of Platyhelminthes (flatworms), Mollusca (aquatic and terrestrial gastropods), Annelida (oligochaetes), Arthropoda (including millipedes, crustaceans and arachnids) and Chordata (amphibians) (Table 11.1) (Morris 1998). This host spectrum represents the dominant terrestrial animal phyla, except for Nematoda. Diptera have acquired the parasitoid habit minimally 21 times (Eggleton and Belshaw 1992), representing about 16,000 species or 20% of all dipteran species (Feener and Brown 1997), although Weinstein and Kuris (2016) list 60 times after an exhaustive literature search. The actual number of instances of dipteran parasitoidism will be shown to be about 100 times, once the life habits and phylogenetic relationships of more obscure lineages, particularly the diverse Asilidae, Phoridae and Sarcophagidae, are known (Eggleton and Belshaw 1992). Notably, in three of the four most diverse families—Bombyliidae, Conopidae and Tachinidae—parasitoidism has evolved only once (Feener and Brown 1997; Yeates and Greathead 1997).

Nine major dipteran lineages at the superfamily level harbor parasitoids. Culicomorpha consists of Chironomidae, some of which are ectoparasitoids of pulmonate gastropod immatures and mayfly naiads. Other minor origins of parasitoidism in Diptera include a single origin from microphagy in Chironomidae (nonbiting midges), and another single origin from mycophagy in Mycetophilidae (fungus gnats). Although Chironomidae extends to the Late Triassic, it is unlikely that these associations are as old. Bibionomorpha (e.g., Poinar Jr 2010) include endoparasitoids of flatworms and the nymphs and adults of sternorrhynchan hemipterans, whose associations most likely are relatively recent and Neogene in origin (Eggleton and Belshaw 1992). Nemestrinimorpha are obligate endoparasitoids of spiders and mites, as well as grasshoppers and beetles, and collectively attack all host life stages. Host associations of Nemestrinimorpha likely extend to the Late Triassic for tanglevein flies (Nemestrinidae). The spider flies (Acroceridae) are a more recently evolved lineage that range from the Late Jurassic to the recent (Mostovski 1998). They are endoparasitoids of spiders and mites and are particularly associated with the parasitoidization of several major clades of araneomorph spiders (Gillung et al. 2018; Fig. 11.14). This lineage originated during the Middle Jurassic and diversified during the Late Cretaceous to Paleogene. The Archisargoidea is the only extinct dipteran superfamily that contains a parasitoid lineage (Fig. 11.15), the highly probable endoparasitoid clade Eremochaetidae, known from the Late Jurassic to earliest Late Cretaceous (Grimaldi and Barden 2016). Although considered typically predaceous (Fig. 11.1f), Asilomorpha also are ectoparasitoids or endoparasitoids,

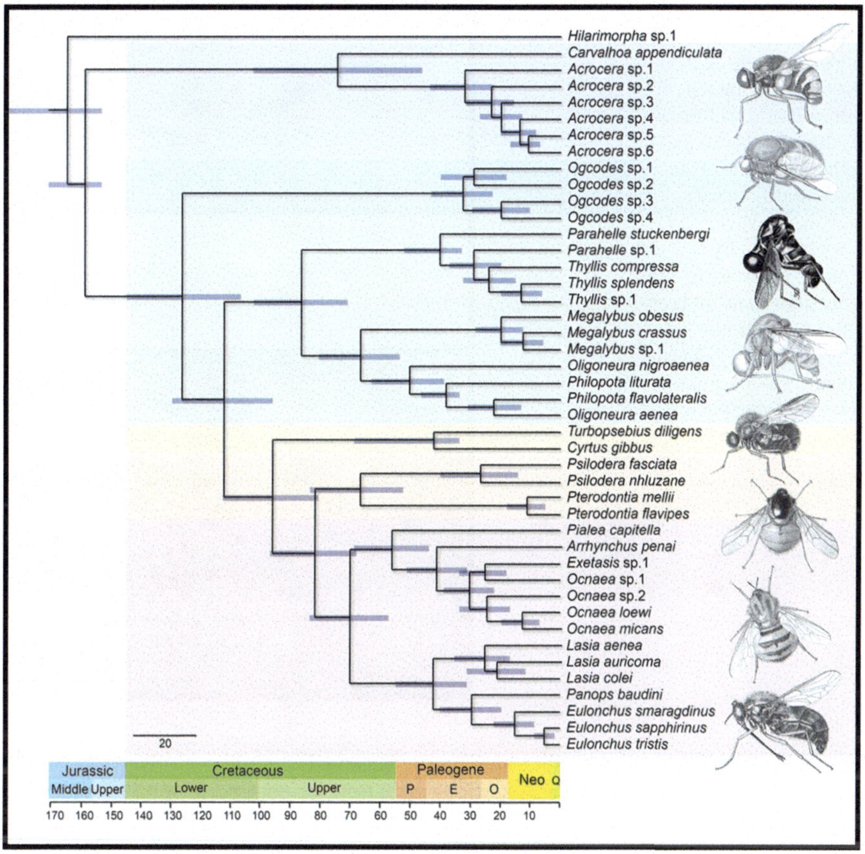

Fig. 11.14 Estimated divergence times among lineages of the spider fly lineage Acroceridae under the fossilized birth-death process. Bars depict the 95% higher posterior probability density of each estimate. Mean ages, their ranges and other details are provided in Gillung et al. (2018). (Reproduced with permission from Fig. 4 in Gillung et al. 2018)

occasionally hyperparasitoids, of beetle and sawfly larval hosts. These hosts are associated with dead wood, or alternatively the eggs, larvae and pupae of grasshoppers, beetles, externally feeding moth larvae, and other parasitoid groups such as Tachinidae and Ichneumonidae (Yeates and Greathead 1997). Some of these associations likely extend to the Middle to Late Jurassic. Empidoidea are ecto- or endoparasitoids of caddisfly pupae, an association that probably originated during the Late Jurassic (Eggleton and Belshaw 1992).

Cyclorrhaphan flies, consisting of the Aschiza and Schizophora sister groups, underwent a Late Cretaceous and Paleogene wave of diversification, with several lineages acquiring parasitoid life habits (Grimaldi and Engel 2005a). Aschiza are endoparasitoids and cleptoparasitoids of the immatures and adults of pulmonate gastropods, oligochaetes, opilionids, spiders and myriapods, as well as all

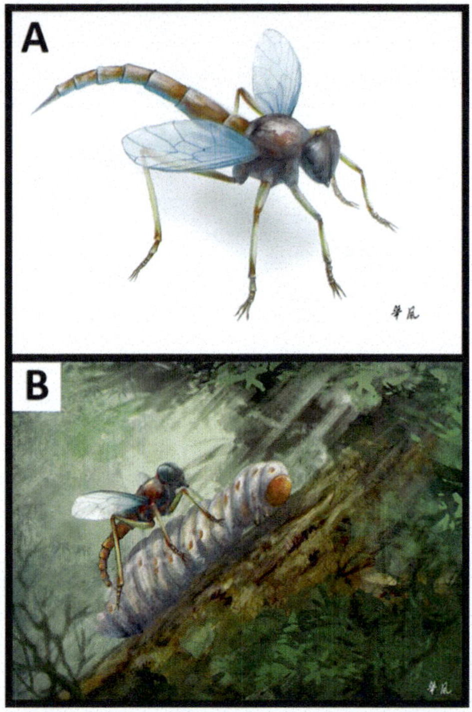

Fig. 11.15 (**a**) Reconstruction of the short-horned fly adult *Zhenia xiai* (Diptera: Eremochaetidae) from Myanmar amber of the Early Cretaceous–Late Cretaceous boundary interval. (**b**) This endoparasitoid fly is depicted in the act of egg laying into its larval host with a hypodermic ovipositor. (Reproduced with permission from Fig. 4 in Zhang et al. 2016)

developmental stages of hemipteran, beetle, fly caddisfly and wasp hosts (Eggleton and Belshaw 1992). Aschizan associations with this varied repertoire of hosts likely originated in the Late Cretaceous and continued throughout the Cenozoic. Schizophora, constituting the Acalyptratae and Calyptratae clades, are endoparasitoids, rarely ectoparasitoids or cleptoparasitoids, of the immature and adult stages of eight, major, non-insectan lineages of invertebrates and vertebrates. These lineages are pulmonate gastropods, freshwater bivalves, oligochaetes, centipedes and millipedes, scorpions, isopod crustaceans, nymphs of cockroaches, termites, orthopterans, mantids, webspinners, earwigs, sternorrhynchan hemipterans, bugs, beetles, moths, sawflies, larvae of aculeate wasps, puparia of calyptrate flies, and mammalian vertebrates (Feener and Brown 1997). The Acalyptratae in particular developed interactions with the most diverse host spectrum of any insect parasitoid clade, and represent relatively recent Cenozoic events compared to phylogenetically more basal, dipteran parasitoid groups. Schizophora parasitoid interactions originated during the Cenozoic initially from a pattern of generalized parasitoid lineages, but continually gave rise to specialized lineages (Stireman III et al. 2006).

The origins of parasitoidism in Diptera have been the most diverse of any order of insects (Eggleton and Belshaw 1992, 1993; Fig. 11.4b). The most common route to parasitoidism in Diptera is from saprophagy, which evolved 13 times at the family level, often in association with social insects. Parasitoidism via saprophagy has been inferred to occur in Phoridae (scuttle flies), Pipunculidae (bigheaded flies),

Conopidae (thickheaded flies), Pyrgotidae (light flies), Cryptochaetidae (scale parasite flies), Chloropidae (frit flies), Anthomyiidae (root maggots), Muscidae (house flies), Calliphoridae (carrion flies), Rhinophoridae (woodlouse flies), Sarcophagidae (flesh flies) and Tachinidae (tachina flies). The second most common route to parasitoidism in Diptera is through predation, which has been documented seven times and frequently has been associated with prey items occurring in or on top of the soil. This route has been inferred for Cecidomyiidae (gall midges), the common ancestor of Nemestrinidae + Acroceridae (tanglevein flies and spider flies), Bombyliidae (bee flies) that are also associated with cleptoparasitoidism, Asilidae (robber flies), Empididae (dance flies), Sciomyzidae (snail-killing flies) and Phacomyiidae (marsh flies). For dipteran parasitoids, it appears that ants are the most common host group (Quevillon and Hughes 2018).

A family-level phylogenetic tree of Diptera indicates that ectoparasitoidism originated 10 times, occurring sporadically throughout the phylogeny, whereas endoparasitoidism originated 17 times, largely confined to the Eremoneura (Wiegmann et al. 2011). This asymmetry in distribution of the two major parasitoid types is attributable to two major morphological features, the piercing ovipositor of some Eremoneura (Yeates and Wiegmann 1999), and compensatory behavioral changes for host seeking in other Eremoneura that lack a penetrative ovipositor (Feener and Brown 1997). Behavioral features promoting parasitoidization that substitute for a rigid, puncturing ovipositor include host consumption of parasitoid eggs that hatch in the gut and invade internal organs (Stireman III et al. 2006), and larvae that aggressively penetrate the host integument (Feener and Brown 1997; Stireman III et al. 2006). These features allow Eremoneura to attack and penetrate hosts that normally would be available only to apocritan Hymenoptera with extensive ovipositor modifications such as lengthening (Eggleton and Belshaw 1992; Feener and Brown 1997). Most Diptera lack a prolonged, robust ovipositor for penetrating long distances through indurated tissues as in Hymenoptera, and instead have a flexible, telescopic ovipositor that is extended by intersegmental membranes, rendering it inefficient for placing parasitoid eggs deep into host tissue (Feener and Brown 1997). Such ovipositor design restricts many dipteran parasitoid-bearing lineages to exposed hosts and a preference for ectoparasitoidism over endoparasitoidism, the latter of which only occasionally occurs across Diptera phylogeny (Eggleton and Belshaw 1993; Wiegmann et al. 2011). Several derived eremoneuran clades—particularly well developed in Tachinidae (tachina flies), Conopidae, Pipunculidae and Phoridae (scuttle flies)—have re-evolved stiffened, piercing, albeit undirected, ovipositors with a sclerotized terminus capable of penetrating their hosts (Feener and Brown 1997; Poinar Jr 2013), allowing for endoparasitoidism. For this reason, there is the dominance of endoparasitoidism over ectoparasoidism particularly among Eremoneura (Table 11.1) (Wiegmann et al. 2011). Nevertheless, dipteran parasitoids have never developed capabilities for attacking wood-boring larvae separated from a bark surface by several centimeters of wood; nor have they been engaged in penetration of thick, hardened tissues such as galls (Eggleton and Belshaw 1993; Quicke 1997). However, a substitute for reaching larvae via an ovipositor is hearing convergence in cricket hosts and their tachina fly parasitoids (Robert et al. 1992).

The second morphological feature is the development of a host-seeking first-instar larva, the planidium, which occur in at least seven dipteran families (Askew 1971) and typically results in endoparasitoidism. Functionally analogous to the triungulin larva of some Neuroptera and Coleoptera and all Strepsiptera, the planidium larva is an active, more or less sclerotized, mobile and host seeking larva that represents a departure from the common practice of the adult female dipteran depositing her eggs directly on, in or adjacent the host (Clausen 1940; Askew 1971). The planidium larva has evolved among Culicomorpha (Chironomidae), Bibionomorpha (Mycetophilidae), Nemestrinimorpha (Acroceridae, Nemestrinidae), Asiloidea (Asilidae), Acalyptratae (Sciomyzidae) and Schizophora (Rhinophoridae) (van Jutting 1938; Greathead 1963; Ferrar 1987; Eggleton and Belshaw 1992), representing a broad swath of Diptera phylogeny and independent originations minimally five or six times.

The trophic origins of the 23 parasitoid families of Diptera are highly diverse but evolved overwhelmingly from saprophagy and predation. For most families it is difficult to ascertain whether parasitoid flies evolved into other trophic modes (Fig. 11.4b). Only two such transitions are known (Eggleton and Belshaw 1992). First, parasitoidism in some Bombyliidae evolved into predation on grasshopper egg pods (Yeates and Greathead 1997). Second, some Sarcophagidae evolved into internal parasites in bees that consume the host only when it dies, a trophic transition that also is suspected in adult scarab beetle and bumble bee hosts. In general, the origin of parasitoidism from predation occurred once in nematocerous clades and several times in brachycerous (non-cyclorrhaphan) clades, whereas parasitoidism via saprophagy is confined to more derived clades among cyclorrhaphan flies. These occurrences are consistent with the predominant detritivore to saprophage diets of cyclorrhaphan flies.

11.5.2.5 Trichoptera (Caddisflies)

Wells (1992) described the life history of *Orthotrichia muscari* a member of Hydroptilidae (purse-case caddisflies) from the Northern Territory of Australia (Table 11.1). This microcaddisfly, in addition to nine other members of the *Orthotrichia aberrans* species group, apparently are parasitoids of similarly aquatic Philopotamidae (fingernet caddisflies) as well as other Hydroptilidae (Wells 2005). This is the only known example of parasitoidism in Trichoptera, and represents the only aquatic mode of parasitoidism that is unique to Insecta.

11.5.2.6 Lepidoptera (Moths)

Members of two families of Lepidoptera are known as ectoparasitoids (Table 11.1). Epipyropidae (planthopper parasite moths) consist of 32 pantropical species that attack fulgoroid planthoppers (Hemiptera: Auchenorrhyncha) and a few lepidopteran larvae (Jeon et al. 2002). The fossil record of planthopper parasite moths is known only for one specimen attached to a leafhopper (Auchenorrhyncha: Cicadellidae) from the early Miocene of the Dominican Republic (Poinar Jr and Poinar 1999). Analogously, Pyralidae (snout moths) contain a single species, *Sthenauge parasitus*, which is an ectoparasitoid on the larvae of the saturniid moth *Aplomerus* (Lepidoptera: Saturniidae), initially feeding on the host's dermal spines but eventually consuming it. Known parasitoidism arose at least twice in Lepidoptera, separately in Epipyropidae and Pyralidae, but likely arose many more times, particularly within lineages of Epipyropidae, given the poor understanding of the life-habits and phylogeny of these moths (Jeon et al. 2002). Parasitoidism in Epipyropidae plausibly evolved from ectoparasitic behavior occurring on the hemipteran host surface, followed by entering the host by piercing its cuticle and then eventually killing it with the exit of the parasite larva (Eggleton and Belshaw 1992). The evolutionary source of parasitoidism in Epipyropidae and Pyralidae is unclear, and there is no fossil record for parasitoid members of either lineage.

11.5.2.7 Hymenoptera (Wasps)

Hymenoptera constitute about 75% of all modern parasitoid species and have the most diverse taxonomic spectrum of parasitoid lineages of any order of insects (Santos and Quicke 2011). The 17 hymenopteran superfamilies of parasitoids are approximately split between ectoparasitoid and endoparasitoid dominated superfamilies; they include 4 superfamilies that contain hyperparasitoids, 4 superfamilies that have predators, and 2 superfamilies with cleptoparasitoids (Gauld and Bolton 1988) (Table 11.1). One superfamily, Chalcidoidea, while overwhelmingly dominated by parasitoid taxa, contains two families, Agaonidae and Chalcididae, which are dominantly represented by taxa such as pollinators, gallers, seed predators and herbivores as well as cleptoparasitoids in plant galls (Eggleton and Belshaw 1992). Family-level diversity per superfamily ranges from 1 to 20. There are 92 families of extant and extinct hymenopteran parasitoids, representing 63% of all insect families with parasitoid life habits (Table 11.2). Of these families, 22 (23.9%) are extinct, some of which represent stem-group lineages. Even though there are 3.75 times more hymenopteran than dipteran parasitoid species, hymenopteran hosts are restricted to one phylum, Arthropoda, and within that phylum only Insecta are attacked (Morris 1998). One possibility for this vast disparity in host utilization is that, whereas parasitoidism arose once or twice in Hymenoptera (Dowton and Austin 1995b). It probably arose at least 10 to 100 times in Diptera (Eggleton and Belshaw 1992). Evidently, competition for the variety of targeted hosts was much less constrained in Diptera.

Sawflies (Tenthredinoidea, Siricoidea) are the sister-group to the parasitoid clades Orussidae and Apocrita, which represent the single (Gauld and Bolton 1988) or dual (Dowton and Austin 1995b) origins of parasitoidism within Hymenoptera. It appears that the parasitoid condition in Hymenoptera is associated with a founder effect of considerably higher content of adenine and thymine, as opposed to cytosine and guanine, in the mitochondrial DNA of Apocrita wasps, when compared to the AT content of non-parasitoid sawflies of Symphyta, including Orussidae (Dowton and Austin 1995a). This extreme codon bias likely occurred during the Early Jurassic, attributable to a founder event reflective of the parasitoid lifestyle (Dowton and Austin 1995b). Because of this causal link between sequence divergence of mitochondrial DNA and parasitoid life habits (Dowton and Austin 1995a), larval sawflies of Symphyta, the basal-most group, would have gone through a predaceous life habit prior to parasitoidism (Morris 1998). Such a life habit could have been present in enclosed feeders such as gallers (Tenthredinidae) or wood borers (Siricidae), in which a feeding strategy included consumption of arthropod cohabitants, including cannibalism, as a trophic prelude to the parasitoid life habit. Facilitating this nutritional transition to the parasitoid condition would be modification of the rigid, cutting ovipositor of a sawfly into a much longer, flexible, piercing or drilling ovipositor of basal Apocrita (Fig. 11.1f), allowing access to larvae through a few centimeters of hard substrates such as wood (Gauld and Bolton 1988). Hymenoptera transformed their sawfly ovipositor into a singular structure that allowed boring and drilling into considerable depths of plant tissue for inserting eggs on or in target larvae. In the process of ovipositor piercing and drilling into plants, chemicals were injected to soften wood, sclerenchyma and other indurated tissues that allowed attack of larval insects within enclosed plant tissues (Gauld and Bolton 1988; Morris 1998). This structural transformation of the ovipositor is considerably different that the condition in Diptera, the other major clade of parasitoid insects, and accounts for the exceptional reach of hymenopteran parasitoids in targeting host larvae.

Orussoidea are a superfamily of sawflies that either have a sister-clade relationship or are paraphyletic to Apocrita. They are ectoparasitoids of wood wasps such as Siricidae or wood-boring beetles such as Buprestidae in dead wood (Gauld and Bolton 1988; Vilhelmsen and Turrisi 2011). A series of basal Apocrita superfamilies of Stephanoidea, "Ephialtitoidea" (Fig. 11.1f), Megalyroidea and Evanioidea are almost all ectoparasitoids of wood-boring larvae that have an overall geochronologic distribution ranging from Early Cretaceous to Recent (Whitfield 2003; Moghaddam and Turrisi 2018; Li et al. 2018a). Within these superfamilies, several extinct, family-level lineages range from late Early Jurassic to mid Cretaceous and are presumed to have similar biologies based on their phylogenetic position within modern lineages of known biologies. One exception to the exclusive ectoparasitoid habits of Stephanoidea, Megalyroidea, Evanioidea and probably "Ephialtitoidea" is the superfamily Trigonalyroidea in which Trigonalidae is an endoparasitoid on external-foliage-feeding, folivorous larvae (Weinstein and Austin 1991; Engel 2016).

The Proctotrupomorpha, the most diverse, monophyletic clade (*sensu* Castro and Dowton 2006) of parasitoid insects, consists of the apocritan superfamilies Proctotrupoidea, Cynipoidea, Platygastroidea, Chalcidoidea, Mymarommatoidea and Serphitoidea (Table 11.1). In many analyses, Proctotrupomorpha also includes the Ceraphronoidea and Ichneumonoidea (Castro and Dowton 2006); by contrast, in other studies these superfamilies are excluded (Dowton et al. 1997; Sharanowski et al. 2010). (The more encompassing view is taken here.) The next three superfamilies are endoparasitoids, the first two of which, Proctotrupoidea and Cynipoidea, attack a variety of larvae occupying principally soil or wood microhabitats, although some families occasionally occur on rotting meat or attack leaf mining larvae (Askew 1971). The third superfamily, Platygastroidea, attack insect eggs, inhabitants of plant galls and scale insects (Whitfield 2003). The next series of superfamilies—Ceraphronoidea, Mymarommatoidea and Serphitoidea—are ectoparasitoids and endoparasitoids on insect eggs, nymphs, larvae, pupae and puparia of hemimetabolous and holometabolous insects (Clausen 1940; Askew 1971). The Ceraphronoidea and Mymarommatoidea have an Early Cretaceous–Holocene fossil record whereas the extinct Serphitoidea, the only extinct superfamily of hymenopteran parasitoids, ranges from Early Cretaceous to mid Cretaceous. The Ceraphronoidea consists of five families, one of which is the Cretaceous Radiophronidae (Fig. 11.16), an inferred ectoparasitoid on larval insects (Ortega-Blanco et al. 2010).

Chalcidoidea and Ichneumonoidea contain a mix of ecto- and endoparasitoids that include hyperparasitoid and predator taxa. At 20 families, Chalcidoidea represents the greatest family-level diversity of any hymenopteran parasitoid superfamily, known for targeting eggs, nymphs, larvae and pupae, especially of concealed insects in cases, galls, leaf mines, seeds, wood borings and insect nests (Clausen 1940; Askew 1971). When compared to other superfamilies, Chalcidoidea display a

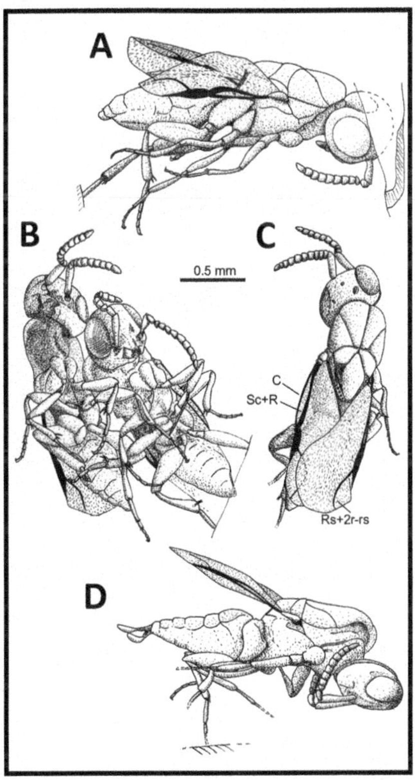

Fig. 11.16 Camera lucida drawings of the ceraphronoid wasps *Radiophron ibericus* and *R.* aff. *ibericus* (Hymenoptera: Radiophronidae) from the Early Cretaceous Peñacerrada 1 locality of northern Spain (Ortega-Blanco et al. 2010). (**a**) Lateral view of holotype MCNA-8754. (**b**) Ventral view of paratypes MCNA-13030.1 and 13030.2. (**c**) Dorsal view of paratype MCNA-13030.1. (**d**) Lateral view of a male of *Radiophron* aff. *ibericus* gen. et sp. nov. (MCNA-8760). (Reproduced with permission from Fig. 3 in Ortega-Blanco et al. 2010)

more recent geochronologic range, from Cenomanian to the present, mostly represented by Cenozoic first occurrences. One interesting clade, Eucharitidae (eucharitid wasps), are the only insect family to exclusively parasitoidize ants (Lachaud and Pérez-Lachaud 2009), a relationship that represents a conserved association extending to the early Paleogene based on a patterns of ancient colonization of novel hosts and later host tracking (Murray et al. 2013). Ichneumonoidea consist of two families that attack eggs, larvae, pupae and adults of wood borers, but occasionally aphid nymphs and ants; their range is Early Cretaceous to Recent (Poinar Jr 1987). A mid-Mesozoic to Recent member of one of these families, Braconidae, is known to have attacked a weevil of Ithyceridae (Poinar Jr and Shaw 2016), as evidenced by a spent cocoon and an exit hole on its host (Fig. 11.17). Braconidae display a diversification pattern accompanied by numerous host shifts among several insect orders (Shaw 1988).

Aculeate Hymenoptera consist of Chrysidoidea, Vespoidea and Apoidea (Fig. 11.1g) that overwhelmingly are ectoparasitoids, collectively, on all developmental stages of external foliage feeding (Orthoptera, Phasmatodea, Lepidoptera) and piercing-and-sucking (Fulgoroidea, Auchenorrhyncha) insects, as well as beetles, ants and bees (e.g., Fig. 11.1n). A few host specialist lineages parasitoidize atypical hosts of cockroaches, webspinners, solifugids and spiders. An example of a specialist lineage is Sclerogibbidae, a family of Chrysidoidea that are ectoparasitoids (Fig. 11.18) on the nymphs and adults of webspinners (Embioptera), a lineage that extends back to the Early Cretaceous (Engel and Grimaldi 2006).

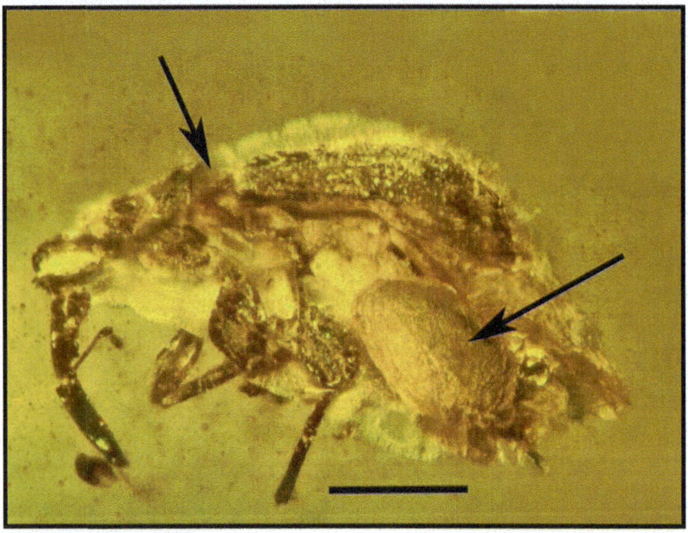

Fig. 11.17 A New York weevil (Coleoptera: Ithyceridae) attacked by a braconid wasp (Hymenoptera: Braconidae) from Myanmar amber, of transitional Early Cretaceous–Late Cretaceous age (99 Ma). The attached braconid wasp cocoon is indicated by the lower-right arrow, and its larval emergence hole is indicated by the upper-left arrow. Horizontal scale bar is 0.8 mm. (Reproduced with permission from Fig. 1 in Poinar Jr and Shaw 2016)

Fig. 11.18 Ventral view of the habitus of the sclerogibbid wasp *Sclerogibbodes embioleia* (Hymenoptera: Sclerogibbidae), ectoparasitoids of webspinners (Embioptera). (Reproduced with permission from Fig. 2 in Engel and Grimaldi 2006)

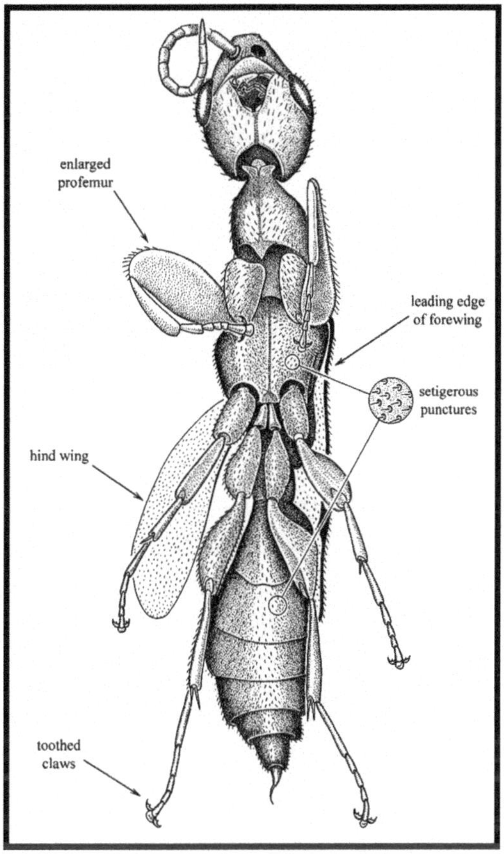

The emergence of parasitoidism in aculeates began during the Late Jurassic to Early Cretaceous boundary interval and continued into the Cenozoic. Evolutionary shifts to the parasitoid life habit in Hymenoptera have involved transitions overwhelmingly from mycophagy and to a lesser extent, phytophagy (Fig. 11.4c) (Dowton and Austin 1995a). The origin of parasitoidism from mycophagy involved Orussidae and most Apocrita lineages whereas the shifts from phytophagy or seed predation (Fig. 11.1m) involved principally the chalcidoids Agaonidae, Eulophidae, Pteromalidae, Tanaostigmatidae and Torymidae, in addition to the nutritionally diverse Cynipidae and Apidae (Eggleton and Belshaw 1992). One significant difference between Hymenoptera and Diptera parasitoids involve searching behavior for the target host (Morris 1998). In Hymenoptera, searching behavior is incumbent on the adult wasp, which places its eggs accurately and directly on the host, resulting in minimal movement of the newly hatched larva as it becomes embedded in the host (Askew 1971). A much different mechanism of host searching is found in Strepsiptera and Coleoptera with a host-seeking, active triungulin larva

(Kathirithamby 2009; Beutel et al. 2016). Similarly, in Diptera the eggs are scattered in the general vicinity of the host. Upon hatching, a special, mobile, first-instar larva of the parasitoid, the planidium, searches for the host via various chemical and physical cues (Askew 1971). Three main nutritional regimes are present to which Hymenoptera have shifted from parasitoid ancestors. First, predation was acquired indirectly via larval provisioning of paralyzed hosts by aculeate bees and wasps, or directly from the parasitoid life habit, such as some ichneumonoid wasps (Eggleton and Belshaw 1992). Second, Hymenoptera parasitoid to phytophage transitions occurred in several families that formerly fed on larvae in highly nutritious habitats such as leaf mines and galls, the Cynipidae being the best example (Fergusson 1990). Third, transitions from egg parasitoidism to egg predation are documented in Evaniidae (Askew 1971). Parasitoidism is a widespread life habit in Hymenoptera that is a rich source of secondary predation and phytophagy.

11.6 Modern Food Webs and the Mid-Mesozoic Parasitoid Revolution (MMPR)

Much of the ensuing discussion references Fig. 11.19, which documents the diversity of major parasitoid clades and groups from the time of their initial appearance during the late Early Jurassic to their representation in the recent record. The fossil diversity of seven major clades or groups of parasitoids—non-proctotrupomorph Hymenoptera, proctotrupomorph Hymenoptera, aculeate Hymenoptera, non-eremoneuran Diptera, eremoneuran Diptera, Strepsiptera and Coleoptera—record first the establishment of the Mid-Mesozoic Parasitoid Revolution (MMPR) and then the subsequent expansion of parasitoid groups and clades through time. Other clades with rare parasitoid members—Neuroptera, Trichoptera and Lepidoptera—each contain one or two families of parasitoid taxa that are not included because of an insufficient or irrelevant fossil record. The MMPR is a time interval during the late Early Jurassic to late Early Cretaceous (phases 1 and 2) in which bottom-to-top regulation of terrestrial food webs dominated by inefficient clades of predators were replaced by top-to-bottom trophic regulation by considerably more efficient parasitoid clades. After the MMPR, these clades subsequently expanded (phases 3 and 4) as parasitoids became trophically entrenched in terrestrial food webs to the present day. The initial pulse of the MMPR consisted of phases 1 and 2 that represented the earliest occurrences and establishment of most parasitoid lineages in terrestrial habitats from the latest Early Jurassic (Toarcian Stage) to the Early Cretaceous (Albian Stage). Commencing with the Late Cretaceous is the subsequent evolutionary diversification and ecological expansion of parasitoids represented by phases 3 and 4 that continues to the present day.

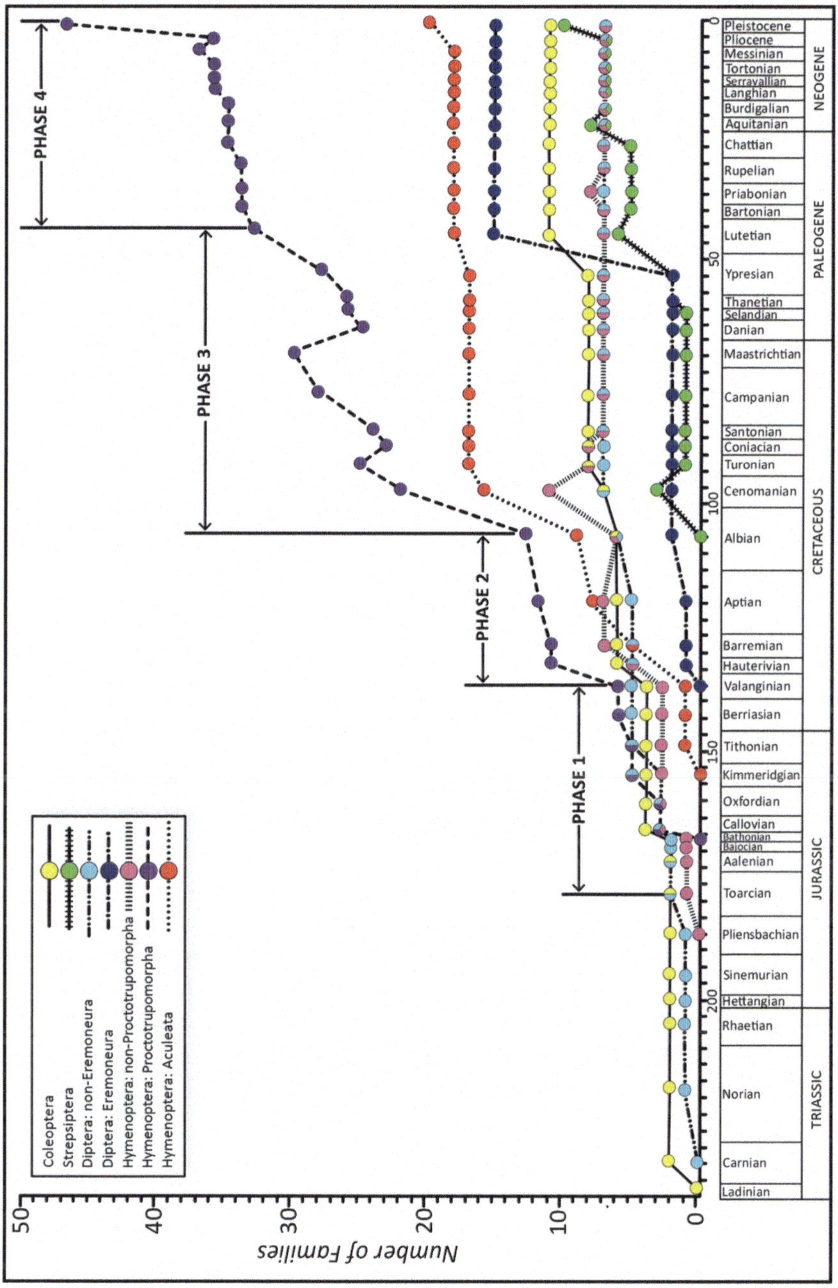

Fig. 11.19 The Mid Mesozoic Parasitoid Revolution (MMPR) represented by phases 1 and 2, and its subsequent expansion represented by phases 3 and 4. The MMPR is the initial expansion of the parasitoid guild during the mid Mesozoic. Data are taken from Table 11.1. Endpoints for the seven groups of parasitoids at 0 million years indicate modern family-level diversities, equivalent to the Holocene Epoch (not labelled). (Geochronologic scale is taken from Walker et al. 2013)

11.6.1 Ovipositors and Host-Seeking First Instar Larvae: Vetting the Parasitoid Taxa

Among the seven parasitoid clades and groups (Fig. 11.19), evidence for parasitoidism is excellent for non-proctotrupomorph and proctotrupomorph Hymenoptera as well as non-eremoneuran and eremoneuran Diptera. For these four groups, the near certainty of their parasitoid status is attributable to presence of a distinctive female ovipositor or oviscapt whose structure is designed to deposit eggs in or on their arthropod or non-arthropod hosts (Askew 1971). The oviposition of eggs often is accomplished by long ovipositors penetrating through considerable distances or thicknesses of intervening tissue, such as wood in the case of ichneumonid wasps targeting wood-boring larvae through several centimeters of wood (Clausen 1940; Gauld 2008). A second type of evidence is the presence of a highly ambulatory, host-seeking triungulin larva that forms the first larval instar of some Neuroptera, Strepsiptera and Coleoptera, particularly Mantispidae, Ripiphoridae, Meloidae, Bothriderinae of Bothrideridae and all strepsipteran clades, a feature that almost assures a parasitoid designation (Crowson 1981; Evans and Steury 2012). An analogous host-seeking larva, the planidium, exists for many parasitoid Diptera, for example, Mycetophilidae, Nemestrinidae, Acroceridae, Asilidae and Rhinophoridae, that also represents parasitoid status (Eggleton and Belshaw 1992). The confidence level is somewhat lowered for aculeate Hymenoptera, as their ovipositor is modified for stinging and paralyzing prey, rather than necessarily for ovipositing eggs in or on their prey (Gauld and Bolton 1988; Whitfield 2003). However, knowledge of the life histories and biology of modern parasitoid aculeate Hymenoptera provides considerable evidence for assigning an aculeate Hymenoptera fossil species to a predator, parasite or parasitoid. In the case of all fossil Rhipiceridae and all Passandridae (Coleoptera), an assignment to parasitoid status is highly probable, owing to the condition that the modern families are depauperate clades that only possess the single life habit of parasitoidism.

The weakest case for a parasitoid assignment are the remaining, highly speciose families of Carabidae, Staphylinidae and Cleridae in Coleoptera, each of which overall is sparsely represented by extant parasitoid taxa. Depending on the subfamily structure of the fossil occurrences and other parasitoid-relevant information for these three beetle families, a judgement was made regarding whether parasitoids were present for a given lineage occurring in a particular geological stage. For example, nearly all modern parasitoid taxa in Staphylinidae are from Subfamily Aleocharinae (Maus et al. 1998), a subgroup that only extends to the Cenomanian (Cai and Huang 2014; Yamamoto et al. 2016). Parasitoid assignments to Staphylinidae were not made in the absence of Aleocharinae for pre-Cenomanian occurrences and in lieu of evidence for the presence of Aleocharinae only in more recent occurrences. While the full range of parasitoid-bearing families listed in Table 11.1 provides the data for Fig. 11.19, determination of parasitoid presence in the fossil record was more circumspect in the narratives for Sects. 11.6.4–11.6.6 below. It is for this lack of trophic specificity, the lack of evidence from the fossil record, and other reasons of uncertainty that the three beetle lineages did not extend to earlier than the late Early Jurassic (Toarcian Stage), and thus were not considered as parasitoids before the MMPR (Fig. 11.19).

11.6.2 The Trophic Cascade and Resource Concentration Hypotheses of Food Webs

A food web consists of a network of trophic interactions among species that specifically identify the consumers and the consumed within a local community (Loreau 2010; Price et al. 2011). With very few exceptions (e.g., Dunne et al. 2014), well-preserved biotas in the fossil record have not been rigorously analyzed for their food-web structure. For well-preserved biotas, modern or fossil, two competing hypotheses would be important for understanding major changes in food-web trophic structure through geologic time (Matson and Hunter 1992). The first option that describes a food web is the trophic cascade hypothesis (Carpenter et al. 1985; but see Polis and Strong 1996) in which carnivores control herbivores in a local biota, and in so doing regulate plants for consumption by herbivores. The trophic cascade concept often is described as top–down control by higher trophic levels of lower trophic levels. The alternative to the trophic cascade is the resource concentration hypothesis (MacArthur and Levins 1964; Schmitz et al. 2000), in which the trophic structure of plants, herbivores and carnivores is competitively dependent on access of plants to resources that allow for growth. The resource concentration hypothesis frequently is mentioned as bottom–up control by lower trophic levels (autotrophs) of higher trophic levels (Oksanen et al. 1981; Hunter and Price 1992; Power 1992). There is evidence that changes in plant traits, induced by shifts in parasitoid attack, can have major and lasting effects on the food web (Bukovinszky et al. 2008). Both hypotheses have major ecological consequences for the beginnings of parasitoidism during the mid Mesozoic (Fig. 11.19).

The trophic structure of food webs can be affected significantly depending on whether consumption of arthropod prey is conducted by predators or by parasitoids. Compared to predators, parasitoids typically are more speciose, possess greater trophic complexity, have higher host specificities, and their arthropod hosts often remain capable of accommodating additional parasitoid individuals (Lafferty et al. 2006; Price et al. 2011; Dunne et al. 2013). These characteristics indicate that parasitoids are much more efficient, on average, than predators in top–down regulation of the food webs. By contrast, resource limitation to plants often has a negative effect on food-web structure by limiting bottom–up energy available to plants, their herbivores and their consumers (Bukovinszky et al. 2008). A relevant study of this latter phenomenon is the Brussels sprouts (primary producer) to aphid (herbivore) to primary parasitoid (primary consumer) to secondary parasitoid (secondary consumer) food web (Bukovinszky et al. 2008). In this food web (Fig. 11.20), source-plant quality has major, bottom–up, cascading effects across trophic levels that is modulated by the quality and quantity of the resource items for plants, such as nutrients, minerals and sunlight (Bukovinszky et al. 2008). In the study, the particular plant type at the bottom of the food chain was either a wild or a domesticated variety of Brussels sprouts, each variety of which differed considerably in secondary chemistry and morphology. These two varieties of plants affected two species of herbivorous aphids, which were attacked and eventually killed (mummified) by five species of primary parasitoid wasps. These primary parasitoid wasps in turn were attacked and later killed (mummified) by ten species of secondary parasitoid wasps (hyperparasitoids) that belonged to two feeding guilds that differ in how the primary parasitoid host was attacked. The secondary parasitoids also left pupal mummies after emergence from their primary parasitoid host.

The variability in plant quality had major, cascading effects on density-mediated and size-mediated effects of the food web, based on the quality and quantity of resources available to the plant (Fig. 11.20). The wild and the domesticated varieties of the Brussels sprouts plant each had major and different effects on food-web trophic levels. These effects were density (number of individuals), size traits (body size, body architecture, secondary chemistry and leaf thickness), and food-web

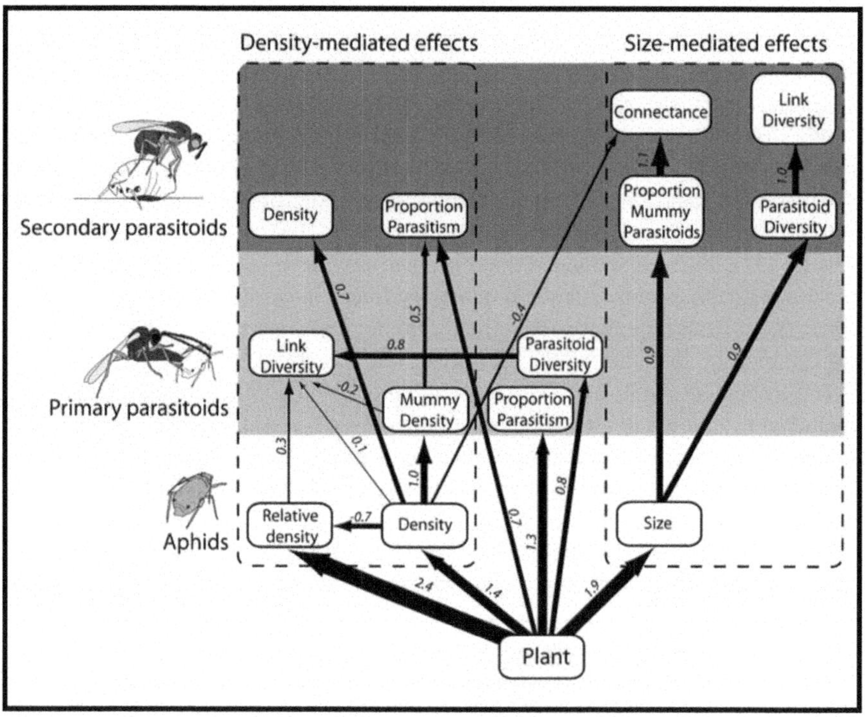

Fig. 11.20 Summary diagram of direct and indirect effects of plant quality on the structure of aphid-parasitoid communities. Arrow thickness is scaled to standardized coefficients from path analysis to illustrate the relative strength of effects. (Reproduced with permission from Fig. 2 of Bukovinszky et al. 2008)

indices such as connectance (proportion of possible links among all species that are realized) and linkage diversity (feeding links among the species of aphid hosts and the species of their parasitoids) (Bukovinszky et al. 2008). One notable effect was aphid body size on the nonadjacent trophic level of secondary parasitoids. This effect began with larger aphids forming larger mummies that resulted in a greater number and size of mummies in the parasitoid guild, and elevated connectance and linkage diversity, as indicated by arrow widths (Fig. 11.20). This study emphasizes that an increase in plant quality causes bottom-to-up cascades across trophic levels that increase the size and number of herbivores and an abundance of primary and secondary parasitoids. An opposite condition is where a resource limitation, such as scarce availability of plant biomass promotes top-to-bottom regulation of trophic levels within a food web (Hunter and Matson 1992).

11.6.3 The Importance of Parasites and Parasitoids in Food Webs

Rather than examine the trophic structure of food webs as they exist in nature, discussed above, another approach is examination of the effect that the addition of parasites have on food web structure. Whereas insect parasitoids, probably are second only to herbivores as the most common lifestyle frequently included as primary data in food webs, parasites generally are not included (Lafferty et al. 2006). One reason for this is that parasites do not cause the death of their hosts, and their effects would be based on interaction strengths, or measures of the intensity or degree of interactions between two ecologically connected species, which are difficult to assess (Ings et al. 2009). However, within the last decade, parasites increasingly have been included explicitly in food webs (Dunne et al. 2013; Lafferty et al. 2008). Notably, incorporation of parasites in food webs that include predator–parasite and parasite–parasite links approximately doubles the connectance in food webs (Fig. 11.21) and changes other food-web indices. These and other, increased food-web indices include the number of links, nestedness (asymmetry of interactions), chain length (arithmetic average of the lengths of all chains in a food web) and linkage density (Lafferty et al. 2006). These data show that the increase of parasites in food webs may have had a supportive and parallel role in propelling the MMPR that largely was attributable to the evolutionary and ecological diversification of several, related parasitoid clades (Fig. 11.19). Because of the positive effect that parasites (and parasitoids) both have for increased connectance and nestedness in ecosystem stability (Lafferty et al. 2006, 2008), it is quite likely that the MMPR buffered local food-web structure by inserting top–down control of trophic cascades that balanced bottom–up control through resource limitation (Labandeira 2015). One relevant effect of the MMPR shown by the Messel food web study (Fig. 11.22) is that for the Messel lake part of the web, the highest trophic level consumer was a large crocodile. By contrast, the highest trophic level for the Messel forest part of the web was

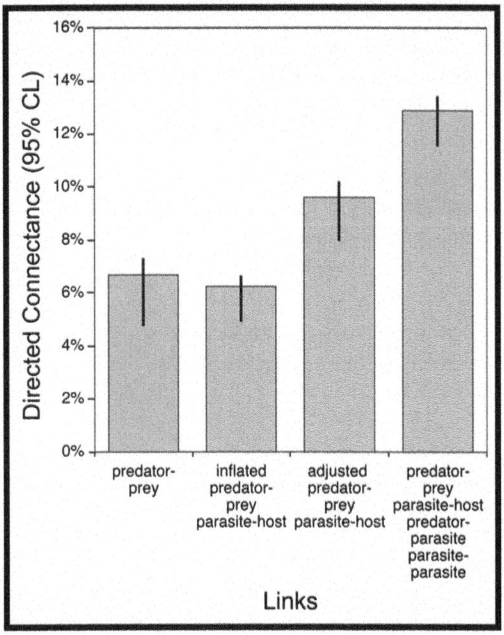

Fig. 11.21 A comparison of directed connectance (number of potential unidirectional links in a food web that have been realized) with and without parasite links in the Carpinteria Salt Marsh along the California coast. The first bar includes only predator–prey links. The second bar adds observed parasite–host links that may have been incorporated in an inappropriate manner (see Lafferty et al. (2006) for details). The third and fourth bars provide two, different methods for determining how parasites affect directed connectance. The third bar excludes parasite–parasite links. Comparison of the third and fourth bars with the first bar indicates that parasites increase directed connectance in food webs. Error bars represent 95% confidence limits. (Reproduced with permission from Fig. 2 of Lafferty et al. 2006)

a parasitoid fly, which incidentally was one additional trophic level higher than that of the crocodile (Dunne et al. 2014).

11.6.4 Top–Down Control of Food Webs by Parasitoids in Modern Ecosystems

During the past 25 years, an increasing number of studies have demonstrated the consequences that parasitoids have in modern ecosystems (Schowalter 2016). Much of this work has been done with the intricate effects in microcosms of hymenopteran and dipteran parasitoids and their hyperparasitoids on leaf mining and external foliage feeding larvae in tropical food webs (Thompson 1984). One study evaluating host specificity in parasitoids involved a community of 66 parasitoids, 60 predators, 19 herbivores, 5 omnivores, and 3 pathogens associated with broom, *Cytisus*

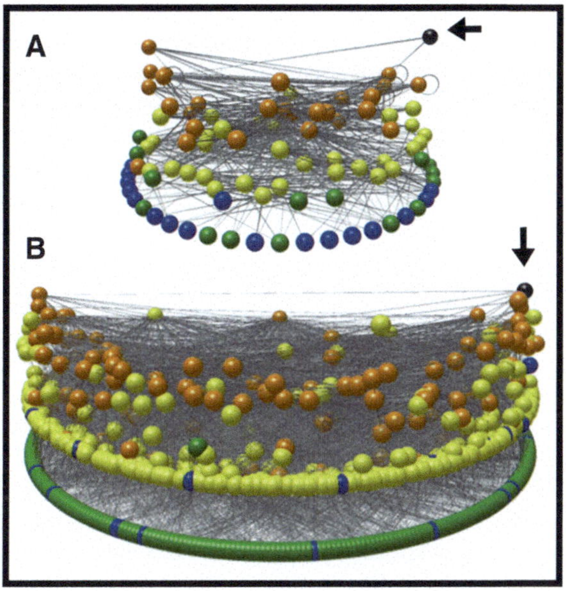

Fig. 11.22 The Messel lake and forest food webs. (**a**) Lake food web and (**b**) adjacent forest food web. Spheres represent taxa; lines represent feeding links. Links that loop indicate cannibalism. The vertical axis corresponds to short-weighted trophic level (the average of one plus the shortest chain length from a consumer to a basal species and the average of one plus the mean trophic level of all the consumer's trophic resources), with autotrophic taxa and detritus at the bottom level (Williams and Martinez 2004). Images produced with Network3D (Yoon et al. 2004; Williams 2010). Colors of nodes correspond to taxonomic affiliation of species. Green, plants including algae and diatoms; blue, bacteria, fungi and detritus; yellow, invertebrates; orange, vertebrates. The upper arrows refer to the highest trophic level in the lake web, a crocodile (black node); the lower arrow refers to the highest trophic level in the terrestrial web, a hyperparasitoid fly (black node). (Reproduced with permission from Fig. 1 of Dunne et al. 2014)

scoparius, at a single site in Berkshire, England (Memmott et al. 2000). This study showed that predators consumed a median of two host species, whereas the median for parasitoids was one host species, indicating greater host specificity for the parasitoids (Memmott et al. 2000). External foliage feeders were more vulnerable than concealed feeders such as leaf miners. The parasitoid sub-webs had considerably lower connectance (higher host specificity) than the predator sub-webs, a feature seen in other studies when parasitoid sub-webs were compared to predator webs (Van Veen et al. 2008). Another measure of parasitoid efficiency concerned a study of a host–parasitoid community in Guanacaste, Costa Rica. In this study (Memmott et al. 1994), leaf-miner host mortality due to parasitoidism varied greatly, but resulted in an overall value of 32.1% of leaf miners succumbing to parasitoids (Memmott et al. 1994).

One important avenue of research has been documenting apparent competition, in which species can interact, or compete, through shared natural enemies such as parasitoids (Holt 1977; Rott and Godfray 2000). In one study of a hyperdiverse community of insects in Belize (Morris et al. 2004), two species of leaf miners, a dipteran and a coleopteran, were removed from a diverse and speciose suite of leaf-mining insects (Fig. 11.23). After the removal of the two species, other species that shared the parasitoids of the removed species experienced lower parasitoid attack rates and increased population densities (Morris et al. 2004), suggesting that apparent competition involving parasitoids is an important feature in structuring tropical insect communities. In another study using a similar ecological context (Bukovinszky et al. 2008), all trophic relationships were examined among a community of aphids, parasitoids and secondary parasitoids in rural England (Müller et al. 1999). The experiment involved 26 species of plants, their 25 species of aphid herbivores, attacked by 18 species of primary parasitoids, and the 28 species of secondary parasitoids—who pursued two different feeding strategies—that attacked them in turn (Müller et al. 1999). The results of this study indicated that, for the 11 webs during the examined period, the ratios of the number of aphid species to number of primary parasitoid species and to the number of secondary parasitoid species were relatively the same across the webs (Fig. 11.24). The ratio of the number of links involving secondary parasitoids and primary aphid parasitoids also was constant across the webs. Quantitative parasitoid overlap graphs for understanding the apparent competition interactions among aphids revealed the robust nature of the indirect links,

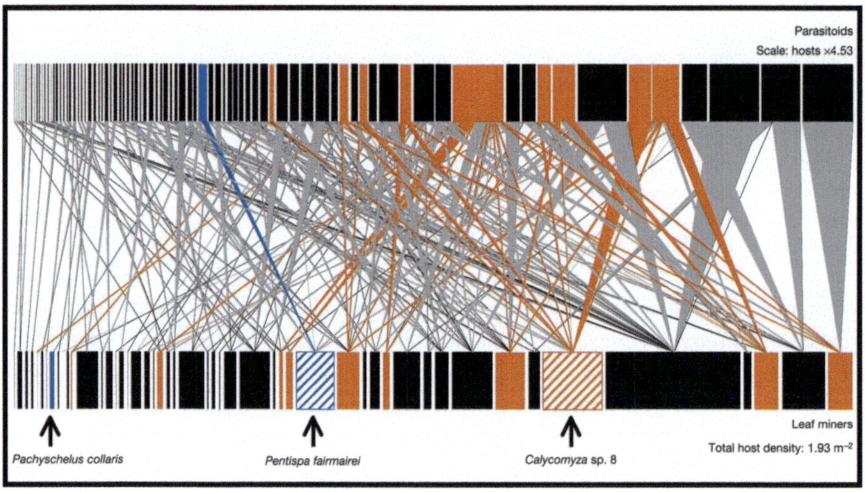

Fig. 11.23 Quantitative food web (Lewis et al. 2002) showing parasitoid species (top bars), leaf-miner species (bottom bars), trophic links among them, and the species predicted to be affected by a species-removal manipulation. The leafmining fly *Calcomyza* sp. 8 and leaf beetle *Pentispa fairmairei* were herbivore species that were removed. Dipteran leaf-miner species present during the sampling period and predicted to be affected indirectly via parasitoids shared with *Calcomyza* sp. 8 are shown in red. The metallic wood-boring beetle *Pachyschelus collaris* (blue) was also predicted to be affected indirectly by the manipulation through parasitoids that it shares with *P. fairmairei*. Only hosts from which the parasitoids were reared are shown in the web. Bar widths are proportional to species abundance at the study site. (Reproduced with permission from Fig. 1 of Morris et al. 2004)

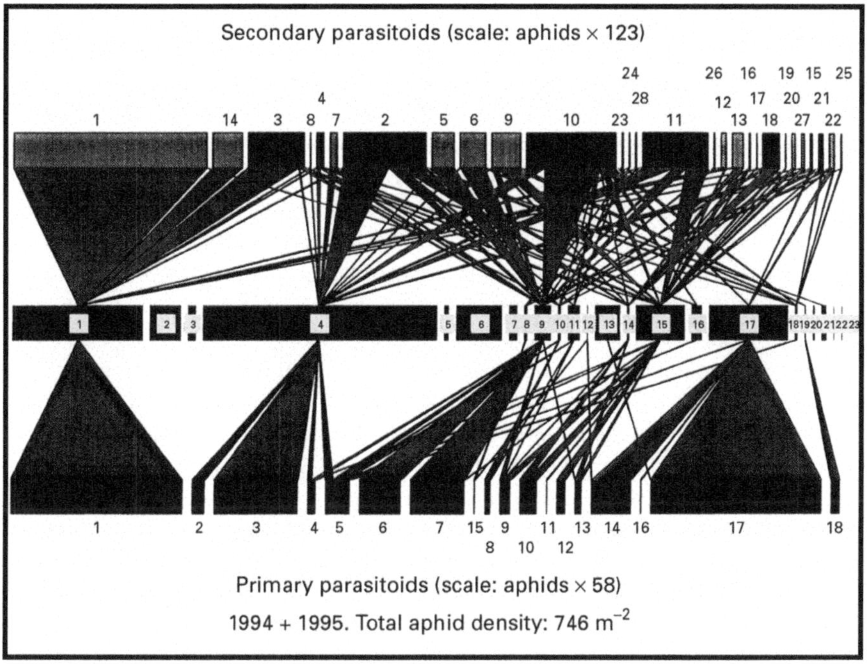

Fig. 11.24 Summary diagram for describing a complete parasitoid web sampled over 2 years. Relative aphid abundances are shown in the center with primary parasitoids below and secondary parasitoids above. (Hyperparasitoids are in grey and primary parasitoids are in black.) The numbers are the species codes (see Müller et al. 1999). Species densities are shown to scale within each month for aphids and the two categories of parasitoids. See Müller et al. (1999) for a fuller description and interpretation of the web diagrams. (Reproduced with permission from Fig. 4 of Müller et al. 1999)

such that common aphid species shared a few, strong, indirect and mostly asymmetrical links through common primary parasitoids and hyperparasitoids. Both studies (Müller et al. 1999; Morris et al. 2004) and other similar studies (Hirao and Mukrakami 2008; Peralta et al. 2014), indicate that the appearance of parasitoid and hyperparasitoid feeding guilds increases the incidence, extent and efficiency of attack on herbivores, resulting in greater top–down regulation of food webs.

11.6.5 Insect Faunas Before the Mid-Mesozoic Parasitoid Revolution

Below, four of the most diverse deposits that soon preceded the beginning of the MMPR are discussed, which range from the Middle Triassic–Late Triassic boundary interval (Ladinian–Carnian stages) to the Early Jurassic (Pliensbachian Stage). These deposits originate from disparate locales in Central Asia, South Africa, China and eastern United States. An examination of insect taxa from these deposits reveal the presence of three insect orders—Hymenoptera, Diptera and Coleoptera—that later in the fossil record would contribute the overwhelming bulk of parasitoid taxa (Fig. 11.19). However, it is evident from the taxa in each of the four most diverse deposits of this time interval that parasitoids are absent.

The first deposit is the Madygen Formation, of Ladinian to early Carnian in age, consistent with a date of 237–220 Ma (Walker et al. 2013). Productive sites of the Madygen Formation are located in the Fergana Valley where the frontiers of Tajikistan, Kyrgyzstan and Uzbekistan complexly intersect (Shcherbakov 2008). The principal localities are from the Dzhailyoucho area that consisted of an extensive Triassic lake deposit. No parasitoid insects are known. The relevant taxa are Hymenoptera containing sawflies of the Xyeloidea (Xyelidae), and Diptera that are represented only by the nematocerous taxa of Tipulomorpha (Limoniidae, †Vladipteridae), Psychodomorpha (†Nadipteridae, †Hennigmatidae) and Bibionomorpha (†Protorhyphidae) (Shcherbakov 2008). The second deposit is the Molteno Formation, which is slightly younger and is centered in the early Late Triassic Carnian Stage, approximately 237–228 Ma (Walker et al. 2013), although accurate age dates have not been determined from radioisotopic age dating. All 106 documented Molteno localities (Labandeira et al. 2018) originate from outcrops of the Karoo Basin, mostly in South Africa that surround Lesotho. The major localities are lake deposits such as Aasvoëlberg 411 and Birds River 111 (Anderson and Anderson 2003). No parasitoid taxa are known (Riek 1974). The only Hymenoptera known is a probable sawfly of Xyelidae (Schlüter 2000), and Diptera are absent from the Molteno Formation (Anderson and Anderson 1993; Labandeira et al. 2018).

The second and third localities, from the Late Triassic localities to Triassic–Jurassic boundary interval display no evidence for Hymenoptera, an increased diversity of non-parasitoid Diptera, and the presence of non-parasitoid Coleoptera, all of which indicate that the MMPR was in the future. The third deposit are the related Beishan Formation and Shangtu Formation of Jilin and Hebei provinces, respectively, which are of Late Triassic age (Rhaetian Stage), corresponding to an age date of 209–201 Ma (Walker et al. 2013), although historically the position of these deposits in the latest Triassic to earliest Jurassic continuum have been contentious (Lin 1982, 1986). There are several localities within each of these two formations have produced fossil insects from moderately diverse assemblages (Lin 1982, 1986; Grimaldi and Engel 2005a). Hymenoptera is absent; Diptera is represented by the nematocerous groups of Tipulomorpha (†Eolimnobiidae, Limoniidae) and Bibionomorpha (†Pleciofungivoridae), and Coleoptera consist of taxa that lack known parasitoid members in the later Mesozoic and Cenozoic faunas (Lin 1982, 1986). Parasitoid taxa have not been discovered.

The fourth deposit is the Cow Branch Formation of the Martinsburg area along the North Carolina–Virginia state border in the eastern United States. The Cow Branch Formation is Late Triassic (Rhaetian Stage) to Early Jurassic (Hettangian Stage) in age, equivalent to 209–199 Ma (Walker et al. 2013). The specific, major locality yielding the fossils is the Solite Quarry, which is one segment of a series of rift basins throughout the east coast of North America that are represented by very fine-grained shales (Grimaldi and Engel 2005a). The Solite Quarry has not yielded Hymenoptera. However, Diptera are diverse compared to other Late Triassic to earliest Jurassic intervals, consisting of the nematocerous Tipulomorpha (†Vladipteridae, Limoniidae), Psychodomorpha (Psychodidae, †Eoptychopteridae, an undetermined family), Culicomorpha (†undetermined family), Bibionomorpha (†Procramptonomyiidae, †Protorhyphidae, †Paraxymyiidae, †Crossaphididae), and stem-group Brachycera (†Prosechamyiidae; Blagoderov et al. 2007). Although the undetermined family of Culicomorpha could have been a parasite as a blood feeder (Fig. 11.8; Blagoderov et al. 2007), all other evidence from subsequent fossil lineages and modern biology would disallow a parasitoid interpretation for this culicomorph. The Solite Coleoptera includes Staphylinidae (Fraser et al. 1996), a rove beetle, which almost certainly did not belong to a subclade such as the Aleocharinae that includes modern parasitoid members (Klimaszewski 1984).

11.6.6 Insect Faunas During the Mid-Mesozoic Parasitoid Revolution

Four of the most informative insect faunas for documenting the initial expansion interval of the MMPR during phases 1 and 2 range from the latest Middle Jurassic (late Callovian Stage) to the mid Early Cretaceous (mid Albian Stage), before the shift from phase 2 to phase 3. The deposits, from oldest to youngest, originate from northeastern China, south-central Kazakhstan, southern England in the United Kingdom, and back to northeastern China. Insect taxa from these four deposits collectively exhibit (1) the first appearances and diversification and ecological expansion of non-proctotrupomorph and proctotrupomorph hymenopteran parasitoids; (2) to a lesser extent, the expansion of non-Eremoneura Diptera and Coleoptera; and (3) the modest beginnings of the Aculeata Hymenoptera and perhaps Strepsiptera diversifications. The oldest Early Cretaceous amber deposits such as Lebanese and Álava amber are undoubtedly extensions of Phase 2 of the MMPR, but their parasitoid faunas remain largely unstudied. By the end of phases 1 and 2 of the parasitoid diversification events, six of the seven major clades of parasitoids were established; the exception was Strepsiptera.

The first two insect faunas are Jurassic in age and they provide documentation for diversification of the earliest lineages of parasitoids. The first fauna is from the Jiulongshan Formation, representing the Yanliao Biota, and located in the Daohugou area where the three provinces of Liaoning, Hebei and Inner Mongolia come together (Ren et al. 2010a; Huang 2016). The age of the Yanliao Biota is 165 Ma, based on secure radioisotopic age dates (Ren et al. 2010b; Huang et al. 2016), and equivalent to the late part of the Callovian Stage (Walker et al. 2013). Yanliao fossils occur mostly in medium to dark hued tuffaceous siltstones that were deposited in fluvial and geographically extensive lacustrine environments (Ren et al. 2010b). By the time the Jiulongshan Formation was deposited, there is evidence for the commencement of a major evolutionary expansion of hymenopteran parasitoid lineages. Probable parasitoids of Neuroptera consisted of one family, Mantispidae, with two genera and two species. Possible parasitoid taxa of Coleoptera may have been present, representing two families, Staphylinidae and Ripiphoridae, four genera and four species (Tan et al. 2010), indicating that beetles were a very minor component of the parasitoid guild. For Staphylinidae, there is no evidence that any subfamilies known to house parasitoid taxa, such as Aleocharinae, were present during the late Middle Jurassic. Dipteran parasitoid taxa were less diverse, consisting of Nemestrinimorpha (Nemestrinidae) and Eremochaetidae representing two families, five genera and eight species (Zhang et al. 2017; Ren et al. 2019), marking the early presence of dipteran parasitoids in the MMPR. Well-represented hymenopteran taxa include a diversity of plant-associated sawflies ("symphytans") that are placed in the superfamilies Xyeloidea (Xyelidae, †Daohugoidae), Tenthredinoidea (†Xyelotomidae), Pamphilioidea (†Xyelydidae), Cephoidea (†Sepulcidae), Siricoidea (Anaxyelidae, Siricidae), and the parasitoid Orussoidea (†Karatavitidae) that shares a sister-group relationship with parasitoid Apocrita. Apocritan parasitoids occur in the three superfamilies Evanioidea (†Anomopterellidae, †Praeaulacidae), Ephialtitoidea (†Ephialtitidae) and Proctotrupoidea (Heloridae, †Mesoserphidae, Pelecinidae, Roproniidae) that collectively account for approximately 15 genera (Rasnitsyn and Zhang 2004, 2010; Gao et al. 2010; Huang and Cai 2016; Li et al. 2018a; Wang et al. 2019; Table 11.1). This distribution of hymenopteran parasitoid families indicates that eight major, family-level lineages of orussoid sawflies and apocritan parasitoid wasps were present, extending into the basal Proctotrupomorpha superclade during Yanliao time (Fig. 11.19). The Jiulongshan Formation probably represents the earliest, most extensive sample of Phase 1 of the MMPR of any biota worldwide.

The second major insect fauna is the Karabastau Formation, representing the Karatau Biota, located along the Karatau Range, near the towns of Aulie (formerly Mikhailovka) and Uspenovka (formerly Galkino) in southern Kazakhstan, and considered Late Jurassic in age. Because of the multiple localities representing multiple stratigraphic horizons at Karatau, the deposits are assigned to a time range that includes the Oxfordian and Kimmeridgian stages (Rasnitsyn and Zherikhin 2002), corresponding to 164–152 Ma (Walker et al. 2013), making the Karatau Biota seven million years younger, on average, than the Yanliao Biota. The insect fossils occur in dark grey shales that preserve detail such as wing eyespots, delicate surface ornamentation and fine hairs on body surfaces (Rohdendorf 1968a; Grimaldi and Engel 2005a). Neuroptera consists of only one genus and species of Mantispidae. Concerning the parasitoid taxa, Coleoptera consisted of 2 families, Staphylinidae and Anthribidae, consisting of 10 genera and 12 species, although Staphylinidae at Karatau have not been assigned to the extant parasitoid subfamily Aleocharinae (Tichomirova 1968; Yu et al. 2019), which contains many modern parasitoid species (Klimaszewski 1984). Parasitoid Diptera include Nemestrinimorpha (Nemestrinidae and Acroceridae) and Archisargoidea (Eremochaetidae), consisting of 3 families, 8 genera and 16 species of potentially parasitoidic flies (Rohdendorf 1968b; Mostovski 1998; Zhang et al. 2016) (Table 11.1) comparable in diversity to that of the Yanliao

Biota. No evidence exists for parasitoid Strepsiptera, Trichoptera or Lepidoptera. Hymenoptera included a similar spectrum of sawflies (Rasnitsyn 1968) and parasitoid lineages that were more diverse than those in the Yanliao Biota. The hymenopteran parasitoid lineages include the parasitoid sawfly Orussoidea (†Karatavitidae), and parasitoid wasps Ephialtitoidea (†Ephialtitidae), Megalyroidea (Megalyridae), Evanioidea (†Anomopterellidae, †Praeaulacidae), Proctotrupoidea (Heloridae, †Mesoserphidae) and the basal aculeate lineage Chrysidoidea (†"Bethylonymidae") (Kozlov 1968; Rasnitsyn 2002) (Table 11.1). The same number of eight, family-level lineages of hymenopteran parasitoids was present in the Karatau Biota as in the Yanliao Biota, but with two family substitutions and a doubling of species to approximately 30, indicating evolutionary stability at the family level but an increased proliferation of species. In general, it appears that parasitoids of the Karatau Biota represent continued stability of major parasitoid lineages when compared with the Yanliao Biota, but increased generation of genera and species at lower taxonomic levels.

The third insect fauna considered during the initial expansion of the MMPR comes from the Lulworth Formation, constituting the lower unit of the Purbeck Limestone Group, and its upper unit, the Durlston Formation, which crop out along the Vale of Wardour area in Dorset, southern England (Rasnitsyn et al. 1998; Coram et al. 2000; Coram and Jepson 2012). The age of the Purbeck strata is earliest Cretaceous (Berriasian Stage), with a corresponding age date of 145–139 Ma (Walker et al. 2013), and is approximately 16 million years younger than the Karatau Biota. The strata of the Purbeck Biota containing the insect fossils consist of fine-grained, mostly thinly bedded limestone occasionally interrupted by algal mat layers (Coram 2003). Parasitoid-containing lineages of Neuroptera, Coleoptera, Strepsiptera, Diptera, Trichoptera, and Lepidoptera are absent. However, as in the preceding two fossil units, a near full complement of sawflies is present, excluding the Orussoidea—the only parasitoid sawfly clade. Apocritan wasps consist of Megalyroidea (Megalyridae), Evanioidea (cf. Aulacidae, †Baissidae, †incertae sedis), Proctotrupoidea (Diapriidae, Proctotrupidae), Ichneumonoidea (Ichneumonidae), the aculeate lineages Chrysidoidea (†"Bethylonymidae") and Apoidea (Sphecidae), and the undetermined apocritan *Apocrites* (Coram and Jepson 2012) (Table 11.1). Including unassigned but distinctive lineages, there are 10 family-level lineages of hymenopteran parasitoids representing about 20 species in the Purbeck Biota.

The Yixian Formation is the fourth insect fauna that highlights Phase 2 of the MMPR (Fig. 11.19). Fossils from the Yixian Formation formed the distinctive Jehol Biota that originate from seven or eight major, fossil-yielding localities, mostly in Liaoning Province of northeastern China, west of Beipiao City and near Liaodong Bay, an arm of the Yellow Sea (Zhang et al. 2010). Although once controversial (Ren et al. 2010b), the date currently is established as mid Early Cretaceous, from 126 to 122 Ma (Swisher III et al. 1999), equivalent to late Barremian to early Aptian in age (Walker et al. 2013), and approximately 18 million years younger than the Purbeck Biota. The fossils occur in broad outcrops of lake deposits with light-colored, highly oxidized, fine-grained mudstone and siltstone strata of buff-colored tuffaceous shales (Ren et al. 2010b). Neuroptera consists of seven genera and nine species of Mantispidae. Potentially parasitoid Coleoptera consisted of 2 families, Staphylinidae and Carabidae, accounting for 21 genera and 37 species (Yu et al. 2019). The overwhelming majority of the beetles were Staphylinidae, although subfamily placement is tricky (Tan et al. 2010). Parasitoid Diptera include Nemestrinimorpha (Nemestrinidae) and Eremochaetidae, consisting of two families, six genera and ten species (Han et al. 2019), generally consistent with the values from the earlier Karatau and Yanliao biotas. For Hymenoptera, the several, major, symphytan lineages detailed in the previous three biotas are repeated in the Jehol Biota. Hymenopteran parasitoid lineages are Ephialtitoidea (†Ephialtitidae), Evanioidea (Aulacidae, †Baissidae, Evaniidae, †Praeaulacidae), Proctotrupoidea (†Mesoserphidae, Heloridae, Pelecinidae, Roproniidae, Serphidae), Ichneumonoidea (Ichneumonidae), and the aculeate Chrysidoidea (†"Bethylonymidae") and Vespoidea (Scoliidae) (Table 11.1). The increase to 13 hymenopteran families of the Jehol Biota from the 8 or 9 occurrences from the previous three biotas is significant. In addition, the presence of approximately 60 Jehol species assigned to the 13 families triples the species from the Purbeck Biota, which consist of about 20 species allocated to 9 families. The Jehol data indicate that there was a quantitative increase in the numbers of parasitoid families and species from Phase 1 to Phase 2 of the MMPR.

11.6.7 Insect Faunas After the Mid-Mesozoic Parasitoid Revolution

After phases 1 and 2 established the MMPR, Phase 3 increased the upward trend by adding a considerable number of families of two lineages: proctotrupomorph Hymenoptera and aculeate Hymenoptera (Fig. 11.19). During Phase 3, the trend lines for all other lineages remained at the same relative levels as they did for Phase 2. In Phase 4, there were increases of families in proctotrupomorph Hymenoptera, eremoneuran Diptera, Coleoptera and the earliest occurrence of valid Strepsiptera; the levels of other lineages essentially held flat. These differences in representation of the seven parasitoid lineages in phases 3 and 4 can be gleaned from four diverse insect faunas that strategically sample every few tens of millions of years the Late Cretaceous and Paleogene.

Some of these trends may be conditioned by taphonomic style, in which the history of the parasitoid fossil record is very asymmetric. Because of the relative hard boundary of no biologically significant amber deposits occurring before approximately 135 million years ago (Labandeira 2014a), deposits before, during and after the MMPR have a differing taphonomic cast, from which the primary data originates. Deposits prior to the MMPR—Madygen (237–220 Ma), Molteno (237–228 Ma), Beishan–Shangtu (209–201 Ma) and Cow Branch biotas (209–199 Ma)—consist only of compression-impression fossils. Deposits during the MMPR—the Yanliao (165 Ma), Karatau (approximately 158–156 Ma), Purbeck

(154–139 Ma) and Jehol (125 Ma) biotas—also contain compression fossils, although during this time the first appearance of major amber deposits such as Lebanese amber (130–120 Ma) (Maksoud et al. 2017) and Spanish Álava Amber (112–104 Ma) occur (Peñalver and Delclòs 2010; Azar et al. 2010). By contrast, insect faunas deposited after the MMPR—Myanmar Amber (99 Ma), Canadian Amber (79–78 Ma), Messel (48 Ma) and Dominican Amber (20.5 Ma) biotas—incorporated more amber material. Consequently, this difference between deposits before and after the MMPR imparted a distinctive preservational trend. Because of this distribution of compression–impression versus amber deposits, there are different taphonomic qualities imparted to each of these 12 deposits. Advantages of compression deposits are: (1) a greater temporal completeness compared to amber deposits that are absent from deposits older than about 135 million years; (2) fossils occurring on rock slabs typically with expansive two-dimensional surfaces; and (3) representation of a range of ecosystems, such as those from fluvial, lacustrine, deltaic and swamp environments (Labandeira 2014a). Advantages of amber deposits are: (1) typically a significantly higher quality or preservation; (2) good availability of trophic data to understand inter-organismic relationships and food webs; and (3) elevated documentation of pathogens, parasites, parasitoids and evidence for disease that rarely are found in compression–impression deposits (Labandeira 2014a; Poinar 2021). Although these two, major modes of preservation are different, they are nevertheless complimentary; for example, amber deposits preserve small to miniscule insects that would rarely be preserved in compression deposits. Compression and amber deposits jointly provide a much more accurate of the fossil record than would their individual representations.

The first major insect fauna of Phase 3 is Myanmar Amber originating in Kachin State, in the Myitkyina and Upper Chindwin Districts along the Hukawng Valley of northern Burma (Cruikshank and Ko 2003). The lithostratigraphic context of Myanmar Amber is still poorly known, although the amber comes from lignite layers interbedded with thin strata of sandstones, siltstones, shale and micritic limestone (Zherikhin and Ross 2000). Source trees of most of Myanmar Amber are the gymnospermous Araucariaceae (*Agathis*) and angiospermous Dipterocarpaceae, which have yielded a lowermost Late Cretaceous radioisotopic date of approximately 99 Ma, equivalent to the early interval of the Cenomanian Stage (Shi et al. 2012) and approximately 23 million years more recent than the Jehol Biota. Myanmar Amber contains a diverse insect biota (Rasnitsyn and Ross 2000; Ross et al. 2010; Ross 2018 and updates). The single neuropteran parasitoid taxon of Myanmar Amber is Hemerobioformia (Mantispidae). Coleopteran parasitoid taxa were Caraboidea (Carabidae), Staphylinoidea (Staphylinidae), Cleroidea (Cleridae), Cucujoidea (Passandridae), Tenebrionoidea (Meloidae, Ripiphoridae) and Curculionoidea (Anthribidae). Strepsipteran parasitoids were Protostrepsiptera (†Phthanoxenidae, †Cretostylopidae), Eleostrepsiptera (†?Mengeidae) and Neostrepsiptera (†Kinzelbachillidae). The single, possible, trichopteran parasitoid taxon is Hydroptilidae. Dipteran parasitoid taxa consisted of Nemestrinimorpha (Acroceridae, Nemestrinidae), Archisargoidea (†Eremochaetidae), Asiloidea (Asilidae, Bombyliidae), Empidoidea (Empididae) and Aschiza (Phoridae,

Pipunculidae). Also diverse were hymenopteran parasitoid taxa that consisted of Stephanoidea (†Aptenoperissidae, †Myanmarinidae, Stephanidae), Megalyroidea (Megalyridae), Trigonalyroidea (†Mametshidae), Evanioidea (Aulacidae, Evaniidae, Gasteruptiidae, †Othniodellithidae, †Praeaulacidae), Proctotrupoidea (Austroniidae, Diapriidae, Heloridae, Pelecinidae, †Peleserphidae, †Spathiopterygidae), Platygastroidea (Platygastridae, Scelionidae), Ceraphronoidea (Ceraphronidae, Megaspilidae, †Stigmaphronidae), Mymarommatoidea (†Gallorommatidae, Mymarommatidae), Serphitoidea (†Serphitidae), Chalcidoidea (Chalcidae, indeterminate family, Mymaridae), Ichneumonoidea (Braconidae, Ichneumonidae), and the aculeate Chrysidoidea (Bethylidae, Chrysididae, Dryinidae, Embolemidae, Scolebythidae), Vespoidea (Pompilidae, Rhopalosomatidae, Sapygidae, Sierolomorphidae, Tiphiidae, Vespidae) and Apoidea (Crabronidae, Sphecidae) (Grimaldi et al. 2002; Ross et al. 2010; Engel et al. 2012a; Cai and Huang 2014; Cai et al. 2017, 2018; Li et al. 2017a, b, 2018a, b; Ross 2018 and updates) (Table 11.1). Parasitoid Hymenoptera in the Myanmar Biota numbers 42 families, representing a third more than that of the earlier Jehol Biota. There are approximately 90 hymenopteran species, about an increase of three times that of the Jehol Biota. The Myanmar Biota represents a considerable expansion in the number of parasitoid species, indicating that these deposits record a major parasitoid diversification event following Phase 2 but before deposition of the Myanmar Biota (Fig. 11.19). However, some of this increase in Myanmar parasitoid diversity could be attributable to exceptional preservation and a concerted effort to mine, process and expand tonnages of material sold abroad (Sokol 2019).

The second major deposit of note, also representing Phase 3, is Canadian Amber, originating from Grassy Lake in southern Alberta, but also occurring at Cedar Lake in western Manitoba where it is secondarily deposited (McKellar and Wolfe 2010). The stratigraphic source of the amber at Grassy Lake are six sub-bituminous coal seams of the Foremost Formation (Pike 1995) that have a late Campanian age, approximately equivalent to 78–72 Ma (Walker et al. 2013), about 24 million years after deposition of Myanmar Amber. The source of the amber initially was thought to be araucariaceous, but plant anatomical and spectroscopic analyses indicate a cupressaceous origin, in particular the tree *Parataxodium* (McKellar et al. 2008). Potential parasitoid taxa of Coleoptera are Caraboidea (Carabidae), Staphylinoidea (Staphylinidae), and Cleroidea (Cleridae). One, unidentified, triungulin larva unassignable to family (Skidmore 2018) represents a parasitoid Strepsiptera. Dipteran parasitoid taxa are still relatively modest compared to earlier occurrences, consisting of taxa in Bibionomorpha (Cecidomyiidae, Mycetophilidae), Asiloidea (Bombyliidae), Empidoidea (Empididae) and Aschiza (Phoridae). No parasitoid taxa of Neuroptera, Trichoptera or Lepidoptera are present. For Hymenoptera, the parasitoid lineages are Evanioidea (Aulacidae), Trigonalyroidea (Mametshidae), Proctotrupoidea (Diapriidae, Proctotrupidae), Cynipoidea (Figitidae, Liopteridae), Platygastroidea (Platygastridae, Scelionidae), Ceraphronoidea (Ceraphronidae, Megaspilidae, Stigmaphronidae), Mymarommatoidea (Mymarommatidae),

Serphitoidea (†Serphitidae), Chalcidoidea (Eulophidae, Eupelmidae, Mymaridae, Rotoitidae, Tetracampidae, Torymidae, Trichogrammatidae), Ichneumonoidea (Braconidae, Ichneumonidae) and the aculeate Chrysidoidea (Bethylidae, Chrysididae, Dryinidae, Scolebythidae) and Apoidea (Sphecidae) (Carpenter et al. 1937; Evans 1969; Yoshimoto 1975; Poinar Jr and Huber 2011; Perrichot et al. 2011; McKellar and Engel 2011a, b, 2012, 2014; McKellar et al. 2013; Engel et al. 2013c; Skidmore 2018) (Table 11.1). Parasitoid Hymenoptera consists of 27 families and 112 described species in Canadian Amber, representing somewhat fewer families but a greater number of species than that of Myanmar Amber. The parasitoid wasp fauna from Canadian Amber houses very few extinct lineages and the first documented occurrences of some modern parasitoid lineages, reflecting a relatively flat level of diversity during the middle of Phase 3, after the MMPR (Fig. 11.19), but with evident taxonomic turnover.

The third deposit toward the end of Phase 3 is Messel, in Hesse, western Germany, and consists of a maar lake resulting from a deep, explosive eruption of rhyolitic magma that formed a deep depression subsequently infilled by sediment (Lorenz and Kurzlaukis 2007). The resulting small lake trapped microorganisms, plants, insects and vertebrates that were excellently preserved (Dunne et al. 2014). The sediments consist of an oil shale that contain fossils entombing a wealth of micromorphological detail, including leaf cuticle, differential hues representing original color patterns, insect setae and other delicate features (Felder and Harms 2004). The Messel Biota was ecologically characterized in a food-web study, consisting of approximately 700 biological species or trophic groups, and the resulting, highly resolved food web (Fig. 11.22) was constructed for the full ecosystem and the separate lake and terrestrial sub-ecosystems (Wedmann 2005; Dunne et al. 2014). The parasitoid community of the Messel Biota is well established (Dunne et al. 2014; Labandeira and Dunne 2014) from both the primary literature and its ecological context in the associated DRYAD data (Labandeira and Dunne 2014). A single parasitoid species represents Hemerobioformia (Mantispidae), and coleopteran parasitoids consisted of Caraboidea (Carabidae), Staphylinoidea (Staphylinidae), and Scarabaeoidea (Scarabaeidae). The sole Strepsiptera parasitoid is Neostrepsiptera (Myrmecolacidae). The Diptera parasitoids were Culicomorpha (Chironomidae), Bibionomorpha (Cecidomyiidae), Nemestrinimorpha (Nemestrinidae) and Muscomorpha–Asiloidea (Asilidae). The Hymenoptera provided the overwhelming bulk of parasitoid taxa, consisting of Proctotrupoidea (family indeterminate), Chalcidoidea (Chalcididae, Eucharitidae, Eulophidae, Torymidae), Ichneumonoidea (Braconidae, Ichneumonidae), the aculeate Vespoidea (Pompilidae, Scoliidae, Tiphiidae), and Apoidea (Sphecidae) (Dunne et al. 2014). The hymenopteran parasitoids of the Messel Biota consisted of 11 families and 12 species. For such a diverse ecosystem, this is a modest account of parasitoidism for the Messel Biota towards the end of Phase 3 (Fig. 11.19). Nevertheless, in the food web analysis

(Dunne et al. 2014; Labandeira and Dunne 2014), the highest trophic level in the forest web was a parasitoid fly (Wedmann 2007) (Fig. 11.22).

Dominican Amber is the fourth and last deposit after the MMPR to be examined. Dominican Amber originates from the Cordillera Septentrional, the northwest–southeast trending mountain axis along the Cordillera Oriental that parallels the coast of the Dominican Republic (Penney 2010). The amber is associated with lignite seams that are interspersed among sandstone and occasionally siltstone in the La Toca and Yanigua Formations that represent environments close to sea level that subsequently have been uplifted and deformed (Iturralde-Vinent and MacPhee 1996). The age of Dominican Amber has been controversial for the past 35 years. Based on a variety of techniques, the age of Dominican Amber is now usually considered as lower Miocene (Grimaldi 1996; Iturralde-Vinent 2001; Penney 2010), equivalent to 16–23 Ma (Walker et al. 2013). The midpoint of this range is about 21 Ma, which is the often-used age of Dominican Amber. The source of the amber is the extinct species of the tree *Hymenaea protera* (Fabaceae), whose leaves, stipules, buds, flowers and pollen are often found dispersed within the amber. The Dominican Amber Biota constitutes one of the most diverse and abundant amber biotas known, and contains a broad spectrum of insect taxa (Arillo and Ortuño 2005; Penney 2010; Poinar Jr 2010). A parasitoid-bearing lineage of Neuroptera is Hemerobioformia (Mantispidae) and Coleoptera that contributed two parasitoid-bearing groups: Staphylinoidea (Staphylinidae) and Tenebrionoidea (Ripiphoridae). Strepsipteran parasitoids were Neostrepsiptera consisting of Bohartillidae, Elenchidae, Myrmecolacidae and Protelencholacidae. Parasitoid-bearing dipteran lineages are Culicomorpha (Chironomidae), Bibionomorpha (Mycetophilidae), Nemestrinimorpha (Acroceridae), and in Muscomorpha: Asiloidea (Asilidae), Empidoidea (Empididae), Aschiza (Phoridae, Pipunculidae) and Schizophora (Muscidae, Tachinidae). For Trichoptera, the sole parasitoid-bearing lineage, Hydroptilidae, is present. Similarly, the obligately parasitoid lineage of Lepidoptera, Zygaenoidea (Epipyropidae), has been recorded. Hymenopteran parasitoid-bearing lineages were the most abundant compared to the seven preceding biotas and contained many major elements of the MMPR. These elements were Orussoidea (Orussidae), Evanioidea (Evaniidae), Platygastroidea (Platygastridae, Scelionidae), Ceraphronoidea (Ceraphronidae), Chalcidoidea (Encyrtidae, Eulophidae, Eupelmidae, Leucospidae, Mymaridae, Pteromalidae, Torymidae), Ichneumonoidea (Braconidae), and the aculeate Chrysidoidea (Bethylidae, Chrysididae, Dryinidae, Sclerogibbidae, Scolebythidae), Vespoidea (Mutillidae, Pompilidae) and Apoidea (Crabronidae, Sphecidae) (Arillo and Ortuño 2005; Engel 2008b; Penney 2010; Poinar Jr 2010) (Table 11.1). The diversity of parasitoid-associated hymenopteran families in the Dominican Amber Biota is 22 families and 94 species, a considerable increase over the Messel Biota but very roughly comparable to the earlier Canadian and Myanmar amber biotas. It appears that the Dominican Amber Biota is positioned in the middle of a plateau of parasitoid diversity representing the 40 million-year-long interval from the middle Eocene (Lutetian Stage) to the late Miocene (Messinian Stage).The parasitoid guilds have a very modern cast, and notably none of the family-level lineages are extinct.

11.7 Parasitoid Clade Diversification in the Early Mid-Mesozoic Parasitoid Revolution

The beginning of the MMPR was mentioned informally as long ago as the mid twentieth century by Carpenter (1954) who recognized the significance of Middle Jurassic wasps with very long ovipositors. The Mid-Mesozoic Parasitoid Revolution (MMPR) is defined as consisting of phases 1 and 2, from the Early Jurassic (Toarcian Stage) to the Early Cretaceous (Albian Stage), which constituted the core of the parasitoid expansion. This period is the MMPR. Phases 3 and 4, from the Cenomanian Stage to the present, is considered the subsequent, post-MMPR expansion, whose diversities for some clades or groups remained flat (Hymenoptera–Aculeata, non-proctotrupomorph Hymenoptera, non-eremoneuran Diptera) whereas for others there were substantial diversity increases (Hymenoptera–Proctotrupomorpha, Diptera–Eremoneura, Coleoptera, Strepsiptera) (Fig. 11.19). This latter period is the post-MMPR expansion. (Another major biological event is the Mesozoic Marine Revolution (Vermeij 1977). This event also had an initial pulse and a subsequent period of expansion.)

Four major insect lineages participated in the MMPR: Hymenoptera, representing 63% of fossil and modern family-level occurrences, Diptera (15.8%), Strepsiptera (11.6%) and Coleoptera (6.8%), and the other lineages of Neuroptera, Trichoptera and Lepidoptera having minor effects (2.8%) (Table 11.2). However, subclades of these lineages played different roles chronologically during the four phases of the ascendancy of parasitoids in continental ecosystems (Table 11.3). For the MMPR, consisting of phases 1 and 2 that range from the late Early Jurassic to the late Early Cretaceous, the overwhelming majority of parasitoid families were members of non-proctotrupomorph Hymenoptera, Proctotrupomorpha–Hymenoptera,

Table 11.3 The four phases of the Mid-Mesozoic parasitoid revolution[a]

Phase	Time interval of expansion	Major clades or groups represented[b]
1	(late) Early Jurassic–(early) Early Cretaceous (Toarcian–Valanginian)	Hymenoptera: Proctotrupomorpha; Diptera: Non-Eremoneura; Coleoptera; Hymenoptera: non-Proctotrupomorpha
2	(early) Early Cretaceous–(late) Early Cretaceous (Hauterivian–Albian)	Hymenoptera: Proctotrupomorpha; Hymenoptera: Aculeata; Hymenoptera; non-Proctotrupomorpha; Coleoptera; Diptera: non-Eremoneura
3	(early) Late Cretaceous–mid Paleogene (Cenomanian–Ypresian)	Hymenoptera: Proctotrupomorpha; Hymenoptera: Aculeata; Coleoptera; Diptera: non-Eremoneura
4	mid Paleogene–Neogene (Lutetian–Recent)	Hymenoptera: Proctotrupomorpha; Hymenoptera: Aculeata; Diptera: Eremoneura; Coleoptera; Hymenoptera: non-Proctotrupomorpha; Strepsiptera

[a]This table is a summary of Fig. 11.19

[b]Listed in order of family-level abundance. This list excludes families of Neuroptera, Trichoptera and Lepidoptera (N = 4) for insufficient numbers to demonstrate a valid pattern

non-Eremoneura–Diptera and Coleoptera (Fig. 11.19). Lineages such as Strepsiptera, Eremoneura–Diptera and Aculeata–Hymenoptera were largely dominant during Phase 3 and especially in Phase 4 that span the earliest Late Cretaceous (Albian Stage) to the Holocene. It is for this reason that there is a focus on two of the most important lineages of the MMPR, the non-proctotrupomorph Stephanoidea and Evanioidea.

11.7.1 Stephanoidea (Stephanid Wasps)

Stephanidae (stephanid wasps) are a relatively nonspeciose family of parasitoid wasps important during phases 1 and 2 of the MMPR in that recent studies have considered Stephanidae as the sister group to all other extant Apocrita, the parasitoid wasps (Sharkey et al. 2012). This enigmatic lineage possesses peculiar characters that distinguish it from all other families of extant and extinct Hymenoptera. Examples of the distinctiveness of Stephanidae include a distinct tuberculate crown that occurs on the head capsule; and a hind femur usually swollen with two or three large, ventral, tooth-like processes and several denticles of lesser size. Additional features are a propodeum (first abdominal segment) with a dorsal profile that is continuously rectilinear throughout, and a metasomal base (the abdomen excluding the propodeum) of the thorax that nearly contacts the metacoxa (Hong et al. 2011; Rasnitsyn and Zhang 2010). Although stephanid wasps are morphologically unusual and rarely encountered in modern habitats, the clade is comprised of ten extant genera with nearly 350 species (van Achterberg and Yang 2004; Aguiar 2004, 2006; Aguiar and Jennings 2005; van Achterberg and Quicke 2006; Aguiar et al. 2010; Hong et al. 2010; Hong and Xu 2011). The geographic distribution of Stephanidae principally is among subtropical and tropical forests worldwide (van Achterberg 2002; Aguiar 2004; Hong et al. 2011). Consistent with their modern, species-poor occurrence, stephanids have a poor fossil record consisting of six genera. Three of these genera are of amber provenance, and are *Archaeostephanus* from the Late Cretaceous of New Jersey, and *Kronostephanus* and *Lagenostephanus* from the Late Cretaceous of Myanmar (Engel and Grimaldi 2004; Engel et al. 2013a, b; Li et al. 2017b). The three Eocene genera are *Protostephanus* from the Florissant Formation of Colorado, USA, and *Electrostephanus* and *Denaeostephanus* from the Baltic Region of northern Europe (Cockerell 1906; Brues 1933; Engel and Grimaldi 2004; Engel 2005b; Engel and Ortega-Blanco 2008). The amber fossil species are widely distributed geographically from Myanmar, the Baltic Region and to New Jersey, USA, while the sole compression fossil comes from Colorado, USA. A second stephanid wasp, *Lagenostephanus lii,* was described from mid-Cretaceous Myanmar Amber (Li et al. 2017b). *Lagenostephanus* is an early apocritan lineage based on overall habitus (Fig. 11.25), particularly head, leg and wing features (Fig. 11.26). It likely resembled morphologically early members of Phase 1 of the MMPR.

A phylogeny of Stephanidae was provided based on character scoring of morphological features from all extinct and extant genera (Li et al. 2017b). Phylogenetic

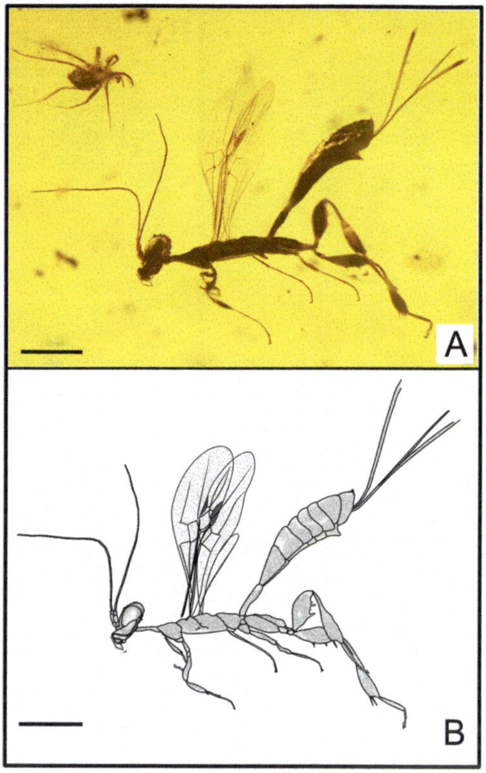

Fig. 11.25 *Lagenostephanus lii.* Holotype CNU-HYM-MA-2014010. (**a**) Photograph of specimen. (**b**) Line drawing of habitus. Scale bars = 1 mm. (Reproduced with permission from Fig. 2 of Li et al. 2017b)

relationships among genera within Stephanidae were presented in a geochronological context consistent with their localities and paleogeographic distributions plotted along a strict consensus tree (Fig. 11.27a). As the two earliest amber stephanids are represented by *Kronostephanus* in the basal subfamily Schlettereriinae and the more derived *Lagenostephanus* in the subfamily Stephaninae, there has been the suggestion that Stephanidae likely were more diverse during the Late Cretaceous (Li et al. 2017b). The earliest Late Cretaceous occurrences and diversification events imply that the origin of Stephanidae occurred geochronologically earlier, perhaps significantly so, than earliest Late Cretaceous. It is notable that in the more basal subfamily Schlettereriinae, the amber genus *Kronostephanus* belongs to a Eurasian distributed clade, and the other amber genus, *Archaeostephanus* from New Jersey amber, occurs in North America during the Late Cretaceous, a biogeographical pattern exhibiting a cosmopolitan distribution for the earliest occurring lineages. The other amber genus, *Lagenostephanus*, also originates from Myanmar. Two middle Eocene amber genera, *Electrostephanus* and *Denaeostephanus* come from the Baltic Region, while a compression fossil genus of late Eocene age, Protostephanus, is from Florissant, Colorado, and has a North American Eocene provenance. As shown (Fig. 11.27b), extant stephanid genera are biogeographically widely distributed. An extant basal genus, *Schlettererius*, is distributed in the Palearctic and

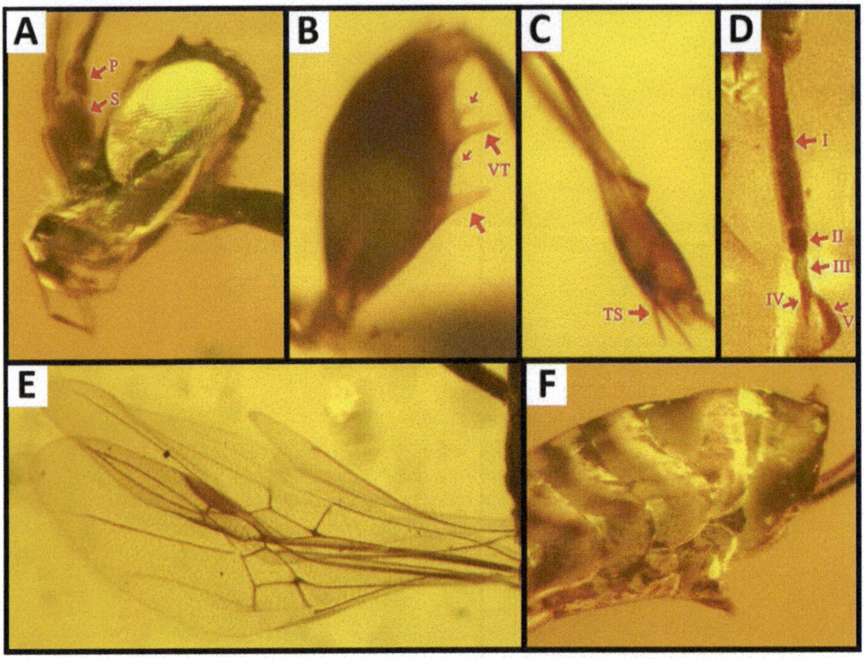

Fig. 11.26 The holotype (CNU-HYM-MA-2014010) of the stephanid wasp *Lagenostephanus lii* (Hymenoptera: Stephanidae) shown in Fig. 11.25, a probable endoparasitoid of wood-boring larvae (Li et al. 2017a, b, c). (**a**) Head in lateral view. (**b**) Hind femur. (**c**) Hind tibia. (**d**) Hind tarsus. (**e**) Hind tarsus. (**f**) Portion of the metasoma in lateral view. Abbreviations: *P* pedicel, *S* scape, *TS* tibial spurs, *VT* ventral tooth, *I–V* five segments of the tarsus. (Reproduced with permission from Fig. 3 of Li et al. 2017b)

Nearctic regions. In addition, *Stephanus* (Stephaninae) is mainly distributed in Eurasia, inhabiting the Oriental and Palearctic regions. As northeastern Asia and northwestern North America became increasingly interconnected during the middle to Late Cretaceous from 80–100 million years ago (Sanmartin et al. 2001; Shih et al. 2009, 2010), a biogeographical connection might have been present for stephanid taxa to migrate from the Palearctic to the Nearctic. Moreover, the seven genera of *Afromegischus*, *Foenatopus*, *Megischus*, *Pseudomegischus*, *Parastephanellus*, *Stephanus* and *Schlettererius* are distributed in Eurasia, which share three of the five genera of *Afromegischus*, *Foenatopus*, *Madegafoenus*, *Megischus* and *Profoenatopus* that are distributed in the Afrotropical Region. Four other modern biogeographic regions are more depauperate, each harboring less than four extant genera. These deep-time and modern biogeographic data clearly indicate that Stephanidae historically have been most diverse in Eurasia but have been widely distributed biogeographically during the past 100 million years and probably a two or three tens of millions of years earlier.

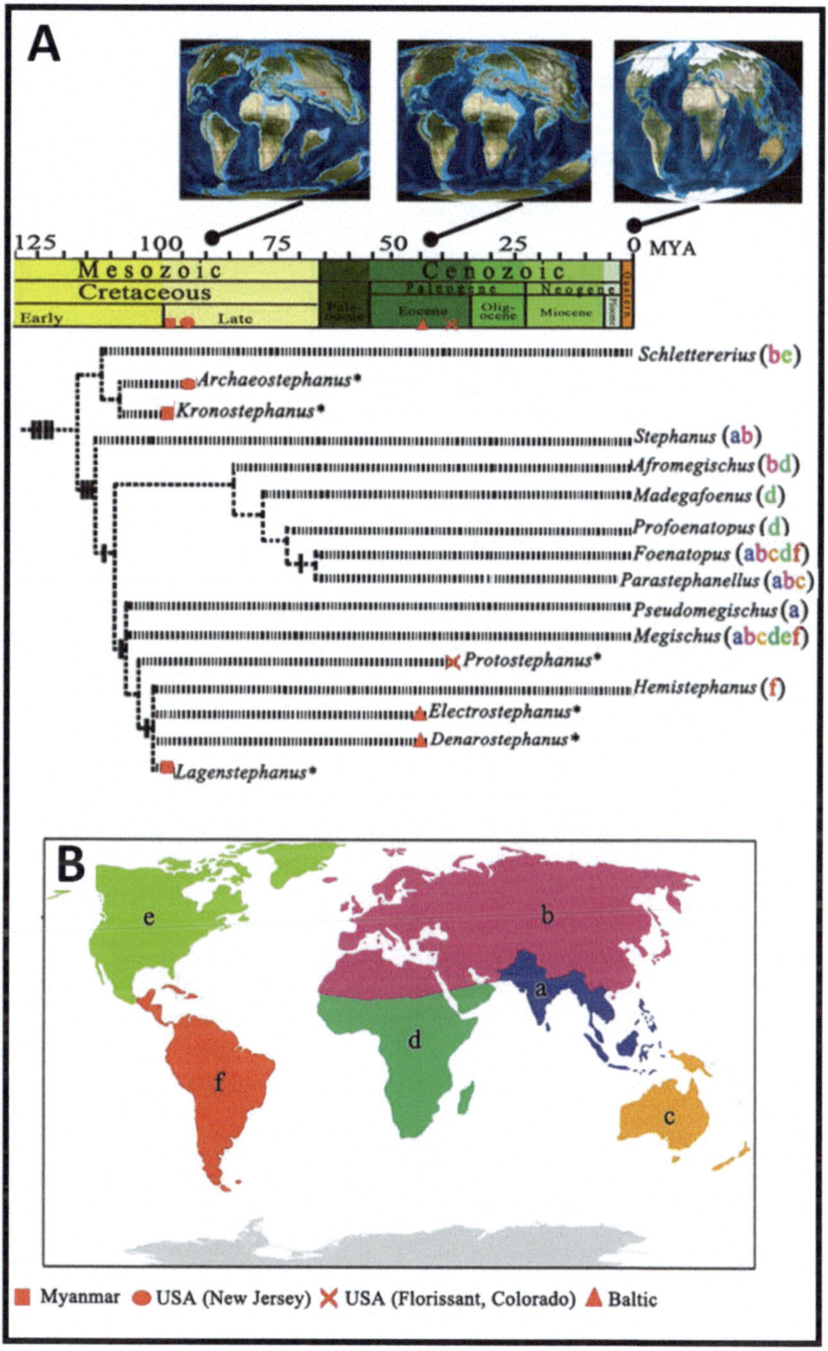

Fig. 11.27 (**a**) Oriental (including India and Myanmar): *Foenatopus, Megischus, Parastephanellus, Pseudomegischus, Stephanus*. (**b**) Palearctic (including all of China and Japan): *Afromegischus, Foenatopus, Megischus, Parastephanellus, Schlettererius and Stephanus*. (**c**) Australasian and Oceanian (including New Guinea and islands east): *Foenatopus, Megischus and Parastephanellus*. (**d**) Afrotropical: *Afromegischus, Foenatopus, Madegafoenus, Megischus and Profoenatopus* (**e**) Nearctic: *Megischus, and Schlettererius*. (**f**) Neotropical (including all of Mexico and the Caribbean): *Foenatopus, Hemistephanus and Megischus*

11.7.2 Evanioidea (Ensign Wasps)

Evanioidea (ensign wasps) also illustrate the diversification of Hymenoptera during phases 1 and 2 of the MMPR, reaching successive stepwise increases in species richness during phases 1 and 2 of the Middle Jurassic to Early Cretaceous as the major non-proctotrupomorph lineage (Table 11.1; Fig. 11.19, pink dot trajectory). Evanioidea are a moderately diverse superfamily of parasitoid wasps that are characterized by two apomorphies (Li et al. 2018a). First, the metasoma is attached high on the propodeum (Goulet and Huber 1993; Grimaldi and Engel 2005a), hence designation of the group as ensign wasps. Second, all functional metasomal spiracles are lost except on the seventh segment (Goulet and Huber 1993; Grimaldi and Engel 2005a). Historically, the superfamily Evanioidea included the three extant families of Evaniidae, Gasteruptiidae and Aulacidae, and later five extinct families from the Mesozoic were added, the Andreneliidae, Anomopterellidae, Baissidae, Othniodellithidae and Praeaulacidae (Rasnitsyn 1972, 1975; Rasnitsyn and Martínez-Delclòs 2000; Engel et al. 2016b). Subsequently, additional ensign wasps were reported, including *Exilaulacus loculatus* (Li et al. 2018a) from mid-Cretaceous Myanmar amber (Fig. 11.28), which was followed in the same report by a preliminary phylogeny of Evanioidea. This phylogeny resulted from morphology and DNA sequence data of selected fossil and extant genera that employed two phylogenetic analytical methods: maximum parsimony and Bayesian inference (Li et al. 2018a). Several distinctive relationships within Evanioidea resulted from the phylogenetic analyses (Fig. 11.29). First, the extinct family Praeaulacidae is paraphyletic and occurs at the base of Evanioidea in-groups. Second, Anomopterellidae is a monophyletic clade and is the sister clade to the remaining families. Third, Aulacidae, Baissidae and Gasteruptiidae do not form a monophyletic clade. Fourth, Othniodellithidae is a monophyletic clade in a position that is more basal to the Andreneliidae + Evaniidae clade. Fifth, Andreneliidae is the sister clade of Evaniidae, and both lineages are monophyletic clades. These results provided clarity to previous, mostly ambiguous, results regarding Evanioidea phylogeny. The inclusion of all evanioid genera, especially fossil taxa, provided a straightforward perspective of Evanioidea phylogenetic events accompanying phases 1 and 2 of the MMPR.

Based on 59 genera and 171 described fossil species of Evanioidea (Zhang and Rasnitsyn 2008; Li et al. 2013a, 2018a), histograms show the frequency of Evanioidea species, genera and family richness during a 169 million-year-long interval from Middle Jurassic to Miocene (Li et al. 2018a; Fig. 11.30a). From these geochronologic, epoch-level data (Fig. 11.30a), a relatively flat level of genus-level richness existed throughout the Mesozoic from 174 to 5 Ma, followed by a considerable decline in the transition to the Cenozoic, and ending in a flat, low level of richness from the mid Eocene to the mid Miocene spanning approximately 40–14 million years ago. Although species-level richness during this time interval is considerably more variable and accentuated, the general pattern is similar to that of generic richness, with both reaching a peak during the Early Cretaceous, followed by a distinct downturn and low levels of occurrences thereafter during the Cenozoic.

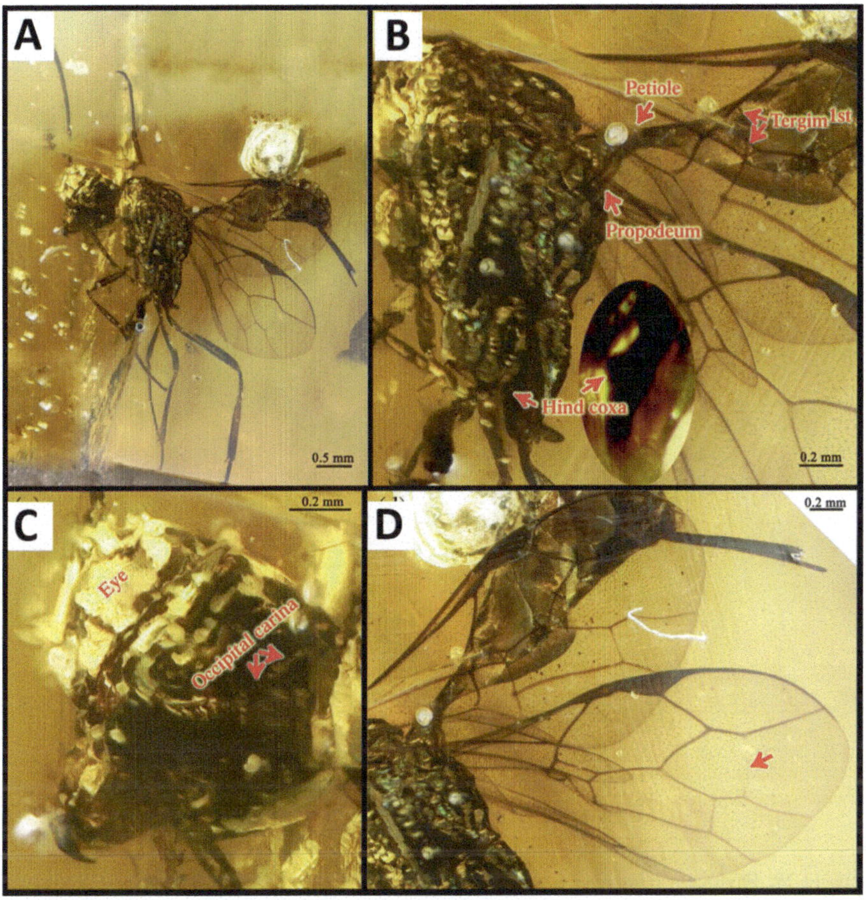

Fig. 11.28 Holotype (CNU-HYM-MA-2014008) of the ensign wasp *Exilaulacus loculatus* (Li et al. 2018a, b) (Hymenoptera: Evaniidae). (**a**) Photograph of the overall habitus. (**b**) Mesosoma and metasoma. (**c**) Head. (**d**) Wings. (Reproduced with permission from Fig. 16 of Li et al. 2018a)

However, the internal composition of families within Evanioidea does vary substantially and displays distinctive patterns (Fig. 11.30b). The species richness of Anomopterellidae peaked during the Middle Jurassic, decreased in the Late Jurassic, after which the Anomopterellidae record disappears, presumably attributable to extinction. A similar pattern is present for Praeaulacidae, a lineage with high richness during the Middle and Late Jurassic that decreased considerably during the Late Cretaceous, after which the Praeaulacidae record ceases, again attributable to extinction. By contrast, the species richness of Baissidae peaks during the Early Cretaceous, apparently becoming extinct by the Cenozoic. Andreneliidae and Othniodellithidae are present, respectively, solely during the Early Cretaceous and Late Cretaceous. The earliest known fossils of the extant family Evaniidae are from the Early Cretaceous, and the majority of species are recorded during the Cretaceous,

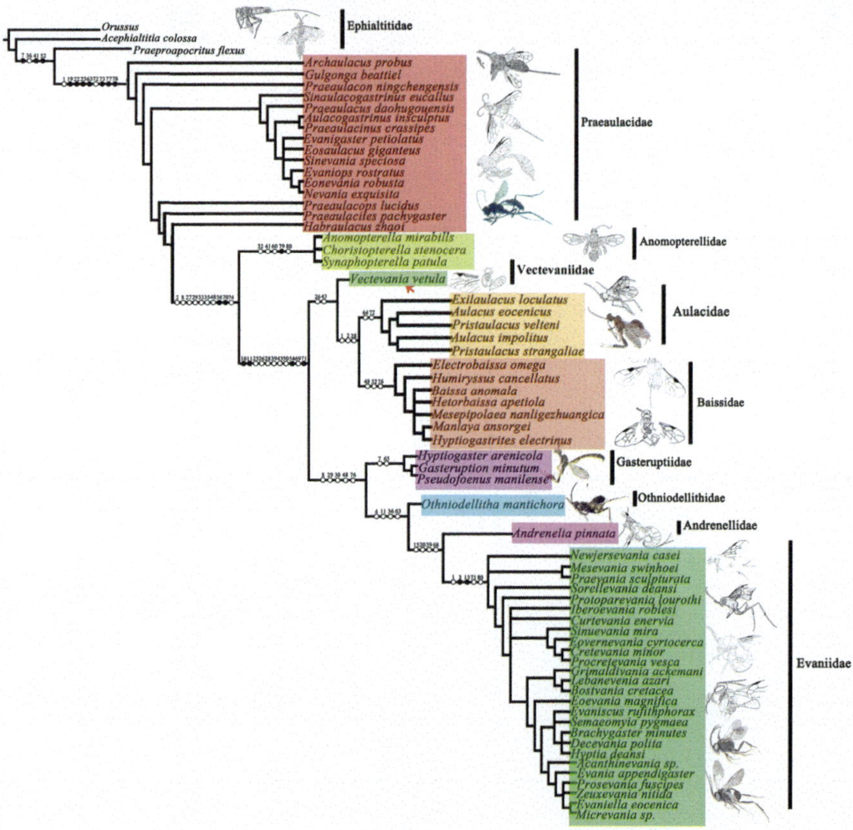

Fig. 11.29 Phylogeny of extant and extinct Evanioidea, based on a strict consensus tree recovered from parsimony analyses of morphological characters, with 582 steps, a consistency index of 0.21 and a retention index of 0.63. Solid circles indicate nonhomoplastic changes and open circles indicate homoplastic changes. (Reproduced with permission from Fig. 24 of Li et al. 2018a)

but a substantial decrease is documented for the Cenozoic. Evaniidae currently are evolutionary relicts. The two, other extant families of Gasteruptiidae and Aulacidae are first documented, respectively, during the Early Cretaceous and Late Cretaceous, both of which persist to the present also as evolutionary relicts. Based on these patterns of species and genera richness through time, there are four families with more than four species occurrences within an epoch that have a distinct pattern of occurrence during the MMPR (phases 1 and 2) and continuing to the post MMPR (phases 3 and 4). Praeaulacidae and Anomopterellidae dominate phases 1 and 2; Baissidae and Evaniidae have elevated occurrences during phases 2 and 3; and Aulacidae has the greatest number of occurrences in Phase 4 (Fig. 11.19). This suggests that Evanioidea, as non-proctotrupomorph Hymenoptera (Fig. 11.19), was one of the earliest participants in the MMPR, and was the greatest contributor of Hymenoptera

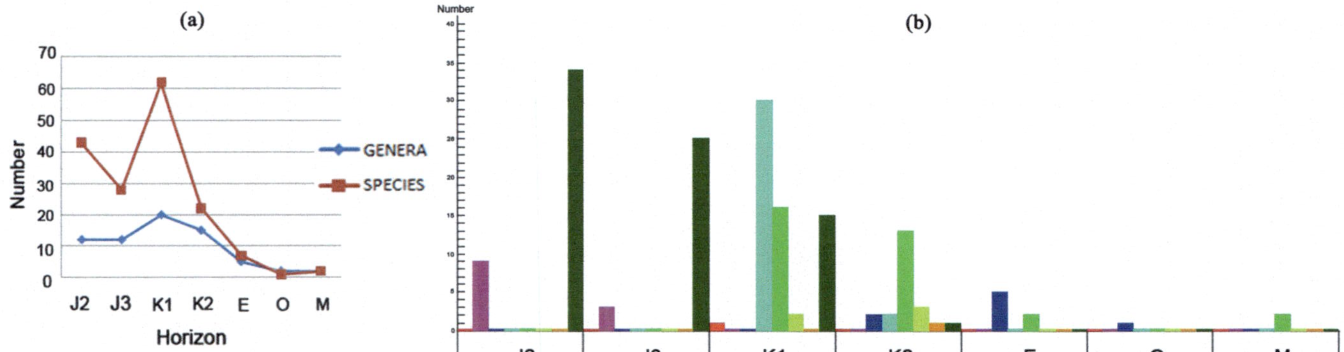

	J2	J3	K1	K2	E	O	M
Andreneliidae	0	0	1	0	0	0	0
Anomopterellidae	9	3	0	0	0	0	0
Aulacidae	0	0	0	2	5	1	0
Baissidae	0	0	30	2	0	0	0
Evaniidae	0	0	16	13	2	0	2
Gasteruptiidae	0	0	2	3	0	0	0
Othniodellithidae	0	0	0	1	0	0	0
Praeaulacidae	34	25	15	1	0	0	0

Fig. 11.30 Taxonomic richness of Evanioidea during the Middle Jurassic to Miocene, showing the contribution of an early parasitoid clade to the Mid-Mesozoic Parasitoid Revolution. (**a**) Total number of Evanioidea species (N = 171) and genera (N = 59) resolved to epoch time intervals during the Middle Jurassic to Miocene. (**b**) The genus and species richness of families (N = 171 occurrences) within the Evanioidea resolved to epoch time intervals during the Middle Jurassic to Miocene. Abbreviations: *J2* Middle Jurassic, *J3* Late Jurassic, *K1* Early Cretaceous, *K2* Late Cretaceous, *E* Eocene, *O* Oligocene, *M* Miocene. Data for the Paleocene, Pliocene and Pleistocene epochs are not reported. (Reproduced with permission from Fig. 25 of Li et al. 2018a)

to Phase 1 and Phase 2, proportionately less so in subsequent phases. This early evolutionary expansion of a major parasitoid group likely was involved in the ecological transformation of terrestrial food webs.

Although divergence-time estimation has played an important role in evolutionary biology (De Baets and Littlewood 2015; Warnock and Engelstädter 2021), it is relevant also in evaluating historical ecological processes such as the MMPR. As a prelude, a study based on total-evidence analysis concluded that the time-of-origin for the order Hymenoptera was at 309 million years ago (Ronquist et al. 2012), or the middle of the Pennsylvanian Period, with node dating providing a very similar result of 311 Ma. Such an age date would be consistent with a major diversification of basal Hymenoptera during the Permian. Nevertheless, the results of that study for both the time of origin and time interval of early diversification of Hymenoptera are considerably older than most previous estimates and the relevant fossil record. A subsequent analysis of the same dataset using the joint and complementary dating of clades (nodes) and terminal lineages (tips) (O'Reilly et al. 2015, O'Reilly and Donoghue 2016) produced a time-calibrated phylogeny of Hymenoptera congruent with the fossil record (Rasnitsyn 1969, 1975, 1988, 2002). These latter estimates acknowledge the origin and early diversification of Hymenoptera as a Middle Triassic to Early Jurassic event (Grimaldi and Engel 2005a), also consistent with the fossil record. However, because of the sparseness of the relevant fossil record it is difficult to evaluate a 74 million-year-long gap. This gap exists between the earliest Late Triassic fossils of Hymenoptera at 235 Ma—*Triassoxyela foveolata* and *Leioxyela antiqua* from the Madygen Biota (Rasnitsyn 1964)—and the presumptive origin of Hymenoptera at 309 Ma (Rasnitsyn 1969; Rasnitsyn and Quicke 2002; Ronquist et al. 2012). A similar pattern exists in Evanioidea (Ronquist et al. 2012), which shows that the earliest divergence time of Evanioidea based on total-evidence dating was the Late Triassic at about 221 Ma (Late Triassic, Norian Stage), 43 million years earlier than a node-dating result of 178 Ma (Early Jurassic, Toarcian Stage) under an internal growth rate model. Currently the earliest record of Evanioidea is several occurrences of Praeaulacidae dated as 165 Ma, of latest Middle Jurassic age (Callovian Stage). These occurrences are *Archaulacus* (Li et al. 2014c), *Aulacogastrinus* (Rasnitsyn 1983), *Eosaulacus* (Zhang and Rasnitsyn 2008), *Nevania* (Zhang and Rasnitsyn 2007), *Praeaulacus* (Rasnitsyn 2008), *Praeaulacon* (Zhang and Rasnitsyn 2008), *Sinaulacogastrinus* (Zhang and Rasnitsyn 2008), the anomopterellid *Anomopterella* (Rasnitsyn 1975) and *Synaphopterella* (Li et al. 2013a). This timing from fossil occurrence data indicates that diversification within Evanioidea would have appeared no later and no earlier than Middle Jurassic. Although combining the divergence time estimation (Ronquist et al. 2012) with the origin age of Evanioidea may push this lineage to the Early Jurassic, or conceivably Late Triassic, currently there is no fossil evidence for such an early origination. A Middle Jurassic origination, indicated by the fossil record, is consistent with the parasitoid habit as a fundamental feature of Evanioidea, and its initial appearance during early Phase 1 of the MMPR, coincident with the Toarcian to Callovian stages of the late Early to Middle Jurassic.

11.8 Discussion: Trophic Specialization and the Mid Mesozoic Parasitoid Diversification

Several recent studies have sought to account for the emergence of parasitoids during the mid Mesozoic. One study reconciled the insect fossil record that shows an upward, rather constant increase in family-level diversity with likely times of considerable evolutionary change resulting from key innovations that should result in a spikier trajectory of insect diversity (Condamine et al. 2016). Two distinct types of diversity analyses were used to reconcile both patterns of insect diversity through time. One approach, using the entire fossil record of family-level data, recorded distinct bursts of diversification that occurred early in insect evolution and subsequently declined gradually to a modern level, interrupted only by occasional extinction events. The second approach employed molecular phylogenetic data that contained 82% of extant insect families and identified surges of diversification, but only for the four, hyperdiverse holometabolous orders. Both approaches did not detect any effect from the origin of angiosperms on insect diversity, a consequence that has been borne out previously from several fossil diversity studies (Dmitriev and Zherikhin 1988; Labandeira and Sepkoski Jr 1993; Jarzembowski and Ross 1996; Labandeira 2014b; but see Wilson et al. 2013). The lack of an effect of angiosperm diversity on insects also is borne out from long-term analyses of major mouthpart types through time (Labandeira 1997, 2019; also see Nel et al. 2018). Rather, the authors concluded that clade-specific innovations were responsible for major diversification events that should be captured by the insect fossil record (Condamine et al. 2016). Such innovations would have included the highly elongate, valved ovipositor of Hymenoptera; the extensible, telescoped ovipositor of Diptera; and the host-seeking, mobile, triungulin larva of Coleoptera and planidium larva of Diptera. Specifically, one of these events involved "… shifts within Diptera and Hymenoptera [that] may be consistent with the development of trophically specialized habits (i.e. parasitoid) …" (Condamine et al. 2016, p. 8). A specific example may be the combination of small size, koinobiont, endoparasitoid and superparasitoid life habits that are associated with a high rate of diversification in particular lineages of wasps such as microgastrine Braconidae (Mardulyn and Whitfield 1999). Given that Diptera and Hymenoptera constitute about 79% of all parasitoid taxa in the fossil record (Table 11.2), there is reason to indicate that the high diversification rate was an indirect reference to the MMPR.

In a separate study (Rainford and Mayhew 2015), a recent phylogeny of Hexapoda, with age dates, was used to ascertain whether specific patterns existed between diet and associated patterns of insect diversity such as clade richness. Two indices of phylogenetic clustering, the net relatedness index and the nearest taxon index, provided the total phylogenetic distance of an insect family with a particular diet (for details see Rainford and Mayhew 2015). (The total phylogenetic distance is the number of all pairwise differences in character states between two phylogenies.) The results of the study indicated that for the diets of detritivory, fungivory, phytophagy, predation, parasitoidism and ectoparasitism, there were no associations between particular dietary substrates and clade richness. Moreover, there was no evidence that clade richness promoted the evolution of antagonisms such as ectoparasitism and parasitoidism. A major conclusion of the study was that taxa with specialized feeding ecologies such as ectoparasitism and parasitoidism exhibited significant phylogenetic clustering and thus were closely related to other taxa with the same diet than to other such taxa with different diets. The results of the study are consistent with previous evaluations (Wiegmann et al. 1993) that failed to demonstrate a stable relationship between parasitoidism and clade richness. These conclusions also highlight the strong dietary conservatism for parasitoid insect families that originated during the mid Mesozoic.

The issue of what caused the triggering of the MMPR is a difficult issue to address. Nevertheless, there are several preconditions that are germane to the issue. The MMPR required multiple steps of biological organization that produced a cascade of events beginning in the late Early Jurassic that was entrenched by the late Early Cretaceous. First, an essential prerequisite was the establishment of Holometabola, which already was present during the Late Carboniferous (Haug et al. 2015). Second, was the development of several key innovations, particularly: (1) the specialized drilling ovipositor of apocritan Hymenoptera; (2) the development of the host-seeking triungulin and planidium first-instar stages of Neuroptera, Coleoptera, Strepsiptera and Diptera; and (3) the telescoped ovipositor of Diptera (Feener and Brown 1997; Gauld 2008; Evans and Steury 2012). Third, was the ecological restructuring of terrestrial ecosystems such that bottom–up, resource-driven food-web structure was replaced by top–down regulation with the emergence of the parasitoid guild that more efficiently regulated primary consumers such as herbivores (Labandeira 2015). Whether the accumulation of these phylogenetic, morphological and ecological aspects caused the separation of the early MMPR into phases 1 and 2, and what propelled phases 3 and 4 during the Late Cretaceous through the Cenozoic, remains a question for further analyses of more finely resolved data.

A related, albeit vexing, issue involves the evolution of the parasitoid community and its component guilds (Mills 1994). The particular issue of concern is whether host evolution of parasitoid lineages proceed from generalist to specialist, the traditional view, or alternatively from specialist to generalist, the uncommon perspective. The traditional version of parasitoid host breadth is that the parasitoid penchant for high animal host specificity in resources results from high extinction rates and a low rate of diversification (Stireman III 2005). Such a view would indicate that host-range evolution proceeds from generalist to specialist and thus should preferentially occur at the terminal lineages of clades. A test of this hypothesis used tachina flies (Diptera: Tachinidae) and it was found, surprisingly, that generalist taxa were iteratively derived from specialist taxa (Gauld et al. 1992; Stireman III 2005). This result highlighted problems in ancestral state reconstruction in previous phylogenetic trees and the need for additional evidence in establishing parasitoid host specificities. Although it is unclear if the specialist-to-generalist pattern in Tachinidae is typical of most parasitoid insects (Stireman 2003; Stireman III 2005), other studies, albeit more limited, have displayed an opposite pattern (Eggleton and Gaston 1992;

Feener and Brown 1997). These data may indicate that for parasitoids, host breadth of intermediate selectivity would be favored (Ferns and Jervis 2016), generating specialists and generalists on opposite sides of the host specificity continuum.

11.9 Summary and Conclusions

Many extant and extinct insects are predators, parasites or parasitoids. However, explicit recognition of parasitoidism as a distinct ecological process on par with predation and parasitism has been a relatively recent development. Parasitoidism historically was categorized by its location on the host (ectoparasitoidism versus endoparasitoidism), by presence on the same host individual of multiple conspecifics (superparasitoidism) or by multiple non-conspecifics (multiparasitoidism). Other descriptive designations common in the parasitoid literature are hyperparasitoidism, the condition of parasitoids living on other parasitoids, and cleptoparasitoidism, the killing of a host individual resulting from a parasitoid absconding food or other vital resources. Direct and indirect evidence for predation, parasitism and parasitoidism in the fossil record consists of biomolecular data, taxonomic affiliations, morphological and functional attributes, gut contents and coprolites of body fossils, in addition to host-tissue damage, plant–insect interactions and sedimentary structures of trace fossils. Parasite and parasitoid insect clades overwhelmingly have targeted 84 clades of holometabolous insects and minimally targeted 4 clades of hemimetabolous insects. The accumulation curve of originations for parasitoid larval dietary substrates through geologic time records a major upward trend, in contrast to more gently increasing trends for fungivory, phytophagy and predation. Contrary to parasitoid evolutionary trajectories, the larval dietary trajectory for ectoparasites resulted in evolutionary cul-de-sacs that did not lead to major diversification events. Parasitoids from three orders of insects—beetles (Coleoptera), flies (Diptera) and wasps (Hymenoptera)—document the multiple and complex paths that various lineages transit into and out of the parasitoid life habit.

Parasites and parasitoids have fossil records ranging from poor to fair, although their modern diversities can be very elevated. Ten groups of parasites occur among hemimetabolous and holometabolous insects. Hemimetabolous parasites are cockroaches (Blattodea) consisting of an extinct fossil lineage; earwigs (Dermaptera), with two lineages parasitic on bats and rodents; bark lice (Psocoptera) containing a sole lineage inhabiting mammal nests; chewing lice and sucking lice (Phthiraptera), composed of four major clades parasitic on birds and principally mammals; and bugs (Hemiptera), consisting of three blood-feeding lineages. Holometabolous parasites are beetles (Coleoptera), of diverse parasitic life habits; fleas (Siphonaptera), whose modern lineages likely form a clade with older mid-Mesozoic giant fleas; flies (Diptera), with blood feeding possibly originating in the Triassic; erebid moths (Lepidoptera) that convergently evolved stylate mouthparts for blood feeding; and wasps (Hymenoptera), with few parasites but an inordinate proliferation of hyperdiverse apocritan parasitoid lineages. The ten parasite taxa differ from the seven parasitoid taxa in five basic ways that involve life-history features and disposition of their hosts. Of parasitoids, the major groups are mantispids (Neuroptera), with 1 origination; beetles, with 10 originations; twisted-wing parasites (Strepsiptera), with a single origination; flies, with approximately 60 originations; caddisflies (Trichoptera) and moths (Lepidoptera), with 1 and 2 originations respectively; and wasps with either 1 or 2 originations, depending on the authority. The bulk of parasitoid diversity is Hymenoptera, accounting for 75% of all extant parasitoid species, consisting of 92 families in 17 superfamilies, and containing 63% of all extinct and extant families (Tables 11.1 and 11.2). Hymenoptera were the major driver of the Mid Mesozoic Parasitoid Revolution (MMPR), resulting in a dramatic expansion of parasitoidic lineages during the Middle Jurassic to Early Cretaceous.

Modern terrestrial food webs are important for understanding the MMPR. Bottom–up food webs explained by the resource concentration hypothesis, account for the pre-MMPR interval leading up to the latest Early Jurassic. Four biotas typical of pre-MMPR time are the Madygen, Molteno, Beishan–Shangtu and Solite biotas, of Middle to Late Triassic age. Once parasitoids originated during the late Early Jurassic, the body-fossil record indicates that their subsequent family-level diversity is subdivided into four temporal phases, each phase of which is characterized by a stepwise increase from the previous diversity level of particular parasitoid clades or groups (Fig. 11.19). During MMPR Phase 1 (Toarcian to Valanginian stages) and Phase 2 (Hauterivian to Aptian stages), a shift ensued from pre-MMPR bottom–up regulation of food webs to MMPR top–down regulation of food webs. This shift is explained by the trophic cascade hypothesis and the trophic efficiency of parasitoids compared to predators (Slansky 1986; Godfray 1994; Harvey et al. 2009). Four biotas typical of the MMPR time interval are the Yanliao, Karatau, Purbeck and Jehol biotas. Two case studies involving early hymenopteran parasitoid clades, Stephanoidea and Evanioidea, document the initial radiation of MMPR lineages that contain lineages that currently are mostly extinct or relict. The post-MMPR interval consisted of Phase 3 (Cenomanian to Lutetian stages) and Phase 4 (Bartonian Stage to recent), during which there was further consolidation of insect parasitoid taxa in food webs. Four biotas illustrative of the post-MMPR interval are Myanmar Amber, Canadian Amber, Messel and Dominican Amber biotas.

There appears to be no association between the MMPR and angiosperm diversity. Rather, three clade-specific innovations are indicated: (1) notably the host-seeking, triungulin larva in Neuroptera, Coleoptera and Strepsiptera, and planidium larva in Diptera; (2) the extrudable, telescoped ovipositor in Diptera; and (3) the long, valved and flexible ovipositor in Hymenoptera. The likely cause of the MMPR required multiple steps of biological organization that produced a series of events from the late Early Jurassic to the late Early Cretaceous. First, was the necessity of the holometabolous condition. Second, was development of the three key innovations of a drilling ovipositor in apocritan Hymenoptera, the host-seeking triungulin and planidium first-instar stages in several holometabolan lineages, and the telescoped ovipositor of Diptera. Third, was reformatting the ecological structure of terrestrial ecosystems from resource driven to trophic-cascade driven food webs resulting from appearance of the parasitoid guild. One outcome of the MMPR, host specialization, is not necessarily associated with clade diversification among parasitoids, as the evolution of host breadth proceeded from specialist to generalist in Tachinidae, the most diverse clade of dipteran parasitoids.

11.10 An Outlook Toward the Future

Hopefully, this contribution will set the stage for further exploration into the paleobiology and evolutionary biology of parasites and parasitoids. Six major questions are posed to spark future work in this fascinating field.

1. What specific feature or features render the parasite life habit an evolutionary cul-de-sac when compared to the evolutionarily more successful parasitoid life habit?
2. Can other evidence be marshalled to understand the vertebrate host identities of the several "giant" flea lineages from the mid Mesozoic?
3. How does the evolutionary transformation from parasitism to parasitoidism occur? Are there modes in which nonparasitic modes of feeding evolve directly into parasitoidic modes of feeding without going through a parasite stage?
4. Are there biological factors determining why some orders undergo one or two originations of the parasitoid life habit (e.g., Strepsiptera, Hymenoptera), whereas other orders undergo many more such originations (e.g., Coleoptera, Diptera)?
5. What accounts for the spectacular increase in taxonomic diversity of parasitoid clades such as Strepsiptera, Eremoneura, Proctotrupomorpha and Aculeata, during the past 170–120 million years? Are such increases explained by innovations such as triungulin or planidium larval stage, a telescopic ovipositor or an elongate drilling ovipositor?
6. Does the postulated transformation of mid-Mesozoic food webs from those initially driven by primary-producers to subsequent ones driven by efficient parasitoid consumers leave other ecological effects on terrestrial ecosystems?

Much of the deep-time history of predation, parasitism and parasitoidism remains unknown. These six questions will be best answered through interdisciplinary collaboration by paleoecologists, entomologists knowledgeable in fossil and modern insect groups, taphonomists, food-web specialists and others that can pool their knowledge in solving issues of common interest.

Acknowledgements We are grateful to Kenneth De Baets and John Huntley for the invitation to provide this review. We thank two reviewers for constructive evaluations of this contribution. Jennifer Wood assembled the figures; Jon Eizyk ably secured copyright permissions for reproduction of the figures. The Smithsonian Institution Libraries provided facilities and interlibrary loan articles essential for the completion of this review. Kevin Johnson and Sandra Schachat provided valuable feedback. We thank David Smith and Matthew Buffington for access to Hymenoptera specimens that were examined for the Stephanoidea and Evanioidea studies. The Paleobiology

Data Base was used in inquiries regarding the fossil records of fossil taxa mentioned in this report. This is contribution 374 of the Evolution of Terrestrial Ecosystems Consortium at the National Museum of Natural History, in Washington, D.C.

References

Aguiar AP (2004) World catalog of the Stephanidae (Hymenoptera: Stephanoidea). Zootaxa 753:1–120

Aguiar AP (2006) The Stephanidae (Hymenoptera) of Mexico, with description of six new species and key to western *Foenatopus* Smith. Zootaxa 1186:1–56

Aguiar AP, Jennings JT (2005) First record of Stephanidae (Hymenoptera) from New Caledonia, with descriptions of four new species of *Parastephanellus* Enderlein. Zootaxa 1001:1–16

Aguiar AP, Jennings JT, Turrisi GF (2010) Three new Middle-Eastern species of *Foenatopus* Smith (Hymenoptera: Stephanidae) with a new host record and key to species with two spots on the metasoma. Zootaxa 2714:40–58

Alekseyev VN, Rasnitsyn AP (1981) Late Cretaceous Megaspilidae (Hymenoptera) from amber of the Taimyr. Paleontol Zh 1981:127–130

Alpert GD (1994) A comparative study of the symbiotic relationship between beetles of the genus *Cremastocheilus* (Coleoptera: Scarabaeidae) and their host ants (Hymenoptera: Formicidae). Sociobiology 25:1–276

Anderson JM, Anderson HM (1993) Terrestrial flora and fauna of the Gondwana Triassic: Part 2 – Co-evolution. In: Lucas SG, Morales M (eds) The nonmarine Triassic, New Mexico Museum of Natural History and Science Bulletin, vol 3. New Mexico Museum of Natural History and Science, Albuquerque, pp 13–25

Anderson JM, Anderson HM (2003) Heyday of the gymnosperms: systematics and biodiversity of the Late Triassic Molteno fructifications. Strelitzia 14:1–398

Ansorge J, Mostovski MB (2002) Redescription of *Prohirmoneura jurassica* Handlirsch 1906 (Diptera: Nemestrinidae) from the Lower Tithonian lithographic limestone of Eichstätt (Bavaria). Neues Jb Paläont Geol Abh 2000:235–243

Antropov AV (2000) Digger wasps (Hymenoptera: Sphecidae) in Burmese amber. Bull Nat Hist Mus Lond Geol 56:59–77

Araújo A, Jansen AM, Bouchet F, Reinhard K, Ferreira LF (2003) Parasitism, the diversity of life, and paleoparasitology. Mem Inst Oswaldo Cruz 98(Suppl 1):5–11

Arillo A, Mostovski MB (1999) A new genus of Prioriphorinae (Diptera, Phoridae) from the Lower Cretaceous amber of Álava (Spain). Stud Dipterol 6:251–255

Arillo A, Nel A (2000) Two new cecidomyiids flies from the Lower Cretaceous amber of Álava (Spain) (Diptera, Cecidomyiidae). Bull Soc Entomol France 105:285–288

Arillo A, Ortuño VM (2005) Catalog of fossil insect species described from Dominican amber (Miocene). Stutt Beit Naturk 352:1–68

Arndt E, Beutel RG, Will K (2005) Carabidae Latreille, 1802. In: Beutel RG, Leschen RAB (eds) Handbuch der Zoologie; Band IV: Arthropoda: Insecta; Teilband 38: Coleoptera, Beetles, Morphology and systematics (Archostemata, Adephaga, Myxophaga, Polyphaga partim), vol 1. Walter de Gruyter, Berlin, pp 119–152

Arnett RH Jr, Thomas MC (2001) American Beetles, Archostemata, Myxophaga, Adephaga, Polyphaga: Staphyliniformia, vol 1. CRC Press, Boca Raton, FL

Arnett RH Jr, Thomas MC, Skelley PE, Frank JH (2002) American Beetles, Polyphaga: Scarabaeoidea through Curculionoidea, vol 2. CRC Press, Boca Raton, FL

Asbakk K, Kumpula J, Oksanen A, Laaksonen S (2014) Infestation by *Hypoderma tarandi* in reindeer calves northern Finland – prevalence and risk factors. Vet Parasitol 200:172–178

Ashford RW (1970) Observations on the biology of *Hemimerus talpoides* (Insecta: Dermaptera). J Zool 162:413–418

Askew RR (1961) On the biology of the inhabitants of oak galls of Cynipidae (Hymenoptera) in Britain. Trans Soc Br Entomol 14:237–268
Askew RR (1971) Parasitic insects. Heinemann, London
Azar D, Nel A (2012) Evolution of hematophagy in "non-biting midges" (Diptera: Chironomidae). Terr Arth Rev 5:15–34
Azar D, Gèze R, Acra F (2010) Lebanese amber. In: Penney D (ed) Biodiversity of fossils in amber from the major world deposits. Siri Scientific Press, Manchester, UK, pp 271–295
Bagheri M, Feiznia S, Arian M, Shabanian R, Mahari R (2013) Continental trace fossils in the Semnan Area (northern Iran). Open J Geol 3:54–61
Bai M, Ahrens D, Yang XK, Ren D (2012) New fossil evidence of the early diversification of scarabs: *Alloioscarabaeus cheni* (Coleoptera: Scarabaeoidea) from the Middle Jurassic of Inner Mongolia, China. Insect Sci 19:159–171
Bailey PT (1989) The millipede parasitoid *Pelidnoptera nigripennis* (F.) (Diptera: Sciomyzidae) for the biological control of the millipede *Ommatoiulus moreleti* (Lucas) (Diplopoda: Julida: Julidae) in Australia. Bull Entomol Res 79:381–391
Ballesteros M, Bardsen BJ, Langeland K, Fauchald P, Stein A, Tveraa T (2011) The effect of warble flies on reindeer fitness: a parasite removal experiment. J Zool 287:34–40
Bänziger H (1968) Preliminary observations on a skin-piercing blood-sucking moth (*Calyptra eustrigata* (Hmps.) (Lep., Noctuidae)) in Malaya. Bull Entomol Res 58:159–164, pls. IX–X
Bänziger H (1971) Bloodsucking moths of Malaya. Fauna 1:4–16
Bänziger H (1996) The proboscis of eye-frequenting and piercing Lepidoptera (Insecta). Zoomorphology 116:77–83
Barling N, Heads SW, Martill DM (2013) A new parasitoid wasp (Hymenoptera: Chalcidoidea) from the Lower Cretaceous Crato Formation of Brazil: the first Mesozoic Pteromalidae. Cretac Res 45:258–264
Basibuyuk HH, Quicke DLJ, Rasnitsyn AP (2000) A new genus of Orussidae (Insecta: Hymenoptera) from the Late Cretaceous New Jersey amber. In: Studies on fossils in amber, with particular reference to the Cretaceous of New Jersey. Backhuys, Leiden, pp 305–311
Basibuyuk HH, Rasnitsyn AP, Fitton MG, Quicke DLJ (2002) The limits of the family Evaniidae (Insecta: Hymenoptera) and a new genus from Lebanese amber. Insect Syst Evol 33:23–34
Batelka J, Prokop J, Engel MS (2016) New ripiphorid beetles in mid-Cretaceous amber from Myanmar (Coleoptera: Ripiphoridae): first Pelecotominae and possible Mesozoic aggregative behaviour in male Ripidiinae. Cretac Res 68:70–78
Batelka J, Prokop J, Pohl H, Bai M, Zhang W, Beutel RG (2019) Highly specialized Cretaceous beetle parasitoids (Ripiphoridae) identified with optimized visualization of microstructures. Syst Entomol 44:396–407
Bedding RA (1965) Parasites of British terrestrial Isopoda. PhD dissertaion, Imperial College of Science and Technology, London
Bell WJ, Roth LM, Nalepa CA (2007) Cockroaches: ecology, behavior, and natural history. Johns Hopkins University Press, Baltimore
Bennett DJ, Engel MS (2005) A primitive sapygid wasp in Burmese amber (Hymenoptera: Sapygidae). Acta Zool Cracov 48B:1–9
Beutel RG, Zhang WW, Pohl J, Wappler T, Bai M (2016) A miniaturized beetle larva in Cretaceous Burmese amber: reinterpretation of a fossil "strepsipteran triungulin". Insect Syst Evol 47:83–91
Blagoderov V, Grimaldi DA, Fraser NC (2007) How time flies for flies: diverse Diptera from the Triassic of Virginia and early radiation of the order. Am Mus Novit 3572:1–39
Blumberg D, Luck RF (1990) Differences in the rates of superparasitism between two strains of *Comperiella bifasciata* (Howard) (Hymenoptera: Encyrtidae) parasitizing California Red Scale (Homoptera: Diaspididae): an adaptation to circumvent encapsulation? Ann Entomol Soc Am 83:591–597
Bohart RM, Menke AS (1976) Sphecid wasps of the World, a generic revision. University of California Press, Berkeley

Bologna MA, Oliverio M, Pitzalis M, Marioittini P (2008) Phylogeny and evolutionary history of the blister beetles (Coleoptera, Meloidae). Mol Phylogenet Evol 48:679–693

Bologna MA, Turco F, Pinto JD (2010) Meloidae Gyllenhal, 1810. In: Leschen RAB, Beutel RG, Lawrence JF (eds) Handbuch der Zoologie; Band IV: Arthropoda: Insecta; Teilband 38: Coleoptera, Beetles, Morphology and systematics (Elateroidea, Bostrichiformia, Cucujiformia partim), vol 2. Walter de Gruyter, Berlin, pp 681–693

Bologna MA, Di Giulio AD (2011) Biological and morphological adaptations in the pro-imaginal phases of the beetle family Meloidae. Atti Accademia Nazionale Italiana di Entomologia 59:141–152

Borkent A (2001) *Leptoconops* (Diptera: Ceratopogonidae), the earliest extant lineage of biting midge, discovered in 120–122 million-year-old Lebanese amber. Am Mus Novit 3328:1–11

Borkent A, Grimaldi DA (2004) The earliest fossil mosquito (Diptera: Culicidae), in mid-Cretaceous Burmese amber. Ann Entomol Soc Am 97(5):882–888

Bouček Z, Noyes JS (1987) Rotoitidae, a curious new family of Chalcidoidea (Hymenoptera) from New Zealand. Syst Entomol 12:407–412

Bouchet F, Lavaud F (1999) Solenophagy and telmophagy: biting mechanisms among various hematophagous insects. Allerg Immun 31:346–350

Boucot AJ, Poinar GO Jr (2010) Fossil behavior compendium. CRC Press, Boca Raton, FL

Bown TM, Hasiotis ST, Genise JF, Maldonado F, Brouwers EM (1997) Trace fossils of Hymenoptera and other insects and paleoenvironments of the Claron Formation (Paleocene and Eocene), Southwestern Utah, U.S. Geol Surv Bull 2153:42–58

Bradley F, Landini W (1984) I fossili del "tripoli" messiano di Gabbro (Livorno). Palaeontogr Ital 73:5–33

Brandão CR, Martins-Neto RG, Vulcano MA (1989) The earliest known fossil ant (first Southern Hemisphere Mesozoic record) (Hymenoptera: Formicidae: Myrmecinae). Psyche 95:195–208

Brandt M, Foitzik S, Fischer-Blass B, Heinze J (2005) The coevolutionary dynamics of obligate ant social parasite systems–between prudence and antagonism. Biological Reviews, 80(2):251–267

Bravo F, Pohl H, Silva-Neto A, Beutel RG (2009) Bahiaxenidae, a "living fossil" and a new family of Strepsiptera (Hexapoda) discovered in Brazil. Cladistics 25(6):614–623

Briggs DE, Summons RE (2014) Ancient biomolecules: their origins, fossilization, and role in revealing the history of life. BioEssays 36:482–490

Brothers D (1995) Bradynobaeniidae. In: Hanson PE, Gauld ID (eds) The Hymenoptera of Costa Rica. Oxford University Press, Oxford, UK, pp 541–548

Brown VK (1973) The biology and development of *Brachygaster minutus* Olivier (Hymenoptera: Evaniidae), a parasite of the oothecae of *Ectobius* spp. (Dictyoptera: Blattidae). J Nat Hist 7:665–674

Brown BV (1997) Systematics and fossil evidence of host-parasitoid relationships of *Calamiscus* Borgmeier (Diptera: Phoridae). J Nat Hist 31:1253–1259

Brown BV (1998) Re-evaluation of the fossil Phoridae (Diptera). J Nat Hist 33:1561–1573

Brues CT (1933) The parasitic Hymenoptera of the Baltic amber. Part 1. Bernstein Forschungen 3:4–178

Brues CT (1937) Superfamilies Ichneumonoidea, Serphoidea and Chalcidoidea. In: Carpenter FM, Folsom JW, Essig EO, Kinsey AC, Brues CT et al (eds) Insects and Arachnids from Canadian amber, Univ Toronto Geol Ser, vol 40. University of Toronto Press, Canada, pp 1–62

Brues CT (1940) Fossil parasitic Hymenoptera of the family Scelionidae from Baltic amber. Proc Am Acad Arts Sci 74:69–90

Budd GE, Mann RP (2018) History is written by the victors: the effect of the push of the past on the fossil record. Evolution 72:2276–2291

Buffington ML, Perkovsky EE, Brady SG (2014) The description of *Rovnoeucoila tympanomorpha* Buffington and Perkovsky, a new genus and species of fossil eucoiline, with observations on the asynchronous evolution of Diglyphosematini (Hymenoptera: Figitidae: Eucoilinae). Proc Entomol Soc Wash 116(3):243–255

Bukovinszky T, van Veen FJF, Jongema Y, Dicke M (2008) Direct and indirect effects of resource quality on food web structure. Science 319:804–807

Burkhardt D, Ślipiński A (2003) Phylogeny and taxonomy of the world Passandridae (Coleoptera). In: Cuccodoro G, Leschen RAB (eds) Systematics of Coleoptera: papers celebrating the retirement of Ivan Löbl, Mem Entomol, vol 17. Associated Publishers, Gainesville, FL, pp 753–883

Burkhardt D, Ślipiński A (2010) Passandridae Blanchard 1845/Erichson, 1845. In: Leschen RAB, Beutel RG, Lawrence JF (eds) Handbuch der Zoologie; Band IV: Arthropoda: Insecta; Teilband 38: Coleoptera, Beetles, Morphology and systematics (Elateroidea, Bostrichiformia, Cucujiformia partim), vol 2. Walter de Gruyter, Berlin, pp 363–368

Burks RA, Heraty JM, Pinto JD, Grimaldi D (2015) Small but not ephemeral: newly discovered species of Aphelinidae and Trichogrammatidae (Insecta: Hymenoptera: Chalcidoidea) from Eocene amber. Syst Entomol 40:592–605

Bush AO, Fernández JC, Esch GW, Seed RJ (2001) Parasitism: the diversity and ecology of animal parasites. Cambridge University Press, Cambridge, UK

Byers GW (1996) More on the origin of Siphonaptera. J Kansas Entomol Soc 69(3):274–277

Cai CY, Huang DY (2014) The oldest micropepline beetle from Cretaceous Burmese amber and its phylogenetic implications (Coleoptera: Staphylinidae). Naturwissenschaften 202:813–817

Cai CY, Huang DY, Newton AF, Eldredge KT, Engel MS (2017) Early evolution of specialized termitophily in Cretaceous rove beetles. Curr Biol 27:1229–1235

Cai CY, Yin ZW, Huang DY (2018) A new ripiphorid beetle from Upper Cretaceous Burmese amber sheds light on early evolution of the extant subfamily Ripidiinae (Coleoptera: Ripiphoridae). C R Palevol 17:351–356

Carlson RW (1979) Superfamily Evanioidea. In: Krombein KC, Hurd PD Jr, Smith DR, Burks BD (eds) Catalog of Hymenoptera in America North of Mexico. Smithsonian Institution Press, Washington, DC, pp 1109–1118

Carpenter FM (1954) The geological history and evolution of insects. Annu Rep Smithson Inst 1953:339–350, pls 1–3

Carpenter FM (1971) Adaptations among Paleozoic insects. In: Yochelson E (ed) Proceedings of the First North American Paleontological Convention. Allen Press, Lawrence, KS, pp 1236–1251

Carpenter FM, Folsom JW, Essig EO, Kinsey AC, Brues CT, Boesel MW, Ewing HE (1937) Insects and arachnids from Canadian amber. Univ Toronto Stud Geol Ser 40:7–62

Carpenter SR, Kitchell JF, Hodgson JR (1985) Cascading trophic interactions and lake productivity. Bioscience 35:634–639

Castro LR, Dowton M (2006) Molecular analyses of the Apocrita (Insecta: Hymenoptera) suggest that the Chalcidoidea are sister to the diaprioid complex. Invert Syst 20:603–614

Cerretti P, O'Hara JE, Wood DM, Shima H, Inclan DJ, Stireman JO III (2014) Signal through the noise? Phylogeny of the Tachinidae (Diptera) as inferred from morphological evidence. Syst Entomol 39:335–353

Cerretti P, Stireman JO III, Pape T, O'Hara JE, Marinho MAT, Rognes K, Grimaldi DA (2017) First fossil of an oestroid fly (Diptera: Calyptratae: Oestroidea) and the dating of oestroid divergences. PLoS One 12(8):e0182108. https://doi.org/10.1371/journal.pone.0182101

Chatzimanolis S, Grimaldi DA, Engel MS, Fraser NC (2012) *Leehermania prorova*, the earliest staphyiniform beetle, from the Late Triassic of Virginia (Coleoptera: Staphylinidae). Am Mus Novit 3761:1–28

Chen J, Wang B, Engel MS, Wappler T, Jarzembowski EA, Zhang HC, Wang XL, Zheng XT, Rust J (2014) Extreme adaptations for aquatic ectoparasitism in a Jurassic fly larva. eLife 2014(3):e02844. https://doi.org/10.7554/eLife.02844

Chen ZZ, Huang S, Chang Y, Wang YG, Shin CK, Ren D (2019) Neuroptera – lacewings. In: Ren D, Shih CK, Gao TP, Wang YG, Yao YZ (eds) Rhythms of insect evolution from the Jurassic and Cretaceous in Northern China. Wiley, Hoboken, NJ, pp 285–336

Chin K (2021) Gastrointestinal parasites of ancient non-human vertebrates: evidence from coprolites and other materials. In: De Baets K, Huntley JW (eds) The evolution and fossil record of parasitism: Coevolution and paleoparasitological techniques. Topics in Geobiology 50

Clausen CP (1940) Entomophagous insects. McGraw Hill, New York

Clayton DH, Bush SE, Johnson KP (2016) Coevolution of life on hosts. The University of Chicago Press, Chicago

Cockerell TDA (1906) Fossil Hymenoptera from Florissant, Colorado. Bull Mus Comp Zool 50:33–58

Cockerell TDA (1917) Arthropods in Burmese amber. Am J Sci 44:360–368

Colwell DD, Hall MJR, Scholl PJ (2006) Introduction. In: Colwell DD, Hall MJR, Scholl PJ (eds) The Oestrid flies: biology, host–parasite relationships, impact and management. CAB International, Wallingford, UK, pp 1–7

Compton SG, Ball AD, Collinson ME, Hayes P, Rasnitsyn AP, Ross AJ (2010) Ancient fig wasps indicate at least 34 Myr of stasis in their mutualism with fig trees. Biol Lett 6:838–842

Condamine FL, Clapham ME, Kergoat GJ (2016) Global patterns of insect diversification: towards a reconciliation of fossil and molecular evidence? Sci Rep 6:19208. https://doi.org/10.1038/srep19208

Cook JL (2014) Review of the biology of parasitic insects in the Order Strepsiptera. Comp Parasitol 81:134–151

Copeland RS, Kirk-Spriggs AH, Muteti S, Booth W, Wiegmann BM (2011) Rediscovery of the "terrible hairy fly", *Mormotomyia hirsuta* Austen (Diptera: Mormotomyiidae), in eastern Kenya, with notes on biology, natural history, and genetic variation of the Ukasi Hill population. Afr Invert 52:363–390

Coram RA (2003) Taphonomy and ecology of Purbeck fossil insects. Acta Zool Cracov 46(Suppl):311–318

Coram RA, Jepson JE (2012) Fossil insects of the Purbeck Limestone Group of Southern England – palaeoentomology from the dawn of the Cretaceous. Siri Scientific Press, Manchester, UK

Coram RA, Jarzembowski EA, Mostovski MB (2000) Two rare eremoneuran flies (Diptera: Empididae and Opetidae) from the Purbeck Limestone Group. Paleontol J 34:370–373

Coty D, Lebon M, Nel A (2016) When phylogeny meets geology and chemistry: doubts on the dating of Ethiopian amber. Ann Soc Entomol France 52:161–166

Coupland JB, Barker GM (2004) Diptera as predators and parasitoids of terrestrial gastropods, with emphasis on Phoridae, Calliphoridae, Sarcophagidae, Muscidae and Fanniidae. In: Barker GM (ed) Natural enemies of terrestrial molluscs. CAB International, London, pp 85–158

Crosskey RW (1980) Family Tachinidae. In: Crosskey RW (ed) Catalog of the Diptera of the Afrotropical region. British Museum of Natural History, London, pp 822–882

Crowson RA (1981) The biology of Coleoptera. Academic, London

Cruikshank RD, Ko K (2003) Geology of an amber locality in the Hukawng Valley, northern Myanmar. J Asian Earth Sci 21:441–455

Cupp EW, Cupp MS, Ribeiro JMC, Kunz SE (1998) Blood-feeding strategy of *Haematobia irritans* (Diptera: Muscidae). J Med Entomol 35:591–595

Currie DC, Grimaldi DA (2000) A new black fly (Diptera: Simuliidae) genus from mid Cretaceous (Turonian) amber of New Jersey. In: Grimaldi DA (ed) Studies on fossils in amber with particular reference to the Cretaceous of New Jersey. Backhuys, Leiden, pp 473–484

Dalgleish RC, Palma RL, Price RD, Smith VS (2006) Fossil lice (Insecta: Phthiraptera) reconsidered. Syst Entomol 31:648–651

Dampf A (1910) *Palaeopsylla klebsiana* n. sp., ein fossiler Flöh aus dem baltischen Bernstein. Schr Phys Ökon Ges Königsberg Pruss 51:248–259

Darling DC, Sharkey MJ (1990) Hymenoptera. Bull Am Mus Natl Mus 195:123–153

De Baets K, Littlewood DTJ (2015) The importance of fossils in understanding the evolution of parasites and their vectors. Adv Parasitol 90:1–51

Deans AR, Basibuyuk HH, Azar D, Nel A (2004) Descriptions of two new Early Cretaceous (Hauterivian) ensign wasp genera (Hymenoptera: Evaniidae) from Lebanese amber. Cretac Res 25:509–516

DeBach P (1964) Biological control of insect pests and weeds. Chapman and Hall, London

Dehon M, Perrard A, Engel MS, Nel A, Michez D (2017) Antiquity of cleptoparasitism among bees revealed by morphometric and phylogenetic analysis of a Paleocene fossil nomadine (Hymenoptera: Apidae). Syst Entomol 42:543–554

Deloya C (1988) Coleopteros lamelicornios asociados a depositos de detritos de *Atta mexicana* (Smith) (Hymenoptera: Formicidae) in el sur del estado de Morelos, México. Folia Entomol Mex 75:77–91

de Moya RS (2019) Implications of a dating analysis of Hippoboscoidea (Diptera) for the origins of phoresis in feather lice (Psocodea: Phthiraptera: Philopteridae). Insect Systematics and Diversity, 3(4):1. https://doi.org/10.1093/isd/ixz008

Dittmar K, Porter ML, Murray S, Whiting MF (2006) Molecular phylogenetic analysis of nycteribiid and streblid bat flies (Diptera: Brachycera, Calyptratae): implications for host associations and phylogeographic origins. Mol Phylogenet Evol 38:155–170

Dittmar K, Zhu Q, Hastriter MW, Whiting MF (2016) On the probability of dinosaur fleas. BMC Evol Biol 16(1):9. https://doi.org/10.1186/s12862-015-0568

Dmitriev VY, Zherikhin VV (1988) Changes in the familial diversity of insects and demonstration of a method of data analysis. In: Ponomarenko AG (ed) The Mesozoic–Cenozoic crisis in the evolution of insects. USSR Academy of Sciences, Moscow, pp 208–215. (in Russian)

Dowton M, Austin AD (1995a) Increased genetic diversity in mitochondrial genes is correlated with the evolution of parasitism in the Hymenoptera. J Mol Evol 41(6):958–965

Dowton M, Austin AD (1995b) The origin of parasitism in the wasps, as inferred from mitochondrial DNA sequences. J Mol Evol 41:958–965

Dowton M, Austin AD, Dillon N, Bartowski E (1997) Molecular phylogeny of the apocritan wasps: the Proctotrupomorpha and Evaniomorpha. Syst Entomol 22:245–255

Drake JM (2003) The paradox of parasites: implications for biological invasion. Proc R Soc B Biol Sci 270:S133–S135

Dunlop J (2021) Chelicerates as parasites. In: De Baets K, Huntley JW (eds) The evolution and fossil Record of parasitism: identification and macroevolution of parasites. Topics in Geobiology 49

Dunne JA, Lafferty KD, Dobson AP, Hechinger RE, Kuris AM, Martinez ND, McLaughlin JP, Muritsen KN, Poulin R, Reise K, Stouffer DB, Thieltges DW, Williams RJ, Zander CD (2013) Parasites affect food web structure primarily through increased diversity and complexity. PLoS Biol 11(6):e1001579

Dunne JA, Labandeira CC, Williams RJ (2014) Highly resolved early Eocene food webs show development of modern trophic structure after the end-Cretaceous extinction. Proc R Soc B 181:20133280. https://doi.org/10.1098/rspb.2013.3280

Durden LA, Musser GG (1994) The sucking lice (Insecta: Anoplura) of the world: a taxonomic checklist with records of mammalian hosts and geographic distributions. Bull Am Mus Nat Hist 218:1–90

Early JW, Masner L, Naumann ID, Austin AD (2001) Maamingidae, a new family of Proctotrupoid wasp (Insecta: Hymenoptera) from New Zealand. Invert Taxon 15:341–352

Edwards D, Selden PA, Richardson JB, Axe L (1995) Coprolites as evidence for plant–animal interaction in Siluro-Devonian terrestrial ecosystems. Nature 377:329–331

Eggleton P, Belshaw R (1992) Insect parasitoids: an evolutionary overview. Philos Trans R Soc Lond B 337:1–20

Eggleton P, Belshaw R (1993) Comparison of dipteran, hymenopteran and coleopteran parasitoids: provisional phylogenetic explanations. Biol J Linn Soc 48:213–226

Eggleton P, Gaston KJ (1990) "Parasitoid" species and assemblages: convenient definitions or misleading compromises? Oikos 59(3):417–421

Eggleton P, Gaston KJ (1992) Tachinid host ranges: a reappraisal (Diptera: Tachinidae). Entomol Gaz 43:139–143

Ellis WN, Ellis-Adam AC (1993) Fossil brood cells of solitary bees on Fuerteventura and Lanzarote, Canary Islands (Hymenoptera: Apoidea). Entomol Ber Amsterdam 53:161–173

Elzinga RJ (1977) Observations on *Sandalus niger* (Sandalidae) with a description of the triungulin larvae. J Kansas Entomol Soc 50:324–328

Engel MS (2002) The first leucospid wasp from the fossil record (Hymenoptera: Leucospidae). J Nat Hist 36:435–441

Engel MS (2003) An anteonine wasp in Cenomanian-Albian amber from Myanmar (Hymenoptera: Dryinidae). J Kansas Entomol Soc 76:616–621

Engel MS (2005a) An Eocene ectoparasite of bees: the oldest definitive record of phoretic meloid triungulins (Coleoptera: Meloidae; Hymenoptera: Megachilidae). Acta Zool Cracov 48B:43–48
Engel MS (2005b) The crown wasp genus *Electrostephanus* (Hymenoptera: Stephanidae): discovery of the female and a new species. Pol Pismo Entomol 74:317–332
Engel MS (2008a) A stem-group cimicid in mid-Cretaceous amber from Myanmar (Hemiptera: Cimicoidea). Alavesia 2:233–237
Engel MS (2008b) An orussid wood wasp in amber from the Dominican Republic (Hymenoptera: Orussidae). Trans Kansas Acad Sci 111(1–2):39–44
Engel MS (2013) A new genus and species of Baissidae in Late Cretaceous amber from New Jersey (Hymenoptera: Evanioidea). Novit Paleoentomol 3:1–8
Engel MS (2016) A new genus and species of maimetshid wasps in Lebanese Early Cretaceous amber (Hymenoptera: Maimetshidae). Novit Paleoentomol 18:1–14
Engel MS, Grimaldi DA (2004) The first Mesozoic stephanid wasp (Hymenoptera: Stephanidae). J Paleontol 78:1192–1197
Engel MS, Grimaldi DA (2006) The first Cretaceous sclerogibbid wasp (Hymenoptera: Sclerogibbidae). Am Mus Novit 3515:1–7
Engel MS, Grimaldi DA (2007a) New false fairy wasps in Cretaceous amber from New Jersey and Myanmar (Hymenoptera: Mymarommatoidea). Trans Kansas Acad Sci 110:159–168
Engel MS, Grimaldi DA (2007b) Cretaceous Scolebythidae and phylogeny of the family (Hymenoptera: Chrysidoidea). Am Mus Novit 3568:1–16
Engel MS, Huang DY (2016) A new twisted-wing parasitoid from mid-Cretaceous amber of Myanmar (Strepsiptera). Cretac Res 58:160–167
Engel MS, Huang DY (2017) A new crown wasp in Cretaceous amber from Myanmar (Hymenoptera: Stephanidae). Cretac Res 69:56–61
Engel MS, Ortega-Blanco J (2008) The fossil crown wasp *Electrostephanus petiolatus* Brues in Baltic Amber (Hymenoptera, Stephanidae): designation of a neotype, revised classification, and a key to amber Stephanidae. ZooKeys 4:55–64
Engel MS, Ortega-Blanco J (2013) A new lineage of enigmatic diaprioid wasps in Cretaceous amber (Hymenoptera: Diaprioidea). Am Mus Novit 3771:1–23
Engel MS, Grimaldi DA, Ortega-Blanco J (2011) Serphitid wasps in Cretaceous amber from New Jersey (Hymenoptera: Serphitidae). Insect Syst Evol 42:197–204
Engel MS, Grimaldi DA, Ortega-Blanco J (2013a) A stephanid wasp in mid-Cretaceous Burmese amber (Hymenoptera: Stephanidae), with comments on the antiquity of the hymenopteran radiation. J Kansas Entomol Soc 86:244–252
Engel MS, McKellar R, Huber J (2013b) A fossil species of the primitive mymarid genus *Borneomymar* (Hymenoptera: Mymaridae) in Eocene Baltic amber. Novit Paleoentomol 5:41647. https://journals.ku.edu/index.php/paleoent/article/view/4651
Engel MS, Ortega-Blanco J, McKellar RC (2013c) New scolebythid wasps in Cretaceous amber from Spain and Canada, with implications for the phylogeny of the family (Hymenoptera: Scolebythidae). Cretac Res 46:31–42
Engel MS, Huang DY, Azar D, Nel A, Davis SR, Alvarado M, Breitkreuz LCV (2015) The wasp family Spathiopterygidae in mid-Cretaceous amber from Myanmar (Hymenoptera: Diaprioidea). C R Palevol 14:95–100
Engel MS, Huang D, Breitkreuz LCV, Azar D, Cai CY, Alvarado M (2016a) A new twisted-wing parasitoid from mid-Cretaceous amber of Myanmar (Strepsiptera). Cretac Res 58:160–167
Engel MS, Huang DY, Alqarni AS, Cai CY (2016b) A remarkable evanioid wasp in mid-Cretaceous amber from northern Myanmar (Hymenoptera: Evanioidea). Cretac Res 60:121–127
English-Loeb GM, Karban R, Brody AK (1990) Arctiid larvae survive attack by a tachinid parasitoid and produce viable offspring. Ecol Entomol 15:361–362
Erwin TL (1979) A review of the natural history and evolution of ectoparasitoid relationships in carabid beetles. In: Erwin TL, Ball GE, Whitehead DR, Halpern AL (eds) Carabid Beetles, their evolution, natural history, and classification. W Junk, The Hague, pp 479–484
Evans HE (1961) A preliminary review of the Nearctic species of *Sierolomorpha* (Hymenoptera). Breviora 140:1–12

Evans HE (1969) Three new Cretaceous aculeate wasps (Hymenoptera). Psyche 76:251–261
Evans AV, Nel A (1989) Notes on *Macrocyphonistus kolbeanus* Ohaus and *Rhizoplatys auriculatus* (Burmeister) with comments on their melittophilous habits (Coleoptera: Melolonthinae: Dynastinae: Phileurini). J Entomol Soc S Afr 52:45–50
Evans AV, Steury VW (2012) The cicada parasite beetles (Coleoptera: Rhipiceridae) of Virginia. Banisteria 39:65–70
Evans HC, Elliot SL, Hughes DP (2011) Hidden diversity behind the zombie-ant fungus *Ophiocordyceps unilateris*: four new species described from carpenter ants in Minas Gerais, Brazil. PLoS One 6(3):e17024. https://doi.org/10.1371/journal.pone.0017024
Evenhuis NL (1994) Catalog of the fossil flies of the world (Insecta: Diptera). Backhuys, Leiden
Falin ZH (2002) Ripiphoridae Gemminger and Harold 1870 (1853). In: Arnett RH Jr, Thomas MC, Skelley PE, Frank JH (eds) American Beetles, Polyphaga: Scarabaeoidea through Curculionoidea, vol 2. CRC Press, Boca Raton, FL, pp 431–444
Farache FH, Rasplus JY, Azar D, Pereira RA, Compton SG (2016) First record of a non-pollinating fig wasp (Hymenoptera: Sycophaginae) from Dominican amber, with estimation of the size of its host figs. J Nat Hist 50:2237–2247
Feener DH Jr, Brown BV (1997) Diptera as parasitoids. Annu Rev Entomol 42:73–97
Felder M, Harms F-J (2004) Lithologie und genetische Interpretation der vulkano-sedimentären Ablagerungen aus der Grube Messel an Hand der Forschungs bohrung Messel 2001 und weiterer Bohrungen. Cour Forsch Senck 252:151–203
Feng Z, Wang J, Rößler R, Ślipiński A, Labandeira CC (2017) Late Permian wood-borings reveal an intricate network of ecological relationships. Nat Commun 18:556. https://doi.org/10.1038/s41467-017-00696-0
Fergusson NDM (1990) A phylogenetic study of the Cynipoidea (Hymenoptera). PhD thesis, Polytechnic University, London
Ferns PN, Jervis MA (2016) Ordinal species richness in insects—a preliminary study of the influence of morphology, life history, and ecology. Entomol Exp Appl 159:270–284
Ferrar P (1987) A guide to the breeding habits and immature stages of Diptera Cyclorrhapha. Brill and Scandinavian Science Press, Leiden
Fisher RC (1961) A study in insect multiparasitism. I. Host selection and oviposition. J Exp Biol 38:267–275
Foster DE (1976) Revision of North American *Trichodes* (Herbst) (Coleoptera: Cleridae). Spec Publ Mus Texas Tech Univ 11:1–86
Frank JH, Gillett-Kaufman JL (2006) Glossary of expressions in biological control. University of Florida IPM-143:1–11
Fraser NC, Grimaldi DA, Olsen PE, Axsmith B (1996) A Triassic Lagerstätte from eastern North America. Nature 380:615–619
Freeland WJ, Boulton WJ (1992) Coevolution of food webs: parasites, predators and plant secondary compounds. Biotropica 24:309–327
Galloway TD (2018) Biodiversity of ectoparasites: lice (Phthiraptera) and fleas (Siphonaptera). In: Foottit RG, Adler PH (eds) Insect biodiversity: science and society, vol II, 1st edn. John Wiley, New York, pp 457–482
Gao TP, Shih CK, Ren D, Feng H (2010) Hymenoptera – pollinators' contribution. In: Dong R, Shih CK, Gao TP, Yao YZ, Zhao YY (eds) Silent stories – insect fossil treasures from Dinosaur era of the Northeastern China. Science Press, Beijing, pp 268–288
Gao TP, Shih CK, Xu X, Wang S, Ren D (2012) Mid-Mesozoic flea-like ectoparasites of feathered or haired vertebrates. Curr Biol 22(8):732–735
Gao TP, Shih CK, Rasnitsyn AP, Xu S, Wang S, Ren D (2014) The first flea with fully distended abdomen from the Early Cretaceous of China. BMC Evol Biol 14:168. https://doi.org/10.1186/s12862-014-0168-1
Garrouste R, Pouillon J-M, Nel A (2016) The first Cenozoic roproniid wasp from the Paleocene of Menat, France (Hymenoptera: Proctotrupoidea). Eur J Taxon 239:1–9. https://doi.org/10.5852/ejt.2016.239
Gaston KJ (1991) The magnitude of global insect species richness. Conserv Biol 5:283–296

Gauld ID (2008) Evolutionary patterns of host utilization by ichneumonoid parasitoids (Hymenoptera: Ichneumonidae and Braconidae). Biol J Linn Soc 35:351–377

Gauld ID, Bolton B (1988) The Hymenoptera. British Museum (Natural History) and Oxford University Press, London

Gauld ID, Gaston KJ, Janzen DH (1992) Plant allelochemicals, tritrophic interactions and the anomalous diversity of tropical parasitoids: the 'nasty' host hypothesis. Oikos 65:353–357

Genise JF, Cladera G (2004) *Chubutolithes gaimanensis* and other wasp trace fossils: breaking through the taphonomic barrier. J Kansas Entomol Soc 77:626–638

Gibson GAP (2009) Description of three new genera and four new species of Nenastatinae (Hymenoptera, Eupelmidae) from Baltic amber, with discussion of their relationships to extant taxa. ZooKeys 20:175–214

Gibson JF, Skevington JH (2013) Phylogeny and taxonomic revision of all genera of Conopidae (Diptera) based on morphological data. Zool J Linn Soc Lond 167:43–81

Gibson GAP, Read J, Huber JT (2007) Diversity, higher classification and higher relationships of Mymarommatoidea (Hymenoptera). J Hymen Res 16:51–146

Gilbert C, Rayor LS (1983) First record of mantisfly (Neuroptera: Mantispidae) parasitizing a spitting spider (Scytodidae). J Kansas Entomol Soc 56:578–580

Gillung JP, Winterton SL, Bayless KM, Khouri Z, Borowiec ML, Yeates D, Kimsey LS, Misof B, Shin S, Zhou X, Mayer C, Peterson M, Wiegmann BM (2018) Anchored Phylogenomics unravels the evolution of spider flies (Diptera, Acroceridae) and reveals discordance between nucleotides and amino acids. Mol Phylogenet Evol 128:233–245

Glaser JD (1976) The biology of *Dynastes tityus* (Linn.) in Maryland (Coleoptera: Scarabaeidae). Coleopt Bull 30:133–138

Gleeson DM, Howitt RLJ, Newcomb RD (2000) The phylogenetic position of the New Zealand batfly, *Mystacinobia zelandica* (Mystacinobiidae; Oestroidea) inferred from mitochondrial 16S ribosomal DNA sequence data. J R Soc N Z 30:155–168

Godefroit P, Sinitsa SM, Dhouailly D, Bolotsky YL, Sizov AV, McNamara ME, Benton MJ, Spagna P (2014) A Jurassic ornithischian dinosaur from Siberia with both feathers and scales. Science 345:451–455

Godfray HCJ (1994) Parasitoids: behavioral and evolutionary ecology. Princeton University Press, Princeton, NJ

Goulet H, Huber JT (1993) Hymenoptera of the world: an identification guide to families. Centre for Land and Biological Resource Research, Ottawa

Gratshev VG, Zherikhin VV (2003) The fossil record of weevils and related beetle families (Coleoptera, Curculionoidea). Acta Zool Cracov 46(Suppl):126–138

Greathead DJ (1963) A review of the insect enemies of Acridoidea (Orthoptera). Trans R Entomol Soc Lond 114:437–517

Green PWC, Turner BD (2005) Food-selection by the booklouse, *Liposcelis bostrychophila* Badonnel (Psocoptera: Liposcelididae). J Stored Prod Res 41:103–113

Greenwalt D (2021) Blood to Molecules: The Fossil Record of Blood and its Constituents. In: De Baets K, Huntley JW (eds) The evolution and fossil record of parasitism: Coevolution and paleoparasitological techniques. Topics in Geobiology 50

Greenwalt DE, Goreva YS, Siljeström SM, Rose T, Harbach RE (2013) Hemoglobin-derived porphyrins preserved in a middle Eocene blood-engorged mosquito. Proc Natl Acad Sci U S A. https://doi.org/10.1073/pnas.1310885110

Griffiths GCD (1972) The phylogenetic classification of Diptera Cyclorrhapha with special reference to the structure of the male postabdomen. W Junk, The Hague

Grimaldi DA (1992) Vicariance biogeography, geographic extinctions, and the North American Oligocene tsetse flies. In: Novacek MJ, Wheeler QD (eds) Extinction and phylogeny. Columbia University Press, New York, pp 178–204

Grimaldi DA (1995) A remarkable new species of *Ogcodes* (Diptera: Acroceridae) in Dominican amber. Am Mus Novit 3127:1–8

Grimaldi DA (1996) Amber: window to the past. Abrams, New York

Grimaldi DA, Barden P (2016) The Mesozoic family Eremochaetidae (Diptera: Brachycera) in Burmese amber and relationships of Archisargoidea: Brachycera in Cretaceous amber, Part VIII. Am Mus Novit 3865:1–29

Grimaldi DS, Engel MS (2005a) Evolution of the insects. Cambridge University Press, New York

Grimaldi DS, Engel MS (2005b) Fossil Liposcelididae and the lice ages (Insecta: Psocodea). Proc R Soc B 273:265–633

Grimaldi DA, Engel MS, Nascimbene PC (2002) Fossiliferous Cretaceous amber from Myanmar (Burma): its rediscovery, biotic diversity, and paleontological significance. Am Mus Novit 3361:1–72

Grimaldi DA, Kathirithamby J, Schawaroch V (2005) Strepsiptera and triungula in Cretaceous amber. Insect Syst Evol 36:1–20

Grissell EE (1980) New Torymidae from Tertiary amber of the Dominican Republic and a world list of fossil torymids (Hymenoptera: Chalcidoidea). Proc Entomol Soc Wash 82:252–259

Gumovsky AV, Perkovsky EE (2005) Taxonomic notes on Tetracampidae (Hymenoptera: Chalcidoidea) with description of a new fossil species of *Dipricocampe* from Rovno amber. Entomol Probl 35:123–130

Gumovsky A, Perkovsky E, Rasnitsyn A (2018) Laurasian ancestors and "Gondwanan" descendants of Rotoitidae (Hymenoptera: Chalcidoidea): what a review of late Cretaceous *Baeomorpha* revealed. Cretac Res 84:286–322

Gurney AB (1953) Notes on the biology and immature stages of a cricket parasite of the genus *Rhopalosoma*. Proc US Natl Mus 103:19–34

Haas F (2018) Biodiversity of Dermaptera. In: Foottit RG, Adler PH (eds) Insect biodiversity: science and society, vol 2, 1st edn. Wiley, New York, pp 315–334

Haas M, Burks RA, Krogmann L (2018) A new lineage of Cretaceous jewel wasps (Chalcidoidea: Diversinitidae). PeerJ 6:e4633. https://doi.org/10.7717/peerj.4633

Haeselbarth E (1979) Zur Parasitierung der Puppen von Forleule (*Panolis flammea* [Schiff.]), Kiefernspanner (*Bupalus piniarius* [L.]) und Heidelbeerspanner (*Boarmia bistortana* [Goeze]) in bayerischen Keifernwäldern. Z Angew Entomol 87:186–202

Halffter G, Edmonds WD (1982) The nesting behavior of dung beetles (Scarabaeinae) – an ecological and evolutive approach. Instituto de Ecología, Mexico City

Halffter G, Matthews EG (1999) The natural history of dung beetles of the subfamily Scarabaeinae. Medical Books, Palermo

Hammond PM (1976) Kleptoparasitic behaviour of *Onthophagus suturalis* Peringuey (Coleoptera: Scarabaeidae) and other dung beetles. Coleopt Bull 30:245–249

Han Y, Ye X, Feng CP, Zhang KY, Shih CK, Ren D (2019) Diptera – true flies with two wings. In: Ren D, Shih CK, Gao TP, Wang YG, Yao YZ (eds) Rhythms of insect evolution: evidence from the Jurassic and Cretaceous of Northern China. Wiley, Hoboken, NJ, pp 497–553

Hanson P (1992) The Nearctic species of *Ormyius* Westwood (Hymenoptera: Chalcidoidea: Ormyidae). J Nat Hist 26:1333–1365

Harper CJ, Krings M (2021) Fungi as parasites: a conspectus of the fossil record. In: De Baets K, Huntley JW (eds) The evolution and fossil Record of parasitism: identification and macroevolution of parasites. Topics in Geobiology 49

Harvey JA, Wagenaar R, Bezemer TM (2009) Interactions to the fifth trophic level: secondary and tertiary parasitoid wasps show extraordinary efficiency in utilizing host resources. J Anim Ecol 78:686–692

Haug JT, Labandeira CC, Santiago-Blay JA, Haug C, Brown S (2015) Life habits, hox genes, and affinities of a 311 million-year-old holometabolan larva. BMC Evol Biol 15:208. https://doi.org/10.1186/s12862-015-0428-8

Haug JT, Nagler C, Haug C, Hörnig MK (2017) A group of assassin fly pupae preserved in a single piece of Eocene amber. Bull Geosci 92(3):283–295. https://doi.org/10.3140/bull.geosci.1621

Haug JT, Müller P, Haug C (2018) The ride of the parasite: a 100-million-year old mantis lacewing larva captured while mounting its spider host. Zool Lett 4:31. https://doi.org/10.1186/s40851-018-0116-9

Hayward A, McMahon DP, Kathirithamby J (2011) Cryptic diversity and female host specificity in a parasitoid where the sexes utilize hosts from separate orders. Mol Ecol 20:1508–1528
Henderickx H, Bosselaers J, Pauwels E, van Hoorebeke L, Boone M (2013) X-ray micro-CT reconstruction reveals eight antennomeres in a new fossil taxon that constitutes a sister clade to *Dundoxenos* and *Triozocera* (Strepsiptera: Corioxenidae). Palaeont Elect 16(3):1–16 [29A]
Hennig W (1965) Die Acalyptratae des baltischen Bernsteins und ihre Bedeutung für die Erfoschung der phylogenetischen Entwicklung dieser Dipteren-Gruppe. Stutt Beitr Naturk 145:1–215
Hennig W (1966) Conopidae im baltischen Bernstein (Diptera: Cyclorrhapha). Stutt Beitr Naturk 154:1–24
Heraty JM, Darling DC (2009) Fossil Eucharitidae and Perilampidae (Hymenoptera: Chalcidoidea) from Baltic amber. Zootaxa 2301:1–16
Heraty JM, Burks RA, Cruaud A, Gibson GAP, Liljeblad J, Munro J, Rasplus J-Y, Delvare G, Janšta P, Gumovsky A, Huber J, Woolley JB, Krogman L, Heydon S, Polaszek A, Schmidt S, Darling DC, Gates MW, Mottern J, Murray E, Dal Molin A, Triapitsyn S, Baur H, Pinto JD, Van Noort S, George J, Yoder M (2013) A phylogenetic analysis of the megadiverse Chalcidoidea (Hymenoptera). Cladistics 29:466–542
Hickman V (1965) On *Planarivora insignis* gen. et sp. n. (Diptera: Mycetophilidae), whose larval stages are parasitic on land planarians. Pap Proc R Soc Tasmania 99:1–8
Hirao T, Mukrakami M (2008) Quantitative food webs of lepidopteran leafminers and their parasitoids in a Japanese deciduous forest. Ecol Res 23:159–168
Hochberg ME, Hawkins BA (1992) Refuges as a predictor of parasitoid diversity. Science 255(5047):973–976
Holloway BA (1976) A new bat-fly family from New Zealand (Diptera: Mystacinobiidae). N Z J Zool 3:279–301
Holt RD (1977) Predation, apparent competition and the structure of prey communities. Theor Pop Biol 12:197–229
Hong YC (1984) New fossil insects of Laiying Group from Laiying Basin, Shandong Province. Prof Pap Stratigr Palaeontol 11:31–41. (in Chinese)
Hong YC (2002) Amber insects of China. Science and Technology Press, Beijing
Hong CD, Xu ZF (2011) A newly recorded genus and species of Family Stephanidae (Hymenoptera, Stephanoidea) from China. Entomotaxonomia 33:71–73
Hong CD, van Achterberg C, Xu ZF (2010) A new species of *Megischus* Brullé (Hymenoptera, Stephanidae) from China, with a key to the Chinese species. ZooKeys 69:59–64
Hong CD, van Achterberg C, Xu ZF (2011) A revision of the Chinese Stephanidae (Hymenoptera, Stephanoidea). ZooKeys 110:1–108
Horion A (1961) Faunistik der mitteleuropäischen Käfer. Band VIII: Clavicornia. 2 Teil (Thoricidae bis Ciidae), *Teredilia*, Coccinellidae. A. Fevel, Überlingen
Houston TF (1987) Fossil brood cells of stenotritid bees (Hymenoptera, Apoidea) from the Pleistocene of South Australia. Trans R Soc S Aust 3:93–97
Howden HF, Storey RI (1992) Phylogeny of the Rhyparini and the new tribe Stereomerini, with descriptions of new genera and species. Can J Zool 70:1810–1823
Hsiao Y, Huang C-L (2018) *Spinotoma ruicheni*: a new Late Cretaceous genus and species of wedge-shaped beetle from Burmese amber (Coleoptera, Ripiphoridae, Pelecotominae). Cretac Res 82:29–35
Hsiao Y, Yu YL, Deng CS, Pang H (2017) The first fossil wedge-shaped beetle (Coleoptera, Ripiphoridae) from the middle Jurassic of China. Eur J Taxon 277:1–13
Huang DY (2014) The diversity and host associations of Mesozoic giant fleas. Natl Sci Rev 1:496–497
Huang DY (2015) *Tarwinia australis* (Siphonaptera: Tarwiniidae) from the Lower Cretaceous Koonwarra fossil bed: morphological revision and analysis of its evolutionary relationship. Cretac Res 52:507–515
Huang DY (ed) (2016) The Daohugou biota. Shanghai Scientific and Technical Publishers, Shanghai. (in Chinese)

Huang DY, Cai CY (2016) The arthropods. In: Huang DY (ed) The Daohugou biota. Shanghai Scientific and Technical Publishers, Shanghai, pp 84–211. (in Chinese)
Huang DY, Engel MS, Cai CY, Wu H, Nel A (2012) Diverse transitional giant fleas from the Mesozoic era of China. Nature 483:201–204
Huang DY, Engel MS, Cai CY, Nel A (2013a) Mesozoic giant fleas from northeastern China (Siphonaptera): taxonomy and implications for Palaeodiversity. Chin Sci Bull 54:1682–1690
Huang DY, Nel A, Cai CY, Lin QB, Engel MS (2013b) Amphibious flies and paedomorphism in the Jurassic period. Nature 495:94–97
Huang DY, Wu Q, Su YT, Sha JG (2016) Isotope chronology. In: Huang DY (ed) The Daohugou biota. Shanghai Scientific and Technical Publishers, Shanghai, pp 71–83. (in Chinese)
Huber JT (1986) Systematics, biology, and hosts of the Mymaridae and Mymarommatidae (Insecta: Hymenoptera): 1758–1984. Entomography 4:185–243
Huber JT (2005) The gender and derivation of genus-group names in Mymaridae and Mymarommatidae (Hymenoptera). Acta Soc Zool Bohem 69:167–183
Hughes DP (2014) On the origins of parasite-extended phenotypes. Integr Comp Biol 54:210–217
Hughes DP, Anderson SB, Hywel-Jones NL, Himaman W, Billen J, Boomsma JJ (2011a) Behavioral mechanisms and morphological symptoms of zombie ants dying from fungal infection. BMC Ecol 11:13. https://doi.org/10.1186/1472-6785-11-13
Hughes DP, Wappler T, Labandeira CC (2011b) Ancient death-grip leaf scars reveal ant–fungal parasitism. Biol Lett 7:67–70
Hull FM (1973) Bee flies of the world: the genera of the family Bombyliidae. Smithsonian Institution Press, Washington, DC
Hunter MD, Matson PA (1992) The relative contributions of top–down and bottom–up forces in population and community ecology. Ecology 73:723
Hunter MD, Price PW (1992) Playing chutes and ladders: heterogeneity and the relative roles of bottom–up and top–down forces in natural communities. Ecology 73:724–732
Ings TC, Montoya JM, Bascompte J, Blüthgen N, Brown L, Dormann CF, Edwards F, Figueroa D, Jacob U, Jones JI, Lauridsen RB, Leger ME, Lewis HM, Olesen JM, van Veen FJF, Warren PH, Woodward G (2009) Ecological networks – beyond food webs. J Anim Ecol 78:253–269
Iturralde-Vinent MA (2001) Geology of the amber-bearing deposits of the Greater Antilles. Carib J Sci 37:141–167
Iturralde-Vinent MA, MacPhee RDE (1996) Age and palaeogeographical origin of Dominican amber. Science 273:1850–1852
Janzen DH (1975) Interactions of seeds and their insect predators/parasitoids in a tropical deciduous forest. In: Price PW (ed) Evolutionary strategies of parasitic insects and mites. Plenum Press, New York, pp 154–186
Janzen J-W (2002) Arthropods in Baltic amber. Ampyx, Halle, Germany
Jarzembowski EA, Ross AJ (1996) Insect origination and extinction in the Phanerozoic. Geol Soc Spec Publ 102:65–78
Jell PA, Duncan PM (1986) Invertebrates, mainly insects, from the freshwater, Lower Cretaceous, Koonwarra fossil bed (Korumburra Group), South Gippsland, Victoria. Mem Assoc Aust Paleont 3:111–205
Jennings JT, Austin AD, Stevens NB (2004) *Hyptiogastrites electrinus* Cockerell, 1917, from Myanmar (Burmese) amber: redescription and its placement within the Evanioidea (Insecta: Hymenoptera). J Syst Palaeontol 2:127–132
Jeon JB, Kim TW, Tripotin P, Kim J-I (2002) Notes on a cicada parasitic moth in Korea (Lepidoptera: Epipyropidae). Korean J Entomol 32:239–241
Jepson JE (2015) A review of the current state of knowledge of fossil Mantispidae (Insecta: Neuroptera). Zootaxa 3964:419–432
Jervis MA (1980) Life history studies on *Aphelopus* species (Hymenoptera: Dryinidae) and *Chalarus* species (Diptera: Pipunculidae), primary parasites of typhlocybine leafhoppers (Homoptera, Cicadellidae). J Nat Hist 14:769–780
Ji Q, Luo ZX, Yuan CX, Tabrum AR (2006) A swimming mammaliaform from the Middle Jurassic and ecomorphological diversification of early mammals. Science 311:1123–1127

Jobling B (1926) A comparative study of the structure of the head and mouth parts in the Hippoboscidae (Diptera Pupipara). Parasitology 18:319–349
Jobling B (1928) The structure of the head and mouth-parts in the Nycteribiidae (Diptera Pupipara). Parasitology 20:254–272
Jobling B (1929) A comparative study of the structure of the head and mouth parts in the Streblidae (Diptera Pupipara). Parasitology 21:417–445
Jobling B (1933) A revision of the structure of the head, mouth-parts and salivary glands of *Glossina palpalis* Rob.-Desv. Parasitology 24:449–490
Johnson NF, Musetti L (1999) Revision of the Proctotrupoid genus *Pelecinus* Latreille (Hymenoptera: Pelecinidae). J Nat Hist 33:1513–1543
Johnson NF, Musetti L, Janzen J-W (2001) A new fossil species of the Australian endemic genus *Peradenia* Naumann & Masner (Hymenoptera: Proctotrupoidea, Peradeniidae) from Baltic Amber. Insect Syst Evol 32:191–194
Johnson KP, Yoshizawa K, Smith VS (2004) Multiple origins of parasitism in lice. Proc R Soc Lond B 271:1771–1776
Johnson KP, Dietrich CH, Friedrich F, Beutel RG, Wipfler B, Peters RS, Allen JM, Petersen M, Donath A, Walden KKO, Kozlov AM, Podsiadlowski L, Mayer C, Meusemann K, Vasilikopoulos A, Waterhouse RM, Cameron SL, Weirauch C, Swanson DR, Persey DM, Hardy NB, Terry I, Liu S, Zhou X, Misof B, Robertson HM, Yoshizawa K (2018) Phylogenomics and the evolution of hemipteroid insects. Proc Natl Acad Sci U S A 115:12775–12780
Kaddumi HF (2005) Amber of Jordan, the oldest prehistoric insects in fossilized resin. Eternal River Museum of Natural History, Amman
Kalugina NS (1991) New Mesozoic Simuliidae and Leptoconopidae and the origin of bloodsucking habit in the lower dipteran insects. Paleontol J 1991(1):66–77
Kathirithamby J (2009) Host-parasitoid associations in Strepsiptera. Annu Rev Entomol 54:227–249
Kathirithamby J, Engel MS (2014) A revised key to the living and fossil families of Strepsiptera, with the description of a new family, Cretostylopidae. J Kansas Entomol Soc 87:385–388
Kathirithamby J, Grimaldi DA (1993) Remarkable stasis in some Lower Tertiary parasitoids: descriptions, new records, and review of Strepsiptera in the Oligo–Miocene amber of the Dominican Republic. Entomol Scand 24:31–41
Kerr PH, Winterton SL (2008) Do parasitic flies attack mites? Evidence in Baltic amber. Biol J Linn Soc 93:9–13
Kimsey LS, Bohart RM (1990) The Chrysidid wasps of the world. Oxford University Press, Oxford, UK
Kinzelbach R (1983) Fächerflüger aus dem Dominikanishen Bernstein (Insecta: Strepsiptera: Myrmecolacidae). Verhand Naturwis Ver Hamburg 26:29–36
Kinzelbach R, Lutz H (1985) Stylopid larva from the Eocene – a spotlight on the phylogeny of the stylopids (Strepsiptera). Ann Entomol Soc Am 78:600–602
Kinzelbach R, Pohl H (1994) The fossil Strepsiptera (Insecta: Strepsiptera). Ann Entomol Soc Am 87:59–70
Kirkpatrick TW (1947) Notes on a species of Epipyropidae (Lepidoptera) parasitic on *Metaphaena* species (Hemiptera: Fulgoridae) at Amani, Tanganyika. Proc R Entomol Soc Lond A22:61–64
Kirk-Spriggs AH, Kotrba M, Copeland RS (2011) Further details of the morphology of the enigmatic African fly *Mormotomyia hirsuta* Austen (Diptera: Mormotomyiidae). Afr Invert 52:145–165
Kitching JW (1980) On some fossil Arthropoda from the Limeworks, Makapansgat, Potgietersus. Palaeontol Afr 23:63–68
Klimaszewski J (1984) A revision of the genus *Aleochara* Gravenhorst of America north of Mexico (Coleoptera: Staphylinidae, Aleocharinae). Mem Entomol Soc Canada 129:1–211
Klotz JH, Dorn PL, Logan JL, Stevens L, Pinnas JL, Schmidt JO, Klotz SA (2010) "Kissing bugs": potential disease vectors and cause of anaphylaxis. Clin Infect Dis 50:1629–1634
Knisley CB, Reeves DL, Stephens GT (1989) Behavior and development of the wasp *Pterombrus rufiventris hyalinatus* Krombein (Hymenoptera: Tiphiidae), a parasite of larval tiger beetles (Coleoptera: Cicindelidae). Proc Entomol Soc Wash 91:179–184

Knutson LV (1972) Pupae of *Nemochtherus angustipennis* (Hine), with notes on feeding habits of robber flies and a review of publications on morphology of immature stages (Diptera: Asilidae). Proc Biol Soc Wash 85:163–178

Knutson LV, Berg CO (1966) Parasitoid development in snail-killing sciomyzid flies. Science 156:522–533

Knutson LV, Flint OS (1979) Pupae of Empididae in pupal cocoons of Rhyacophilidae and Glossosomatidae (Diptera–Trichoptera). Proc Entomol Soc Wash 73:314–329

Kocarek P, John V, Hulva P (2013) When the body hides the ancestry: phylogeny of morphologically modified epizoic earwigs based on molecular evidence. PLoS One 8(6):e66900. https://doi.org/10.1371/journal.pone.0066900

Kogan M, Poinar GO Jr (2010) New fossil *Stylops* (Strepsiptera: Stylopidae) from Dominican amber. Neotrop Entomol 39(2):227–234

Kolibáč J (2010) Cleridae Latreille, 1802. In: Leschen RAB, Beutel RG, Lawrence JF (eds) Handbuch der Zoologie; Band IV: Arthropoda: Insecta; Teilband 38: Coleoptera, Beetles, Morphology and systematics (Elateroidea, Bostrichiformia, Cucujiformia partim), vol 2. Walter de Gruyter, Berlin, pp 257–261

Kolibáč J, Huang D (2016) The oldest known clerid fossils from the Middle Jurassic of China, with a review of Cleridae systematics. Syst Entomol 41:808–823

Kovalev VG, Mostovski MB (1997) A new species of snipe flies (Diptera, Rhagionidae) from the Mesozoic deposits of eastern Transbaikalia. Paleontol Zh 1997(5):86–90. (in Russian)

Kozlov MA (1968) Jurassic Proctotrupoidea (Hymenoptera). In: Rohdendorf BB (ed) Jurassic insects of Karatau. USSR Academy of Sciences, Moscow, pp 237–240. (in Russian)

Kozlov MA (1974) An early Cretaceous ichneumon of the family Pelecinidae (Hymenoptera, Pelecinoidea). Paleontol Zh 1974(1):144–146. (in Russian)

Kozlov MA, Rasnitsyn AP (1979) On the limits of the family Serphitidae (Hymenoptera: Proctotrupoidea). Entomol Obozr 58:402–416

Krassilov VA (2008a) Mine and gall predation as top down regulation in the plant–insect systems from the Cretaceous of Negev, Israel. Palaeogeogr Palaeoclimat Palaeoecol 261:261–269

Krassilov VA (2008b) Evidence of temporary mining in the Cretaceous fossil mine assemblage of Negev, Israel. Insect Sci 15:285–290

Krassilov VA, Zherikhin VV, Rasnitsyn AP (1997) *Classopollis* in the guts of Jurassic insects. Palaeontology 40:1095–1101

Krell F-T (2006) Fossil record and evolution of Scarabaeoidea (Coleoptera: Polyphaga). Coleopt Soc Monogr 5:120–143

Krenn HW, Aspöck H (2012) Form, function and evolution of the mouthparts of blood-feeding Arthropoda. Arthropod Struct Dev 41:101–118

Krikken J (1978) Valgine beetles: a preliminary review of the genera, with descriptions of two novelties. Zool Meded 53:153–164

Krogman L (2013) First fossil record of cerocephaline wasps with a description of a new genus and species from Dominican amber (Hymenoptera: Chalcidoidea: Pteromalidae: Cerocephalinae). Hist Biol 25:43–49

Krzeminski W, Jarzembowski EA (1999) *Aenne triassica* sp. n., the oldest representative of the family Chironomidae (Insecta: Diptera). Pol Pismo Entomol 68:445–449

Kulicka R (1979) Mengea mengei sp. n. from the Baltic amber. Prace Muz Ziemi 32:109–112

Kuris AM (1974) Trophic interactions: similarity of parasitic castrators to parasitoids. Q Rev Biol 49:129–148

Kuris AM, Hechinger RF, Shaw JC, Whitney KL, Aguirre-Macedo L, Boch CA, Dobson AP, Dunham EJ, Fredensborg BL, Huspeni TC, Lorda J, Mababa L, Mancini FT, Mora AB, Pickering M, Tallhouk NL, Torchen ME, Lafferti KD (2008) Ecosystem energetic implications of parasite and free-living biomass in three estuaries. Nature 454:515–518

Kutty SN, Pape T, Wiegmann BN, Meier R (2001) Molecular phylogeny of the Calyptratae (Diptera: Cyclorrhapha) with an emphasis on the superfamily Oestroidea and the position of the Mystacinobiidae and McAlpine's fly. Syst Entomol 35:614–635

Labandeira CC (1994) A compendium of fossil insect families. Contrib Biol Geol Milwaukee Publ Mus 88:1–71
Labandeira CC (1997) Insect mouthparts: ascertaining the paleobiology of insect feeding strategies. Annu Rev Ecol Syst 28:153–193
Labandeira CC (1998) Early history of arthropod and vascular plant associations. Annu Rev Earth Planet Sci 26:329–377
Labandeira CC (2002) Paleobiology of predators, parasitoids, and parasites: death and accommodation in the fossil record of continental invertebrates. In: Kowalewski M, Kelley PH (eds) The fossil record of predation, Paleont Soc Pap, vol 8. Cambridge University Press, New York, pp 211–249
Labandeira CC (2005) Fossil history and evolutionary ecology of Diptera and their associations with plants. In: Yeates DK, Wiegmann BM (eds) The evolutionary biology of flies. Columbia University Press, New York, pp 217–273
Labandeira CC (2010) The pollination of mid Mesozoic seed plants and the early history of long-proboscid insects. Ann Mo Bot Gard 97:469–513
Labandeira CC (2014a) Amber. In: Laflamme M, Schiffbauer JD, Darroch SAF (eds) Reading and writing of the fossil record: preservational pathways to exceptional fossilization, Paleont Soc Pap, vol 20. Cambridge University Press, New York, pp 163–216
Labandeira CC (2014b) Why did terrestrial diversity not increase during the angiosperm radiation? Mid-Mesozoic, plant-associated insect lineages harbor clues. In: Pontarotti P (ed) Evolutionary biology: genome evolution, speciation, coevolution and origin of life. Springer, Cham, Switzerland, pp 261–299
Labandeira CC (2015) The fundamental importance of parasitoids in terrestrial ecosystems. Geol Soc Am Abstr Prog 47(7):215
Labandeira CC (2019) The fossil record of insect mouthparts: innovation, functional convergence and associations with other organisms. In: Krenn H (ed) Insect mouthparts—form, function, development and performance. Wiley, New York
Labandeira CC, Dunne J (2014) DRYAD data for "Highly resolved early Eocene food webs show development of modern trophic structure after the end-Cretaceous extinction", (Dunne, Labandeira and Williams). Proc R Soc. https://doi.org/10.5061/dryad.ps0f0
Labandeira CC, Phillips TL (1996) A Carboniferous petiole gall: insight into early ecologic history of the Holometabola. Proc Natl Acad Sci U S A 93:8470–8474
Labandeira CC, Sepkoski JJ Jr (1993) Insect diversity in the fossil record. Science 261:310–315
Labandeira CC, Kvaček J, Mostovski MB (2007a) Pollination drops, pollen, and insect pollination of Mesozoic gymnosperms. Taxon 56:663–695
Labandeira CC, Wilf P, Johnson KR, Marsh F (2007b) Guide to insect (and other) damage types on compressed plant fossils (Version 3.0 – Spring 2007). Smithsonian Institution, Washington, DC
Labandeira CC, Anderson JM, Anderson HM (2018) Arthropod herbivory in Late Triassic South Africa: the Molteno Biota, the Aasvoëlberg 411 locality and the developmental biology of a gall. In: Tanner L (ed) The late Triassic world: earth in a time of transition, Topics in geobiology, vol 46. Springer, Cham, Switzerland, pp 623–719
Lachaud J-P, Pérez-Lachaud G (2009) Impact of natural parasitism by two eucharitid wasps on a potential biocontrol agent ant in southeastern Mexico. Biol Control 48:92–99
Lafferty KD, Dobson AP, Kuris AM (2006) Parasites dominate food web links. Proc Natl Acad Sci U S A 103:11211–11216
Lafferty KD, Allesina S, Arim M, Briggs CJ, De Leo G, Dobson AP, Dunne JA, Johnson PTJ, Kuris AM, Marcogliese DJ, Martinez ND, Memmot J, Marquet PA, McLaughlin JP, Mordecai EA, Pascual M, Poulin R, Thieltges DW (2008) Parasites in food webs: the ultimate missing links. Ecol Lett 11:533–546
Lai PY (1988) Biological control: a positive point of view. Proc Hawaiian Entomol Soc 28:179–190
Lambkin CL (1986) A revision of the Australian Mantispidae (Insecta, Neuroptera) with a contribution to the classification of the family. I. General and Drepanicinae. Aust J Zool 16:1–174
Lambrecht FI (2018) Palaeoecology of tsetse flies and sleeping sickness in Africa. Proc Am Philos Soc 124:367–385

LaSalle J, Gauld ID (1991) Parasitic Hymenoptera and the biodiversity crisis. Redia 74:315–334

Lattin JD (1999) Bionomics of the Anthocoridae. Annu Rev Entomol 44:207–231

Lawrence JF (2005) Rhipiceridae Latreille, 1834. In: Beutel RG, Leschen RAB (eds) Handbuch der Zoologie; Band IV: Arthropoda: Insecta; Teilband 38: Coleoptera, Beetles, Morphology and systematics, vol 1. Walter de Gruyter, Berlin, pp 456–460

Lawrence JF, Ślipiński A (2013) Australian Beetles. Morphology, classification and keys. CSIRO Publishing, Collingwood, Australia

Lawrence JF, Anderson DM, Beal RS Jr, Becker EC, Bell RT, Bosquet Y, Bright DE, Brown HP, Carlson DC, Cooper KW, Dogger JR, Dybas HS, Foster DE, Frank JH, Kavanaugh DH, LaBella DM, Lawson FA, LeSage L, Lloyd JE, Newton AF Jr, Pfaffenberger GS, Reichardt H, Selander RB, Spangler PJ, Spilman TJ, de Viedma MG, Wheeler QE, Young DK (1991) Order Coleoptera. In: Stehr FW (ed) Immature insects, vol 2. Kendall-Hunt, Dubuque, IA

Lehane MJ (1991) Biology of blood-sucking insects. HarperCollins Academic, London

Lehmann GUC (2003) Review of biogeography, host range and evolution of acoustic hunting in Ormiini (Insecta, Diptera, Tachinidae), parasitoids of night-calling bushcrickets and crickets (Insecta, Orthoptera, Ensifera). Zool Anz 242:107–120

Lelej AS (1996) Males of the genus *Protomutilla* (Hymenoptera: Mutillidae) from Baltic amber. Paleontol Zh 1996:104–106. (in Russian)

Lent H, Wygodzinsky P (1979) Revision of the Triatominae (Hemiptera, Reduviidae), and their significance as vectors of Chagas' Disease. Bull Am Mus Nat Hist 163:123–520

Leung TLF (2017) Fossils of parasites: what can the fossil record tell us about the evolution of parasitism? Biol Rev 92:410–430

Leung TLF (2021) Parasites of fossil vertebrates: what we know and what can we expect from the fossil record? In: De Baets K, Huntley JW (eds) The evolution and fossil record of parasitism: Identification and macroevolution of parasites. Topics in Geobiology 49

Lewis RE, Grimaldi DA (1997) A pulicid flea in Miocene amber from the Dominican Republic (Insecta: Siphonaptera: Pulicidae). Am Mus Novit 3205:1–9

Lewis OT, Memmot J, Lasalle J, Lyal CHC, Whitefoord C, Godfray HCJ (2002) Structure of a diverse tropical forest insect–parasitoid community. J Anim Ecol 71:855–873

Li LF, Rasnitsyn AP, Shih CK, Ren D (2013a) Anomopterellidae restored, with two new genera and its phylogeny in Evanioidea (Hymenoptera). PLoS One 8(12):e82587

Li LF, Shih CK, Ren D (2013b) Two new wasps (Hymenoptera: Stephanoidea: Ephialtitidae) from the Middle Jurassic of China. Acta Geol Sin 87:1486–1494

Li LF, Shih CK, Ren D (2014a) New fossil evaniids (Hymenoptera, Evanioidea) from the Yixian Formation of western Liaoning, China. Cretac Res 47:48–55

Li LF, Shih CK, Ren D (2014b) Revision of *Anomopterella* Rasnitsyn, 1975 (Insecta, Hymenoptera, Anomopterellidae) with two new Middle Jurassic species from northeastern China. Geol Carpath 65:365–374

Li LF, Shih CK, Ren D (2014c) New fossil Praeaulacinae wasps (Insecta: Hymenoptera: Evanioidea: Praeaulacidae) from the Middle Jurassic of China. Zootaxa 3814:432–442

Li LF, Shih CK, Ren D (2014d) Two new species of *Nevania* (Hymenoptera: Evanioidea: Praeaulacidae: Nevaniinae) from the Middle Jurassic of China. Alcheringa 38:140–147

Li LF, Rasnitsyn AP, Shih CK, Ren D (2015a) A new genus and species of Praeaulacidae (Hymenoptera: Evanioidea) from Upper Cretaceous Myanmar amber. Cretac Res 55:19–24

Li LF, Shih CK, Rasnitsyn AP, Ren D (2015b) New fossil ephialtitids elucidating the origin and transformation of the propodeal-metasomal articulation in Apocrita (Hymenoptera). BMC Evol Biol 15:45. https://doi.org/10.1186/s12862-015-0317-1

Li LF, Kopylov DS, Shih CK, Ren D (2017a) The first record of Ichneumonidae (Insecta: Hymenoptera) from the Upper Cretaceous of Myanmar. Cretac Res 70:152–162

Li LF, Rasnitsyn AP, Labandeira CC, Shih CK, Ren D (2017b) Phylogeny of Stephanidae (Hymenoptera: Apocrita) with a new genus from Upper Cretaceous Myanmar amber. Syst Entomol 42:194–203

Li LF, Shih CK, Ren D (2017c) New fossil helorid wasps (Hymenoptera, Proctotrupoidea) from the Early Cretaceous of China. Alcheringa 41:474–486

Li LF, Rasnitsyn AP, Shih CK, Labandeira CC, Buffington M, Li DQ, Ren D (2018a) Phylogeny of Evanioidea (Hymenoptera, Apocrita), with descriptions of new Mesozoic species from China and Myanmar. Syst Entomol 43:810–842
Li LF, Shih CK, Rasnitsyn AP, Li DQ, Ren D (2018b) A new wasp of Myanmarinidae (Hymenoptera: Stephanoidea) from the mid-Cretaceous Myanmar amber. Cretac Res 86:33–40
Light JE, Smith VS, Allen JM, Durden LA, Reed DI (2010) Evolutionary history of mammalian sucking lice (Phthiraptera: Anoplura). BMC Evol Biol 10:292. http://www.biomedcentral.com/1471-2148/10/292
Lin QB (1982) Insecta. In: Paleontological atlas of East China. Part 3, Volume of Mesozoic and Caenozoic. Geological Publishing House, Beijing. (in Chinese)
Lin QB (1986) Early Mesozoic fossil insects from South China. Science Press, Beijing. (in Chinese with English summary)
Lin YC, Chan ML, Ko CW, Hsieh MY (2004) Nail infestation by *Liposcelis bostrychophila* Badonnel. Clin Exp Dermatol 29:620–621
Lin XD, Labandeira CC, Shih CK, Hotton CL, Ren D (2019) Life habits and evolutionary biology of new two-winged long-proboscid scorpionflies from mid-Cretaceous Myanmar amber. Nat Commun 10:1235. https://doi.org/10.1038/s41467-019-09236
Linsley EG, McSwain JW, Smith RF (1952) The life history and development of *Rhipiphorus smithi* with notes on their phylogenic significance (Coleoptera, Rhipiphoridae). Univ Calif Publ Entomol 9:291–314
Liu ZW, Engel MS (2010) Baltic amber Ibaliidae (Hymenoptera: Cynipoidea): a new genus with implications for the phylogeny and historical biogeography of the family. Syst Entomol 35:164–171
Liu ZW, Engel MS, Grimaldi DA (2007) Phylogeny and geological history of the cynipoid wasps (Hymenoptera: Cynipoidea). Am Mus Novit 3583:1–48
Lohrmann V, Engel MS (2017) The wasp larva's last supper: 100 million years of evolutionary stasis in the larval development of rhopalosomatid wasps (Hymenoptera: Rhopalosomatidae). Foss Rec 20:239–244
Loreau M (2010) From populations to ecosystems. Princeton University Press, Princeton, NJ
Lorenz V, Kurzlaukis S (2007) Root zone processes in the phreatomagmatic pipe emplacement model and consequences for the evolution of maar–diatreme volcanoes. J Volcan Geoth Res 150:4–32
Lukashevich ED, Mostovski MB (2003) Hematophagous insects in the fossil record. Paleontol J 37(2):153–161
Maa TC (1964) A review of old world Polyctenidae (Hemiptera: Cimicoidea). Pac Insect 6:494–516
Maa TC (1966) Redescription of the fossil *Ornithomyia rottensis* (Statz). Pac Insect Mon 10:3–9
MacArthur RH, Levins R (1964) Competition, habitat selection, and character displacement in a patchy environment. Proc Natl Acad Sci U S A 51:1207–1210
Mackauer M, Chau A (2001) Adaptive self superparasitism in a solitary parasitoid wasp: the influence of clutch size on offspring size. Ecology 15:335–343
Maddison DR, Baker MD, Ober KA (1999) Phylogeny of carabid beetles as inferred from 18S ribosomal DNA (Coleoptera: Carabidae). Syst Entomol 24:103–138
Maksoud S, Azar D, Granier B, & Gze R (2017). New data on the age of the Lower Cretaceous amber outcrops of Lebanon. Palaeoworld, 26(2), 331–338
Malyshev SI (1968) Genesis of the Hymenoptera and the phases of their evolution. Methuen, London
Mardulyn P, Whitfield JB (1999) Phylogenetic signal in the COI, 16S, and 28S genes for inferring relationships among genera of Microgastrinae (Hymenoptera; Braconidae): evidence of a high diversification rate in this group of parasitoids. Mol Phylogenet Evol 12:282–294
Marinho MAT, Wolff M, Ramos-Pastrana Y, Azeredo-Espin AML, Amorim DS (2017) The first phylogenetic study of Mesembrinellidae (Diptera: Oestroidea) based on molecular data: clades and congruence with morphological characters. Cladistics 33:134–152
Marshall AG (1981) The ecology of ectoparasitic insects. Academic, London
Marshall SA (2012) Flies – the natural history and diversity of Diptera. Firefly Books, Buffalo

Martins AC, Luz DR, Melo GAR (2018) Palaeocene origin of the Neotropical lineage of cleptoparasitic bees Ericrocidini–Rhathymini (Hymenoptera, Apidae). Syst Entomol 43:510–521
Martin AJ, Varricchio DJ (2011) Paleoecological utility of insect trace fossils in dinosaur nesting sites of the Two Medicine Formation (Campanian), Choteau, Montana. Historical Biology, 23(01):15–25
Martins-Neto RG (2003) The fossil tabanids (Diptera Tabanidae): when they began to appreciate warm blood and when they began to transmit diseases? Mem Inst Oswaldo Cruz 98(Suppl 1):29–34
Matson PA, Hunter MD (1992) Special feature: the relative contributions to top–down and bottom–up forces in population and community ecology. Ecology 73:723
Maus C, Mittmann B, Peschke K (1998) Host records of parasitoid *Aleochara* Gravenhorst species (Coleoptera, Staphylinidae) attacking puparia of cyclorrhaphous Diptera. Deut Entomol Z 45:231–254
May RM (1988) How many species are there on Earth? Science 241:1441–1449
May RM (1992) How many species inhabit the Earth? Sci Am 267(4):42–48
McAlpine JF (1970) First record of calypterate flies in the Mesozoic Era (Diptera: Calliphoridae). Can Entomol 102:342–346
McKellar RC, Engel MS (2011a) New Stigmaphronidae and Megaspilidae (Hymenoptera: Ceraphronoidea) from Canadian Cretaceous amber. Cretac Res 32:794–805
McKellar RC, Engel MS (2011b) The serphitid wasps (Hymenoptera: Proctotrupomorpha: Serphitoidea) of Canadian Cretaceous amber. Syst Entomol 36:192–208
McKellar RC, Engel MS (2012) Hymenoptera in Canadian Cretaceous amber (Insecta). Cretac Res 35:258–279
McKellar RC, Engel MS (2014) New bethylid and chrysidid wasps (Hymenoptera: Chrysidoidea) from Canadian Late Cretaceous amber. Paläontol Z 88:433–451
McKellar RC, Wolfe AP (2010) Canadian amber. In: Penney D (ed) Biodiversity of fossils in amber from the major world deposits. Siri Scientific Press, Manchester, UK, pp 149–166
McKellar RC, Wolfe AP, Tappert R, Muehlenbachs K (2008) Correlation of Grassy Lake and Cedar Lake ambers using infrared spectroscopy, stable isotopes, and palaeoentomology. Canad J Earth Sci 45:1061–1082
McKellar RC, Kopylov DS, Engel MS (2013) Ichneumonidae (Insecta: Hymenoptera) in Canadian Late Cretaceous amber. Foss Rec 16:217–227
Melo GAR (2000) Biology of an extant species of the scolebythid genus *Dominibythus* (Hymenoptera: Chrysidoidea: Scolebythidae), with description of its mature larva. In: Austin AD, Dowton M (eds) Hymenoptera evolution, biodiversity and biological control. CSIRO, Collingwood, Australia, pp 281–284
Memmott J, Godfray HCJ, Gauld ID (1994) The structure of a tropical host-parasitoid community. J Anim Ecol 63(3):521–540
Memmott J, Martinez ND, Cohen JE (2000) Predators, parasitoids and pathogens: species richness, trophic generality and body sizes in a natural food web. J Anim Ecol 69:1–15
Mermudes JRM, Leschen RAB (2014) Anthribidae Billberg, 1820. In: Leschen RAB, Beutel RG, Lawrence JF (eds) Handbuch der Zoologie; Band IV: Arthropoda: Insecta; Teilband 38: Coleoptera, Beetles, Morphology and systematics (Phytophaga), vol 3. Walter de Gruyter, Berlin, pp 309–315
Michelson V (2000) Oldest authentic record of a fossil calyptrate fly (Diptera): a species of Anthomyiidae from early Coenozoic Baltic amber. Stud Dipterol 7:11–18
Mikulás R, Genise JF (2003) Traces within traces: holes, pits and galleries in walls and fillings of insect trace fossils in paleosols. Geol Acta 1:339–348
Mills NJ (1994) Parasitoid guilds: defining the structure of the parasitoid communities of endopterygote insect hosts. Environ Entomol 23:1066–1083
Misof B, Liu SL, Meusemann K, Peters RS, Donath A, Mayer C, Frandsen PB, Ware J, Flouri T, Beutel RG, Niehuis O, Petersen M, Izquierdo-Carrasco F, Wappler T, Rust J, Aberer AJ, Aspöck U, Aspöck H, Bartel D, Blanke A, Berger S, Böhm A, Buckley TR, Calcott B, Chen JQ,

Friedrich F, Fukui M, Fujita M, Greve C, Grove P, Gu SC, Huang Y, Jermiin LS, Kawahara AY, Krogmann L, Kubiak M, Lanfear R, Letsch H, Li Y, Li Z, Li JG, Lu HR, Machida R, Mashimo Y, Kapli P, McKenna DD, Meng GL, Nakagaki Y, Navarrete-Heredia JL, Ott M, Ou YX, Pass G, Podsiadlowski L, Pohl H, von Reumont BM, Schütte K, Sekiya K, Shimizu S, Slipiński A, Stamatakis A, Song WH, Su X, Szucsich NU, Tan MH, Tan XM, Tang M, Tang JB, Timelthaler G, Tomizuka S, Trautwein M, Tong XL, Uchifume T, Walzl MG, Wiegmann BM, Wilbrandt J, Wipfler B, Wong TKF, Wu Q, Wu GX, Xie YL, Yang SZ, Yang Q, Yeates DK, Yoshizawa K, Zhang Q, Zhang R, Zhang WW, Zhang YH, Zhao J, Zhou CJ, Zhou LL, Ziesmann T, Zou SJ, Li YG, Xu X, Zhang Y, Yang HM, Wang J, Wang J, Kjer KM, Zhou X (2014) Phylogenomics resolves the timing and pattern of insect evolution. Science 346:763–767

Mockford EL (1971) Psocoptera from sleeping nests of the dusky-faced wood rat in southern California (Psocoptera: Atropidae; Psoquillidae, Liposcelidae). Pan Pac Entomol 47:127–140

Mockford EL (2018) Biodiversity of Psocoptera. In: Foottit RG, Adler PH (eds) Insect biodiversity: science and society, vol 2, 1st edn. Wiley, New York, pp 417–456

Moghaddam MG, Turrisi GF (2018) Taxonomic and faunistic study of Aulacidae (Hymenoptera, Evanioidea) from Iran, with illustrated key to species. Zoosyst Evol 94:95–108

Morris AK (1998) A model of trophic evolutionary pathways. MS Thesis, Oregon State University, Corvallis

Morris RJ, Lewis OT, Godfray CJ (2004) Experimental evidence for apparent competition in a tropical forest food web. Nature 428:310–313

Mostovski MB (1998) A revision of the nemestrinid flies (Diptera: Nemestrinidae) described by Rohdendorf, and a description of new taxa of the Nemestrinidae from the Upper Jurassic of Kazakhstan. Paleontol J 32:369–375

Mostovski MB, Jarzembowski EA, Coram R (2003) Tabanids and athericids (Diptera: Tabanidae, Athericidae) from the Lower Cretaceous of England and Transbaikalia. Paleontol Zh 2003(2):57–64

Muldrew JA (1953) The natural immunity of the larch sawfly (*Pristophora erichsonii* (Htg.)) to the introduced parasite *Mesoleius tenthredinis* Morley in Manitoba and Saskatchewan. Can J Zool 31:313–332

Müller CB, Adriaanse ICT, Belshaw R, Godfray HCJ (1999) The structure of an aphid–parasitoid community. J Anim Ecol 68:346–370

Murray EA, Carmichael AE, Heraty JM (2013) Ancient host shifts followed by host conservatism in a group of ant parasitoids. Proc R Soc B 280:20130495. https://doi.org/10.1098/rspb.2013.0495

Nagatomi A, Soroida K (1985) The structure of the mouthparts of the orthorrhaphous Brachycera (Diptera) with special reference to blood-sucking. Beitr Entomol 35:263–308

Nagler C, Haug JT (2015) From fossil parasitoids to vectors: insects as parasites and hosts. Adv Parasitol 90:137–200

Nakata S, Maa TC (1974) A review of the parasitic earwigs (Dermaptera: Arixeniina; Hemimerina). Pac Insect 16:307–374

Narchuk EP (1972) Predation and parasitism in the evolution of chloropid flies. Zool Zh 51:1342–1352. (in Russian)

Naumann ID, Masner L (1985) Parasitic wasps of the Proctotrupoid complex: a new family from Australia with a key to world families (Hymenoptera: Proctotrupoidea sensu lato). Austr J Zool 33:761–783

Nel A, Azar D (2005) The oldest parasitic Scelionidae: Teleasinae (Hymenoptera: Platygastroidea). Pol Pismo Entomol 74:333–338

Nel A, De Ploëg G (2004) New fossil bee flies (Diptera: Bombylioidea) in the lowermost Eocene amber of the Paris Basin. Geol Acta 2:57–65

Nel A, Perrichot V, Néraudeau D (2003) The oldest trigonalid wasp in the late Albian amber of Charente-Maritime (SW France) (Hymenoptera: Trigonalidae). Ecol Geol Helv 96:503–508

Nel A, Waller A, De Ploëg G (2004) An aulacid wasp in the Lowermost Eocene amber from the Paris Basin (Hymenoptera: Aulacidae). Geol Acta 2:67–74

Nel P, Bertrand S, Nel A (2018) Diversification of insects since the Devonian: a new approach based on morphological disparity of mouthparts. Sci Rep 8:3516. https://doi.org/10.1038/s41598-018-21938-1

Newton AF (2005) Leiodidae Fleming, 1821. In: Beutel RG, Leschen RAB (eds) Handbuch der Zoologie; Band IV: Arthropoda: Insecta; Teilband 38: Coleoptera, Beetles, Morphology and systematics (Archostemata, Adephaga, Myxophaga, Polyphaga partim), vol 1. Walter de Gruyter, Berlin, pp 269–280

Olmi, M. (1998) New Embolemidae and Dryinidae (Hymenoptera Chrysidoidea). Frust Entomol 20:30–118

O'Neill KM (2001) Solitary wasps: behavior and natural history. Cornell University Press, Ithaca, NY

O'Reilly J, Donoghue PC (2016) Tips and nodes are complementary, not competing approaches to the calibration of molecular clocks. Biol Lett 12:20150975. https://doi.org/10.1098/rsbl.2015.0975

O'Reilly JE, dos Reis M, Donoghue PC (2015) Dating tips for divergence-time estimation. Trends Genet 31:637–650

Obame-Nkoghe J, Rahola N, Bourgarel M, Yangari P, Prugnolle F, Maganga GD, Leroy E-M, Fontenille D, Ayala D, Paupy C (2016) Bat flies (Diptera: Nyceribiidae and Streblidae) infesting cave-dwelling bats in Gabon: diversity, dynamics and potential role in *Polychromophilus melanipherus* transmission. Parasit Vect 9:333. https://doi.org/10.1186/s13071-016-1625-z

Oberprieler SK, Rasnitsyn AP, Brothers DJ (2012) The first wasps from the Upper Jurassic of Australia (Hymenoptera: Evanioidea, Praeaulacidae). Zootaxa 3503:47–54

Oksanen L, Fretwell SD, Arruda J, Niemalä P (1981) Exploitation ecosystems in gradients of primary productivity. Am Nat 118:240–261

Oldroyd H (1964) The natural history of flies. W.W. Norton, New York

Ortega-Blanco J, Rasnitsyn AP, Delclòs X (2010) A new family of ceraphronoid wasps from Early Cretaceous Álava amber, Spain. Acta Palaentol Pol 55:265–276

Ortega-Blanco J, Delclòs X, Engel MS (2011a) The wasp family Embolemidae in Early Cretaceous amber from Spain (Hymenoptera: Chrysidoidea). J Kansas Entomol Soc 84:36–42

Ortega-Blanco J, Delclòs X, Engel MS (2011b) Diverse stigmaphronid wasps in Early Cretaceous amber from Spain (Hymenoptera: Ceraphronoidea: Stigmaphronidae). Cretac Res 32:762–773

Ortega-Blanco J, Delclòs X, Engel MS (2011c) A protorhyssaline wasp in Early Cretaceous amber from Spain (Hymenoptera: Braconidae). J Kansas Entomol Soc 84:51–57

Oryan A, Razavi SM, Bahrami S (2009) Occurrence and biology of goat warble fly infestation by Przhevalskiana silenus (Diptera: Oestridae) in Iran. Vet Parasitol 166:178-181

Parfin S (1958) Notes on the bionomics of the Mantispidae (Neuroptera: Planipennia). Entomol News 69:203–207

Pape T (2001) Phylogeny of Oestridae (Insecta: Diptera). Syst Entomol 26:131-171

Paulian R (1988) Biologie des Coléoptères. Lechevalier, Paris

Peck SB (2006) Distribution and biology of the ectoparasitic beaver beetle *Platypsyllus castoris* Ritsema in North America (Coleoptera: Leiodidae: Platypsyllinae). Insect Mundi 20:85–94

Peñalver E, Delclòs X (2010) Spanish amber. In: Penney D (ed) Biodiversity of fossils in amber from the major world deposits. Siri Scientific Press, Manchester, UK, pp 236–270

Peñalver E, Engel MS (2006) Two wasp families rare in the fossil record (Hymenoptera): Perilampidae and Megaspilidae from the Miocene of Spain. Am Mus Novit 3540:1–12

Peñalver E, Fontal-Cazalla FM, Pujade-Villar J (2013) *Palaeogronotoma* n. gen. from the Miocene of Spain, the first Tertiary fossil record of the subfamily Eucoilinae (Hymenoptera: Figitidae). Geodiversitas 35:643–653

Penney D (2010) Dominican amber. In: Penney D (ed) Biodiversity of fossils in amber from the major world deposits. Siri Scientific Press, Manchester, UK, pp 22–39

Penteado-Dias AM, van Achterberg C (2002) First record of the genus *Probethylus* Ashmead (Sclerogibbidae: Probethylinae) from Brazil, with a description of a new species. Zool Meded 76:105–107

Peralta G, Frost CM, Rand TA, Didham RK, Tylianakis JM (2014) Complementarity and redundancy of interactions enhance attack rates and spatial stability in host–parasitoid food webs. Ecology 95:1888–1896
Perkovsky EE (1999) Novy podnejurskiy rod I vid Leiodinae (Coleoptera, Leiodidae) iz Mongolii. Vest Zool 33:77–79
Perkovsky EE, Zosimovich VY, Vlaskin AP (2010) Rovno amber. In: Penney D (ed) Biodiversity of fossils in amber from the major world deposits. Siri Scientific Press, Manchester, UK, pp 116–136
Perrichot V (2009) Long-tailed wasps (Hymenoptera: Megalyridae) from Cretaceous and Paleogene European amber. Paleont Contrib 2009(1):1–19
Perrichot V, Nel A (2008) Eocene bethylid wasps from French amber (Hymenoptera: Bethylidae). Neues Jb Paläont Geol Abh 248:91–101
Perrichot V, Nel A, Quicke DLJ (2008) New braconid wasps from French Cretaceous amber (Hymenoptera, Braconidae): synonymization with Eoichneumonidae and implications for the phylogeny of Ichneumonoidea. Zool Scr 38:79–88
Perrichot V, Ortega-Blanco J, McKellar RC, Delclòs X, Azar D, Nel A, Tafforeau P, Engel MS (2011) New and revised maimetshid wasps from Cretaceous ambers (Hymenoptera, Maimetshidae). ZooKeys 130:421–453
Perrichot V, Beaucournu J-C, Velten J (2012) First extinct genus of a flea (Siphonaptera: Pulicidae) in Miocene amber from the Dominican Republic. Zootaxa 3438:54–61
Perrichot V, Engel MS, Nel A, Ortega-Blanco J, Delclòs X, Soriano C, Lohrmann V, Tafforeau P (2014) "False ants" from the Cretaceous (Hymenoptera: Chrysidoidea: Falsiformicidae). Mitt Entomol Ver Stutt 19:26
Peters RS, Krogmann L, Mayer C, Donath A, Gunkel S, Meusemann K, Kozlov A, Podsiadlowski L, Peterson M, Lanfear R, Diez PA, Heraty J, Kjer KM, Klopfstein S, Meier R, Polidori C, Schmitt T, Liu S, Zhou X, Wappler T, Rust J, Misof B, Nehuis O (2017) Evolutionary history of the Hymenoptera. Curr Biol 27:1013–1018
Philips TK, Ivie MA (2002) Bothrideridae Erickson 1845. In: Arnett RH Jr, Thomas MC, Skelley PE, Frank JH (eds) American Beetles, Polyphaga: Scarabaeoidea through Curculionoidea, vol 2. CRC Press, Boca Raton, FL, pp 358–362
Pielowska A, Sontag E, Szadziewski R (2018) Haematophagous arthropods in Baltic amber. Ann Zool 68:237–249
Pike EM (1995) Amber Taphonomy and the Grassy Lake, Alberta, Amber Fauna. PhD thesis, Department of Biological Sciences, University of Calgary, Calgary
Pinto JD, Selander RB (1970) The bionomics of the blister beetles of the genus *Meloe* and a classification of the New World species. Illinois Biol Mon 42:1–122
Pohl H, Beutel RG (2005) The phylogeny of the Strepsiptera (Hexapoda). Cladistics 21:328–374
Pohl H, Beutel RG (2008) The evolution of the Strepsiptera (Hexapoda). Zoology 111:318–338
Pohl H, Beutel RG (2016) †*Kinzelbachilia ellenbergi* – a new ancestral species, genus and family of Strepsiptera (Insecta). Syst Entomol 41:287–297
Pohl H, Beutel RG (2019) Effects of miniaturization in primary larvae of Strepsiptera (Insecta). Arthropod Struct Dev 48:49–55
Pohl H, Kinzelbach R (1995) First record of a female stylopid (Strepsiptera: ?Myrmecolacidae) parasite of a prionomyrmecine ant (Hymenoptera: Formicidae) in Baltic amber. Insect Syst Evol 32:143–146
Pohl H, Kinzelbach R (2001) Neufunde von Fächerflügern aus dem Baltischen und Dominikanischen Bernstein (Strepsiptera: Bohartillidae & Myrmecolacidae). Mitt Geol Paläont Inst Mus Univ Hamburg 78:197–209
Pohl H, Beutel RG, Kinzelbach R (2005) Protoxenidae fam. nov. (Insecta, Strepsiptera) from Baltic amber – a "missing link" in strepsipteran phylogeny. Zool Scr 35:57–69
Pohl H, Batelka J, Prokop J, Müller P, Yavorskaya MI, Beutel RG (2018) A needle in a haystack: Mesozoic origin of parasitism in Strepsiptera revealed by first definite Cretaceous primary larva (Insecta). PeerJ 6:e5943. https://doi.org/10.7717/peerj.5943
Pohl H, Hammel JU, Richter A, Beutel RG (2019) The first fossil free-living late instar larva of Strepsiptera (Insecta). Arthropod Syst Phylogeny 77:125–140

Poinar GO Jr (1987) Fossil evidence of spider parasitism by Ichneumonidae. J Arachnol 14:399–400
Poinar GO Jr (2003) Trends in the evolution of insect parasitism by nematodes as inferred from fossil evidence. J Nematol 35:129–132
Poinar GO Jr (2004a) Palaeomyia burmitis (Diptera: Phlebotominae), a new genus and species of Cretaceous sand flies with evidence of blood-sucking habits. Proc Entomol Soc Wash 106:598–605
Poinar GO Jr (2004b) Evidence of parasitism by Strepsiptera in Dominican amber. BioControl 49:239–234
Poinar GO Jr (2005) *Triatoma dominicana* sp. n. (Hemiptera: Reduviidae: Triatominae), and *Trypanosoma antiquus* sp. n. (Stercoraria: Trypanosomatidae), the first fossil evidence of a triatomine–trypanosomatid vector association. Vect Borne Zoonot Dis 5(1):72–81
Poinar GO Jr (2009) *Cascoplecia insolitis* (Diptera: Cascopleciidae), a new family, genus and species of flower-visiting, unicorn fly (Bibionomorpha) in Early Cretaceous Burmese amber. Cretac Res 31:71–76
Poinar GO Jr (2010) Palaeoecological perspectives in Dominican amber. Ann Soc Entomol France 46:23–52
Poinar GO Jr (2011) *Vetufebrus ovatus* n. gen., n. sp. (Haemospororida: Plasmodiidae) vectored by a streblid bat fly (Diptera: Streblidae) in Dominican amber. Parasit Vect 4:229. https://doi.org/10.1186/1756-3305-4-229
Poinar GO Jr (2013) *Stenaspidiotus microptilus* n. gen., n. sp. (Coleoptera: Chrysomelidae: Chrysomelinae) in Dominican amber with evidence of tachinid (Diptera: Tachinidae) oviposition. Hist Biol 25:101–105
Poinar GO Jr (2015) A new genus of fleas with associated microorganisms in Dominican amber. J Med Entomol 52:1234–1240
Poinar GO Jr (2018) Vertebrate pathogens vectored by ancient hematophagous arthropods. Hist Biol 32(7):888-901. https://doi.org/10.1080/0891263.2018.1545018
Poinar GO Jr (2019) Fossil record of viruses and parasitic bacteria and protozoa. In: De Baets K, Huntley JW (eds) The evolution and fossil record of parasitism – identification and macroevolution of parasites. Springer, Berlin
Poinar GO Jr, Brown A (2012) The first fossil streblid bat fly, *Enischnomyia stegosoma* n. g., n. sp. (Diptera: Hippoboscoidea: Streblidae). Syst Parasitol 81:79–86
Poinar GO Jr, Brown A (2014) New genera and species of jumping ground bugs (Hemiptera: Schizopteridae) in Dominican and Burmese amber, with a description of a meloid (Coleoptera: Meloidae) triungulin on a Burmese specimen. Ann Soc Entomol France 50:372–381
Poinar GO Jr, Buckley R (2010) *Doratomantispa burmanica* n. gen., n. sp. (Neuroptera: Mantispidae), a new genus of mantidflies in Burmese amber. Hist Biol 23:169–176
Poinar GO Jr, Huber JT (2011) A new genus of fossil Mymaridae (Hymenoptera) from Cretaceous amber and key to Cretaceous mymarid genera. ZooKeys 130:461–472
Poinar GO Jr, Miller JC (2002) First fossil record of endoparasitism of adult ants (Formicidae: Hymenoptera) by Braconidae (Hymenoptera). Ann Entomol Soc Am 95:41–43
Poinar GO Jr, Poinar R (1999) The amber forest: a reconstruction of a vanished world. Princeton University Press, Princeton, NJ
Poinar GO Jr, Poinar R (2004a) *Paleoleishmania proterus* n. gen., n. sp. (Trypanosomatidae: Kinetoplastida) from Cretaceous Burmese amber. Protist 155:305–310
Poinar GO Jr, Poinar R (2004b) Evidence of vector-borne disease of Early Cretaceous reptiles. Vector Borne Zoonotic Dis 4:281–284
Poinar GO Jr, Shaw SR (2007) *Megalyra baltica* Poinar and Shaw n. sp. (Hymenoptera: Megalyridae), a long-tailed wasp from Baltic amber. Zootaxa 1478:65–68
Poinar GO Jr, Shaw SR (2016) Endoparasitism of a Cretaceous adult weevil by a euphorine wasp (Hymenoptera: Braconidae). Neues Jb Geol Paläont Abh 282:109–113
Poinar GO Jr, Telford SR Jr (2005) Paleohaemoproteus burmacis gen. n., sp. n. (Haemosporida: Plasmodiidae) from an Early Cretaceous biting midge (Diptera: Ceratopogonidae). Parasitology 131:79–84
Poinar GO Jr (2021) Fossil record of viruses, parasitic bacteria and parasitic protozoa. In: De Baets K, Huntley JW (eds) The evolution and fossil Record of parasitism: identification and macroevolution of parasites. Topics in Geobiology 49

Polis GA, Strong DR (1996) Food web complexity and community dynamics. Am Nat 147:813–846
Ponomarenko AG (1976) A new insect from the Cretaceous of Transbaikalia, a possible parasite of pterosaurians. Paleontol Zh 1976:102–106
Ponomarenko AG (1995) The geological history of beetles. In: Pakaluk J, Ślipiński SA (eds) Biology, phylogeny, and classification of coleoptera: papers celebrating the 80th birthday of Roy A. Crowson. Museum and Institute of Zoology, Warsaw, pp 155–171
Popham EJ (1984) The genus *Hemimerus*, insect parasites of the giant rat. Nyala 10:39–42
Poulin R (2011) The many roads to parasitism: a tale of convergence. Adv Parasitol 74:1–40
Poulin R, Morand S (2000) The diversity of parasites. Q Rev Biol 75:277–293
Power ME (1992) Top–down and bottom–up forces in food webs: do plants have primacy. Ecology 73:733–746
Prentice M, Poinar GO Jr, Milki R (1996) Fossil scolebythids (Hymenoptera: Scolebythidae) from Lebanese and Dominican amber. Proc Entomol Soc Wash 98:802–811
Price PW (1980) Evolutionary biology of parasites. Princeton University Press, Princeton, NJ
Price PW (1984) Insect ecology. Wiley, New York
Price PW, Denno RF, Eubanks MD, Finke DL, Kaplan I (2011) Insect ecology, 4th edn. Cambridge University Press, Cambridge, UK
Putz FE, Holbrook NM (1989) Strangler fig rooting habits and nutrient relations in the Llanos of Venezuela. Am J Bot 76:781–788
Quevillon LE, Hughes DP (2018) Pathogens, parasites, and parasitoids of ants: a synthesis of parasite biodiversity and epidemiological traits. bioRxiv. https://doi.org/10.1101/384495
Quicke DLJ (1997) Parasitic wasps. Chapman and Hall, New York
Qvarnström M, Niedźwiedzki G, Žigaitė Ž (2016) Vertebrate coprolites (fossil faeces): an underexplored *Konservat-Lagerstätte*. Earth Sci Rev 162:44–57
Rainford JL, Mayhew PJ (2015) Diet evolution and clade richness in Hexapoda: a phylogenetic study of higher taxa. Am Nat 186:777–791
Rainford JL, Hofreiter M, Nicholson DB, Mayhew PJ (2014) Phylogenetic distribution of extant richness suggests metamorphosis is a key innovation driving diversification in insects. PLoS One 9(10):e10985. https://doi.org/10.1371/journal.pone.0109085
Ramírez LC, Corsolino J, Di Iorio O (2016) First fossil record of parasitic flat-bark beetle (Coleoptera: Passandrididae) from the Eocene of Patagonia, Argentina. Ameghiniana 53:160–169
Rasnitsyn AP (1964) New Triassic hymenopterans from Central Asia. Paleontol Zh 1964(1):88–96
Rasnitsyn AP (1968) New Mesozoic sawflies (Hymenoptera, Symphyta). In: Rohdendorf BB (ed) Jurassic insects of Karatau. USSR Academy of Sciences, Moscow, pp 190–236. (in Russian)
Rasnitsyn AP (1969) The origin and evolution of lower Hymenoptera. Trans Paleontol Inst 123:1–196. (in Russian)
Rasnitsyn AP (1972) Late Jurassic hymenopterous insects (Praeaulacidae) of Karatau. Paleontol Zh 1972:70–87. (in Russian)
Rasnitsyn AP (1975) Hymenoptera Apocrita of the Mesozoic. Trans Paleontol Inst 147:1–134. (in Russian)
Rasnitsyn AP (1980) On the system of the family Aulacidae (Hymenoptera) in connection with a new finding in the Cretaceous of Manlay. Trans Joint Soviet-Mongolian Paleontol Exped 13:65–67. (in Russian)
Rasnitsyn AP (1983) New names for fossil insects. Paleontol Zh 3:103. (in Russian)
Rasnitsyn AP (1985) Jurassic insects of Siberia and Mongolia. Trudy Paleontol Inst Akad Nauk SSSR 211:1–192. (in Russian)
Rasnitsyn AP (1988) An outline of evolution of the hymenopterous insects (order Vespida). Orient Insects 22:115–145
Rasnitsyn AP (1991a) Early Cretaceous evaniomorphous hymenopterans. Vest Zool 1991:128–132. (in Russian)
Rasnitsyn AP (1991b) Early members of Evaniomorphous hymenopteran families Stigmaphronidae, Cretevaniidae and the subfamily Kotujellitinae (Gasteruptiidae). Paleontol J 1991:128–132. (in Russian)

Rasnitsyn AP (1992) *Strashila incredibilis*, a new enigmatic mecopteroid insect with possible siphonapteran affinities from the Upper Jurassic of Siberia. Psyche 99:323–333
Rasnitsyn AP (1996) New Cretaceous Embolemidae (Vespida = Hymenoptera; Chrysidoidea). Mem Entomol Soc Wash 17:183–187
Rasnitsyn AP (2000) An extremely primitive aculeate wasp in the Cretaceous amber from New Jersey (Vespida: ?Sierolomorpidae). In: Grimaldi DA (ed) Studies on fossils in amber, with particular reference to the Cretaceous of New Jersey. Backhuys, Leiden, pp 327–332
Rasnitsyn AP (2002) Superorder Vespidea Laicharting, 1781. Order Hymenoptera Linné, 1758 (=Vespida Laicharting, 1781). In: Rasnitsyn AP, Quicke DLJ (eds) History of insects. Kluwer, Dordrecht, pp 242–254
Rasnitsyn AP (2008) Hymenopterous insects (Insecta: Vespida) in the Upper Jurassic deposits of Shar Teg, SW Mongolia. Russ Entomol J 17:299–310
Rasnitsyn AP (2013) *Vectevania vetula* Cockerell, 1922 from the uppermost Eocene of Bembridge Marls, England, and the system of the family Gasteruptiidae s. l. [Vespida (= Hymenoptera), Evanioidea]. Proc Russ Entomol J 84:98–106
Rasnitsyn AP, Brothers DJ (2009) New genera and species of Maimetshidae (Hymenoptera: Stephanoidea) *s.l.* from the Turonian of Botswana, with comments on the status of the family. Afr Invert 50:191–205
Rasnitsyn AP, Kovalev OV (1988) The oldest Cynipoidea (Hymenoptera, Archaeocynipidae fam. n.) from the Early Cretaceous Transbaikalia. Vest Zool 1988:18–21
Rasnitsyn AP, Krassilov VA (2000) The first documented occurrence of phyllophagy in Pre-Cretaceous insects: leaf tissues in the gut of Upper Jurassic insects from southern Kazakhstan. Paleontol J 34:301–309
Rasnitsyn AP, Martínez-Delclòs X (1999) New Cretaceous Scoliidae (Vespida = Hymenoptera) from the Lower Cretaceous of Spain and Brazil. Cretac Res 20:767–772
Rasnitsyn AP, Martínez-Delclòs X (2000) Wasps (Insecta: Vespida = Hymenoptera) from the Early Cretaceous of Spain. Acta Geol Hisp 35:65–95
Rasnitsyn AP, Öhm-Kühnle C (2018) Three new female *Aptenoperissus* from mid-Cretaceous Burmese amber (Hymenoptera: Stephanoidea: Aptenoperissidae): unexpected diversity of paradoxical wasp suggest insular features of source biome. Cretac Res 91:168–175
Rasnitsyn AP, Quicke DLJ (2002) History of insects. Kluwer Academic, Dordrecht
Rasnitsyn AP, Ross AJ (2000) A preliminary list of arthropod families present in the Burmese amber collection at the Natural History Museum, London. Bull Nat Hist Mus Lond Geol 56:21–24
Rasnitsyn AP, Sharkey MJ (1988) New Eoichneumonidae from Early Cretaceous of Siberia and Mongolia (Hymenoptera: Ichneumonoidea). In: Gupta VK (ed) Advances in parasitic Hymenoptera research. E.J. Brill, Leiden, pp 169–197
Rasnitsyn AP, Strelnikova OD (2017) Tracheal system and biology of the Early Cretaceous *Saurophthirus longipes* Ponomarenko, 1976 (Insecta, ?Aphaniptera, Saurophthiroidea stat. nov.). Paleontol J 51(2):171–182
Rasnitsyn AP, Zhang HC (2004) Composition and age of the Daohugou hymenopteran (Insecta, Hymenoptera = Vespida) assemblage from Inner Mongolia, China. Palaeontology 47:1507–1517
Rasnitsyn AP, Zhang HC (2010) Early evolution of Apocrita (Insecta, Hymenoptera) as indicated by new findings in the Middle Jurassic of Daohugou, northeast China. Acta Geol Sin 84:834–873
Rasnitsyn AP, Zherikhin VV (1999) First fossil chewing louse from the Lower Cretaceous of Baissa, Transbaikalia (Insecta, Pediculida = Phthiriaptera, Saurodectidae fam. n.). Russ Entomol J 8:253–255
Rasnitsyn AP, Zherikhin VV (2002) 4: Appendix: Alphabetic list of selected insect fossil sites. 4.1: Impression fossils. In: Rasnitsyn AP, Quicke DLJ (eds) History of insects. Kluwer, Dordrecht, pp 437–444
Rasnitsyn AP, Jarzembowski EA, Ross AJ (1998) Wasps (Insecta: Vespida = Hymenoptera) from the Purbeck and Wealden (Lower Cretaceous) of southern England and their biostratigraphical and palaeoenvironmental significance. Cretac Res 19:329–391

Rasnitsyn AP, Ansorge J, Zessin W (2003) New hymenopterous insects (Insecta: Hymenoptera) from the Lower Toarcian (Lower Jurassic) of Germany. Neues Jb Geol Paläontol Abh 227:321–342

Rasnitsyn AP, Poinar G Jr, Brown AE (2017) Bizarre wingless parasitic wasp from mid-Cretaceous Burmese amber (Hymenoptera, Ceraphronoidea, Aptenoperissidae fam. nov.). Cretac Res 69:113–118

Rehn JAG, Rehn JWH (1935) A study of the genus *Hemimerus* (Dermaptera, Hemimerina, Hemimeridae). Proc Acad Nat Sci Philadelphia 87:457–508

Rehn JAG, Rehn JWH (1937) Notes on the genus *Hemimerus* (Dermaptera, Hemimerina, Hemimeridae). Proc Acad Nat Sci Philadelphia 89:331–335

Reinhardt K, Siva-Jothy MT (2007) Biology of the bed bugs (Cimicidae). Annu Rev Entomol 52:351–274

Ren D (1998) Flower-associated Brachycera flies as fossil evidence for Jurassic angiosperm origins. Science 280:85–88

Ren D, Shih CK, Gao TP, Yao YZ, Zhao YY (2010a) Silent stories – insect fossil treasures from Dinosaur Era of the Northeastern China. Science Press, Beijing

Ren D, Tan JJ, Shih CK, Gao TP (2010b) The upper Mesozoic stratigraphic characteristics in northeastern China. In: Ren D, Shih CK, Gao TP, Yao YZ, Zhao YY (eds) Silent stories – insect fossil treasures from Dinosaur Era of the Northwestern China. Science Press, Beijing, pp 12–21

Ren D, Shih CK, Labandeira CC (2011) A well-preserved aneuretopsychid from the Jehol Biota of China (Insecta: Mecoptera: Aneuretopsychidae). ZooKeys 121:17–28

Ren D, Shih CK, Gao T, Wang Y, Yao Y. 2019. Rhythms of Insect Evolution. Wiley

Reutter OM (1913) Lebensgewohenheiter und Instinkte der Insekter. Friedlander, Berlin

Richter R, Storch G (1980) Beiträge zur Ernährungsbiologie eozäner Fledermause aus der "Grübe Messel". Nat Mus 110:353–367

Riek EF (1974) Upper Triassic insects from the Molteno "formation", South Africa. Palaeontol Afr 17:19–31

Robert D, Amoroso J, Hoy RR (1992) The evolutionary convergence of hearing in a parasitoid fly and its cricket host. Science 258:1135–1137

Roberts H (1980) Descriptions of the developmental stages of *Sosylus* spp. (Coleoptera: Colydiidae) from New Guinea, parasites and predators of ambrosia beetles (Coleoptera: Platypodidae). Bull Entomol Res 70:245–252

Robin N (2021) Importance of data on paleosymbioses for parasite–host evolution. In: De Baets K, Huntley JW (eds) The evolution and fossil record of parasitism – coevolution, pathologies and paleoparasitological techniques. Springer, Berlin

Robin N, Foldi I, Godinot M, Petit G (2016) Scale insect larvae preserved in vertebrate coprolites (Le Quesnoy, France, Lower Eocene): paleoecological insights. Sci Nat 103:85. https://doi.org/10.1007/s00114-016-1412-x

Robinson WH (1971) Old and new biologies of *Megaselia* species (Diptera: Phoridae). Stud Entomol 14:321–368

Rocha LSG, Burt TO, de Mello-Patiu CA, Skevington JH (2015) The first *Stylogaster* Macquart, 1835 (Diptera: Conopidae) fossil, from Oligo-Miocene Dominican amber, and some phylogenetic and biogeographic considerations. Foss Rec 18:119–125

Rohdendorf BB (1968a) Jurassic insects of Karatau. USSR Academy of Sciences, Moscow

Rohdendorf BB (1968b) New Mesozoic nemestrinids (Diptera, Nemestrinidae). In: Rohdendorf BB (ed) Jurassic insects of Karatau. USSR Academy of Sciences, Moscow, pp 180–189. (in Russian)

Roig-Alsina A (1994) A new genus of Plumariidae, with notes on the relationships among the genera of the family (Hymenoptera, Chrysidoidea, Plumariidae). Mitt Mün Entomol Ges 84:91–96

Ronquist F, Klopfstein S, Vilhelmsen L, Schulmeister S, Murray DL, Rasnitsyn AP (2012) A total-evidence approach to dating with fossils, applied to the early radiation of the Hymenoptera. Syst Biol 61:13–50

Ross AJ (2018) Burmese (Myanmar) amber taxa, on-line checklist v.2017.4. https://www.researchgate.net/profile/Andrew_Ross27/publication/322332066_Burmese_Myanmar_amber_taxa_on-line_checklist_v20174/links/5a54aa8f0f7e9bbc105a691b/Burmese-Myanmar-amber-taxa-on-line-checklist-v20174.pdf. Accessed May 2019

Ross AJ, Mellish C, York P, Crighton B (2010) Burmese amber. In: Penney D (ed) Biodiversity of fossils in amber from the major world deposits. Siri Scientific Press, Manchester, UK, pp 208–235
Rott AS, Godfray HCJ (2000) The structure of the leafminer–parasitoid community. J Anim Ecol 69:274–289
Salem A, Franc M, Jacquiet P, Bouhsira E, Liénard E (2012) Feeding and breeding aspects of *Stomoxys calcitrans* (Diptera: Muscidae) under laboratory conditions. Parasite 19:309–317
Sánchez MV, Genise JF (2009) Cleptoparasitism and detritivory in dung beetle fossil brood balls from Patagonia, Argentina. Palaeontology 52:837–848
Sanmartin I, Enghoff H, Ronquist F (2001) Patterns of animal dispersal, Vicariance and diversification in the Holarctic. Biol J Linn Soc 73:345–390
Santos AM, Quicke DLJ (2011) Large-scale diversity patterns of parasitoid insects. Entomol Sci 14:371–382
Schär S, Vorburger C (2013) Host specialization of parasitoids and their hyperparasitoids on a pair of syntopic aphid species. Bull Entomol Res 103:530–537
Schlüter TE (2000) *Moltenia rieki* n. gen., n. sp. (Hymenoptera: Xyelidae?), a tentative sawfly from the Molteno Formation (Upper Triassic), South Africa. Paläontol Z 74:75–78
Schmidt W, Schurmann M, Teichmüller M (1958) Biß-Spuren an Früchten des Miozän-Waldes der niederrheinischen Braunkohlen-formation. Fortsch Geol Rhein Westfalen 2:563–572
Schmidt AR, Perrichot V, Svojtka M, Anderson KB, Belete KH, Bussert R, Döfelt H, Jancke S, Mohr B, Mohrmann E, Nascimbene PC, Nel A, Nel P, Ragazzi F, Roghi G, Saupe E, Schmidt K, Schneider H, Selden PA, Vávra N (2010) Cretaceous African life captured in amber. Proc Natl Acad Sci U S A 107:7329–7334
Schmitz OJ, Hambäck PA, Beckerman AP (2000) Trophic cascades in terrestrial systems: a review of the effects of carnivore removals on plants. Am Nat 155:141–153
Scholtz CH, Chown SL (1995) The evolution of habitat use and diet in the Scarabaeoidea: a phylogenetic approach. In: Pakaluk J, Ślipiński SA (eds) Biology, phylogeny, and classification of coleoptera: papers celebrating the 80th birthday of Roy A. Crowson. Museum and Institute of Zoology, Warsaw
Scholtz CH, Grebennikov VV (2005) Family Scarabaeidae. In: Beutel RG, Leschen RAB (eds) Handbuch der Zoologie; Band IV: Arthropoda: Insecta; Teilband 38: Coleoptera, Beetles, Morphology and systematics, vol 1. Walter de Gruyter, Berlin, pp 386–425
Schowalter TD (2016) Insect ecology: an ecosystem approach, 4th edn. Elsevier, Amsterdam
Schremmer F (1961) Morphologische Anpassungen von Tieren – insbesonders Insekten – an die Gewinnung von Blumennahrung. Verhand Deut Zool Ges Saarbrücken 1961:375–401
Schuh RT, Weirauch C, Wheeler WC (2009) Phylogenetic relationships within the Cimicomorpha (Hemiptera: Heteroptera): a total-evidence analysis. Syst Entomol 34:15–48
Selander RB, Mathieu JM (1964) The ontogeny of blister beetles (Coleoptera, Meloidae). I. A study of three species of the genus *Pyrota*. Ann Entomol Soc Am 57:711–732
Sharanowski BJ, Robbertse B, Walker J, Voss SR, Yoder R, Spatafora J, Sharkey MJ (2010) Expressed sequence tags reveal Proctotrupomorpha (minus Chalcidoidea) as sister to Aculeata (Hymenoptera: Insecta). Mol Phylogenet Evol 57:101–112
Sharkey MJ, Carpenter JM, Vilhelmsen L, Heraty J, Liljeblad J, Dowling APG, Schulmeister S, Murray D, Deans AR, Ronquist F, Krogmann L, Wheeler WC (2012) Phylogenetic relationships among superfamilies of Hymenoptera. Cladistics 28:80–112
Shaw SR (1988) Euphorine phylogeny: the evolution of diversity in host-utilization by parasitoid wasps (Hymenoptera: Braconidae). Syst Entomol 13:323–335
Shaw SR (2004) Essay on the evolution of adult-parasitism in the subfamily Euphorinae (Hymenoptera: Braconidae). Proc Russ Entomol Soc St Petersburg 75:82–95
Shcherbakov DE (2008) Madygen, Triassic Lagerstätte number one, before and after Sharov. Alavesia 2:113–124
Shcherbakov DE (2017) Cretaceous Saurophthiridae (Aphaniptera) as pupiparous pre-fleas of diving pterosaurs. Paleontol J 51:183–185
Shear WA, Palmer JM, Coddington JA, Bonamo PM (1989) A Devonian spinneret: early evidence of spiders and silk use. Science 246:479–481

Shi GS, Grimaldi DA, Harlow GE, Wang J, Wang J, Yang MC, Lei WY, Li QL, Li XH (2012) Age constraint on Burmese amber based on U–Pb dating of zircons. Cretac Res 37:155–163

Shi XQ, Zhao YY, Shih CK, Ren D (2013) New fossil mesoserphid wasps (Insecta, Hymenoptera, Proctotrupoidea) from the Jehol Biota, China. Zootaxa 3710:591–599

Shi XQ, Zhao YY, Shih CK, Ren D (2014) Two new species of *Archaeohelorus* (Hymenoptera, Proctotrupoidea, Heloridae) from the Middle Jurassic of China. ZooKeys 369:49–59. https://doi.org/10.3897/zookeys.369.6561

Shih CK, Lin CX, Ren D (2009) The earliest fossil record of pelecinid wasps (Insecta: Hymenoptera: Proctotrupoidea: Pelecinidae) from Inner Mongolia, China. Ann Entomol Soc Am 102:20–38

Shih CK, Feng H, Liu CX, Zhao YY, Ren D (2010) Morphology, phylogeny, evolution, and dispersal of pelecinid wasps (Hymenoptera: Pelecinidae) over 165 million years. Ann Entomol Soc Am 103:875–885

Shih CK, Feng H, Ren D (2011) New fossil Heloridae and Mesoserphidae wasps (Insecta, Hymenoptera, Proctotrupoidea) mesoserphid wasps (Insecta, Hymenoptera, Proctotrupoidea) from the Middle Jurassic of China. Ann Entomol Soc Am 104:1334–1348

Silvestri F (1906) Contribuzioni alla conoscenza biologia degli imenotteri parassiti. I. Biologia del *Litomastix truncatellus* (Dalm.). Boll Lab Zool Gen Agr Portici 1:17–64

Simutnik SA (2002) A new genus of encyrtid wasp (Hymenoptera, Chalcidoidea, Encyrtidae) from late Eocene Rovno amber (Ukraine). Vest Zool 36:99–102

Simutnik SA, Perkovsky EE (2006) A description of the encyrtid male (Hymenoptera: Chalcidoidea: Encyrtidae) with archaic structure of metasoma from Rovno amber. Vest Zool 40:283–286

Skidmore RE (2018) Checklist of Canadian amber inclusions in the Canadian National Collection of Insects. Research Branch, Agriculture and Agri-Food, Canada. http://www.biology.ualberta.ca/facilities/strickland/SKIDMORECNCCanadianAmberInclusions.pdf

Slansky F (1986) Nutritional ecology of endoparasitic insects and their hosts: an overview. J Insect Physiol 32:255–261

Ślipiński A, Lord N, Lawrence JF (2010) Bothrideridae Erichson, 1845. In: Leschen RAB, Beutel RG, Lawrence JF (eds) Handbuch der Zoologie; Band IV: Arthropoda: Insecta; Teilband 38: Coleoptera, Beetles, Morphology and systematics (Elateroidea, Bostrichiformia, Cucujiformia partim), vol 2. Walter de Gruyter, Berlin, pp 411–422

Smith KGV (1966) The larva of *Thecophora occidensis*, with comments upon the biology of Conopidae. J Zool Lond 149:263–267

Smith KGV, Cunningham-von Someren GR (1985) The larva of *Stylogaster varifrons* Malloch (Dipt., Stylogastridae). Entomol Month Mag 121:81–85

Smith JA, Tierney SM, Park YC, Fuller S, Schwarz MP (2007) Origins of social parasitism: the importance of divergence ages in phylogenetic studies. Mol Phylogenet Evol 43:1131–1137

Sohn J-C, Labandeira C, Davis D, Mitter C (2012) An annotated catalog of fossil and subfossil Lepidoptera (Insecta: Holometabola) of the world. Zootaxa 3286:1–132

Sokol J (2019) Troubled treasure. Science 364:722–729

Spahr U (1987) Ergänsungen und Berichtigungen zu R. Keilbach's Bibliographie und Liste der Bernsteinfossilen-Ordnung Hymenoptera. Stutt Beit Naturk Ser B Geol Paläontol 127:1–121

Spasojevic T, Wedmann S, Klopfstein S (2017) Seven remarkable new fossil species of parasitoid wasps (Hymenoptera, Ichneumonidae) from the Eocene Messel Pit. PLoS One 13(6):e0197477

Statz G (1940) Neue Dipteren (Brachycera et Cyclorhapha) aus dem Oberoligocän von Rott. Palaeontographica Abt A 91:120–174

Stebnicka Z (1999) A new genus and species of termitophilous Eupariini from Ecuador with checklist of the Neotropical genera (Coleoptera: Scarabaeidae: Aphodiinae). Acta Zool Cracov 42:289–295

Stevens JR (2003) The evolution of myiasis in blowflies (Calliphoridae). Int J Parasitol 33:1105–1113

Stevens JR, Willman JF, Otranto D, Wall R, Pape T (2006) The evolution of myiasis in humans and other animals in the Old and New Worlds (part II): biological studies. Trends Parasitol 22:181–188

Stireman III (2003) What determines host range in parasitoids? An analysis of a tachinid parasitoid community. Oecologia 135:629–638

Stireman JO III (2005) The evolution of generalization? Parasitoid flies and the perils of inferring host range evolution from phylogenies. J Evol Biol 18:325–336

Stireman JO III, O'Hara JE, Wood M (2006) Tachinidae: evolution, behavior, and ecology. Annu Rev Entomol 51:525–555

Strelnikova OD, Rasnitsyn AP (2016) A preliminary study of the respiratory and alimentary systems of the Early Cretaceous "flea" *Saurophthirus longipes* Ponomarenko, 1976 (Insecta, ?Aphaniptera, Saurophthiridae). Far East Entomol 327:1–7

Stuckenberg BR (1973) The Athericidae: a new family in the lower Brachycera (Diptera). Ann Natl Mus 21:649–673

Sung G-H, Hywel-Jones NL, Sung J-M, Luangsa-ard JJ, Shrestha B, Spatafora JW (2007) Phylogenetic classification of *Cordyceps* and the clavicipitaceous fungi. Stud Mycol 57:5–59

Swift J (1733) On Poetry: a Rhapsody. J. Huggonson, London

Swisher CC III, Wang YQ, Wang XL, Xu X, Wang Y (1999) Cretaceous age for the feathered dinosaurs of Liaoning, China. Nature 400:58–61

Szadziewski R (1998) New mosquitoes from Baltic amber (Diptera: Culicidae). Pol Pismo Entomol 67:233–244

Talamas EJ, Johnson NF, Buffington ML, Ren D (2017) *Archaeoteleia* Masner in the Cretaceous and a new species of *Proteroscelio* Brues (Hymenoptera, Platygastroidea). J Hymenoptera Res 56:241–261

Tan JJ, Chang HL, Yue YL, Shih CK, Ren D (2010) Coleoptera – sacred, precious and strong. In: Ren D, Shih CK, Gao TP, Yao YZ, Zhao YY (eds) Silent stories – insect fossil treasures from Dinosaur Era of the Northeastern China. Science Press, Beijing, pp 180–215

Tangelder IRM, Krikken J (1982) Termitophilous scarabs of the tribe Corythoderini: a taxonomic review (Coleoptera: Aphodiinae). Zool Verhandl 194:3–114

Teeling EG, Springer MS, Madsen O, Bates P, O'Brien SJ, Murphy WJ (2005) A molecular phylogeny for bats illuminates biogeography and the fossil record. Science 307:580–584

Thayer MK (2005) Staphylinidae Latreille, 1802. In: Beutel RG, Leschen RAB (eds) Handbuch der Zoologie; Band IV: Arthropoda: Insecta; Teilband 38: Coleoptera, Beetles, Morphology and systematics (Archostemata, Adephaga, Myxophaga, Polyphaga partim), vol 1. Walter de Gruyter, Berlin, pp 296–344

Thomas MC (2002) Passandridae Erichson 1845. In: Arnett RH Jr, Thomas MC, Skelley PE, Frank JH (eds) American Beetles, Polyphaga: Scarabaeoidea through Curculionoidea, vol 2. CRC Press, Boca Raton, FL, pp 327–328

Thomas DB, Mangan RL (1989) Oviposition and wound-visiting behavior of the screwworm fly, *Cochliomyia hominivorax* (Diptera: Calliphoridae). Ann Entomol Soc Am 82:526–534

Thompson JN (1984) Insect diversity and the trophic structure of communities. In: Huffaker CB, Rabb RL (eds) Ecological entomology. Wiley, New York, pp 591–606

Thorpe WH (1941) A description of six new species of the genus *Cryptochaetum* (Diptera–Agromyzidae) parasitic on *Icerya purchasi* (Coccidae, Monophlebini). Proc Zool Soc Lond 1930:929–971

Tichomirova AL (1968) Staphylinid beetles from Jurassic Karatau. In: Rohdendorf BB (ed) Jurassic insects of Karatau. USSR Academy of Sciences, Moscow, pp 138–154. (in Russian)

Todd K (2011) Maria Sibylla Merian (1647–1717): an early investigator of parasitoids and phenotypic plasticity. Terr Arthropod Rev 4:131–144

Tomassini RL, Montalvo CI, Ezquiaga MC (2016) The oldest record of flea/armadillos interaction as example of bio erosion on osteoderms from the late Miocene of the Argentine Pampas. Int J Paleopathol 15:65–68

Tompkins DM, Clayton DH (1999) Host resources govern the specificity of swiftlet lice: size matters. J Anim Ecol 68:489–500

Torchio PF (1972) *Sapyga pumila* Cresson, a parasite of *Megachile rotundata* (F.) (Hymenoptera: Sapygidae; Megachilidae). I: Biology and description of immature stages. Melanderia 10:1–22

Townes HK (1977) A revision of the Rhopalosomatidae (Hymenoptera). Contrib Am Entomol Inst 15:1–34
Trjapitsyn VA (1963) A new hymenopteran genus from Rovno amber. Paleontol Zh 1963(3):89–95
Turrisi FG, Vilhelmsen L (2010) Into the wood and back: morphological adaptations to the wood-boring parasitoid lifestyle in adult aulacid wasps (Hymenoptera: Aulacidae). J. Hymen Res 19:244–258
Ülgentürk S (2001) Parasitoids and predators of Coccidae (Homoptera: Coccoidea) species on ornamental plants in Ankara, Turkey. Acta Phytopath Entomol Hung 36:369–375
Ülgentürk S, Toros S (1996) Ankara ilinde *Anthribus fasciatus* Först (Cole: Anthribidae) üzerinde ön çalişmalar. Türkiye III Entomol Kongr, Ankara
Vala J-C, Bailey PT, Gasc C (1990) Immature stages of the fly *Pelidnoptera nigripennis* (Fabricius) (Diptera: Phaeomyiidae), a parasitoid of millipedes. Syst Entomol 15:391–399
Valentine BD (2002) Anthribidae Billberg 1820. In: Arnett RH Jr, Thomas MC, Skelley PE, Frank JH (eds) American Beetles, Polyphaga: Scarabaeoidea through Curculionoidea, vol 2. CRC Press, Boca Raton, FL, pp 695–700
van Achterberg C (2002) Apanteles (Choeras) gielisi spec. nov. (Hymenoptera: Braconidae: Microgastrinae) from The Netherlands and the first report of Trichoptera as host of Braconidae. Zool Meded Leiden 76:53–60
van Achterberg C, Quicke DLJ (2006) Taxonomic notes on Old World Stephanidae (Hymenoptera): description of *Parastephanellus matsumotoi* sp. n. from Japan, redescription to *Commatopus xanthocephalus* (Cameron) and keys to the genera *Profoenatopus* van Achterberg and *Megischus* Brullé. Tijd Entomol 149:215–225
van Achterberg C, Yang ZQ (2004) New species of the genera *Megischus* Brullé and *Stephanus* Jurine from China (Hymenoptera: Stephanoidea: Stephanidae), with a key to world species of the genus *Stephanus*. Zool Meded Leiden 78:101–117
van de Kamp T, Schwermann AH, dos Santos RT, Lösel PD, Engler T, Etter W, Farago T, Göttlicher J, Heuveline V, Kopmann A, Mähler B, Mörs T, Odar J, Rust J, Jerome NT, Vogelgesang M, Baumbach T, Krogmann L (2018) Parasitoid preserved in mineralized fossils. Nat Commun 9:3325. https://doi.org/10.1038/s41467-018-05654-y
van Dijk J, De Baets K (2021) Biodiversity and host-parasite (co)extinction. In: De Baets K, Huntley JW (eds) The evolution and fossil record of parasitism: Coevolution and paleoparasitological techniques. Topics in Geobiology 50
van Jutting BT (1938) A freshwater pulmonate (*Physa fontinalis* (L.)) inhabited by the larva of a non-biting midge, (*Tendipes* (*Parachironomus*) *varus* Gtgh). Arch Hydrobiol 32:693–699
Van Veen FJF, Müller CB, Pell JK, Godfray HCJ (2008) Food web structure of three guilds of natural enemies: predators, parasitoids and pathogens of aphids. J Anim Ecol 77:191–200
Vermeij GJ (1977) The Mesozoic marine revolution: evidence from snails, predators and grazers. Paleobiology 3:245–258
Vetter RS, Vincent LS, Intyre AA, Clarke DE, Reinker KI, Danielsen DWR, Robinson LJ, Kabashima JN, Rust MK (2012) Predators and parasitoids of egg sacs of the widow spiders, *Latrodectus geometricus* and *Latrodectus hesperus* (Araneae: Theridiidae) in southern California. J Arachnol 40:209–214
Vilhelmsen L, Turrisi GF (2011) Per arborem ad astra: morphological adaptations to exploiting the woody habitat in the early evolution of Hymenoptera. Arthropod Struct Dev 40:2–20
Vilhelmsen L, Zimmermann D (2014) *Baltorussus* total makeover: rejuvenation and sex change in an ancient parasitoid wasp lineage. PLoS One 9(6):e98412. https://doi.org/10.1371/journal.pone.0098412
Vinikour WS, Anderson RV (1981) Diptera larvae (Empididae and Chironomidae) in Trichoptera pupal cases (Glossosomatidae and Limnephilidae). Entomol News 92:69–74
Voigt E (1952) Ein Haareinschluss mit Phthirapteren-Eiern im Bernstein. Mitt Geol Staat Hamburg 21:59–74
Vršanský P, Ren D, Shih CK (2010) Nakridletia ord nov. – enigmatic insect parasites support sociality and endothermy of pterosaurs. AMBA Projekly 8:1–16

Vršanský P, Šmídová L, Sendi H, Barna P, Müller P, Ellenberger S, Wu H, Ren X, Lei X, Azar D, Šurka J, Su T, Deng W, Shen X, Lv J, Bao T, Bechly G (2019) Parasitic cockroaches indicate complex states of earliest proved ants. Biologia 74:65–89

Waage JK, Greathead DJ (1986) Insect parasitoids. Academic, London

Walker JD, Geissman JW, Bowring SA, Babcock LE (2013) The Geological Society of America geologic time scale. Geol Soc Am Bull 125:259–272

Wang B, Kathirithamby J, Engel MS (2015) The first twisted-wing parasitoid in Eocene amber from northeastern China (Strepsiptera: Myrmecolacidae). J Nat Hist 50:1305–1313

Wang M, Rasnitsyn AP, Zhang HC, Shih CK, Ren D (2019) Revising the systematic position of the extinct family Daohugoidae (basal Hymenoptera). J Syst Palaeontol 17:1025–1035

Wappler T, Smith VS, Dalgleish RC (2004) Scratching an ancient itch: an Eocene bird louse fossil. Proc R Soc Lond B 271(Suppl):S255–S258

Warnock RCM, Engelstädter J (2021) The molecular clock as tool for understanding host-parasite evolution. In: De Baets K, Huntley JW (eds) The evolution and fossil record of parasitism: Coevolution and paleoparasitological techniques. Topics in Geobiology 50

Wedmann S (2000) Die Insekten der oberoligozänen Fossillagerstätte Enspel (Westerwald, Deutschland): Systematik, Biostrationomie und Paläoökologie. Mainz Naturwiss Arch 23:1–154

Wedmann S (2005) Annotated taxon-list of the invertebrate animals from the Eocene fossil site Grube Messel near Darmstadt, Germany. Cour Forsch Senck 255:103–110

Wedmann S (2007) A nemestrinid fly (Insecta: Diptera: Nemestrinidae: cf *Hirmoneura*) from the Eocene Messel pit (Germany). J Paleontol 81:1114–1117

Wedmann S, Yeates DK (2008) Eocene records of bee flies (Insecta, Diptera, Bombyliidae, *Comptosia*): their palaeobiogeographic implications and remarks on the evolution history of bombyliids. Palaeontology 51:231–240

Weiblen GD (2002) How to be a fig wasp. Annu Rev Entomol 47:299–330

Weinstein SB, Austin AD (1991) The host relationships of trigonalyid wasps (Hymenoptera: Trigonalyidae) with a review of their biology and catalogue to world species. J Nat Hist 25:399–433

Weinstein SB, Kuris AM (2016) Independent origins of parasitism in Animalia. Biol Lett 12:20160324. https://doi.org/10.1098/rsbl.2016.0324

Weirauch C, Munro JB (2009) Molecular phylogeny of the assassin bugs (Hemiptera: Reduviidae), based on mitochondrial and nuclear ribosomal genes. Mol Phylogenet Evol 53:287–299

Weitschat W, Wichard W (2010) Baltic amber. In: Penney D (ed) Biodiversity of fossils in amber from the major world deposits. Siri Scientific Press, Manchester, UK, pp 80–115

Wellnhofer P (2008) Archaeopteryx. Der Urvogel von Solnhofen. Verlag Friedrich Pfeil, Munich

Wells A (1992) The first parasitic Trichoptera. Ecol Entomol 17:299–302

Wells A (2005) Parasitism by hydroptilid caddisflies (Trichoptera) and seven new species of Hydroptilidae from northern Queensland. Aust J Entomol 44:385–391

Whitfield JB (1998) Phylogeny and evolution of host–parasitoid interactions in Hymenoptera. Annu Rev Entomol 43:129–151

Whitfield JB (2003) Phylogenetic insights into the evolution of parasitism in Hymenoptera. Adv Parasitol 54:69–99

Whiting MF, Whiting AS, Hastriter MW, Dittmar K (2008) A molecular phylogeny of fleas (Insecta: Siphonaptera): origins and host associations. Cladistics 24:677–707

Wiegmann BM, Mitter C, Farrell B (1993) Diversification of carnivorous parasitic insects: extraordinary radiation or specialized dead end? Am Nat 142:737–754

Wiegmann BM, Trautwein MD, Winkler IS, Barr NB, Kim J-W, Lambkin C, Bertone MA, Casel BK, Bayless KM, Heimberg AM, Wheeler BM, Peterson KJ, Pape T, Sinclair BJ, Skevington JH, Blagoderov V, Caravas J, Kutty SN, Schmitt-Ott U, Kampmeier GE, Thomson FC, Grimaldi DA, Beckenbach AT, Courtney GW, Friedrich M, Meier R, Yates DK (2011) Episodic radiations in the fly tree of life. Proc Natl Acad Sci U S A 108:5690–5695

Williams RJ (2010) Network3D Software. Microsoft Research, Cambridge, UK

Williams RJ, Martinez ND (2004) Limits to trophic levels and omnivory in complex food webs: theory and data. Am Nat 163:458–468
Wilson EO (1971) The insect societies. Harvard University Press, Cambridge MA
Wilson JS, von Dohlen CD, Forister ML, Pitts JP (2013) Family-level divergences in the stinging wasps (Hymenoptera: Aculeata), with correlations to angiosperm diversification. Evol Biol 40:101–107
Windsor DA (1998) Most species on Earth are parasites. Int J Parasitol 28:1939–1941
Winkler IS, Blaschke JD, Davis DJ, Stireman JO III, O'Hara JE, Cerretti P, Moulton JK (2015) Explosive radiation or uninformative genes? Origin and early diversification of tachinid flies (Diptera: Tachinidae). Mol Phylogenet Evol 88:38–54
Winterton SL, Wiegmann BM, Schlinger EI (2007) Phylogeny and Bayesian divergence time estimations of small-headed flies (Diptera: Acroceridae) using multiple molecular markers. Mol Phylogenet Evol 43:808–832
Wood DM (2006) Morphology of adult Oestridae. In: Colwell DD, Hall MJR, Scholl PJ (eds) The Oestrid flies: biology, host–parasite relationships, impact and management. CAB International, Wallingford, UK, pp 78–92
Yamamoto S, Maruyarna M, Parker J (2016) Evidence for social parasitism of early insect societies by Cretaceous rove beetles. Nat Commun 7:13658. https://doi.org/10.1038/ncomms13658
Yao YZ, Cai WZ, Cai WZ, Xu X, Shih CK, Engel MS, Zheng XT, Zhao YY, Ren D (2014) Blood-feeding true bugs in the Early Cretaceous. Curr Biol 24:1786–1792
Yeates DK, Greathead D (1997) The evolutionary pattern of host use in the Bombyliidae (Diptera): a diverse family of parasitoid flies. Biol J Linn Soc 60:149–185
Yeates DK, Wiegmann BM (1999) Congruence and controversy: Toward a higher-level phylogeny of Diptera. Annu Rev Entomol 44:397–428
Yoon I, Williams RJ, Levine E, Yoon S, Dunne JA, Martinez ND (2004) Webs on the Web (WoW): 3D visualization of ecological networks on the WWW for collaborative research and education. Electron Imag Conf Proc IS&T/SPIE Symp Electr Imag Visual Data Anal 5295:124–132
Yoshimoto CM (1975) Cretaceous chalcidoid fossils from Canadian amber. Can Entomol 107:499–528
Yu YL, Liu ZH, Shih CK, Ren D (2019) Coleoptera – beetles. In: Ren D, Shih CK, Gao TP, Wang YG, Yao YZ (eds) Rhythms of insect evolution: evidence from the Jurassic and Cretaceous in Northern China. Wiley, Hoboken, NJ, pp 336–428
Zaspel JM, Kononenko VS, Goldstein PZ (2007) Another blood feeder? Experimental feeding of a fruit-piercing moth species on human blood in the Primorye territory of Far Eastern Russia (Lepidoptera: Noctuidae: Calpinae). J Insect Behav 20:437–451
Zaspel JM, Zahiri R, Hoy MA, Janzen D, Weller SJ, Wahlberg N (2012) A molecular phylogenetic analysis of the vampire moths and their fruit-piercing relatives (Lepidoptera: Erebidae: Calpinae). Mol Phylogenet Evol 65:786–791
Zenker MM, Botton M, Teston JA, Specht A (2010) Noctuidae moths occurring in grape orchards in Serra Gaúcha, Brazil and their relation to fruit-piercing. Rev Brasil Entomol 54:288–297
Zhang JF (2017) New findings of *Flagellisargus* J Zhang, 2012 (Diptera, Brachycera, Archisargidae), with discussion of the placements of some controversial taxa. Deut Entomol Z 64:111–122
Zhang HC, Rasnitsyn AP (2007) Nevaniinae subfam. n., a new fossil taxon (Insecta: Hymenoptera: Evanioidea: Praeaulacidae) from the Middle Jurassic of Daohugou in Inner Mongolia, China. Insect Syst Evol 38:149–166
Zhang HC, Rasnitsyn AP (2008) Middle Jurassic Praeaulacidae (Insecta: Hymenoptera: Evanioidea) of Inner Mongolia and Kazakhstan. J Syst Palaeontol 6:463–487
Zhang HC, Zhang JF (2001) Proctotrupoid wasps (Insecta, Hymenoptera) from the Yixian Formation of western Liaoning Province. Acta Micropaleont Sin 18:11–28
Zhang HC, Rasnitsyn AP, Zhang JF (2002a) Two ephialtitid wasps (Insecta, Hymenoptera, Ephialtitoidea) from the Yixian Formation of western Liaoning, China. Cretac Res 23:401–407

Zhang HC, Rasnitsyn AP, Zhang JF (2002b) Pelecinid wasps (Insecta: Hymenoptera: Proctotrupoidea) from the Yixian Formation of western Liaoning, China. Cretac Res 23:87–98

Zhang HC, Rasnitsyn AP, Zhang JF (2002c) The oldest known scoliid wasps (Insecta, Hymenoptera, Scoliidae) from the Jehol biota of western Liaoning, China. Cretac Res 23:77–86

Zhang HC, Wang B, Fang Y (2010) Evolution of insect diversity in the Jehol Biota. Sci China Earth Sci 53:1908–1917

Zhang QQ, Zhang JF, Feng YT, Zhang HC, Wang B (2016) An endoparasitoid Cretaceous fly and the evolution of parasitoidism. Sci Nat 103:2. https://doi.org/10.1007/s00114-015-1327-y

Zhang QQ, Zhang JF, Wang B (2017) First record of the subfamily Archinemestriniinae in the family Nemestrinidae (Diptera: Brachycera) from Upper Cretaceous Burmese amber. Cretac Res 75:141–145

Zhang Q, Rasnitsyn AP, Wang B, Zhang HC (2018a) Hymenoptera (wasps, bees and ants) in mid-Cretaceous Burmese amber: a review of the fauna. Proc Geol Assoc 129:736–747

Zhang Q, Rasnitsyn AP, Wang B, Zhang HC (2018b) Peleserphidae, a new family of basal proctotrupomorphs (Hymenoptera: Proctotrupoidea) from mid-Cretaceous Burmese amber. Cretac Res 86:66–72

Zhang Q, Rasnitsyn AP, Wang B, Zhang HC (2018c) Myanmarinidae, a new family of basal Apocrita (Hymenoptera: Stephanoidea) from upper-Cretaceous Burmese amber. Cretac Res 81:86–92

Zherikhin VV, Ross AJ (2000) A review of the history, geology and age of Burmese amber (Burmite). Bull Nat Hist Mus Geol Ser 56:3–10

Zhu QY, Hastriter MW, Whiting MF, Dittmar K (2015) Fleas (Siphonaptera) are Cretaceous, and evolved with Theria. Mol Phylogenet Evol 90:129–139

Index

© The Editor(s) (if applicable) and The Author(s) 2021

K. De Baets, J. Huntley (eds.), *The Evolution and Fossil Record of Parasitism*, Topics in Geobiology 49, https://doi.org/10.1007/978-3-030-42484-8

C

E

G

I

The manufacturer's authorised representative in the EU is Springer Nature Customer Service Centre GmbH, Europaplatz 3, 69115 Heidelberg, Germany. If you have any concerns regarding our products, please contact ProductSafety@springernature.com

Printed and bound by CPI Group (UK) Ltd, Croydon, CR0 4YY
23/09/2026
02221471-0010